GEL ELECTROPHORESIS – ADVANCED TECHNIQUES

Edited by **Sameh Magdeldin**

Gel Electrophoresis – Advanced Techniques
Edited by Sameh Magdeldin

Published by InTech
Janeza Trdine 9, 51000 Rijeka, Croatia

Publishing Process Manager Martina Durovic

Technical Editor Teodora Smiljanic

Cover Designer InTech Design Team

Additional hard copies can be obtained from orders@intechopen.com

Gel Electrophoresis – Advanced Techniques, Edited by Sameh Magdeldin
p. cm.
ISBN 978-953-51-0457-5

For EU product safety concerns:
IN TECH d.o.o., Prolaz Marije Krucifikse Kozulić 3, 51000 Rijeka, Croatia,
info@intechopen.com or visit our website at intechopen.com.

Meet the editor

Dr Sameh Magdeldin is a senior postdoctoral researcher at the Medical School, Niigata University, Japan, and an academic associate professor (lecturer) at the Physiology Department, SCU, Egypt. He received his MVSc and PhD in Physiology and second PhD in Proteomics, and is an expert in two–dimensional poly acrylamide gel electrophoresis of protein fractionation, a well-known gel-based fractionation approach. Dr Magdeldin is also interested in shotgun proteomics analysis, reversed-phase chromatography and label-free comparative proteomic approaches. He created the outstanding "All and None" methodology for analyzing large throughput proteomic data, published in a highly respected proteomics journal. He currently serves as a guest editor, associate editor and a peer-reviewer in several international journals. Dr Magdeldin received several grants and awards such as the 8th HUPO Congress young investigator award, JSN award, Grant-in-aid for young scientist, and the young researcher overseas grant from the Japan Society for the Promotion of Science (JSPS).

Contents

Preface XIII

Part 1 Electrophoresis Application in Ecological and Biotechnological Aspects 1

Chapter 1 **Application of Gel Electrophoresis Techniques to the Study of Wine Yeast and to Improve Winemaking 3**
María Esther Rodríguez, Laureana Rebordinos, Eugenia Muñoz-Bernal, Francisco Javier Fernández-Acero and Jesús Manuel Cantoral

Chapter 2 **Proteomics in Seaweeds: Ecological Interpretations 21**
Loretto Contreras-Porcia and Camilo López-Cristoffanini

Chapter 3 **Gel Electrophoresis Based Genetic Fingerprinting Techniques on Environmental Ecology 51**
Zeynep Cetecioglu, Orhan Ince and Bahar Ince

Chapter 4 **Gel Electrophoresis of Grapevine (*Vitis vinifera* L.) Isozymes – A Review 67**
Gizella Jahnke, János Májer and János Remete

Chapter 5 **Molecular Electrophoretic Technique for Authentication of the Fish Genetic Diversity 83**
Tsai-Hsin Chiu, Yi-Cheng Su, Hui-Chiu Lin and Chung-Kang Hsu

Chapter 6 **Gel Electrophoresis for Investigating Enzymes with Biotechnological Application 97**
Maria de Lourdes T. M. Polizeli, Simone C. Peixoto-Nogueira, Tony M. da Silva, Alexandre Maller and Hamilton Cabral

Part 2 Electrophoresis Application in Bacteriology, Parasitology, Mycology and Public Health 111

Chapter 7 **Application of Molecular Typing Methods to the Study of Medically Relevant Gram-Positive Cocci 113**
Laura Bonofiglio, Noella Gardella and Marta Mollerach

Chapter 8 **Molecular Microbiology Applied to the Study of Phytopathogenic Fungi 139**
Carlos Garrido, Francisco J. Fernández-Acero, María Carbú, Victoria E. González-Rodríguez, Eva Liñeiro and Jesús M. Cantoral

Chapter 9 **Molecular and Proteolytic Profiles of *Trypanosoma cruzi* Sylvatic Isolates from Rio de Janeiro-Brazil 157**
Suzete A. O. Gomes, Danielle Misael, Cristina S. Silva, Denise Feder, Alice H. Ricardo-Silva, André L. S. Santos, Jacenir R. Santos-Mallet and Teresa Cristina M. Gonçalves

Chapter 10 **Pulsed Field Gel Electrophoresis in Molecular Typing and Epidemiological Detection of Methicillin Resistant *Staphylococcus aureus* (MRSA) 179**
Velazquez-Meza Maria Elena, Vázquez-Larios Rosario, Hernández Dueñas Ana Maria and Rivera Martínez Eduardo

Chapter 11 **Usefulness of Pulsed Field Gel Electrophoresis Assay in the Molecular Epidemiological Study of Extended Spectrum Beta Lactamase Producers 193**
Patrick Eberechi Akpaka and Padman Jayaratne

Part 3 Electrophoresis Application in the Analysis of Protein-Nucleic Acid Interaction and Chromosomal Replication 203

Chapter 12 **Electrophoretic Mobility Shift Assay: Analyzing Protein – Nucleic Acid Interactions 205**
Carolina Alves and Celso Cunha

Chapter 13 **Analysis of Chromosomal Replication Progression by Gel Electrophoresis 229**
Elena C. Guzmán and Enrique Viguera

Part 4 Electrophoresis Application in Enzymology 245

Chapter 14 **Polyacrylamide Gel Electrophoresis an Important Tool for the Detection and Analysis of Enzymatic Activities by Electrophoretic Zymograms 247**
Reyna Lucero Camacho Morales, Vanesa Zazueta-Novoa, Carlos A. Leal-Morales, Alberto Flores Martínez, Patricia Ponce Noyola and Roberto Zazueta-Sandoval

Chapter 15 **Applications of Zymography (Substrate-SDS-PAGE) for Peptidase Screening in a Post-Genomic Era 265**
Claudia M. d'Avila-Levy, André L. S. Santos, Patrícia Cuervo, José Batista de Jesus and Marta H. Branquinha

Part 5 Temporal Temperature Gel Electrophoresis 289

Chapter 16 **Temporal Temperature Gel Electrophoresis to Survey Pathogenic Bacterial Communities: The Case of Surgical Site Infections 291**
Romano-Bertrand Sara, Parer Sylvie, Lotthé Anne, Colson Pascal, Albat Bernard and Jumas-Bilak Estelle

Part 6 Two-Dimensional Gel Electrophoresis (2-DE) 313

Chapter 17 **Two-Dimensional Gel Electrophoresis Reveals Differential Protein Expression Between Individual *Daphnia* 315**
Darren J. Bauer, Gary B. Smejkal and W. Kelley Thomas

Chapter 18 **Two-Dimensional Gel Electrophoresis and Mass Spectrometry in Studies of Nanoparticle-Protein Interactions 327**
Helen Karlsson, Stefan Ljunggren, Maria Ahrén, Bijar Ghafouri, Kajsa Uvdal, Mats Lindahl and Anders Ljungman

Chapter 19 **Two Dimensional Gel Electrophoresis in Cancer Proteomics 359**
Soundarapandian Kannan, Mohanan V. Sujitha, Shenbagamoorthy Sundarraj and Ramasamy Thirumurugan

Part 7 Other Applications of Gel Electrophoresis Technique 391

Chapter 20 **Enzymatic Staining for Detection of Phenol-Oxidizing Isozymes Involved in Lignin-Degradation by *Lentinula edodes* on Native-PAGE 393**
Eiji Tanesaka, Naomi Saeki, Akinori Kochi and Motonobu Yoshida

Chapter 21 **Protection Studies by Antioxidants Using Single Cell Gel Electrophoresis (Comet Assay) 413**
Pınar Erkekoglu

Chapter 22 **Gel Electrophoresis as Quality Control Method of the Radiolabeled Monoclonal Antibodies 447**
Veronika Kocurová

Chapter 23 **Gel Electrophoresis as a Tool to Study Polymorphism and Nutritive Value of the Seed Storage Proteins in the Grain Sorghum 463**
Lev Elkonin, Julia Italianskaya and Irina Fadeeva

Chapter 24 **Extraction and Electrophoresis of DNA from the Remains of Mexican Ancient Populations 479**
Maria de Lourdes Muñoz, Mauro Lopez-Armenta, Miguel Moreno-Galeana, Alvaro Díaz-Badillo, Gerardo Pérez-Ramirez, Alma Herrera-Salazar, Elizabeth Mejia-Pérez-Campos, Sergio Gómez-Chávez and Adrián Martínez-Meza

Preface

"Gel electrophoresis is finding a wide application because its unique separatory power"
Oliver Smithies

Since the first report about gel electrophoresis, released in 1930, and despite the alternative fractionation technologies that have recently emerged, gel electrophoresis remains the most popular and most powerful technique that provides a perfect fractionation of macromolecules (DNA, RNA and Proteins). Hundreds of techniques could be set according to experimental atmosphere, including various substrates (agarose or polyacrylamide), gel conditions (native or denaturating), buffers (TAE, TBE, etc) and others. This broad variability enabled further progression and development of gel-based fractionation such as isoelectric focusing, zymograms and 2D-PAGE. With wide application ranging from restriction mapping and genetic fingerprinting, to forensic application and co-application with blotting system, there is no doubt that gel electrophoresis made an overwhelming progress in the *"OMICS"* era.

It is my pleasure to introduce the book "Gel Electrophoresis - Advanced Techniques". This book presents a wide research application of this sophisticated technique with wider scope to better cover this popular topic.

Sameh Magdeldin, MVSc, PhD (Physiology), PhD (Proteomics)
Senior post doc researcher and Proteomics team leader
Medical School, Niigata University, Japan
Assistant Professor (Lecturer), Physiology Department
Suez Canal University
Egypt

Part 1

Electrophoresis Application in Ecological and Biotechnological Aspects

Application of Gel Electrophoresis Techniques to the Study of Wine Yeast and to Improve Winemaking

María Esther Rodríguez, Laureana Rebordinos, Eugenia Muñoz-Bernal, Francisco Javier Fernández-Acero and Jesús Manuel Cantoral
Microbiology Laboratory, Faculty of Marine and Environmental Sciences University of Cadiz, Puerto Real Spain

1. Introduction

Yeasts are unicellular fungi that are frequently used as a model and tools in basic science studies. This is the case of the laboratory yeasts *Saccharomyces cerevisiae*, which were introduced in the laboratory for genetics and molecular studies in about 1935. There is, however, a second type of yeast comprising those used in industrial processes, for example, in brewing, baking and winemaking. Wine yeast and its properties have been known to humans for as long as civilizations have existed, and the earliest evidence of this yeast has been dated to Neolithic times (Mortimer, 2000).

Most wine yeast strains are diploid and have a low frequency of sporulation. Another important characteristic of wine yeasts, and those used in other industries, is their highly polymorphic chromosomes: their genetic constitution is affected by the frequent and extensive mutation they undergo. These effects include (i) aneuploidy, (ii) polyploidy, (iii) amplification and deletion of chromosomal region or single gene, and (iv) the presence of hybrid chromosomes. The chromosomal polymorphism obtained by applying the technique known as pulsed field gel electrophoresis (PFGE) has been used to characterize and to classify strains that belong to the same species.

In the wine industry, knowledge of the yeast species responsible for the alcoholic fermentation is important because these yeasts with their metabolism contribute significantly to the organoleptic characteristics of the finished wine (Fleet, 2008). The diverse range of yeasts associated with the vinification process can be classified in two groups. The first group is formed principally by the genera *Hanseniaspora, Torulaspora, Metschnikowia, Candida, Zygosaccharomyces*, etc. These yeasts initiate spontaneous alcoholic fermentation of the must, but they are soon replaced by the second group, formed by *Saccharomyces* yeasts, which are present during the subsequent phases of the fermentation until it is completed. Within the genus *Saccharomyces* the species most relevant for the fermentation process are *S. cerevisiae* and *S. bayanus* var. *uvarum*; this is because they have become of interest for their biotechnological properties. However, there is currently increasing interest in the non-*Saccharomyces* yeasts for the development of innovative new styles of wine (Viana et al.,

2009). In the industry, knowledge of specific strains of these microorganism species is important for (i) their selection; (ii) their use as starter cultures; and (iii) improving the fermentation process.

During the 1990's the development of molecular techniques has enabled the identification and characterization of different strains belonging to the same species of yeast, and it has been possible to establish the ecology of spontaneous fermentations in many of the world's winemaking regions (Fleet, 2008). These techniques also constitute a powerful tool not only for the selection of the most suitable yeast, since they tell us which yeasts are the most representative in the fermentation process, but also for obtaining information on the addition to the must of particular strains of yeast in the case of inoculated fermentations (Rodríguez et al., 2010).

Two of the approaches most often used for the molecular characterization of industrial yeast are analysis of the electrophoretic karyotypes by pulsed-field gel electrophoresis (PFGE) and analysis of the restriction fragment length polymorphism of the mitochondrial DNA (mtDNA-RFLP). We have used PFGE in winemaking to analyse the diversity of wild yeasts in spontaneous fermentation of a white wine produced in a winery in SW Spain with the object of selecting the most suitable autochthonous starter yeast; and from the results of the inoculation, we were able to make decisions for improving the efficiency of the process and to establish procedures for the proper performance of the inoculation (Rodríguez et al., 2010). We have also applied the analysis of the karyotypes to characterize natural yeasts in biodynamic red wines in another region of Spain. In this chapter we also evaluate the use of the mtDNA-RFLP technique for quick monitoring of the dominance of inoculated strains in industrial fermentation, without any need for the prior isolation of yeast colonies (Rodríguez et al., 2011).

Another electrophoretic technique has been used to show substantial changes in protein levels in selected wine yeasts under specific growth conditions. It has recently been stated that the proteome is "the relevant level of analysis to understand the adaptations of wine yeasts for fermentation" (Rossignol et al., 2009). Following this, in-depth studies are now being made of the proteome of wine yeast strains and the relationship between the proteome and wine quality and winery processes. We are now exploring more generally the relevance of proteomics to wine improvement. In this chapter, we will summarize the efforts being made by the proteomics research community to obtain the knowledge needed on proteins in the post-genomics era

2. Pulsed-field gel electrophoresis (PFGE) for the study of yeast population

PFGE as a system encompasses a series of techniques in which the intact chromosomes of microorganisms like yeasts and filamentous fungi are submitted to the action of a pulsing electric field in two orientations that is changing direction, in a matrix of agarose. The best-known PFGE modality is the CHEF system (Contour-clamped Homogeneous Electric Fields); this consists of a hexagon of 24 electrodes surrounding the gel that produce a homogeneous electric field alternating between two directions orientated at 120° with respect to each other. Using this system, Chu et al. (1986) resolved the electrophoretic karyotype of *Saccharomyces cerevisiae* in 15 bands in a size range of 200-2200 kb. Before performing the electrophoresis the yeast cells must be suitably treated, avoiding the direct manipulation of the genetic material to prevent possible rupture of the chromosomes. The cells are then embedded in blocks of agarose which are subsequently treated with a reducing agent and K proteinase to destabilize the wall and cytoplasmatic membranes, respectively (Figure 1), thus facilitating the release of the DNA when submitted to the action of an electric field.

This methodology for correctly obtaining the karyotype of *S. cerevisiae* is based on the procedure described by Carle & Olson (1985) and optimized by Rodríguez et al. (2010). It also depends on the concentration of the agarose gel (1%), buffer (0.5 x TBE), initial and final switch (60-120 seconds respectively), run time (24 hours), voltage (6 V/cm) and buffer temperature (14 °C).

The analytical results given by this technique are the number and size of the yeast chromosomes, and it allows specific strains of *Saccharomyces* to be differentiated because their karyotypes show distinct bands running below the 500-kb marker. It also allows the differentiation between *S. cerevisiae* and *Saccharomyces bayanus* var. *uvarum* (*S. uvarum*) species (Naumov et al., 2000, 2002).

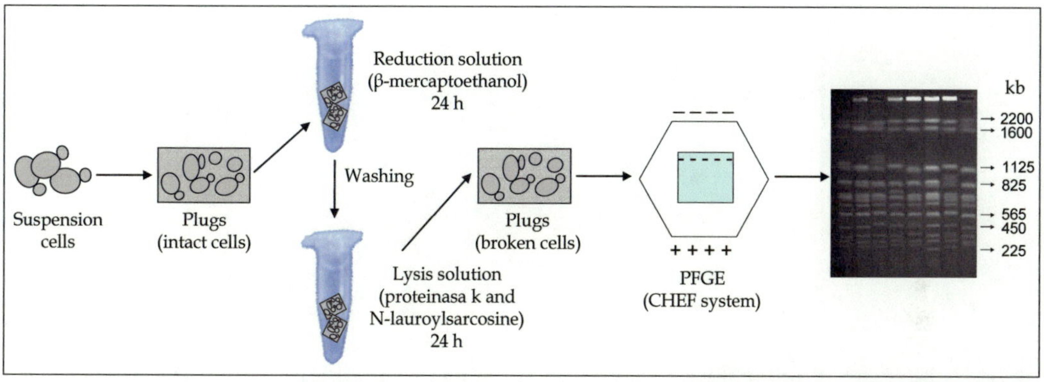

Fig. 1. Methodology for characterizing yeast strains, using pulsed field gel electrophoresis to obtain the karyotype.

In previous research the PFGE technique has been used to analyse the dynamics of the yeast population during the spontaneous fermentations of wine (Demuyter et al., 2004; Martínez et al., 2004; Naumov et al., 2002; Raspor et al., 2002; Rodríguez et al., 2010), and it has also been used to characterize other industrial yeasts including baker's and brewer's yeast (Codón et al., 1998). Another relevant application of PFGE has been to characterize the yeast population which is present in the *flor velum* that grows on the surface of fino-type sherry wines in the barrel, during their biological ageing process (Mesa et al., 1999, 2000). The results revealed an interesting correlation between the yeast genotypes and the different blending stages.

One disadvantage of the technique is that it is laborious, expensive and requires specialized personnel; increasingly, therefore, analysts are resorting to other simpler and faster techniques to discriminate between yeast clones, like, for example, interdelta analysis of sequences or microsatellite analysis (Cordero-Bueso et al., 2011; Le Jeune et al., 2007; Schuller et al., 2007). However, the methodology proposed in Figure 1 enables a large number of yeast isolates to be processed, and PFGE is considered a most suitable technique for discriminating between yeast clones (Schuller et al., 2004).

In our laboratory, this technique has been used to characterize the wine yeast population responsible for the spontaneous fermentation of a white wine produced in a winery in SW Spain (Rodríguez et al., 2010). Analyses of industrial-scale fermentations (in 400 000-l fermentation vessels) were carried out during two consecutive vintages. In 1999 and 2000 a total of 211 and 228 yeast colonies, respectively, from different vessels, were characterised by karyotyping. The degree of polymorphism observed was high, and 17 different karyotypic patterns were detected in 1999, and 21 patterns in 2000. In the two campaigns, we also found patterns belonging to non-*Saccharomyces* yeasts, the karyotypes of which did not show the four bands running below 500-kb. During the fermentation, this population was displaced by *S. cerevisiae* strains; patterns I, II, III and V were predominant during entire fermentation process in 1999, whereas in 2000 patterns II and V were predominant (see Figure 2). Those were the yeast strains selected for inoculating the industrial fermentations, as will be explained bellow.

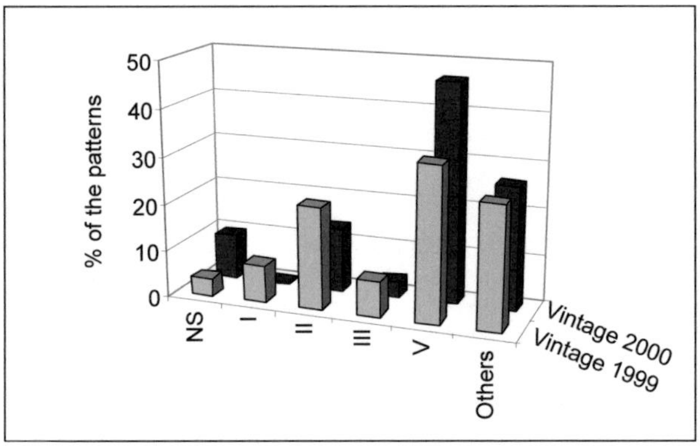

Fig. 2. Frequencies of the majority karyotype patterns (%) obtained in the spontaneous fermentations of 1999 and 2000. NS corresponds to non-*Saccharomyces* yeasts.

The results of the characterization of the yeasts also showed that the different strains changed their proportion, and there was a sequential substitution of strains during the fermentation; this gave a valuable indication of the dynamics of the yeasts population throughout the process. Some of these changes were specific to a particular fermentation phase, suggesting that the yeast strains with different electrophoretic karyotypes also differ in their adaptation to the evolving environment at different phases of the fermentation process.

Although the diversity of wild yeast can contribute to high-quality and unique flavour in the finished wine, spontaneous fermentation is often unpredictable and might introduce less desirable traits to the product, sometimes even spoiling a production batch. Other risks associated with spontaneous fermentation include either slow or arrested fermentation. To avoid these problems, winemakers often add cultures of selected yeasts, in the form of active dried yeasts or autochthonous yeasts. Nevertheless, in some cases, these yeasts used as starters are not able to displace the wild yeasts present in the must, since the wild yeasts can be very competitive (Esteve-Zarzoso et al., 2000; Lopes et al., 2007).

In our work on the analysis of the karyotype, it has been possible to monitor the yeast population under industrial conditions for several years when fermentations of the white wine were inoculated with selected autochthonous yeast strains. This has allowed inoculation strategies to be designed for the correct development of inoculated yeast while retaining the unique regional character of the finished wine (Rodríguez et al., 2010). The strains with patterns I, II, III and V were the most representative during the spontaneous fermentation process (Figure 2) and they could be isolated at the late fermentation phase. These autochthonous yeasts show valuable traits of enological interest, such as high fermentative capacity, ethanol tolerance, and they had a killer phenotype. The capacity of each strain to compete within the mixed population was also tested under semi-industrial conditions by PFGE. The results show that the strains with karyotypes II and V were the most vigorous competitors, followed by the strain with patterns III and I, which were detected in lower proportions (Rodríguez et al., 2010). Therefore, the strains with patterns II, III and V were used to inoculate the fermentations in the year 2001; strains with patterns II and V were used in 2002, 2003 and 2004; and from the year 2005 until the present (2011), only the strain with pattern V has been used.

The inoculation of industrial vessels of the winery of this study presented several peculiarities. For example: (i) in each vintage year several vessels with a total capacity of 400 000-l were inoculated; (ii) the inoculums comprising the selected autochthonous yeast strains were prepared from fresh YEPD plates (1% yeast extract, 2% glucose, 2% peptone and 2% agar) by preparing a starter in which each scaling-up round was performed when the °bé reached a value between 1-2, and in each round, the fermentation volume was increased tenfold to give high initial levels of inoculum (> 60×10^6 viable cells/ml) and ensure the correct development of the inoculated strain; (iii) once the starter cultures were scaled-up and added to a 400 000-l container, partial volumes were withdrawn and used for the inoculation of other 400 000-l vessels of the winery; and (iv) the 400 000-l vessels received random additions of fresh must until reaching the final volume. The frequency and timing of these additions depended on the production yield of fresh must during the vintage campaign.

In our results (Table 1) only the strain with pattern V, called P5, was dominant under industrial conditions and, for this reason, the number of the starter strains were reduced over the years and currently only the strain with karyotype V is used for the inoculation of the industrial fermentations. In spite of this strain's good capacity for achieving dominance,, in some years a high degree of polymorphism was detected in the fermentations; and the cause of the unexpected predominance of wild yeast karyotype was linked to several factors, including: a sudden decrease in the temperature of one of the vessels during the scaling-up process of the inoculation in 2002; the method of inoculation and the scaling-up process, which were changed in 2003, whereby the inoculums of pure culture did not represent 10% and each scaling-up round was performed when the inoculums had a high sugar content (around 3-5 °Bé); and the storage of must in the vessels in which spontaneous fermentation had occurred.

When karyotype V was dominant in the fermentations, the wine obtained had fruity characteristics, with well-balanced acidity, that satisfied the wine producer. Although we did not obtain a comprehensive aromatic characterization of the wine, the panel of wine tasters (Figure 3) considered the wine produced in these fermentations better than the wine obtained either in the spontaneous fermentations of 1999 and 2000 or in the vintage years of 2002 and 2003, when the karyotype V (strain P5) was not detected in high proportion in the yeast population.

Patterns	2001	2002	2003	2004	2005	2007	2008
II	7.6	9.2	3.6	0.4	0	10.7	0
III	8.5	0	0	0	0	0	0
V	83	0.6	19.3	99.6	79.8	50	100
(NS)	0.2	0	0.4	0	0.4	0	0
Others	0.7	90.2	76.7	0	19.8	39.3	0
Total yeast analysed	423	174	223	235	278	142	240

Table 1. Frequencies (%) of the inoculated strains in the industrial fermentations for seven vintage years. This analysis was not performed in the vintage year of 2006. NS: non-*Saccharomyces.*

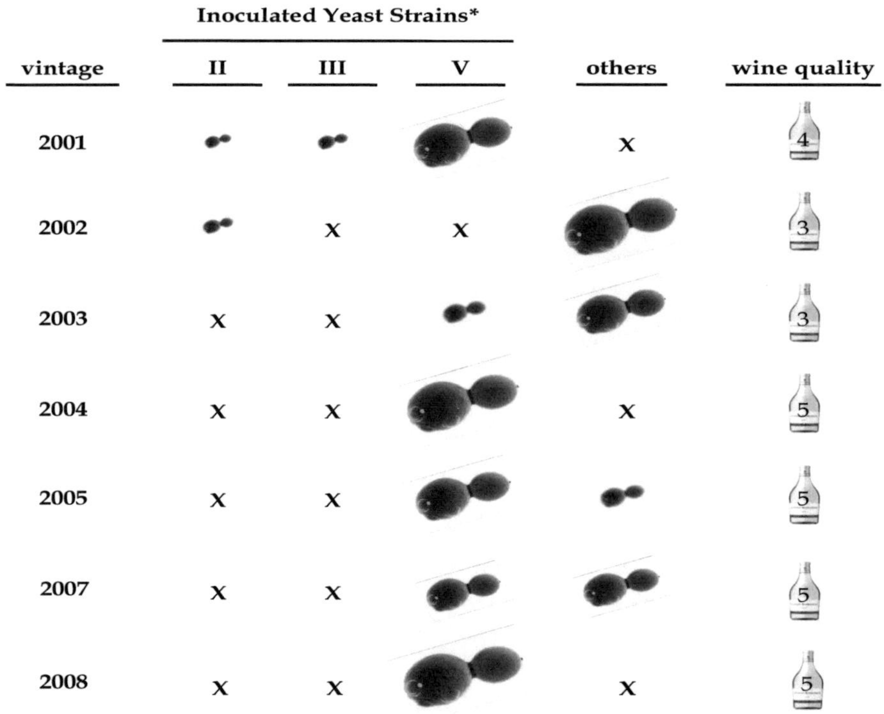

*Strain group II was inoculated in the years 2001 to 2004; the strain with pattern III was inoculated only in 2001; and strain group V was inoculated in all vintages

Fig. 3. Composition of the wine yeast population in each vintage, and its relationship with the quality of the final product. For each year the size of the yeast cell shown is proportional to the contribution of each strain to the total wine yeast population of the winery. The x symbol indicates that the proportion of the strain(s) within the population was below 5%. Wine quality was evaluated by a panel of expert wine-testers from the winery, who graded the final product on a scale from 1 to 5 based on fruity wine with well-balanced acidity desired by the producer (5 indicates highest quality and 1 lowest quality). Predominance of the strain with pattern V corresponded to a better quality of the wine.

By using PFGE to study the yeast population of the inoculated fermentations, the producer was able to make informed decisions for improving the process; the common factors in the vintages of 2001, 2004, 2005, 2007 and 2008, in which the inoculated strain was dominant, can be highlighted. These factors were the following: (i) the culture was not scaled-up to the next volume until the yeast had fully depleted the sugar to less than 1 °Bé (one degree is equivalent to 18 g/l of fermentable sugars in the must); therefore all cultures reached a high alcohol content before the addition of fresh must; (ii) the inoculum was always diluted less than 10-fold in each scaling-up round; (iii) the temperature of the fermentation was kept at 17 °C.

We think that these criteria favoured the adaptation of the inoculums to the conditions of the must obtained in each vintage and to the final conditions within the 400 000-l industrial vessels. In addition, these criteria favoured the predominance of the inoculated strain with pattern V.

In another study with biodynamic red wines, carried out in the Ribera del Duero D. O. Region (Valladolid, Spain), spontaneous fermentations were also analysed applying PFGE. We studied seven fermentations in three phases during the fermentation process: initial (IF), middle (MF) and final (EF), and 20 isolated strains per sample were characterized by applying PFGE (417 strains in 2008, and 412 strains in 2009). The results for two consecutive vintages studied showed the presence of different types of the yeast during the fermentations that were grouped in three populations. The first population was formed by non-*Saccharomyces* yeast, whose strains showed patterns with the absence of bands running below the region of 500 kb, which are specific to *S. cerevisiae* strains as reported above. The second population comprised *Saccharomyces bayanus* var. *uvarum* (*S. uvarum*); and the third population included *Saccharomyces cerevisiae* yeast. The strains of *S. uvarum* were differentiated from the *S. cerevisiae* strains by the presence of two small chromosomes in the region of 245-370 kb, instead of three as for S. *Cerevisiae*, as reported by Naumov et al. (2000, 2002). Non-*Saccharomyces* (NS) yeasts were dominant in the initial phase of fermentation but were displaced in the subsequent and final phases of the process by another population of yeasts. *S. uvarum* yeasts were present mainly in the phase mid-way through the fermentation; then the population of *S. cerevisiae* yeasts displaced the NS and *S. uvarum* yeasts, and remained dominant until the end of the fermentation, in the majority of the deposits analysed. The low frequency of detection of *S. uvarum* at the end of the fermentation could be indicative of its lower ethanol tolerance compared to *S. cerevisiae*. Within each population yeast strains were also found with different karyotyping patterns, and the distribution (by %) of these varied in the seven deposits analysed during the two consecutive years studied. Thus, for *S. uvarum*, considerable variability of strains and a total of 12 different electrophoretic patterns were detected (Figure 4): uI-uVII for vintage 2008; and uI-uIII, uV, uVIII-uXII in 2009. The strains with patterns uI, uII, uIII and uV, followed by uIV (in 2008) and uIX (in 2009) were the most representative in two years studied. Within the population of the *S. cerevisiae* yeasts, the variability of the patterns was higher than in *S. uvarum* ; 29 (cI-cXXIX) and 27 (cI-cVII, cX-cXII, cXV, cXVII, cXIX, cXXII, cXXIV, cXXX-cXLI) electrophoretic karyotype patterns were detected for 2008 and 2009 respectively. The *S. cerevisiae* yeast strains most representative of the fermentation process in these years were those that showed the karyotypes cIII, cVI, cXI and cXII.

The yeast population dynamics presented in this biodynamic red wine were different from those observed in other studies of white wines in which *S. uvarum* was dominant during spontaneous alcoholic fermentation (Demuyter et al., 2004).

Although *S. uvarum* has been found in other producing regions of the world, such as Alsace (Demuyter et al., 2004), at the moment there are no studies about the population dynamics of *S. uvarum* in Ribera del Duero, Spain.

The use of the PFGE technique allows analysts to detect a high degree of polymorphism in the population of the yeast and to monitor the dynamics of yeast ecology during the fermentation; this is because it is able to show the occurrence of gross chromosomal rearrangements, which is the phenomenon that mainly accounts for the rapid evolution of yeast clones subjected to industrial conditions (Infante et al., 2003). The technique also shows the most representative yeast strains in the industrial winemaking process, which are partially but significantly responsible for the finished wine's quality. Knowledge of these main strains can be used as a criterion for making a first selection of the autochthonous yeast. Later, with these previously selected strains, PFGE can be applied to study the features of these strains that are of enological interest, as described in recent years, which fall into three main categories: (i) properties that affect the performance of the fermentation process; (ii) properties that determine the quality of the wine; and (iii) properties associated with the commercial production of yeast (reviewed in Fleet, 2008). The yeasts selected by these means can then be used for inoculation in the fermentation process, thus improving winemaking.

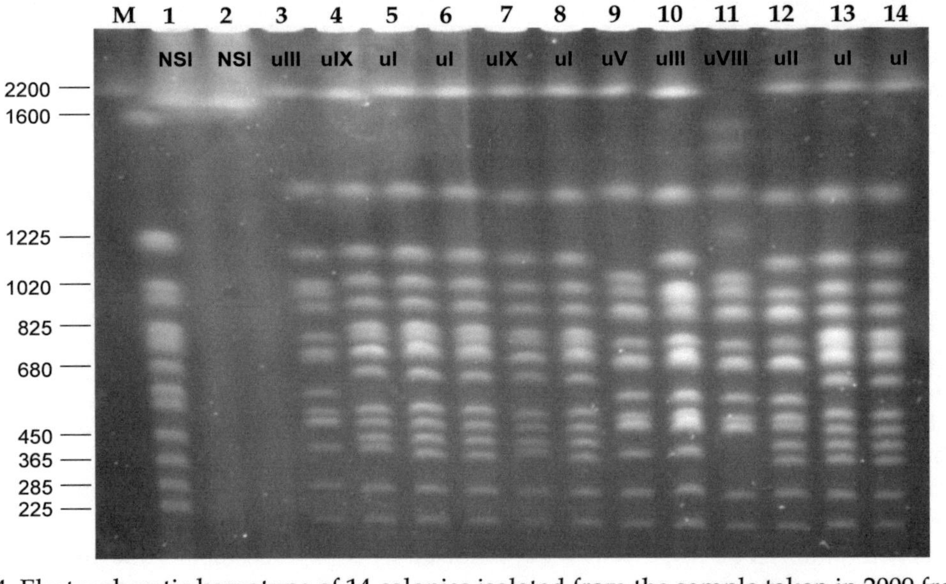

Fig. 4. Electrophoretic karyotype of 14 colonies isolated from the sample taken in 2009 from a vessel in the phase mid-way through the fermentation process. Colonies 3-14 correspond to different karyotype patterns (uI-uIX) of yeast strains found in the *S. uvarum* species. Isolates 1-2 correspond to non-*Saccharomyces* strains which show the same pattern (NSI), with absence of bands running below 1225 kb. The chromosomes of the *S. cerevisiae* YNN295 strain were used as reference (M).

3. Application of mtDNA-RFLP as a rapid method for monitoring the inoculated yeast strains in wine fermentations

Although PFGE has been reported to be the most efficient in discriminating between different strains of *S. cerevisiae,* the mtDNA-RFLP technique is frequently used to differentiate between yeast isolates of the same species (González et al., 2007) because it enables a larger number of strains to be analyzed in a shorter time; it is a fast, simple, reliable and economic method, which does not require sophisticated material or specialized personnel (Fernández-Espinar et al., 2006). For these reasons, it is a very suitable technique for use by industry.

Most of the mitochondrial DNA in yeasts does not code proteins, and contains a high proportion of AT bases. Analysts can take advantage of this characteristic to characterize yeasts; it involves measuring variation in sequences in the mtDNA affecting the restriction sites of several endonucleases. Endonucleases such as *Alu*I, *Hinf*I or *Rsa*I, recognise the very frequent restrictions in the chromosomal DNA but not in the mitochondrial DNA, leading to a total cleavage of the chromosomal DNA in small pieces. These pieces can be easily differentiated from the mitochondrial fragments, which appear as bands with an electrophoretic mobility corresponding to molecules greater than 2 kb, generating polymorphisms that allow the characterization between yeast strains.

In previous studies applying this technique it has been demonstrated that the population of a fermentation vessel is "taken over" by wild yeasts, which displace the inoculated yeast strain, reducing it to a minority presence (Esteve-Zarzoso et al., 2000; Lopes et al., 2007; Raspor et al., 2002). In our research, when we have analyzed the inoculated fermentation of white wine as described above, we have found several examples of real situations that led to a significant decrease in the proportion of the inoculated strain (pattern V) and, in consequence, the quality of the wine was reduced. In order to minimize the impact of unwanted yeasts, wineries need a simple method for rapid diagnosis of the degree of dominance of inoculated strains, a method that could be performed routinely during the fermentation process (Ambrona et al., 2006, López et al., 2003). With this object we have used RFLP analysis of mtDNA for the rapid monitoring of the dominance, or otherwise, of inoculated yeast strains in industrial fermentations of white and red wines in a winery in southern Spain (Rodriguez et al., 2011).

We apply this technique directly to samples of fermenting wine without previously isolating yeast colonies. For white wine fermentations, a rapid assay is performed consisting of taking a sample of fermenting must, purifying the DNA from harvested cells, and obtaining the restriction patterns by digestion with endonuclease *Hinf*I. The same protocol is applied to red wine fermentation, but an overnight cultivation step is added before purification of the DNA (Figure 5).

The criterion for considering the result of the rapid test to be positive was obtaining restriction patterns of mtDNA that were identical for the total cells and the inoculated strain; when this is the case, the starter yeast can be taken as being dominant in the fermentation. The result was considered negative when additional bands, or absence of bands, were observed in the patterns; in this case neither the dominance, nor even the presence, of the inoculated yeast strain can be assured.

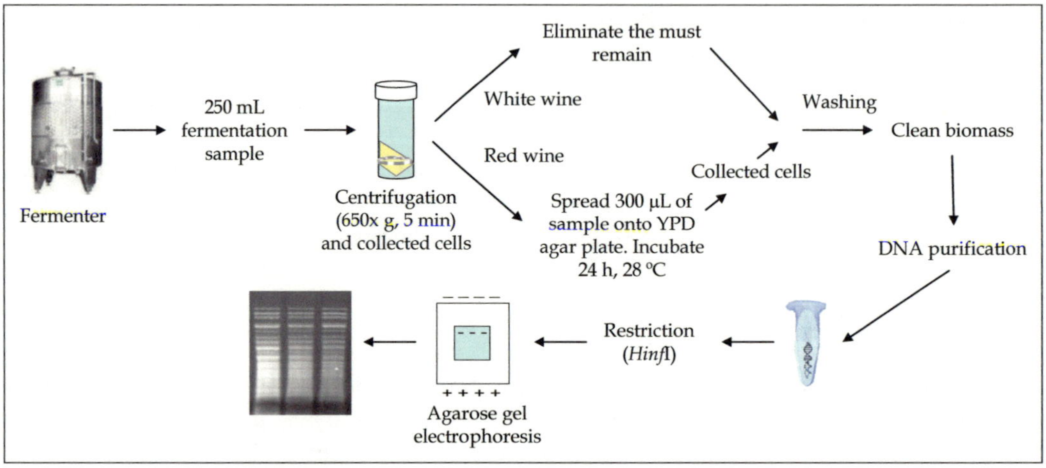

Fig. 5. Methodology for the rapid test assays by restriction analysis of the mtDNA.

This methodology of the RFLP test has been applied since 2005 in a winery of SW Spain and the results were obtained 11 and 23 hours after sampling, for white and red wine respectively (Rodríguez et al., 2011). If the wine-producer knows whether or not the presence of the inoculated yeasts has suffered a sudden decrease, and the inoculated strain is no longer dominant, in any phase of the fermentation process, a rapid intervention can be made. Since this year the winery has only used the selected autochthonous strain P5 for the inoculation of its white wine fermentations; this strain was in a clear majority in spontaneous fermentations, as reported above, and it shows karyotype pattern V. The peculiarities of the inoculation of industrial vessels for this wine have been described above (section 2). After applying the RFLP test, the correct course of the fermentation of all starters was assured before the inoculation of the industrial vessels. Generally, all the fermentations are tested in at least two different phases of the process: first after refills with fresh must and again when the fermentations are finished.

In addition, the results were checked using PFGE to validate the previous results obtained from the RFLP test. For this validation, 34 samples tested by RFLP were analyzed by electrophoretic karyotype of 323 colonies for white wine. The results indicated that when RFLP test was positive, the inoculated strain was present in the fermentation at 64% or more. When the RFLP tests were negative it was confirmed by PFGE that the starter yeast was present at only 60% or less.

The red wine of the winery was fermented in stainless steel vessels of 27.000 l and no refills of must were carried out. The fermentations were inoculated with several commercial active dried wine yeasts (ADWY) by hydration following manufacturer's instructions. For this wine, for 331 colonies analyzed by PFGE, the results of the RFLP test correctly predicted the results obtained later by applying PFGE. However, in this case, the limit found for white wine cannot be established because, from the positive results obtained by applying the RFLP test, the presence of the inoculated strain was greater than 75%, and all the negative results were at 55% or less (Rodríguez et al., 2011). Nevertheless, further experiments will be necessary to confirm these correlations because the RFLP test shows qualitative results and the actual percentage implantation of the starter yeast cannot be known when the results are positive or negative. Figure 6 shows examples of positive and negative results for the last two years (2009 and 2010) for the dominance and non-dominance respectively of the inoculated strain P5 in white wine of the same winery described above.

For the 2009 vintage, we tested several different vessels at the initial phase of the fermentation, for white wine (Panel A). The results were positive for all cases after applying the rapid test, i.e. the restriction patterns of the samples and the inoculated strain were identical, and the strain P5 was responsible at the beginning of the fermentation process displacing other wild yeasts.

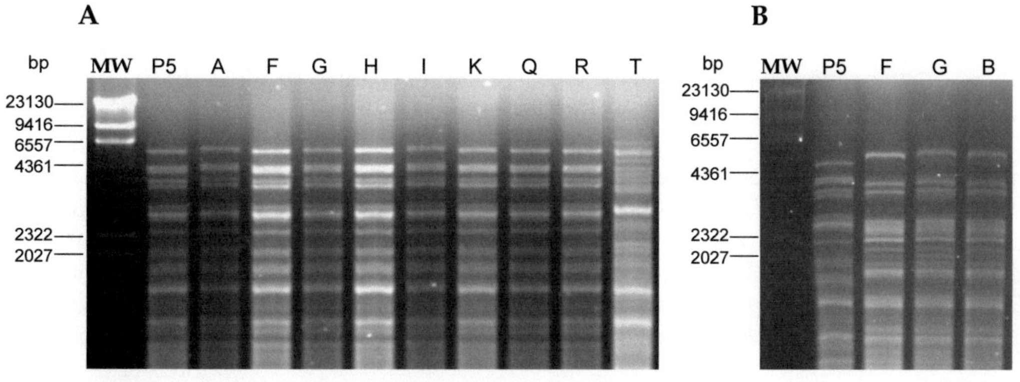

Fig. 6. Rapid test based on mtDNA-RFLP with *HinfI* of samples from white wine fermentations which were inoculated with autochthonous yeast strain P5. Panel A shows results of samples taken in 2009 from vessels A, F, G, H, I, K, Q, R and T at the initial fermentation phase (11-7 °Bé). Panel B shows results of samples taken in the 2010 vintage from vessels F, G and B during the main phase of the fermentation (5-3 °Bé). MW is the lambda-*Hind*III molecular marker.

Only the fermentation vessel reference T was considered negative for the RFLP test at the beginning of the process. In this vessel evidence was observed of spontaneous fermentation before the inoculation, due to the conditions in which the must was stored. When this must was inoculated, the strain P5 did not implant successfully and it was concluded that another wild yeast population was dominant.

In situations like this, the winemaker can take the decision not to use the fermenting must for the inoculation of other vessels.

Panel B (Figure 6) shows examples of the negative results of the RFLP test in three vessels mid-way during the fermentation process. In these deposits the population was perhaps similar because the restriction pattern of the total cells in each sample was similar. After the vintage, 20 colonies were isolated from the same sample previously analyzed by rapid test in the vessel G (Figure 6, panel B) in order to confirm the composition of the yeast population by karyotype. Surprisingly, all the clones show the same pattern as a commercial yeast strain used for the inoculation in the fermentation of another type of white wine in the previous year. It is assumed that this commercial yeast was also dominant in the fermentations sampled in vessels F and B. In previous studies researchers have reported the risks in using commercial yeast, because they can become part of the microbiota of the winery, effectively creating their own ecosystem, and can subsequently be predominant in the fermentations (Santamaría et al., 2005). In our study, we think that this commercial yeast was present in equipment which was not properly cleaned. When the wine-producer was more careful in the next vintage, there were no problems of contaminations by commercial yeast, and the dominant yeast in the fermentations was the inoculated autochthonous yeast P5 (data not shown). Therefore, it was confirmed that the commercial yeast had not acquired an ecological niche because it presumably did not adapt well to the ecosystem of a properly-cleaned winery.

As stated, the results of the RFLP can be obtained 11 and 23 hours after taking the sample for white and red wine respectively. However, this time can be shortened further, because it depends on the method used to rupture cells, on the number of samples analyzed per day, and on whether the samples contain a greater amount of must residues. In the case of red wine, there was another problem in shortening the test time, because the residues were difficult to clear by centrifugation; we think that some compounds remaining in the digested DNA samples were inhibitory for the endonuclease. Therefore, a step has been added in the protocol of the rapid test (Figure 5) in which the sample of the red must is plated on YPD-agar and incubated overnight at 28 °C. Nevertheless, we think that, for red wine, the time taken to obtain the results could also be shortened further, like that for white wine, if the clean biomass can be separated from the must residues in a few minutes. To achieve this, further experiments will need to be carried out.

4. Relevance of proteomic analysis in the winemaking process

In brief, proteomics can be described as a set of techniques for unravelling complex mixtures of proteins. In spite of it being a relatively recent technique, most of the systems used are widely known by the research community. However, the crucial work for its final "take-off" as a viable technique has been the modifications made to the mass spectrometry system, to allow the analysis of peptides and proteins. The exponential growth in the number of entries for genes and/or proteins in the databases now makes protein analysis and identification much easier, as well. This, combined with the use of powerful methods of fractionation and separation of peptides and proteins, such as 2D-PAGE (two dimensional polyacrylamide electrophoresis) and high resolution liquid chromatography, proteomics has been consolidated since the mid-90's, as the science for massive protein analysis; it is now the main methodology for unravelling biological processes, leading some authors to describe the current period as the "post-genomic era".

Proteomics has been defined as the set of techniques for studying the complex mixture of proteins, named the Proteome, that exists in any specific cell, microorganism, tissue, etc, used in specific experimental conditions, culture, sampling, etc. It is a highly dynamic system, and is more complex than genomics because, while the genome of an organism is more or less constant, the number of proteomes obtained from a specific genome is infinite. It depends on the assayed cell, tissue, culture conditions, etc., because each change produces a modification in the observed proteome. An additional factor of complexity derives from the fact that there are changes that occur in proteome that are not encoded in the genome. These changes mainly originate from two sources: (i) the editing of the mRNA; and (ii) post-translational modifications (PTMs) that normally serve to modify or modulate the activity, function or location of a protein in different physiological or metabolic contexts. More than

200 different PTMs have been described (including phosphorylation, methylation, acetylation, etc.) that transform each single gene into tens or hundreds of different biological functions. Before the advances made in proteomics, the differential analysis of the genes that were expressed in different cell types and tissues in different physiological contexts was done mainly through analysis of mRNA. However, for wine yeast, it has been proved that there is no direct correlation between mRNA transcripts and protein content (Rossignol et al., 2009). It is known that mRNA is not always translated into protein, and the amount of protein produced by a given amount of mRNA depends on the physiological state of the cell. Proteomics confirms the presence of the protein and provides a direct measure of its abundance and diversity.

In terms of methodology, proteomics approaches are classified in two groups: (i) gel-free systems based on the use of various chromatography methods; and (ii) gel-based methods that use mainly two-dimensional polyacrylamide gel electrophoresis (2DE). This latter approach will form the focus of our discussion here, given the subject matter of this book. As a succinct summary, the typical workflow of a proteomic experiment begins with the experimental design. This must be studied in depth, and it will delimit the conclusion obtained, even more so when comparisons are made between two strains, cultures or physiological stages, among others. As an optimum, only one factor among the various different assayed conditions must change (Fernandez-Acero et al., 2007). Several biological replicates, usually from 3 to 5, will be required depending on the strategy adopted. The next key step is to obtain a protein extract of high enough quality to separate complex mixtures of proteins. Usually, the protein extraction is done in sequential steps. First the tissue, cells, etc. are ruptured using mechanical or chemical techniques. Then, proteins are precipitated and cleaned. Most existing protocols use acetone and trichloroacetic acid. In the next step the proteome is defined and visualized using electrophoretic techniques. 2DE has been widely used for this purpose. Using this technique, proteins are separated using two different parameters. In the first dimension, proteins from the purified extract are separated by their iso-electric point using an iso-electrofocusing (IEF) device. Then, the focused strips are loaded in a polyacrylamide gel where the proteins are separated by their molecular weight. This system allows the separation of hundreds of proteins from one complex mixture. The gels are visualized with unspecific (Comassie, Sypro, etc.) or specific (e.g. Phospho ProQ diamond) protein stains. The gels are digitalized and analyzed with specific software to reveal the significant spots. These spots are identified using mass spectrometry; commonly, for 2DE approaches, MALDI TOF/TOF is used. The huge list of identified proteins obtained is studied to discover the biological relevance of each identification.

In spite of the many achievements of proteomics, only a few proteomic studies have been carried out on wine yeast, whereas mRNA expression has been widely used to study a broad range of industrial conditions. However, Rossignol et al. (2009) show that substantial changes in protein levels during alcoholic fermentation are not directly associated with changes in the transcriptome; this suggests that the mRNA is selectively processed, degraded and/or translated. This conclusion is important: it is the proteome, not the genome nor the transcriptome, that is the relevant level of analysis for understanding the adaptations of wine yeasts during alcoholic fermentation, since these are responsible for the phenotype.

The usual strategy for wine production is the inoculation of selected yeast strains into the must, decreasing the lag phase, a quick and complete fermentation of the must, and a high degree of reproducibility of the final product. The development of global analysis methodology has allowed a detailed analysis to be made of changes in gene expression and protein levels at various time-points during vinification. Zuzuarregui et al. (2006) presented a comparison between the mRNA and protein profiles of two yeast strains with different fermentation behaviours, which correlates with divergence in the fermentation profiles. The results indicate changes in the mRNA and protein levels and, probably, post-translational modifications of several proteins, some of them involved in stress response and metabolism.

Another proteomic approach was aimed at studying the adaptation of a wild-type wine yeast strain, isolated from a natural grape must, to physiological stresses during spontaneous fermentation (Trabalzini et al., 2003). Using 2DE, changes in the yeast proteome were monitored during glucose exhaustion, before the cells begin their stationary phase. The proteome adaptation of *S. cerevisiae* seems to be directed or caused by the effects of ethanol, leading to both hyperosmolarity and oxidative responses. Through the use of a wild-type *S. cerevisiae* strain and PMSF, which is a specific inhibitor of vacuolar proteinase B, it was also possible to distinguish the specific contributions of the vacuole and the proteasome autoproteolytic process. This is the first study that follows the adaptation of a physiologically wild wine yeast strain progressively to the exhaustion of an essential nutrient, glucose.

To monitor yeast stress Salvadó et al. (2008), using ADWY (active dried wine yeast) inoculated into the must, have observed its behaviour in different stress situations, i.e. high sugar concentration or low pH. The main responses after inoculation in a fermentable medium were the activation of several genes of the fermentation pathway and the monoxidative branch of the pentose pathway, and the induction of a huge cluster of genes related to ribosomal biogenesis and protein synthesis. The changes that occur during the lag phase are characterized by an overall change in the protein synthesis and reflect the physiological conditions of the yeast, which affects the fermentative capacity and fermentation performance. Certain enological practices increase these stressful conditions for ADWY. This is the case of low-temperature fermentation, which improves taste by restructuring flavour profiles, with potential enological applications. This study focuses on changes that occur in ADWY after inoculation in a synthetic wine. These changes reflect adaptation to a new medium.

Previous reports have shown that proteomic analysis of wine yeast is the most relevant tool for understanding the physiological changes involved in winery processes. The information obtained may improve the quality of the final product. Our group has been a pioneer in fungal proteomic approaches (Fernandez-Acero et al., 2007, 2011; Garrido et al., 2010), and in line with this, we are now exploring the relevance of proteomics in wine improvement (Muñoz-Bernal et al., 2011). Our group has developed new protocols for obtaining the proteome and subproteomes of yeast, and the results to date suggest that there is a lot of biological information to be studied and analyzed from the proteomic perspective. The relevance of this achievement for winery processes could be significant.

5. Conclusions

Application of the PFGE technique allows the yeast population in the wine fermentation process to be characterized. The technique has been reported to be the most efficient for discriminating between *S. cerevisiae* yeast clones (Schuller et al., 2004), and it differentiates these from the specie *S. bayanus* var. *uvarum* (Naumov et al., 2002). It is also able to reveal the occurrence of gross chromosomal rearrangements, which account for the rapid evolution shown by yeast in industrial environments (Infante et al., 2003). Using PFGE, we have detected a high degree of polymorphism in the population of spontaneous fermentations of different types of wine produced in different regions of Spain, and it was observed that there were yeast strains that were specific to a particular phase of the fermentation process. This suggests that yeast strains with different karyotypes also differ in their adaptation to the evolving environment at different phases of the fermentation process. Studies for the molecular characterization of wine yeast represent a first step for selecting autochthonous yeast strains which are better adapted to specific conditions of a particular wine-making region. Moreover, such knowledge in respect of yeast populations may lead to the identification of a new natural source of wine yeast that could be used by the industry in the future as a new commercial starter (Fleet, 2008).

Studies by PFGE of the yeast population in inoculated fermentations also allow producers to understand and make informed decisions for improving their processes. Our results suggest that the success of the inoculation protocol is highly dependent on adequate preparation of the inoculums, which must facilitate the adaptation of the inoculated strains to the final conditions of the fermentation.

The RFLP test designed to monitor and confirm that the population of the inoculated yeast has reached and maintained predominance, in white or red wines, is proposed as a response to one of the major challenges for microbiological control in the wine industry. In our results real situations are shown taking place during actual wine fermentations; for example spontaneous fermentations sometimes occur before the inoculation. We offer a test which the winemaker can use to obtain a reliable indication of whether or not wild yeasts are displacing the inoculated strains. If the strategy presented is followed, the wine producer would be able to identify and correct in time the unwanted evolution of the yeast population - usually by re-inoculating the selected strains and/or correcting a deviation in temperature or change in some other parameter of the vessel that might have caused the unwanted situation.

Our studies are among the first examples carried out at the industrial scale showing how molecular techniques can be successfully applied to improve quality and efficiency in the winemaking process.

Despite the achievements already made, we are also exploring the potential use of the latest molecular proteomics techniques to unravel the biological component of the complex winemaking processes. Proteomics data collected to date strongly suggest that these techniques are potentially very useful for controlling the fermentation process and for assuring the quality of the finished wine; they offer excellent prospects for improving these processes in the near future.

6. Acknowledgements

This work was supported by grants PETRI 95-0855 OP from the DGICYT of the Ministry of Science and Innovation, and OT 054/174/015/020/114/136/104 from Bodegas Barbadillo S.L. of Sanlúcar de Barrameda, Spain, and CDTI-IDI-20101408.

7. References

Ambrona, J., Vinagre, A., Maqueda, M., Álvarez, M. L., Ramirez, M. (2006). Rhodamine-pink as a genetic marker for yeast populations in wine fermentation. *Journal of Agricultural and Food Chemistry*, Vol. 54, pp. 2977-2984, ISSN 0021-8561

Carle, G. F., Olson, M. V. (1985). An electrophoretic karyotype for yeast. *Proceedings of the National Academy of Sciences of the United States of America*, Vol. 82, pp. 3756-3760, ISSN 0027-8424

Chu, G., Vollarath, D., Davis, R. W. (1986). Separation of large DNA molecules by contour-clamped homogeneous electric field. *Science*, Vol. 234, pp. 1582-1585, ISSN 0036-8075

Codón, J. M., Benítez, T., Korhola, M. (1998). Chromosomal polymorphism and adaptation to specific industrial environments of *Saccharomyces* strains. *Applied Microbiology and Biotechnology*, Vol. 49, pp. 154-163, ISSN 0175-7598

Cordero-Bueso, G., Arroyo, T., Serrano, A., Tello, J., Aporta, I., Vélez, M. D., Valero, E. (2011). Influence of the farming system and vine variety on yeast communities associated with grape berries. *International Journal of Food Microbiology*, Vol. 145, pp. 132-139, ISSN 0168-1605

Demuyter, C., Lollier, M., Legras, J. L., Le Jeune, C. (2004). Predominance of *Saccharomyces uvarum* during spontaneous alcoholic fermentation, for three consecutive years, in an Alsatian winery. *Journal of Applied Microbiology*, Vol. 97, pp. 1140-1148, ISSN 1364-5072

Esteve-Zarzoso, B., Gostíncar, A., Bobet, R., Uruburu, F., Querol, A. (2000). Selection and molecular characterization of wine yeast isolated from the `El Penedès´ area (Spain). *Food Microbiology*, Vol. 17, pp. 553-562, ISSN 0740-0020

Fernández-Acero, F. J., Carbú, M., Garrido, C., Vallejo, I., Cantoral, J. M. (2007). Proteomic Advances in Phytopathogenic Fungi. *Current Proteomics*, Vol. 4, No. 2, pp. 79-88, ISSN 1570-1646

Fernández-Acero, F. J., Carbú, M., El-Akhal, M. R., Garrido, C., González-Rodríguez, V. E., Cantoral, J. M. (2011). Development of Proteomics-Based Fungicides: New Strategies for Environmentally-Friendly Control of Fungal Plant Diseases. *International Journal of Molecular Sciences*, Vol. 12, pp. 795-816, ISSN 1422-0067

Fernández-Espinar, M. T., Martorell, P., De Llanos, R., Querol, A. (2006). Molecular methods to identify and characterize yeasts in foods and beverages. In: Querol, A., Fleet, G. H. (Eds), The Yeast Handbook: Yeast in Food and Beverages, Springer, pp. 55-82

Fleet, G.H. (2008). Wine yeasts for the future. *FEMS Yeast Research*, Vol. 8, pp. 979-995, ISSN 1567-1356

Garrido, C., Cantoral, J. M., Carbú, M., González-Rodríguez, V. E., Fernández-Acero, F. J. (2010). New Proteomic Approaches to Plant Pathogenic Fungi. *Current Proteomics*, Vol. 7, pp. 306-315, ISSN 1570-1646

González, S. S., Barrio, E., Querol, A. (2007). Molecular identification and characterization of wine yeast isolated from Tenerife (Canary Island, Spain). *Journal of Applied Microbiology*, Vol. 102, pp. 10185-1025, ISSN 1364-5072

Infante, J.J., Dombek, K.M., Rebordinos, L., Cantoral, J.M. & Young, E.T. (2003). Genome-wide caused chromosomal rearrangements play a major role in the adaptive evolution of natural yeast. *Genetics*, Vol. 165, pp. 1745-1759, ISSN 0016-6731

Le Jeune, C., Lollier, M., Demuyter, C., Erny, C., Legras, J. L., Aigle, M., Masneuf-Pomarède, I. (2007). Characterization of natural hybrids of *Saccharomyces cerevisiae* and *Saccharomyces bayanus* var. *uvarum*. FEMS Yeast Research, Vol. 7, pp. 540-549, ISSN 1567-1356

Lopes, C. A., Rodríguez, M. E., Sangorrín, M., Querol, A., Caballero, A. C. (2007). Patagonian wines: implantation of an indigenous strain of Saccharomyces cerevisiae in fermentations conducted in traditional and modern cellars. *Journal of Industrial Microbiology & Biotechnology*, Vol. 34, pp. 139-149, ISSN 1367-5435

Lopez, V., Frenández-Espinar, M. T., Barrio, E., Ramón, D., Querol, A. (2003). A new PCR-based method for monitoring inoculated wine fermentations. *International Journal of Food Microbiology*, Vol. 81, pp. 63-71, ISSN 0168-1605

Martínez, M., Gac, S., Lavin, A., Ganga, M. (2004). Genomic characterization of *Saccharomyces cerevisiae* Straits isolated from wine-producing areas in South America. Journal of Applied Microbiology, Vol. 96, pp. 1161-1168, ISSN 1364-5072

Mesa, J. J., Infante, J. J., Rebordinos, L., Cantoral, J. M. (1999). Characterization of yeast involved in the biological ageing of Sherry wines. *LWT Food Science and Technology*, Vol. 32, pp. 114-120, ISSN 0023-6438

Mesa, J. J., Infante, J. J., Rebordinos, L., Sánchez, J. A., Cantoral, J. M. (2000). Influence of the yeast genotypes on enological characteristics of Sherry wines. American Journal of Enology and Viticulture, Vol. 51, 1, pp. 15-21, ISSN 0002-9254

Mortimer, R.K. (2000). Evolution and variation of the yeast (*Saccharomyces*) genome. *Genome Research*, Vol. 10, pp. 403-409, ISSN 1088-9051

Muñoz-Bernal, E., Fernández-Acero, F. J., Rodríguez, M. E., Cantoral, J. M. (2011). Variación en el perfil 2DE en *Saccaromyces bayanus* var. *uvarum* inducido por temperatura. Análisis diferencial de proteoma. XXIII Congreso Nacional de Microbiología, Salamanca (Spain), 11-14 July 2011

Naumov, G. I., Masneuf, I., Naumova, E. S., Aigle, M., Dubourdieu, D. (2000). Association of *Saccharomyces bayanus* var. *uvarum* with some French wines: genetics analysis of yeast populations. *Research in Microbiology*, Vol. 151, pp. 683-691, ISSN 0923-2508

Naumov, G. I., Naumova, E. S., Antunovics, Z., Sipiczki, M. (2002). *Saccharomyces bayanus* var. *uvarum* in Tokaj wine-making of Slovakia and Hungary. *Applied Microbiology and Biotechnology*, Vol. 59, pp. 727-730, ISSN 0175-7598

Raspor, P., Cus, F., Povhe Jemec, K., Zagorc, T., Cadez, N., Nemanic, J. (2002). Yeast population dynamics in spontaneous and inoculated alcoholic fermentations of *Zametovka* must. *Food Technology and Biotechnology*, Vol. 40, pp. 95-102, ISSN 1330-9862

Rodríguez, M.E., Infante, J.J., Molina, M., Domínguez, M., Rebordinos, L. & Cantoral, J.M. (2010). Genomic characterization and selection of wine yeast to conduct industrial fermentations of a white wine produced in a SW Spain winery. *Journal of Applied Microbiology*, Vol. 108, pp. 1292-1302, ISSN 1364-5072

Rodríguez, M.E., Infante, J.J., Molina, M., Rebordinos, L. & Cantoral J.M. (2011). Using RFLP-mtDNA for the rapid monitoring of the dominant inoculated yeast strain in industrial wine fermentations. *International Journal of Food Microbiology*, Vol. 145, pp. 331-335, ISSN 0168-1605

Rossignol, T., Kobi, D., Jacquet-Gutfreund, L. & Blondin, B. (2009). The proteome of a wine yeast strain during fermentation: correlation with the transcriptome. *Journal of Applied Microbiology*, Vol. 107, pp. 47-55, ISSN 1364-5072

Salvadó, Z., Chiva, R., Rodríguez-Vargas, S., Rández-Gil, F., Mas, A., Guillamón, J. M. (2008). Proteomic evolution of a wine yeast during the first hours of fermentation. *FEMS Yeast Research*, Vol. 8, pp. 1137-1146, ISSN 1567-1356

Santamaría, P., Garijo, P., López, R., Tenorio, C., Gutiérrez, A. R. (2005). Analysis of yeast population during spontaneous alcoholic fermentation: Effect of the age of the cellar and the practice of inoculation. *International Journal of Food Microbiology*, Vol. 103, pp. 49-56, ISSN 0168-1605

Schuller, D., Valero, E., Dequin, S. & Casal, M. (2004). Survey of molecular methods for the typing of wine yeast strains. *FEMS Microbiology Letters*, Vol. 231, pp. 19-26, ISSN 1567-1356.

Schuller, D., Casal, M. (2007). The genetic structure of fermentative vineyard-associated *Saccharomyces cerevisiae* populations revealed by microsatellite analysis. *Antonie van Leeuwenhoek*, Vol. 91, pp. 137-150, ISSN 0003-6072

Trabalzini, L., Paffetti, A., Scaloni, A., Talamo, F., Ferro, E., Coratz, G., Bovalini, P., Santucci, A. (2003). Proteomic response to physiological fermentation stresses in a wild-type wine strain of *Saccharomyces cerevisiae*. *Biochemistry Journal*, Vol. 370, pp. 35-46

Viana, F., Gil, J. V., Vallés, S., Manzanares, P. (2009). Increasing the levels of 2-phenyl acetate in wine through the use of a mixed culture of *Hanseniaspora osmophila* and *Saccharomyces cerevisiae*. *International Journal of Food Microbiology*, Vol. 135, pp. 68-74, ISSN 0168-1605

Zuzuarregui, A., Monteoliva, L., Gil, C., del Olmo, M. (2006). Transcriptomic and proteomic approach for understanding the molecular basis of adaptation of *Saccharomyces cerevisiae* to wine fermentation. *Applied and Environmental Microbiology*, Vol. 72, No. 1, pp. 836-847, ISSN 0099-2240

2

Proteomics in Seaweeds: Ecological Interpretations

Loretto Contreras-Porcia and Camilo López-Cristoffanini
Universidad Andrés Bello, Faculty of Ecology and Natural Resources
Department of Ecology and Biodiversity, Santiago
Chile

1. Introduction

Macro and micro-algae are fundamental components of coastal benthic ecosystems and are responsible for a large part of the coastal primary production (Lobban & Harrison, 1994). Adverse effects on these groups caused by natural or anthropogenic phenomena, can affect directly or indirectly organisms of higher trophic levels and the integrity of entire ecosystems. In this context, both the ecological and economic importance of many algal species justifies the need to expand our knowledge on the molecular biology of these organisms.

The distribution and abundance of algal species occurring in the marine zone results from the interplay of biotic (i.e. competition and herbivore pressure) and abiotic (i.e. tolerance to extreme and fluctuating environments) factors (Abe et al., 2001; Burritt et al., 2002; Davison & Pearson, 1996; Pinto et al., 2003; van Tamelen, 1996). For example, the distribution of macroalgal species at the upper limit of the rocky intertidal zone is principally determined by abiotic factors such as UV radiation, light, salinity, temperature changes, nutrient availability and desiccation (e.g. Aguilera et al., 2002; Burritt et al., 2002; Cabello-Pasini et al., 2000; Contreras-Porcia et al., 2011a; Véliz et al., 2006). On the other hand, the microalgae diversity is maintained by a combination of variable forces - environmental oscillations (e.g. habitat instability), more severe disturbances and recovery from catastrophic forcing - backed by the powerful dispersive mobility of this group (Reynolds, 2006). The richness, relative abundance and occasional dominances of the phytoplankton in successive years, depends on water movements, thermal stress and carbon fluxes, but mainly on nutrient enrichment of the sea (Hodgkiss & Lu, 2004; Holm-Hansen et al., 2004; Reynolds, 2006; Wang et al., 2006; Zurek & Bucka, 2004).

Superimposed on the natural abiotic oscillations, algae are also exposed to various other sources of stress, particularly those resulting from human industrial, urban and agricultural activities. Among these is copper mining, whose wastes have reportedly caused severe and negative effects on the coasts of England (Bryan & Langston, 1992), Canada (Grout & Levings, 2001; Marsden & DeWreede, 2000), Australia (Stauber et al., 2001) and Chile (Correa et al., 1999). Although copper is a micronutrient for plants and animals, occurring naturally in coastal seawater at levels at or below 1 μg L^{-1} (Apte & Day, 1998; Batley, 1995; Sunda, 1989), at higher concentrations it becomes highly toxic. The phenomenon of toxicity in algae is strongly influenced by the speciation of this metal (Gledhill et al., 1997), and within the cell it likely operates through the Haber-Weiss reaction, characterized by a heavy metal-catalyzed production of hydroxyl radicals from hydrogen peroxide (Baker & Orlandi, 1995). For example, in northern Chile, mine wastes originated at a copper mine pit are disposed of directly into the sea. The rocky intertidal zone along the impacted coasts shows a severe reduction in species richness, and the macroalgal assemblage is reduced to the opportunistic algae *Ulva compressa* (Plantae, Chlorophyta) and *Scytosiphon lomentaria* (Chromista, Ochrophyta) (Medina et al., 2005). This negative effect on the biota has been widely recognized as the result of the persistent high levels of copper in the water, by far the most important metal brought into the system by mine wastes (Medina et al., 2005). Many macroalgae species are absent, such as *Lessonia nigrescens* complex (Chromista, Ochrophyta), which are key components in structuring the intertidal zone (Ojeda & Santelices, 1984). As for microalgae, an example is a mine effluent that contained high levels of copper, which was disposed in a reservoir named Venda Nova in northern Portugal. There, a phytoplankton survey was carried out between the years 1981-1982. A shift in the dominant species was demonstrated when compared with an uncontaminated area, Alto Rabagão. More than 50% of the algal species developed lower populations. Also, at the most polluted zone, phytoplankton density, biomass and richness were strongly reduced (Oliveira, 1985).

In macro and micro-algae it is possible to determine that under natural abiotic factors, a common cellular response could involve the over-production of reactive oxygen species (ROS) (Andrade et al., 2006; Contreras et al., 2005, 2007b, 2009; Contreras-Porcia et al., 2011a; Kumar et al., 2010; Lee & Shin, 2003; Liu et al., 2007; Rijstenbil, 2001). ROS are ubiquitous by-products of oxidative metabolism that are also involved in intracellular signalling processes (e g. Blokhina & Fagerstedt, 2010; Rhee, 2006). ROS are produced directly by the excitation of O_2 and the subsequent formation of singlet oxygen, or by the transfer of one, two or three electrons to O_2. This results in the formation of superoxide radicals, hydrogen peroxide or hydroxyl radicals, respectively (Baker & Orlandi, 1995). Oxidative damage to cellular constituents such as DNA/RNA, proteins and lipids may occur (e g. Contreras et al., 2009; Vranová et al., 2002) when ROS levels increase above the physiological tolerance range. However, a coordinated attenuation system can be activated in order to eliminate this ROS over-production, and therefore, the oxidative stress condition (e. g. Burritt et al., 2002; Ratkevicius et al., 2003; Rijstenbil, 2001). For example, in the coastal zones of northern Chile it has been demonstrated that the high copper levels in the seawater generate in sensitive species a high oxidative stress condition, which appears as the starting point for a series of molecular defense responses. In first place, the condition of oxidative stress has been demonstrated by the direct production of ROS and oxidized lipid in individuals living at an impacted site as well as in those transplanted from control sites to the impacted site (Contreras et al., 2005; Ratckevicius et al., 2003). Compared with high tolerant species such as *Ulva* and *Scytosiphon,* in low tolerant species such as *L. nigrescens* the ROS production by copper, specifically superoxide anions, is poorly attenuated, which is reflected in i) higher levels of oxidized lipids, ii) the generation of cellular alterations and iii) negative effects on early developmental stages of the life cycle (Andrade et al., 2006; Contreras et al., 2007a; 2009). Thus, ecophysiological differences are evident between diverse algal species. This is also true for microalgal species since there are species-specific responses to oxidative stress caused by high levels of copper. For example, it was demonstrated that 4 species of phytoplankton under high concentrations of copper only grew up to 80-95% of that observed in the control condition (Bilgrami & Kumar, 1997). Furthermore, a study including two microalgae species exposed to copper stress showed significant differences between them. In the high tolerant species, *Scenedesmus vacuolatus,* in comparison to the low tolerant species, *Chlorella kessleri,* the chlorophyll a/chlorophyll b ratio was partially reduced. Likewise, both the antioxidant enzyme activity and protein content were progressively increased (Sabatini et al., 2009).

Another environmental factor that affects the abundance and distribution in macroalgae is desiccation. It is an important stress factor faced by living organisms because, as cells lose water, essential macromolecules are induced to form non-functional aggregates and organelles collapse (Alpert, 2006). Some animals (Clegg, 2005) and plants are well adapted to significant water losses, displaying full physiological recovery during rehydration (Alpert, 2006; Farrant, 2000). Compared to vascular plants or animals, in macroalgae the effects of desiccation on the physiology and the molecular mechanisms involved in its tolerance are poorly understood. For example, in one of the few reports available, the activation of different antioxidant enzymes, such as ascorbate peroxidase (AP) and glutathione reductase (GR) was recorded in the upper intertidal macroalga *Stictosiphonia arbuscula* (Plantae, Rhodophyta) (Burritt et al., 2002) as a response to desiccation-mediated oxidative stress. The remaining studies have focused on assessing the capacity to tolerate desiccation displayed by measuring the photosynthetic apparatus activity in *Porphyra, Gracilaria, Chondrus,* and *Ulva* species among others (Abe et al., 2001; Ji & Tanaka, 2002; Smith et al., 1986; Zou & Gao, 2002). Presently, the only study using molecular approaches to unravel the desiccation tolerance responses, found that genes encoding for photosynthetic and ribosomal proteins are up-regulated in *Fucus vesiculosus* (Chromista, Ochrophyta) (Pearson et al., 2001, 2010). Additionally, independent studies have shown that diverse physiological parameters are altered by desiccation including the lipid and protein levels (Abe et al., 2001), photosynthetic alterations (*Fv/Fm*) as well as cellular morphology and ontogenetic changes (e.g. Contreras-Porcia et al., 2011b; Varela et al., 2006). Moreover, in microalgae it has been shown that salt (i.e. changes in water osmolarity) and temperature stress can be highly stressful and may finally trigger a programmed cell death (PCD) (Kobayashi et al., 1997; Lesser, 1997; Takagi et al., 2006; Zuppini et al., 2010). In these species the effects of both types of stress have been widely studied, and have been reported to provoke photosynthetic alterations, ROS production and ultimately cell death (Liu et al. 2007; Lesser, 1996; Mishra & Jha, 2011; Vega et al., 2006).

Recently, the red species *Porphyra columbina* Montagne (Plantae, Rhodophyta) was recognised among the macroalgae that are highly tolerant to natural desiccation stress. *P. columbina* is highly seasonal and grows abundantly along the upper intertidal zone (Hoffmann & Santelices, 1997; Santelices, 1989). This alga is well adapted to the extreme fluctuating regimes of water/air exposure, as demonstrated by the formation of sporophytic thalli from monoecious fronds (*n*) during long daily periods of desiccation stress due to its position in the intertidal zone (Contreras-Porcia et al., 2012). Additionally, desiccation in *P. columbina* induces morphological and cellular alterations accompanied by a loss of ca. 96 % of the water content (Contreras-Porcia et al., 2011b). Specifically, under natural desiccation stress, the production of ROS (i.e. H_2O_2 and O_2) in *P. columbina* is significantly induced (Contreras-Porcia et al., 2011b). However, during the high tide, ROS quickly returned to basal levels because *P. columbina* displays an efficient antioxidant system. In addition, at biomolecular level, only a low production of oxidized proteins is recorded during desiccation, due to the efficient antioxidant system of this alga.

The results mentioned above, indicate that desiccation in *P. columbina* causes an over-production of ROS, which is efficiently attenuated. Morphological and photosynthetic changes could be operating as tolerance mechanisms, due to the fact that these responses principally prevent biomolecular alterations, protein aggregation and cellular collapse. For example, it has been proposed that cell wall folding is a cellular strategy used to prevent tearing the plasmalemma from the cell wall during desiccation, ensuring cell integrity (Contreras-Porcia et al., 2011b). The activation of antioxidant enzymes and the photoinhibition of the photosynthetic apparatus help to explain the attenuation of ROS. Thus, ROS excess is buffered by the activation of several physiological and biochemical responses, which suggest a mechanism allowing this plant to tolerate desiccation (Contreras-Porcia et al., 2011b). The ecophysiological responses in this species help, in part, to account for its position and dominance at the highest level in the intertidal zone, and thereby, suggesting desiccation stress tolerance as a determinant trait for explaining that situation. In fact, our recent results demonstrate that the magnitude of the effects generated by desiccation in algae is related to the position of the species in the intertidal zone. Additionally, this work demonstrated the exceptional metabolism of *P. columbina* used to buffer this stress condition. Thus, the determinations of novel metabolic pathways are necessaries in order to fully understand the high desiccation tolerance in this species, for example at the proteomic level. In fact, in this time our forces are concentrated in resolving the proteomic profile of this species under natural hydration and desiccation stress.

Finally, the need to unravel the mechanisms associated with tolerance to different environmental factors by algal species opens the electrophoretic and proteomic approximations as important tools in comprehending and explaining the observed tolerances. However, little information regarding electrophoretic and proteomic analysis is available in algal species. Compared with other group of organisms (e.g. vascular plant or animals) protein extraction in macroalgae has been extraordinary difficult, due principally to the limited knowledge at biochemical and molecular levels. In this context, the present chapter aims to understand the different proteomic approaches utilized in this group of organisms in order to comprehend their ecophysiological behaviour.

2. Proteomic methodology in micro and macroalgae

Sample preparation, in particular the quality of protein extraction, is critical to the successful resolution of 2-DE patterns. In fact, when protein extraction protocols from higher plants are applied to algae, the 2-DE resolution is reduced (Contreras et al., 2008; Hippler et al., 2011). Due to the large variation in cellular biochemical composition among diverse organisms, which affects solubility and recovery of a complex mixture from the sample, there are no 2-DE sample preparation protocols accurate for all organisms. In macro and microalgae the protein extraction protocol must be optimized, due to the high concentration of photosynthetic pigments that are known to interfere with the resolution of the 2-DE gels (e.g. Contreras et al., 2008; Wang et al., 2003; Wong et al., 2006). Particularly, in macroalgae protein extraction is difficult due to a low concentration and the co-extraction of contaminants such as anionic polysaccharides, polyphenols and salts, which are highly concentrated in the tissue (Chinnasamy & Rampitsch, 2006; Cremer & Van de Walle, 1985;

Flengsrub & Kobro, 1989; Mechin et al., 2003). These contaminants pose a significant difficulty for 2-DE, as they cause horizontal and vertical streaking, smearing and a reduction in the number of distinctly resolved protein spots. Thus, the selection of the most appropriate protein extraction method is necessary in order to obtain high quality extracts, and therefore, a high quality 2-DE pattern. For a better understanding and explanation of the current techniques and methodology in algae proteomic, this chapter has been divided in two sections: microalgae and macroalgae methodology.

2.1 Microalgae methodology

It is important to highlight that due to the small size of microalgae, all of the protein extraction protocols for these organisms begin with a centrifugation step in order to pellet cells. This helps to concentrate cells, and consequently allows a correct extraction of the desired proteins.

2.1.1 Early proteomic studies

One of the first proteomics studies on microalgae dates from the year 1972, in which Mets & Bogorad showed alterations in the chloroplast ribosomes proteins of erythromycin-resistant mutants of *Chlamydomonas reinhardtii* (Plantae, Chlorophyta) compared to the wild-type. The ribosomal protein extraction performed on this work was the LiCl-urea method described by Leboy et al. (1971) that was developed for *Escherichia coli* (as cited in Mets & Bogorad, 1972). Thus, the Mets & Bogorad work was a precursor to microalgae proteomic studies. Here, ribosomes are disrupted and freed of RNA by adding LiCl. Then, the samples are centrifuged to precipitate total RNA and the supernatant, which contains the proteins, is retained.

Several studies in the same decade also focused their attention on characterizing ribosomal proteins (e.g. Götz & Arnold, 1980; Hanson et al., 1974). The Hanson et al. (1974) work based their protocols on the Mets & Bogorad (1972) research paper and also used *C. reinhardtii* as model species. Instead, in 1980 Götz & Arnold used a different ribosomal protein extraction after testing several protocols. The procedure chosen was the acetic-acid method in presence of $MgCl_2$ according to Kaltschmidt & Wittmann (1972), method that was also first developed for *E. coli*. In this method, $MgCl_2$ and glacial acetic acid are added to the ribosome suspension, and then the mixture is centrifuged to pellet RNA. For better mixture cleaning, the pellet can be extracted a second time in the same way.

Not all studies from this decade focused their attention on ribosomal proteins as was the case of the work of Piperno et al. (1977), in which the protein mixture came from *Chlamydomona* flagella and axonemes. It is important to highlight this research since the extraction method used was very rustic. After the flagella and axoneme separation, the proteins were dissolved only in SDS and kept for 2-DE analysis.

2.1.2 Current proteomic studies

Recent studies evaluate more complex protein mixtures, so the method chosen must be more accurate in extracting proteins with minimum contaminants and interferents. In fact, a work in *C. reinhardtii* that performed an analysis of all the thylakoid membranes proteins used a more complex protocol (Hippler et al., 2001) than the ones previously discussed in this chapter. This method uses methanol in order to precipitate cell debris and retains proteins in the supernatant. Then, chloroform is added and the sample is vortexed and centrifuged. The upper phase containing DNA is discarded. Afterwards, methanol is added to the sample in order to pellet proteins and leave the RNA in the aqueous phase. Finally, the pellet is washed with methanol in order to remove contaminants.

In 2003, a study tested different protein extraction protocols in the microalga *Haematococcus pluvialis* (Plantae, Chlorophyta) in order to determine which ones yielded better results (Wang et al., 2003). After cell disruption, the samples were dialysed to remove any salt left in the samples, which are known to interfere in the IEF step. After the dyalisis, each sample was treated in three different ways: i) proteins were left to precipitate in a non-denaturing preparation, ii) a mixing of dialysate with acetone was kept at -20 °C o/n to allow complete precipitation and iii) a mixing of dialysate with TCA in acetone containing β-mercaptoethanol also kept at -20°C o/n. Methods ii) and iii) were denaturing procedures but it was procedure iii) the one that yielded 2-DE gels with higher resolution (detailed protocol in Appendix A).

The work by Kim et al. (2005) is interesting since the protein extraction protocol used is relatively simple when compared to others (e.g. Wang et al., 2003; Contreras et al., 2008) (detailed protocol in Appendix A). Proteins of *Nannochloropsis oculata* (Chromista, Ochrophyta) are obtained in very short time compared to the other microalgae protocols, however, not with the same quality as the more complex protocols. In another *C. reinhardtii* work, but this time conducting a whole cell proteomic study (Förster et al., 2006), a protocol described by Mathesius et al. (2001) that is suited for root proteins was used (detailed protocol in Appendix A). This procedure is denaturing and relatively simple, but includes washing steps that help to improve the quality of the protein extracts compared to the one used on *N. oculata* (Kim et al., 2005). A work from 2009 in the microalga *Haematococcus lacustris* also had a denaturing protocol in which pelleted cells were grounded to a fine powder in liquid nitrogen (Tran et al., 2009). Then, they are disrupted with a lysis buffer containing urea, thiourea, DTT, CHAPS, Tris-base and a plant protease inhibitor cocktail tablet. Samples are centrifuged to separate cell debris, and then the pellet is resuspended in acetone to precipitate proteins and remove contaminants. Finally, the samples are again centrifuged, acetone is removed by air-drying and pellet is clean and ready for 2-DE gels.

Chlamydomonas reinhardtii is one of the most studied microalgae worldwide and as noted in this chapter, proteomics studies are no exception. Another protocol for this algae dates from 2011, in this one the cells are disrupted with a lysis buffer containing urea, CHAPS and thiourea (Mahong et al., 2012). The sample is centrifuged and the supernatant retained. To eliminate possible photosynthetic pigments and other hydrophobic compounds, the samples are washed with ice-cold acetone. Then, each sample is centrifuged, and the pellet is ready for electrophoretic processes.

2.1.3 Gel loading: From proteins to gels

Another key step in obtaining 2-DE gels is gel loading and gel running. After protein extraction, the pellet must be resuspended in a rehydration buffer, which is generally the same in all works. Then, proteins are loaded in order to perform the IEF step for their correct horizontal migration, however, the protocols varied according both to the biological model and the protein type extracted (i.e. soluble or membrane proteins). Finally, proteins separated in the IEF step are loaded in to the second dimension (SDS-PAGE). Thus, in this section rehydration buffers, IEF steps and second dimension gels will be analyzed.

2.1.3.1 Early proteomic studies

In the work of Mets & Bogorad (1972), ribosomal proteins were only run in the IEF step at 1.5 mA for 4 h but it was enough to separate them due to the low quantity of proteins that were obtained in this extraction. The second dimension was run at 25 mA, enough time to allow the protein migration, since the 2-DE gel patterns are very clear and well resolved. Also, no vertical or horizontal streaking is present, thereby, permitting clear protein detection. It is not astonishing to observe similar 2-DE patterns in quality terms in the work by Hanson et al. (1974), since both the ribosomal protein extraction and the two-dimensional gel electrophoresis were performed essentially as described by Mets & Bogorad (1972). Therefore, no vertical or horizontal streaking was found, resulting in gels with high resolution. Both protein extraction and gel electrophoresis proved to be very efficient and adequate for protein separation. However, it should be emphasized that the patterns from both works are easier to obtain, since the protein mixture is very simple since it only came from ribosome structures.

Unlike the ribosomal protein mixture, others do not generate 2-DE patterns with the same resolution. One case may be flagella and axonemes of *C. reinhardtii* in which a larger number of proteins are founded. Piperno et al. (1977) compared proteins of this structure from both wild-type and paralyzed mutants strains of this species. The IEF step was performed at 300 V for 18-19 h and followed by 400 V for 1.5 h. The second dimension was first run at 25 mA (initial voltage: 60 V) for 1 h and then it was raised to 50 mA. The run continued until the dye in the molecular weight standard had reached the bottom according to Ames and Nikaido (1976) (as cited in Piperno et al., 1977). The 2-DE gels had minimum vertical streaking, but lot of horizontal streaking and big stains regardless of the sample. The horizontal streaking could be due to a more complex protein mixture; however, the protein extraction protocol of this work is very deficient since it only uses SDS. Regardless of this, some spots were easily detected in the gels allowing for comparison between wild-types and mutant strains. Finally, in the work of Götz & Arnold (1980) ribosomal proteins from eight species were evaluated with two gels showing clear and well-resolved 2-DE patterns. The protein extraction was well suited for all species. Therefore, the $MgCl_2$-acetic acid method proved efficient in a large number of species, but again it was used to extract only ribosomal protein, so minimum contaminants are present.

2.1.3.2 Current proteomic studies

In more recent papers, such as those described in the previous section, the rehydration buffer used to resuspend the proteins prior to gel loading is key for the proper migration of proteins. The most commonly used buffer contains the reagents thiourea, urea, CHAPS, DTT, ampholytes and bromophenol blue. However, the concentrations of the reagents vary among the different works, so choosing the most accurate one is no easy task. As an example, we chose the protocol described by Wang et al. (2003) in which several reagents were tested to determine which one that yielded the best 2-DE pattern (i.e. no streaking and more defined spots) (see Appendix A). The majority of researchers state in their works that after resuspending the proteins, the mixture must be left at room temperature for at least 1 h (e.g. Hippler et al., 2001; Kim et al., 2005; Tran et al., 2009). Likewise, the amount of proteins normally loaded is 500 μg, concentration enough to yield well resolved gels (e.g. Förster et al., 2006; Mahong et al., 2012; Wang et al., 2003).

The IEF profile contains several steps, which vary between the different works, so making comparisons is complicated and not very productive. Nowadays, researchers worldwide use IPG gel strips for a better protein migration, which leads to a better 2-DE pattern (e.g. Mahong et al., 2012; Wang et al., 2003). Having said that, all IPG gel strips must be first rehydrated for at least 10 h before setting the IEF profile. As an example we chose the IEF profile of Wang et al. (2003) which was initiated at 250 V for 15 min, and gradually ramped to 10,000 V over 5 h, and remained at 10,000 V for an additional 6 h.

After the IEF steps and prior to the second dimension, IPG gel strips must be incubated twice in an equilibration buffer containing Tris-HCl, urea, glycerol and SDS. The first time DTT is added to the equilibration buffer in order to denaturate proteins, whereas the second time iodoacetamide is added to alkylate the reduced cysteines and inhibit protein refolding. After equilibration, IPG gel strips are ready to be loaded on to the second dimensional SDS-PAGE for the vertical protein separation (i.e. according to their molecular weight). Gel thickness will vary in each experiment in order to allow the desired protein separation. Regardless of this, gels are run until the bromophenol blue reaches the bottom of the gel since it migrates faster than the proteins. The last step for obtaining the 2-DE gel is gel staining in which two principal stains are used: blue Coomassie and silver nitrate. Regardless of this, generally prior to staining, the gels are washed with deionised water. After staining, the excess of dye is removed with deionised water to obtain well-defined gels with minimum background noise.

Now with the gels stained, we are able to determine which protocol(s) yielded the best 2-DE gel(s) in terms of patterns quality (i.e. minimum or none streaking, spots with defined circles, a maximum spot number and high spot intensity). In the work by Kim et al. (2005) 2-DE gel images show smearing, some vertical streaking and high horizontal streaking specifically in the acidic side of the gel. Also, spots are not well-defined circles and are overlapped among them. Similar were the image gels by Tran et al. (2009), because smearing as well as vertical and horizontal streaking are present in the acidic part of the 2-DE gel. Also several spots were overlapped among them; nevertheless a few of them were well defined. These were the two protocols that yielded the worst results (e.g. poor gel resolution quality) and this must be to the simplicity of the protein extraction protocols used. The two protocols that follow in terms of 2-DE gel quality are those of Hippler et al. (2001) and Förster et al. (2006). In both works 2-DE gels are of high quality, which obviously obey more complex protein extraction protocols. In the oldest work, there are several traits that give this images high quality: i) minimum horizontal streaking, ii) well defined spots (i.e circle shaped), iii) highly stained spots and iv) high number of spots (since only thylakoid membrane proteins were extracted) (Hippler et al., 2001). The high quality of 2-DE gels is probably due to that only a portion of the cell proteins was extracted having less contaminants interfering in both IEF and second dimension. Förster et al. (2006) 2-DE gel images show a high number of spots and most of them are well define with almost no smearing. However, a lot of vertical streaking is observed in the gels, thus the problems must be found in the second dimension since minimum horizontal streaking is present.

Finally, the protocols that yielded the 2-DE images of higher quality were those developed by Wang et al. (2003) and Mahong et al. (2012). In both works, total proteins were extracted from two different microalgae, *H. pluvialis* and *C. reinhardtii* respectively. Highlight that both protocols are the most complex ones among all six analyzed. Gels from both works succeeded in having reduced streaking as well as defined, highly stained and high number of spots. Nevertheless, if one must choose between both, it is Mahong et al. (2012) protocol the one with the best results since gels in this work have minimum background allowing an easier spot detection.

2.2 Macroalgae methodology

The difficulty in obtaining high quality 2-DE gels from macroalgae was first highlighted by Wong et al. (2006), who obtained algal proteins from *Gracilaria changii* (Plantae, Rhodophyta) using four different extraction methods: 1) direct precipitation by trichloroacetic acid/acetone, 2) direct lysis using urea buffer, 3) tris buffer and 4) phenol/chloroform. However, only methods 3) and 4) were compared for their suitability to generate *G. changii* proteins for two-dimensional gel electrophoresis. It was stated in this work that the phenol/chloroform method (detailed protocol in Appendix B) was Ideal for obtaining well resolved 2-DE patterns. Nevertheless, the quality of the 2-DE profiles was poor due to the presence of high amounts of interfering substances accompanied by low protein yield and horizontal and vertical streaking along gels regardless the pH gradient. Thus, this method is not fully accurate for this algal species.

As part of an on-going work focused on unravelling the metabolic processes occurring in physiologically stressed brown macroalgae, a new method for protein extraction that minimizes the co-extraction of non-protein compounds using two structurally distinct brown algal species *Scytosiphon gracilis* (Chromista, Ochrophyta) (Contreras et al., 2007b) and *Ectocarpus siliculosus* (Chromista, Ochrophyta) (Contreras et al., 2008) was developed. In order to do this, several protein extraction methods available in the literature were tested. However, neither of the previous protocols was ideal for obtaining a good quality algal protein extraction, due to high background noise, band distortion, and more importantly, very low protein dissolution. The protocol developed in this work allowed the use of a highly resolving 2-DE protein analyses, providing the opportunity to unravel potentially novel physiological processes unique to this group of marine organisms (see Table 1 and Results section). Specifically, the protocol uses an initial desalting step with Milli Q water - phosphate buffer in order to remove the salt from the algal tissues. Afterwards, the tissue is pulverized using liquid nitrogen and homogenized with sucrose, EDTA and CHAPS. The proteins are extracted using phenol and washed with ammonium acetate. Finally, the quality of the extracted proteins is improved by using the 2-D clean-Up Kit (GE Healthcare).

In another important proteomic work with macroalgae developed by Kim et al. (2008) and published contemporarily with the Contreras et al. (2008) work, using as models the red algae *Bostrychia radicans* and *B. moritziana* (Plantae, Rhodophyta), used a lysis buffer comprised principally by urea and thiourea (detailed protocol in Appendix B). Although these species belong to the same group of red algae like *G. changii*, the simplicity of this method utilized in comparison with the phenol one (Wong et al., 2006) is due to the morphological characteristics of this species (see image in Appendix B).

The work described by Yotsukura et al. (2010) presents a similar protocol to the one described by Contreras et al. (2008). Here, proteins are extracted from the brown alga *Saccharina japonica* (Chromista, Ochrophyta), important kelp described principally on the coastal areas of northern Japan. In this protocol, the protein extraction was improved by using phenol as the principal component in the lysis buffer (detailed protocol in Appendix B). This protocol was also used in *Ecklonia cava* (Chromista, Ochrophyta), other important kelp found on the coast along the Sea of Japan, and also good quality 2-D patterns were obtained (Yotsukura et al., 2012). The use of phenol in the protein extraction described by Contreras et al. (2008) has been recently used in the red alga *Porphyra columbina* in order to identify the proteins that are over-induced during desiccation stress tolerance responses. A highly resolved 2-DE protein was obtained using this method (Fig. 1), with minor modifications (detailed protocol in Appendix B), such as an important rinse of the protein pellet due principally to the over-production of phycocyanin and phycoerythrin. Thus, the phenol protocol developed by Contreras et al. (2008) could be used in macroalgae species from different taxonomic groups.

The first dimension of the 2-DE in the works mentioned above used approximately 200-500 µg of extracted proteins. However, for the isoelectric focusing (IEF) the protocols varied depending on the algal species used. For example, in *Bostrychia radicans* and *B. moritziana*, the voltage was linearly increased from 150-3,500 V during 3 h, followed by a constant 3,500 V, with focusing complete after 96 V. In *Scytosiphon gracilis* and *Ectocarpus siliculosus*, on the other hand, the strips are actively rehydrated for 15 h in IEF buffer containing the proteins and focused at 20°C with the following successive steps: a liner increase from 0 to 250 V for 15 min, a gradient phase from 250 V to 10,000 V for 4 h, and the a hold phase at 10,000 V for a total of 60 kVh. Using this protocol, the IEF for *Porphyra columbina* has some modifications, principally in a total operational voltage of 70 kVh.

3. Results and discussion: From gel to molecular/ecological interpretation

Proteomic analyses have proved to be an important molecular approximation that enables comparisons between species and/or cell variants, and understanding of cell function and stress tolerance (e.g. metals, high salinity, high temperatures, among others) (e.g. Contreras et al., 2010; Kim et al., 2005; Ritter et al., 2010). Due to the particularity of the cellular components (e.g. high content of polysaccharides) of this group of organisms, protein extraction has been the principal problem. However, as stated in the previous sections, some protocols have proved capable of producing high quality protein extracts for 2-DE electrophoresis (microalgae: Mahong et al., 2012; Wang et al., 2003 and macroalgae: Contreras et al., 2008). A high quality protein extract will yield high-resolved 2-DE patterns. Therefore, with a suitable protocol the use of a proteomic approximation appears to be of high importance for understanding various physiological responses in this group of organisms. However, it is imperative to highlight that proteomic works in micro and principally in macroalgae, are considerably lower in comparison with vascular plants and animals. Then, our effort in this chapter was concentrated in describing those important works utilizing as model the algal assemblage.

One of the first proteomic studies in microalgae characterized the chloroplastic ribosomal proteins of wild-type and erythromycin-resistant mutants of *Chlamydomonas reinhardtii* (Mets & Bogorad, 1972). In the mutant *ery-M2d* a protein of the 52S subunit was missing when compared to the wild-type. Nevertheless, low intensity proteins spots with the same

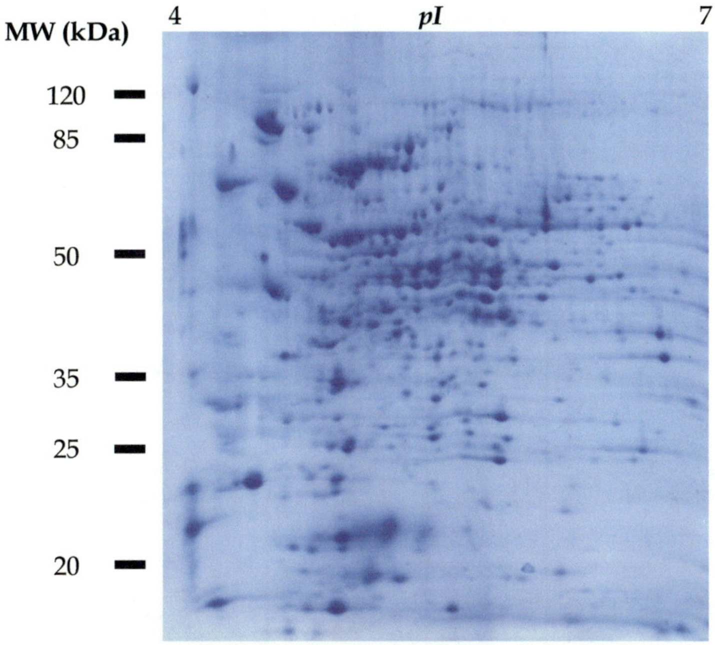

Fig. 1. 2-D proteome of *Porphyra columbina* under natural desiccation. First dimension was performed on a linear gradient IPG strip of pH 4-7 using 600 µg of total proteins. The 12.5% SDS-PAGE gel was stained with colloidal Coomassie blue.

pI but different molecular weight were found, indicating that the *ery-M2* gene is involved in determining the properties of that protein. Hanson et al. (1974) performed 2-DE gels in order to characterize cytoplasmic and chloroplastic ribosomal proteins. Their results showed that the number of proteins in both small and large subunits was higher in cytoplasmic than in chloroplastic ribosomes, indicating that cytoplasmic ribosomes are more complex. Another study did a comparison between the ribosomal proteins of 8 species including *C. reinhardtii* (Götz & Arnold, 1980). The results showed that the number of proteins in both subunits was similar among all species, and that the *Polytoma papillatun* (Plantae, Chlorophyta) proteins were the most similar to those of *C. reinhardtii* in terms of protein homology.

Piperno et al. (1977) analysed the flagella proteins of wild-type and paralyzed mutants of *C. reinhardtii* in order to identify the mutated protein that incapacitates the mobility in mutant strains. In the flagella of *pf* 14, which completely lack radial spokes and associated spokeheads, 12 polypeptides were missing. Also in *pf* 1 flagella, where spokes are clearly present but spokeheads appear to be absent, 6 polypeptides were missing. Then, protein electrophoretic studies confirmed the phenotypical characteristics displayed by both paralyzed mutants, where the missing proteins may be involved in spokes and spokeheads correct morphology. Another work in *C. reinhardtii,* used a proteomic approach to analyse photosynthetic thylakoid membrane proteins isolated from wild-type and mutant strains (Hippler et al., 2001). The two mutant strains were Δ*ycf4* (PSI-deficient) and *crd1* (which is conditionally reduced in PSI and LHCI under copper-deficiency). In this work more than 30 different LHCP spots were identified using a tandem quadrupole mass spectrometer,

Protein	Expression level[a]	n° of peptides analized[b]	Species, n° access[c]	*pI*, Mw (KDa)
Transferase	over	13	*Dechloromonas aromatica* (Q47F82)	5.6, 65
tRNA synthetase	over	24	*Helicobacter pylori* (P56126)	5.9, 60
Phosphomannomutase	over	24	*Schizosaccharomyces pombe* (Q9UTJ2)	5.3, 53
Proteosome, subunit α	over	24	*Oryza sativa* (Q10KF0)	5.6, 53
ATP synthase, subunit α	over	13	*Syntrophus aciditrophicus* (Q2LQZ7)	6.0, 67
Ribulose biphosphate carboxylase large chain	over	22	*Porphyra yezoensis* (Q760T5)	9.6, 65
Glyceraldehyde 3-phosphate dehydrogenase 1	over	22	*Gracilaria verrucosa* (P30724)	6.2, 43
Peptidase/Protease	over	19	*Methanothermobacter* (O27355)	6.3, 42
tRNA binding protein	over	23	*Anaplasma* (Q2GJX4)	6.2, 38
ATP binding protein	over	24	*Methanocaldococcus jannaschi* (Q58049)	6.4, 38
Transcriptional regulator	over	24	*Mesorhizobium loti* (CAD31581.1)	8.7, 29
Carbohydrate kinase	over	35	*Salmonella enterica* (YP_152740.1)	3, 18.2
RNA binding protein	over	28	*Bacillus phage* (P06953)	6.4, 25
ABC transporter subunit	over	40	*Theileria parva* (XP_764551.1)	8, 20.6
RNA polymerase, subunit α	over	19	*Francisella tularensis* (Q5NHU3)	9.2, 17.5
Peroxiredoxin	over	20	*Porphyra purpurea* (P51272)	9.5, 18
Chaperonine	over	10	*Caulobacter crescentus* (P48222)	5.6, 8
ABC transporter subunit	over	45	*Desulfitobacterium hafniense* (ZP_01371968.1)	6.2, 8.2
ABC transporter subunit	over	35	*Janibacter* sp. (ZP_00996449.1)	8.6, 7.3

Table 1. Proteins differentially expressed in *S. gracilis* exposed to copper excess. The analysis by MSMS allowed to obtain various protein peptides which were identified by BLASTP (NCBI). (a) Changes in expression level compared with controls. (b) Number of peptides analyzed by LC/MS/MS. (c) NCBI access number of the species with the highest identity obtained by BLASP.

thereby, permitting proteins with transmembrane domains to be separated with high resolution. Here, the results showed that LHCI spots were present on Δ*ycf4* and absent on *crd1* mutants.

Proteomics approaches have been helpful in understanding tolerance to naturally or anthropologically occurring environmental factors (e.g. high light, thermal stress and heavy metals respectively) in different species. Due to anthropological activities (e.g. industry and mining), heavy metals such as cadmium (Cd) and copper (Cu) are accumulating in the environment (Vermeer & Castilla, 1991; Medina et al., 2005). At high concentrations these metals are a source of abiotic stress, and can be highly toxic to organisms. In this matter, proteomic approaches are of high utility because they may provide new information regarding mechanisms to cope with the stress induced by the high concentration of metals. For example, in the work developed by Contreras et al. 2010, the copper-tolerance capacity of the brown algae species *Scyotsiphon gracilis* was evaluated by means of the 2-DE approximation. In this work, using the protocol previously described by Contreras et al. 2008 in the Appendix B, 19 over-expressed proteins were identified, including a chloroplast peroxiredoxin, a cytosolic phosphomannomutase, a cytosolic glyceraldehyde-3-phosphate dehydrogenase, 3 ABC transporters, a chaperonine, a subunit of the proteasome and a tRNA synthetase, among others (Table 1). The possible involvement of these over-expressed proteins in buffering oxidative stress and avoiding metal uptake in *S. gracilis* exposed to copper excess is discussed considering this proteomic information. For example, the peroxiredoxine (PRX) is an enzyme involved in the detoxification of hydrogen peroxide and fatty acid hydroperoxides (Dietz et al., 2006). In plants, *prx* transcripts increase in response to different abiotic stresses such as salinity, drought and metals (Dietz, 2003; Wood et al., 2003). Furthermore, PRX in the microalga *C. reinhardtii* and the red macroalga *Porphyra purpurea* (Plantae, Rhodophyta) have shown high similarity with plant PRXs (Baier and Dietz, 1997; Goyer et al., 2002). The expression of PRX in *C. reinhardtii* seems regulated by light, oxygen and redox state (Goyer et al., 2002). Thus, the PRX identified in *S. gracilis* may play an important role in oxidative stress buffering and in lipoperoxides detoxification. In fact, we have recently demonstrated the active participation of this enzyme in copper tolerant species in comparison with sensitive ones, where the over-expression of this enzyme is localized in the cortical cells (Lovazzano et al., personal communication). The proteomic information obtained by Contreras et al. 2010 in *S. gracilis* opens the opportunity of understanding many biological/physiological processes in algae. Using this information and those obtained using a biochemistry approximation, it is possible to strongly suggest a cross-talk between different pathways to re-establish the cellular homeostasis distorted by copper-associated oxidative stress in this species as well as in other tolerant ones (Fig. 2). Thus, the differential ability of each species to deal with oxidative stress resulting from the high copper levels, explains the persistence of tolerant species and the absence of sensitive ones at copper contaminated zones.

Using the method described by Contreras et al. 2008, it was also possible to evaluate differential tolerance in *Ectocarpus siliculosus* strains, originated from habitats with contrasting histories of copper levels (Ritter et al., 2010). Here, the authors showed a differential stress tolerance between 50 and 250 $\mu g\ L^{-1}$ of copper. This difference was also observed at the level of the 2-DE proteome profile. For example, in the tolerant strains from a copper contaminated site (i.e. Chañaral, Chile) a specific expression of PSII Mn-stabilizing protein, fucoxanthine chlorophyll a-c binding protein and vanadium-dependent bromoperoxidase proteins, among others, was observed. Thus, the occurrence of the differential proteome profile among the strains could be strongly suggested by the persistence copper driving force in the evolution of *Ectocarpus siliculosus* from the copper contaminated sites (Ritter et al., 2010). In other brown macroalgae such as *Ecklonia cava,* it was possible to observe the effects of temperature on the proteomic profile (Yotsukura et al., 2012). Here, the authors define that the differential protein expression induced by temperature could be considered as an important biomarker of the health individuals in the culture conditions.

In *Saccharina japonica* it was possible to observe differences at the level of the proteome under seasonal variation and pH conditions (Yotsukura et al., 2010; 2012). Under seasonal variation, the specific expression of different proteins was identified, among them the vanadium-dependent bromoperoxidase (Yotsukura et al., 2010). Comparatively, under different pH culture conditions the over-expression of several proteins was described such as: glyceraldehyde-3-phosphate dehydrogenase, actin, phosphoglycerate kinase, elongation factor Tu and ATP synthase subunit β, among others. Thus, different metabolic pathways could be induced in brown macroalgae according to the type of stress factor. In this context, the utilization of the 2-DE approximation has been extraordinarily important in unravelling the tolerance mechanisms associated with environmental variables from natural and anthropogenic sources. In fact, the identification of important enzymes, never before described in algae (i.e. Peroxiredoxine (Contreras et al., 2010) and vanadium-dependent bromoperoxidase (Ritter et al., 2010)), opens the opportunity to further understanding the biology of this group of organisms.

In microalgae, several works have also been reported. For example, Wang et al. (2004) studied the proteome changes of *Haematococcus pluvialis* under oxidative stress induced by the addition of acetate and Fe^{2+} and exposure to excess of high light intensity. About 70 proteins were identified in which 19 were up-regulated (e.g. antioxidant enzymes and sugar synthesis proteins) and 13 were down-regulated (e.g. metabolism and cell growth proteins). Also, transient regulated proteins were identified in which 31 were up-regulated (e.g. antioxidant enzymes) and only 8 were down-regulated (e.g. chloroplastic proteins). In 2006, Förster et al. performed a proteome comparison among wild-type and two very high light-resistant mutants (*VHL^R-S4* and *VHL^R-S9*) under different high light stress. About 1500 proteins were detected in the gel and 83 proteins from various metabolic pathways were identified by peptide mass fingerprinting. The results revealed complex alterations in response to the stress, where total proteins varied drastically in the wild-type compared to the mutants. Nevertheless, the mutant *VHL^R-S4* proved to have better adaptation to high light stress since a more controlled protein regulation was conducted (e.g. up-regulation of several chaperonins and down-regulation of energy metabolism proteins). Another work conducted in *H. pluvialis* analyzed the proteome under high irradiance, but combined with nitrogen starvation (Tran et al., 2009). In the gels, about 900 protein spots were detected of which 13 were down-regulated and 36 up-regulated. Among the up-regulated proteins, a glutathione peroxidase and a translocase from the outer mitochondrial membrane were matched to *C. reinhardtii*; therefore, these stress responses may be common among these microalgae. A study assessing a proteomic analysis on *C. reinhardtii* under a short-term exposure to irradiance revealed significant down regulation of several heat-shock proteins (HSPs) (Mahong et al., 2012) under differential times of exposition to this stress (0 h, 1.5 h, 3 h and 6 h of high light). Spot densities allowed the determination that early rearrangement of the light-harvesting antenna proteins occurs, where this was manifested by the up- and down-regulation of several protein spots identified as LHC-II polypeptides. Moreover, increased expression of proteins involved in carbohydrate metabolism was found, which could help accelerate the utilization of electrons generated, in order to minimize the risk of superoxide formation. Surprisingly, after 6 hours of high light several molecular chaperones were down-regulated and this could result in drastic effects on cell structure and function. Nevertheless, *C. reinhardtii* is normally light-sensitive which could be explained by the down-regulation of molecular chaperones.

In microalgae, the response of species to heavy metal contamination has also been evaluated. A proteomic analysis conducted on *N. oculata* showed differences between protein expression

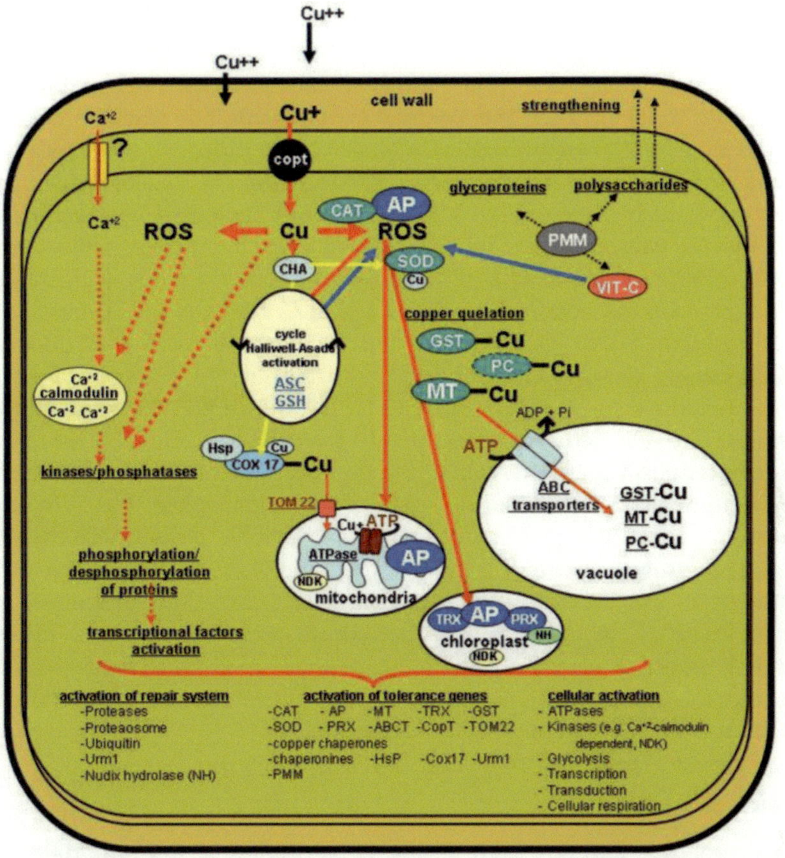

Fig. 2. Cellular events involved in the mechanisms of copper stress tolerance in algae. Dotted arrows indicate routes not directly evidenced in brown algae. The alteration of the state redox, cell damage, and the metal may trigger the antioxidant machine [i.e. compounds and antioxidant enzymes (activation of cycle Halliwell-Asada (MDHAR, DHAR and GP), CAT, SOD, AP, TRX, PRX)] as the activations of protein/genes that form part of various metabolic pathways. Proteins such as HSP or CHA may be involved in the protein protection as in the transport of the metal to proteins that use it as a cofactor, respectively. The sequestration of the metal by different proteins (i.e. MT, GST and PC) is an important homeostatic pathway of tolerance to the metal. The strengthening of the cell wall can increase the resistance to the entry of the metal to the cell. Copts, copper transporter; ROS, reactive oxygen species; MDHAR, monodehydroascorbate reductase; DHAR, dehydroascorbate reducatase; GP, gluthathione peroxidase; AP, ascorbate peroxidase; SOD, superoxide dismutase; CAT, catalase; TRX, tioredoxine; PRX, peroxiredoxine; PMM, phosphomannomutase; VIT-C, vitamin C or ascorbic acid; CHA, copper chaperone; GST, glutathione-s-transferase; PC, phytochelatin; MT, metallothionein; ASC, ascorbate; GSH, glutathione; HSP, heat shock protein; TOM 22; cox 17 transporter; Urm1 (modifier protein type ubiquitin); NDK, nucleoside diphosphate kinase.

of treated (10 μM Cd for 4 days) and untreated (control) cells (Kim et al., 2005). The protocol used in this work, as was discussed before, yielded deficient 2-DE gels, resulting in few proteins detected with only 11 of them with significant changes. Also, the Cd concentration was far from toxic levels suggesting that changes in the protein expression were not needed. This is a non-sequenced species, and therefore, cross-species protein identification was conducted in order to identify those expressed in *N. oculata*. The results showed that malate dehydrogenase and NADH-dehydrogenase were newly induced, whereas glyceraldehyde 3-phosphate dehydrogenase was suppressed. The induction of malate dehydrogenase could be a defense mechanism against Cd toxicity, since at least in *C. reinhardtii* this enzyme controls the malate valve system, which exports reducing power from the chloroplast. Another work assessing Cd toxicity evaluated the proteomic profiles of treated (150 µM Cd) and untreated (control) mutants lacking cell walls of *C. reinhardtii* (Gillet et al., 2006). These mutants are more sensitive to heavy metals due to the lack of a cell wall (Macfie et al., 1994 as cited in Gillet et al., 2006). It was observed that cadmium slowed down the growth rate, and furthermore, induced a 30-50% of growth inhibition. In this work, an elevated number of protein spots were detected and subsequently identified. In fact, 20 proteins were down-regulated in response to Cd stress. Among the down-regulated proteins were those that are involved in amino acid and nitrogen metabolism, chloroplast function and molecule biosynthesis to minimize ROS production. The most variable protein was the RubisCo large subunit, where the protein spot in the control treatment was 15.3 times more intense than in the Cd treatment. It was observed that enzymes with antioxidant properties, chaperonins, and enzymes involved in ATP and carbohydrate metabolism were up-regulated. In addition, in both works chloroplast proteins were found to be down-regulated and proteins involved in antioxidant response to be up-regulated. Therefore, the Cd tolerance mechanism may be similar among different species of microalgae.

4. Conclusions

Micro and macroalgae contain high levels of compounds that interfere with protein extraction. These compounds lead to precipitation of insoluble polymers where the proteome obtainment is almost impossible. However, many efforts have been made in the last years to minimize the coprecipitation of those compounds, and thus now important proteomic protocols are available. For example in macroalgae, it is highlighted the use of phenol during the protein extraction, resulting in consistent electrophoresis runs in several species, conciliating suitable quality and reliability for 2-DE gels and its downstream analysis. The advantage of using phenol as an extracting agent resides in its capacity to disrupt membranes, leaving most of the water-soluble molecules totality in the aqueous phase.

Compared to animals and vascular plants, there is limited information about the use of 2-DE in either micro or macroalgae, both at technical and proteomic level. In fact, low number of published information in the proteomic context can be registered in micro and macroalgae. For example, in this group of organisms only about 42 works can be founded in the www.ncbi.nlm.nih.gov data base, using as search the concepts proteomic or proteome. On the other hand, in vascular plants it is possible to find about 3,100 works in the same data base and ca. 13,100 in animals. Thus, insignificant information exists nowadays in the proteome involvement in algal species, independent of the taxonomic status, ecological importance and economic value of this group of organisms.

2-DE in algal species has allowed the identification of several pathways involved in tolerance mechanisms, associated principally to different abiotic factors. For example, under copper stress the identification of proteins such as peroxiredoxine, enzyme involved in the detoxification of hydrogen peroxide and fatty acid hydroperoxide has Allowed to understand the differential degree of tolerance between Copper tolerant and sensitive species. In fact, using the proteomic protocol described in these species, which uses phenol in the protein extraction, a differential proteome profile in algal individuals between desiccation stress and normal hydration was founded. In this context, new tolerance mechanisms will be revealed using this approximation in order to understand the high desiccation tolerance that exists in this species in comparison with many others, including that from the same phylum. Thus, 2-DE approximation is an important tool that can be interconnected with those obtained to ecological level in order to understand mechanisms of stress tolerance, and therefore explanation of distribution patterns at local and latitudinal scale.

5. Appendix A: Microalgae methodology

Chlamydomonas reinhardtii **(Plantae, Chlorophyta)** (Förster et al., 2006). The figure correspond to the species mentioned.

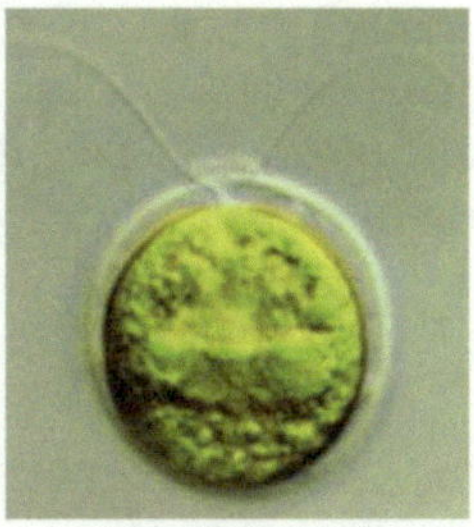

This protocol is an adaption of the one described by Mathesius and co-workers (Grotewold, 2003) that are suited for plant material.

1. Collect *Chlamydomonas* cells by centrifugation at 5,000 x *g* for 5 min at 20° C. Determine fresh weight of cell pellets. Samples can be stored at -80°C for later use.
2. Grind pelleted *Chlamydomonas* cells to a fine powder in liquid nitrogen using a mortar and pestle after addition of 0.5 g of glass powder per 1 g fresh weight of pelleted cells.
3. Suspend the ground material in -20°C cold acetone containing 10% w/v TCA and 0.007% w/v DTT. Sonicate this suspension on ice six times for 10 s each with intermittent 1-min breaks using an ultrasonicator. Centrifuge samples at 35,000 x *g* for 15 min at 4°C.
4. Wash the pellet twice by resuspension in -20°C acetone containing 0.07% w/v DTT, placing it at -20°C for 30 min and centrifuging at 12,000 x *g* for 15 min at 4°C.
5. Lyophilize the pellet for 3 min and resuspend in sample buffer containing 9 M urea, 4% w/v CHAPS, 1% w/v DTT, 0.8% v/v ampholytes (ones suited for the desired 2D-gel), 25 mM Tris base, 1 mM PMSF and 5 mM EDTA.
6. Sonicate samples twice in a sonic bath in an ice-water mixture for 5 min and centrifuge them at 19,000 x *g* for 15 min at 20°C.
7. Determine protein concentration of the sample (e.g. with a Bradford assay or a BCA assay) and keep at –80°C until used for isoelectric focusing.

Haematococcus pluvialis **(Plantae, Chlorophyta)** (Wang et al., 2003). The figure correspond to the species mentioned.

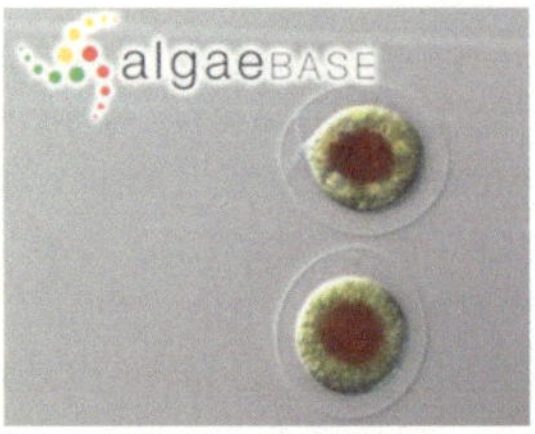

In this study a number of key chemical reagents were evaluated, the protocol that yielded the best 2-DE results is detailed below.

1. Collect *H. pluvialis* cells by centrifugation at 3,000 x *g* for 5 min and wash the pellet three times with cold deionized water.
2. Resuspend cell pellet in one volume of 50 mM Tris-HCl buffer pH 8.0, 3 mM DTT, 5 mM MgCl2, 10% glycerol, 0.5% PVP, 5 mM Na_2-EDTA, 1 mM PMSF, 5 mM benzamidin, 5 mM acoproic acid and 1% v/v plant protease inhibitor cocktail.
3. Disrupt cells by one passage through a pre-cooled French Press Cell at a pressure of 20,000 psi. Centrifuge cell lysate at 3,000 *g* for 10 min to pellet cell debris.
4. Collect the supernatant and centrifuge at 100,000 *g* for 1 h.
5. Dialyze the supernatant from the previous centrifugation step against 250 mL of 85% w/v sucrose solution at 4°C for 2 h. Precipitate the dialysate with 9 volumes of ice-cold 10% w/v TCA in acetone containing 0.07% w/v β-mercaptoethanol at -20°C overnight.
6. Centrifuge samples at 15,000 x *g*. Discard supernatant and wash the pellet with acetone containing 0.07% w/v β-mercaptoethanol to remove TCA.
7. Then, remove residual acetone by air-drying.
8. Resuspend pellet in solubilization buffer containing 2 M thiourea, 8 M urea, 4% CHAPS, 2 mM TBP, and 0.2% ampholytes (ones suited for the desired 2-DE gel).
9. Determine protein concentration of the sample (e.g. with a Bradford assay or a BCA assay) and keep at -80°C until used for isoelectric focusing.

Nannochloropsis oculata **(Chromista, Ochrophyta)** (Kim et al., 2005). The figure correspond to the species mentioned.

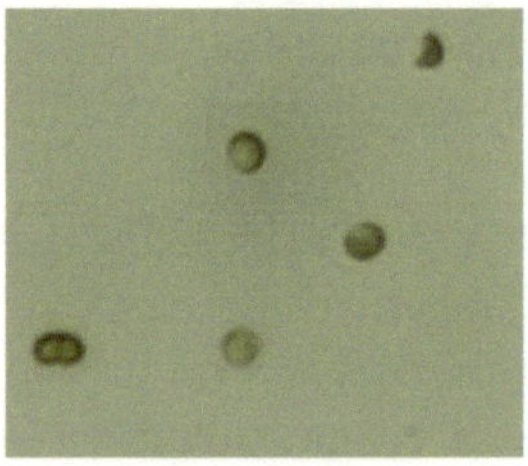

In this study only one method for protein extraction was performed, the details are shown below.

1. Collect *N. oculata* cells by centrifugation at 12,000 x *g* for 10 min. Suspend cell pellet in PBS buffer pH 7.2.

2. Mix suspension with the same volume of sample buffer containing 0.3% w/v SDS, 1% w/v β-mercaptoethanol and 0.05 M Tris-HCl pH 8.0.
3. Denature solution at 100°C for 3 min, cool on ice and treat with DNase/RNase. Precipitate proteins with 10% TCA in 100% Acetone at -70°C for 3 h.
4. Wash the pellet with 100% acetone several times and then air-dry it at room temperature for 5 min.
5. Determine protein concentration of the sample (e.g. with a Bradford assay or a BCA assay) and keep at -80°C until used for isoelectric focusing.

6. Appendix B: Macroalgae methodology

Gracilaria changii **(Plantae, Rhodophyta)** (Wong et al., 2006). The figure correspond to the species mentioned.

In this study two protein extraction methods were analysed in two-dimensional gels. The best results were yielded by the phenol/chloroform method, which is detailed below.

1. Grind frozen seaweeds at -70°C into a fine powder with a mortar and pestle in liquid nitrogen. Put approximately 100 mg of the resulting powder into a 1.5 mL tube for a single extraction.
2. Add 1 mL of TRI reagent (containing phenol and guanidine-isothiocyanate) to 100 mg of seaweed powder and homogenize the mixture.
3. Store the homogenate for 5 min at room temperature to clarify phases. Reserve the phenolic phase and add 200 µL of chloroform per 1 mL of TRI reagent.
4. Cover the samples and shake vigorously for 15 seconds. Store the resulting mixture at room temperature for 2-15 min. Centrifuge mixture at 12,000 *g* for 15 min at 4°C.
5. Discard upper aqueous phase containing RNA, and retain interphase and lower red phenol-chloroform phase containing DNA and proteins.
6. Add ethanol to the reserved phases in order to precipitate DNA.
7. Retain phenol/ethanol supernatant and add 3 volumes of acetone to precipitate proteins, mix by inversion for 10-15 sec to obtain a homogenous solution.
8. Store sample for 10 min at room temperature and sediment the protein precipitate at 12,000 *g* for 10 min at 4° C.
9. Discard the phenol/ethanol supernatant and disperse the protein pellet in 0.5 mL of 0.3 M guanidine hydrochloride in 95% ethanol + 2.5% v/v glycerol.
10. Add another 0.5 mL aliquot of the guanidine hydrochloride/ethanol/glycerol solution to the sample and store for 10 min at room temperature. Centrifuge the proteins at 8,000 *g* for 5 min.

11. Discard the wash solution and perform two more washes in 1 mL each of the guanidine Hydrochloride/ethanol/glycerol wash solution.
12. Perform a final wash in 1 mL of ethanol containing 2.5% glycerol v/v. At the end of the 10 min of ethanol wash, centrifuge the proteins at 8,000 *g* for 5 min.
13. Discard the alcohol and air-dry the pellet for 7-10 min at room temperature.
14. Resolubilize protein pellet in 40 mM Tris buffer pH 8.8 containing 8 M urea, 4% CHAPS and 2 mM TBP.
15. Determine protein concentration of the sample (e.g., with a Bradford assay or a BCA assay) and keep at -80°C until used for isoelectric focusing.

Scytosiphon gracilis and *Ectocarpus siliculosus* **(Chromista, Ochrophyta)** (Contreras et al., 2008). The figures correspond to the species mentioned.

In this method, a major extraction of proteins was obtained in comparison with pervious macroalgae methods described.

1. Remove the excess salt by rinsing in Milli Q water and 50 mM Tris-HCl pH 8.8.
2. Freeze seaweed material at -80°C before pulverization.
3. Homogenize seaweed material using a mortar-driven homogenizer in sample lysis solution composed of 1.5% PVP, 0.7 M sucrose, 0.1 M KCl, 0.5 M Tris-HCl pH 7.5, 250 mM EDTA, protease inhibitor cocktail, 2% v/v β-mercaptoethanol and 0.5% w/v CHAPS
4. Equal volume of Tris-HCl pH 7.5-satured phenol is added and the mixture homogenized at 4°C. Then, by centrifugation the upper phase is removed and the lower phase is re-extracted using the same volume of phenol.
5. The proteins in the phenol phase are precipitated by means of ammonium acetate (0.1 M in methanol). The protein pellet obtained by centrifugation is washed in 80% ice-cold acetone and cold acetone containing 20 mM DTT.
6. Determine protein concentration of the sample and keep at -20°C until used for isoelectric focusing.

Bostrychia radicans / B. moritziana **(Plantae, Rhodophyta)** (Kim et al., 2008). The figure correspond to *B. moritziana*.

In this study a very simple method for protein extraction was performed, the details are shown below.

1. Freeze seaweed material at -80°C before pulverization.
2. Homogenize seaweed material using a mortar-driven homogenizer in sample lysis solution composed of 7 M urea, 2 M thiourea, containing 4% w/v CHAPS, 1% w/v DTT, 2% v/v ampholytes and 1 mM benzamidine.
3. Perform freezing and thawing steps for five times for 1 day*.
4. Extract proteins for 1 h at room temperature with vortexing.
5. Centrifuge mixture at 15,000 *g* for 1 h at 18°C.
6. Retain soluble fraction and discard insoluble material.
7. Determine protein concentration of the sample (e.g., with a Bradford assay or a BCA assay) and keep at -80°C until used for isoelectric focusing.

* If cell lysing is found to be difficult, used a bead beater in order to facilitate the process.

Saccharina japonica and *Ecklonia cava* **(Chromista, Ochrophyta)** (Yotsukura et al., 2010; 2012). The figures correspond to the species mentioned.

1. Remove the salt excess in Milli Q water.
2. Freeze seaweed material at -80°C before pulverization. Homogenize the tissue in 99.5% cold ethanol and centrifuged.
3. The protein pellet is rinsed in 99.5% ethanol and 100% acetone and resuspended in 0.1 M Tris-HCl buffer pH 8.0, 30% sucrose, 2% SDS, 5% β-mercaptoethanol and phenol.

4. The solution is vortexed and centrifuged at room temperature and the upper phase collected. Agitate solution in 0.1 M ammonium acetate and kept at -20°C.
5. The protein pellet obtained by centrifugation is rinsed in 0.1 M ammonium acetate in methanol and 80% acetone, subsequently dried in evaporator and preserved at –80°C until protein quantification.

Porphyra columbina **(Plantae, Rhodophyta).** The figure correspond to the species mentioned.

This method is an adaptation of the method performed by Contreras et al. 2008, where the details of it are shown below*.

1. Pulverize 3-5 g of frozen seaweeds at -80°C to a fine powder with a mortar and pestle in liquid nitrogen.

2. Resuspend pulverized tissue in 5-10 mL of buffer lysis containing 0.5 M Tris-HCl pH 7.5, 0.7 M sucrose, 0.5 M KCl, 250 mM EDTA, 1.5% w/v PVP, 0.5% w/v CHAPS and 2% v/v β-mercaptoethanol and homogenize for 15 min at 4°C.
3. Add an equal volume of Tris–HCl pH 7.5-saturated phenol and homogenize for 15 min at 4° C. Centrifuge the homogenate at 2,000 *g* for 30 min.
4. Retain only upper phenol phase containing proteins being careful not to remove the interphase.
5. Add ½ volumes of Tris–HCl pH 7.5-saturated phenol and mix well by inversion.
6. Centrifuge the homogenate at 2,000 *g* for 20 min.
7. Retain newly only upper phenol phase containing proteins being careful not to remove the interphase and mix with the previously retained upper phase.
8. Add 5 volumes of 0.1 M ammonium acetate on methanol ice-cold.
9. Shake vigorously to mix the solution and leave to precipitate for 4 h at -20°C.
10. Centrifuge at 2,000 *g* for 40 min. Discard supernatant and wash pellet with 8 volumes of 0.1 M ammonium acetate on methanol ice-cold.
11. Shake vigorously to mix the solution and leave to precipitate for 30 min at -20°C.
12. Centrifuge at 2,000 *g* for 30 min.
13. The proteins pellet is washed in 80% ice-cold acetone and cold acetone containing 20 mM DTT.
14. The protein pellet is then washed in ice-cold acetone 80% and ice-acetone 60% in methanol to remove the majority of contaminants.
15. Determine protein concentration of the sample and keep at -20°C until used for isoelectric focusing.

* Method not yet published.

7. Acknowledgment

This work was supported by FONDECYT 11085019 to LC. Additional funding cames from FONDAP 1501-0001 (CONICYT) to the Center for Advanced Studies in Ecology and Biodiversity (CASEB) Program 7. Finally, we are especially grateful to Nicole Ehrenfeld and Javier Tapia for image acquisition and Daniela Thomas, Alejandra Nuñez and Aaron Mann for text editing.

8. References

Abe, S., Kurashima, A., Yokohama, Y. & Tanaka, K. (2001). The cellular ability of desiccation tolerance in Japanese intertidal seaweeds. *Botanica Marina*, Vol. 44, No.4, pp.125-131, ISSN 0006-8055

Aguilera, J., Bischof, K., Karsten, U., Hanelt, D. & Wiencke, C. (2002). Seasonal variation in ecophysiological patterns in macroalgae from an Arctic fjord. II. Pigment accumulation and biochemical defence systems against high light stress. *Marine Biology*, Vol. 140, pp. 1087-1095, ISSN: 0025-3162

Alpert, P. (2006). Constraints of tolerance: why are desiccation-tolerant organisms so small or rare?. *The Journal of Experimental Biology*, Vol. 209, 1575-1584, ISSN: 1477-9145

Andrade, S., Contreras, L., Moffett, J.W. & Correa, J.A. (2006). Kinetics of copper accumulation in *Lessonia nigrescens* (Phaeophyceae) under conditions of environmental oxidative stress. *Aquatic Toxicology*, Vol. 78, pp. 398-401, ISSN: 0166-445X

Apte, S.C. & Day, G.M. (1998). Dissolved metal concentrations in the Torres Strait and Gulf of Papua. *Marine Pollution Bulletin*, Vol. 30, pp. 298-304, ISSN: 0025-326X

Baier, M. & Dietz, J. (1997). The plant 2-Cys peroxiredoxin BAS1 is a nuclear-encoded chloroplast protein, its expressional regulation, phylogenetic origin, and implications for its specific physiological function in plants. *The Plant Journal* Vol. 12, pp. 179–190, ISSN: 1365-313X

Baker, C.J. & Orlandi, E.W. (1995). Active oxygen in plant pathogenesis. *Annu. Rev. Phytopathology*, Vol. 33, pp. 299-321, ISSN: 0066-4286

Batley, G.E. (1995). Heavy metals and tributyltin in Australian coastal and estuarine waters. In Zann, L.P. & Sutton ,D. [Eds.] Pollution State of the Marine Environment Report for Australia. Technical Annex 2, Great Barrier Reef Marine Park Authority, Canberra, pp. 63–72.

Bilgrami, K.S. & Kumar, S. (1997). Effects of copper, lead and zinc on phytoplankton growth. *Biologia Plantarum*, Vol. 39, pp. 315-317, ISSN: 0006-3134

Blokhina, O. & Fagerstedt, K.W. (2010). Reactive oxygen species and nitric oxide in plant mitochondria: origin and redundant regulatory systems. *Physiologia Plantarum*, Vol. 138, pp. 447-462, ISSN: 1399-3054

Bryan, N.L. & Langston, W.J. (1992) Bioavailability, accumulation and effects of heavy metals in sediments with special reference to United Kingdom estuaries: a review. *Environmental Pollution*, Vol. 76, pp. 89-131, ISSN: 0269-7491

Burritt, D. J., Larkindale, J. & Hurd, C.L. (2002). Antioxidant metabolism in the intertidal red seaweed *Stictosiphonia arbuscula* following desiccation. *Planta*, Vol. 215, pp. 829-838, ISSN: 0032-0935

Cabello-Pasini, A., Diaz-Martín, M.A., Muñiz-Salazar, R., Zertuche-Gonzalez, J.A. & Pacheco-Ruiz, I. (2000). Effect of temperature and desiccation on the photosynthetic performance of *Porphyra perforata*. *Journal of Phycology*, Vol. 36, pp. 10-14, ISSN: 1529-8817

Chinnasamy, G. & Rampitsch, C. (2006). Efficient solubilization buffers for two-dimensional gel electrophoresis of acidic and basic proteins extracted from wheat seeds. *Biochimica et Biophysica Acta - Proteins and Proteomics*, Vol. 1764, pp. 641–4, ISSN: 1570-9639

Clegg, J.S. (2005). Desiccation Tolerance in Encysted Embryos of the Animal Extremophile, Artemia. *Integrative and Comparative Biology*, Vol. 45, pp. 715–724, ISSN 1557-7023

Contreras, L., Moenne, A. & Correa, J.A. (2005). Antioxidant responses in *Scytosiphon lomentaria* (Phaeophyceae) inhabiting copper-enriched coastal environments. *Journal of Phycology*, Vol. 41, pp. 1184-1195, ISSN: 1529-8817

Contreras, L., Medina, M.H., Andrade, S., Oppliger, V. & Correa J.A. (2007a). Effects of copper on early developmental stages of *Lessonia nigrescens* Bory (Phaeophyceae). *Environmental Pollution*, Vol. 145, pp. 75-83, ISSN: 0269-7491

Contreras, L., Dennett, G., Moenne, A., Palma, E. & Correa, J.A. (2007b). Molecular and morphologically distinct *Scytosiphon* species (Scytosiphonales, Phaeophyceae) display similar antioxidant capacities. *Journal of Phycology*, Vol. 43, pp. 1320-1328, ISSN: 1529-8817

Contreras, L., Ritter, A., Denett, G., Boehmwald, F., Guitton, N., Pineau, C., Moenne, A., Potin, P., & Correa, JA. (2008). Two-dimensional gel electrophoresis analysis of brown algal protein extracts. *Journal of Phycology*, Vol. 44, pp. 1315-1321, ISSN: 1529-8817

Contreras, L., Mella, D., Moenne, A. & Correa, J.A. (2009). Differential responses to copper-induced oxidative stress in the marine macroalgae *Lessonia nigrescens* and *Scytosiphon lomentaria* (Phaeophyceae). *Aquatic Toxicology*, Vol. 94, pp. 94-102, ISSN: 0166-445X

Contreras, L., Moenne, A., Gaillard, F., Potin, P. & Correa, J.A. (2010) Proteomic analysis and identification of copper stress-regulated proteins in the marine alga *Scytosiphon gracilis* (Phaeophyceae). *Aquatic Toxicology*, Vol. 96, pp. 85-89, ISSN: 0166-445X

Contreras-Porcia, L., Denett, G., González, A., Vergara, E., Medina, C., Correa, J.A. & Moenne, A. (2011a). Identification of copper-induces genes in the marine alga *Ulva compressa* (Chlorophyta). *Marine Biotechnology*, Vol. 13, pp. 544-556, ISSN: 1436-2228

Contreras-Porcia, L., Thomas, D., Flores, V. & Correa, J.A. (2011b). Tolerance to oxidative stress induced by desiccation in *Porphya columbina* (Bangiales, Rhodophyta). *Journal of Experimental Botany*, Vol. 62, pp. 1815–1829, ISSN 0022-0957

Contreras-Porcia, L., Callejas, S., Thomas, D., Sordet, C., Pohnert, G., Contreras, A., Lafuente, A., Flores-Molina, M.R. & Correa, J.A. (2012). Seaweeds early development: detrimental effects of desiccation and attenuation by algal extracts. *Planta*, Vol. 235, pp. 337-348, ISSN: 0032-0935

Correa, J.A., Castilla, J.C., Ramírez, M., Varas, M., Lagos, N., Vergara, S., Moenne, A., Román, D. & Brown, M.T. (1999). Copper, copper-mine tailings and their effect on marine algae in northern Chile. *Journal of Applied Phycology*, Vol. 11, pp. 57-67, ISSN: 0921-8971

Cremer, F. & Van de Walle, C. (1985). Method for extraction of proteins from green plant tissues for two-dimensional polyacrylamide gel electrophoresis. *Analytical Biochemistry*, Vol. 147, pp. 22–6, ISSN: 0003-2697

Davison, R.I. & Pearson, G.A. (1996). Stress tolerance in intertidal seaweeds. *Journal of Phycology*, Vol. 32, pp. 197-211, ISSN: 1529-8817

Dietz, K.J. (2003). Plant peroxiredoxins. *Annual Review of Plant Biology*, Vol. 54, 93-107, ISSN: 1543-5008

Dietz, K.J, Jacob, S., Oelze, M.L., Laxa, M., Tognetti, V., Nunes de Miranda, S.M., Baire, M. & Finkenmeier, I. (2006). The function of peroxiredoxins in plant organelle redox metabolism. *Journal of Experimental Botany*, Vol. 57, pp. 1697–1709, ISSN 0022-0957

Farrant, J.M. (2000). A comparison of mechanisms of desiccation tolerance among three angiosperm resurrection plant species. *Plant Ecology*, Vol. 151, pp. 29-39, ISSN: 1385-0237

Flengsrub, R. & Kobro, G. (1989). A method for two-dimensional electrophoresis of proteins from green plant tissues. *Analytical Biochemestry*, Vol. 177, pp. 33–36, ISSN: 0003-2697

Förster, B., Mathesius, U. & Pogson, B.J. (2006). Comparative proteomics of high light stress in the model alga *Chlamydomonas reinhardtii*. *Proteomics*, Vol. 6, pp. 4309-4320, ISSN: 1615-9861

Gillet, S., Decottignies, P., Chardonnet, S. & Le Maréchal, P. (2006). Cadmium response and redoxin targets in *Chlamydomonas reinhardtii*: a proteomic approach. *Photosynthesis Research*, Vol. 89, pp. 201-211, ISSN: 0166-8595

Gledhill, M., Nimmo, M., Hill, S.J. & Brown, M.T. (1997). The toxicity of copper II species to marine algae, with particular reference to macroalgae. *Journal of Phycology*, Vol. 33, pp. 2-11, ISSN: 1529-8817

Goyer, A., Haslekas, C., Miginiac-Maslow, M., Klein, U., Le Maréchal, P., Jacquot, J.P. & Decottignies, P. (2002). Isolation and characterization of a thioredoxin-dependent peroxidase from *Chlamydomonas reinhardtii*. *European Journal Biochemistry*, Vol. 269, pp.272–282, ISSN: 1742-4658

Götz, H. & Arnold, C.G. (1980). Comparative electrophoretic study on ribosomal proteins from algae. *Planta*, Vol. 149, pp. 19-26, ISSN: 0032-0935

Grotewold, E. (2003). Methods in Molecular Biology. Plant Functional Genomics. Humana Press, New York.

Grout, J.A. & Levings, C.D. (2001). Effects of acid mine drainage from an abandoned copper mine, Britannia Mines. Howe Sound, British Columbia, Canada, on transplanted blue mussels (*Mytilus edulis*). *Marine Environmental Research*, Vol. 51, pp. 265-288, ISSN: 0141-1136

Hanson, M.R., Davidson, J.N., Mets, L.J. & Bogorad, L. (1974). Characterization of chloroplast and cytoplasmic ribosomal proteins of *Chlamydomonas reinhardtii* by two-dimensional gel electrophoresis. *Molecular and General Genetics*, Vol. 132, pp. 105-118, ISSN: 1432-1874

Hippler, M., Klein, J., Fink, A, Allinger, T. & Hoerth, P. (2001). Towards functional proteomics of membrane protein complexes: analysis of thylakoid membranes from *Chlamydomonas reinhardtii*. *The Plant Journal*, Vol. 28, pp. 595- 606, ISSN: 1365-313X

Hodgkiss, L.J. & Lu, S. (2004). The effects of nutrients and their ratios on phytoplankton abundance in Junk Bay, Hong Kong. *Hydrobiologia*, Vol. 512, pp. 215-229, ISSN: 0018-8158

Hoffmann, A.J. & Santelices, B. (1997). Flora marina de Chile central. Ediciones Universidad Católica de Chile, 434 pp.

Holm-Hansen, O., Naganobu, M., Kawaguchi, S., Kameda, T., Krasovski, I., Tchernyshkov, P., Priddle, J., Korb, R., Brandon, M., Demer, D., Hewitt, R.P., Kahru, M. & Hewes, C.D. (2004). Factors influencing the distribution, biomass, and productivity of phytoplankton in the Scotia Sea and adjoining waters. *Deep-Sea Research II*, Vol. 51, pp. 1333-1350, ISSN: 0967-0645

Ji, Y. & Tanaka, J. (2002). Effect of desiccation on the photosynthesis of seaweeds from the intertidal zone in Honshu, Japan. *Phycological Research*, Vol. 50, pp. 145-153, ISSN: 1440-1835

Kaltschmidt, E. & Wittmann, H.G. (1972). Ribosomal proteins: XXXII: comparison of several extraction methods for proteins from *Escherichia coli* ribosomes. *Biochimie*, Vol. 54, pp. 167-175, ISSN: 0300-9084

Kim, H.K., Shim, J.B., Klochkova, T.A., West, J.A. & Zuccarello, G.C. (2008). The utility of proteomics in algal taxonomy: *Bostrychia radicans/B. moritziana* (Rhodomelaceae, Rhodophyta) as a model study. *Journal of Phycology*, Vol. 44, pp. 1519-1528, ISSN: 1529-8817

Kim, Y.K., Yoo, W.I., Lee, S.H. & Lee, M.Y. (2005). Proteomic analysis of cadmium-induced protein profile alterations from marine alga *Nannochloropsis oculata*. *Ecotoxicology*, Vol. 14, pp. 589-596, ISSN: 0963-9292

Kobayashi, M., Kurimura, Y. & Tsuji Y. (1997). Light-independent, astaxanthin production by the Green microalga *Haematococcus pluvialis* under salt stress. *Biotechnology letter*, Vol. 19, pp. 507-509, ISSN: 0141-5492

Kumar, M., Kumari, P., Gupta, V., Reddy, C.R.K. & Jha, B. (2010). Biochemical responses of red alga *Gracilaria corticata* (Gracilariales, Rhodophyta) to salinity induced oxidative stress. *Journal of Experimental Marine Biology and Ecology*, Vol. 391, pp. 27-34, ISSN: 0022-0981

Lee, M.Y. & Shin, H.W. (2003). Cadmium-induced changes in antioxidant enzymes from the marine alga *Nannochloropsis oculata*. *Journal of Applied Phycology*, Vol. 15, pp. 13-19, ISSN: 0921-8971

Lesser, M.P. (1996). Elevated temperatures and ultraviolet radiation cause oxidative stress and inhibit photosynthesis in symbiotic dinoflagellates. *Limnology and oceanography*, Vol. 41, pp. 271-283, ISSN: 0024-3590

Lesser, M.P. (1997). Oxidative stress causes coral bleaching during exposure to elevated temperatures. *Coral reefs*, Vol. 16, pp. 187-192, ISSN: 1432-0975

Liu, W., Au, D.W.T., Anderson, D.M., Lam, P.K.S. & Wu, R.S.S. (2007). Effects of nutrients, salinity, pH and light:dark cycle on the production of reactive oxygen species in the alga *Chattonella marina*. *Journal of Experimental Marine Biology and Ecology*, Vol. 346, 76-86, ISSN: 0022-0981

Lobban , C.S. & Harrison, P.J. (1994). Seaweed Ecology and Physiology. Cambridge University Press, New York.

Mahong, B., Roytrakul, S., Phaonaklop, N., Wongratana, J. & Yokthongwattana, K. (2012). Proteomic analysis of a model unicellular green alga, *Chlamydomonas reinhardtii*, during short-term exposure to irradiance stress reveals significant down regulation of several heat-shock proteins. *Planta* Vol. 235, pp. 499-511, ISSN: 1432-2048

Marsden, A.D. & DeWreede, R.E. (2000). Marine macroalgal community, structure, metal content and reproductive function near an acid mine outflow. *Environmental Pollution*, Vol. 110, pp. 431-440, ISSN: 0269-7491

Mathesius, U., Keijzers, G., Natera, S.H.A., Weinman, J.J., Djordjevic, M.A. & Rolfe, B.G. (2001). Establishment of a root proteome reference map for the model legume *Medicago truncatula* using the expressed sequence tag database for peptide mass fingerprinting. *Proteomics,* Vol. 1, pp. 1424-1440, ISSN: 1615-9861

Medina, M., Andrade, S., Faugeron, S., Lagos, N., Mella, D. & Correa, J.A. (2005). Biodiversity of rocky intertidal benthic communities associated with copper mine tailing discharges in northern Chile. *Marine Pollution Bulletin,* Vol. 50, pp. 396-409, ISSN: 0025-326X

Mechin, V., Consoli, L., Guilloux, M. L. & Damerval, C. (2003). An efficient solubilization buffer for plant proteins focused in IPGs. *Proteomics,* Vol. 3, pp. 1299–302, ISSN: 1615-9861

Mets, L. & Bogorad, L. (1972). Altered chloroplast ribosomal proteins associated with erythromycin-resistant mutant in two genetic systems of *Chlamydomonas reinhardtii*. *Proceedings of the National Academy of Sciences of the USA,* Vol. 69, pp. 3779-3783, ISSN: 1091-6490

Mishra, A. & Jha, B. (2011). Antioxidant response of the microalga *Dunaliella salina* under salt stress. *Botanica Marina,* Vol. 54, pp. 195-199, ISSN: 1437-4323

Ojeda, F.P. & Santelices, B. (1984) Ecological dominance of *Lessonia nigrescens* (Phaeophyta) in central Chile. *Marine Ecology Progress Series,* Vol. 19, pp. 83-89. ISSN: 0171-8630

Oliveira, R. (1985). Phytoplankton communities response to a mine effluent rich in copper. *Hydrobiologia,* Vol. 128, pp. 61-69, ISSN: 0018-8158

Pearson, G.A., Serrao, E.A. & Cancela, M.L. (2001). Suppression subtractive hybridization for studying gene expression during aerial exposure and desiccation in fucoid algae. *European Journal of Phycology,* Vol. 36, pp. 359-366, ISSN: 0967-0262

Pearson, G.A., Hoarau, G., Lago-Leston, A., Coyer, J.A., Kube, M., Reinhardt ,R., Henckel, K., Serrao, E.T.A., Corre, E. & Olsen, J.L. (2010). An expressed sequence tag analysis of the intertidal brown seaweeds *Fucus serratus* (L.) and *F. vesiculosus* (L.) (Heterokontophyta, Phaeophyceae) in response to abiotic stressors. *Marine Biotechnology,* Vol. 12, pp. 195-213, ISSN: 1436-2228

Pinto, E., Sigaud-Kutner, T.C.S., Leitai, M.A., Okamoto, O.K., Morse, D., & Colepicolo, P. (2003). Heavy metal-induced oxidative stress in algae. *Journal of Phycology,* Vol. 39, pp. 1008-1018, ISSN: 1529-8817

Piperno, G., Huang, B. & Luck, D.J.L. (1977). Two-dimensional analysis of flagellar proteins from wild-type and paralyzed mutants of *Chlamydomonas reinhardtii. Proceedings of the National Academy of Sciences of the USA,* Vol. 74, pp. 1600-1604, ISSN: 1091-6490

Ratkevicius, N., Correa, J.A. & Moenne, A. (2003). Copper accumulation, synthesis of ascorbate and activation of ascorbate peroxidase in *Enteromorpha compressa* (L.) Grev. (Chlorophyta) from heavy-metal enriched environments in northern Chile. *Plant Cell and Environment,* Vol. 26, pp. 1599-1608, ISSN: 1365-3040

Reynolds, C.S. (2006) Ecology of Phytoplankton. Cambridge University Press, New York.

Rhee, S.G. (2006). Cell signaling. H_2O_2, a necessary evil for cell signaling. *Science,* Vol. 312, pp. 1882-1883, ISSN: 1095-9203

Rijstenbil, J.W. (2001). Effects of periodic, low UVA radiation on cell characteristics and oxidative stress in the marine planktonic diatom *Ditylum brightwellii. European Journal of Phycology,* Vol. 36, pp.1-8, ISSN: 0967-0262

Ritter, A., Ubertini, M., Romac, S., Gaillard, F., Delage, L., Mann, A., Cock, J.M., Tonon, T., Correa J.A. & Potin P. (2010). Copper stress proteomics highlights local adaptation of two strains of the model brown alga *Ectocarpus siliculosus. Proteomics,* Vol. 10, pp. 2074-2088, ISSN: 1615-9861

Sabatini, S.E., Juárez, A. B., Eppis, M.R., Bianchi, L., Luquet, C.M. & Ríos de Molina, M.C. (2009). Oxidative stress and antioxidant defenses in two green microalgae exposed to copper. *Ecotoxicology and Environmental Safety,* Vol. 72, pp. 1200-1206, ISSN: 0147-6513

Santelices, B. (1989). Algas marinas de Chile. Distribución Ecológica. Utilización y Diversidad. Pontificia Universidad Católica de Chile.

Smith, C.M., Satoh, K. & Fork, D.C. (1986). The effects of osmotic tissue dehydration and air drying on morphology and energy transfer in two species of *Porphyra. Plant Physiology,* Vol. 80, pp. 843-847, ISSN: 0032-0889

Stauber,J.L., Benning,R.J., Hales,L.T. ,Eriksen,R. & Nowak,B. (2001) Copper bioavailability and amelioration of toxicity in Macquarie Harbour, Tasmania, Australia. *Marine & Freshwater Research,* Vol. 51, pp. 1-10 , ISSN: 1323-1650

Sunda, W.G. (1989). Trace metal interactions with marine phytoplankton. *Biological Oceanography,* Vol. 6, pp. 411-442, ISSN: 0196-5581

Takagi, M., Karseno & Yoshida T. (2006). Effect of salt concentration on intracelular accumulation of lipids and triacylglyceride in marine microalgae *Dunaliella* cells. *Journal of Bioscience and Bioengineering,* Vol. 101, pp. 223-226, ISSN: 1389-1723

Tran, N.P., Park, J.K., Hong, S.J. & Lee, C.G. (2009). Proteomics of proteins associated with astaxanthin accumulation in the green algae *Haematococcus lacustris* under the influence of sodium orthovanadate. *Biotechnology Letters,* Vol. 31, pp. 1917-1922, ISSN: 0141-5492 van Tamelen, P.G. (1996). Algal zonation in tidepools, experimental evaluation of the roles of physical disturbance, herbivory and competition. *Journal of Experimental Marine Biology and Ecology*, Vol. 201, pp. 197-231, ISSN: 0022-0981

Varela, D.A., Santelices, B., Correa, J.A. & Arroyo, M.K. (2006). Spatial and temporal variation of photosynthesis in intertidal *Mazzaella laminarioides* (Bory) Fredericq (Rhodophyta, Gigartinales). *Journal of Applied Phycology*, Vol. 18, pp. 827-838, ISSN: 0921-8971

Vega, J.M., Garbayo, I., Domínguez, M.J. & Vigara, J. (2006). Effect of abiotic stress on photosynthesis and respiration in *Chlamydomonas reinhardtii* induction of oxidative stress. *Enzyme and Microbial Technology*, Vol. 40, pp. 163-167, ISSN: 0141-0229

Véliz, K., Edding, M., Tala, F. & Gómez, I. (2006). Effects of ultraviolet radiation on different life cycle stages of the south Pacific Kelps, *Lessonia nigrescens* and *Lessonia trabeculata* (Laminariales, Phaeophyceae). *Marine Biology*, Vol. 149, pp.1015-1024, ISSN: 0025-3162

Vermeer, K. & Castilla, J.C. (1991). High cadmium residues observed during a pilot study in shorebirds and their prey downstream from the El Salvador copper mine, Chile. *Bulletin of Environmental Contamination and Toxicology*, Vol. 46, pp. 242-248, ISSN: 0007-4861

Vranová, E., Inzé, D. & van Breusegen, F. (2002). Signal transduction during oxidative stress. *Journal of Experimental Botany*, Vol. 53, pp. 1227–1236, ISSN 0022-0957

Wang, S.B., Hu, Q., Sommerfeld, M. & Chen, F. (2003). An optimized protocol for isolation of soluble proteins from microalgae for two-dimensional gel electrophoresis analysis. *Journal of Applied Phycology*, Vol. 15, pp. 485-496, ISSN: 0921-8971

Wang, S.B., Chen , F., Sommerfeld, M. & Hu, S. (2004). Proteomic analysis of molecular response to oxidative stress by the green alga *Haematococcus pluvialis* (Chlorophyceae). *Planta*, Vol. 220, pp. 17-29, ISSN: 0032-0935

Wang, Z., Qi, Y., Chen, J., Xu, N. & Yang, Y. (2006). Phytoplankton abundance, community structure and nutrients in cultural areas of Daya Bay, South China Sea. *Journal of Marine Systems*, Vol. 62, pp. 85-94, ISSN: 0924-7963

Wood, Z.A., Poole, L.B. & Karplus, B.A. (2003). Peroxiredoxin evolution and the regulation of hydrogen peroxide signaling. *Science*, Vol. 300, pp. 650–653, ISSN: 1095-9203

Wong, P.F., Tan, L.J., Nawi, H. & AbuBakar, S. (2006). Proteomics of the red alga, *Gracilaria changii* (Gracilariales, Rhodophyta). *Journal of Phycology*, Vol. 42, pp. 113-120, ISSN: 1529-8817

Yotsukura, N., Nagai, K., Kimura, H. & Morimoto, K. (2010). Seasonal changes in proteomic profiles of Japanese kelp: *Saccharina japonica* (Laminariales, Phaeophyceae). *Journal of Applied Phycology*, Vol. 22, pp. 443-451, ISSN: 0921-8971

Yotsukura, N., Nagai, K., Tanaka, T., Kimura, H. & Morimoto, K. (2011). Temperature stress-induced changes in the proteomic profiles of *Ecklonia cava* (Laminariales, Phaeophyceae). *Journal of Applied Phycology*. Vol. 24, pp. 163-171. ISSN: 1573-5176

Zou, D. & Gao, K. (2002). Effects of desiccation and CO_2 concentration on emersed photosynthesis in *Porphyra haitanensis* (Bangiales, Rhodophyta), a species farmed in China. *European Journal of Phycology*, Vol. 37, pp. 587-592, ISSN: 0967-0262

Zuppini, A., Gerotto, C. & Baldan, B. (2010). Programmed cell death and adaptation: two different types of abiotic stress response in a unicellular chlorophyte. *Plant and Cell Physiology*, Vol. 51, pp. 884-895, ISSN 0032-0781

Zurek, R. & Bucka, H. (2004). Horizontal distribution of phytoplankton and zooplankton from the littoral towards open waters under wind stress. *Oceanological and Hydrobiological Studies*, Vol. 33, pp.69-81, ISSN: 1730-413X

3

Gel Electrophoresis Based Genetic Fingerprinting Techniques on Environmental Ecology

Zeynep Cetecioglu[1], Orhan Ince[1] and Bahar Ince[2]
[1]*Istanbul Technical University, Environmental Engineering Department, Maslak, Istanbul*
[2]*Bogazici University, Institute of Environmental Science, Rumelihisarustu-Bebek, Istanbul*
Turkey

1. Introduction

Molecular tools in environmental microbiology have been applied extensively in last decades because of the limitations in culture-dependent methods (Amann *et al.*, 1995; Muyzer *et al.*, 1996; Head *et al.*, 1998). Despite isolation techniques are provided detailed knowledge about the single species in terms of morphology, biochemistry, and also genetic (Bitton, 2005), they have important drawbacks. The first one is to find the selective media favoring the desired microbial group. Additionally, isolated species cannot reflect their behaviors in the natural environment. Until today, 19.000 microbial species have been isolated (DSMZ, 2011; http://www.dsmz.de), however it is accepted that this number is only a small portion of real diversity (Amann *et al.*, 1995). Besides, using the molecular tools in natural and engineering systems, we can find the answer to the questions such as 'which species do exist?', 'which species are active?', 'how many microorganisms are there?', which species do utilize the specific compounds?'.

Microbial ecology studies need identification of species based on a comprehensive classification system that perfectly reflects the evolutionary relations between the microorganisms (Pace, 1996). Zuckerkandl and Pauling (1965) indicated that nucleic acids could document evolutionary history. Due to the pioneering studies, nucleic acids, especially 16S rRNA, are the ultimate biomarkers and hereditary molecules probably because of their essential role in protein synthesis, making them one of the earliest evolutionary functions in all cellular life-forms (Olsen *et al.*, 1986; Pace *et al.*, 1986; Woese, 1987; Stahl *et al.*, 1988). In particular, 16S rRNA and 16S rDNA have been used in phylogenetic analysis and accepted as ideal evolutionary chronometer.

Genetic fingerprinting techniques are one of the most applied molecular tools based on 16S rRNA in microbial ecology studies. These techniques such as denaturing gradient gel electrophoresis (DGGE), temperature gradient gel electrophoresis (TGGE), amplified ribosomal DNA restriction (ARDRA) or restriction fragment length polymorphism (RFLP), terminal restriction fragment length polymorphism (T-RFLP), and single strand conformation polymorphism (SSCP), have been developed for estimation of diversity in ecosystems, screening clone libraries, following the diversity changes with respect to time and location and also identification of species (Hofman-Bang, 2003). This approach comes into prominence because of fast, less labor-intensive features (Muyzer and Smalla, 1998).

These methods have been used to characterize the microbial diversity in different environments such as activated sludge (Liu *et al.*, 1997; Curtis and Craine, 1998), anaerobic reactors (Leclerc *et al.*, 2004), sediments (Muyzer and De Wall, 1993, Cetecioglu *et al.*, 2009), lake water (Ovreas *et al.*, 1997), hot springs (Santegoeds *et al.*, 1996), biofilm (Santegoeds *et al.*, 1998). The method can be used for as both qualitative and semi-quantitative approaches on biodiversity estimations.

In this chapter, these genetic fingerprinting techniques based on gel electrophoresis are discussed. Also exemplarily applications are presented.

2. Microbial ecology and characterization of microbial community via molecular tools

Biochemical conversions occurred in environment are determined by black box model because of limitations to identify microbial communities which are responsible of these (un)known processes (Amann *et al.*, 1995). Acquisition of pure cultures is necessary to obtain an insight into the physiology, biochemistry and genetics of isolated microorganisms. In spite of developments on cultivation methods everyday, still a small portion of the microbial species within the nature can be isolated by culture-dependent techniques (Giovannoni *et al.*, 1990). Another problem in microbial ecology is the complications on identification and classification of the species based on their morphological features. Since the morphological features of the microorganisms cannot give the detailed information about their evaluation relationships. In order to determine the role of microbial diversity in natural or engineered systems, the questions about microbial population including 'Who is there? How many microorganisms are there them? Where are they located? What are they doing? How do populations respond to changes in environmental conditions? What is the relationship between diversity and community stability?' have to be answered. Accordingly, culture independent methods, which give information about microbial ecosystem in terms of diversity, function, etc., are more reliable (Muyzer *et al.*, 1998; Head *et al.*, 1998).

To increase our knowledge about microbial communities and our understanding of their composition, dynamics and interactions within microbial ecosystems, nucleic acid analysis give a wide range opportunity nowadays. Molecular phylogeny not only employs nucleic acid documentation and evolutionary history but also provides a motivation for identification and quantification of microbial species (Olsen and Woese, 1993). The phylogenetic tree of all living organisms is represented in Figure 1. Ribosomal RNA and its gene are the main biomarkers and hereditary molecules for prokaryotes because of their essential role in protein synthesis making them one of the earliest evolutionary functions in all cellular life-forms (Woese, 1987). Therefore prokaryotes can be detected, identified and enumareted by the analysis of 16S rRNA and 16S rDNA.

16S rRNAs and 16S rDNAs, which encode them, are ideal biomarker because they exist in all prokaryotes, they have conserved and their variable regions give the opportunity to identify species even strains as seen in Figure 2. While the conserved regions of 16S rRNA make this molecule as an evolutionary clock instead of their selectively neutral mutational changes (Woese, 1987, Amann *et al.*, 1995), their variable regions allow phylogenetic determination on different taxonomic level (Amann *et al.*, 1995; Head *et al.*, 1998).

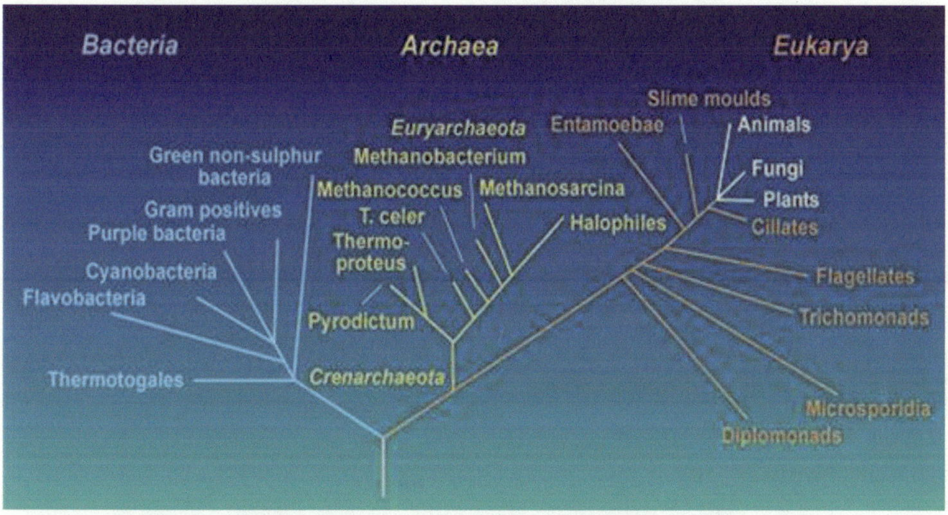

Fig. 1. The rRNA phylogenetic tree of life (Madigan *et al.*, 2009).

As a result, to design general or specific primers and probes for 16S rDNAs and 16S rRNAs provides study options about identification and evolution of microorganisms because this molecule is fairly large (≈1500 nucleotides) including sufficient sequence information. Also the abundance is high within most cells (10^3 to 10^5 copies) and they can be detected easily (Amann *et al.*, 1995). While even secondary structure of 16S rRNA molecule is highly conserved, many variable regions randomly change during evolution. This differential variation explains the relationship between microorganisms evolutionarily. Data obtained from this analysis are adequate to compare statistically significant phylogenetic relations (Olsen *et al.*, 1986). Therefore 16S rRNA and its encoding gene have been widely used to investigate community diversity. The rapidly growing 16S rDNA sequence data bank, accessible (http://www.ebi.ac.uk/) provides the opportunity to get information about 16S rDNA sequences of the determined cultured and uncultured species (Dahllöf, 2002).

In spite of the advantages of using 16S rRNA molecule for phylogenetic analysis, the main limitations are that the heterogeneity between multiple copies of this molecule in one species interferes pattern analysis, confuses the explanation of diversity obtained from clone libraries and sequences retrieved from banding patterns (Dahllöf, 2002).

3. Fingerprinting techniques and their application areas

Fingerprinting techniques provide a separation in microbial community according to their genetic pattern or profile (Muyzer, 1998). A variety of fingerprinting techniques such as *denaturing/temperature gradient gel electrophoresis, amplified ribosomal DNA restriction analysis, terminal restriction fragment length polymorphism, and single strand confirmation polymorphism* has been developed to assess diversity and dynamics in the ecosystem (Hofman-Bang, 2003). The first fingerprinting technique was used in 1980's, which based on the electrophoretic separation in high-resolution polyacrylamide gels of 5S rRNA and tRNA

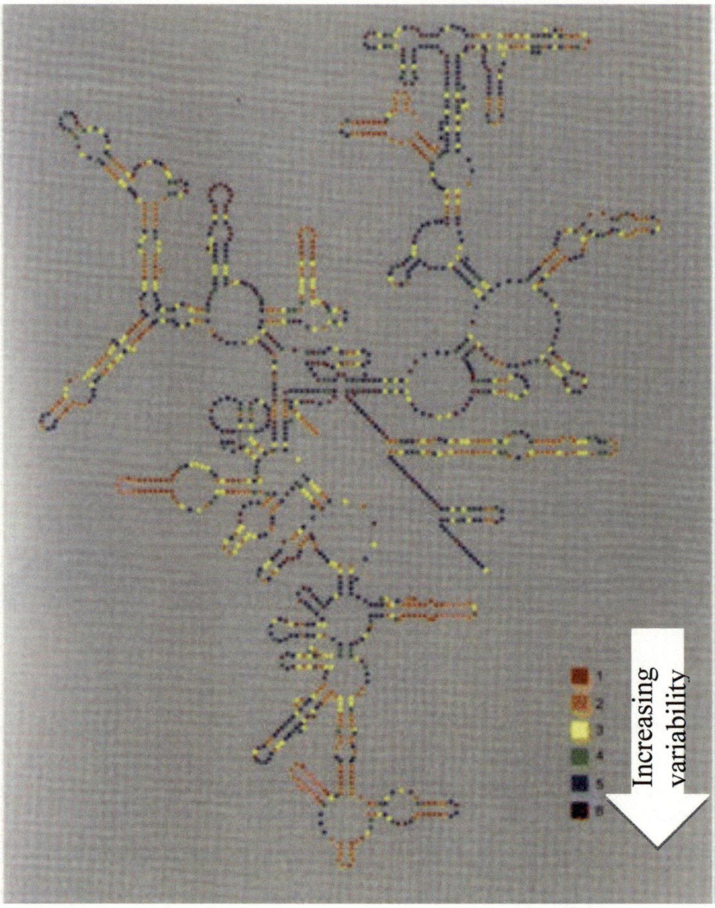

Fig. 2. Secondary structure of the 16S rRNA of E. coli, showing conserved and variable regions (Van de Peer *et al.*, 1996).

obtained from natural samples (Hofle, 1988 and 1990). In 1993, Muyzer *et al.* introduced a new fingerprinting technique to apply on microbial ecology, *denaturing gradient gel electrophoresis* (DGGE). In this method, PCR amplified DNA fragments can be separated according to their nucleic acid pattern. This method has become widespread in a short time. Then another similar technique has been developed, *temperature gradient gel electrophoresis* (TGGE). These methods provide not only analysis of the structure and species composition of microbial communities but also identification of several uncultured microorganisms (Heuer *et al.*, 1997 and Cetecioglu *et al.*, 2009).

3.1 Denaturing/Temperature Gradient Gel Electrophoresis (DGGE/TGGE)

DGGE is a gel electrophoresis technique to separate same length-DNA fragments based on their base sequence differences. In theory, it is sensitive to observe even one base difference on sequence because of melting patterns of the fragments (Muyzer *et al.*, 1993). This method provides a fast, and labor-intensive approach to determine the diversity and the microbial community within an ecosystem, to monitor the changes on dynamics and also to screen the clone libraries (Muyzer and Smalla, 1998). Furthermore, DGGE can be used as qualitative and semi-quantitative approach for biodiversity estimations.

3.1.1 Principles of the experiment

The optimal gradient is the main concern for DGGE/TGGE experiments since the main purpose is separation of DNA fragments according to their melting behaviours. Perpendicular polyacrylamide gels are used according to incremental gradients of denaturants or temperature. The sample including same-length DNA fragment mixtures is loaded to gel for running by electrophoresis. After completing electrophoresis, the gel is stained by a dye such as ethidium bromide, SYBR gold, SYBR green, etc. for obtaining sample pattern. While linear gradient is created by chemical denaturants as urea and formamide for DGGE, temporal temperature gradient is used to separate the DNA fragments in TGGE. Melting pattern of double strand DNA fragments is based on their hydrogen bond content: GC rich DNA fragments melts at higher denaturant/temperature region of the gradient. Complete separation of the double strand DNA is prevented by using GC-clamp primer during the amplification of target DNA region (Dorigo *et al.*, 2005). The schematic explanation of DGGE is given in Figure 3.

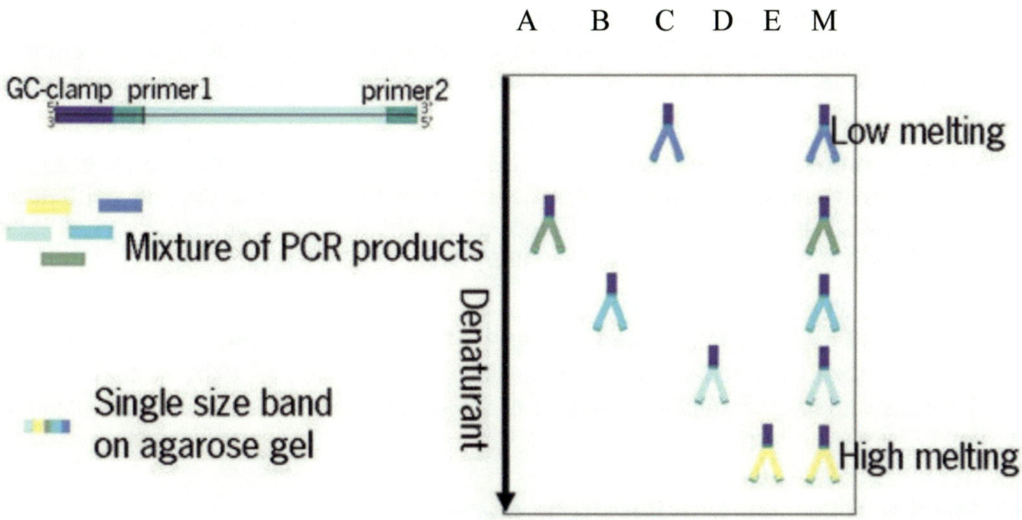

Fig. 3. Principle of DGGE (A: organism a, B: organism b, C: organism c, D: organism d, E: organism E, M: mix sample) (Plant Research International, 2011).

The main difficulties and limitations of the DGGE/TGGE can be listed as:

1. Proper primer selection to represent whole community
2. Optimization of electrophoresis conditions (Muyzer *et al.*, 1993)
3. Limitations on sensitivity for detection of rare community members (Vallaeys *et al.*, 1997)
4. Separation of only small DNA fragments up to 500 bp (Muyzer and Smalla, 1998)
5. Biases coming from PCR amplification such as chimeric products or fidelity errors
6. Heteroduplex formations, multiple bands or due to resolution of the gel, or different fragments resulting from existence of several rRNA coding regions, (Curtis and Craine, 1998).

3.1.2 Application area

DGGE/TGGE is used for several purposes in microbial ecology. The first and the most common application is to reveal and to compare community complex of the microbial diversity within different environments. Curtis and Craine (1998) used this technique to show the bacterial complexity of different activated sludge samples. Connaughton *et al.* (2006) used PCR-DGGE method to find out bacterial and archaeal community structure in a high-rate anaerobic reactor operated at 18 °C. This technique was used to reveal the microbial community in a lab-scale thermophilic trickling biofilter producing hydrogen (Ahn *et al.*, 2005). Another biofilm study showed the bacterial diversity in a river by 16S rDNA PCR-DGGE method (Lyautey *et al.*, 2005). In another study, the authors showed that the different bacterial and archaeal profiles within the highly polluted anoxic marine sediments in the different locations from the Marmara Sea (Cetecioglu *et al.*, 2009). Ye *et al.* (2011) showed the temporal variability of cyanobacteria in the water and sediment of a lake.

Furthermore the scientists use these techniques, mostly DGGE, to analyse the community changes over time. Santagoeds *et al.* (1998) used PCR-DGGE method to monitor the changes in sulphate reducing bacteria in biofilm. Ferris and Ward (1997) also performed similar approach to reveal seasonal changes in bacterial community from hot spring microbial mat. Kolukirik *et al.* (2011) used 16S rDNA PCR-DGGE technique to represent the local and seasonal bacterial and archaeal shifts in hydrocarbon polluted anoxic marine sediments.

These fingerprinting techniques are widely used to monitor simple communities instead of complex environments. It is one of the detection methods to analyse the cultivation/ isolation approaches and to determine the enrichment cultures (Santagoeds *et al.*, 1996; Ward *et al.*, 1996; Teske *et al.*, 1996; Muyzer, 1997; Bucholz-Cleven *et al.*, 1997).

Also DGGE/TGGE are commonly chosen for comparison of the efficiency of the DNA extraction protocols (Heuer and Smalla, 1997; Lieasack *et al.*, 1997) and the screening of the clone libraries (Heuer and Smalla, 1997; Lieasack *et al.*, 1997, Kolukirik *et al.*, 2011) because rapid and reliable results are caused to perform less time (Kowalchuk *et al.*, 1997).

3.2 Amplified Ribosomal DNA Restriction Analysis (ARDRA)

Recognition site of restriction enzymes are changed for different microbial species. The principle of amplified ribosomal DNA restriction analysis (ARDRA), also called as restriction fragment length polymorphism (RFLP), is based on this knowledge. The combination of PCR and restriction can, for example, be used for enhanced amplification of minor DNA templates (Green and Minz, 2005).

In the first step of this technique, ribosomal DNA is amplified by PCR to avoid undesired and/or dominant DNA templates. Then, the 16S rDNA PCR products are digested into specific DNA fragments by restriction enzymes. At the final step, the fragments are loaded to high-resolution gel for electrophoresis. The schematic representation of the principle of ARDRA is given in Figure 4. The main advantage of this technique is to provide rapid comparison of rRNA genes (Moyer *et al.*, 1994).

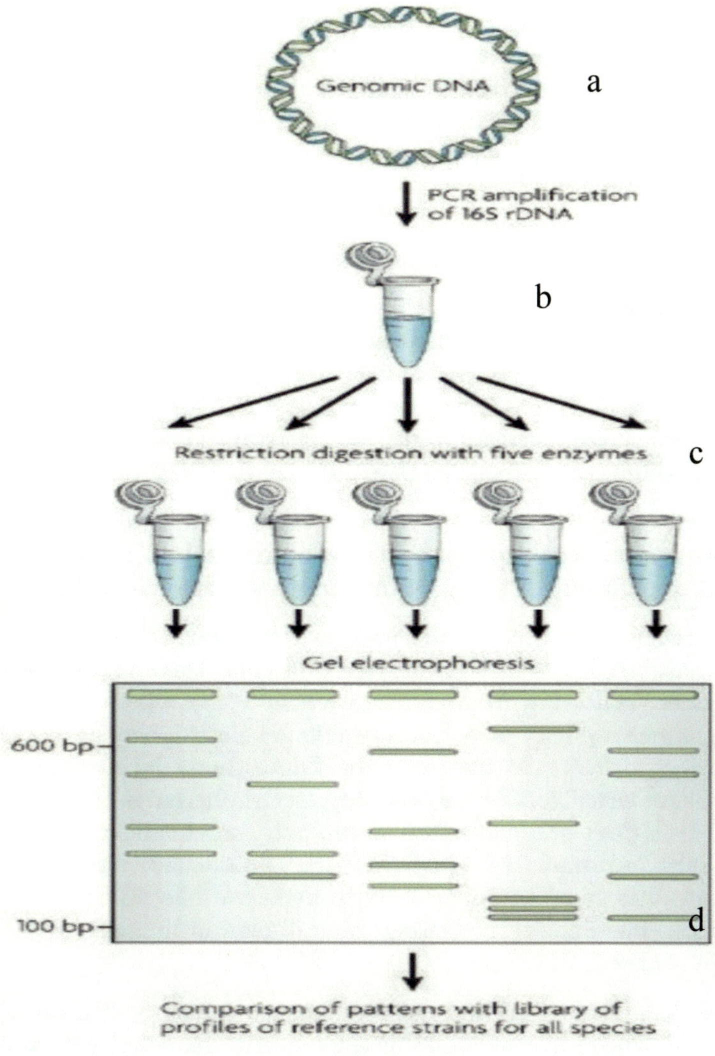

Fig. 4. Steps of ARDRA (a: Genomic DNA extraction, b: PCR reaction for specific region, c: restriction digestion, d: gel electrophoresis) (Dijkshoorn *et al.*, 2007).

The application areas of this technique are also similar to DGGE. It is varied from detection isolates or clones to determination of whole community in an environment. For these different purposes, different gel types can be used. While agarose gel is sufficient to detect isolates or clones, polyacrylamide gels are necessary for better resolution in the community analysis (Martinez-Murcia *et al.*, 1995).

In the literature, there are different studies performed by ARDRA. Lagace *et al.* (2004) identified the bacterial community of maple trees. A wide variety of the organisms were detected from different groups. Barbeiro and Fani used this technique to investigate more specific bacterial group, Acinetobacteria, within 3 sewage treatment plants (1998). In 1995, Vaneechoutte and his colleagues performed similar study for Acinetobacter strains. They showed that this technique is less prone to contamination problems for detection. In another study, ARDRA was used to screen bacterial and archaeal clone libraries to detect the microbial community within an anaerobic reactor to treat fodder beta silage (Klocke *et al.*, 2007). Also there are some studies to investigate the microbial community in soil (Smith *et al.*, 1997; Viti and Giovannetti, 2005).

3.3 Terminal Restriction Length Polymorphism (T-RFLP)

Terminal Restriction Fragment Length Polymorphism (T-RFLP) is another fingerprinting technique to obtain profiles of microbial communities. The principle of this method is to separate the genes according to position of their restriction site closest to a labelled end of an amplified gene (Figure 5). The main difference from ARDRA is that the restriction enzymes using in T-RFLP only detect terminal restriction fragments (T-RF). Also this method is used qualitative and quantitative analysis like DGGE (Liu *et al.*, 1997).

The method is carried out in a series of steps including PCR, restriction enzyme digestion, gel electrophoresis and recognition of labelled fragments. Like most other fingerprinting techniques, PCR amplification of a target gene is the first step of T-RFLP.

After DNA extraction, target gene amplification is carried out using one or both the primers having their 5′ end labelled with a fluorescent molecule. Then amplicons are digested by restriction enzymes. Following the restriction reaction, the digested DNA fragments are separated using either capillary or polyacrylamide gel electrophoresis in a DNA sequencer with a fluorescence detector so that only the fluorescently labelled terminal restriction fragments (TRFs) are visualized. At the final step, electropherom is obtained as a result of T-RFLP profiling. Using this graph, electropherom, only target restricted DNA fragments are detected and also satisfactorily quantified by automated electrophoresis systems. Quantification analysis gives an opportunity to make various statistical methods, such as similarity indices, hierarchical clustering algorithms, ordination methods, and self-organizing maps (Liu *et al.*, 1997).

In the literature, T-RFLP was carried out for different purposes like other fingerprinting techniques. In 1997, while Liu *et al.* used this technique to characterize microbial diversity in different environments such as activated sludge, enriched sludge from lab-scale bioreactor, aquifer sand, termite, Moeseneder and his colleagues (1999) optimized T-RFLP to determine marine bacterioplankton communities and to compare this technique to DGGE. In 2000, Horz and his colleagues reported major sub-groups of ammonia oxidizing bacteria by using amoA functional gene. Methane-oxidizing bacteria from landfill site cover soil were detected by T-RFLP combined with RNA dot-blot hybridization (Stralis-Pavese *et al.*, 2006). Also in the same study, RFLP method is used to screen clone libraries. Lueders and Friedrich tried to determine PCR amplification bias by T-RFLP in 2003. Blackwood and his colleagues used T-RFLP for quantitative comparison of microbial communities from different environments such as soil and bioreactors (2003). Additionally this technique was used to screen clone libraries (Moeseneder *et al.*, 2001). Liu *et al.* (2011) performed T-RFLP to determine the microbial shift during bioremediation of petroleum hydrocarbon contaminated soil.

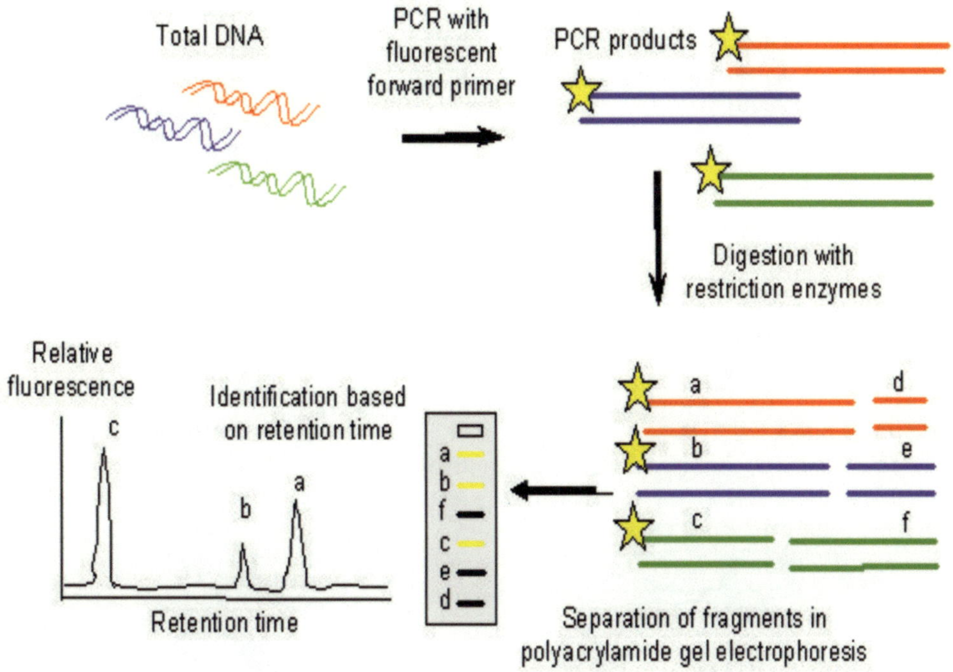

Fig. 5. Steps of T-RFLP (Kaksonen, 2011).

3.4 Single Strand Conformation Polymorphism (SSCP)

Single Strand Conformation Polymorphism (SSCP) is also a fingerprinting technique to separate same-length DNA fragments according to their differences in mobility caused by the secondary structure. The principle of this technique is represented in Figure 6. None of denaturant is used in this method to detect the mobility of the secondary structure of DNA fragments. Each band on SSCP gel corresponds to a distinct microbial sequence, indicating the presence of a microbial strain or species retrieved from the sample (Leclerc *et al.*, 2001; Lee *et al.*, 1996). The main limitation of SSCP, which is similar to DGGE/TGGE, is that one single strand DNA sequence can form more than one stable conformation and this fragment can be represented by multiple bands (Tiedje *et al.*, 1999). The advantage of this technique compared to other fingerprinting methods is that it does not require GC-clamp and gradient gel. SSCP is easier and more straightforward.

SSCP is mostly performed to determine the microbial community profile in different environments such as bioreactor and natural ecosystems. Firstly Lee *et al.* (1996) applied this method to obtain genetic profile of microbial communities. Then Schwieger and Tebbe (1998) used SSCP to determine the community profile including up to 10 bacterial strains. In another study, this method was combined with colony PCR to determine population levels of single and multiple species within plant and environmental samples (Kong *et al.*, 2005). Schmalenberger *et al.* (2008) investigated bacterial communities in an acidic fen by SSCP following by sequencing analysis. In this study, each representative band was cut, then cloned and sequenced to identify species. Also SSCP was carried out to determine the bacterial profile in an aerobic continuous stirred tank reactor (CSTR) treating textile wastewater (Khelifi *et al.*, 2009). Also this technique was applied for determination of *Clostiridum* sp. based on difference their [Fe-Fe]-hydrogenase gene (Quemeneur *et al.*, 2010).

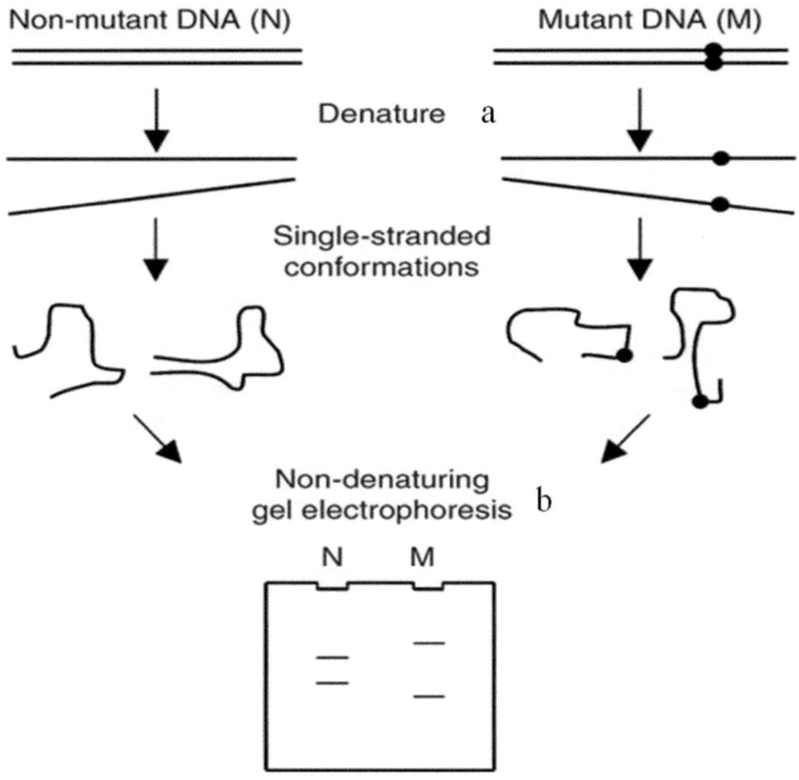

Fig. 6. Steps of SSCP (a: denaturation of ds DNA, b: electrophoresis) (Gasser *et al.*, 2007).

4. Conclusion

The principles of all fingerprinting techniques are similar. DGGE/TGGE, ARDRA, T-RFLP and (SSCP) have been developed to screen clone libraries, to estimate the level of diversity in environmental samples, to follow changes in community structure, to compare diversity and community characteristics in various samples and simply to identify differences between communities. While some of the scientists have showed that sensibilities and resolution of all these techniques are similar, DGGE is still more common application compared to other mentioned techniques. The main reasons of it are that the application of

DGGE is easier and more effective and also less equipment is necessary for it.

5. References

Ahn, Y., Park, E-J., Oh, Y-K., Park, S., Webster, G., Weightman, A.J. 2005. Biofilm microbial community of a thermophilic trickling biofilter used for continuous biohydrogen production. *FEMS Microbiology Letters*, 249: 31-38.

Amann, R.I., Ludwig, W., and Schleifer, K.H., 1995: Phylogenetic identification and in situ detection of individual microbial cells without cultivation. *Microbiol Rev*, 59: 143–169.

Barberio, C., fani, R., 1998. Biodiversity of an Acinetobacter population isolated from activated sludge. *Res. Microbiol.*, 149: 665-673.

Bitton, G., 2005. *Wastewater Microbiology*. 2nd ed., New York, Wiley-Liss.

Blackwood, C.B., Marsh, T., Kim, S-H., Paul, E.A., 2003. Terminal Restriction Fragment Length Polymorphism Data Analysis for Quantitative Comparison of Microbial Communities. *Appl. Environ. Microb.*, 69(2): 926-932.

Buchholz-Cleven, B.E.E., Rattunde, B., Straub, K.L., 1997. Screening for genetic diversity of isolates of anaerobic Fe(II)oxidizing bacteria using DGGE and wholecell hybridization. *Systematic and Applied Microbiology*, 20, 301–309.

Cetecioglu, Z., Ince, B. K., Kolukirik, M., and Ince, O. (2009). Biogeographical Distribution and Diversity of Bacterial and Archaeal Communities within Highly Polluted Anoxic Marine Sediments from the Marmara Sea. *Marine Pollution Bulletin*, 3(58): 384-395.

Connaughton, S., Collins, G., and O'Flaherty, V. 2006. Development of microbial community structure and activity in a high-rate anaerobic bioreactor at 18 ^{0}C. *Water Research*, 40: 1009 - 1017

Curtis, T.P., Craine, N.G., 1998. The comparison of the bacterial diversity of activated sludge plants. *Water Science and Technology*, 37, 71-78.

Dahllöf, I., 2002. Molecular community analysis of microbial diversity. *Current Opinion in Biotechnology*, 13, 213–217.

Dijkshoorn, L., Nemec, A., Seifert, H., 2007. An increasing threat in hospitals: multidrug-resistant Acinetobacter baumannii. *Nature Reviews Microbiology*, 5, 939–951

Dorigo,U., Volatier, L., Humbert, J. F., 2005. Molecular approaches to the assessment of biodiversity in aquatic microbial communities. *Water Research*, 39, 2207-2218.

Ferris, M.J., Ward, D.M., 1997. Seasonal distributions of dominant 16S rRNA defined populations in a hot spring microbial mat examined by denaturing gradient gel electrophoresis. *Applied and Environmental Microbiology*, 63, 1375–1381.

Giovannoni, S.J., Britschgi, T.B., Moyer, C.L., Field, K.G. 1990. Genetic diversity in Sargasso Sea bacterioplankton, *Nature*, 345, 60–63.

Green SJ & Minz D (2005) Suicide polymerase endonuclease restriction, a novel technique for enhancing PCR amplification of minor DNA templates. *Appl. Environ. Microbiol.*, 71: 4721-4727

Head, I.M., Saunders, J.R., Pickup R.W., 1998. Microbial evolution, diversity, and ecology: A decade of ribosomal RNA analysis of uncultivated microorganisms. *Microbiology and Ecology*, 35, 1–21.

Heuer, H., Smalla, K., 1997. Application of denaturing gradient gel electrophoresis (DGGE) and temperature gradient gel electrophoresis (TGGE) for studying soil microbial communities. In: van Elsas JD, Trevors JT & Wellington EMH (Eds) *Modern Soil Microbiology*, Marcel Dekker, New York. 353–373.

Höfle MG (1988) Identifcation of bacteria by low molecular weight RNA profiles: a new chemotaxonomic approach. *J. Microbiol. Meth.*, 8: 235–248

HöfleMG (1990) Transfer RNAs as genotypic fingerprints of eubacteria. *Arch. Microbiol.*, 153: 299–304

Hofman-Bang, J., Zheng, D., Westermann, P., Ahring, B. K., Raskin, L., 2003. Molecular ecology of anaerobic reactor systems. *Advances in Biochemical Engineering and Biotechnology*, 81, 151-203.

Horz, H-P., Rotthauwe, J-H., Lukow, T., Liesack, W., 2000. Identification of major subgroups of ammonia-oxidizing bacteria in environmental samples by T-RFLP analysis of amoA PCR products. *Journal of Microbiological Methods*, 39: 197-204.

Kaksonen, A. Molecular approaches for microbial community analysis. 08/11/2011. http://wiki.biomine.skelleftea.se/biomine/molecular/index_12.htm

Khelifi, E., Bouallagui, H., Touhami, Y., Godon, J-J., Hamdi, M., 2009. Bacterial monitoring by molecular tools of a continuous stirred tank reactor treating textile wastewater. *Bioresource Technology*, 100: 629-633.

Klocke, M., Mahnert, P., Mundt, K., Souidi, K., Linke, B., 2007. Microbial community analysis of a biogas-producing completely stirred tank reactor fed continuously with fodder beet silage as mono-substrate. *System. Appl. Microb.*, 30: 139-151.

Kolukirik, M., Ince, O., Cetecioglu, Z., Celikkol, S., Ince, B.K., 2011: Local and Seasonal Changes in Microbial Diversity of the Marmara Sea Sediments, *Marine Pollution Bulletin*, 62 (11): 2384-2394.

Kong, P., Richardson, P.A., Hong, C., 2005. Direct colony PCR-SSCP for detection of multiple pythiaceous oomycetes in environmental samples. *Journal of Microbiological Methods*, 61: 25-32.

Kowalchuk, G.A., Stephen, J.R., de Boer, W., Prosser, J.I., Embley, T.M., Woldendorp, J.W., 1997. Analysis of ammoniaoxidizing bacteria of the subdivision of the class Proteobacteria in coastal sand dunes by denaturing gradient gel electrophoresis and sequencing of PCR-amplified 16S ribosomal DNA fragments. *Applied and Environmental Microbiology*, 63, 1489–1497.

Lagace, L., Pitre, M., Jacques, M., Roy, D., 2004. Identification of the Bacterial Community of Maple Sap by Using Amplified Ribosomal DNA (rDNA) Restriction Analysis and rDNA Sequencing. *Appl. Environ. Microb.*, 70(4): 2052-2060.

Leclerc, M., Delbes, C., Moletta, R. and Godon J-J., 2001. Single strand conformation polymorphism monitoring of 16S rDNA Archaea during start-up of an anaerobic digester. *FEMS Microbiology Ecology*, 34, 213–20.

Leclerc, M., Delgenes, J.P., Godon, J.J., 2004. Diversity of the Archaeal community in 44 anaerobic digesters as determined by single strand conformation polymorphism analysis and 16S rDNA sequencing. *Environmental Microbiology,* 6, 809–819.

Lee, D-H., Zo, Y-G., Kim, S-J., 1996. Nonradioactive Method To Study Genetic Profiles of Natural Bacterial Communities by PCR–Single Strand-Conformation Polymorphism. Appl. *Microb. Environ.*, 62(9): 3112-3120.

Liesack, W., Janssen, P.H., Rainey, F.A., WardRainey, N.L., Stackebrandt, E. 1997. Microbial diversity in soil: The need for a combined approach using molecular and cultivation techniques. In: van Elsas JD, Trevors JT & Wellington EMH (Eds) *Modern Soil Microbiology*. Marcel Dekker, New York. pp. 375–439

Liu, P-W.G., Chang, T.C., Whang, L-M., Kao, C-H., Pan, P-T, Cheng, S-S., 2011. Bioremediation of petroleum hydrocarbon contaminated soil: Effects of strategies and microbial community shift. *International Biodeterioration&Biodegradation,* 65: 1119-1127.

Liu, W.T., Marsh, T.L., Cheung, H., Forney, L.J., 1997. Characterization of microbial diversity by determining terminal restriction fragment length polymorphisms of genes encoding 16S rRNA. *Applied and Environmental Microbiology,* 63, 4516–4522.

Lueders, T. and Friedrich, M.W., 2003. Evaluation of PCR Amplification Bias by Terminal Restriction Fragment Length Polymorphism Analysis of Small-Subunit rRNA and mcrA Genes by Using Defined Template Mixtures of Methanogenic Pure Cultures and Soil DNA Extracts. *Appl. Environ. Microb.*, 69(1): 320-326.

Lyautey, E., Lacoste, B., Ten-Hage, L., Rols, J-L., Garabetian, F. 2005. Analysis of bacterial diversity in river biofilms using 16S rDNA PCR-DGGE: methodological settings and fingerprints interpretation. *Water Research,* 39: 380-388.

Madigan, M.T., Martinko, J.M., Parker, J., 2009. *Brock Biology of Microorganisms,* (12th edition), Prentice Hall, Inc., New Jersey.

Martinez-Murcia, A.J., Acinas, S.G., Rodriguez Valera, F., 1995. Evaluation of prokaryotic diversity by restrictase digestion of 16S rDNA directly amplified from hypersaline environments. *FEMS Microbiology Ecology,* 17, 247–256.

Moeseneder, M.M., Arrieta, J.M., Muyzer, G., Winter, C., Herndl, G.J., 1999. Optimization of Terminal-Restriction Fragment Length Polymorphism Analysis for Complex Marine Bacterioplankton Communities and Comparison with Denaturing Gradient Gel Electrophoresis. *Appl. Environ. Microb.*, 65(8): 3518-3525.

Moeseneder, M.M., Winter, C., Arrieta, J.M., Herndl, G.J., 2001. Terminal-restriction fragment length polymorphism _T-RFLP/ screening of a marine archaeal clone library to determine the different phylotypes. *Journal of Microbiological Methods,* 44: 159-172.

Moyer, C., Dobbs, F. C., Karl, D.M. 1994. Estimation of diversity and community structure through restriction fragment length polymorphism distribution analysis of bacterial 16S rRNA genes from a microbial mat at an active, hydrothermal vent system, Loihi Seamount, Hawaii. *Applied and Environmental Microbiology*, 60, 871-879.

Muyzer, G., Brinkhoff, T., Nübel, U., Santegoeds, C., Schafer, H., Wawer, C., 1997. Denaturing gradient gel electrophoresis (DGGE) in microbial ecology. In: Akkermans, A.D.L, van Elsas, J.D., de Bruijn, F.J., (Eds) *Molecular Microbial Ecology Manual*, 3.4.4, 1–27. Kluwer Academic Publishers, Dordrecht, The Netherlands.

Muyzer, G., De Waal, E.C., and Utterlinden, A.G., 1993: Profiling of complex microbial populations by denaturing Gradient Gel Electrophoresis analysis of polymerase chain reaction-amplified genes coding for 16S rRNA. *Appl Environ Microbiol*, 59: 695–700.

Muyzer, G., Hottentrager, S., Teske, A., Wawer, C., 1996. Denaturing gradient gel electrophoresis of PCR amplified 16S rDNA - A new molecular approach to analyse the genetic diversity of mixed microbial communities. In: Akkermans ADL, van Elsas JD & de Bruijn FJ (Eds) *Molecular Microbial Ecology Manual*, 3.4.4, 1–23. Kluwer Academic Publishers, Dordrecht, The Netherlands.

Muyzer, G., Smalla, K., 1998. Application of denaturing gradient gel electrophoresis (DGGE) and temperature gradient gel electrophoresis (TGGE) in microbial ecology. *Antonie Van Leeuwenhoek*, 73, 127–141.

Olsen, G.J., D.J. Lane, S.J. Giovannoni, D.A. Stahl and N.R. Pace. (1986). Microbial ecology and evolution: a ribosomal RNA approach. *Ann. Rev. Microbiol.*, 40:337-365.

Olsen, G.J., Woese, C.R., 1993. Ribosomal RNA: a key to phylogeny. *The FASE B Journal*, 7, 113-123.

Øvreas, L., Forney, L., Daae, F.L., Torsvik, V., 1997: Distribution of bacterioplankton in meromictic Lake Saelevanet, as determined by denaturant gradient gel electrophoresis of PCR-amplified gene fragments coding for 16S rRNA. *Appl. Environ. Microbiol.*, 63, 3367–3373.

Pace, N. R., 1996. New perspective on the natural microbial world: Molecular microbial ecology. *ASM News*, 62:463-470

Pace, N.R., G.J. Olsen, and C.R. Woese. (1986). Ribosomal RNA phylogeny and the primary lines of evolutionary descent. *Cell*, 45:325-326

Plant Research International, Wageningen University, Characterization identification and detection, 12.12.2011, Available from:
<http://documents.plant.wur.nl/pri/biointeractions/images/slide14.jpg, 2011>

Quemeneur, M., Hamelin, J., Latrille, E., Steyer, J-P., Trably, E., 2010. Development and application of a functional CE-SSCP fingerprinting method based on [FeeFe]-hydrogenase genes for monitoring hydrogen-producing Clostridium in mixed cultures. *Intern. Jour. Hydrogen Energy.*, 35: 13158-13167.

Santegoeds, C.M., Ferdelman, T.G., Muyzer, G., de Beer, D., 1998. Structural and functional dynamics of sulfate-reducing populations in bacterial biofilms. *Applied and Environmental Microbiology*, 64, 3731–3739.

Santegoeds, C.M., Nold, S.C., Ward, D.M., 1996. Denaturing gradient gel electrophoresis used to monitor the enrichment culture of aerobic chemoorganotrophic bacteria from a hot spring cyanobacterial mat. *Applied and Environmental Microbiology*, 62, 3922–3928.

Schmalenberger, A., Tebbe, C.C., Kertesz, M.A., Drake, H.L., Küsel, K., 2008. Two-dimensional single strand conformation polymorphism (SSCP) of 16S rRNA gene fragments reveals highly dissimilar bacterial communities in an acidic fen. *European Journal of Soil Biology*, 44: 495-500.

Schwieger, F and Tebbe, C.C., 1998. A New Approach To Utilize PCR-Single-Strand-Conformation Polymorphism for 16S rRNA Gene-Based Microbial Community Analysis. *Appl. Microb. Environ.*, 64(12): 4870-4876.

Smit, E. Leeflang, P., Wernars, K., 1997. Detection of shifts in microbial community structure and diversity in soil caused by copper contamination using amplified ribosomal DNA restriction analysis. *FEMS Microbiology Ecology*, 23: 249-261.

Stahl, D.A., Flesher, B., Mansfield, H.R., Montgomery, L., 1988: Use of phylogenetically based hybridization probes for studies of ruminal microbial ecology. *Applied and Environmental Microbiology*, 54, 1079–1084.

Stralis-Pavese, N., Bodrossy, L., Reichenauer, T.G., Weilharter, A., Sessitsch, A., 2006. 16S rRNA based T-RFLP analysis of methane oxidizing bacteria—Assessment, critical evaluation of methodology performance and application for landfill site cover soils. *Applied Soil Microbiology*, 31: 251-266.

Teske, A., Wawer, C., Muyzer, G., Ramsing, N. B., 1996. Distribution of sulfate-reducing bacteria in a stratified fjord (Mariager Fjord, Denmark) as evaluated by most-probable-number counts and denaturing gradient gel electrophoresis of PCR-amplified ribosomal DNA fragments. *Applied and Environmental Microbiology*, 62, 1405–1415.

Tiedje, J.M., Asuming-Brempong, S., Nusslein, K., Marsh, T.L., Flynn, S.J., 1999. Opening the black box of soil microbial diversity. *Applied Soil Ecology*, 13, 109-122.

Vallaeys, T., Topp, E., Muyzer, G., Macheret, V., Laguerre, G., Soulas, G., 1997. Evaluation of denaturing gradient gel electrophoresis in the detection of 16S rDNA sequence variation in rhizobia and methanotrophs. *FEMS Microbiology Ecology*, 24, 279–285.

Van de Peer, Y., van der Auwera G., de Wachter R., 1996. The evolution of stramenopiles and alveolates as derived by "substitution rate calibration" of small ribosomal subunit RNA. *Journal of molecular evolution*, 42, 201-210.

Vaneechoutte, M., Dijkshoorn, L., Tjernberg, I., Elaichouni, A., De Vos, P., Claeys, G., Verschraegen, G., 1995. Identification of Acinetobacter Genomic Species by Amplified Ribosomal DNA Restriction Analysis. *Jour. Of Clinical Microb.*, 33(1): 11-15.

Vanhoutte, T., Huys, G., Cranenbrouck, S., 2005: Exploring microbial ecosystems with denaturing gardient gel electrophoresis (DGGE). BCCM Newletter: http://bccm.belspo.be/newsletter/17-05/bccm02.htm

Viti, C. and Giovannetti, L., 2005. Characterization of cultivable heterotrophic bacterial communities in Cr-polluted and unpolluted soils using Biolog and ARDRA approaches. *Applied Soil Ecology*, 28: 101-102.

Ward, D.M., Santegoeds, C.M., Nold, S.C., Ramsing, N.B., Ferris, M.J., Bateson, M.M., 1996. Biodiversity within hot spring microbial mat communities: molecular monitoring of enrichment cultures. *Antonie van Leeuwenhoek*, 71, 143–150.

Woese, C.R., 1987. Bacterial evolution. Microbiology Reviews, 51, 221–271.

Ye, W., Tan, J., Liu, X., Lin, S., Pan, J., Li, D., Yang, H., 2011. Temporal variability of cyanobacterial populations in the water and sediment samples of Lake Taihu as determined by DGGE and real-time PCR. *Harmful Algae,* 10: 472-479.

4

Gel Electrophoresis of Grapevine (*Vitis vinifera* L.) Isozymes – A Review

Gizella Jahnke, János Májer and János Remete
University of Pannonia Centre of Agricultural Sciences
Research Institute for Viticulture and Enology
Badacsony
Hungary

1. Introduction

Several articles were written from the beginning of the fifties about the presence of plant enzymes in multiple forms. The major discussion was questioning whether these forms are artifacts that rose during the purification or not. To show that these forms are not artifacts in 1952 Jermyn divided his original peroxidase juice into two parts by acidic precipitation. The precipitate contained the A and B, while the supernatant the C and D points. Two components were found in the purified peroxidase solution; one migrated to the anode, the other to the cathode (Jermyn and Thomas, 1954).

The first major step for the starting up of isozyme analysis was the development of starch gel electrophoresis by Smithies (1955). The second major step was the demonstration of the direct visualization of isozymes in the stach gel by specific histochemical stains by Hunter and Markert in 1957 (McMillin in Tanksley and Orton, 1983).

The term isozyme was formed by Market and Moller (1959), using this word for different molecular forms of enzymes with the same substrate specifity.

Proteins - as the primary products of structural genes - are very alluring for the direct genetic studies. Variation in the DNA coding sequences frequently (but not all the cases) causes variation in the primary conformation of the proteins. In un-natural environments the detection of this variation is very difficult, because in such conditions the base of the separation is only the size of the protein (molecular weight). In natural environments the change of a single amino acid can detectably modify the migration. The extraction from a single tissue can contain a lot of proteins, which - in the case of non-specific (e.g. Comassie blue) staining - can result in a complex pattern, that makes it difficult to identify the homolog (allelic) and non-homolog enzymes. This problem can be solved by the application of enzyme-specific staining after the electrophoresis (Shields et al. in Tanksley and Orton, 1983).

The analysis of isozymes and their functions is the subject of functional genomics. The study of the gene expression in the level of RNA and proteins can give answers to a lot of open questions (Bernardi, 2004).

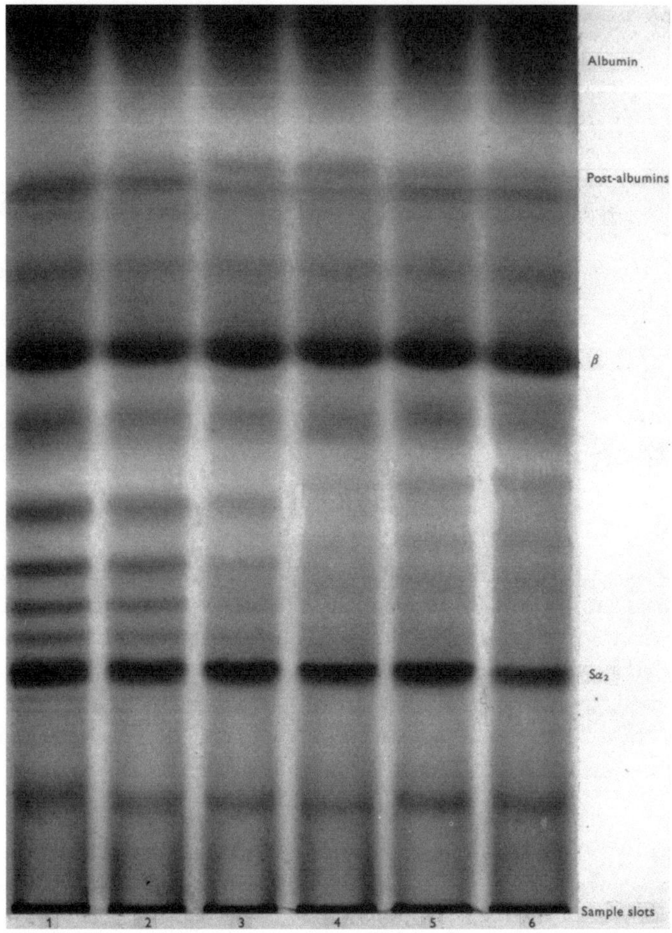

Fig. 1. A photograph of the result obtained by vertical starch-gel electrophoresis (approx. 19 hr. at 5v/cm.) with serum samples from six healthy individuals. Only the section of gel from the sample slots to the albumin is included in the photograph. Samples 1 and 2 are from female identical twins, 45 years old. Samples 3-5 are from 9-year-old female non-identical quadruplets and sample 6 is from their mother. (Smithies, 1958)

2. Grouping of isozymes

Isozymes are divided into three categories depending on the way they are biosynthesized:

1. isoenzymes (multilocus isozymes) arise from multiple gene loci, which code for structurally distinct polypeptide chains of the enzyme;
2. allozymes (or alleloenzymes), are structurally distinct variants of a particular polypeptide chain, coded by multiple alleles at a single locus;
3. secondary isozymes result from post-translational modifications of the enzyme structure;

The distinction between multiple alleles and multiple gene loci as causes of isozyme formation is that multiple alleles are the result of differences between individual members of a certain species, whereas multiple loci are common to all members of a species (Markert, 1975.).

The most probable reason for the presence of multilocus enzyme forms is the gene duplication. The gene duplication – the multiplication of genes in the genom –can come into existence by, for example, not equal crossing over. The frequency of mutation of various structural genes can be different, as a result of which some genes only rarely present in different allelic variant, as more alleles present in the population of other isoforms. This difference can be accepted as the evidence of a separate locus.

For the formation of multilocus isoforms a different evolutionary way can be imagined. It is probable, that the variation of the structural genes of originally different enzymes can cause the formation of similar catalytic functions. (H. Nagy, 1999).

Enzymes with variable substrates generally show higher variability itseves (catechol oxidase, acid phosphatase, peroxidase, esterase), but the amount of allozymic polymorphism is an increasing function of environmental variation. "Observations on natural pupulatians are cited which substantiate the claim that allozymic polymorphism is primarily due to selection acting on environmental variation in gene function. ...a large portion of the observed allozymic variation is due to a rather specific type of phenomenon: substrate variability" (Gillespie and Langley, 1974).

Enzymes with a single, special substrate show lower variability (glucose phosphate isomerase, phosphoglucomutase, glutamate-oxalacetate transaminase, glucose-6-phosphate transaminase etc.), but the banding patterns are less affected by the environment (Gillespie and Kojima, 1968; Gillespie and Langley, 1974).

3. Separation of isozymes by gel electrophoresis

Isoenzymes can be separated by electrophoresis or isoelectric focusing. The isozymes – under given proper circumstances – show peculiar patterns in the gel, which are called zymogramm (Hunter and Markert 1957).

Electrophoresis is a type of chromathography. The power for the separation of proteins is the difference in voltage between the two ends of the gel. The movement of proteins in the electric field is effected by their weight, shape and charge (Smith, 1960; Bálint and Bíró in Bíró, 1989).

The gel for the separation can be made from starch, agarose or acrilamide (Fig 2.). A standardized method of starch gel electrophoresis is used by UPOV (1996) for the analysis of identity of plant cultivars by isozyme analysis (Baum, 1986).

Advantages of the starch gel are that it is non-toxic, and more isozymes can be analysed by the slicing of a thick gel. More recently polyacrylamide gels are used because of their larger resolution. The porous structure of poliacrilamide gel is formed through a process of polymerization of acrylamide (CH_2=CH-CO-NH_2) and bis-acrylamide (CH_2=CH-CO-NH-CH_2-NH-CO-CH=CH_2). As a result of polimerisation a colourless, diaphanous, flexible, and consistent gel arises, which is resistant to scalding or chilling. The density, viscosity and size of poles are determined by the concentration of acrylamide and bis-acrylamide. (Hajósné Novák és Stefanovitsné Bányai in Hajósné Novák, 1999).

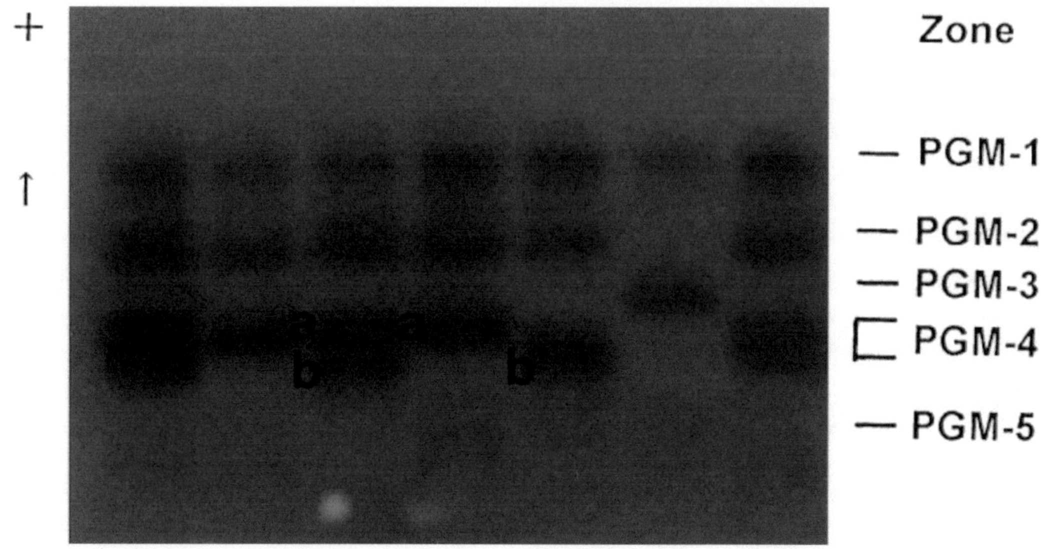

Fig. 2. Segregation of isozyme bands in PGM zone 4. Each column shows representative individuals of a selfed population of Cherry Bell-4. (Nomura et al., 1999.)

4. Separation of isozymes by isoelectric focusing

The charge of proteins is determined by the ratio of acidous and alkaline molecule parts and the rate of their dissociation. The rate of dissociation is determined by the pH of the surroundings of the molecule. The isoelectric point (pI) of the protein is the pH at which the acidous and alkaline molecule parts equally dissociate, the protein's net charge being zero. In a surrounding where the pH is lower than the isoelectric point, the net charge of the protein will be positive, in turn when the pH is higher than the IP, the net charge will be negative (Hames, 1990).

Isoelectric focusing of proteins can be carried out in a gel, in which a pH gradient is generated. Under voltage the proteins migrate to the point of the gel, where their net charge is equal to zero (pI). For this method thin poliacrilamide or agarose gels are used (Fig. 3).

The advantages of isoelectric focusing in opposition of gel electrophoresis are that the isozyme variants can be identified based on their isoelectric points, which results more accurate determination of isoforms, than the identification based on Rf values. On the other hand, the used gels in isoelectric focusing are thinner, so the separation is faster (Patterson and Payne, 1989).

Previously isoelectric focusing had disadvantages, as it required practice and the staining of the gels, because of the wide pH gradient, was difficult (Patterson és Payne, 1989), but nowadays these cause no problem. With the use of ampholites, the preparation of pH gradient gels needs even less practice, and you can even purchase ones. Neither causes problems with the staining of this so-called immobilized pH gradient gels, because the pH gradient can be removed by the washing of the gel.

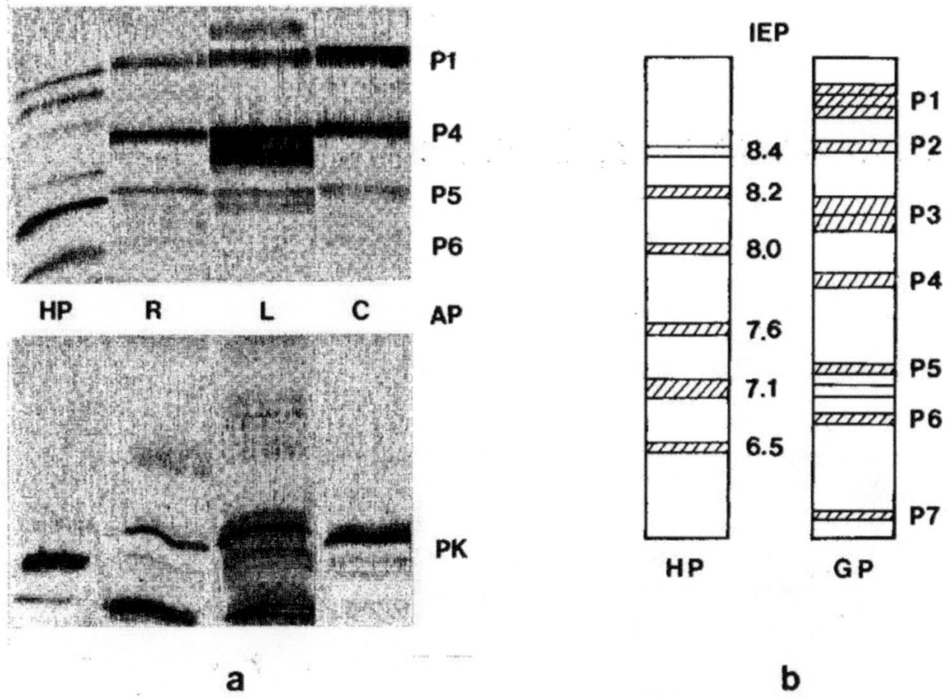

Fig. 3. Isoelectric focusing patterns of horseradish (HP), and grapevine (GP) peroxidases. IEP isoelectric points, P1-P7 names of isoenzyme bands, AP application zone, PK cathodic peroxidase. -a) anodic and cathodic pattern of peroxidases obtained from different organs of *Vitis riparia*. L Leaf, R root, C callus. The bands of horseradish peroxidases are distorted due to a margin effect on the slab. -b) Shematic diagram of the anodic part of peroxidase patterns. (Bachmann and Blaich, 1988)

5. Advantages and disadvantages of isozyme analyses

"The utility of isozymes as genetic markers is generally attributed to their frequent polymorphism, codominance, single gene-Mendelian inheritance, rapid, simple and relatively inexpensive assay and their ubiquity in plant tissues and organs (even in embrios and pollen). Although the selective neutrality of isozymes has been debated, it seems highly probable that they are adaptive under certain circumstances." (Bretting and Widrlechner 1995a).

Other advantages of isozyme analysis are the rapid analyses of samples, a small amount of plant material is sufficient. Young plants can be tested and selected based on their genotypes for features, which morphologically appear later. These can mean significant temporal and financial savings in the case of the breeding of annual plants. Now the best cost-efficient markers are the isozymes (Bretting and Widrlechner, 1995b).

Disadvantages of isozyme analysis as against the DNA markers, that they are organ-, tissue- and developmental stage-specific (Fig 4.). They often go through post-transcriptional modifications, which limit their usage (Staub et al., 1996).

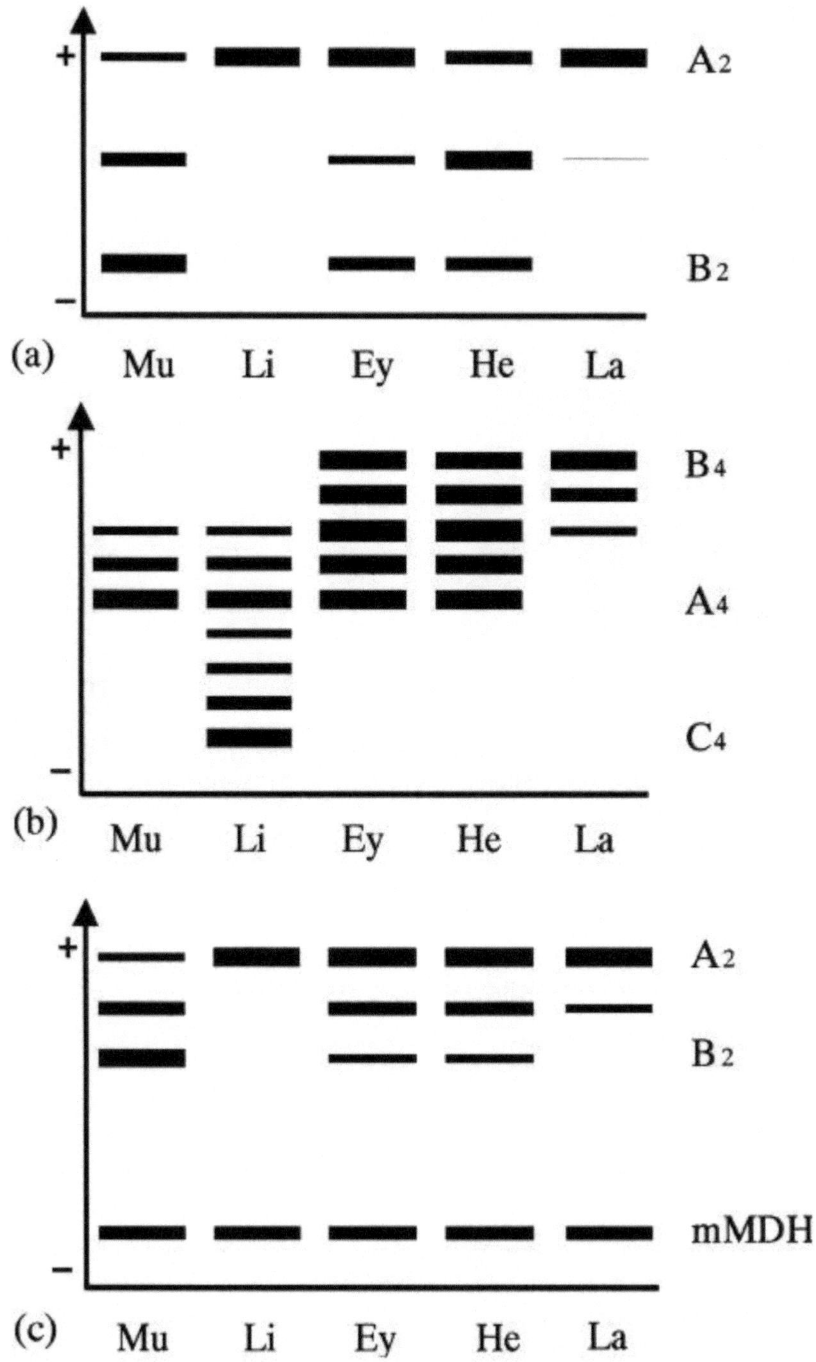

Fig. 4. Schematic representation of the differential expression of GPI (a), LDH (b) and MDH (c) isozymes in adult tissues (Mu, white muscle; Li, liver; Ey, eye; He, heart) and larvae (La) of L. cephalus. The egg pattern is identical to the larval one. Differences in line thickness refer to different staining intensities. (Manaresi et al. 1998)

6. Isozyme analysis in grape

Enzyme banding patterns for over 60 varieties of wine and table grapes were determined by gel electrophoresis by Wolfe (1976). Enzymes were extracted from ripe berries of each variety and separated by electrophoresis in a starch gel. Enzyme bands were detected by developing the gels in a buffered solution that produced an insoluble dye at the site of enzyme activity. The varieties were assayed for leucine aminopeptidase, indophenol oxidase, acid phosphatase, catechol oxidase, alcohol dehydrogenase, esterase, and peroxidase. The first four enzymes listed were found the most useful for distinguishing varieties.

Enzyme-banding patterns of catechol oxidase, acid phosphatase, esterase, alcohol dehydrogenase, indophenol oxidase, and leucine aminopeptidase obtained by enzyme staining of starch gel electropherograms allow the distinction of berries of the grape cultivars Perlette, Thompson Seedless, Superior Seedless and an early ripening sport of Superior Seedless (Schwennesen et al., 1982).

Twenty-seven cultivars and feral accessions from four Vitis species were examined by SUBDEN et al. (1987) for 12 isozyme systems exhibiting polymorphism. Using extracts from woody tissue and a protocol to avoid isozyme inactivation by polyphenolics and other materials, 27 of 29 strains exhibited unique isozyme banding patterns for glucose-6-phosphate isomerase, peptidase, and acid phosphatase. Implications for genetic homogeneity screening of nursery stock or identifying unknown samples are discussed.

German researchers analysed the isozymes of peroxidase by isoelectric focusing. Purified internodal phloem extracts from dormant wood were used. In the 6-11 pH range 8 bands were found, 71 Vitis species and varieties were identified (Bachmann and Blaich, 1988).

Genetic analysis of 11 allozyme polymorphisms was performed by Weeden (1988) on the progeny of 'Cayuga White' x 'Aurore', two complex interspecific grape (*Vitis*) hybrids. Segregation for most of the polymorphisms closely approximated monogenic Mendelian ratios, and eight new isozyme loci were defined for grape. Joint segregation analysis among the isozyme loci revealed three multilocus linkage groups (ACP-1—PGM-c; ACP-2—AAT-c; GPI-c—LAP-1). These results demonstrate that sufficient allozyme polymorphism exists in grape to establish many multilocus linkage groups and that this genetic analysis can be accomplished using extant progeny or progeny readily produced from highly heterozygous clones.

The pattern of the systems PER and ACP from 8 vines of *Vitis vinifera* L. has been studied in 1988 by Royo et al. (1989). A method to differentiate and characterize 6 clones of *Vitis vinifera* L. has been established by gaining the variability of the PER pattern from the band pattern constantly present in any vine, and the total band pattern from another vine (not only amongst the vines but also along the vegetative cycle). In the vines investigated no difference has been found for the ACP system.

Three enzymes in 5 cultivars of *Vitis vinifera* L. are analyzed by PAGE in young leaves. With acid phosphatase, arylesterase and glutamat-oxalat transaminase more or less different isoenzyme patterns of the different cultivars were obtained. There were no interclonal differences. The most polymorph enzyme was the arylesterase. The best results were obtained with young leaves from sprouting buds (Eiras-Dias et al., 1989).

Starch-gel electrophoresis was used by Walters et al (1989) for the analysis of *Vitis vinifera* L. cultivars, interspecific *Vitis* hybrids and wild individuals of *Vitis riparia* Michx. They suggest a simple and inexpensive procedure for the extraction of active enzymes from grape, which is rapid and efficient. Starch-gel electrophoresis with different optimized gel-electrode buffer systems is used for 40 different isoenzymes, 14 of which were consistently resolvable and showed variation among different cultivars.

Isozyme analysis is one of the means suitable to characterize clonally propagated cultivars. Isoelectric focusing was used to reveal differences in isozyme patterns between tissue-cultured plants and mother plants, for the cultivars Barbera, Queen of the Vineyards, Dolcetto and Delight. In cultivar Barbera both 2n and 4n plants were considered. Leaf samples were collected from shoots grown on cuttings under controlled environmental conditions and from plants obtained by tissue culture. The buds used for tissue culture were taken from the same shooted cuttings. Leaf extracts were analyzed by isoelectric focusing considering the following isozymes: AcPH (acid phosphatase), GPI (glucose phosphate isomerase) and PGM (phosphoglucomutase). The banding patterns of GPI and PGM showed differences among the cultivars, while for AcPH there seemed to be no differences among them in the pH range considered. There were no differences between isozyme patterns of the Barbera 2n and Barbera 4n. The main difference between in vitro plants and mother plants was the amount of isozyme evaluated by densitometric measurements. In all the cultivars, the amount of isozymes for AcPH was higher in mother plants than in in vitro ones, while for PGM and GPI it was the opposite. This can be due to the different environmental conditions affecting cellular metabolism (Botta et al., 1990).

The idea of using woody stems during the resting period instead of leaves for the isozyme analysis arose in 1990. Kozma et al. (1990) analysed the esterase isozymes of varieties from different convarietas and interspecific hybrid families by poliacrilamide gel electrophoresis and isoelectric focusing. Based on their results they established, that the phloem extracts from woody stems collected in the resting period or shoots colleting in spring give good reproducible patterns, but the leaf extracts give irreproducible patterns.

Tests were carried out on different types of calli and somatic embryos of V. rupestris using 2-D electrophoresis. The investigation carried out by Martinelli et al. (1993) is focused on the isozyme patterns of AcP (acid phosphatase), ADH (alcohol dehydrogenase), EST (esterase), G6PDH (gluconate-6-phosphate dehydrogenase) and PGM (phosphoglucomutase). A typical variation of isozyme pattern could be observed during the different steps of somatic embryogenesis. De-differentiated callus showed other types of isoenzyme pattern compared to those obtained during the development of somatic embryos.

Similarly to Kozma et al. Bachmann (1994) used extracts from phloem of dormant canes for the isozyme analysis. This comprehensive study has analysed the peroxidase isozyme banding patterns of 313 different cultivars and species of *Vitis* using isoelectric focusing on polyacrylamid gels. The author reports that acidic peroxidases were characteristic for *Vitis vinifera* L. cultivars with only a 5 % frequency of occurrence in other *Vitis* species. Variation in neutral to basic peroxidases could be used to group together similar cultivars independent of berry colour, e.g. Pinot noir, Pinot gris, Pinot blanc and Pinot meunier. However, other examples of colour variants, e.g. Merlot noir and blanc were clearly different using peroxidase banding.

Shiraishi et al. (1994) used GPI and PGM isozyme banding patterns for the detection of hybrid origin of seedlings during their triploide grape breeding. First they analysed 99 diploide cultivars, 20 diploide plants from 8 wild *Vitis* species and populations from the crosses between them . In the GPI-2 locus 13 in the GPM locus 11 allels was found. Data showd high genetical differences between Vitis species . After that, the GPI-2 and PGM-2 genotype of 6 diploide and 4 tetraploide cultivars (used for the crosses) were determined. 15 diploide x tertraploide crosses were made. Trisomy gene expression was detected in 92 out of 98 seedlings, as 6 showed diploide patterns.

Seed proteins and enzymes (AcP, ADH, EST, G-6-PDH, MDH, PGM, POD) from several cultivars and wild ecotypes of *Vitis vinifera* L. have been used to evaluate taxonomic differences between *V. vinifera* sspp. sativa and sylvestris (Scienza et al., 1994). Only total proteins in the pH range of 4.0-5.5 and AcP, EST and G-6-PDH were useful for genotype differentiation. The cluster analysis (UPGMA), based on Jaccard genetic distance and determined on the presence/absence of electrophoretic profiles, reveals 2 distinct groups, supporting the hypothesis of the authors that *V. sativa* and *V. silvestris* should be regarded as 2 separate taxa.

Studies on the induction characteristics and the fine structure of grapevine cells cultured in vitro were undertaken with cultivar Monastrell berry samples of different developmental stages between fruit set and veraison (Zapata et al., 1996). Medium composition, electron microscopy application and protoplast isolation procedures are explained. It could be shown that the intensity of cell development and callus induction percentage depended on the berry growth stage; the de-differentiation process is mainly located in meso-carp tissues. Cultured cells showed to be highly vacuolated with their cytoplasm reduced to a very thin peripheral layer (containing golgi sacks).

Ros Barceló et al. (1996) studied the gene expression of isozymes of providase in downy mildew resistant (*Vitis vinifera* x *Vitis rupestris*) x *Vitis riparia* hybrids and in the susceptible *Vitis vinifera* parent. The peroxidase isoenzyme type B3 (PI=8,9) expressed in the phloem and leaves of resistant hybrids was completely absent in the susceptible parent.

To test whether the basic peroxidase isoenzyme B3 may be considered as a molecular marker of disease resistance in *Vitis* species, suspension cell cultures derived from the downy mildew susceptible *V. vinifera* parent species were treated with an elicitor (cellulase Onoztika R-10) from the soil fungus *Trichoderma viride,* a specific and well-known elicitor of disease resistance reactions in grapevines. The results showed that treatment with the elicitor induces, simultaneously with the activation of the disease resistance mechanism, the appearance of B3 in the cell cultures. These results suggest that the basic peroxidase isoenzyme B3 may be considered as a marker of disease resistance in Vitis species.

Isoenzymes from grapevine woody stems and shoots were evaluated for their use in identification of varieties and clones by Royo et al. (1997). Plant extracts were separated by polyacrylamide gel electrophoresis. Isoenzyme analysis was carried out for esterases, peroxidases, catechol oxidase, glutamate oxalacetate transaminase and acid phosphatase. The plant material was grown and sampled at two localities in Spain, with different climatic conditions. Sampling was carried out bimonthly for two consecutive years in order to find out the influence of the environment and time of the year. Each isozyme system had a pattern defined by 'fixed' bands that were always present at both localities and during the resting period of the plant (autumn —winter).

An evaluation of the genetic diversity of 'Albariño' (*Vitis vinifera* L.) was carried out by Vidal et al. (1998). The 73 isozyme and 308 RAPD markers were common in the samples tested. The results show the existence of a genetic homogeneity within 'Albariño'cultivated in Galicia. Minor ampelographic differences among samples could be due to external factors rather than to genetic differences.

DNA and isoenzyme analyses were used to characterize 20 table grape cultivars including Moscato d'Amburgo, Italia, Sultanina, Bicane and some recently released new cultivars (Crespan et al., 1999). GPI and PGM isoenzyme systems were able to separate the cultivars into 9 groups whereas the 8 microsatellite loci that were analysed revealed a higher discriminating power. Parentage analysis confirmed that the cultivar Italia was obtained from the crossing Bicane x Moscato d'Amburgo.

Hungarian researchers used isoelectric focusing for the peroxidase and esterase isozymes of some grapevine cultivars. Samples were gathered at different times of the year. The leaf samples after blooming were found the best for the identification of varieties, but they found the phloem extracts of woody stems also suitable for cultivar identification (Stefanovits-Bányai et al., 1999; Stefanovits-Bányai et al, 2002).

Sixty-four Muscat flavoured grapevine accessions were analysed in the work of Crespan and Milani (2001). An analysis was performed at two isozymes and 25 microsatellite loci. The 64 accessions were reduced to 20, which were easily distinguishable from each other at the molecular level by as few as two microsatellite loci. The remaining 44 were found to be synonyms. Three mutants with red and pink coloured berries were identified in the Moscato bianco group. Moscato nero encompasses at least two, Moscato rosa three different varieties. It seems that only two of the analysed Muscats are the main progenitors of the Muscat family: Moscato bianco and Muscat of Alexandria, which in turn are joined by a direct parent-offspring link.

Sànchez-Escribano et al. (1998) analysed 43 table grape varieties by 6 isoenzyme systems (PER, CO, GOT SOD, EST, AcP). The last 2 enzymes were found unsuitable for identification, by the combination of the zymogram of the other 4 enzymes, they were able to identify 31 cultivars, as the remaining 12 were clustered to 5 groups.

Protein and esterase isozyme patterns of authentic grapes and wines of 13 white wine cultivars were determined by means of isoelectric focussing (range of pI: 2.5-10) by Paar et al (1999). Esterase staining with grapes showed active zones mainly in the alkaline pI-range, with most of the cultivars, however indicating no qualitative, but only quantitative differences. Staining of the protein patterns of grapes and wines with Coomassie Brilliant Blue proved to be well suitable for the differentiation of cultivars. With grapes as well as wines the most predicative bandings focussed in the acid pI-range of 4. With the cultivars Grüner Veltliner, Rotgipfler and Riesling Italico the protein banding patterns were so characteristic, that these cultivars were easily identified, whereas with the other cultivars detailed comparisons of the phenogrammes were necessary.

Isozyme and RAPD markers were used for the characterization of Hungarian grapevine varieties and their parents (HAJÓS-NOVÁK And HAJDÚ 2003). The cathecol-oxidase system was found the most suitable for identification proposes.

Isozyme and SSR analysis were carried out for the differentiation of the grapevine cultivars Kéknyelű and Picolit. The name of the grapevine cultivar 'Kéknyelű' has become inseparable from the name of the Badacsony vine region (Hungary), whose fame is also well known beyond the Hungarian frontier. In the Vitis International Variety Catalogue (http://www.genres.de/idb/vitis/) 'Kéknyelű' is reported, as the synonym of the Italian grapevine cultivar 'Picolit'. Vertical poliacrylamide-gel electrophoresis was used for the investigation isoenzymes of catechol-oxidase (CO) and acid phosphatase (AcP). Microsatellite analyses were carried out at 6 loci (VVS2, VVS16, VrZag79, VVMD7, VMC4A1, VMC4G6). The results of the isoenzymatic and microsatellite analyses confirmed, that this two cultivars are different (Jahnke et al., 2007).

Jahnke et al. (2009) investigated the genetic diversity of Hungarian grapevine cultivars with biochemical and molecular markers (isoenzyme and SSR). The isoenzyme patterns of 4 enzyme systems (catechol-oxidase, glutamate-oxalacetate-transaminase, acid phosphatase and peroxidase) and the microsatellite profile in 6 loci (VVS2, VVS16, VVMD7, VMC4A1, VMC4G6, VrZag79) of 48 grapevine varieties were analysed.

The results with CO, GOT, AcP and PER enzymes were reproducible and the zymograms obtained from the woody stems were independent from the time of sampling during the dormant period of the grape (Fig 5.).

Based on the isoenzyme patterns of these 4 enzymes most of the investigated varieties (40/48) were identified. A correlation was found between the isoenzyme patterns and the classification to convarietas of the varieties.

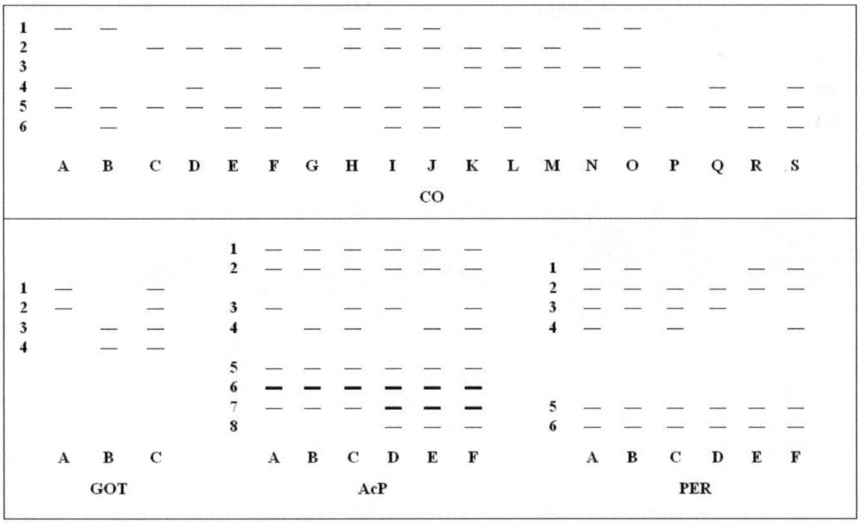

Fig. 5. Characteristic interpretative zymograms observed for CO, GOT, AcP and PER enzymes. The letters mark the different types of isoenzyme patterns, while numbers refer to the number of different isoenzyme bands (Jahnke et al. 2009.)

It was established, that while the varieties of the convarietas pontica differed from those of the convarietas orientalis and occidentalis, the two latter groups could have not been differentiated from each other. Based on the SSR (simple sequence repeat) analyses 46 of the

48 investigated varieties were identified. Even 'Pinot blanc' and 'Pinot gris' cultivars belonging to the same conculta (Pinot) could be differentiated in their VMC4A1 locus.

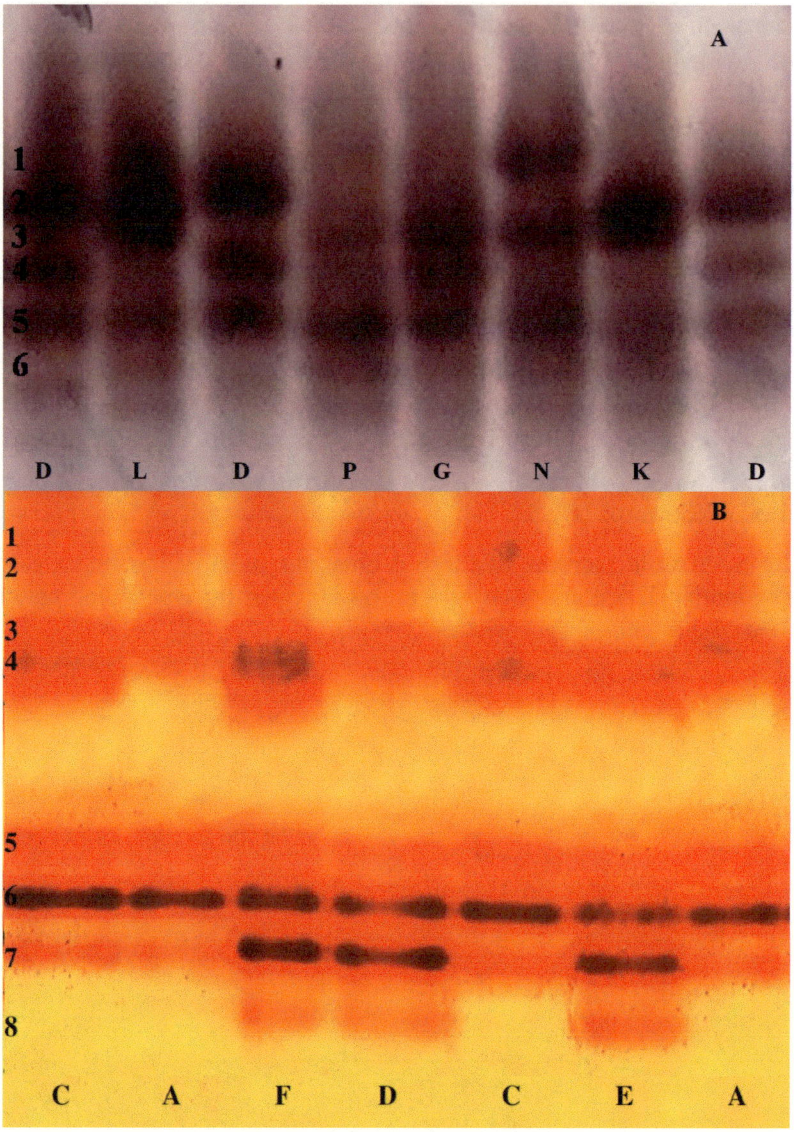

Fig. 6. Isoenzyme gel photos for CO (A) and AcP (B) respectively. The numbers show the band numbers, and the capital letters the banding pattern types shown in Figure 5. (Jahnke et al. 2009)

7. References

Bachmann O. (1994): Peroxidase isoenzyme patterns in Vitaceae. Vitis 33: 151-153.

Bachmann O., Blaich R. (1988): Isoelectric focusing of grapevine peroxidases as a tool for ampelography. Vitis 27: 147-155.

Bálint M., Bíró E. (1989): A fehérjekutatás fizikai-kémiai, preparatív és analitikai módszerei. 129-166. p. In: Bíró E. (Eds.) Biokémia I. Budapest: Tankönyv K.

Baum B. R. (1986): International Registration of Cultivars with Emphasis on Barley: Procedures and Methods of Producing a Register. Acta Horticulturae (182):237-250.

Bernardi G. (2004): Structural and Evolutionary Genomics. Natural Selection in Genome Evolution. Elsevier, Amsterdam

Botta R., Vallania R., Miaja M. L., Vergano G., Me G. (1990): Isozyme pattern comparison between tissue-cultured grapevines and mother plants. Vitis Special Issue: 88-92.

Bretting P. K., Widrechner M. P. (1995b): Genetic Markers and Horticultural Germplasm Management. HortScience 30(7) 1349-1356.

BrettingP. K., Widrlechner M. P (1995a): Genetic markers and plant genetic resource management. Plant Breeding Reviews (13): 11-86.

Crespan M., Botta R., Milani N. (1999): Molecular characterisation of twenty seedless table grape cultivars (Vitis vinifera L.). Vitis 38: 87-92.

Crespan M., Milani N. (2001): The Muscats: A molecular analysis of synonyms, homonyms and genetic relationship within a large family of grapevine cultivars. Vitis 40: 23-30.

Eiras-Dias J. E. J., Sousa B., Cabral F., Carcalho I. (1989): Isozymatic characterisation of portuguese vine varieties of Vitis vinifera L. Rivista di Viticoltura e di Enologia 1: 23-26.

Gillespie J. H., Kojima K. (1968): The degree of polymorphisms in enzymes involved in energy production compared to that in nonspecific enzymes in 2 Drosophila-ananassae populations. Proceedings of the National Academy of Science of the USA 61: 582-585.

Gillespie J. H., Langleyc. H. (1974): A general model to account of enzyme variation of natural populations. Genetics 76: 837-846.

H. Nagy A. (1999): A genetikai variabilitás vizsgálata izoenzimekkel. 25-39 p. In: HAJÓSNÉ NOVÁK M. (Eds.): Genetikai variabilitás a növénynemesítésben. Budapest: Mezőgazda Kiadó, 142 p.

Hajósné Novák M., Stefanovitsné Bányai É. (1999): Az izoenzimek elválasztása gélelektroforézissel. 99-104 p. In: HAJÓSNÉ NOVÁK M. (Szerk.): Genetikai variabilitás a növénynemesítésben. Budapest: Mezőgazda Kiadó, 142 p.

Hajós-Novák M., Hajdu E. (2003): Isozyme and DNA fingerprinting characterization of two hungarian wine grape hybrids and their parents. Acta Horticulturae 603: 211-216.

Hames B. D. (1990): One-dimensional polyacrylamide gel electrophoresis. 1.-139. p. In: Hames B.D., Rickwood D. (Eds.): Gel electrophoresis of proteins. ILR press at Oxford University press, Oxford- New York- Tokyo, Sec. Ed.

Hunter R. L., Markert C. L. (1957): Histochemical Demonstration of Enzymes Separated by Zone Electrophoresis in Starch Gels. *Science* (125): 1294-1295

Hunter R. L., Markert C. L. (1957): Histochemical demonstration of enzymes separated by zone electrophoresis in starch gels. Science 125: 124-1295.

Jahnke G., Korbuly J., Májer J., Györffyné Molnár Júlia (2007): Discrimination of the grapevine cultivars 'Picolit' and 'Kéknyelű ' with molecular markers. Scientia Horticulturae (114): 71-73.

Jahnke G., Májer J., Lakatos A., Györffyné Molnár J., Deák E., Stevanovits-Bányai É., Varga P. (2009): Isoenzyme and microsatellite analysis of Vitis vinifera L. varieties from the Hungarian grape germplasm. Scientia Horticulturae (120): 213–221.

Jermyn M. A., Thomas R. (1954): Multiple components in horse-radish peroxidase. Biochemical Journal. 56: 631-639.

Kozma P., H. Nagy A., Juhász O. (1990): Inheritence of isoenzymes and soluable proteins in grape vareties and F1 hybrids. Proceedings of the 5th International Symposium on Grape Breeding ; Vitis Special Issue: 134-141

Li, B.; Xu, Y. & Choi, J. (1996). Applying Machine Learning Techniques, *Proceedings of ASME 2010 4th International Conference on Energy Sustainability*, pp. 14-17, ISBN 842-6508-23-3, Phoenix, Arizona, USA, May 17-22, 2010

Lima, P.; Bonarini, A. & Mataric, M. (2004). *Application of Machine Learning*, InTech, ISBN 978-953-7619-34-3, Vienna, Austria

Manaresi S., Mantovani B., Zaccanti F. (1998): Comparative Analysis of Glucosephosphate Isomerase, Lactate Dehydrogenase and Malate Dehydrogenase Isozymes in 9 Cyprinid Species from Italy. Zoological Science 15:461-467.

Markert C., Moller L. F. (1959): Multiple forms of enzymes: tissue, ontogenetic, and species specific patterns. Proceedings of the National Academy of Sciences of the USA. 45: 753-763.

Markert C. (Ed.). (1975): Isozymes. Vol 1–4, Academic Press, New York.

Martinelli L., Scienza A., Villa P., De Ponti P., Gianazza E. (1993): Enzyme Markers for Somatic Embriogenezis in Vitis. Journal of Plant Phisiology 141: 476-481. p.

Mcmillin D. E. (1983): Plant isozymes: a historical perspective. 3-13. p. In: TANKSLEY S. D., ORTON T. J. (Szerk.): Isozymes in Plant Genetics and Breeding. Amsterdam: Elsevier Science Publishers B.V.

Nomura K., Yoneda K., Inoue H., Tateishi A., Kanehira T., Isobe K. (1999): Identification and linkage analysis of isozymes loci in open-pollinated varieties of radish (*Raphanus sativus* L.). Scientia Horticulturae 82: 1-8.

Paar E., Doubek S., Eder R. (1999): Differenzierung von Wiessweinsorten mittels isoelektischer Fokussierung. Mitteilungen Klosterneuburg 49:176-185.

Patterson B. A., Payne L. A. (1989): Zymograms of plant extracts using isoelectric focusing on ultra thin layers. Acta Horticulturae 247:163-169.

Ros Barceló A., Zapata J. M., Calderón A. A. (1996): A Basic Peroxidase Isoenzyme, Marker of Resistance Against Plasmopara viticola in Grapevines, is Induced by an Elicitor from Trichoderma viride in Susceptible Grapevines. Journal of Phytopathology 144: 309-313

Royo B., Gonzalez J., Laquidain M. J., Larumbe M. P. (1989): Caracterization mediante analisi isoenzimatico de clones de la vinifera "Garnacha" (Vitis vinifera L.). Investigación agraria. Producción y protección vegetales. 4: 343-354.

Royo J. B., Cabello F., Miranda S., Gogorcena Y., Gonzalez J., Moreno S., Itoiz R., Ortiz J. M. (1997): The use of isoenzymes in characterisation of grapevines (Vitis vinifera L.). Influence of the environment and time of sampling. Scientia Horticulturae 69: 145-155.

Sànchez-Escribano E., Ortiz J. M., Cenis J. L. (1998): Identification of table grape cultivars (Vitis vinifera L.) by the isozymes from the woody stems. Genetic Resources and Crop Evolution 45: 173-179.

Schwennesen J., Mielke E. A., Wolfe W. H. (1982): Identification if Seedless Table Grape Cultivars and Bud Sport with Berry Isozymes. HortScience 17: 366-368.

Scienza A., Villa P., Tedesco G., Parini L., Ettort C., Magenes S., Gianazza E. (1994): A chemotaxonomic investigation on Vitis vinifera L. II. Comparison among ssp. sativa traditional cultivars and wild biotypes of ssp. silvestris from various Italian regions. Vitis 33: 217-224.
Scienza A., Villa P., Tedesco G., Parini L., Ettort C., Magenes S., Gianazza E. (1994): A chemotaxonomic investigation on Vitis vinifera L. II. Comparison among ssp. sativa traditional cultivars and wild biotypes of ssp. silvestris from various Italian regions. Vitis 33: 217-224.
Shields C. R., Orton T. J., Stuber W. (1983): An outline of general resource needs and procedures for the electrophoretic separation of active enzymes from plant tissue. 443-467 p. In: Tanksley S. D., Orton T. J. (Eds.): Isozymes in Plant Genetics and Breeding. Amsterdam: Elsevier Science Publishers

Shiraishi S, Ohmia Wakama C., Hiramatsu M. (1994): Variation of Gluceosephosphate Isomerase and Phosphoglucomutase Isozymes in Vitis and Their use in Grape Breeding. J. Fac. Agr., Kyushu Univ., 38 (3.4), 255-272.

Siegwart, R. (2001). Indirect Manipulation of a Sphere on a Flat Disk Using Force Information. *International Journal of Advanced Robotic Systems,* Vol.6, No.4, (December 2009), pp. 12-16, ISSN 1729-8806

Smith I. (1960). Chromatographic and electrophoretic techniques. Vol. II Zone electrophoresis. Heinemann, London.

Smithies O. (1955): Zone Electrophoresis in Starch Gels.: Group Variations in the Serum Proteins of Normal Human Adults. Biochemical Journal 61: 629-641.

Smithies O. (1958): An Improved Procedure for Starch-gel Electrophoresis: Further Variations in the Serum Proteins of Normal Individuals. Biochemical Journal 71: 585-587.

Staub J. E., Serquen F. C., Gupta M. (1996): Genetic Markers, Map Construction, and Their Application in Plant Breeeding. HortScience 31: 729-740.

Stefanovits-Bányai É, Lakatos Zs., Kerepesi I., Balogh I. (1999): Esterase and peroxidase isozyme patterns of some Vitis vinifera L. species. Publicationes Uniersitate Horticulturae Alimentariae LIX: 42-46.

Stefanovits-Bányai É., Lakatos S., Hajós-Novák M., Hajdu E., Balogh I. (2002): Recent developments in biochemical characterisation of Vitis vinifera L. varieties in Hungary. International Journal of Horticultural Science 8: 57-61.

Subden R. E., Krizus A., Lougheed S. C., Carey K. (1987): Isozyme Characterisation of Vitis Species and Some Cultivars. American Journal of Enology and Viticulture 38:176-181.

UPOV (1996): Guidelines for the conduct of tests for distinctness, uniformity and stability. Rape seed (Brassica napus L. oleifera).

UPOV, Geneva, Switzerland TG/36/6

Vidal J. R., Moreno S., Masa A., Ortiz J. M. (1998): Study of the genetic homogeneity of Albariño (Vitis vinifera L.) growing in Galicia (Spain) using isozyme and RAPD markers. Vitis 37: 145-146.

Walters T. W., Posluszny U., Kevan P. G. (1989): Isozyme analysis of the grape (Vitis). I. A practical solution. Canadian Journal of Botany 67: 2894-2899.

Weeden N. F., Reisch B. I., Martens M. H. E. (1988): Genetic Analysis of Isozyme Polimorohism in Grape. Journal of the American Society for Horticultural Sciences 113 (5): 765-769.

Wolfe W. H (1976): Identification of grape varieties by isozyme banding patterns. American Journal of Enology and Viticulture 27: 68-73.

5

Molecular Electrophoretic Technique for Authentication of the Fish Genetic Diversity

Tsai-Hsin Chiu, Yi-Cheng Su, Hui-Chiu Lin and Chung-Kang Hsu
[1]*Department of Food Science, National PengHu University of Science and Technology*
[2]*Seafood Research and Education Center, Oregon State University*
[3]*Penghu Marine Biology Research Center, Fisheries Research Institute, COA, EY*
[1,3]*Taiwan*
[2]*USA*

1. Introduction

Cobia (*Rachycentron canadum*) is the sole representative of their family, the Rachycentridae They are distributed worldwide in tropical and subtropical seas, as the Atlantic and Pacific Oceans (Miao, et al., 2009). There are several species, including cobia, seabream, red progy, snappers, scads and groupers that are raised by cage culture in Taiwan. Among these cage-cultured fishes, cobia certainly takes a leading distribution in both annual total production (81.9%) and total value production (75.4%) as compared to the rest in Taiwan (Fisheries Agency 2006).

Giant grouper (*Epinephelus lanceolatus*) are also found in tropical and subtropical waters from the Indo-Western Pacific Ocean. It is one of the two largest species of groupers in the world. Due to its fast growth and high price, giant grouper currently is regarded as a favorite species for marine culture in Taiwan (Hseu, et al., 2004).

Red coral trout (*Plectropomus leopardus*) a reef-associated fish in Western Pacific, distributed from southern Japan to Australia and eastward to the Caroline Islands (Zhang, et al., 2010). Only few studies concerning population genetics of *Plectropomus leopardus* has been reported.

All of cobia, giant grouper, and red coral trout are high-valued fish market in Taiwan and neighboring countries, including China, Japan, and Vietnam. For the globalization of the seafood industry, seafood authentication and food safety are very important. We must know that the source of fish or accurately species of the fish. Traditional method to distinguish the fish species was observed the external traits. It can cause the error judgment. Today, DNA-based methods are also more frequently employed for food authentication (Lockley and Bardsley, 2000). It has proven to be reliable, sensitive and fast for many aspects of fish species and food authentication. Asensio et al. (2009) were suggesting that the species-specific PCR method could be potentially used by regulatory agencies as routine control assay for the commercial grouper fillets authentication. PCR-based methods commonly used for fish species identification include PCR-sequencing, random amplified polymorphic DNA (RAPD), inter simple sequence repeat (ISSR). Those methods are simplicity, specificity and sensitivity.

Recently, many researchers have reported for the assessment of genetic structure of aquaculture species such as red mullet (Mullus barbatus), Tropical abalone (Haliotis asinina), and suminoe oyster (Crassostrea ariakensis) using several kinds of molecular markers (Garoia et al., 2004; Tang et al., 2004; Zhang et al., 2005; Maltagliati, et al., 2006), including RAPD, ISSR, AFLP, and RFLP methods. Many molecular methods are available for studying various aspects of wild populations, captive broodstocks and interactions between wild cultured stocks of fish and other aquatic species (Okumus and Ciftci, 2003). Among those methods, RAPD and ISSR technology were cheaper, simple, and fast. And just only one primer could obtain the different profiles for genomic analysis (Welsh and McClelland, 1990). RAPD is simple, rapid and cheap, it have high polymorphism. RAPD analysis has been used to evaluate genetic diversity for species, subspecies and population identification in common carp (Bártfai, et al., 2003), Indian major carps (Barman, et al., 2003). The microsatellite method was already used to study of genetic diversity of other grouper (Antoro, et al., 2006; Ramirez, et al., 2006; Wang, et al., 2007). Zeng, et al. (2008) have report that genetic analysis of Malaysia and Taiwan wild populations of giant grouper by microsatellite method. Their results were shown polymorphic loci in those populations, but they didn't discriminate the wild and cultivated populations of giant grouper. Beside, genetic markers can be suitable for assessing the differences between culture stocks and wild population and monitoring the changes in the genetic variation (Okumuş, et al., 2003). Monitoring the genetic diversity of natural populations and fish raised in fish hatcheries is fundamentally important for species conservation. Molecular markers can be very useful in this context (Povh et al., 2008).

In our study, we try to identify the seafood products, including cobia, giant grouper, and red coral trout from cultivated and wild populations by molecular markers, and provide the fish population genetic diversity for seafood management and good monitoring for brood stock management.

2. Methods and material

2.1 Fish sampling and genomic DNA extraction

14 giant grouper (*Epinephelus lanceolatus*) and 14 cobia (*Rachycentron canadum*) were collected from Southern Taiwan, as Penghu Island, Kaohsiung, and Pengtung during 2007-2009. Those were selected in different cultured farms and local markets. 14 of red coral trout (*Plectropomus leopardus*) were collected from Penghu Island during 2009-2011. All samples were described in Table 1. All specimens were confirmed in the laboratory.

Approximately 1g of fish muscle tissue samples were cut into small pieces and pulverized in liquid nitrogen. The powdered fish samples were obtained and extracted genomic DNA using the QIAGEN® DNeasy Blood kit (QIAGEN Inc., Valencia, California) according to manufacturer's instructions. The extracted DNA concentration in 200 μl of sterile water and then the quality of DNA were assessed by a Qubit™ Fluorometer (invitrogen, USA). The DNAs were stored at -20°C until PCR amplifications.

2.2 PCR-RAPD method

A total of 95 RAPD primers were used for PCR, which were shown in Table 2. Those sequences were obtained from University of British Columbia Biotechnology Laboratory,

No.	Species	Sampling sources	No.	Species	Sampling sources	No.	Species	Sampling sources
1	*E. lanceolatus*	Wild	R1	*R. canadum*	Wild	71411	*P. leopardus*	Wild
E2	*E. lanceolatus*	Cultivated	R2	*R. canadum*	Wild	71412	*P. leopardus*	Wild
E3	*E. lanceolatus*	Cultivated	R3	*R. canadum*	Cultivated	71413	*P. leopardus*	Wild
E4	*E. lanceolatus*	Cultivated	R4	*R. canadum*	Wild	82011	*P. leopardus*	Wild
E5	*E. lanceolatus*	Wild	R5	*R. canadum*	Cultivated	91321	*P. leopardus*	Wild
E6	*E. lanceolatus*	Cultivated	R6	*R. canadum*	Wild	91711	*P. leopardus*	Wild
E7	*E. lanceolatus*	Wild	R7	*R. canadum*	Cultivated	91712	*P. leopardus*	Wild
E8	*E. lanceolatus*	Wild	R8	*R. canadum*	Cultivated	71421	*P. leopardus*	Cultivated
E9	*E. lanceolatus*	Wild	R9	*R. canadum*	Wild	71422	*P. leopardus*	Cultivated
E10	*E. lanceolatus*	Cultivated	R10	*R. canadum*	Wild	71423	*P. leopardus*	Cultivated
E11	*E. lanceolatus*	Cultivated	R11	*R. canadum*	Wild	81621	*P. leopardus*	Cultivated
E12	*E. lanceolatus*	Cultivated	R12	*R. canadum*	Wild	80622	*P. leopardus*	Cultivated
E13	*E. lanceolatus*	Wild	R13	*R. canadum*	Wild	81623	*P. leopardus*	Cultivated
E14	*E. lanceolatus*	Wild	R14	*R. canadum*	Wild	91322	*P. leopardus*	Cultivated

Table 1. Specimens of *E. lanceolatus, R. canadum, and P. leopardus* fish analyzed and locality where they were collected

RAPD Analysis Kit (Amersham Pharmacia Biotech, Piscataway, NJ), and Operon primer kit (Operon, Advanced Biotechnologies). DNA amplification was performed in a final volume of 25 µl in the "Gene Amp PCR System 2720" thermal cycler (Applied Biosystems Inc., USA). The reaction mix contained 20 mM Tris-HCl, pH8.0, 50 mM KCl, 2 mM $MgCl_2$, 10 mM dNTPs each (dATP, dCTP, dGTP, dTTP), 20 µM of primer, 2.5 U *Taq*-polymerase (Promega, Co., Wisconsin, USA) and 1 µl of the 10 ng extracted DNA. The pre-amplification PCR procedure was: treatment at 94°C for 5 min, followed by 35 cycles of denaturation at 94°C for 30s, annealing at a primer-specific annealing temperature as table 2 for 30s and extension at 72°C for 30s, and final extension at 72°C for 10 min. The annealing temperature of Cobia was 36°C. A 10 µl of the PCR product were analyzed in a 2 % agarose gel in 0.5 X TBE. The electrophoresis was performed at a constant voltage of 150 V for 150 min and 250 V for 1 min. The gel was stained with ethidium bromide and visualized under UV light.

2.3 PCR-ISSR method

ISSR primers of this study were listed in the Table 3. A total 59 primers were screened. Pre-amplification PCR reaction was conducted in 25 µl reaction containing 12.5 µl PCR master mix (Promega, Co., Wisconsin, USA), 1µl each primer, 1 µl of the 10 ng extracted DNA, and 10.5 µl dH_2O. Then, the mixtures were subjected to 94°C for 5 min, followed by 35 cycles of denaturation at 94°C for 30s, 30s at a primer-specific annealing temperature as table 3,

Primer	Sequence of primer (5′-3′)	Tm(℃)	references
RAPD16	GGTGGCGGGA	56	RAPD Primer Set#1, University of British Columbia
RAPD17	CCTGGGCCTC	56	
RAPD22	CCCTTGGGGG	56	
RAPD23	CCCGCCTTCC	56	
RAPD31	CCGGCCTTCC	56	
RAPD34	CCGGCCCCAA	56	
RAPD50	TTCCCCGCGC	56	
RAPD56	TGCCCCGAGC	56	
RAPD63	TTCCCCGCCC	56	
RAPD64	GAGGGCGGGA	56	
RAPD65	AGGGGCGGGA	56	
RAPD67	GAGGGCGAGC	56	
RAPD70	GGGCACGCGA	56	
RAPD71	GAGGGCGAGG	56	
RAPD73	GGGCACGCGA	56	
RAPD81	GAGCACGGGG	56	
RAPD83	GGGCTCGTGG	56	
RAPD84	GGGCGCGAGT	56	
RAPD86	GGGGGGAAGG	56	
RAPD87	GGGGGGAAGC	56	
RAPD88	CGGGGGATGG	56	
RAPD89	GGGGGCTTGG	56	
RAPD94	GGGGGGAACC	56	
RAPD95	GGGGGGTTGG	56	
RAPD96	GGCGGCATGG	56	
RAPD105	CTCGGGTGGG	56	RAPD Primer Set#2
RAPD106	CGTCTGCCCG	56	
RAPD115	TTCCGCGGGC	56	
RAPD149	AGCAGCGTGG	56	
RAPD157	CGTGGGCAGG	56	
RAPD158	TAGCCGTGGC	56	
RAPD173	CAGGCGGCGT	56	
RAPD174	AACGGGCAGC	56	
RAPD190	AGAATCCGCC	56	
RAPD196	CTCCTCCCCC	56	
RAPD198	GCAGGACTGC	56	
RAPD210	GCACCGAGAG	56	Set#3
RAPD211	GAAGCGCGAT	56	
RAPD218	CTCAGCCCAG	56	
RAPD241	GCCCGACGCG	56	
RAPD245	CGCGTGCCAG	56	
RAPD270	TGCGCGCGGG	56	
RAPD286	CGGAGCCGGC	56	
RAPD287	CGAACGGCGG	56	
RAPD319	GTGGCCGCGC	56	Set#4
RAPD480	GGAGGGGGGA	56	Set#5
RAPD534	CACCCCCTGC	56	Set#6
RAPD536	GCCCCTCGTC	56	

Primer	Sequence of primer (5′-3′)	Tm (℃)	references
RAPD540	CGGACCGCGT	56	Set#6
RPAD542	CCCATGGCCC	56	
RAPD563	CGCCGCTCCT	56	
RAPD571	GCGCGGCACT	56	
RAPD584	GCGGGCAGGA	56	
RAPD585	CCCGCGAGTC	56	
RAPD592	GGGCGAGTGC	56	
RAPD595	GTCACCGCGC	56	
RAPD598	ACGGGCGCTC	56	
RAPD601	CCGCCCACTG	56	Set#7
RAPD603	ACCCACCGCG	56	
RAPD606	CGGTCGGCCA	56	
RAPD615	CGTCGAGCGG	56	
RAPD620	TTGCGCCCGG	56	
RAPD625	CCGCTGCAGC	56	
RAPD626	CCAAGCCCGG	56	
RAPD640	CGTGGGGCCT	56	
RAPD647	CCTGTGGGGG	56	
RAPD769	GGGTGGTGGG	56	Set#8
RAPD770	GGGAGGAGGG	56	
RAPD771	CCCTCCTCCC	56	
RAPD772	CCCACCACCC	56	
primer1	GGTGCGGGAA	36	Ready-To-Go. RAPD Analysis Kit (Amersham Pharmacia Biotech, Piscataway, N J)
primer 2	GTTTCGCTCC	36	
primer 3	GTAGACCCGT	36	
primer 4	AAGAGCCCGT	36	
primer 5	AACGCGCAAC	36	
primer 6	CCCGTCAGCA	36	
OPA1	CAGGCCCTTC	56	(Operon, Advanced Biotechnologies)
OPA2	TGCCGAGCTG	56	
OPA3	AGTCAGCCAC	56	
OPA4	AATCGGGCTG	56	
OPA5	AGGGGTCTTG	56	
OPA6	GGTCCCTGAC	56	
OPA7	GAAACGGGTG	56	
OPA8	GTGACGTAGG	56	
OPA9	GGGTAACGCC	56	
OPA10	GTGATCGCAG	56	
OPA11	CAATCGCCGT	56	
OPA12	TCGGCGATAG	56	
OPA13	CAGCACCCAC	56	
OPA14	TCTGTGCTGG	56	
S514	CAGGATTCCC	56	Portman International (China) Limited, Hong Kong
S1036	AAGGCACGAC	56	
S1040	CCTGTTCCCT	56	
S1042	TCGCACAGTC	56	
S1201	CCATTCCGAG	56	

Table 2. RAPD primers of PCR amplification

extension at 72°C for 30s, and final extension at 72°C for 5 min before analysis by the electrophoresis as described previously.

Primer	Sequence of primer(5′-3′)	Tm(℃)	References
ISSR1	$(GGAC)_3A$	48	Pazza et al. (2007)
ISSR2	$(GGAC)_3C$	48	
ISSR3	$(GGAC)_3T$	48	
ISSR4	$(TGTC)_4$	48	
ISSR5	$(GGAC)_4$	48	
ISSR6	$(GGAT)_4$	48	
ISSR7	$(TAGG)_4$	48	
ISSR8	$(GACA)_4$	48	
ISSR801	$(AT)_8T$	50	Liu et al. (2006)
ISSR817	$(CA)_8A$	48	
ISSR825	$(AC)_8T$	48	
ISSR842	$(CA)_8YG$	52	
ISSR848	$(CA)_8RG$	51	
ISSR850	$(GT)_8YC$	51	
ISSR855	$(AC)_8YT$	50	
ISSR856	$(AC)_8YA$	49	
ISSR858	$(TG)_8RT$	48	
ISSR859	$(TG)_8RC$	52	
ISSR860	$(TG)_8RA$	51	
ISSR888	$BDB(CA)_7$	52	
SAS1	$(GTG)_4GC$	55	Maltagliati et al.(2006)
SAS3	$(GAG)_4G$	55	
UBC809	$(AG)_8G$	55	
UBC811	$(GA)_8C$	55	
UBC827	$(AC)_8G$	55	
IT1	$(CA)_8GT$	55	
IT2	$(CA)_8AC$	55	
IT3	$(GAG)_4AG$	55	
PT1	$(GT)_8C$	55	
ISSR807	$(AG)_8T$	48	UBC Primer Set#9, University of British Columbia
ISSR819	$(GT)_8A$	48	
ISSR822	$(TC)_8A$	48	
ISSR831	$(AT)_8YA$	48	
ISSR834	$(AG)_8YT$	48	
ISSR843	$(CT)_8RA$	48	
ISSR852	$(TC)_8RA$	48	
ISSR861	$(ACC)_6$	48	
ISSR862	$(AGC)_6$	48	
ISSR868	$(GAA)_6$	48	
ISSR871	$(TA)_8RG$	48	
ISSR873	$(GACA)_4$	48	
ISSR877	$(TGCA)_4$	48	
ISSR9	$(GAG)_5RY$	55	Hou *et al.*, 2006
ISSR10	$VBV(CA)_8$	54	
ISSR11	$VDV(GT)_8$	51	
ISSR12	$HVHT(GT)_7$	51	
ISSR13	$(CT)_8A$	49	
ISSR14	$(TG)_8GT$	48	
ISSR15	$(AG)_8TG$	54	
ISSR16	$(TC)_8C$	52	
ISSR17	$(TG)_8G$	53	
ISSR18	$(TG)_6R$	45	

ISSR19	$(CA)_6RY$	44
ISSR20	$(GT)_6YR$	44
ISSR21	$(GT)_6AY$	43
ISSR22	$(ACTG)_4$	52
ISSR23	$(GACA)_4$	48
ISSR24	$(CAC)_6$	57

Table 3. ISSR primers of PCR amplification

2.4 Genetic distances and phylogenetic analysis

Patterns from RAPD and ISSR methods were scored for the presence (1) or absence (0) of clear bands to analyze genetic similarities using the Dice coefficient of similarity. Similarity matrix cluster and phylogenetic analysis was used to reveal association among strains based on the unweighted pair group method with arithmetic averages (UPGMA) using the NTSYSpc software (Numerical taxonomy and multivariate analysis system, version 2.01b, State University of New York, Stony Brook, NY, USA) according to Rohlf (1997).

3. Results

3.1 RAPD method of giant group

A total of 14 giant grouper including cultivate and wild were obtained. Analysis on species of giant grouper in cultivate and wild. For RAPD method amplification, the species, For RAPD method, total 95 of RAPD oligonucletide primers were used to screening the genetic diversity of giant grouper. There are 21 RAPD primers (22.1%) have polymorphic bands. Total 279 bands were generated by those primers and 86 polymorphic bands (31%). The primer RAPD 115 (5'-TTCCGCGGGC-3') was got the more diversity than other primers, have 8 polymorphic bands (Fig 1). The RAPD 245 primer was generated the less bands, only

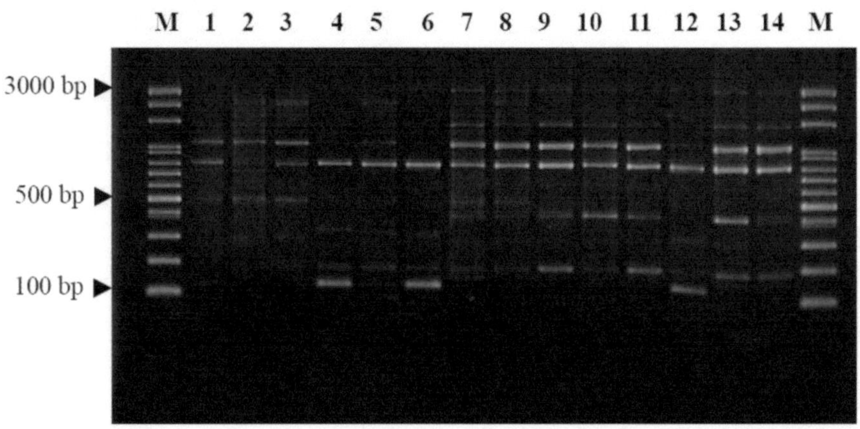

Fig. 1. RAPD profiles of the 14 *E. lanceolatus* fish obtained using the primer 115. Lanes M: Bio-100 bp DNA Ladder. Lanes 1-7: *E. lanceolatus* fish wild population (E1, E5, E7, E8, E9, E13, and E14); Lanes 8-14: *E. lanceolatus* fish cultivate population (E2, E3, E4, E6, E10 E11, and E12).

2 polymorphic bands. The sequence and PCR-RAPD condition were listed in Table 2. All primers were generated bands ranging in size from 100 to 3000 bp. The results shown that the ratio of polymorphic bands were between 13.3~66.7% by 21 RAPD primers. For dendrogram analysis, two groups were identified by RAPD 115 primer (Fig 2). E1, E5, E14, E7, E8, and E9 samples were clustered in group I, which were collected from wild population. Group II, which including E13, E10, E2, E3, E4, E6, E11, and E12 samples. For Group II, all samples were belonged to cultivated populations. For giant grouper, wild (seven samples) and cultivated (seven samples) populations of giant grouper can be discriminated by RAPD method.

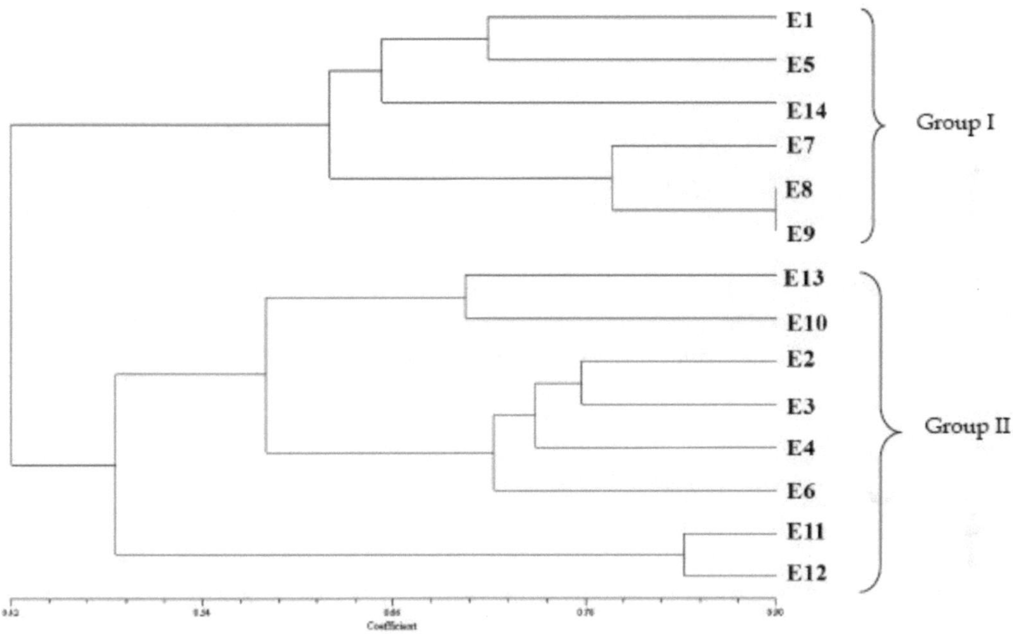

Fig. 2. UPGMA consensus dendrogram of dissimilarity among individuals analyzed using the primer RAPD 115.

3.2 ISSR method of giant group

Results of ISSR analysis, 59 primers were used in this study. According the results of ISSR method, 17 primers (29%) have polymorphic patterns. Total 166 bands were generated, 58 polymorphic bands (34.9%). The primer ISSR IT3 was got the more diversity than other primers, have 20 bands. The ISSR 15 primer was generated the less bands, only 3 bands. All the polymorphic patterns were ranged between 100~3000 bp. ISSR primer 868 (5′-$(GAA)_6$-3′) was better distinguished than other primers. The result was shown in Fig 3. For giant grouper, the patterns of ISSR primer868 could discriminate giant grouper between wild and cultivated populations. For dendrogram analysis, four groups were clustered by ISSR primer868 primer (Fig 4). Among those groups, Group I, Group III, and Group IV were collected from wild population. Samples were clustered in Group II were from cultivated populations. We also found that the results of ISSR method have the same tend to RAPD method. ISSR method was more discriminate ability than RAPD method.

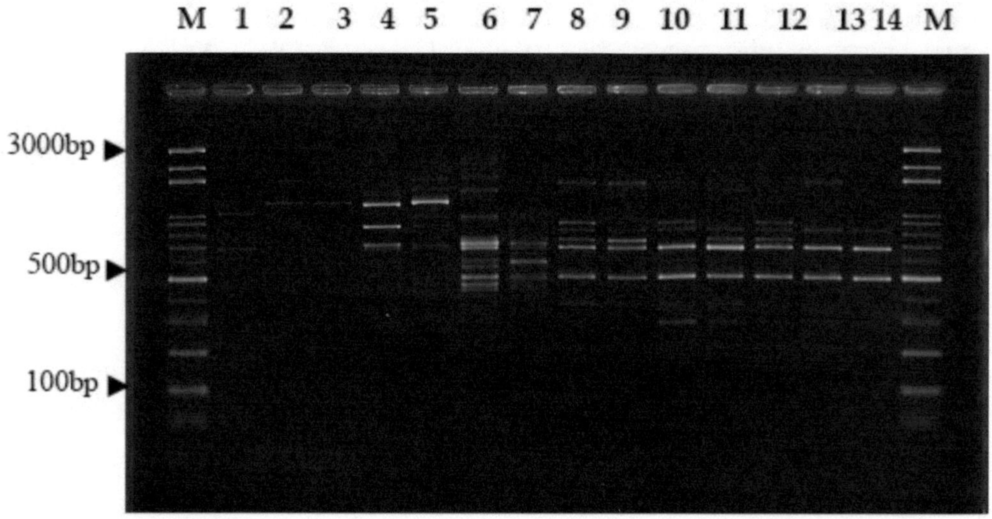

Fig. 3. ISSR profiles of the 14 giant grouper fish obtained using the primer ISSR 868. Lanes M: Bio-100bp DNA Ladder. Lanes 1-7: giant grouper fish wild population (E1, E5, E7, E8, E9, E13, and E14); Lanes 8-14: giant grouper fish cultivated population (E2, E3, E4, E6, E10, E11, and E12).

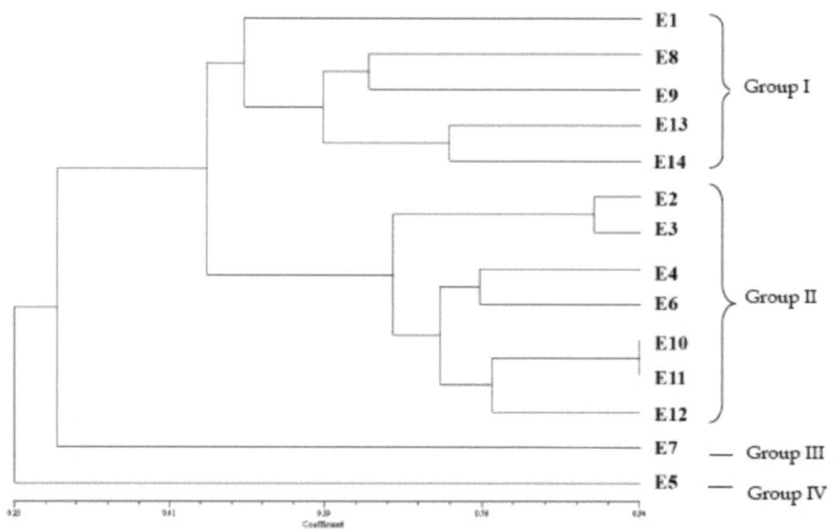

Fig. 4. UPGMA consensus dendrogram of dissimilarity among individuals analysed using the primer ISSR 868.

3.3 RAPD and ISSR methods of cobia

Ninety-five RAPD primers and 59 ISSR primers were used for PCR amplification. The results were shown that all the cobia samples were the same patterns and no polymorphic bands. The primer ISSR UBC809 and RAPD31 were generated 12 to 17 bands and in size from 100 to 2000 bp. The results were also shown in Fig 5 and Fig 6.

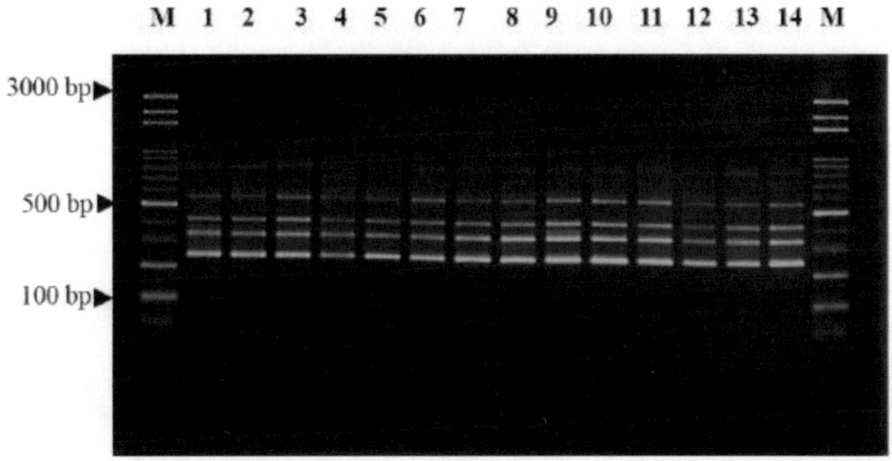

Fig. 5. ISSR profiles of the 14 *R. canadum* fish obtained using the primer UBC 809. Lanes M: Bio-100 bp DNA Ladder. Lanes: 1, 2, 4, 6, 9, 10, 11, 12, 13, and 14 (R1, R2, R4, R6, R9, R10, R11, R12, R13, and R14) *R. canadum* fish wild population; Lanes: 3, 5, 7, and 8 (R3, R5, R7, and R8) *R. canadum* fish cultivate population.

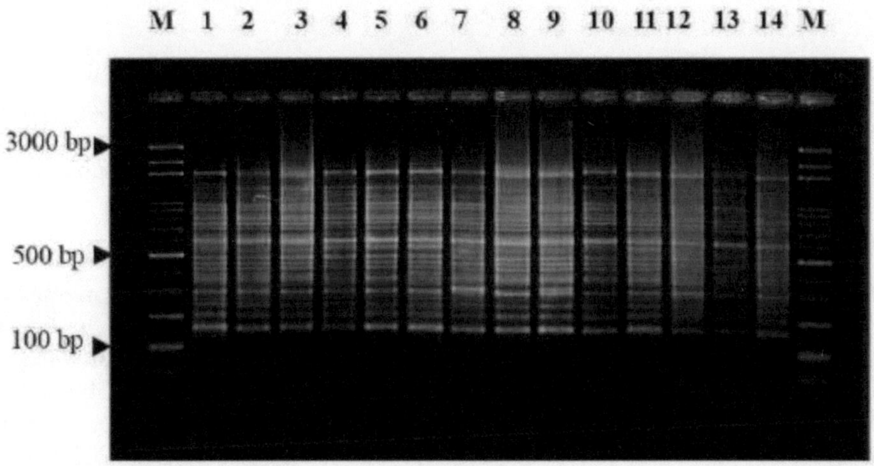

Fig. 6. RAPD profiles of the 14 *R. canadum* fish obtained using the primer 31. Lanes M: Bio-100 bp DNA Ladder. Lanes : 1, 2, 4, 6, 9, 10, 11, 12, 13, and 14 (R1, R2, R4, R6, R9, R10, R11, R12, R13, and R14) *R. canadum* fish wild population; Lanes: 3, 5, 7, and 8 (R3, R5, R7, and R8) *R. canadum* fish cultivate population.

Sequence variability of mitochondrial DNA regions was low between the six cobia (Garnet, et al., 2002). These results were also similar in our study. Both RAPD and ISSR methods have no different patterns. Hence, this could provide more useful information of molecular genetic data in population and stock enhancement studies.

3.4 ISSR methods of red coral trout

For screening ISSR primers, the ISSR primer15 (ISSR15: 5′-$(AG)_8$TG -3′) was better distinguished than other primers. The result was shown in Fig 7. The primer ISSR 15 was generated 10 to 16 bands and in size from 200 to 2000 bp. For dendrogram analysis, three groups were identified. Group I, the 71412, 71413, 71422, and 82011 were clustered in the group. Group II, including 81621, 91321, 71411, and 91712 samples; group III including 71423, 80622, 91322, 71421, and 81623 samples. All the nodes of the dendrograms ranged from 90 to 100%. The result was shown in Fig 8.

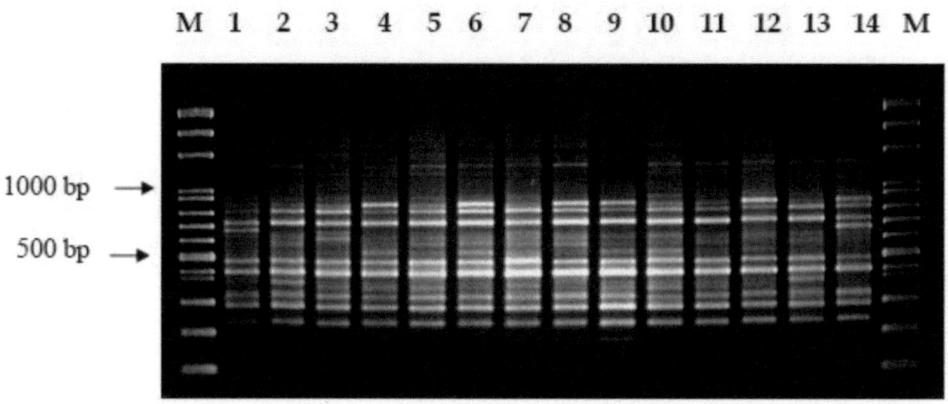

Fig. 7. ISSR profiles of the 14 red coral trout obtained using the primer ISSR 15. Lanes M: Bio-100 bp DNA Ladder. Lanes: 4, 5, 6, 8, 9, 10, and 12 (71411, 71412, 71413, 82011, 91321, 91711, and 91712) red coral trout fish wild population; Lanes: 1, 2, 3, 7, 11, 13, and 14 (71421, 71422, 71423, 81621, 80622, 81623, and 91322) red coral trout fish cultivate population.

4. Discussions

RAPD and ISSR methods were generally used for genetic diversity and populations study; those methods also could be used to analyze the breeding relationship. For species identification and genetic resource/diversity analysis, RAPD and microsatellitsa method were recommended (Liu & Cordes, 2004). The RAPD techniques has been used for discrimination of populations of species of the genus Barbus, grouper, Nile perch and wreck fish, salmonids, among others (Partis & Wells, 1996; Callejas & Ochando, 2001; Asensio et al., 2002; Jin, Cho, Seong, Park, Kong & Hong, 2006). Genetic analysis with RAPD markers is relatively easy, fast, and efficient. RAPD analysis, however, may not be practical for identifying interbreed species (Martinez, Elvevoll & Haug, 1997). SSRs are inherited in a co-dominant fashion. This allows one to discriminate between homo- and heterozygous state, and increases the efficiency of genetic mapping and population genetic studies. ISSR markers have recently been used successfully for genetic analysis in hatchery and wild *Paralichthys olivaceus* strains. It indicates that molecular marker systems contribute greater levels of capability for the detection of polymorphism, and provide a better solution for the assessment of genetic variations (Shikano, 2005; Liu, Chen, Li & Li, 2006).

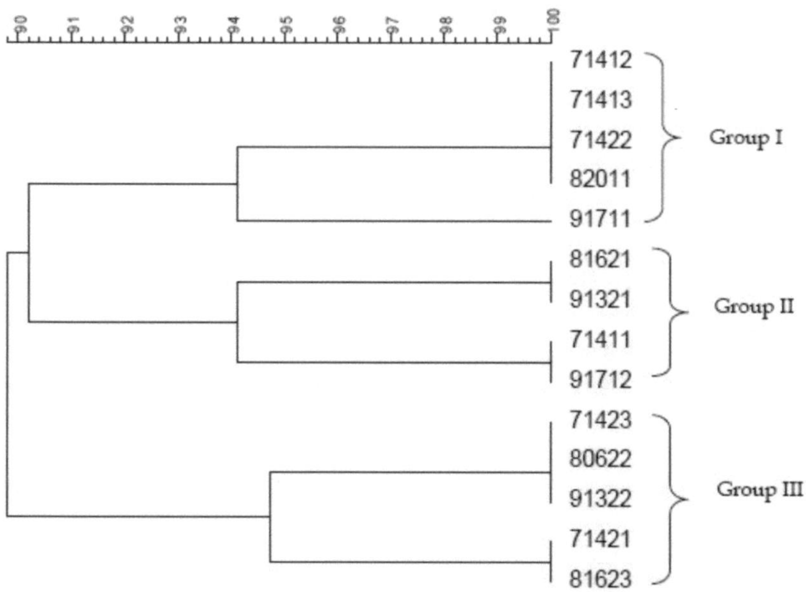

Fig. 8. UPGMA consensus dendrogram of dissimilarity among individuals analysed using the primer ISSR 15.

According to our results, RAPD and ISSR method can be effectively discriminate giant grouper from different sources, such as cultivate and wild, even if fish are from different cultivate farms. But cobia and red coral trout were less discriminate ability. The study found that ISSR and RAPD methods were positively high correlations. Giant grouper species have highly genetic diversity. A comparison of RAPD and ISSR patterns in 14 giant grouper samples, ISSR primers have higher polymorphism and fewer bands than those of RAPD primers. It could provide simple and convenient method to discriminate genetic variation of giant grouper samples. In this study, ISSR method could distinguish genetic variation within specie and different populations. Some reports also have suggested that ISSR may reveal a much higher numbers of polymorphic fragments per primer than those of RAPD (Esselman, et al., 1999). Among these markers, microsatellite DNAs have revolutionized the use of molecular genetic markers in the applications mentioned before, and the markers are destined to dominate this type of studies in the coming years (Asensio, 2007). It also has been revealed as important tools in studies regarding the genetic structure of populations, phylogeographic relations and phylogenetic reconstruction in fish (Antunes, et al., 2010).

5. Conclusion

We developed DNA molecular marker techniques which could be used to generate information for fish genetic diversity, species identification, trace genetic variation between different individuals in aquaculture, authenticate fish, fishery products and provide good reference resources for species sources and relationships.

6. Acknowledgment

This research project was supported by the Council of Agriculture (COA) (97AS-4.1.2-AI-I2).

7. References

Antoro, S.; Na-Nakom, U. & Koedprang, W. (2006). Study of Genetic Diversity of Orange-Spotted, *Epinephelus coioides*, from Thailand and Indonesia Using Microsatellite Markers. *Marine Biotechnology*, Vol.8, pp. 17-26, ISSN 1432-1793

Antunes, R. S. P.; Gomes, V. N.; Prioli, S. M. A. P.; Prioli, R. A.; Júlio Jr.; H. F. ; Prioli, L. M.; Agostinho, C. S. & Prioli, A. J. (2010). Molecular Characterization and Phylogenetic Relationships Among Species of the Genus *Brycon* (Characiformes: Characidae) from Four Hydrographic Basins in Brazil. *Genetics and Molecular Research*, Vol.9, pp. 674-684, ISSN 1676-5680

Asensio, L.; González, I.; Fernández, A.; Rodríguez, M. A.; Lobo, E.; Hernández, P. E.; García, T. & Martín, R. (2002). Application of Random Amplified Polymorphic DNA (RAPD) Analysis for Identification of Grouper (*Epinephelus guaza*), Wreck fish (*Polyprion americanus*), and Nile perch (*Lates niloticus*) Fillets. *Journal of Food Protection*, Vol.65, pp,432-435, ISSN 0362-028x

Asensio, L. (2007). PCR-Based Methods for Fish and Fishery Products Authentication. *Trends in Food Science and Technology*, Vol.18, pp. 558-566, ISSN 0924-2244

Asensio, L.; González, I.; Rojas, M.; García, T. & Martín, R. (2009). PCR-Based Methodology for the Authentication of Grouper (*Epinephelus marginatus*) in Commercial Fish Fillets. *Food control*, Vol.20, pp. 618-622, ISSN 0956-7135

Barman, H. K.; Barat, A.; Yadav, B. M.; Banerjee, S.; Meher P. K.; Reddy, P. V. G. K. & Jana, R. K. (2003). Genetic Variation Between Four Species of Indian Major Carps as Revealed by Random Amplified Polymorphic DNA Assay. *Aquaculture*, Vol. 217, pp. 115-123, ISSN 0044-8486

Bártfai, R.; Egedi, S.; Yue, G. H.; Kovács, B.; Urbányi, B.; Tamás, G.; Horváth, L. & Orbán, L. (2003). Genetic Analysis of Two Common Carp Broodstocks by RAPD and Microsatellite Markers. *Aquaculture*, Vol. 219, pp. 157-167, ISSN 0044-8486

Callejas, C. & Ochando, M. D. (2001). Molecular Identification (RAPD) of the Eight Species of the Genus Barbus (*Cyprinidae*) in the Iberian Peninsula. *Journal of Fish Biology*, Vol. 59, pp. 1589-1599, ISSN 1095-8649

Esselman, E. J.; Jianqiang, L.; Crawford, D. J.; Windus, J. L. & Wolfe, A. D. (1999). Clonal Diversity in the Rare *Calamagrostis porteri* ssp. *insperata* (Poaceae): Comparative Results for Allozymes and Random Amplified Polymorphic DNA and Intersimple Sequence Repeat Markers. *Molecular Ecology*, Vol. 8, pp. 443-453, ISSN 1365-294X

Garber, A. F.; Grater, W. D.; Stuck, K. C. & Franks, J. S. (2002). Characterization of the Mitochondrial DNA Control Region of Cobia, *Rachycentron vanadium*, from Mississippi Coastal Waters. *Gulf and Caribbean Fisheries Institute*, Vol. 53, pp. 570-578, ISSN 0072-9019

Garoia, F.; Guarniero, I.; Piccinetti, C. & Tinti, F. (2004). First Microsatellite Loci of Red Mullet (Mullus barbatus) and Their Application to Genetic Structure Analysis of Adriatic Chared Stock. *Marine biotechnology*, Vol. 6, pp. 446-452, ISSN 1436-2236

Hou, L.; Lü, H.; Zou, X.; Bi, X.; Yan, D. & He, C. (2006). Genetic Characterizations of Mactra Veneriformis (Bivalve) Along the Chinese Coast Using ISSR-PCR Marker. *Aquaculture*, Vol. 261, pp. 865-871, ISSN 0044-8486

Hseu, J.-R.; Hwang, P.-P. & Ting, Y.-Y. (2004). Morphometric Model and Laboratory Analysis of Intracohort Cannibalism in Giant Grouper *Epinephelus lanceolatus* Fry. *Fisheries Science*, Vol. 70, pp. 482-486, ISSN 1444-2906

Jin, L. G.; Cho, J. G.; Seong, K. B.; Park, J. Y.; Kong, I. S. & Hong, Y. K. (2006). 18 rRNA Gene Sequences and Random Amplified Polymorphic DNA Used in Discriminating Manchurian Trout from Other Freshwater Salmonids. *Fisheries Science*, Vol. 72, pp. 903-905, ISSN 1444-2906

Liu, Y.-G.; Chen, S.-L.; Li, J. & Li, B.-F. (2006). Genetic Diversity in Three Japanese Flounder (*Paralichthys olivaceus*) Populations Revealed by ISSR Markers. *Aquaculture*, Vol. 255, pp. 565-572, ISSN 0044-8486

Liu, Z. J. & Cordes, J. F. (2004). DNA Marker Technologies and Their Applications in Aquaculture Genetics. *Aquaculture*, Vol. 238, pp. 1-37, ISSN 0044-8486

Lockley, A. & Bardsley, R. (2000). DNA-Based Methods for Food Authentication. *Trends in Food Science & Technology*, Vol. 11, pp. 67-77, ISSN 0924-2244

Maltagliati, F.; Lai, T.; Casu, M.; Valdesalici, S. & Castelli, A. (2006). Identification of Endangered Mediterranean Cyprinodontiform Fish by Means of DNA Inter-Simple Sequence repeats (ISSRs). *Biochemical systematics and Ecololgy*, Vol. 34, pp. 626-634, ISSN 0305-1978

Martinez, I.; Elvevoll, E. O. & Haug, T. (1997). RAPD Typing of North-East Atlantic Minke Whale (Balaenoptera acutorostrata). *ICES Journal of Marine Science*, Vol. 54, pp. 478-484, ISSN 1095-9289

Miao, S.; Jen, C. C.; Huang, C. T. & Hu S.-H. (2009). Ecological and Economic Analysis for Cobia *Rachycentron canadum* Commercial Cage Culture in Taiwan. *Aquaculture International*, Vol. 17, pp. 125–141, ISSN 0967-6120

Okumus, Í. & Çiftci, Y. (2003). Fish Population Genetics and Molecular Marker: II - Molecular Markers and Their Applications in Fisheries Aquaculture. *Turkish Journal of Fisheries and Aquatic Sciences*, Vol. 3, pp. 51-79, ISSN 1303-2712

Partis, L. & Wells, R. J. (1996). Identification of Fish Species Using Rrandom Amplified Polymorphic DNA (RAPD). *Molecular and Cellular Probes*, Vol. 10, pp. 435-441, ISSN 0890-8508

Pazza, R.; Kavalco K. F.; Prioli So^nia M. A. P.; Prioli, A. J. & Bertollo, L. A. C. (2007). Chromosome Polymorphism in Astyanax Fasciatus (Teleostei, Characidae), Part 3: Analysis of the RAPD and ISSR Molecular Markers. *Biochemical Systematics and Ecololgy*, Vol. 35, pp. 843-851, ISSN 0305-1978

Povh, J. A.; Lopera-Barrero, N. M.; Ribeiro, R. P.; Lupchinski Jr. E.; Gomes, P. C. & Lopes, T. S. (2008). Genetic Monitoring of Fish Repopulation Pprograms Using Molecular Markers. *Ciencia e Investigación Agraria*, Vol. 35, pp. 1-10, ISSN 0718-1620

Ramirez, M. A.; Pareicia-Acevedo, J. & Planas, S. (2006). New Microsatellite Resources for Groupers (Serranidae). *Molecular Ecology Notes*, Vol. 6, pp. 813-817, ISSN 1755-0998

Rohlf, F. J. (1997). "NTSYS-pc v. 2.01b Numerical Taxonomy and Multivariate Analysis System." Applied Biostatistics, Inc., Setauket, NY.

Shikano, T. (2005). Marker-Based Estimation of Heritability for Body Color Variation in Japanese Flounder Paralichthys Olivaceus. *Aquaculture*, Vol. 249, pp. 95-105, ISSN 0044-8486

Tang, S.; Tassanakajon, A.; Klinbunga, S.; Jarayabhand, P. & Menasveta, P. (2004). Population Structure of Tropical Abalone (Haliotis asinine) in Coastal Waters of Thailand Determined Using Microsatellite Markers. *Marine Biotechnology*, Vol. 6, pp. 604-611, ISSN 1432-1793

Wang, S. F.; Du, J. Y. & Wang, J. (2007). Identification of *Epinephelus malabaticus* and *Epinephelus coioides* Using DNA Markers. *Acta Oceanology Sinica*, Vol. 26, pp. 122-129, ISSN 1869-1099

Welsh, J. & McClelland, M. (1990). Fingerprinting Genomes Using PCR Arbitrary Primers. *Nucleic Acids Research*, Vol. 18, pp. 7213-7218, ISSN 1362-4962

Zeng, H.-S.; Ding, S.-X.; Wang, J. & Su, Y.-Q. (2008). Characterization of Eight Polymorphic Microsatellite Loci for the Giant Grouper (*Epinephelus lanceolatus* Bloch). *Molecular Ecology Resources*, Vol. 8, pp. 805-807, ISSN 1755-0998

Zhang, Q.; Allen Jr. S. K. & Reece, K. S. (2005). Genetic Variation in Wild and Hatchery Stocks of Suminoe Oyster (Crassostrea ariakensis) Assessed by PCR-RFLP and Microsatellite Markers. *Marine Biotechnology*, Vol. 7, pp. 588-599, ISSN 1432-1793

Zhang, J.; Liu, H. & Song, Y. (2010). Development and Characterization of Polymorphic Microsatellite Loci for a Threatened Reef Fish *Plectropomus leopardus*. Conservation Genetics Resources, Vol. 2 (Supplement 1), pp. 101-103, ISSN 1877-7260

6

Gel Electrophoresis for Investigating Enzymes with Biotechnological Application

Maria de Lourdes T. M. Polizeli[1*], Simone C. Peixoto-Nogueira[1], Tony M. da Silva[1], Alexandre Maller[2] and Hamilton Cabral[3]
[1]*Biology Department, Faculty of Philosophy Sciences and Letters of Ribeirão Preto, São Paulo University*
[2]*Biochemistry and Immunology Department School of Medicine of Ribeirão Preto, São Paulo, São Paulo University*
[3]*Science Pharmaceutical Department School of Pharmaceutical Science of Ribeirão Preto, São Paulo University*
Brazil

1. Introduction

Enzymes are vital biological catalysts with essential action in the metabolism of all living beings. Moreover, enzymes have a very significant role in various industrial sectors, including baking, brewery, detergents, textile, pharmaceutical, animal feed, cellulose pulp biobleaching, biofuels and others (Polizeli et al., 2005; 2009; 2011). Industrial enzymes are especially produced from microorganisms, such as bacteria, yeasts and filamentous fungi.

Here, we describe some basic knowledge about microbial enzymes with potential application in industry, their properties and some biochemical methods for the detection of amylase, pectinase, proteases, and xylanase activities on polyacrylamide gel electrophoresis. These methods are more qualitative procedures than quantitative, once they may be used to confirm the electrophoretic homogeneity of purified enzymes through chromatographic process. In addition, we described the principle of each method approached in this chapter, which grants a better understanding of each procedure.

1.1 Overall considerations about industrial enzymes

Enzymes are optimum biological catalysts present in all living beings, and they, under adequate conditions, catalyze in their active sites the natural substrates from the metabolic route reactions. Quite often, the metabolic enzymes act in a sequence, and the product generated in a reaction becomes the substrate for the following phase, as diagramed:

$$\text{Substrate}\,1 \xrightarrow{\text{enzyme}\,1} \text{Product}\,1 + \text{Enzyme}\,1 \quad (1)$$

$$\text{Substrate}\,2\,(=\text{Product}\,1) \xrightarrow{\text{enzyme}\,1} \text{Product}\,2 + \text{Enzyme}\,2 \quad (2)$$

Some properties of the enzymes make them excellent competitors against traditional chemical catalysts, due to their great catalytic efficacy (kcat), considering that the main objective of any biological transformation is to obtain a short-term high conversion of substrate into product. Besides this characteristic, enzymes present a high specificity and selectivity, according to their metabolic role, acting in optimum conditions of pH and temperature. Also, they do not pose damage to the environment because they are biodegradable.

Applied biological catalysis had its origins in ancient China and Japan, in the manufacturing of foods and alcoholic beverages in which amylases and proteases of vegetal and microbial origins were employed. It dates back to the end of the XIX century with the introduction of standardized preparations of rennilases in the production of cheese. After this period, the implementation of new industrial applications of enzymes was slow, arising intensively in the last 40 years.

The main reasons for the current importance of enzymes in the industrial scenario are due to the development of application processes of proteases in detergents, pectinases in juices and glucoamylase in the production of glucose from starch. Additionally, the employment of recombinant DNA techniques allowed the obtainment of high productivity and the most suitable design of enzymes.

Following, we present the application of enzymes in several technical sectors justifying the development of methods for the visualization of enzymatic activities in electrophoresis gels.

1.2 Some industrial enzymes and their applications

Proteases, amylases, cellulases and lipases are used in the preparation of detergents and that is the greatest industrial market. Enzymes are added to increase wash efficiency. For so, they require lower usage temperatures, reduce wash periods and agitation costs. They act on both proteic and starchy residuals as well as on fats; they enhance clothes softness and restore color brightness. They require, in general, suitable thermostability and activity in alkaline pH.

Animal feed - the addition of cellulases, xylanases, proteases, lipases, ligninases and phytases in ruminants and monogastrics foods leads to the digestibility of grass and forage, reduces pollutants, decreases the release of carbon dioxide, soluble carbon hydrates and phosphorus. β-glucans increase food viscosity decreasing starch digestibility, but the addition of β-glucanases increases food assimilation, resulting in weight gain (Facchini et al., 2011a, 2011b).

Swine and poultry feed needs the addition of phytase, a phosphatase that acts on acid or alkaline media and dephosphorylates phytic acid releasing phosphorus to the environment (Haefner et al., 2005). That can damage the environments where the soil contains plenty of phosphate, as it is observed in Europe.

Xylanases and ligninases may also take part in the biobleaching of the cellulose pulp for the manufacturing of paper, whereas cellulases are used for the modification of the textile fiber properties, giving them the pre-wash effect. Those three enzymatic systems participate in the sugar cane hydrolysis for the bioethanol manufacturing. Such procedure has been widely adopted in many countries like Brazil, which has a number of flexpower vehicles (Betini et al., 2009; Michelin et al., 2009, 2011).

Proteases and lipases also have a role in the dairy industry, acting in the production of cheese (Gupta et al., 2002). Chemokine, extracted from the stomach of calves, acts in the Milk coagulation, leading to the formation of cheese. Lipases give the aroma and hot flavor in cheese (Hasan et al., 2006). Still in the food industry, pectinases and cellulases increase the extraction of oils through pressing (coconut, sunflower, soybean, olive, etc.).

One sector that has been widely economically explored is the application of amylases in processes of starch saccharification (Silva et al., 2009a, 2009b). For so, there is the need of several enzymes such as α-amylase, which forms maltooligosaccharides; glucoamylases and β-amylases, which hydrolyze starch to glucose and maltose and the glucosyltransferases with the production of cyclodextrins. With synergistic action of the amylolytic system, there is the production of maltose syrup used in breweries, as well as the glucose syrup which is preferably converted by glucose isomerase to fructose syrup, due to the high sweetening power. Such compounds may be used in the manufacturing of sauce, child feeding, gums, candies, ice-cream, pharmaceutical products, canned products etc. Amylases also act in the textile industry to remove starch added to cotton to increase resistance (Gupta et al., 2003).

In the pharmaceutical industry, the use of enzymes is increasingly growing, reflecting the *in vivo* catalysis potential. We can highlight the use of pancreatin, obtained from the swine pancreas, which is used as adjuvant in the digestive process of people who have genetic disorders leading to digestive problems or who, due to surgical removal of the pancreas or precocious aging, present digestive problems. Many enzymes are used as therapeutical agents, such as asparaginase and glutaminase, collagenase, hyaluronidase, ribonuclease, streptokinase, uricase and uroquinase (Prakashan, 2008).

Semi synthetic penicillins (ampicillin and amoxicillin) were launched in the market to replace penicillin, given the acquired resistance of some microorganisms. All kinds of penicillin have the same basic structure: 6-aminopenicillanic acid (6-APA), a thiazolidine ring bound to a beta lactam that takes a free amino group. In the synthesis of semi synthetic penicillin there is the enzymatic hydrolysis of G penicillin with the penicillin G acylase. After purified and concentrated, the 6-APA released in the hydrolysis is used as an intermediary in the synthesis of amoxicillin or ampicillin (Cabral et al., 2003).

The enzymes can be used in the cosmetics industry in creams against skin aging and acne, buffing cream, oral hygiene and hair dying.

Enzymes may be used in analytical applications, due to their high specificity, identifying substances in complex mixtures such as blood, urine and other biological fluids. They participate in tests for glucose (glucose oxidase), urea, amino acids, proteins, ethanol, etc (Godfrey & West, 1996).

In fine chemistry, the list of compounds produced by enzymatic biocatalysis is huge, and we highlight, as an example, vitamin C and several L-amino acids. Acrylamide is used as a monomer in the production of polyacrylamide, widely used as flocculating polymer. Acrylamide was initially produced chemically, but the technological disadvantages such as the formation of toxic residuals and the costs in the purification process made the enzymatic way a viable process.

2. Some industrial enzymes and their applications

2.1 Preparation of polyacrylamide gel electrophoresis for activity enzymatic

The detection of enzymatic activities for industrial use, in electrophoresis gel, happens when PAGE (polyacrylamide gel electrophoresis) is employed, which is an electrophoresis performed in non-denaturing conditions. The Figure 1 illustrates some steps used to preparation of the electrophoresis gel.

In this kind of procedure, there is not preferably the addition of the sodium dodecyl sulfate - SDS detergent, β-mercaptoethanol or another reducing agent, such as dithiothreitol - DTT or urea. Also, the protein samples in its native form (not denaturated) are not boiled before their application in the gel because enzymes will lose its activities, if denaturated. The enzymatic activity may also be detected in gels of the SDS-PAGE type, if the samples were not boiled or added by any reducing agent that denatures the protein. Sodium dodecyl sulfate (SDS) is an anionic surfactant whose role is to bestow the proteins with uniform load density. SDS presents a high negative load and a hydrophobic tail that interacts with the polypeptidic chains in an approximated ratio of 1.4 g of SDS for each gram of protein, making them negatively loaded. In the lack of SDS, the proteins with equal mass may migrate differently in the pores of the gel due to the load differential of their tridimensional structures.

PAGE may be performed in a pH 4.5 or 8.9, depending on the isoelectric point - pI, of the sample under study. In order to accomplished zymograms, it is performed SDS-PAGE; however, the samples generally correspond to a crude extract or a partial purified extract which are either not boiled or added by β-mercaptoethanol, DTT or urea.

2.2 Preparation of the sample for application in electrophoresis

The preparation of proteic solutions for the application in electrophoresis is an important phase. It is important to highlight once again that most enzymes used industrially have microbial origin (fungi, yeasts or bacteria) and also that normally, their synthesis is followed by the elimination of a number of primary and secondary metabolites produced by the very microorganisms, as well as other compounds present in the cultivation medium, such as vitamins, salts, carbohydrates, amino acids and peptides.

In order to avoid the interference of such factors in the electrophoresis, especially when there is the application of enzymatic extracts without previous purification, it is necessary to pay close attention to the type of sample that is being prepared.

Below are some measures and precautions that must be adopted:

i. Dialysis: This procedure aims at the removal of substances with smaller molar mass, such as salts, carbohydrates and amino acids, which may interfere in the electrophoresis quality;
ii. Attention to the concentration of the sample applied in the electrophoresis. In general lines, around 10 μg of proteins is necessary for a good visualization in the electrophoresis gel after the dying phase. For the detection of the enzymatic activity, considerable enzymatic levels are necessary. If the protein solution presents a lower concentration, the application of any procedure for the concentration of proteins is

necessary, such as lyophilization, use of filtering membranes with defined molar mass, ammonium sulfate precipitation and even the use of solvents (ice acetone or ethanol). We must bear in mind that those two last processes need an additional dialysis for the removal of the ammonium sulfate or carbohydrates when the precipitation happens through the action of solvents.

iii. For the detection of activity in electrophoresis gel, the run must take place in low temperatures, such as a refrigerator or a cold chamber.

iv. Attention must also be paid to the native load of proteins and the separation must depend only on its molar mass. For so, proteins may be mixed with SDS, becoming negatively loaded, as it has already been described.

2.3 Electrophoresis separation techniques

2.3.1 Electrophoresis in non-denaturing conditions (PAGE)

Electrophoresis is going to be performed in pH (4.5 or 8.9), in a polyacrylamide gel that may range from 5 to 15%, depending on the size and the load of the protein under study. For proteins loaded negatively, the running buffer will consist of Tris-HCl and glycine, pH 8.9. For proteins loaded positively, there is going to be a buffer with β-alanine and glacial acetic acid, pH 4.5. Both procedures must be performed at 4°C. Table 1 indicates necessary volumes to obtain PAGE gels with different concentrations.

2.3.1.1 PAGE for acid proteins (-), (Davis, 1964)

Solution A	
Tris-HCl	9.75 g
HCl (1 M)	12 mL
TEMED	0.05 mL
Distilled water	25 mL
pH adjusted for pH 8.7	
Solution C	
Acrylamide	9.6 g
Bis-acrylamide	0.32 g
Distilled water	20 mL
pH adjusted for pH 8.9	
Solution G	
Ammonium persulfate	0.007 g
Distilled water	5 mL
Dye	
Bromophenol blue	
Glycerol	
Running buffer	
Tris-HCl 50mM	
glycine 36mM, pH 8.9	

Preparation of the samples and markers:

- In a sterile microtube, put 18 µL of proteic sample and 2 µL of running buffer Tris-HCl 50mM and glycine 36mM, pH 8.9;
- Add 2 µL of the dying solution Bromophenol blue 0.1% and 4 µL of glycerol.

2.3.1.2 PAGE alkaline proteins (+), Reisfeld et al. (1962)

Solution A	
KOH (1 M)	48 mL
Glacial acetic acid	17.2 mL
TEMED	4 mL
Complete with distilled H_2O	100 mL
pH adjusted for 4.3	
Solution C	
Acrylamide	19.2 g
Bis-acrylamide	0.54 g
Complete with distilled H_2O	40 mL
pH adjusted for 4.5	
Solution G	
ammonium persulfate	0.28 g
Complete with distilled H_2O	100 mL
Dye	
Methyl green	
Glycerol	

Sample preparation:

- In a sterile microtube put 18 µL of sample and 2 µL of running buffer consisting of 31.2 g of β-alanine, 8 mL of glacial acetic acid and an amount of distilled water sufficient to reach the volume of 1000 mL;
- Add 2 µL of the solution Methyl green 0.1% and 4 µL of glycerol.

Solution	Polyacrylamide concentration (%)							
	4	5	6	7	8	10	12	14
A (mL)	1.05	1.05	1.05	1.05	1.05	1.05	1.05	1.05
C (mL)	0.71	0.876	1.05	1.22	1.49	1.77	2.10	2.45
H_2O (mL)	6.65	6.47	6.30	6.13	5.97	5.60	5.25	4.95

Table 1. Preparation of 8.4 mL of PAGE in different concentrations.

- Set up the electrophoresis bowl;
- In a Becker, add the solutions A, C and H_2O. Mix;

- Add 50 µL of ammonium persulfate 10% and quickly put the mixture in the electrophoresis bowl;
- Place the comb;
- Wait until the gel solidifies and apply the samples (up to 30 µL) in different lanes;
- Connect the energy cables in the respective jacks of the energy source adjusted to 70 mAmps and 120 Volts;
- Turn on the source and wait for the samples to run throughout the gel before turning it off;
- Remove the gel carefully and process it.

2.3.2 Electrophoresis gel SDS-PAGE, Laemmli (1970)

The electrophoresis must be performed in pH 8.9 and in the presence of SDS (sodium dodecyl sulfate), with the gel concentration ranging from 5 to 15%. Table 2 indicates necessary volumes to obtain SDS-PAGE gels with different concentrations.

Solution A

Tris-HCl	36.5 g
TEMED	230 µL
Distilled water	9 mL

Adjust the pH with HCl concentrated for pH 8.9 and store it at 4-6°C.

Solution C

Acrylamide	28 g
Bis-acrylamide	0.74 g
Distilled water	100 mL

Store at 4-6°C in a glass flask with Amberlite resin due to the degradation of acrylamide in acid and ammonia.

Solution E

SDS	0.21 g
Distilled H_2O	100 mL

Store at room temperature.

Sample buffer

Tris-HCl	0.755 g
Glycerol	1 mL

Dissolve with 17.5 mL of distilled H_2O and adjust the pH to 6.75 with concentrated HCl and add:

SDS	2 g
Bromophenol blue	0.001 g
Distilled water	100 mL

Freeze aliquots for a further use.

Run Buffer

Tris	3.025 g
glycine	14.4 g
SDS	1.0 g

Dissolve in an amount of distilled water sufficient for 1000mL, pH 8.9.

Preparation of samples and markers:

- In a sterile microtube, place a sample buffer and a molar weight marker in the ration of 1:1;
- In a sterile microtube, place the buffer sample and the sample in the ratio of 1:3.

Solution/Reagent	Gel concentration (%)					
	5	6	7	8	10	12
A (mL)	1	1	1	1	1	1
C (mL)	1.4	1.7	2	2.3	2.85	3.45
E (mL)	3.8	3.8	3.8	3.8	3.8	3.8
H_2O (mL)	1.8	1.5	1.2	0.9	0.35	-
Ammonium persulfate(g)	0.00425	0.00425	0.0057	0.006	0.007	0.00708

Table 2. Preparation of SDS-PAGE

- Set up the electrophoresis bowl;
- In a Becker, add the solutions A, C, E and H_2O and mix;
- Add the ammonium persulfate and quickly place the solution in the electrophoresis bowl; Place the comb;
- Wait until the gel solidifies and apply the patterns (0.5-5 µL) and the samples (up to 30 µL) in different lanes;
- Connect the energy cables in the respective jacks of the source regulated for 70 mAmps and 120 Volts;
- Connect the source and wait until the samples run throughout the gel before turning it off;
- Remove the gel carefully and process it.

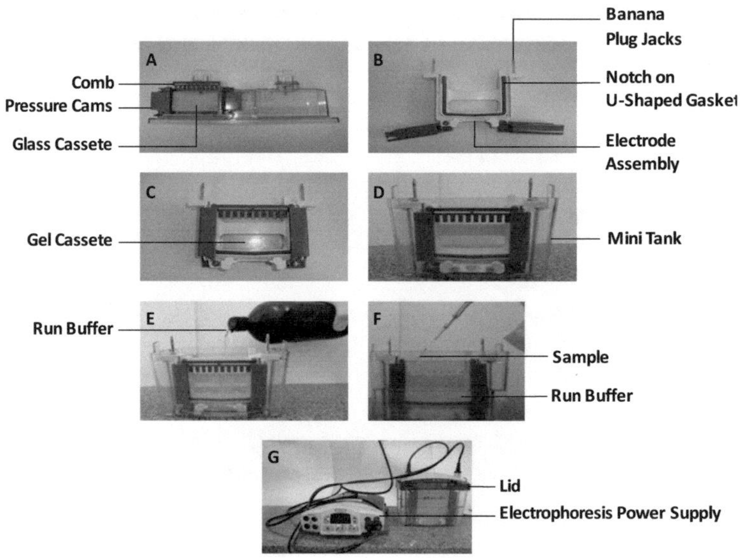

Fig. 1. Preparation and development of vertical polyacrylamide electrophoresis gel using Bio-Rad™ system. (A) and (B) accessories for stacking gel. (C) and (D) preparation of the electrophoresis gel; (E) addition of run buffer; (F) application of the samples; (G) development of the electrophoresis.

3. Methods for specific enzymes detection in electrophoresis gel

3.1 Amylase

The method consists of the conduction of an electrophoresis in polyacrylamide gel polymerized with starch 0.5%. The electrophoretic run is performed in the pH adequated to the amylase isoelectric point. After the end of the run, the gel must be immersed in the suitable temperature and buffer during at least one hour. The gel is going to be revealed with a solution of iodine (I_2 10 mM) and potassium iodide (KI 14 mM) until the appearance of activity bands. Fig. 2A illustrates the activity of α-glucosidase, one of the enzymes of the amylolytic system, which leads to the formation of glucose as end product (Aquino et al., 2001; 2003; Silva et al., 2009a, 2009b).

3.2 Pectinase

Method I - After conducting a PAGE 4.5 or 8.9 (depending on the enzyme pI), the gel containing the enzyme must be incubated with the substrate - a solution containing citric pectin or sodium polypectate 1% in the suitable buffer of the enzyme under study. In Fig. 2B there was the use of 1% of sodium polypectate in a sodium acetate buffer 100mM, pH 4.0 and incubation at 50°C (enzyme optimum temperature), for 2 hours for the dying with 0.02% Ruthenium red $[(Ru_3O_2(NH_3)_{14})Cl_6.4H_2O)]$, a dye capable of interacting with the pectic substates (Sterling, 1970). Thus, in the region where the protein migrated to and hydrolyzed the substrate, there is a halo with a whitened coloration that contrasts against the rest of the red-colored gel.

Method II - The citric pectin must be dissolved in gel buffer with the aid of a magnetic agitator, followed by the addition of acrylamide, bis-acrylamide and TEMED solutions. Crystals of ammonium persulfate are added immediately before the plate gel is overflown. After the run, incubate the gel for 1-2 hours with 100 mL of malic acid 0.1M, at 4°C, in order to cause a gradual change to pH 3.0. Such period allows the enzyme to interact with the pectin polymerized in the acrylamide gel in its suitable pH range. Wash with distilled water and color in Ruthenium red 0.02%, during 30 to 120 min. Wash with distilled water.

Result: against a redish gel, it is possible to notice the polygalacturonase activity due to the formation of clear, opaque or colorless areas.

3.3 Xylanase

In order to detect the xylanase activity in gel, the polyacrylamide must be polymerized with 0.5% xylan dissolved in the buffer of the electrophoresis to be performed (PAGE 4.5 or 8.9, depending on the isoelectric point of the enzyme under study). After the electrophoresis run the gel must be incubated in the temperature and in the reaction buffer which is mostly suitable for the xylanase under study for at least 1 hour. After this period, the gel is going to be stained with 1% Congo red ($C_32H_{22}N_6Na_2O_6S_2$) a sodium salt of benzidinediazo-bis-1-naphtylamine-4-sulfonic acid. Thus, in the region where the protein migrated to and hydrolyzed the substrate, there is a halo with a whitened coloration that contrasts against the rest of the red-colored gel. (Fig. 2C) (Sandrim et al., 2005; Damásio et al., 2011).

Xylanases may also be observed through zymograms (Fig. 2D). For so, the samples must be applied in SDS-PAGE without the addition of β-mercaptoethanol, any other reducing agent and without boiling. The advantage of the activity detection through zymograms is that effectively all active isoforms present there may be detected, once that regardless on the isoelectric point, all proteins will migrate in the gel, because of their molar mass.

In this case, the SDS gel must be performed with the running buffer commonly adopted in the protocols and, after the electrophoresis run, the proteins will have to be transferred to another gel composed by agarose and xylan, a substrate that is specific to the activity to be detected. Such transferring happens when there is a kind of "sandwich" with the polyacrylamide gel and the agarose + substrate gel. The transferring must happen overnight. After this period, the agarose + substrate gel, now also with the proteins to be analyzed, will have to be incubated in the buffer that is suitable for the isoforms under study, during one hour, following with the coloration suitable for the activity detection of the enzyme analyzed.

3.4 Proteases

The zymography may also be applied for proteases. It is a simple, quantitative and functional technique to analyze the activity of proteases (Leber & Balkwill, 1997). It consists basically of two stages, the separation through electrophoresis, followed by the activity detection of the enzyme in polyacrylamide gel, in non-reducing conditions (without treatment with DTT or β-mercaptoethanol) (Dong-Min et al., 2011). This technique has been used to evaluate the level of proteases in tissues or biological fluids, and it bears the advantage of distinguishing different kinds of enzymes due to the characteristic of mobility that each enzyme presents (Raser et al., 1995). The protease activity in zymography is observed as a clear band, indicating the substrate proteolysis after colored with Coomassie Brilliant Blue (Kim et al., 1998).

This methodology is widely employed for the detection of Matrix metalloproteinases (MMPs). However, it can also be employed for other types of proteases, with the need of adjustments in the methodology, such as the substitution of the substrate, generally gelatin for casein. Unfortunately, the zymography with casein is very little sensitive, when compared to the zymography with gelatin. Besides, casein migrates in the gel during the electrophoresis due to its relative low molar mass. That results in two clearly defined areas in the gel: the upper part, which still contains excess casein and the lower part, with less casein (Beurden and Von denHoff, 2005).

For the detection of proteases (Fig. 2E), the sample must be diluted in the sample buffer (5x) of the gel (0.4M Tris-HCl, pH 6.8; 5% SDS; 20% glycerol 0.03% Bromophenol blue). The samples cannot be boiled, because this process denatures the enzyme and it will no longer present activity (Kleiner & Stetler-Stevenson, 1994). The electrophoresis of the samples containing the protease must be performed according to Laemmli (1970). The gel concentration must be prepared according to the molar mass of its protease. The electrophoresis may be performed in constant 100V for 1-2 hours at 4°C.

For the development of the proteolytic activity, the gel must be incubated with 70 mL of buffer with the appropriate reaction pH, for 5 min., 4°C, 100 rpm. Following, the buffer must be removed and the gel must be incubated with 70 mL of Triton X-100 2.5% prepared in the reaction buffer. The gel must be kept at 100 rpm, for 30 min, 4°C. This step is for the removal of the SDS and the activation of the protease. Afterwards, the excess Triton X-100 must be removed. For so, add 70 mL of buffer with the appropriate reaction pH, incubate for 30 min., at 4°C, 100rpm.

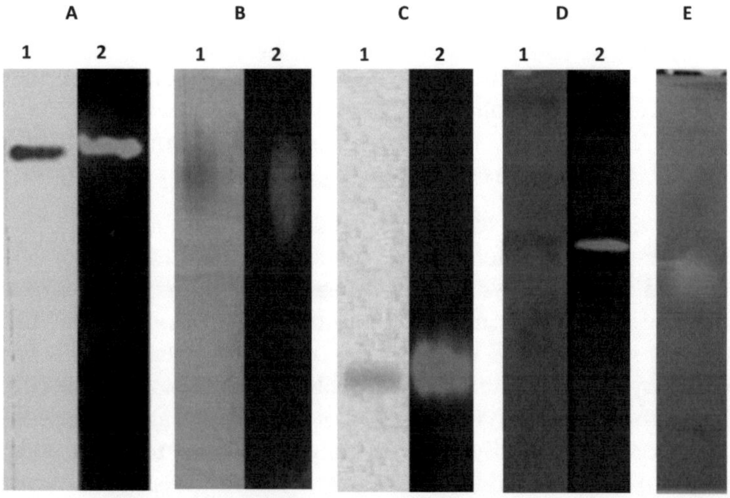

Fig. 2. Activity gels for different enzymes in SDS-PAGE 10%. (A1) α-glucosidase (a type of amylase) revealed with Coomassie Blue; (A2) α-glucosidase revealed with 10 mM Iodine solution and 14 mM potassium iodide; (B1) polygalacturonase revealed with silver solution; (B2) polygalacturonase activity revealed with 0.02% ruthenium red; (C1) xylanase revealed with silver solution; (C2) xylanase activity revealed with Congo red; (D1) Zymogram for xylanase revealed with silver solution and (D2) Congo red; (E) protease activity revealed with Coomassie Blue.

Following, remove the buffer and add the casein solution at 3%, prepared in a buffer with the enzyme reaction pH. The gel must be incubated for 30 min, at 4°C, for the diffusion of the casein to the gel. After that, the gel must be bathed at the enzyme reaction temperature for the period of 1-2 hours, so that the enzymatic reaction occurs (such period may be adjusted according to each enzyme and concentration).

The excess casein must be removed by bathing the gel for 5 times with distilled water at room temperature (García-Carreño et al., 1993 and Kleiner & Stetler-Steveson, 1994) with modifications.

For the coloration, the gel must be stained with a solution containing 40% ethanol, 10% acetic acid and 0.1% Coomassie brilliant blue R-250 (García-Carreño et al., 1993). In this stage, the gel shows a blue bottom and where the hydrolysis took place, there will be a white halo. The gel needs to have the excess Coomassie brilliant blue removed; hence, it will have to be discolored with a discoloring solution composed by 40% ethanol and 10% acetic acid (García-Carreño et al., 1993). The clear zones over the blue bottom indicate the protease activity.

Some enzymes need activators, such as Ca^{2+}, DTT and EDTA to show their activity. Those can be added together with the substrate. Other substrates, such as hemoglobin, bovine serum albumin, gelatin and collagen, may have their coloration improved through the use of other dyes, as for example, Amide black (García-Carreño et al., 1993).

4. Conclusion

The biological catalyzers present several advantages over their chemical similar, particularly, the regio and stereoselectivity that lead to the formation of products which are enantiomerically pure and in conformity with the norms established for the food, pharmaceutical and agriculture industry. The enzymes are efficient under the energetic point of view, operating in controlled pH, temperature and pressure. The development of the recombinant DNA technology enabling the expression of enzymes in different hosts has resulted in the production of more efficient biological catalyzers.

Several methods are attributed to enzymatic determinations. The most widely used are the colorimetric ones, where the reactions occur with specific substrates, generally leading to the formation of colored products, which can be easily quantified in spectrophotometers or through acrylamide gel electrophoresis, non-denaturing conditions. The detection of enzymatic activity through PAGE involves the migration potential of the enzyme in gel, which is influenced by the molar mass of the protein and its loads in specific pH. Thus, the visualization of the enzymatic activity in gels is seen as an advantaging condition, once that it is possible to consider that hardly did that enzyme migrate together with interferents or contaminating substances that could be in a crude extract and that could lead to errors in the enzymatic levels. PAGE for enzymatic activities can be considered as an elegant method that has been increasingly employed in researches.

The zymogram technique demands low enzyme concentrations, which can reach the order of nanograms. Several adaptations of substrates may be made in this technique, because the majority of substrates used are low-cost and yield good results. With this technique, we can infer how many types or isoforms of enzymes of a same class are present in the crude extract, for example, what the molar mass is and even in some cases, the quantification.

Hence, the proposal for the optimization of stages of enzymatic activity detection in electrophoresis must be conducted for each specific extract to be studied, yielding in this way high sensitivity and precision.

5. Acknowledgment

Dr. Maria de Lourdes T.M. Polizeli and Dr Hamilton Cabral are Research Fellows of CNPq. Dr Simone C. Peixoto-Nogueira, Dr Tony M. Silva and Dr Alexandre Maller are recipients of FAPESP Fellowship. This chapter concerns research data of the project National System for Research on Biodiversity (SISBIOTA-Brazil, CNPq 563260/2010-6/FAPESP number 2010/52322-3). We thank Dr. Abilio for the English technical assistance.

6. References

Aquino, A.C.M.M., Jorge, J.A., Terenzi, H.F. & Polizeli, M.L.T.M. (2001).Thermostable glucose-tolerant glucoamylase produced by the thermophilic fungus *Scytalidium thermophilum* 15.1. *Folia Microbiologica*, Vol. 45 (6), pp. 11-16.

Aquino, A.C.M.M., Jorge, J.A., Terenzi, H. F. & Polizeli, M.L.T.M. (2003). Studies on a thermostable alpha-amylase from the thermophilic fungus *Scytalidium thermophilum*. *Applied Microbiology and Biotechnology*, Vol. 61, pp. 323-328.

Betini, J.H.A., Michelin, M., Peixoto-Nogueira, S.C., Jorge, J.A., Terenzi, H.F. & Polizeli, M.L.T.M. (2009). Xylanases from *Aspergillus niger, Aspergillus niveus* and *Aspergillus ochraceus* produced under solid-state fermentation and their application in cellulose pulp bleaching. *Bioprocess and Biosystems Engineering,* Vol. 32, pp. 819-824.

Beurden Sanoek-van, P.A.M. & Von den Hoff, J.W. (2005). Zymographic techniques for the analysis of matrix metalloproteinases and their inhibitors. *BioTechniques,* Vol. 38, pp. 73-82.

Cabral, J.M.S., Aires-Barros, M.R. & Gama, M. (2003). *Engenharia Enzimática,* Lidel edições Técnicas, 972-757-272-3, Lousã.

Damásio, A.R.L., Silva, T.M., Almeida, F.B.R., Squina, F., Ribeiro, D.A. R., Leme, A.F.P., Segato, F., Prade, R.A., Jorge, J.A., Terenzi, H.F. & Polizeli, M.L.T.M. (2011). Heterologous expression of an *Aspergillus niveus* xylanase GH11 in *Aspergillus nidulans* and its characterization and application. *Process Biochemistry* Vol. 46, pp. 1236-1242.

Davis, B.J. (1964). Disc Electrophoresis. II. Method and application to human serum proteins. *Annals of the New York Academy of Sciences.* Vol.28, 121, pp. 404-427.

Dong-Min, C., Kim, K.E., Ahn, K.H., Park, C.S., Kim, D.H., Koh, H.B., Chun, H.K., Yoon, B.D., Kim, H.,J., Kim, M.S. & Choi, N.S. (2011). Silver-stained fibrin zymography: separation of proteases and activity detection using a single substrate-containing gel. *Biotechnology Letters,* Vol. 33, pp. 1663-1666.

Facchini, F.D., Vici, A.C., Reis, V.R., Jorge, J.A., Terenzi, H.F., Reis, R.A. & Polizeli, M.L.T.M. (2011a). Production of fibrolytic enzymes by *Aspergillus japonicus* C03 using agro-industrial residues with potential application as additives in animal feed. *Bioprocess and Biosystems Engineering,* Vol. 34 (3), pp. 347-55.

Facchini, F.D.A., Vici, A.C., Benassi, V. M., Freitas, L.A.P., Reis, R.A., Jorge, J.A., Terenzi, H.F. & Polizeli, M.L.T.M. (2011b). Optimization of fibrolytic enzyme production by *Aspergillus japonicus* C03 with potential application in ruminant feed and their effects on tropical forages hydrolysis. *Bioprocess and Biosystems Engineering,* Vol. 34, pp. 1027–1038.

García-Carreño, F.L.; Dimes, L.E. & Haard, N.F. (1993). Substrate-gel electrophoresis for composition and molecular weight of proteinases or proteinaceous proteinase inhibitors. *Analytical Biochemistry* Vol. 214, pp. 65-69.

Godfrey, T & West, S. (1996). *Industrial Enzymology* (2º ed.), Stokton Press, 1561591637 0333594649, New York.

Gupta, R., Beg, Q.K. & Lorenz, P. (2002). Bacterial alkaline proteases: molecular approaches and industrial applications. *Applied Microbiology and Biotechnology,* Vol. 59, (April 2002), pp. 15–32, 0175-7598.

Gupta R, Gigras, P., Mohapatra, H., Goswam,i V.K. & Chauhan, B. (2003). Microbial α-amylases: a biotechnological perspective. *Process Biochemistry,* Vol 38, (june 2003), pp. (1599-616), 1359-5113.

Haefner, S., Knietsch, A., Scholten, E., Braun, J., Lohscheidt, M. & Zelder, O. (2005). Biotechnological production and applications of phytases. *Applied Microbiology and Biotechnology.* Vol. 68, (5), (September 2005), pp. 588-597, 1432-0614.

Hasan, F., Shah, A.A.A. & Hameed, A. (2006). Industrial applications of microbial lipases. *Enzyme and Microbial Technology.* Vol. 39, 2, (june 2006), pp. 235-251, 0141-0229.

Kim, SH., Choi, N.S. & Lee, W.Y. (1998). Fibrin zymography: a direct analysis of fibrinolytic enzymes on gels. *Analytical Biochemistry,* Vol. 263, pp. 115-116.

Kleiner, D.E. & Stetler-Stevenson, W.G. (1994). Quantitative zymography: detection of picogram quantities of gelatinases. *Analytical Biochemistry*, Vol. 218, pp. 325-329.

Laemmli, U.K. (1970). Cleavage of structural protein during the assembly of the head of bacteriophage T4. *Nature*, Vol. 227, pp. 680-685.

Leber, T.M. & Balkwill, F.R. (1997). Zymography: a single-step staining method for quantification of proteolytic activity on substrate gels. *Analytical Biochemistry*, Vol. 249, pp. 24-28.

Michelin, M., Peixoto-Nogueira, S.C., Betini, J.H,A., Silva, T.M., Jorge, J.A., Terenzi, H.F. & Polizeli, M.L.T.M. (2009). Production and properties of xylanases from *Aspergillus terricola* Marchal and *Aspergillus ochraceus* and their use in cellulose pulp bleaching. *Bioprocess and Biosystems Engineering*, Vol. 33, (7) (December 2009), pp. 813-821, 1615-7591.

Michelin, M., Silva, D. P., Ruzene, D. S., Vicente, A. A., Jorge, J. A., Terenzi, H. F., & Polizeli, M. L. T. M. (2011). Production of xylanolytic enzymes by *Aspergillus terricola* in stirred tank and airlift tower loop bioreactors. *Journal of Industrial Microbiology & Biotechnology*, DOI 10.1007/s10295-011-0987-7.

Polizeli, M.L.T.M., Rizzatti, A.C.S., Monti, R.; Terenzi, H.F., Jorge, J.A. & Amorim, D.S. Xylanases from fungi: properties and industrial applications. Review. (2005). *Applied Microbiology and Biotechnology*, Vol 67, pp. 577-591.

Polizeli, M.L.T.M. (2009). Properties and commercial applications of xylanases from fungi. *In: Mycotechnology: Current Trends and Future Prospects*, M.K.Rai (Editor). I.K. International Publisher, New Delhi., chapter 4, pp. 82-101.

Polizeli, M. L. T. M., Corrêa, E. C., Polizeli, A. M., & Jorge, J. A. (2011). Hydrolases from microorganisms used for degradation of plant cell wall and bioenergy production. *In: Routes to Cellulosic Ethanol*. Editors: Marcos S. Buckeridge & Gustavo H. Goldman. Springer Verlag, chapter 8, pp. 115-134.

Prakashan, N. (2008). *Biochemistry Basics and Applied* (3º edition), Rachana Enterprises, 81-85790-52-3.

Raser, K.J., Posner, A. & Wang, K.K.W. (1995). Casein Zymography: A method to study µ-Calpain, m-Calpain, and their inhibitory agents. *Archives of Biochemistry and Biophysics*, nº 1, v. 319, pp. 211-216.

Reisfeld, R.A., Lewis, U.J. & Williams, D.E. (1962). Disk electrophoresis of basic proteins and peptides on polyacrylamide gels. *Nature* Jul. Vol. 21 (195), pp. 281-283.

Sandrim, V.C., Rizzatti, A.C.S., Terenzi, H.F., Jorge, J.A., Milagres, A.M.F. & Polizeli, M.L.T.M. (2005). Purification and biochemical characterization of two xylanases produced by *Aspergillus caespitosus* and their potential for the bleaching of kraft pulp. *Process Biochem*istry Vol. 40(5), pp. 1823-1828.

Silva, T.M., Maller, A., Damásio, A.R.L., Michelin, M., Ward, R.J., Hirata, I.Y., Jorge, J.A., Terenzi, H.F. & Polizeli, M.L.T.M. (2009). Properties of a purified thermostable glucoamylase from *Aspergillus niveus*. *Journal of Industrial Microbiology and Biotechnology*, Vol. 36, (august 2009a) pp. 1439-1446, 1367-5435.

Silva, T.M., Michelin, M., Damásio, A.R.L., Maller, A., Almeida, F.B.D.R., Ruller, R., Ward, R.J., Rosa, J.C., Jorge, J.A., Terenzi, H.F. & Polizeli, M.L.T.M. (2009b). Purification and biochemical characterization of a novel alpha-glucosidase from *Aspergillus niveus*. *Antonie van Leeuwenhoek*, Vol. 96, pp. 569-578.

Sterling, C. Crystal-structure of ruthenium red and stereochemistry of its pectic stain. (1970). *American Journal of Botany*, Vol. 57, pp. 172–175.

Part 2

Electrophoresis Application in Bacteriology, Parasitology, Mycology and Public Health

7

Application of Molecular Typing Methods to the Study of Medically Relevant Gram-Positive Cocci

Laura Bonofiglio, Noella Gardella and Marta Mollerach
Department of Microbiology, Immunology and Biotechnology, University of Buenos Aires Argentina

1. Introduction

The development of molecular genotyping methods has been a landmark in the possibility of classifying microorganisms below the species level. The ability to differentiate efficiently related bacterial isolates is essential for the control of infectious diseases and has become a necessary technology for clinical microbiology laboratories.

Strain typing is an integral part of epidemiological investigations of bacterial infections. Typing methods fall into two broad categories: phenotypic and genotypic methods. Phenotypic methods are those that characterize the products of gene expression in order to differentiate strains. Properties such as biochemical profiles, antimicrobial susceptibility profiles, bacteriophage types, and antigens present on the cell surface are examples of phenotypic methods that can be used for typing isolates. Since they involve gene expressions, these properties have a tendency to vary, based on changes in growth conditions and growth phase, being often difficult to detect.

Methods for distinguishing among bacterial strains have profoundly changed over the last years mainly due to the introduction of molecular technology. Genotypic strain typing methods are based on the analysis of differences in the chromosomal and extrachromosomal nucleic acid sequences between strains. Molecular epidemiology of infectious diseases integrates practices and principles of molecular biology with those of epidemiology (Tenover et al. 1997).

Investigations of presumed outbreaks of bacterial infections in hospitals often require strain typing data to identify outbreak-related strains and to distinguish epidemic from endemic or sporadic isolates.

All typing systems can be characterized in terms of typeability, reproducibility, discriminatory power, ease of performance, and ease of interpretation. For each isolate, the system should provide an interpretable result, preferably based on objective criteria. Ideally, results should be reproducible from day to day and from laboratory to laboratory and should allow differentiation of unrelated strains. Additionally, the method should be standardized and if possible should be technically simple, cost-effective, and rapid (van Belkum, et al. 2007).

The results of bacterial strain typing have many different applications including outbreak investigation and surveillance in clinical care settings and public health investigations and also within other contexts such as food and pharmaceutical industries and environmental analysis.

The aim of this chapter is to provide an overview of the methods available for analyzing bacterial isolates, focusing on those methods employed for typing *Streptococcus pneumoniae* and *Staphylococcus aureus*. Different molecular approaches have been used to better understand the epidemiology of these medically relevant gram-positive cocci (Willems. et al. 2011).

2. Genotypic methods

The application of molecular biology tools to infectious disease epidemiology is perhaps just as revolutionary in advancing knowledge and concepts in epidemiology. Genotypic typing methods assess genome variation in bacterial isolates.

The advantages of nucleic acid-based typing systems lie in that they are less likely to be affected by growth conditions or the laboratory manipulations to which organisms are subjected. Undoubtedly, genetic materials undergo changes due to natural or artificial selective pressures, but this mechanism is exactly the basis for their typeability.

Compared to the classical phenotypic typing techniques, genotypic typing techniques have several advantages such as general applicability and a high discriminatory power.

A molecular technique must take into consideration the relative accumulation of variation (short or long term) of a targeted set of genes in a pathogen. Nearly all the typing systems can be grouped into variants of just three basic analytical procedures: (i) PCR, (ii) the use of restriction enzymes, and (iii) nucleic acid sequencing. These procedures allow for the use of common equipment and standard reagents to analyze many different types of infectious agents. In addition, genotypic characterization of pathogens facilitates standardization of information storage and data analyses, interpretation, and communication, which are all amenable to computer-assisted manipulations.

2.1 PCR-based typing methods

In the last years, a number of PCR-based strategies have been developed for use as typing tools. PCR can be readily performed with commercially available supplies and there is little variation in the reagents and equipment needed to perform PCR assays from different microorganisms. The major advantages of PCR-based techniques are speed and simplicity.

2.1.1 Repetitive element sequence-based PCR (rep-PCR)

A variety of repetitive DNA sequence elements have been identified in bacterial pathogens, which have been exploited to develop strategies for bacterial typing. Rep-PCR is a simple PCR-based technique that targets multiple copies of repetitive elements in the bacterial genome to generate DNA fingerprints (Versalovic et al. 1991). Primers designed to anneal in the outward direction, near the end of these repetitive elements bind to multiple non-coding, repetitive sequences interspersed throughout the bacterial genome. Multiple DNA fragments between those sites (interrepeat fragments) are amplified. Since the number and location of the repetitive elements are variable, the sizes and number of effectively amplified fragments vary depending on the strain (Figure 1).

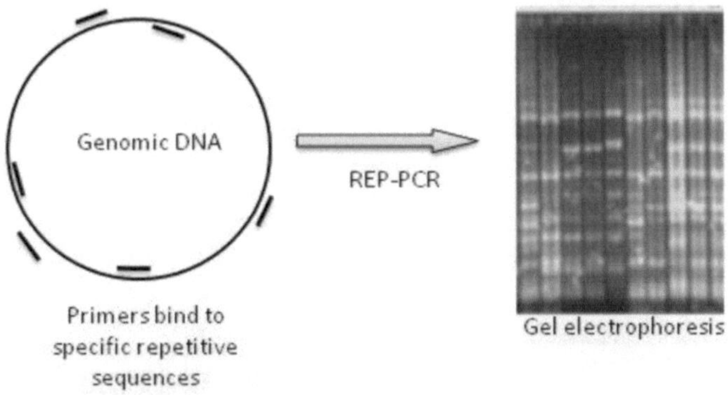

Fig. 1. Schematic representation of REP-PCR assay. On the right BOX-PCR patterns of *S. pneumoniae* isolates using BOXAR1 primer.

Two different Rep-PCR have been used for typing enteric bacteria: a 38-bp repetitive extragenic palindromic element (REP) and a 126-bp enterobacterial repetitive intergenic consensus (ERIC) sequence (Versalovic et al. 1991), whose function has not yet been elucidated.

A BOX repetitive element is a highly conserved repeated DNA element that has been identified in the *Streptococcus pneumoniae* (pneumococcus) chromosome. Although the function of this element has not yet been completely understood, it has been demonstrated that the presence of a BOX element is associated with variation in colony opacity of the pneumococcus (Saluja & Weiser 1995). BOX-PCR has been effectively used for typing *S. pneumoniae* as well as other bacterial species (van Belkum et al. 1996).

Several genetic elements have been used for developing Rep-PCR to type *Staphylococcus aureus*. The element IS*256* occurs in the genome either independently or as part of the composite transposon Tn*4001*, IS*256* insertion position is strain-specific and spaced close enough to allow amplification of polymorphic inter-IS*256* element sequences (Deplano et al. 1997). Another element used for this methodology is RW3A, a repetitive sequence initially found in *Mycoplasma pneumoniae*, which also generates strain-specific DNA fragments when *S. aureus* DNA is used as template (van der Zee et al. 1999).

2.1.2 Randomly Amplified Polymorphic DNA-PCR (RAPD-PCR) or Arbitrarily-Primed PCR (AP-PCR)

Randomly Amplified Polymorphic DNA-PCR (RAPD-PCR), also referred to as Arbitrarily-Primed PCR (AP-PCR), is a variation of the PCR technique employing a single, generally short primer, that is not targeted to amplify a specific bacterial DNA sequence. Low annealing temperatures are used during amplification, allowing imperfect hybridization at multiple chromosomal locations. When the primer binds in two sites on opposite strands, at the proper orientation and with sufficient affinity to allow the initiation of polymerization, the amplification of the fragment between those sites will occur. The amplified products will be various different-sized fragments that can be resolved by conventional agarose gel electrophoresis (Figure 2).

Although the method is much faster than many other typing methods, it is much more susceptible to technical variation. Slight variations in the reaction conditions or the reagents can lead to difficulty in result reproducibility and in the band patterns generated. Therefore, trying to make comparisons among potential outbreak strains can be very problematic (van Belkum et al. 1995). When RAPD-PCR is tightly controlled, it can provide a high level of discrimination, especially when multiple amplifications with different primers are performed.

PCR-based typing methods are simplest and rapid genotyping methods, but is remarkable for its susceptibility to minor variations in experimental conditions.

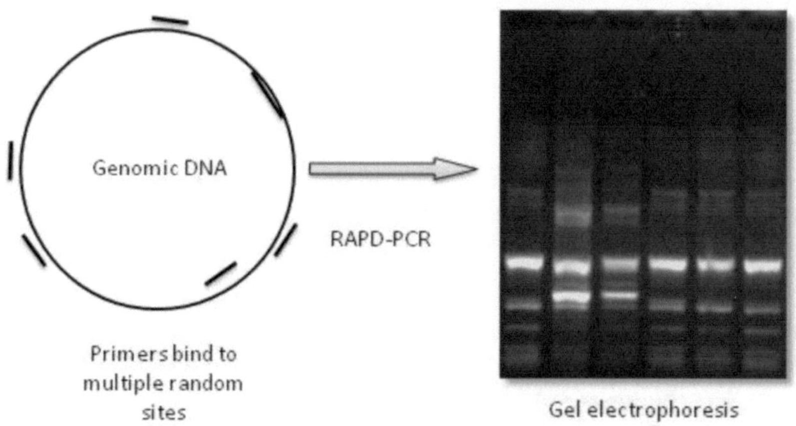

Fig. 2. Schematic representation of RAPD-PCR assay. On the right RAPD-PCR patterns of *Enterococcus faecalis* isolates using D8635 primer.

2.2 Based on enzymatic restriction of chromosomal DNA

2.2.1 Restriction endonuclease analysis of chromosomal DNA by hybridization with a nucleic acid probe (Southern blotting)

Following digestion with high frequency restriction endonucleases, chromosomal DNA is separated into different-sized fragments by conventional agarose gel electrophoresis, but this type of polymorphism is difficult to interpret due to the high number of fragments generated. However, interpretation of these polymorphisms can be facilitated by a Southern blot hybridization technique. By this methodology, fragments are separated by electrophoresis and transferred to a nitrocellulose or nylon membrane and hybridized using specific chemically or radioactively-labeled probes (Figure 3). DNA probes are designed for specific sequences that are found in multiple copies and in different positions of the chromosome. One of the most frequently used probes is ribosomal RNA (16s rRNA) because most species have more than one chromosomal rRNA operon distributed around the chromosome. This particular technique is denominated ribotyping. In recent years, the use of this technique has declined, mainly due to its limited discriminatory power compared to other techniques. Several DNA probes have been employed for the study of methicillin-resistant *S. aureus* (MRSA) outbreaks, including various insertion sequences, such as *IS*431, *IS*256 and *mecA* gene.

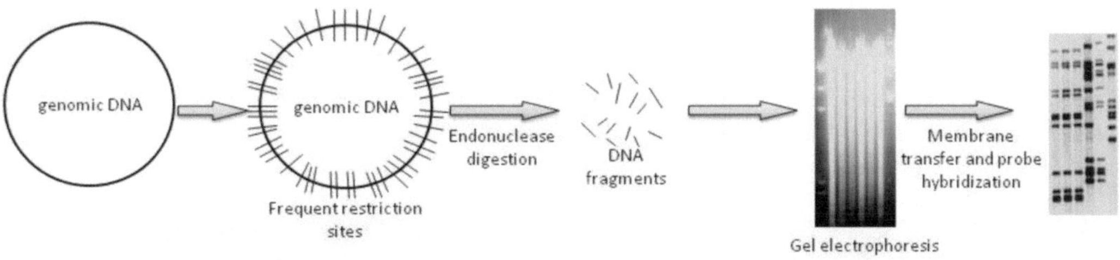

Fig. 3. Schematic representation of restriction endonuclease analysis by Southern-blotting assay. On the right *IS*6110-restriction fragment length of *Mycobacterium tuberculosis* isolates.

2.2.2 Pulsed-field gel electrophoresis (PFGE) of chromosomal DNA

Pulsed-field gel electrophoresis is based on the digestion of bacterial DNA with restriction endonucleases that recognize few sites along the chromosome, generating large DNA fragments (30-800 Kb) that cannot be effectively separated by conventional electrophoresis. The basis for PFGE separation is the size-dependent time-associated reorientation of DNA migration achieved by periodic switching of the electric field in different directions. The DNA fragments will form a distinctive pattern of bands in the gel, which can be analyzed visually and electronically (Figure 4 A). Bacterial isolates with identical or very similar band patterns are more likely to be related genetically than bacterial isolates with more divergent band patterns.

This technique is laborious and includes several steps, requires good standardization and takes at least two days for obtention of results. Procedures will differ to some extent depending on the organism that is being analyzed.

Regarding DNA preparation, PFGE requires intact DNA for restriction endonuclease treatment. The risk of mechanical breakage to DNA molecules during the extraction procedure is avoided by embedding intact organisms into agarose plugs where cells are enzymatically lysed and cellular proteins digested. After endonuclease treatment, the agarose plugs containing the digested DNA are then submitted to PFGE (Figure 4B). The choice of the restriction enzyme for DNA digestion and pulse-time switching parameters for PFGE are critical variables for the obtention of restriction profiles to show well- resolved fragments.

Recent protocols can be completed in as little as two days through shortcuts such as the direct addition of lytic enzymes to the agarose mixture before the blocks are cast and also high temperature short-term washes which facilitate the extraction of unwanted compounds (Goering 2010; Halpin et al. 2010).

Isolates with identical PFGE patterns were considered to represent the same epidemiological type. Isolates differing by one genetic event were considered epidemiologically-related subtypes, expecting that a single genetic event could occur in the chromosome of an organism as it moved from patient to patient. Isolates differing by two genetic events were also deemed to be potentially related, while three or more chromosomal differences were thought to represent an epidemiologically-significant difference (unrelated isolates). Van Belkum suggested a more conservative approach where only nosocomial isolates differing by a single genetic event (up to four differences in the PFGE restriction fragment pattern) were considered related subtypes. The terminology within both proposed formats was left intentionally vague, understanding that molecular typing is only one component of epidemiological evaluation which must include other available clinical data for accurate analysis (Tenover et al. 1995; van Belkum et al. 2007; Goering 2010).

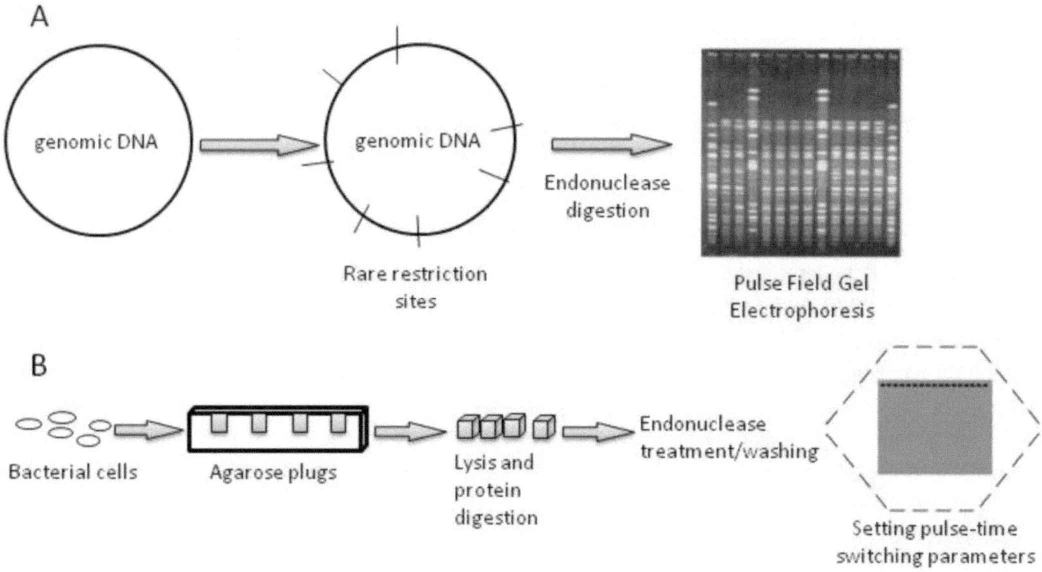

Fig. 4. A. Schematic representation of pulse field gel electrophoresis. On the right PFGE of *Sma*I-digested genomic DNA of *S. aureus* isolates. B. Sequence of steps involved in PFGE.

Furthermore, isolates with more uniform PFGE profiles require more conservative interpretation. The fact that two strains share the same pattern does not prove that they are epidemiologically related. The establishment of an epidemiologic relationship depends on the frequency with which the "indistinguishable" pattern is seen among epidemiologically-unrelated isolates and correlation with clinical and epidemiological information. If common contact between two patients with strains having the same pulsed-field gel electrophoresis (PFGE) type can be established, the chances are greater that an epidemiologic link could be ascribed. Thus, the greatest power of PFGE typing lies in showing strain dissimilarity rather than in proving similarity or relatedness. These considerations must be taken into account for banding pattern analysis from other molecular typing methods.

In some instances, initial unsatisfactory PFGE results may be aided by the use of an alternative restriction enzyme (Kam et al. 2008; Bosch et al.) or, in more difficult situations, the use of one or more additional typing methods (van Belkum et al. 2007).

The intra- and interlaboratory reproducibility of this method depends on understanding and controlling variables (Cookson et al. 1996; van Belkum et al. 1998; te Witt et al. 2010).

This success is due to an emphasis on standardized quality control especially in major areas of potential PFGE variability such as DNA sample preparation, choice of restriction enzyme, and electrophoresis conditions.

PFGE has been applied to a wide range of microorganisms and has remarkable discriminatory power and reproducibility. It is currently considered the strain typing method of choice for many commonly encountered pathogens. However, one of the main notable limitations is the need for specialized and relatively expensive equipment.

2.3 DNA sequencing-based methods

Genotyping methods based on DNA sequencing discriminate among bacterial strains directly from polymorphisms in their DNA considering the original sequence of nucleotides.

2.3.1 Single-locus sequence typing

Sequencing of a single genetic locus has been used for epidemiological studies of many bacterial species, yielding valuable typing results. In this approach, it is essential to select highly variable gene sequences. Valuable typing results have been obtained for *S. pyogenes* by DNA sequencing of 150 nucleotides coding for the N-terminal end of M protein (*emm* typing) (Beall et al. 2000). Another example is spa typing for *S.aureus* that consists in sequencing of the X region of the protein A gene (*spa typing*). This technique is widely used for subtyping methicillin-resistant *S. aureus* (MRSA) strains (Shopsin, 1999, 2000; Shopsin & Kreiswirth 2001; Harmsen et al. 2003), (Figure 5).

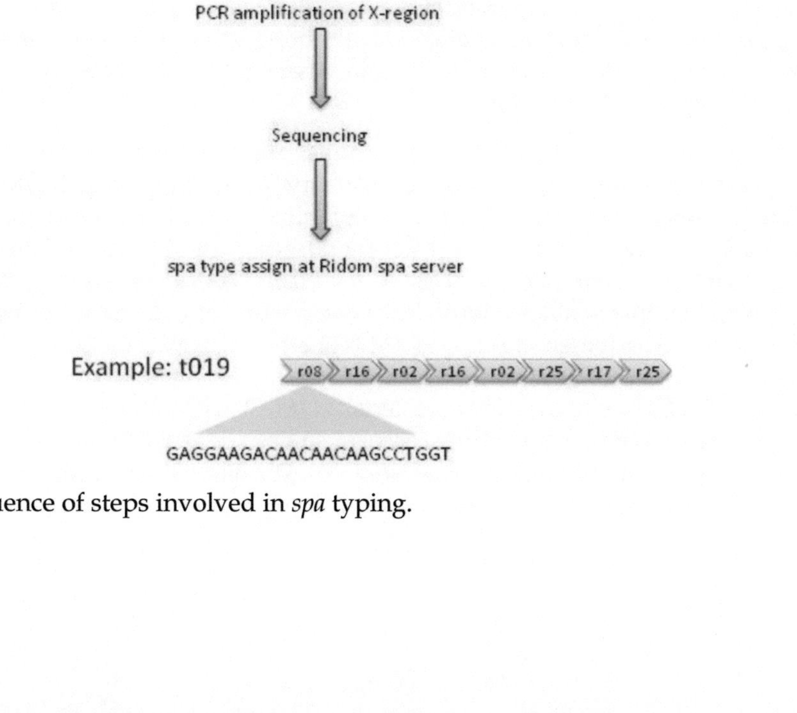

Fig. 5. Sequence of steps involved in *spa* typing.

2.3.2 Multi-locus sequence typing (MLST)

MLST is a genotyping method based on the measurement of DNA sequence variation in a set of housekeeping genes (usually seven genes) whose sequences are constrained because of the essential function of the proteins they encode. This method was proposed in 1998 as a general approach to provide accurate, portable data that were appropriate for the epidemiological investigation of bacterial pathogens and which also reflected their evolutionary and population biology (Maiden et al. 1998).

MLST schemes have been developed for several species and databases containing the allelic profiles of a great number of strain types with corresponding clinical information that can be readily consulted over the Internet (http://www.mlst.net/ and http://pubmlst. org/), (Aanensen & Spratt 2005). Additional information such as date, place of isolation and antibiotype is included in the database when a strain is deposited so this database is continuously expanding as new STs are identified and additional nucleotide sequence data are deposited.

Internal fragments of the seven housekeeping genes are amplified by PCR from chromosomal DNA using the primer pairs described in the web site. The amplified fragments are directly sequenced in each direction. The sequences at each of the seven *loci* are then compared with all the known alleles at that *locus,* and a number representing a previously described allele (or a new one) is assigned to the *locus*. For a given isolate, alleles present at each gene position are combined into an allelic profile and assigned a sequence type (ST) designation (Maiden et al. 1998). Relationships among isolates are assessed by comparisons of allelic profiles: closely related isolates have identical STs, or STs that differ at a few *loci,* whereas unrelated isolates have unrelated STs (Figure 6).

A number of clustering algorithms have been employed to analyze the data in the MLST scheme, including UPGMA (unweighted pair group method with arithmetic mean) and eBURST analyses (Feil et al. 2004).

The original conception of MLST used the allele number as the primary unit of analysis (Enright & Spratt 1998; Maiden et al. 1998) which was appropriate for organisms where horizontal genetic exchange is common. However, MLST data can also be interpreted by tree-building approaches that use nucleotide substitutions rather than allelic changes as the unit of analysis; this is more pertinent to bacteria where mutational change predominates over genetic exchange in the evolution of variants.

An important advantage of MLST is that results are unambiguous and easily and unequivocally exchangeable, much more so than images of agarose gel electrophoresis patterns. MLST drawbacks are practical, including limited accessibility and high cost. It is a relatively expensive technique available for the characterization of bacterial isolates, mainly in reference or research laboratories. However, MLST is increasingly applied as an informative typing tool that enables international comparison of isolates. It has been applied to problems as diverse as the emergence of antibiotic-resistant variants (Crisostomo et al. 2001; Enright et al. 2002), the association of particular genotypes with virulence (Brueggemann et al. 2003) or antigenic characteristics (Meats et al. 2003) and also the global spread of disease caused by novel variants (Albarracin Orio et al. 2008). In addition to these medically-motivated epidemiological analyses, MLST data have been exploited in evolutionary and population analyses (Jolley et al. 2000) that estimate recombination and mutation rates (Feil et al. 2001) and in investigation of the evolutionary relationships among bacteria that are classified as belonging to the same genus (Godoy et al. 2003).

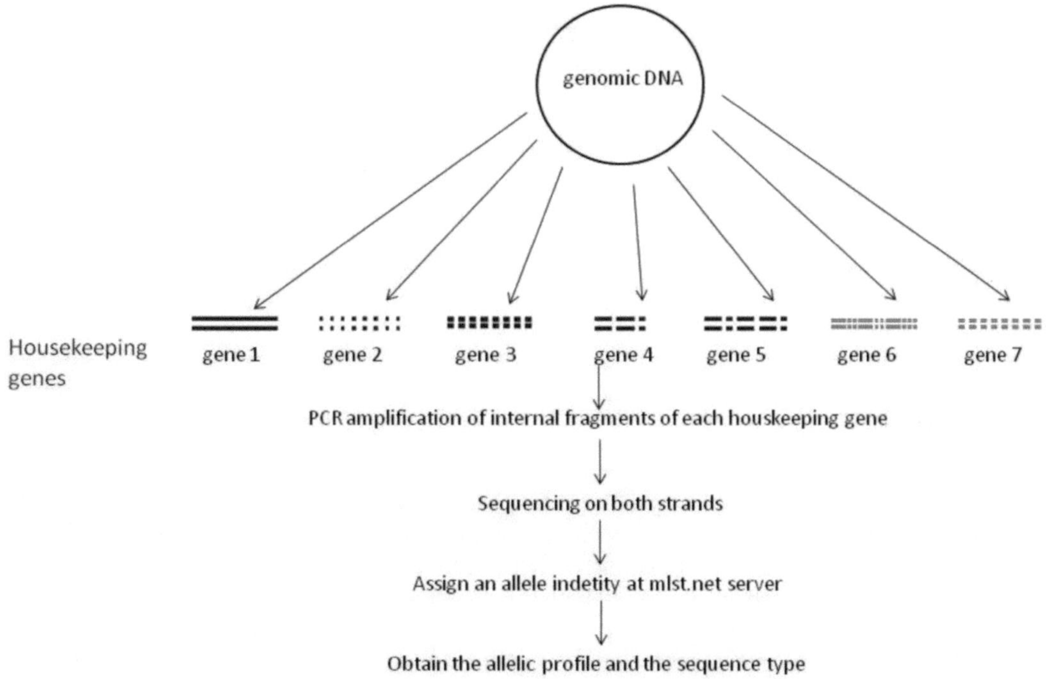

Fig. 6. Sequence of steps involved in MLST scheme. Adapted from Vazquez et al. 2004.

2.4 Analysis of results obtained by molecular epidemiology

Comparison and interpretation of raw data generated by molecular typing methods, such as gel electrophoresis band patterns, sequence alignments, or hybridization matrix patterns could be performed by visual analysis when there are few strains. However, if the analysis includes many strains, the comparison turns out to be very difficult. Therefore computer programs have become indispensable in molecular epidemiological investigations.

Computer programs that compare band sequences or patterns employ clustering algorithms that can generate dendrograms or trees illustrating the arrangement of the clusters produced. For pattern recognition, such as electrophoretic banding patterns or hybridization matrices, additional programs are needed to capture, digitize, and normalize the patterns.

There are different commercially available platforms for databasing and gel analysis that have been developed for computer-assisted analysis such as BioNumerics, GelCompar (Applied Maths, Sint-Martens-Latem, Belgium), Diversity Database Fingerprinting Software (Bio-Rad Laboratories, Hercules, Ca). Treecon (Van de Peer and De Wachter 1994).

3. Epidemiologic applications of bacterial typing techniques

A more comprehensive knowledge of the evolution and the epidemiology of bacterial pathogens had been obtained by combination of genetic, phenotypic, spatial and temporal data.

Multiple techniques have been developed to assess genomic differences among different isolates or clones of the same species, such as PCR-based methods and PFGE. These methods present portability problems and limited comprehension of the processes by which variation occurs. However, DNA sequence-based techniques generate portable differentiation in bacterial populations that can be used to understand their phylogenetic history. Extensive genomic and phenotypic diversity exists within populations of microbial pathogens of the same species. This diversity reflects the evolutionary divergence arising from mutations and gene flux. These distinctive characters are scored by typing systems which are designed to optimize discrimination between epidemiologically related and unrelated isolates of the pathogen of interest (Maslow and Mulligan 1996; Struelens 1996).

Epidemiologic typing systems can be used for outbreak investigations to confirm and delineate the transmission patterns of one or more epidemic clone(s), to test hypotheses about the sources and transmission vehicles of these clones and to monitor the reservoirs of epidemic organisms. Typing also contributes to epidemiologic surveillance and evaluation of control measures by documenting the prevalence over time and the circulation of epidemic clones in infected populations. Clearly, different requirements will be needed for these distinct applications (Maslow and Mulligan 1996; Struelens 1996).

Typing can be undertaken at two different levels, depending on the situation: i) short term or local epidemiology, when organisms are recovered in a defined setting over a short period of time, which is used to study nosocomial outbreaks, local transmission and carriage, and the relationship between isolates associated with carriage and infection in a given geographic area, ii) long term or global epidemiology, when strains are recovered from one geographic area related to those isolated worldwide or strains recovered at different times.

Local epidemiology is applied to study outbreaks with the aim to characterize that the increase in incidence of infection is caused by enhanced transmission of a specific strain. In this framework, typing methods are applied to investigate the sources of contamination and the route of transmission. Accurate application of bacterial typing will support appropriate control measures designed to contain or interrupt the outbreak and prevent further spread of disease. Typing may also be used for isolates cultured from the same patient over time to help define whether a second episode of infection is due to relapse or reinfection. PCR fingerprinting is the simplest and most rapid genotypic method for local application; however, PCR typing is very susceptible to minor variations in experimental conditions and reagents. Therefore, the method is more appropriate for the comparison of a limited number of samples processed simultaneously and run on one gel. By contrast, PFGE has good reproducibility and is highly discriminatory. Therefore, PFGE is considered the current gold standard for outbreak and local epidemiology studies.

At a different level, collaborative studies have been performed to define major internationally disseminated bacterial clones of important human pathogens. Currently, MLST in combination with PFGE is the most appropriate strategy for long term epidemiology and have reached useful conclusions from infectious disease surveillance data. The evaluation of global population genetic structure, genetic evolution, genetic diversity and pathogenicity has been successfully developed within this framework.

For eukaryotes, clones are genetically identical organisms. However, in bacterial epidemiology, the clone concept is of an even more pragmatic nature, denoting isolates obtained during real outbreaks with common features (e.g. multiple antibiotic-resistant isolates) from different geographic locations, the so-called epidemic clones.

The threshold of marker similarity used for definition of a clone need to be adjusted to the species studied, the typing system used, the environmental selective pressure and the time and space scale of the study (Tibayrenc 1995; Struelens 1996). Mutation rate and gene flux vary between species, pathovars and environments. In vivo micro-evolution of most pathogens remains poorly understood. Subclonal evolution and emergence of variants that occur in individual hosts or during prolonged transmission can be recognized by several high resolution molecular typing systems, like, for instance, macrorestriction analysis by pulsed-field gel electrophoresis (Struelens 1996).

4. Strategies applied for surveillance and typing of relevant gram-positive pathogens

4.1 Methicillin-resistant *Staphylococcus aureus*

Staphylococcus aureus is recognized as one of the most important human pathogens. *It* has shown great ability to acquire resistance to different antimicrobial agents. The first isolation of methicillin-resistant *S. aureus* (MRSA) was reported in 1960 and since then, the prevalence of this pathogen has increased.

Methicillin resistance is conferred by the *mecA* gene which codes for an additional penicillin-binding protein named PBP 2a; this protein has reduced affinity to β-lactam agents. This gene is located in a mobile genetic element of variable size known as staphylococcal cassette chromosome *mec* (SCC*mec*). So far, eight types and several subtypes of SCC*mec* have been characterized (Deurenberg & Stobberingh 2008; Chambers & Deleo 2009).

The incidence of MRSA varies geographically throughout the world. MRSA has emerged as an important pathogen among hospitalized patients. Most hospital-acquired infections caused by methicillin-resistant *Staphylococcus aureus* (HA-MRSA) are associated with a relatively small number of epidemic clones that spread over different continents. According to the Sistema Informático de Resistencia (Asociación Argentina de Microbiología, Buenos Aires, Argentina), MRSA strains are among the most prevalent nosocomial pathogens (http://www.aam.org.ar) in Argentina, whereas the Brazilian clone, the pediatric clone and the Cordobés clone have been found to be the main clones associated with HA-MRSA infections (Corso et al. 1998; Sola et al. 2002; Gardella et al. 2005).

However, since 1990, MRSA has been recognized as a cause of infections in people without established risk factors for HA-MRSA, such as recent hospitalization, surgery, residence in a long-term care facility, receipt of dialysis, or presence of invasive medical devices (Fridkin et al. 2005; Chambers & Deleo 2009). These infections are thought to be acquired in the community and are referred to as community-associated MRSA infections (CA-MRSA). This term has also been used to refer to MRSA strains with bacteriological characteristics considered typical of isolates recovered from patients with CA-MRSA infections (Salgado et al. 2003). HA-MRSA strains are generally resistant to antibiotics other than β-lactams, whereas typical CA-MRSA strains are only resistant to methicillin. HA-MRSA isolates frequently harbor SSC*mec* types-I, II and III whereas CA-MRSA strains carry types IV and V (Ma et al. 2002; Naimi et al. 2003).

The Panton-Valentine leukocidin (PVL) toxin has been described as a genetic marker of CA-MRSA isolates, rarely identified in HA-MRSA isolates (Ma et al. 2002; Naimi et al. 2003). Several studies have demonstrated that the presence of PVL genes is associated with *S. aureus* recovered from patients suffering from primary skin infections (Lina et al. 1999), severe necrotizing pneumonia, and increased complications of hematogenous osteomyelitis; however, the role of PVL in the pathogenesis of *S. aureus* infections has not yet been fully elucidated.

The spectrum of disease caused by CA-MRSA appears to be similar to that of methicillin-susceptible *Staphylococcus aureus* (MSSA) in the community. Skin and soft tissue infections (SSTIs), specifically furuncles (abscessed hair follicles or "boils"), carbuncles (coalesced masses of furuncles), and abscesses, are the most frequently reported clinical manifestations (Fergie & Purcell 2001; Baggett et al. 2003; Fridkin et al. 2005). Less commonly, MRSA has been associated with severe and invasive staphylococcal infections in the community, including necrotizing pneumonia, bacteremia, osteomyelitis, toxic shock syndrome, and meningitis (Deurenberg & Stobberingh 2008).The rapid emergence of these infections has been one of the most unexpected events in bacterial infectious diseases in the recent years.

Distinct genetic lineages associated with CA-MRSA infections have been determined by typing and their geographic dissemination evaluated in different countries. In Latin America, CA-MRSA has been described several times (Ma et al. 2005; Ribeiro et al. 2005; Alvarez et al. 2006; Gardella et al. 2008).

Typing of MRSA strains is necessary for proper epidemiological investigations of sources and modes of transmission of these strains in hospitals, and the design of appropriate control measures and the application of different typing methods have contributed to understanding the emergence of MRSA in the community. Phenotyping methods generally have limited discriminatory power and poor typeability; therefore, a number of molecular techniques have been developed for *S. aureus* typing, namely restriction fragment length polymorphism (RFLP) analysis techniques, including ribotyping and Southern blot analysis with probes for mobile elements present in multiple copies in the staphylococcal genome, like insertion sequences (IS*256*,IS*257*, IS*431* and IS*1181*) and transposons (Tn*554* and Tn*4001*) (Wei et al. 1992; Tenover et al. 1994; Kreiswirth et al. 1995).

Among PCR methods, rep-PCR and RAPD-PCR analysis were found to be epidemiologically useful, but interlaboratory studies showed that reproducibility is an important drawback for these techniques (Saulnier et al. 1993; van Belkum et al. 1995; Deplano et al. 1997; van der Zee et al. 1999). Pulsed-field gel electrophoresis (PFGE) analysis is an accurate and discriminating method which is now used as the reference method for *S. aureus* typing in some reference centers (Bannerman et al. 1995). However, PFGE analysis is costly and technically demanding and still requires interlaboratory standardization (Cookson et al. 1996). PFGE proved to be a highly discriminatory and sensitive technique in microepidemiological (local or short term) and macroepidemiological (national, continental, or long term) surveys (Struelens et al. 1993; McDougal et al. 2003). Nevertheless, some authors have argued that the stabilities of PFGE markers may be insufficient for the reliable application of PFGE to long-term or macroepidemiological studies (Blanc et al. 2002).

Sequence-based methods such as multilocus sequence typing (MLST) has proved to be adequate for long-term global epidemiology and the study of the recent evolution of *S. aureus*

(Enright, 2000, 2002). Another useful technique is the Staphylococcal cassette chromosome *mec* (SCC*mec*) typing, based on the molecular characterization by multiplex PCR of the mobile genetic element carrying the methicillin-resistant gene (*mecA*) (Oliveira and de Lencastre 2002). The combination of the MLST type and the SCC*mec* type, defined as the "clonal type," is now used for the international nomenclature of MRSA clones (Enright et al. 2002).

Moreover, single-locus DNA sequencing of repeat regions of the *coa* (coagulase) gene and the *spa* gene (protein A), respectively, could be used for reliable and accurate MRSA typing (Shopsin, 1999 2000; Tang et al. 2000; Shopsin & Kreiswirth 2001; Harmsen et al. 2003). Spa typing is especially interesting for rapid typing of MRSA in a hospital setting since it offers higher resolution than coa typing (Shopsin et al. 2000). The repeat region of the *spa* gene is subject to spontaneous mutations, as well as to loss and gain of repeats. Repeats are assigned an alpha-numerical code, and the spa type is deduced from the order of specific repeats. There is a good correlation between clonal groupings determined by MLST and the respective spa types (Harmsen et al. 2003).

We performed different studies to characterize MRSA clones within diverse scenarios of Argentina. In 2005, we demonstrated the replacement of the multiresistant MRSA "Brazilian" clone (SCC*mec* III, ST239) by the "Cordobes" clone (SCC*mec* I, ST5), a MRSA clone susceptible to rifampin, minocycline and trimethoprim/sulfamethoxazole in two university hospitals. Isolates were characterized by using RAPD-PCR and PFGE and SCCmec typing (Gardella et al. 2005).

Later, we analyzed community-associated methicillin-resistant *Staphylococcus aureus* (CA-MRSA) isolates recovered from patients suffering from different types of infections. All CA-MRSA isolates carried SCC*mec* type IV. Four major clones were detected in Argentina by PFGE. The largest cluster was named CAA clone: Pulsotype A, spa type 311, ST 5, LPV (+) (Gardella et al. 2008) and two isolates of this clone were recovered from two cases of acute bacterial meningitis (von Specht et al. 2006), (Figure 7).

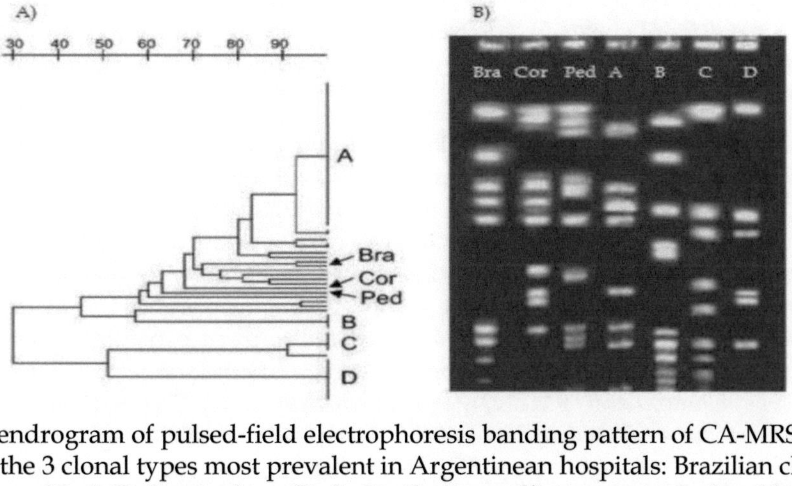

Fig. 7. (A) Dendrogram of pulsed-field electrophoresis banding pattern of CA-MRSA isolates and the 3 clonal types most prevalent in Argentinean hospitals: Brazilian clone (Bra), Cordobes clone (Cor), Pediatric clone (Ped). Similarity coefficient was calculated by using Dice coefficient, and cluster analysis was performed by the unweighted pair-group method. Four major pulsotypes were coded from A, B, C and D and representative HA-MRSA strains of prevalent clones in Argentina.

We also characterized CA-MRSA strains isolated from skin and soft-tissue infections in isolates recovered from Uruguay in 2005. In that study, we identified three major groups of CA-MRSA strains (1, 2, and 4) that were defined according to phenotypic and genotypic characteristics. The most frequent group, G1, showed a PFGE pattern identical to that of CA-MRSA strains previously isolated in Uruguay and Brazil; these strains are still producing SSTI, illustrating the stability of this emergent pathogen over time, as well as its excellent adaptation to the community environment (Pardo et al. 2009).

During the 2008 school- year period we conducted the first epidemiological study of *S. aureus* carriage in Argentina. Carriage was investigated in all children attending the last year of kindergarten in a city of Buenos Aires province, Argentina. Of 316 healthy children, 31.0% carried *S. aureus*, including 14 MRSA carriers (4.4%). All MRSA isolates carried the SCC*mec* type IV cassette. Eight of those 14 carriers were closely related to the CAA clone, which was responsible for the most severe community-acquired MRSA infections caused in our country (PFGE A, SCC*mec* IV, *spa* t311, ST5), (Gardella et al.).

Our results should serve as a warning for the health system since the main clone circulating in the community presents epidemic characteristics and also possesses a genetic background (ST5) of demonstrated plasticity and efficiency to be established as prevalent in the hospital environment.

4.2 Molecular epidemiology of *Streptococcus pneumoniae*

Streptococcus pneumoniae is a human pathogen of increasing clinical relevance causing important diseases such as meningitis, pneumonia, bacteremia and otitis media. Surveillance has become progressively more important because of the worldwide distribution of penicillin-resistant and multidrug-resistant pneumococci clones in the last 15 years.

S. pneumoniae with resistance to one or more antibiotics has been isolated since 1990. Penicillin is the drug of choice for the treatment of pneumococcal infections. The resistance mechanism includes the modification of penicillin-binding proteins (PBP), which in general is associated to cephalosporin resistance.

The clinical relevance of multiple antibiotic-resistant pneumococcal strains has led to the creation of a network. The Pneumococcal Molecular Epidemiology Network (PMEN) was established in 1997 with the aim of global surveillance of antibiotic-resistant *S.pneumoniae* and the standardization of nomenclature and classification of resistant clones (http://www.sph.emory.edu/PMEN/index.html). The PMEN also includes major invasive antibiotic-susceptible clones that have a wide geographic spread. Up to date, there are currently 43 clones described by the PMEN. Of those, three penicillin- resistant and two penicillin- susceptible PMEN clones have been detected in Argentina http://www.sph.emory.edu/PMEN/(Zemlickova et al. 2005). Clones to be included in the Network have to be subjected to PFGE, MLST and PBP fingerprinting to confirm that they differ from previously accepted ones. PBP fingerprinting is a typing technique that includes PCR amplification of the *pbp1a*, *pbp2b* and *pbp2x* genes with previously described primers (Gherardi et al. 2000). Amplified genes are digested with *HaeIII+DdeI* (*pbp1a*) and *HaeIII+RsaI* (*pbp2b* and *pbp2x*) restriction enzymes and electrophoresed on 3% gels. This technique has been used in many reports over the last 20 years to study the molecular epidemiology of *S. pneumoniae* resistance to β-lactams (Zhang et al. 1990; Munoz et al. 1991). For macrolide-resistant strains, tests for *erm* and *mef* genes have to be performed (http://www.sph.emory.edu/PMEN). The PMEN network has included BOX-PCR typing in the guidelines for the recognition of pneumococcal clones (McGee et al. 2001).

In addition, the use of modern typing methods, mainly MLST, has greatly helped track the geographical spread of specific *S. pneumoniae* strains and follow the dynamics of microbial populations over time. The application of all these techniques have shown that the spread of multiresistant international clones defined by the PMEN is the main cause of increase in pneumococcal resistance to β-lactams and other drugs (Munoz et al. 1991; Klugman 2002; Smith et al. 2006; Sadowy et al. 2007; Soriano et al. 2008). In pneumococcus, each serotype may typically be made up of a number of clones, which are not closely related and are not equivalent in terms of antibiotic resistance.

Furthermore, molecular methods showed that the evolution of penicillin-resistance and multiresistance is a phenomenon in which the acquisition and/or alteration of molecular targets is mainly a consequence of intergenic change and that *S. pneumoniae* diversity has largely been driven by recombination (Hermans et al. 1997; Enright & Spratt 1998; Enright et al. 1999; Descheemaeker et al. 2000; McGee et al. 2001).

As multiresistance has increased the difficulties of treating this serious bacterial infection, prevention through vaccination has become even more important. There are at least 91 known pneumococcus capsular types, with 23 capsular types included in the current pneumococcal polysaccharide (adult) vaccine and 13 types included in the current conjugate (child) vaccine. To overcome serotype specificity of actual vaccines, upcoming pneumococcal vaccines should offer a different approach to the prevention of pneumococcal disease and the decrease in carriage. Several proteins have been identified as possible candidates to develop more appropriate vaccines. One of them , the pneumococcal surface protein A (PspA), is a surface virulence factor, antigenically variable yet cross-reactive that interferes with complement-mediated clearance of pneumococci (McDaniel et al. 1991; Tu et al. 1999).Since 1993, six Latin-American countries have been participating in an epidemiological surveillance study conducted by the Pan American Health Organization (PAHO) in order to determine the relative prevalence of capsular types and antimicrobial resistant patterns of *S. pneumoniae* causing invasive infections in children <5 years of age. One of these studies showed that, the prevalence of penicillin resistant *S. pneumoniae* (PRSP) in Argentina was 24.4%, which was significantly associated with the expansion of serotype 14 clone that had been previously described in Europe expressing serotype 9V (Rossi et al. 1998). A similar situation was encountered in Uruguay in the same period (Camou et al. 1998).

Ongoing surveillance programs for invasive pneumococcal disease also monitor the appropriateness of existing vaccine formulations and provide valuable information on which to base the formulation and application of new vaccines that are currently under development. In this framework, from 1993 to 2000, with the participation of the Argentinean *Streptococcus pneumoniae* Working Group, 1293 invasive isolates were studied to determine capsular type distribution and antimicrobial susceptibility. We selected a sample of 149 strains, having the same serotype distribution as in the total collection, in order to characterize the distribution of PspA variants among Argentinean invasive isolates recovered from children less than 6 years of age. The genetic relatedness among the isolates of the major serotypes was also evaluated by BOX-PCR because it is a quick molecular method that is suitable for investigation of genetic relatedness of pneumococcal strains and provides results whose interpretation is relatively unambiguous (Hermans et al. 1995; van Belkum et al. 1996). This study provided epidemiological information about the PspA family distribution and the genetic diversity of Argentinean *S. pneumoniae* isolates and informed of the potential coverage of a PspA- based vaccine. It was the first insight into diversity of PspA within strains circulating in Argentina (Mollerach et al. 2004). Family 1 PspA was detected in 54.4% of the isolates, 41.6% of which were family 2 and 4.0% expressed both family 1 and family 2 PspAs. This observation indicates that a PspA vaccine containing only family 1 and family 2 PspAs should be able to cover the bulk of the strains in this region. Box typing revealed the Argentinian strains were from at least 10 clonally related groups.

In some cases, a strong association between one PspA type and a certain capsular type was found. For example, serotype 1 and 5 and the majority of isolates of penicillin-susceptible serotype 14 isolates exhibited PspA family 1. On the other hand, serotypes 7 F, serotype14 PRSP and the majority of type 9V isolates were assigned to PspA family 2. BOX-PCR analysis revealed genetic homogeneity of serotype 14 PRSP and serotype 5 isolates. Antibiotype suggests correlation with the Spain^{9V}-3 clone and Colombia^{5}-19 clone, respectively. These clones had been previously described in the region (Gamboa et al. 2002; Brandileone et al. 2004; Zemlickova et al. 2005).

Nowadays, the sequencing of DNA allows to compare the results between laboratories and to obtain a global look for the situation of the circulating multidrug-resistant clones. This effort includes a database that contains information concerning the clones that are currently widespread in different parts of the world (http://spneumoniae.mlst.net).

In the framework of PAHO in Latin American countries, surveillance data revealed that penicillin-nonsusceptible *S. pneumoniae* (PNSP) type 6B increased from 15.8 % in the period between 1993-1997, to 67.3 % in 1998-2002 ($p<0.001$). Serotype 6 ranks fourth among capsular types causing invasive diseases in Argentinean patients under 6 years of age, and it has been included in the heptavalent conjugate polysaccharide vaccine licensed in Argentina in 2001 and also in the 13-valent introduced in 2010 (Organización Panamericana de la Salud 2007; Ruvinsky et al. 2008). This serotype is a frequent cause of invasive diseases (Riedel et al. 2007; Gabastou et al. 2008; Darabi et al.). We characterized the population of penicillin non-susceptible *S. pneumoniae* type 6B strains isolated from pediatric patients in Argentina between 1993-2002 with the use of molecular typing methods including BOX-PCR, PFGE and MLST (Bonofiglio et al. 2011) (Figure 8). The results of the study showed that the increase in penicillin resistance in serotype 6B may be partly explained by the entrance of the Poland^{6B}-20 clone, which is a PMEN clone not previously described in Argentina. Our findings showed that the Poland^{6B}-20 clone established in 1999; and the use of BOX-PCR and PFGE subtypes suggested that horizontal transfer or differentiation events had occurred after the common lineage became established. Dissemination of this clone could be traced through demographic data, as isolates representative of the clone had been recovered in different regions of Argentina and its expansion is also responsible for the emergence of erythromycin-resistance in *S. pneumoniae* serotype 6B. The pneumococcal MLST database currently contains information of 81 strains of the Poland^{6B}-20 clone.

Other similar studies were carried out by Sadowy et al, who analyzed isolates recovered in Poland in the period 2003-2005 using serotyping, MLST and sequencing of *murM* and *pspA* alleles. They demonstrated that the vast majority of the isolates (90.7%) belong to international multiresistant clones whereas, the Spain 9V -ST156 clonal complex being the most prevalent. Moreover, this clone has evolved rapidly, as demonstrated by the observed number of STs , the use of another approach of MLST (multiple locus variable-number-tandem repeat analysis) and the polymorphism of *pbp* and *pspA* genes (coding for penicillin-binding proteins and the pneumococcal surface protein A, respectively) (Sadowy et al.).

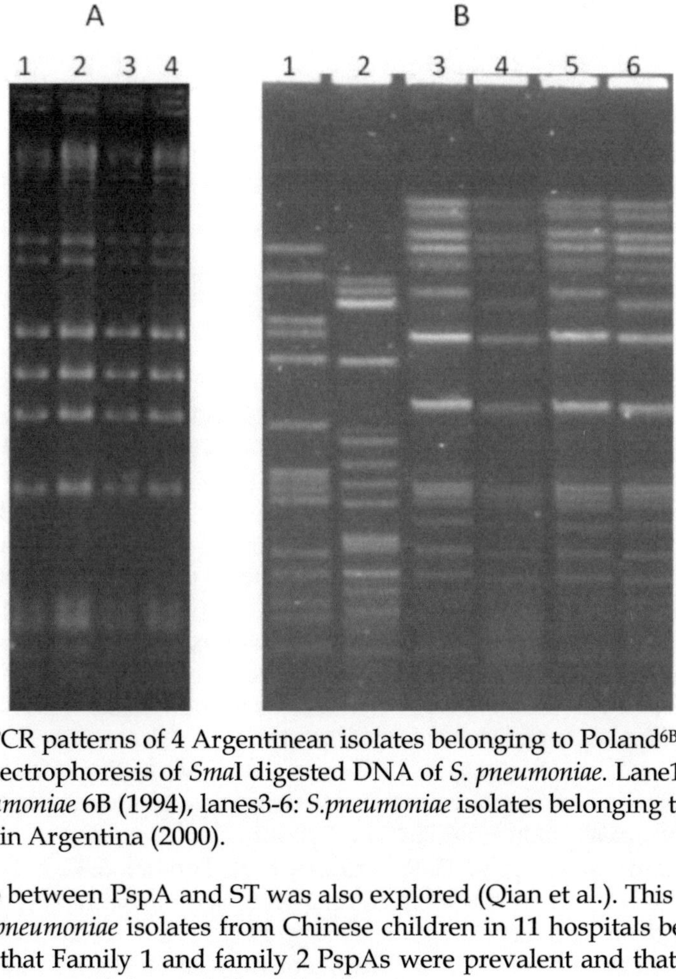

Fig. 8. A) BOX-PCR patterns of 4 Argentinean isolates belonging to Poland^{6B}-20 clone. B). Pulse field gel electrophoresis of *Sma*I digested DNA of *S. pneumoniae*. Lane1: *S.pneumoniae* R6, lane2: *S.pneumoniae* 6B (1994), lanes3-6: *S.pneumoniae* isolates belonging to Poland^{6B}-20 clone recovered in Argentina (2000).

The relationship between PspA and ST was also explored (Qian et al.). This author analyzed 171 invasive *S. pneumoniae* isolates from Chinese children in 11 hospitals between 2006 and 2008. He found that Family 1 and family 2 PspAs were prevalent and that strains with the same ST always presented the same PspA family.

5. Conclusion

This chapter has reviewed some of the most popular molecular methods for the epidemiological typing of two medically relevant gram-positive cocci, discussing their principles, strengths and weaknesses. We have described several examples of our recent work showing the application of molecular typing techniques to the study of two relevant pathogens. The examples we have considered herein include a relative clonal species such as *S.aureus*, and on the other hand, a pathogen showing a high recombination rate, such as *S. pneumoniae*.

Molecular epidemiology has enormous potential in understanding the evolution of bacterial populations and can help establish appropriate control measures and interventions, including the use of vaccines, therapeutics, public health actions and ongoing pathogen surveillance.

6. Acknowledgments

This work was supported in part by grants from the University of Buenos Aires, Buenos Aires, Argentina, and Agencia Nacional de Promoción Científica y Tecnológica (Buenos Aires, Argentina; PICT 1634) to MM. MM and LB are members of "Carrera del Investigador", CONICET, Argentina.

The authors thank clinical microbiologists for providing isolates and clinical data. In particular, Martha von Spetch and Mabel Regueira; as well as Alejandra Corso for providing the pediatric clone isolate.

7. References

Aanensen, D. M. & B. G. Spratt (2005).The multilocus sequence typing network: mlst.net. *Nucleic. Acids Res.* 33(Web Server issue): W728-33.1362-4962 (Electronic)

Albarracin Orio, A. G., P. R. Cortes, M. Tregnaghi, G. E. Pinas & J. R. Echenique (2008).A new serotype 14 variant of the pneumococcal Spain^{9V}-3 international clone detected in the central region of Argentina. *J. Med. Microbiol.* 57(Pt 8): 992-9.0022-2615 (Print)

Alvarez, C. A., O. J. Barrientes, A. L. Leal, G. A. Contreras, L. Barrero, S. Rincon, L. Diaz, N. Vanegas & C. A. Arias (2006).Community-associated methicillin-resistant *Staphylococcus aureus*, Colombia. *Emerg. Infect. Dis.* 12(12): 2000-1.1080-6040 (Print)

Baggett, H. C., T. W. Hennessy, R. Leman, C. Hamlin, D. Bruden, A. Reasonover, P. Martinez & J. C. Butler (2003).An outbreak of community-onset methicillin-resistant *Staphylococcus aureus* skin infections in southwestern Alaska. *Infect. Control Hosp. Epidemiol.* 24(6): 397-402.0899-823X (Print)

Bannerman, T. L., G. A. Hancock, F. C. Tenover & J. M. Miller (1995).Pulsed-field gel electrophoresis as a replacement for bacteriophage typing of *Staphylococcus aureus*. *J. Clin. Microbiol.* 33(3): 551-5.0095-1137 (Print)

Beall, B., G. Gherardi, M. Lovgren, R. R. Facklam, B. A. Forwick & G. J. Tyrrell (2000).*emm* and *sof* gene sequence variation in relation to serological typing of opacity-factor-positive group A streptococci. *Microbiology* 146 (Pt 5): 1195-209.1350-0872 (Print)

Blanc, D. S., P. Francioli & P. M. Hauser (2002).Poor value of pulsed-field gel electrophoresis to investigate long-term scale epidemiology of methicillin-resistant *Staphylococcus aureus*. *Infect. Genet. Evol.* 2(2): 145-8.1567-1348 (Print)

Bonofiglio, L., M. Regueira, J. Pace, A. Corso, E. Garcia & M. Mollerach (2011).Dissemination of an erythromycin-resistant penicillin-nonsusceptible *Streptococcus pneumoniae* Poland(6B)-20 clone in Argentina. *Microb. Drug Resist.* 17(1): 75-81.1931-8448 (Electronic)

Bosch, T., A. J. de Neeling, L. M. Schouls, K. W. van der Zwaluw, J. A. Kluytmans, H. Grundmann & X. W. Huijsdens (2010).PFGE diversity within the methicillin-resistant *Staphylococcus aureus* clonal lineage ST398. *BMC Microbiol* 10: 40.1471-2180 (Electronic)

Brandileone, M. C., A. L. Andrade, E. M. Teles, R. C. Zanella, T. I. Yara, J. L. Di Fabio & S. K. Hollingshead (2004).Typing of pneumococcal surface protein A (PspA) in *Streptococcus pneumoniae* isolated during epidemiological surveillance in Brazil: towards novel pneumococcal protein vaccines. *Vaccine* 22(29-30): 3890-6.0264-410X (Print)

Brueggemann, A. B., D. T. Griffiths, E. Meats, T. Peto, D. W. Crook & B. G. Spratt (2003).Clonal relationships between invasive and carriage *Streptococcus pneumoniae* and serotype- and clone-specific differences in invasive disease potential. *J. Infect. Dis.* 187(9): 1424-32.0022-1899 (Print)

Camou, T., M. Hortal & A. Tomasz (1998).The apparent importation of penicillin-resistant capsular type 14 Spanish/French clone of *Streptococcus pneumoniae* into Uruguay in the early 1990s. *Microb. Drug Resist.* 4(3): 219-24.1076-6294 (Print)

Cookson, B. D., P. Aparicio, A. Deplano, M. Struelens, R. Goering & R. Marples (1996).Inter-centre comparison of pulsed-field gel electrophoresis for the typing of methicillin-resistant *Staphylococcus aureus*. *J. Med. Microbiol.* 44(3): 179-84.0022-2615 (Print)

Corso, A., I. Santos Sanches, M. Aires de Sousa, A. Rossi & H. de Lencastre (1998).Spread of a methicillin-resistant and multiresistant epidemic clone of *Staphylococcus aureus* in Argentina. *Microb Drug Resist* 4(4): 277-88.1076-6294 (Print)

Crisostomo, M. I., H. Westh, A. Tomasz, M. Chung, D. C. Oliveira & H. de Lencastre (2001).The evolution of methicillin resistance in *Staphylococcus aureus*: similarity of genetic backgrounds in historically early methicillin-susceptible and -resistant isolates and contemporary epidemic clones. *Proc. Natl. Acad. Sci. U. S. A.* 98(17): 9865-70.0027-8424 (Print)

Chambers, H. F. & F. R. Deleo (2009).Waves of resistance: *Staphylococcus aureus* in the antibiotic era. *Nat. Rev. Microbiol.* 7(9): 629-41.1740-1534 (Electronic)

Darabi, A., D. Hocquet & M. J. Dowzicky (2010).Antimicrobial activity against *Streptococcus pneumoniae* and *Haemophilus influenzae* collected globally between 2004 and 2008 as part of the Tigecycline Evaluation and Surveillance Trial. *Diagn. Microbiol. Infect. Dis.* 67(1): 78-86.1879-0070 (Electronic)

Deplano, A., M. Vaneechoutte, G. Verschraegen & M. J. Struelens (1997).Typing of *Staphylococcus aureus* and *Staphylococcus epidermidis* strains by PCR analysis of inter-IS256 spacer length polymorphisms. *J Clin Microbiol* 35(10): 2580-7.0095-1137 (Print)

Descheemaeker, P., S. Chapelle, C. Lammens, M. Hauchecorne, M. Wijdooghe, P. Vandamme, M. Ieven & H. Goossens (2000).Macrolide resistance and erythromycin resistance determinants among Belgian *Streptococcus pyogenes* and *Streptococcus pneumoniae* isolates. *J. Antimicrob. Chemother.* 45(2): 167-73.0305-7453 (Print)

Deurenberg, R. H. & E. E. Stobberingh (2008).The evolution of *Staphylococcus aureus*. *Infect Genet Evol* 8(6): 747-63.1567-1348 (Print)

Enright, M. C., N. P. Day, C. E. Davies, S. J. Peacock & B. G. Spratt (2000).Multilocus sequence typing for characterization of methicillin-resistant and methicillin-susceptible clones of *Staphylococcus aureus*. *J. Clin. Microbiol.* 38(3): 1008-15.0095-1137 (Print)

Enright, M. C., A. Fenoll, D. Griffiths & B. G. Spratt (1999).The three major Spanish clones of penicillin-resistant *Streptococcus pneumoniae* are the most common clones recovered in recent cases of meningitis in Spain. *J. Clin. Microbiol.* 37(10): 3210-6.0095-1137 (Print)

Enright, M. C., D. A. Robinson, G. Randle, E. J. Feil, H. Grundmann & B. G. Spratt (2002).The evolutionary history of methicillin-resistant *Staphylococcus aureus* (MRSA). *Proc. Natl. Acad. Sci. U. S. A.* 99(11): 7687-92.0027-8424 (Print)

Enright, M. C. & B. G. Spratt (1998).A multilocus sequence typing scheme for *Streptococcus pneumoniae*: identification of clones associated with serious invasive disease. *Microbiology* 144 (Pt 11): 3049-60.1350-0872 (Print)

Feil, E. J., E. C. Holmes, D. E. Bessen, M. S. Chan, N. P. Day, M. C. Enright, R. Goldstein, D. W. Hood, A. Kalia, C. E. Moore, J. Zhou & B. G. Spratt (2001).Recombination within natural populations of pathogenic bacteria: short-term empirical estimates and long-term phylogenetic consequences. *Proc. Natl. Acad. Sci. U. S. A.* 98(1): 182-7.0027-8424 (Print)

Feil, E. J., B. C. Li, D. M. Aanensen, W. P. Hanage & B. G. Spratt (2004).eBURST: inferring patterns of evolutionary descent among clusters of related bacterial genotypes from multilocus sequence typing data. *J. Bacteriol.* 186(5): 1518-30.0021-9193 (Print)

Fergie, J. E. & K. Purcell (2001).Community-acquired methicillin-resistant *Staphylococcus aureus* infections in south Texas children. *Pediatr. Infect. Dis. J.* 20(9): 860-3.0891-3668 (Print)

Fridkin, S. K., J. C. Hageman, M. Morrison, L. T. Sanza, K. Como-Sabetti, J. A. Jernigan, K. Harriman, L. H. Harrison, R. Lynfield & M. M. Farley (2005).Methicillin-resistant *Staphylococcus aureus* disease in three communities. *N. Engl. J. Med.* 352(14): 1436-44.1533-4406 (Electronic)

Gabastou, J. M., C. I. Agudelo, M. C. Brandileone, E. Castaneda, A. P. de Lemos & J. L. Di Fabio (2008).Characterization of invasive isolates of *S. pneumoniae, H. influenzae,* and *N. meningitidis* in Latin America and the Caribbean: SIREVA II, 2000-2005. *Rev. Panam. Salud Publica* 24(1): 1-15.1020-4989 (Print)

Gamboa, L., T. Camou, M. Hortal & E. Castaneda (2002).Dissemination of *Streptococcus pneumoniae* clone Colombia(5)-19 in Latin America. *J. Clin. Microbiol.* 40(11): 3942-50.0095-1137 (Print)

Gardella, N., S. Murzicato, S. Di Gregorio, A. Cuirolo, J. Desse, F. Crudo, G. Gutkind & M. Mollerach (2011).Prevalence and characterization of methicillin-resistant *Staphylococcus aureus* among healthy children in a city of Argentina. *Infect. Genet. Evol.* 11(5): 1066-71.1567-7257 (Electronic)

Gardella, N., R. Picasso, S. C. Predari, M. Lasala, M. Foccoli, G. Benchetrit, A. Famiglietti, M. Catalano, M. Mollerach & G. Gutkind (2005).Methicillin-resistant *Staphylococcus aureus* strains in Buenos Aires teaching hospitals: replacement of the multidrug resistant South American clone by another susceptible to rifampin, minocycline and trimethoprim-sulfamethoxazole. *Rev. Argent. Microbiol.* 37(3): 156-60.0325-7541 (Print)

Gardella, N., M. von Specht, A. Cuirolo, A. Rosato, G. Gutkind & M. Mollerach (2008).Community-associated methicillin-resistant *Staphylococcus aureus*, eastern Argentina. *Diagn. Microbiol. Infect. Dis.* 62(3): 343-7.0732-8893 (Print)

Gherardi, G., C. G. Whitney, R. R. Facklam & B. Beall (2000).Major related sets of antibiotic-resistant Pneumococci in the United States as determined by pulsed-field gel electrophoresis and pbp1a-pbp2b-pbp2x-dhf restriction profiles. *J. Infect. Dis.* 181(1): 216-29.0022-1899 (Print)

Godoy, D., G. Randle, A. J. Simpson, D. M. Aanensen, T. L. Pitt, R. Kinoshita & B. G. Spratt (2003).Multilocus sequence typing and evolutionary relationships among the causative agents of melioidosis and glanders, *Burkholderia pseudomallei* and *Burkholderia mallei*. *J. Clin. Microbiol.* 41(5): 2068-79.0095-1137 (Print)

Goering, R. V. (2010).Pulsed field gel electrophoresis: a review of application and interpretation in the molecular epidemiology of infectious disease. *Infect. Genet. Evol.* 10(7): 866-75.1567-7257 (Electronic)

Halpin, J. L., N. M. Garrett, E. M. Ribot, L. M. Graves & K. L. Cooper (2010).Re-evaluation, optimization, and multilaboratory validation of the PulseNet-standardized pulsed-field gel electrophoresis protocol for *Listeria monocytogenes*. *Foodborne Pathog. Dis.* 7(3): 293-8.1556-7125 (Electronic)

Harmsen, D., H. Claus, W. Witte, J. Rothganger, D. Turnwald & U. Vogel (2003).Typing of methicillin-resistant *Staphylococcus aureus* in a university hospital setting by using novel software for spa repeat determination and database management. *J. Clin. Microbiol.* 41(12): 5442-8.0095-1137 (Print)

Hermans, P., M. Sluijter, T. Hoogenboezem, H. Heersma, A. v. Belkum & R. d. Groot (1995).Comparative Study of Five Different DNA Fingerprint Techniques for Molecular Typing of *Streptococcus pneumoniae* Strains. *J. Clin. Microbiol.* 33: 1606-1612

Hermans, P. W., M. Sluijter, S. Dejsirilert, N. Lemmens, K. Elzenaar, A. van Veen, W. H. Goessens and R. de Groot (1997).Molecular epidemiology of drug-resistant pneumococci: toward an international approach. *Microb. Drug Resist.* 3(3): 243-51.1076-6294 (Print)

Jolley, K. A., J. Kalmusova, E. J. Feil, S. Gupta, M. Musilek, P. Kriz & M. C. Maiden (2000).Carried meningococci in the Czech Republic: a diverse recombining population. *J. Clin. Microbiol.* 38(12): 4492-8.0095-1137 (Print)

Kam, K. M., C. K. Luey, M. B. Parsons, K. L. Cooper, G. B. Nair, M. Alam, M. A. Islam, D. T. Cheung, Y. W. Chu, T. Ramamurthy, G. P. Pazhani, S. K. Bhattacharya, H. Watanabe, J. Terajima, E. Arakawa, O. A. Ratchtrachenchai, S. Huttayananont, E. M. Ribot, P. Gerner-Smidt & B. Swaminathan (2008).Evaluation and validation of a PulseNet standardized pulsed-field gel electrophoresis protocol for subtyping *Vibrio parahaemolyticus*: an international multicenter collaborative study. *J. Clin. Microbiol.* 46(8): 2766-73.1098-660X (Electronic)

Klugman, K. P. (2002).The successful clone: the vector of dissemination of resistance in *Streptococcus pneumoniae*. *J. Antimicrob. Chemother.* 50 Suppl S2: 1-5.0305-7453 (Print)

Kreiswirth, B. N., S. M. Lutwick, E. K. Chapnick, J. D. Gradon, L. I. Lutwick, D. V. Sepkowitz, W. Eisner & M. H. Levi (1995).Tracing the spread of methicillin-resistant *Staphylococcus aureus* by Southern blot hybridization using gene-specific probes of *mec* and Tn554. *Microb Drug Resist* 1(4): 307-13.1076-6294 (Print)

Lina, G., Y. Piemont, F. Godail-Gamot, M. Bes, M. O. Peter, V. Gauduchon, F. Vandenesch & J. Etienne (1999).Involvement of Panton-Valentine leukocidin-producing *Staphylococcus aureus* in primary skin infections and pneumonia. *Clin. Infect. Dis.* 29(5): 1128-32.1058-4838 (Print)

Ma, X. X., A. Galiana, W. Pedreira, M. Mowszowicz, I. Christophersen, S. Machiavello, L. Lope, S. Benaderet, F. Buela, W. Vincentino, M. Albini, O. Bertaux, I. Constenla, H. Bagnulo, L. Llosa, T. Ito & K. Hiramatsu (2005).Community-acquired methicillin-resistant *Staphylococcus aureus*, Uruguay. *Emerg. Infect. Dis.* 11(6): 973-6.1080-6040 (Print)

Ma, X. X., T. Ito, C. Tiensasitorn, M. Jamklang, P. Chongtrakool, S. Boyle-Vavra, R. S. Daum & K. Hiramatsu (2002).Novel type of staphylococcal cassette chromosome *mec* identified in community-acquired methicillin-resistant *Staphylococcus aureus* strains. *Antimicrob. Agents Chemother.* 46(4): 1147-52.0066-4804 (Print)

Maiden, M. C., J. A. Bygraves, E. Feil, G. Morelli, J. E. Russell, R. Urwin, Q. Zhang, J. Zhou, K. Zurth, D. A. Caugant, I. M. Feavers, M. Achtman & B. G. Spratt (1998).Multilocus sequence typing: a portable approach to the identification of clones within populations of pathogenic microorganisms. *Proc. Natl. Acad. Sci. U. S. A.* 95(6): 3140-5.0027-8424 (Print)

Maslow, J. & M. E. Mulligan (1996).Epidemiologic typing systems. *Infect. Control Hosp. Epidemiol.* 17(9): 595-604.0899-823X (Print)

McDaniel, L. S., J. S. Sheffield, P. Delucchi & D. E. Briles (1991).PspA, a surface protein of *Streptococcus pneumoniae,* is capable of eliciting protection against pneumococci of more than one capsular type. *Infect. Immun.* 59(1): 222-8.0019-9567 (Print)

McDougal, L. K., C. D. Steward, G. E. Killgore, J. M. Chaitram, S. K. McAllister & F. C. Tenover (2003).Pulsed-field gel electrophoresis typing of oxacillin-resistant *Staphylococcus aureus* isolates from the United States: establishing a national database. *J. Clin. Microbiol.* 41(11): 5113-20.0095-1137 (Print)

McGee, L., L. McDougal, J. Zhou, B. G. Spratt, F. C. Tenover, R. George, R. Hakenbeck, W. Hryniewicz, J. C. Lefevre, A. Tomasz & K. P. Klugman (2001).Nomenclature of major antimicrobial-resistant clones of *Streptococcus pneumoniae* defined by the pneumococcal molecular epidemiology network. *J. Clin. Microbiol.* 39(7): 2565-71.0095-1137 (Print)

Meats, E., A. B. Brueggemann, M. C. Enright, K. Sleeman, D. T. Griffiths, D. W. Crook & B. G. Spratt (2003).Stability of serotypes during nasopharyngeal carriage of *Streptococcus pneumoniae. J. Clin. Microbiol.* 41(1): 386-92.0095-1137 (Print)

Mollerach, M., M. Regueira, L. Bonofiglio, R. Callejo, J. Pace, J. L. Di Fabio, S. Hollingshead & D. Briles (2004).Invasive *Streptococcus pneumoniae* isolates from Argentinian children: serotypes, families of pneumococcal surface protein A (PspA) and genetic diversity. *Epidemiol. Infect.* 132(2): 177-84.0950-2688 (Print)

Munoz, R., T. J. Coffey, M. Daniels, C. G. Dowson, G. Laible, J. Casal, R. Hakenbeck, M. Jacobs, J. M. Musser, B. G. Spratt and Tomasz, A. (1991) Intercontinental spread of a multiresistant clone of serotype 23F *Streptococcus pneumoniae. J. Infect. Dis.* 164(2): 302-6.0022-1899 (Print)

Naimi, T. S., K. H. LeDell, K. Como-Sabetti, S. M. Borchardt, D. J. Boxrud, J. Etienne, S. K. Johnson, F. Vandenesch, S. Fridkin, C. O'Boyle, R. N. Danila & R. Lynfield (2003).Comparison of community- and health care-associated methicillin-resistant *Staphylococcus aureus* infection. *J. A. M. A.* 290(22): 2976-84.1538-3598 (Electronic)

Oliveira, D. C. & H. de Lencastre (2002).Multiplex PCR strategy for rapid identification of structural types and variants of the mec element in methicillin-resistant *Staphylococcus aureus. Antimicrob. Agents Chemother.* 46(7): 2155-61.0066-4804 (Print)

Organización Panamericana de la Salud (2007). Informe regional de SIREVA II: datos por país y por grupos de edad sobre las características de los aislamientos de

Streptococcus pneumoniae,Haemophilus influenzae y *Neisseria meningitidis*, en procesos invasivos, 2000–2005. Washington, D.C.

Pardo, L., V. Machado, M. Mollerach, M. I. Mota, L. P. Tuchscherr, P. Gadea, N. Gardella, D. O. Sordelli, M. Vola, F. Schelotto & G. Varela (2009).Characteristics of Community-Associated Methicillin-Resistant *Staphylococcus aureus* (CA-MRSA) Strains Isolated from Skin and Soft-Tissue Infections in Uruguay. *Int. J. Microbiol.* 2009: 472126.1687-9198 (Electronic)

Qian, J., K. Yao, L. Xue, G. Xie, Y. Zheng, C. Wang, Y. Shang, H. Wang, L. Wan, L. Liu, C. Li, W. Ji, Y. Wang, P. Xu, S. Yu, Y. W. Tang & Y. Yang (2011).Diversity of pneumococcal surface protein A (PspA) and relation to sequence typing in *Streptococcus pneumoniae* causing invasive disease in Chinese children. *Eur. J. Clin. Microbiol. Infect. Dis.*1435-4373 (Electronic)

Ribeiro, A., C. Dias, M. C. Silva-Carvalho, L. Berquo, F. A. Ferreira, R. N. Santos, B. T. Ferreira-Carvalho & A. M. Figueiredo (2005).First report of infection with community-acquired methicillin-resistant *Staphylococcus aureus* in South America. *J. Clin. Microbiol.* 43(4): 1985-8.0095-1137 (Print)

Riedel, S., S. E. Beekmann, K. P. Heilmann, S. S. Richter, J. Garcia-de-Lomas, M. Ferech, H. Goosens & G. V. Doern (2007).Antimicrobial use in Europe and antimicrobial resistance in *Streptococcus pneumoniae*. *Eur. J. Clin. Microbiol. Infect. Dis.* 26(7): 485-90.0934-9723 (Print)

Rossi, A., A. Corso, J. Pace, M. Regueira & A. Tomasz (1998).Penicillin-resistant *Streptococcus pneumoniae* in Argentina: frequent occurrence of an internationally spread serotype 14 clone. *Microb. Drug Resist.* 4(3): 225-31.1076-6294 (Print)

Ruvinsky, R., M. Regueirea, A. Corso, S. Fossari, M. Moscoloni, S. García & L. Di-Fabio (2008).*Streptococcus pneumoniae* Invasive Diseases:Surveillance of serotypes in children in Argentina. *Abstract presented at the 46th Annual Interscience Conference on Antimicrobial Agents and Chemotherapy (ICAAC).* Washington DC, USA. October 25-28. Abstract no. G1-2087.

Sadowy, E., R. Izdebski, A. Skoczynska, P. Grzesiowski, M. Gniadkowski & W. Hryniewicz (2007).Phenotypic and molecular analysis of penicillin-nonsusceptible *Streptococcus pneumoniae* isolates in Poland. *Antimicrob. Agents Chemother.* 51(1): 40-7.0066-4804 (Print)

Sadowy, E., A. Kuch, M. Gniadkowski & W. Hryniewicz (2010).Expansion and evolution of the *Streptococcus pneumoniae* Spain^{9V}-ST156 clonal complex in Poland. *Antimicrob. Agents Chemother.* 54(5): 1720-7.1098-6596 (Electronic)

Salgado, C. D., B. M. Farr & D. P. Calfee (2003).Community-acquired methicillin-resistant *Staphylococcus aureus*: a meta-analysis of prevalence and risk factors. *Clin. Infect. Dis.* 36(2): 131-9.1537-6591 (Electronic)

Saluja, S. K. & J. N. Weiser (1995).The genetic basis of colony opacity in *Streptococcus pneumoniae*: evidence for the effect of box elements on the frequency of phenotypic variation. *Mol. Microbiol.* 16(2): 215-27.0950-382X (Print)

Saulnier, P., C. Bourneix, G. Prevost & A. Andremont (1993).Random amplified polymorphic DNA assay is less discriminant than pulsed-field gel electrophoresis for typing strains of methicillin-resistant *Staphylococcus aureus*. *J. Clin. Microbiol.* 31(4): 982-5.0095-1137 (Print)

Shopsin, B., M. Gomez, S. O. Montgomery, D. H. Smith, M. Waddington, D. E. Dodge, D. A. Bost, M. Riehman, S. Naidich & B. N. Kreiswirth (1999).Evaluation of protein A gene polymorphic region DNA sequencing for typing of *Staphylococcus aureus* strains. *J. Clin. Microbiol.* 37(11): 3556-63.0095-1137 (Print)

Shopsin, B., M. Gomez, M. Waddington, M. Riehman & B. N. Kreiswirth (2000).Use of coagulase gene (*coa*) repeat region nucleotide sequences for typing of methicillin-resistant *Staphylococcus aureus* strains. *J. Clin. Microbiol.* 38(9): 3453-6.0095-1137 (Print)

Shopsin, B. & B. N. Kreiswirth (2001).Molecular epidemiology of methicillin-resistant *Staphylococcus aureus*. *Emerg. Infect. Dis.* 7(2): 323-6.1080-6040 (Print)

Smith, A. J., J. Jefferies, S. C. Clarke, C. Dowson, G. F. Edwards & T. J. Mitchell (2006).Distribution of epidemic antibiotic-resistant pneumococcal clones in Scottish pneumococcal isolates analysed by multilocus sequence typing. *Microbiology* 152(Pt 2): 361-5.1350-0872 (Print)

Sola, C., G. Gribaudo, C. M. S. Groups, A. Vindel, L. Patrito & J. L. Bocco (2002).Identification of a novel methicillin-resistant *Staphylococcus aureus* epidemic clone in Córdoba, Argentina, involved in nosocomial infections. *Journal of Antimicrobial Microbiology* 40(4): 1427-1435

Soriano, F., F. Cafini, L. Aguilar, D. Tarrago, L. Alou, M. J. Gimenez, M. Gracia, M. C. Ponte, D. Leu, M. Pana, I. Letowska & A. Fenoll (2008).Breakthrough in penicillin resistance? *Streptococcus pneumoniae* isolates with penicillin/cefotaxime MICs of 16 mg/L and their genotypic and geographical relatedness. *J. Antimicrob. Chemother.* 62(6): 1234-40.1460-2091 (Electronic)

Struelens, M. J. (1996).Consensus guidelines for appropriate use and evaluation of microbial epidemiologic typing systems. *Clin. Microbiol. Infect.* 2(1): 2-11.1469-0691 (Electronic)

Struelens, M. J., R. Bax, A. Deplano, W. G. Quint & A. Van Belkum (1993).Concordant clonal delineation of methicillin-resistant *Staphylococcus aureus* by macrorestriction analysis and polymerase chain reaction genome fingerprinting. *J. Clin. Microbiol.* 31(8): 1964-70.0095-1137 (Print)

Tang, Y. W., M. G. Waddington, D. H. Smith, J. M. Manahan, P. C. Kohner, L. M. Highsmith, H. Li, F. R. Cockerill, 3rd, R. L. Thompson, S. O. Montgomery & D. H. Persing (2000).Comparison of protein A gene sequencing with pulsed-field gel electrophoresis and epidemiologic data for molecular typing of methicillin-resistant *Staphylococcus aureus*. *J. Clin. Microbiol.* 38(4): 1347-51.0095-1137 (Print)

te Witt, R., A. van Belkum, W. G. MacKay, P. S. Wallace & W. B. van Leeuwen (2010).External quality assessment of the molecular diagnostics and genotyping of meticillin-resistant Staphylococcus aureus. *Eur. J. Clin. Microbiol. Infect. Dis.* 29(3): 295-300.1435-4373 (Electronic)

Tenover, F. C., R. Arbeit, G. Archer, J. Biddle, S. Byrne, R. Goering, G. Hancock, G. A. Hebert, B. Hill, R. Hollis, W. R. Javis, B. Kreiswirth, W. Eisner, J. Maslow, L. K. McDougal, J. M. Miller, M. Mulligan & M. A. Pfaller.(1994).Comparison of traditional and molecular methods of typing isolates of *Staphylococcus aureus*. *J. Clin. Microbiol.* 32(2): 407-15.0095-1137 (Print)

Tenover, F. C., R. D. Arbeit, R. V. Goering, P. A. Mickelsen, B. E. Murray, D. H. Persing & B. Swaminathan (1995).Interpreting chromosomal DNA restriction patterns produced by pulsed-field gel electrophoresis: criteria for bacterial strain typing. *J. Clin. Microbiol.* 33(9): 2233-9.0095-1137 (Print).

Tenover, F.C., Arbeit, R.D., Goering, R. V. (1997).How to select and interpret molecular strain typing methods for epidemiological studies of bacterial infections: a review for healthcare epidemiologists. Molecular Typing Working Group of the Society for Healthcare Epidemiology of America. Infect. Control Hosp. Epidemiol.18(6):426-39.0899-823X (print).

Tibayrenc, M. (1995).Population genetics and strain typing of microorganisms: how to detect departures from panmixia without individualizing alleles and loci. *C. R. Acad. Sci. III.* 318(1): 135-9.0764-4469 (Print)

Tu, A. H., R. L. Fulgham, M. A. McCrory, D. E. Briles & A. J. Szalai (1999).Pneumococcal surface protein A inhibits complement activation by *Streptococcus pneumoniae. Infect. Immun.* 67(9): 4720-4.0019-9567 (Print)

van Belkum, A., J. Kluytmans, W. van Leeuwen, R. Bax, W. Quint, E. Peters, A. Fluit, C. Vandenbroucke-Grauls, A. van den Brule, H. Koeleman, W. Melchers, J. Meis, A.Elaichouni, M. Vaneechoutte, F. Moonens, N. Maes, M. Struelens, F. Tenover & H. Verbrugh (1995).Multicenter evaluation of arbitrarily primed PCR for typing of *Staphylococcus aureus* strains. *J. Clin. Microbiol.* 33(6): 1537-47.0095-1137 (Print)

van Belkum, A., M. Sluijuter, R. de Groot, H. Verbrugh & P. W. Hermans (1996).Novel BOX repeat PCR assay for high-resolution typing of *Streptococcus pneumoniae* strains. *J. Clin. Microbiol.* 34(5): 1176-9.0095-1137 (Print)

van Belkum, A., P. T. Tassios, L. Dijkshoorn, S. Haeggman, B. Cookson, N. K. Fry, V. Fussing, J. Green, E. Feil, P. Gerner-Smidt, S. Brisse & M. Struelens (2007).Guidelines for the validation and application of typing methods for use in bacterial epidemiology. *Clin. Microbiol. Infect.* 13 Suppl 3: 1-46.1198-743X (Print)

van Belkum, A., W. van Leeuwen, M. E. Kaufmann, B. Cookson, F. Forey, J. Etienne, R. Goering, F. Tenover, C. Steward, F. O'Brien, W. Grubb, P. Tassios, N. Legakis, A. Morvan, N. El Solh, R. de Ryck, M. Struelens, S. Salmenlinna, J. Vuopio-Varkila, M. Kooistra, A. Talens, W. Witte & H. Verbrugh (1998).Assessment of resolution and intercenter reproducibility of results of genotyping *Staphylococcus aureus* by pulsed-field gel electrophoresis of SmaI macrorestriction fragments: a multicenter study. *J Clin Microbiol* 36(6): 1653-9.0095-1137 (Print)

van der Zee, A., H. Verbakel, J. C. van Zon, I. Frenay, A. van Belkum, M. Peeters, A. Buiting & A. Bergmans (1999).Molecular genotyping of *Staphylococcus aureus* strains: comparison of repetitive element sequence-based PCR with various typing methods and isolation of a novel epidemicity marker. *J. Clin. Microbiol.* 37(2): 342-9.0095-1137 (Print)

Vázquez,J. A., Berrón, S.(2004).Multilocus sequence typing: el marcador molecular de la era de Internet.Enfermedades Infecciosas y Microbiología Clínica (2004). Enfermedades infecciosas y microbiología clínica.22 (2): 113-120.0213005X

Versalovic, J., T. Koeuth & J. R. Lupski (1991).Distribution of repetitive DNA sequences in eubacteria and application to fingerprinting of bacterial genomes. *Nucleic Acids. Res.* 19(24): 6823-31.0305-1048 (Print)

von Specht, M., N. Gardella, P. Tagliaferri, G. Gutkind & M. Mollerach (2006).Methicillin-resistant *Staphylococcus aureus* in community-acquired meningitis. *Eur. J. Clin. Microbiol. Infect. Dis.* 25(4): 267-9.0934-9723 (Print)

Wei, M. Q., E. E. Udo & W. B. Grubb (1992).Typing of methicillin-resistant *Staphylococcus aureus* with IS256. *FEMS Microbiol. Lett.* 78(2-3): 175-80.0378-1097 (Print).

Willems, R. J., Hanage, W.P., Bessen, D. E., Feil, E. J. (2011).Population biology of Gram-positive pathogens: high-risk clones for dissemination of antibiotic resistance.FEMS Microbiol. Rev.35(5):872-900.0168-6445 (print).

Zemlickova, H., M. I. Crisostomo, M. C. Brandileone, T. Camou, E. Castaneda, A. Corso, G. Echaniz-Aviles, M. Pasztor & A. Tomasz (2005).Serotypes and clonal types of penicillin-susceptible *Streptococcus pneumoniae* causing invasive disease in children in five Latin American countries. *Microb. Drug Resist.* 11(3): 195-204.1076-6294 (Print)

Zhang, Q. Y., D. M. Jones, J. A. Saez Nieto, E. Perez Trallero & B. G. Spratt (1990).Genetic diversity of penicillin-binding protein 2 genes of penicillin-resistant strains of *Neisseria meningitidis* revealed by fingerprinting of amplified DNA. *Antimicrob Agents Chemother* 34(8): 1523-8.0066-4804 (Print).

8

Molecular Microbiology Applied to the Study of Phytopathogenic Fungi

Carlos Garrido, Francisco J. Fernández-Acero, María Carbú,
Victoria E. González-Rodríguez, Eva Liñeiro, and Jesús M. Cantoral
Microbiology Laboratory, Faculty of Marine and Environmental Sciences
University of Cádiz, Puerto Real
Spain

1. Introduction

Fungi is an extensive group of eukaryotic microorganisms, generally they are microscopic and usually filamentous. It is estimated that there are between 70,000 and 1.5 millions species of fungi, most of them are being discovering and describing (Agrios, 2005). Most of the known hundred thousand fungal species are strictly saprophytic, living on decomposing dead organic matter. About fifty species cause disease in human, and more than ten thousand species can cause disease in one (obligate parasite) or many kinds of plants (non-obligate parasites) (Fernández-Acero et al., 2007a).

Phytopathogenic fungi are able to infect any tissue at any stage of plant growth. Plant pathogenic fungi show a complex life cycles, including both sexual and asexual reproduction stages (Agrios, 2005). Moreover, complex infection cycles and carbon assimilation is displayed (Garrido et al., 2010). These biological variability give them the possibility to develop its biological role from very climatologically different environments, since dry and desert zones until wet and hot regions in the tropic and equatorial area to the capacity to attack all plant tissues, from leaves to roots (Agrios, 2005).

During the last decades, the development of molecular methods has lead the Scientifics community to accumulate a high quantity of information from different molecular approaches (Fernández-Acero et al., 2011; Garrido et al., 2009b). Advances into Genomics, Transcriptomics, Proteomics, and more recently, Metabolomics are transforming research into fungal plant pathology, providing better and more accurate knowledge about the molecular biology and infection mechanisms showed by these fungi (Garrido et al., 2010).

Since 1992, our research group has been working with two of the most aggressive plant pathogens, which have been established such a model organisms for molecular and phytaphology studies: *Botrytis cinerea* and *Colletotrichum acutatum* (Fernández-Acero et al., 2006b, 2007a; Garrido et al., 2009b, 2010; Perfect et al., 1999). These genera include some of the most destructive plant pathogen species known. They induce worldwide diseases as, between others, the grey mould on grapes and the anthracnose on strawberries, respectively (Coley-Smith et al., 1980; Elad et al., 2004; Sutton, 1992). The losses caused by the phytopathogenic fungi *Botrytis cinerea* and *Colletotrichum acutatum* have been quantified between 10 and 100 million of Euros per year in Europe (Fernández-Acero et al., 2007a). Losses caused by *B. cinerea* in French vineyards oscillate between 15% and 40%; in Holland, *B. cinerea* generates losses of about 20% of the flower crop; and, in Spain, the losses fluctuate between 20% and 25% of the strawberry crops (Fernández-Acero et al., 2007a). *Colletotrichum* spp. causes up to 80% plant death in nurseries and yield losses of >50%, being a major disease of cultivated strawberry (Denoyes-Rothan et al., 2004; Garrido et al., 2009a).

Our group has carried out an intense research activity of the molecular microbiology of these plant pathogens. These studies involve several molecular approaches in which the gel electrophoresis plays an important role. In this chapter, we will summarize the results obtained, and the molecular methods used for the study and characterization of the phytopathogen fungi *Botrytis cinerea* and *Colletotrichum* spp., all of them strongly related with different types of gel electrophoresis approaches and downstream protocols, including, between others, Pulse Field Gel Electrophoresis, agarose gel electrophoresis of DNA, Restriction Fragment Polymorphism Analyses, Southern-blot, Polyacrylamide Gel Electrophoresis and Two dimensional gel electrophoresis of proteins. These electrophoretic methods will be used to structure the development of chapter, describing the technical bases of each method and showing the approaches carried out and the results obtained.

2. Chromosomal polymorphism and genome organization in *Botrytis cinerea* and *Colletotrichum* spp.

Botrytis cinerea and *Colletotrichum acutatum* are two species of phytopathogenic fungi that show a very high level of phenotypic diversity among isolates. These fungi show complex cycles of life and infection, including both sexual and asexual forms (Garrido et al., 2008; Vallejo et al., 2002). Also high levels of somatic variability appear when the fungi are grown "*in vitro*", depending on the medium, temperature, light and other factors, which even determine differences in cultural characteristics, production of reproductive structures and pathogenicty between others (Bailey & Jeger, 1992; Carbu, 2006; Garrido et al., 2009b; Rebordinos et al., 1997, 2000; Vallejo et al., 1996, 2001). These fungi do not show a high level of host specificity and they infect many different genera of hosts, adapting their infection strategy and metabolism to the environment conditions and kind of plant colonized. They are notoriously variable genera about which many fundamental questions relating to taxonomy, evolution, origin of variation, host specificity and mechanisms of pathogenesis remain to be answered (Bailey & Jeger, 1992; Elad et al., 2004).

Many research projects are aimed to study the genome organization and chromosomal polymorphism trying to find the origin of phenotypic variability showed by these fungi. In the past decades, several strategies have been tested on lower fungi such as *B. cinerea* and *Colletotrichum* spp., i.e. cytological karyotyping, analysis of progeny from crosses between strains, sexual hybridizations, etc. These assays looked for a relation between molecular and phenotypic variability (Carbu, 2006; Faretra & Antonacci, 1987; Faretra et al., 1988; Vallejo et al., 1996). Cytological studies showed a very high level of difficulty in this group of microorganisms due to small size and/or the difficulty to condense sufficiently the chromosomes to make them visible by microscope. These characteristics made difficult to obtain reliable information about the genome organization of these fungi, and the obtaining of conclusive results about their biological mechanisms of recombination and chromosomal polymorphisms.

The development of Pulse- field gel electrophoresis (PFGE) resolved many problems found with cytogenetic studies in filamentous fungi. This technique has been widely used since the 90s for genomic characterization into fungal plant pathogens. PFGE allows the separation of large DNA molecules (DNAs from 100 bases to over 10 megabases (Mb) may be effectively resolved) which would all co-migrate in conventional agarose gels. This technique has proved to be a very useful tool to study aspects of genome organization in several yeast and fungi. It has led to the discovery that most species exhibit chromosome-length polymorphisms (CLPs), revealing a high level of intraspecific, and even population-level variability (Vallejo et al., 2002).

Technically, PFGE resolves chromosome-sized DNAs by alternating the electric field between spatially distinct pairs of electrodes. The electrophoresis cell consists of an array with 24 horizontal electrodes arranged in a hexagon. Agarose gels are electrophoresed horizontally, submerged under recirculated buffer. The system (CHEF-"Clamped Homogeneous Electric Field" and PACE "Programmable Autonomously Controlled Electrodes", from BIO-RAD) provides highly uniform, or homogenous, electric fields within the gel, using an array of 24 electrodes, which are held to intermediate potentials to eliminate lane distortion. Thus, lanes are straight. The system maintains uniform field using patented Dynamic Regulation. The electrodes sense changes in local buffer conductivity due to buffer breakdown, change in buffer type, gel thickness, or temperature, and potentials.

The preparation of samples for resolving chromosomal karyotypes by PFGE is not exempt of difficulty due to the biological characteristic of fungal cells. Fungus has to be growth in an optimal culture medium and mycelium harvest after determinate time which depends of the fungal species. This time is very important because is necessary to obtain the highest number of fungal cells in metaphase stage (Carbu, 2006; Garrido et al., 2009b). Chromosomes are condensed and highly coiled in metaphase, which makes them most suitable for visual analysis. After young mycelium is harvested, it is necessary to produce protoplasts using different mixes of lysing enzymes, which digest the fungal cell wall after incubation. Protoplast suspensions are mixed with low melting point agarose, adjusted to final concentration of 1×10^8 protoplast ml^{-1}, and solidified plugs of agarose containing protoplast are digested with proteinase K. The digestion produces pores in the plasma membrane, providing the possibility to extract the chromosomal by PFGE (Garrido et al., 2009b).

Gels are prepared with a special type of agarose. It depends of the DNA molecules sizes because there are different commercial preparations, some of them for DNA molecules higher than 10 Mb, i.e. PFGE ™Megabase agarose (Bio-Rad). Plugs are cast in the gel, and this is placed in the center of the hexagon formed by the 24 electrodes. Many parameters of the electrophoresis have to be optimized, since the type and concentration of running buffer, temperature of buffer, voltage and time of pulses, angles of electric fields. Depending of instrument setup, we can resolve the electrophoretic karyotype (EK) only with one experiment, like in the case of *Botrytis cinerea*; or even it could be necessary two different steps/running conditions, due to the high differences in sizes of the chromosomal DNA molecules. After electrophoresis, gels are stained using i.e. ethidium bromide and visualized using a UV light system.

PFGE has been widely used by our group to study the genome organization and Chromosomal Polymorphisms (CPL) in *B. cinerea* and *C. acutatum*. We have determined the number and sizes of chromosomes in both species, and therefore we have estimated the genome size for these fungi; the high level of CPL displayed by them, represented in the different EK profiles showed by the strains; and PFGE has made possible downstream applications such as Southern-blot analysis using different probes. All the results accumulated during the last years have provided a better understanding about the genome organization and the molecular bases of asexual and sexual reproduction of these fungi. They proved that polymorphism has been observed in both asexual and sexual fungi and most likely results from both mitotic and meiotic processes, especially in the case of *Botrytis cinerea* (Vallejo et al., 2002).

When a study of PFGE has made, it is usual to find chromosomal bands of different intensity and therefore it is important to consider several technical aspects that can have influence in the interpretation of the final results, and the conclusions obtained: i) a double band could be composed of two coumpounds of a couple of homologous chromosomes or of two heterologous chromosomes of similar size, and ii) two homologous chromosomes can differ in size and appear like two heterologous ones. Due to this fact, depending on the aims of the study, sometimes further hybridization studies are necessary in order to determine the linkage groups of each of the bands (Carbu, 2006; Vallejo et al., 1996).

Botrytis cinerea strains studied by our group were isolated from different hosts and geographical origins. We found different EK profiles between isolates, which did not follow any correlation with the host, year of isolation, or phenotypical characteristics. We have found that the number of chromosomal bands varied between 5 and 12, and they ranged between 1.80 and 3.8 Mb. These results made possible to estimate the minimal genome size of *B. cinerea* genome, found between 14.5 and 22.7 Mbp (Carbu, 2006; Vallejo et al., 1996, 2002) (Fig. 1a).

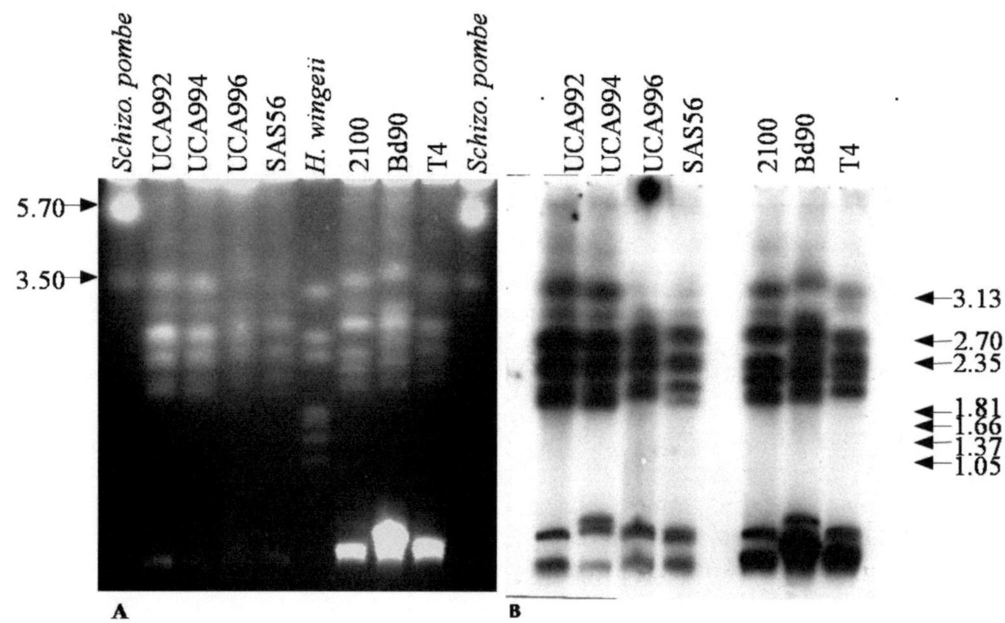

Fig. 1. A.- PFGE chromosomal separation of selected *B. cinerea* isolates. The molecular sizes were estimated using *Schizosaccharomyces pombe* (line 1 and 10), and *Hansenula wingeii* (line 6) chromosomes as reference molecular markers (Bio-Rad). B.- Southern-blot hybridization using a telomeric DNA probe to hybridise the PFGE separated chromosomal bands.

The *B. cinerea* strains showed a high level of CLPs, revealing the facility to support chromosomal rearrangements in this species, and could be the basis of the high degree of adaptability to the environmental conditions. Our group has also studied crosses between strains with different EK profiles. This study had as main aim to analyze the chromosomal rearrangements and chromosomal segregation in the crossed strains, in order to clarify the controversy appeared about the possibility that a high level of CLPs between strains, could inhibit meiosis (Zeigler, 1998), and therefore to be one possible reason to explain the low level of sexual reproduction that take place in *B. cinerea* under natural conditions (Carbu, 2006; Giraud et al., 1997).

The crosses between strains produced fertile strains (more than 100 ascopores studied) and our results demonstrated that chromosomal rearrangements did not affect the capacity to reproduce sexually in *B. cinerea*. It was observed than only several isolates recovered the parental EKs. New chromosomes sizes were identified and some bands were lost from the parental to descendants EKs. All these results, along with a segregation analyses carried out in the decendants, represented strong evidence that some strains might not be haploid, and that aneuploidy and differences in ploidy levels are present in this species (Vallejo et al., 2002). Our group has also studied how during a long period of time, reproducing the fungus "*in vitro*", there were not detected changes in the EK of a given strain. All results together, proved that mitotic growth does not provide EK variability in this fungus, being the chromosomal rearrangements generated after meiotic recombination the causal agent of EK variability in *B. cinerea* (Carbu, 2006).

In the case of the species *C. acutatum*, there were not data published about the EKs and CLPs among isolates until the last 2009 (Garrido et al., 2009b). PFGE had been used with other species of this genus, like *C. gloeosporioides* (Masel et al., 1993) and *C. lindemouthianum* (O´Sullivan et al., 1998). *Colletotrichum* spp. displayed an estimated genome sizes higher than *B. cinerea*. Protocols to separate the chromosomes molecules were carried out in two different experimental setups, including variations in the pulse of electric field, percentage of agarose gels and duration of the assays (Masel et al., 1993), i.e. for separation of larger chromosomal molecules in *C. gloeosporioides*, Masel et al. (1993) optimized an PFGE approach running a electrophoresis of seven days long. During this experiment, it was necessary to replace the running buffer each two days to obtain a better resolution in the final image. Similar protocols were used to resolve EK from *C. lindemouthianum* strains(O´Sullivan et al., 1998).

The karyotype of *C. acutatum* was studied by our group in several strains isolated from different geographical origins. They had showed differences in the morphological characteristics in relation to the color and texture of mycelium, ratio of growth in different medium, pathogenicity and level of conidia production (Garrido et al., 2008, 2009b). Protocol to obtain *C. acutatum* protoplasts and the PFGE conditions to separate chromosomes were optimized based on our previous experience with *B. cinerea*. We optimized a PFGE running conditions to separate chromosomes between approximatele 0.1 and 9 Mb after only 72 h. of running. This protocol improved substantially those previously described for *Colletotrichum* spp., which took longer due to the two steps needed to resolve the complete karytopyes. Those longer protocols (Masel et al., 1993) were also tested, and we got the same number and sizes of chromosomal bands, proving the improvement of our optimized 72-hours protocol (Garrido et al., 2009b).

C. acutatum strains showed EK profiles containing between six and nine chromosomal bands with different sizes ranging from 0.1 and 8 Mb. The total minimal genome size estimated for *C. acutatum* ranged between 29 and 36 Mb, which is similar to that previously described for other species of *Colletotrichum* (Masel et al., 1993; O´Sullivan et al., 1998). We observed CLPs between strains studies but further analyses with a high number of isolates could be necessary in order to obtain strong conclusions about the CLPs showed by the species and how this variability could affect the sexual and asexual reproduction of this species in the environment (Garrido et al., 2009b).

PFGE gels from *B. cinerea* and *C. acutatum* were used in downstream applications, like Southern-blot analyses. Gels were transferred to Hybond-N membranes and they hybridised with a telomeric probe confirming that all the bands represented chromosomes. The description of Southern-blot analyses will be described in the next section, but it proved how PFGE, not only provides the possibility to obtain interesting conclusions about the biology and genome organization of these fungi, but also gel electrophoresis techniques are often the starting point for interesting downstream applications that provide more information in the researches of these fungi (Fig. 1b).

In our PFGE studies in *B. cinerea* and *C. acutatum*, it has not been observed a higher EKs variability that showed by phenotypic characteristics among strains (Carbu, 2006; Garrido et al., 2008, 2009a, 2009b; Rebordinos et al., 2000; Vallejo et al., 1996). Phenotypic features were very highly variable between strains with the same EKs. Therefore, we cannot conclude that there is a direct relation between morphological, physiological and pathogenic variability directly related with heterokaryosis, aneuploidy and a variable level of ploidy among strains. New proteomics approaches to *B. cinerea* and *Colletotrichum* spp., which will be described during next pages, is contributing with very interesting data, that in conjunction with genomic information, disclose that phenotypic variation is more related with the synthesis of proteins and their post-transductional modifications, and not only by genotypes encoding them (Fernández-Acero et al., 2011).

3. Phylogenetic relationships between strains of *Colletotrichum* spp. using telomeric fingerprinting

Colletotrichum acutatum is a widely spread species that can be found throughout the world (Whitelaw-Weckert et al., 2007). *C. acutatum* causes anthracnose on a number of economically important crops, including woody and herbaceous crops, ornamentals, fruits, conifers and forage plants (Sreenivasaprasad & Talhinhas, 2005). It was classified as an organism of quarantine significance in Canada from 1991 to 1997, in the UK and the EU since 1993, and it can be found widely spread in the southwest region of USA (EPPO/CABI, 1997; Garrido et al., 2009a; Mertely and Legard, 2004). Investigations of *C. acutatum* were focused in two main aspects of the pathogen: i) cultural and morphological studies (Afanador-Kafuri et al., 2003; Denoyes-Rothan & Baudry, 1995; Garrido et al., 2008;) and ii) molecular approaches using molecular techniques including isoenzyme comparisons, Restriction Fragment Length Polymorphism (RFLP) analyses of mitochondrial DNA, Amplified Fragment Length Polymorphism (AFLP), AT rich analyses, Random Amplified Polymorphic DNA (RAPD), and ITS sequences analyses for specific PCR sequencing and identification (Buddie et al., 1999; Freeman et al., 1993; Garrido et al., 2009a, 2009b; Sreenivasaprasad et al., 1996; Talhinhas et al., 2005).

Sreenivasaprasad & Talhinhas (2005) studied *C. acutatum* populations from several hosts and different geographical origins. They established molecular groups based on sequences analyses of the internal transcribed spacers (ITS) of ribosomal DNA polymorphic regions (Sreenivasaprasad & Talhinhas, 2005). ITS regions have been widely used on molecular approaches for studying relationship between microorganisms, and it is also very useful regions for designing molecular approaches to identification and diagnostic protocols, due to the high variability showed by the sequences among species and even strains (Garrido et al., 2009a). The classification carried out by Sreenivasaprasad & Talhinhas (2005), established eight molecular groups for *C. acutatum* species. These molecular groups have been widely used to study the genotypic and phenotypic diversity of this fungus, and to classify isolates from different origin (Whitelaw-Weckert et al., 2007).

During the last years, we carried out a study to classify a worldwide collection of *C. acutatum* strains isolated from thirteen countries (Australia, Canada, France, Germany, Japan, The Netherlands, New Zealand, Norway, Portugal, Spain, Switzerland, USA and UK). For this purpose we used two different molecular approaches in order to study the phylogenetic relationship between strains: i) a sequencing analysis of the internal transcribed spacers (ITS) of the 5.8S ribosomal DNA polymorphic regions; ii) a telomeric fingerprinting study by Southern-blot hybridization, using a telomeric probe after RFLP digestions of genomic DNA (Garrido et al., 2009b).

In total, eighty-one 5.8S-ITS sequences were studied, several strains were sequenced by our group, and other ones used from databases such as reference sequences for allocating our strains in the previously established molecular groups for *C. acutatum*. ITS regions, including 5.8S rDNA, were amplified by conventional PCR using universal primers ITS1 and ITS4 (White et al., 1990). After PCR amplification, products were loaded in a conventional 1% agarose gel for conventional DNA electrophoresis. Products were cut from the gels using a purification kit, DNA was quantified, and subsequently sequenced in both directions (Garrido et al., 2009b).

The phylogenetic study carried out with the sequences allowed us to allocate the strains into *C. acutatum* molecular groups described by Sreenivasaprasad & Talhinhas (2005), but the analysis of bootstrap in the neighbout-joining phylogenetic tree, published by Garrido et al. (2009), showed interesting data about the molecular groups. In base of that analyses, the nine molecular groups previously described (Whitelaw-Weckert et al., 2007), could be grouped in only four groups. Our results proved that A1, A2, A5 A8 and A9 subgroups showed a bootstrap support of 90%, and therefore could be considered such as large group in base to the analyses of the sequences of ITS regions (Garrido et al., 2009b). The same result was observed for subgroups A6 and A4, since these subgroups clustered together with a strong bootstrap support of 91% (Garrido et al. 2009). Our results supported a new classification into four molecular groups instead the nine previously described for this species in base to the ITS sequences (Garrido et al., 2009b).

The phylogenetic analyses showed that the majority of the strains studied grouped in the group A2. This happened because many strains from Spain were included in the analyses. The results proved the high level of similarity between *C. acutatum* strains isolated from Spain. It is also interesting that the A2 group included, principally, isolates from Spain, Portugal, France, UK and USA. *C. acutatum* was first described in the southwest region of the USA, and then it was observed in France and UK at the beginning of the 80s. It is not clear how the pathogen was introduced into production fields in Europe, but it is thought that the pathogen could have arrived since the American nurseries to the EU (Freeman & Katan, 1997). It should have arrived to France, UK and the Iberian Peninsula fields. The arrival of the pathogen was facilitated by the intense international trade between these countries related with strawberry crop. Therefore the fungus could be introduced by infected plants, contaminated soil associated with strawberry crowns at planting, and quiescent infections on strawberry leaves or fruits (Garrido et al., 2008, 2009b; Leandro et al., 2001, 2003).

In order to complete the phylogenetic classification of our *C. acutatum* strain collection, a different molecular approach was carried out. The results obtained were compared with those from the ITS sequences analyses. We used the profiles obtained after restriction enzymes digestions of genomic DNA, and then hybridized with a telomeric probe by Southern-blot hybrisization. Genomic DNA of *C. acutatum* strains were digested to completion with several restriction enzymes in independent experiments (*Bam*HI, *Eco*RI, *Hin*dIII and *Pst*I). Gel electrophoresis is an intermediate point of the complete protocol. It make possible to separate the DNA fragments obtained after the restriction enzymes digestions. In this case, we used a 1.5% agarose gel, and electrophoresis was carried out in a conventional horizontal tray for DNA electrophoresis (Bio-Rad). After separation of digested fragments, gels were blotting to Hybond-N membrans, being ready for subsequently hybridization (Garrido et al., 2009b).

For Southern-blot hybridization we used a telomeric probe to get hybridization in the telomeric regions. These regions are located at the end of the lineal chromosomes of most eukaryotic organisms, and they are named telomeres. Telomeres are regions of repetitive DNA sequences that protect the end of the chromosome from deterioration or from fusion with neighboring chromosomes. The repeated sequences is dependent of the species. For *C. acutatum* telomeres, we produced our telomeric probe, (TTAGGG)n, by PCR in the absence of a template using (TTAGGG)5 and (CCCTAA)5 primers as it was described by Ijdo et al. (Ijdo et al., 1991). The Hybond-N membranes were allowed to hybridize with the telomeric probe; films images were digitalized and telomeric profiles were analysed using Fingerprinting II software v3.0 (Bio-Rad).

The experimental setup described provided the possibility to obtain two different kinds of results/conclusions from the study: I) Selected restriction enzymes used for RFLP did not produce any cut in the telomeric regions of *C. acutatum* strains. Each band represents a physically distinct telomere extremity. Therefore, taking into consideration the higher number of telomeric extremities and then divided into two, we can determine the number of chromosomes among strains studied. II) The fingerprinting analyses of the telomeric profiles, carried out using Fingerprinting II software, make possible to produce phylogenetic trees based in the similarity of the profiles showed among the strains. Therefore, these results could be compared with those obtained from phylogenetic groups based on ITS sequences.

Among the fifty-two isolates analysed by telomeric fingerprinting, the number of band or telomeres oscillated between twelve and eighteen. Therefore, the minimum number of estimated chromosomes was from six to nine among *C. acutatum* isolates (Garrido et al.,

2009b). In this study the number of strains studied was higher than those studied by FPGE, and although fingerprinting analyses did not make possible to study the chromosomal length polymorphisms among the isolates, the minimum numbers of estimated chromosomes are coincident with those obtained from FPGE analyses, showed in the last section of this chapter.

The telomeric profiles obtained for each isolate of *C. acutatum* were analysed. UPGMA dendogram showed a representative grouping among the isolates, which was coincident with the grouping in the neighbuor-joining phylogenetic tree based on sequences of rDNA ITS regions (Garrido et al., 2009b). All the strains previously classified in the A2 molecular groups, also clustered in a large group with more than 70% of similarity based in this case in the telomeric fingerprinting profiles. These results proved the high level of similarity shows by these isolated, not only based in sequence similarity of one specific region but also in their genotypes and genome organization among *C. acutatum* strains, which suggests a common origin of the strains among the different molecular groups (Garrido et al., 2009b; Talhinhas et al., 2005).

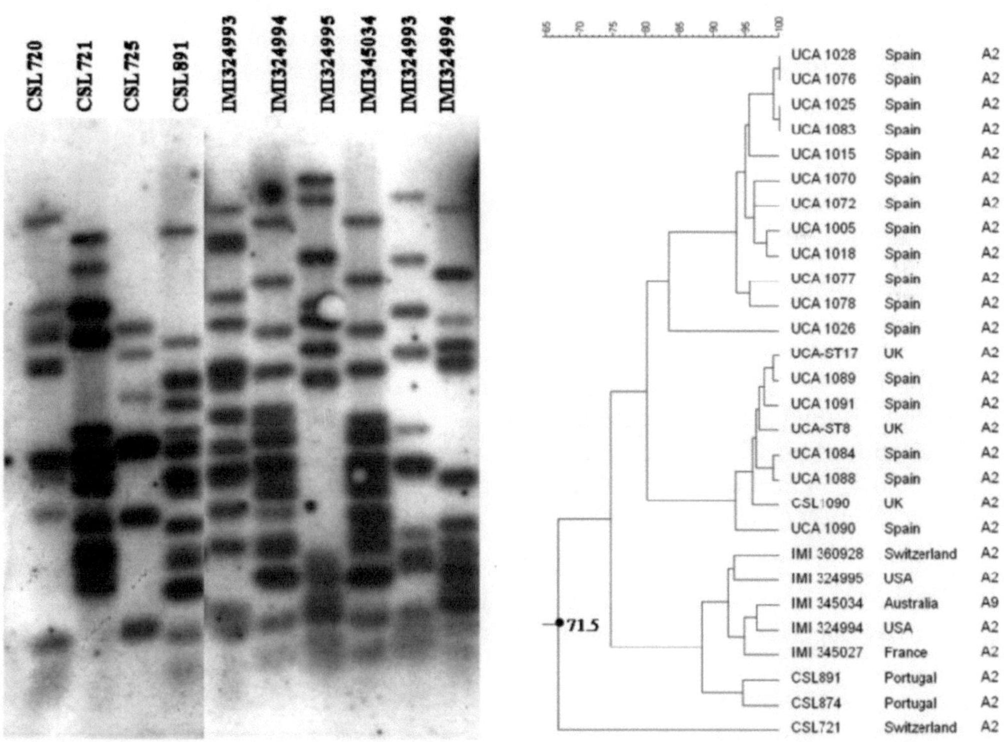

Fig. 2. Left.- Telomeric fingerprinting patterns obtained by telomeric hybridisation of Southern blots from HindIII-DNA digestions. Right.- Combined UPGMA dendograms with the *C. acutatum* isolates belonging to A2 group, based on Dice coefficients generated using a composite data set from individual experiments of each enzyme digestion (*Bam*HI, *Eco*RI, *Hin*dIII and *Pst*I) hybridised with a telomeric probe.

4. Develop of molecular methods for detection and identification of phytophatogenic fungi - Monitoring of the diseases causing by *Colletotrichum* spp.

Many fungal plant pathogens produce similar symptoms when they develop diseases among different hosts. Currently, the ability to detect, identify and quantify plant pathogens accurately is the cornerstone of plant pathology (Garrido et al., 2011). The reliable identification of the organism(s) responsible for a crop disease is an essential prerequisite to apply the correct disease management strategies and the most appropriate control measures to take. Besides, many pathogens are subjected to special regulation through quarantine programs agreed among producer countries. For all these reasons, pathogen identification is crucial to all aspect of fungal diagnostics and epidemiology in the field of plant pathology, but also in medical science, environmental studies and biological control (Alastair McCartney et al., 2003; Atkins et al., 2003).

Since the 1990's, new methods based on molecular biology have provided new tools for more accurate and reliable detection, identification and quantification of plant pathogens. These methods are based on immunological and DNA/RNA study strategies, including, amongst others: RFLP analyses of mitochondrial DNA (Garrido et al., 2008; Sreenivasaprasad et al., 1992), AFLP, AT-rich analyses (Freeman et al., 2000a, 2000b), RAPD-DNA (Whitelaw-Weckert et al., 2007), genus-specific and species-specific PCR primers (Garrido et al., 2008; Martínez-Culebras et al., 2003; Mills et al., 1992; Sreenivasaprasad et al., 1996), real-time PCR studies (Garrido et al., 2009a), and ELISA assays (Hughes et al., 1997). Diagnosis time can be reduced from a period of weeks, typically experienced with culture plating, to only a few days, thus allowing the appropriate control methods to implemented much sooner and more effectively (Atkins et al., 2003).

Advances in polymerase chain reaction technology have opened alternative approaches to the detection and identification of fungal pathogens. The development of PCR technology relies on three fundamental steps: i) the selection of a specific target region of DNA/RNA to identify the fungus; ii) extraction of total community DNA/RNA from the environmental sample; iii) a method for identifying the presence of the target DNA/RNA region in the sample (Garrido et al., 2011). Our group have optimised a very high sensitive protocol for diagnosis and identification of the fungal genus *Colletotrichum*, and the species *C. acutatum* and *C. gloeosporioides* (Garrido et al., 2009a).

The sensitivity of PCR-based protocols depends mainly on the instrumentation and technique used (i.e. conventional PCR *vs.* real-time PCR), but in a high proportion of cases this sensitivity depends on the quality of the total community DNA/RNA extracted from the environmental samples. Garrido et al. (2009) optimized a DNA extraction protocol that can be used for samples of strawberry plant material directly, or from fungal colonies removed from an agar plate. This method uses sample material physically ground using a grinding machine, in the presence of CTAB lysis buffer. The lysated samples are washed in various chemical products (chloroform, isopropanol, ethanol, etc.) and then the final step involves using Magnesil® beads and GITC lysis buffer (guanidinium thiocyanate buffer) in a Kingfisher robotic processor (Kingfisher ML, Thermo Scientific). The new method was tested with roots, crowns, petioles, leaves and fruits and the extraction methods always showed very high yields of DNA in both quantity and quality. Although, a wide range of commercial kits are available for extraction of fungal DNA, they can represent a high cost per sample analysed, and they are not always totally reliable in not co-extracting PCR inhibitors, needed a dilution of samples prior to PCR reactions. The optimised protocol did not co-extracted PCR-inhibitors from any samples, and therefore, the sensitivity of the detection protocol is improved using this DNA extraction protocols (Garrido et al., 2011).

To date, conventional PCR has been a fundamental part of fungal molecular diagnosis, but it shows several limitations: i.e. gel-based methods, possibility of quantification, sensitivity, etc. The development of real-time PCR has been a valuable response to these limitations (Garrido et al., 2011). This technology improve the sensitivity, accuracy and it is less time-consuming that conventional end-point PCR. For development and optimization of *Colletotrichum* diagnosis protocols, the commonly-used ribosomal RNA genes were used, because of the highly variable sequences of the internal transcribed spacers ITS1 and ITS2, which separate the 18S/5,8S and 5,8S/28S ribosomal RNA genes, respectively (Garrido et al., 2009a). Specific genus and species sets of primers and probes were designed for real-time PCR amplifications using TaqMan® chemistry technology. This system consists of a fluorogenic probe specific to the DNA target, which anneals to the target between the PCR primers; TaqMan® tends to be the most sensitive and simply methods for real-time PCR detection (Garrido et al., 2009a, 2011).

The specificity of all assays was tested using DNA from isolates of six species of *Colletotrichum* and from DNA of another nine fungal species commonly found associated with strawberry material. All the new assays were highly specific for *Colletotrichum* spp., *C. acutatum* and *C. gloeosporioides*, no cross-reactions were observed with either related plant pathogens or healthy strawberry plant material. The sensitivity of the new real-time PCR assays was compared with that of previously published conventional PCR assays; they were confirmed to be 100 times more sensitive than the latter. The *C. acutatum*-specific real-time PCR assay was also compared with an existing ELISA assay for the diagnosis of this pathogen. Real-time PCR permitted the detection of the pathogen in samples that gave negative results for *C. acutatum* using ELISA. The real-time PCR assay detected the equivalent of 7.2 conidia per plant inoculated with a serial dilution of *C. acutatum* spores, demonstrating the high degree of sensitivity of the method (Garrido et al., 2009a).

The new protocols were tested for monitoring the development of anthracnose disease in strawberry in the field in the south of Spain. The real-time PCR results showed a progressive increase of target DNA between January and June. The results showed that an increase in lesion development was accompanied by an increase in the amount and incidence of the pathogen as the season progressed. These results showed that new methods are suitable for diagnosis, identification and monitoring of the disease using field samples of strawberry and also, they permitted the detection of the pathogens from artificially infected symptomless plant material. Therefore, the methods described, based on real-time PCR, proved useful for studying the epidemiological routes of these strawberry pathogens in fields and nurseries (Garrido et al., 2009a, 2011).

5. Proteomics approaches of phytopathogenic fungi

In spite of the advances done by the described techniques above, nowadays proteomics is the most realistic and effective set of tools to unravel complex mixtures of proteins, describing the current molecular biology age as "post-genomic era". The term proteome was coined in 1995 by Wilkins et al (Wilkins et al., 1995), later the term proteomics appeared by James et al. (James, 1997). Proteome is defined as the complete set of proteins expressed by an organism, in a particular biological state. Proteomics may be introduced as a set of techniques that allow to study and to describe the proteome. The impact of the proteomic approaches is mainly based in a group of widely used techniques such as liquid chromatography or two dimensional gel electrophoresis, to separate complex protein mixtures, defining the proteome. However, the increasing relevance of these studies has been pushed by the improvements done in mass spectrometry system, allowing the analysis of peptides and proteins and/or by the increase number of proteins entries in the databases, making easier protein analysis and identification.

Main proteome characteristic is that it is a high dynamic system. It is even more complex than genomics, due to while the genome of an organism is more or less constant, the number of obtained proteomes from a specific genome is almost infinite. It depends of the assayed cell, tissue, culture conditions, etc. Each change produces a modification in the observed proteome. An additional factor of complexity is that there are changes that occur in proteome that are not encoded in the genome. These changes are mainly based on two sources, (i) the editing of the mRNA and (ii) post-translational modifications (PTMs) that normally serve to modify or modulate the activity, function or location of a protein in different contexts physiological or metabolic. There are more than 200 different described PTMs (phosphorylation, methylation, acetylation, etc.). They transform each single gene into tens or hundreds of different biological functions. Before proteomics achievements, the differential analysis of the genes, that were expressed in different cell types and tissues in different physiological contexts, was done mainly through analysis of mRNA. However, for wine yeast it has been proved that there is no direct correlation between mRNA transcripts and protein content (Rossignol et al., 2006). It is known that mRNA is not always translated into protein, and the amount of protein produced by a given amount of mRNA depends on the physiological state of the cell. Proteomics confirms the presence of the protein and provides a direct measure of its abundance and diversity.

In terms of methodology, proteomics approaches are classified in two groups, (i) gel free systems, based in the use of different chromatography methods, and (ii) gel based methods, using mainly two dimensional polyacrylamide gel electrophoresis (2DE), that will be the core of our discussion. As a schematic summary, the typical workflow of a proteomic experiment begins with the experimental design. It must be deeply studied, and it will delimit the obtained conclusions, even more when comparison between two strains, cultures or physiological stages between others, are done. From an optimal point of view, only one factor must change between the different assayed conditions (Fernández-Acero et al., 2007a, 2007b). It must contain the use of different biological replicates depending of the used strategy, usually from 3 to 5. The next key step is to obtain a protein extract with enough quality to separate the complex mixture of proteins. Usually, the protein extraction is done in sequential steps (Garrido et al., 2010). First, the biological sample is disrupted using mechanical or chemical techniques. Then, proteins are precipitated and cleaned. Most of the protocol use acetone and trichloroacetic acid. During the next step the proteome is defined and visualized using electrophoretic techniques. 2DE has been widely used for this purpose. Using this technique proteins are separated using two different parameters. During the first dimension, proteins are separated by their isoelectric point using an isoelectrofocusing (IEF) device. Then, the focused strips are used to load in a polyacrylamide gel, where the proteins are separated by their molecular weight. This system allows the separation of hundreds of proteins from a complex mixture. The gels are visualized with unspecific protein stains (those that stain total proteins, such as Coomassie, Sypro, Silver, etc.), or specific ones (those staining solution prepared to detect specific groups of proteins, mainly post-translationals modifications, i.e. Phospho ProQ diamond). The gels are digitalized and analyzed with specific software to reveal the significant spots. Those spots are identified using mass spectrometry. MALDI TOF/TOF is commonly used for 2DE approaches. The huge list of identified proteins obtained is studied to reveal the biological relevance of each identification.

Unfortunately, the number of papers related to fungal proteomics is still poor compared with the application of this technology to other biological sources. As an example, a simple search in WOK website (web of knowledge, http://www.accesowok.fecyt.es/) get 809 entries when the terms "proteom*"and "fung*" are used, whereas 51237 entries are displayed when "proteom*" is used alone. In spite of the numerical results obtained may vary depending of the used keywords and web resource, the fact is that there is a lot of work to do to bring fungal proteomic information at the same level that is obtained with other biological sources. This lack is mainly caused by (i) the difficulties to obtain proteins with enough quality to 2DE separations and (ii) the lack of protein sequences listed in the databases. Our research group was pioneer solving these problems and preparing the first proteomic approaches to the phytopathogenic fungi *Botrytis cinerea* (Fernández-Acero et al. 2006).

Fungi posse strong cell walls. This makes difficult the cell breakage using standard protocols. Moreover, fungal proteins extract are characterised by its high concentration of glycosylated proteins that produces dense extracts, dragging a lot of impurities that disturb protein electrofocusing. We optimized a protocol based on a first phosphate buffer solubilisation followed by a typical TCA/Acetone precipitation. Using this protocol we developed the first proteomic map of *Botrytis cinerea* (Fernández-Acero et al., 2006b). Using this optimized approach we prepared a differential proteomics approach based on 2DE, comparing the proteomes of two *B. cinerea* strains differing in virulence (Fernández-Acero et al., 2007b). In spite of this protocol has been widely cited and used (Cobos et al., 2010; Fernández-Acero et al., 2010, 2011; Michielse et al., 2011; Moreira et al., 2011; Sharma et al., 2010; Yang et al., 2011), our recent data suggest that the phosphate buffer solubilisation produces an artificial enrichment of soluble proteins in our assayed extracts. For this reason, we improved our method using a phenol based protocol preparing a *Botrytis cinerea* map during cellulose degradation (Fernández-Acero et al., 2010). Based on this protocol, adding a previous step of precipitation with DOC, we developed the analysis of the main fungal subproteome, the secretome. We identified 76 secreted proteins from cultures where the virulence was induced with different plant-based elicitors (Fernández-Acero et al., 2010). New projects to unravel proteome content of *Botrytis cinerea* and *Colletotrichum acutatum* are running.

All the proteomic approaches developed on *B. cinerea* has been facilitated by the availability of fungal genome sequence (Amselem et al., 2011) (http://urgi.versailles.inra.fr/Species/Botrytis, and http://www.broadinstitute.org/annotation/genome/botrytis_cinerea/Home.html). Summarizing all our identified spots, we do not get the 3% of the predicted genome. The method to capture new fungal proteins, its identification by mass spectrometry and to determine their biological relevance needs to be determined yet. By using our previous experience with *B. cinerea*, we are developing proteomic approaches to *C. acutatum*. Its conidial germination, mycelia dataset and secretome are characterized by 2DE. The key challenge is in our opinion, the use of the collected information to develop new methodologies to fight against plants pathogens. As a future prospect, the development of new environmental friendly proteomics-based fungicides has been discussed (Fernández-Acero et al., 2011).

6. Acknowledgements

This research has been financed by the Spanish Government DGICYT - AGL2009-13359-CO2/AGR, by the Andalusian Government: Junta de Andalucía, PO7-FQM-002689, http://www.juntadeandalucia.es/innovacioncienciayempresa; Programa Operativo 2007-2013 (FEDER-FSE) (18INSV2407, 18INSV2610), and by the CeiA3 International Campus of excellence in Agrifood (18INACO177.002AA, http://www.uco.es/cei-A3/). Victoria E. González-Rodríguez was financed by the grant FPU of the Ministerio de Educación, Government of Spain, Ref. AP2009-1309; Eva Liñeiro was financed by the grant of the University of Cádiz Ref. 2010-152.

7. References

Afanador-Kafuri, L., Minz, D., Maymon, M., Freeman, S. (2003). Characterization of *Colletotrichum* isolates from Tamarillo, Passiflora, and Mango in Colombia and identification of a unique species from the genus. *Phytopathology*, Vol. 93, pp. 579-587, 0031-949X

Agrios, G.N. (2005). *Plant Pathology*. (5th Ed). Elsevier Academic Press, ISBN: 13: 978-0120445653, San Diego, CA

Alastair McCartney, H., Foster, S.J., Fraaije, B.A., Ward, E. (2003). Molecular diagnostics for fungal plant pathogens. *Pest Management Science*, Vol. 59, pp. 129-142, 1526-498X

Amselem, J., Cuomo, C.A., van Kan, J.A.L., Viaud, M., Benito, E.P., Couloux, A., Coutinho, P.M., de Vries, R.P., Dyer, P.S., Fillinger, S., Fournier, E., Gout, L., Hahn, M., Kohn, L., Lapalu, N., Plummer, K.M., Pradier, J.-., Quévillon, E., Sharon, A., Simon, A., Have, A., Tudzynski, B., Tudzynski, P., Wincker, P., Andrew, M., Anthouard, V., Beever, R.E., Beffa, R., Benoit, I., Bouzid, O., Brault, B., Chen, Z., Choquer, M., Collémare, J., Cotton, P., Danchin, E.G., Da Silva, C., Gautier, A., Giraud, C., Giraud, T., Gonzalez, C., Grossetete, S., Güldener, U., Henrissat, B., Howlett, B.J., Kodira, C., Kretschmer, M., Lappartient, A., Leroch, M., Levis, C., Mauceli, E., Neuvéglise, C., Oeser, B., Pearson, M., Poulain, J., Poussereau, N., Quesneville, H., Rascle, C., Schumacher, J., Ségurens, B., Sexton, A., Silva, E., Sirven, C., Soanes, D.M., Talbot, N.J., Templeton, M., Yandava, C., Yarden, O., Zeng, Q., Rollins, J.A., Lebrun, M.-., Dickman, M. (2011). Genomic analysis of the necrotrophic fungal pathogens *Sclerotinia sclerotiorum* and *Botrytis cinerea*. *PLoS Genetics*, Vol. 7, No. 8, e1002230. doi:10.1371/journal.pgen.1002230, 1553-7390

Atkins, S.D., Clark, I.M., Sosnowska, D., Hirsch, P.R., Kerry, B.R. (2003). Detection and quantification of *Plectosphaerella cucumerina*, a potential biological control agent of potato cyst nematodes, by using conventional PCR, Real-Time PCR, Selective Media, and Baiting. *Applied and Environmental Microbiology*, Vol. 69, pp. 4788-4793, 0099-2240

Bailey, J.A., Jeger, M.J. (1992). *Colletotrichum: biology, pathology and control*. CAB international, ISBN: 0-85198-756-7, Wallingford

Buddie, A.G., Martínez-Culebras, P.V., Bridge, P., García-López, M.D., Querol, A., Cannon, P.F., Monte, E. (1999). Molecular characterization of *Colletotrichum* strains derived from strawberry. *Mycological Research*, Vol. 103, pp. 385-394, 0953-7562

Carbu, M. (2006). *Estudio de la variabilidad genetica y organizacion cromosomica en el hongo fitopatogeno "Botrytis cinerea". Thesis doctoral*. University of Cádiz, Cádiz

Cobos, R., Barreiro, C., Mateos, R.M., Coque, J.R. (2010). Cytoplasmic- and extracellular-proteome analysis of *Diplodia seriata*: A phytopathogenic fungus involved in grapevine decline. *Proteome Sciences*, Vol. 8, doi: 10.1186/1477-5956-8-46, 1477-5956

Coley-Smith, J.R., Verhoheff, K., Jarvis, W.R. (1980). *The Biology of Botrytis*. Academic Press, ISBN 0-1217 98508X, London, UK

Denoyes-Rothan, B., Guerin, G., Lerceteau-Kohler, E., Risser, G. (2004). Inheritance of resistance to *Colletotrichum acutatum* in *Fragaria x ananassa* . *Phytopathology*, Vol. 95, pp. 405-412, 0031-949X

Denoyes-Rothan, B., Baudry, A. (1995). Species identification and pathogenicity study of French *Colletotrichum* strains isolated from strawberry using mophological and cultural characteristics. *Phytopathology*, Vol. 85, pp. 53-57, 0031-949X

Elad, Y., Williamson, B., Tudzynski, P., Delen, N. (2004). *Botrytis: Biology, Pathology and Control*. Springer, ISBN: 978-1-4020-6586-1, Dordrecht, The Netherland.

EPPO/CABI (1997). *Quarantine Pests for Europe* (2nd edition). Edited by Smith IM, McNamara DG, Scott PR, Holderness M. CABI International, Wallingford, UK, 1425 pp.

Faretra, F., Antonacci, E., Pollastro, S. (1988). Sexual bahaviour and matting system of *Botryotinia fuckeliana* (*Botrytis cinerea*). *Microbiology*, Vol. 134, No. 9, pp. 2543-2550, 1350-0872

Faretra, F., Antonacci, E. (1987). Production of apothecia of *Botryotinia fuckeliana* (de Bary) Whetz under controled environmental conditions. *Phytopathologia Mediterranea*, Vol. 26, pp. 29-35, 0031-9465

Fernández-Acero, F.J., Colby, T., Harzen, A., Carbú, M., Wieneke, U., Cantoral, J.M., Schmidt, J. (2010). 2-DE proteomic approach to the *Botrytis cinerea* secretome induced with different carbon sources and plant-based elicitors. *Proteomics*, Vol. 10, pp. 2270-2280, 1615-9853

Fernández-Acero, F.J., Carbú, M., Garrido, C., Vallejo, I., Cantoral, J.M. (2007a). Proteomic Advances in Phytopathogenic Fungi. *Current Proteomics*, Vol. 4, pp. 79-88, 1570-1646

Fernández-Acero, F.J., Jorge, I., Calvo, E., Vallejo, I., Carbú, M., Camafeita, L.E., Garrido, C., López, J.A., Cantoral, J.M., Jorrin, J. (2007b). Proteomic analysis of phytopathogenic fungus *Botrytis cinerea* as a potential tool for identifying pathogenicity factors, therapeutic targets and for basic research. *Archives of Microbiology*, Vol. 187, pp. 207-215, 1432-072X

Fernández-Acero, F.J., Carbú, M., Garrido, C., Collado, I.G., Cantoral, J.M., Vallejo, I. (2006a). Screening Study of Potential Lead Compounds for Natural Product-based Fungicides Against *Phytophthora* Species. *Journal of Phytopathology*, Vol. 154, pp. 616-621, 0931-1785

Fernández-Acero, F.J., Jorge, I., Calvo, E., Vallejo, I., Carbú, M., Camafeita, E., Lopez, J.A., Cantoral, J.M., Jorrin, J. (2006b). Two-dimensional electrophoresis protein profile of the phytopathogenic fungus *Botrytis cinerea*. *Proteomics*, Vol. 6, No. 1, pp. 88-96, 1615-9853

Fernández-Acero, F.J., Carbú, M., El-Akhal, M.R., Garrido, C., González-Rodríguez, V.E., Cantoral, J.M. (2011). Development of proteomics-based fungicides: New strategies for environmentally friendly control of fungal plant diseases. *International Journal of Molecular Sciences*, Vol. 12, pp. 795-816, 1422-0067

Freeman, S., Katan, T. (1997). Identification of *Colletotrichum* species responsible for anthracnose and root necrosis of strawberry in Israel. *Phytopathology*, Vol. 87, pp. 516-521, 0031-949X

Freeman, S., Minz, D., Jurkevitch, E., Maymon, M., Shabi, E. (2000a). Molecular analyses of *Colletotrichum* species from Almond and other fruit. *Phytopathology*, Vol. 90, pp. 608–614, 0031-949X

Freeman, S., Shabi, E., Katan, T. (2000b). Characterization of *Colletotrichum acutatum* causing anthracnose of Anemone (*Anemone coronaria* L.). *Phytopathology*, Vol. 66, No. 12, pp. 5267-5272, 0031-949X

Freeman, S., Pham, M., Rodríguez, R.J. (1993). Molecular genotyping of *Colletotrichum* species based on arbitrarily primed PCR, A+T-rich DNA, and nuclear DNA analyses. *Experimental Mycology*, Vol. 17, pp. 309-322, 0147-5975

Garrido, C., Carbú, M., Fernández-Acero, F.J., González-Rodríguez, V.E., Cantoral, J.M. (2011). New Insight in the Study of Strawberry Fungal Pathogens. *Genes, Genomes and Genomics*, Vol. 5, No. 1, pp. 24-39, 1749-0383

Garrido, C., Cantoral, J.M., Carbú, M., González-Rodríguez, V.E., Fernández-Acero, F.J. (2010). New proteomic approaches to plant pathogenic fungi. *Current Proteomics*, Vol. 7, pp. 306-315, 1570-1646

Garrido, C., Carbú, M., Fernández-Acero, F.J., Boonham, N., Colyer, A., Cantoral, J.M., Budge, G. (2009a). Development of protocols for detection of *Colletotrichum acutatum* and monitoring of strawberry anthracnose using real-time PCR. *Plant Pathology*, Vol. 58, pp. 43-51, 1365-3059

Garrido, C., Carbú, M., Fernández-Acero, F.J., Vallejo, I., Cantoral, J.M. (2009b). Phylogenetic relationships and genome organization of *Colletotrichum acutatum* causing anthracnose in strawberry. *European Journal of Plant Pathology*, Vol. 125, pp. 397-411, 0929-1873

Garrido, C., Carbú, M., Fernández-Acero, F.J., Budge, G., Vallejo, I., Colyer, A., Cantoral, J.M. (2008). Isolation and pathogenicity of *Colletotrichum* spp. causing anthracnose of strawberry in south west Spain. *European Journal of Plant Pathology*, Vol. 120, pp. 409-415, 0929-1873

Giraud, T., Fortini, D., Levis, C., Leroux, P., Brygoo, Y. (1997). RFLP markers show genetic recombination in *Botryotinia fuckeliana* (*Botrytis cinerea*) and transposable elements reveal two sympatric species. *Molecular Biology and Evolution*, Vol. 14, pp. 1177-1185, 1537-1719

Hughes, K.J.D., Lane, C.R., Cook, R.T.A. (1997). Development of a rapid method for the detection and identification of *Colletotrichum acutatum*, In: *Diagnosis and Identification of Plant Pathogens*, Dehne HW, Adam G, Diekmann M, Frahm J, Mauler-Machnik A & van Halteren P), pp. 113-116. Kluwer Academic, ISBN: 0792347714, Dordrecht (NL).

Ijdo, J.W., Wells, R.A., Baldini, A., Reeders, S.T. (1991). Improved telomere detection using a telomere repeat probe (TTAGGG)n generated by PCR. *Nucleic Acids Research*, Vol. 19, pp. 4780, 0305-1048

James, P. (1997). Protein identification in the post-genome era: the rapid rise of proteomics. *Quarterly Reviews of Biophysics*, Vol. 30, pp. 279-331, 0033-5835

Kim, Y., Nandakumar, M.P., Marten, M.R. (2007). Proteomics of filamentous fungi. *Trends in Biotechnology*, Vol. 25, pp. 395-400, 0167-7799

Leandro, L.F.S., Gleason, M.L., Nutter, F.W., Wegulo, S.N., Dixon, P.M. (2003). Strawberry plant extracts stimulate secondary conidiation by *Colletotrichum acutatum* on symptomless leaves. *Phytopathology*, Vol. 93, pp. 1285-1291, 0031-949X

Leandro, L.F.S., Gleason, M.L., Nutter, F.W., Wegulo, S.N., Dixon, P.M. (2001). Germination and Sporulation of *Colletotrichum acutatum* on symptomless strawberry leaves. *Phytopathology*, Vol. 91, pp. 659-664, 0031-949X

Martínez-Culebras, P.V., Querol, A., Suarez-Fernández, M.B., García-López, M.D., Barrio, E. (2003). Phylogenetic relationships among *Colletotrichum* pathogens of strawberry and design of PCR primers for their identification. *Journal of Phytopathology*, Vol. 151, pp. 135-143, 0931-1785

Masel, A., Irwin, J.A.G., Manners, J.M. (1993). DNA addition or deletion is associated with a major karyotype polymorphism in the fungal phytopathogen *Colletotrichum gloeosporioides*. *Molecular Genetics and Genomics*, Vol. 237, pp. 73-80, 1617-4615

Mertely, J.C., Legard, D.E. (2004). Detection, isolation, and pathogenicity of *Colletotrichum* spp. from strawberry petioles. *Plant Disease*, Vol. 88, pp. 407-412, 0191-2917

Michielse, C.B., Becker, M., Heller, J., Moraga, J., Collado, I.G., Tudzynski, P. (2011). The *Botrytis cinerea* Reg1 protein, a putative transcriptional regulator, is required for pathogenicity, conidiogenesis and for the production of secondary metabolites. *Molecular Plant-Microbe Interaction*, Vol 24, No. 9, pp. 1074-1085, 0894-0282

Mills, P.R., Sreenivasaprasad, S., Brown, A.E. (1992). Detection and differentiation of *Colletotrichum gloeosporioides* isolates using PCR. *FEMS Microbiology Letters*, Vol. 98, pp. 137-144, 0378-1097

Moreira, J., Almeida, R., Tavares, L., Santos, M., Viccini, L., Vasconcelos, I., Oliveira, J., Raposo, N., Dias, S., Franco, O. (2011). Identification of Botryticidal Proteins with Similarity to NBS–LRR Proteins in Rosemary Pepper (*Lippia sidoides* Cham.) Flowers. *The Protein Journal*, Vol. 30, pp. 32-38, 1572-3887

O´Sullivan, D., Tosi, P., Creusot, F., Cooke, M., Phan, T., Dron, M., Langin, T. (1998). Variation in genome organization of the plant pathogenic fungus *Colletotrichum lindemuthianum*. *Current Genetic*, Vol. 33, pp. 291-298, 0172-8083

Perfect, S.E., Bleddyn Hughes, H., O'Connell, R.J., Green, J.R. (1999). REVIEW: *Colletotrichum*: A model genus for studies on pathology and fungal-plant interactions. *Fungal Genetics and Biology*, Vol. 27, pp. 186-198, 1087-1845

Rebordinos, L., Vallejo, I., Santos, M., Collado, I.G., Carbú, M., Cantoral, J.M. (2000). Análisis genético y relación con patogenicidad en *Botrytis cinerea*. *Revista Iberoamericana de Micología*, Vol. 17, pp. 37-42, 1130-1406

Rebordinos, L., Santos, M., Vallejo, I., Collado, I.G., Cantoral, J.M. (1997). Molecular characterization of the phytopathogenic fungus *Botrytis cinerea*. *Recent Research Developments in Phytochemistry*, Vol. 1, pp. 293-307, 81-308-0052-7

Rossignol, M., Peltier, J., Mock, H., Matros, A., Maldonado, A.M., Jorrín, J.V. (2006). Plant proteome analysis: A 2004-2006 update. *Proteomics*, Vol. 6, pp. 5529-5548, 1615-9853

Sharma, N., Rahman, M.H., Kav, N.N.V. (2010). A possible proteome-level explanation for differences in virulence of two isolates of a fungal pathogen *Alternaria brassicae*. *Journal of Plant Biochemistry and Biotechnology*, Vol. 19, pp. 40-49, 0971-7811

Sreenivasaprasad, S., Talhinhas, P. (2005). Genotypic and phenotypic diversity in *Colletotrichum acutatum*, a cosmopolitan pathogen causing anthracnose on a wide range of hosts. *Molecular Plant Pathology*, Vol. 6, pp. 361-378, 0885-5765

Sreenivasaprasad, S., Sharada, K., Brown, A.E., Mills, P.R. (1996). PCR-based detection of *Colletotrichum acutatum* on strawberry. *Plant Pathology*, Vol. 45, pp. 650-655, 1365-3059

Sreenivasaprasad, S., Brown, A.E., Mills, P.R. (1992). DNA sequence variation and interrelationships among *Colletotrichum* species causing strawberry anthracnose. *Physiological and Molecular Plant Pathology*, Vol. 41, pp. 265-281, 0885-5765

Sutton, B.C. (1992). The genus *Glomerella* and its anamorph *Colletotrichum* , In: *Colletotrichum: Biology, Pathology and Control,* Bailey, J.A. & Jeger, M.J., pp. 1-26, CBA International, ISBN: 0-85198-756-7, Wallingford, UK.

Talhinhas, P., Sreenivasaprasad, S., Neves-Martins, J., Oliveira, H. (2005). Molecular and phenotypic analyses reveal association of diverse *Colletotrichum acutatum* groups and a low level of *C. gloeosporioides* with olive anthracnose. *Applied and Environmental Microbiology*, Vol. 71, pp. 2987-2998, 0099-2240

Vallejo, I., Carbú, M., Muñoz, F., Rebordinos, L., Cantoral, J.M. (2002). Inheritance of chromosome-length polymorphisms in the phytopathogenic ascomycete *Botryotinia fuckeliana* (anam. *Botrytis cinerea*). *Mycological Research*, Vol. 106, pp. 1075-1085, 0953-7562

Vallejo, I., Rebordinos, L., Collado, I.G., Cantoral, J.M. (2001). Differential behaviour of mycelial growth of several *Botrytis cinerea* strains on either Patchoulol- or Globulol-amended media. *Journal of Phytopathology*, Vol. 149, pp. 113-118, 0931-1785

Vallejo, I., Santos, M., Cantoral, J.M., Collado, I.G., Rebordinos, L. (1996). Chromosomal polymorphism in *Botrytis cinerea* strains. *Hereditas*, Vol. 124, pp. 31-38, 0018-0661

White, T.J., Bruns, S.L., Taylor, J.W. (1990). Amplification and direct sequencing of fungal ribosomal RNA genes for phylogenetics, In: *PCR Protocols: A Guide to Methods and Applications*, Innis, M. A., Gelfand, D. H., Sninsky, J. J., and White, T. J., pp. 315, Academic Press, Inc., New York

Whitelaw-Weckert, M.A., Curtin, S.J., Huang, R., Steel, C.C., Blanchard, C.L., Roffey, P.E. (2007). Phylogenetic relationships and pathogenicity of *Colletotrichum acutatum* isolates from grape in subtropical Australia. *Plant Pathology,* Vol. 56, pp. 448-463, 1365-3059

Wilkins, M.R., Sanchez, J.C., Gooley, A.A., Appel, R.D., Humphery-Smith, I., Hochstrasser, D.F., Williams, K.L. (1995). Progress with proteome projects: why all proteins expressed by a genome should be identified and how to do it. *Biotechnology & Genetic Engineering Reviews,* Vol. 13, pp. 19-50, 0264-8725

Yang, J., - Wang, L., - Ji, X., - Feng, Y., - Li, X., - Zou, C., - Xu, J., - Ren, Y., - Mi, Q., - Wu, J., - Liu, S., - Liu, Y., - Huang, X., - Wang, H., - Niu, X., - Li, J., - Liang, L., - Luo, Y., - Ji, K., - Zhou, W., - Yu, Z., - Li, G., - Liu, Y., - Li, L., - Qiao, M., - Feng, L., - Zhang, K. (2011). Genomic and Proteomic Analyses of the Fungus *Arthrobotrys oligospora* Provide Insights into Nematode-Trap Formation. *PLoS Pathogens,* Vol. 7, e1002179, 1553-7366.

Zeigler, R.S. (1998). Recombination in *Magnaporte grisea. Annual review of phytpathology,* Vol. 36, pp. 249-275, 0066-4286

9

Molecular and Proteolytic Profiles of *Trypanosoma cruzi* Sylvatic Isolates from Rio de Janeiro-Brazil

Suzete A. O. Gomes[1,2] et al.*
[1]*Laboratório de Biologia de Insetos, GBG*
Universidade Federal Fluminense-UFF, Rio de Janeiro, RJ
[2]*Laboratório de Transmissores de Leishmanioses*
Setor de Entomologia Médica e Forense IOC-FIOCRUZ-Rio de Janeiro, RJ
Brazil

1. Introduction

Chagas disease, also known as American trypanosomiasis, has its epidemiology conditioned to the (i) triatominae vectors, (ii) etiologic agent, *Trypanosoma cruzi*, and (iii) sylvatic and sinantropic reservoirs, the mammals. Social factors associated with economic factors, such as industry development, population growth and rural area colonization, which lead directly to ecological imbalance, provide favorable conditions for the disease establishment (Barretto, 1967; Ávila-Pires, 1976).

In 1909, Carlos Chagas releases his discovery on a new human disease, the American trypanosomiasis, subsequently known as Chagas disease. Carlos Chagas described the etiologic agent, the protozoan belonging to the Trypanosomatidae family *Trypanosoma cruzi*, and its insect vector belonging to the Hemiptera order, Triatominae subfamily, the so-called kissing bug (Chagas, 1909; Lent & Wygodzinsky, 1979).

The natural history of the Chagas disease probably started millions of years ago probably as a sylvatic enzooty, and it is still present in different areas from Brazilian territory. The arrival of men in these areas, as well as comprehensive deforestation caused by extensive farming during the past 300 years has caused triatomine insects, formerly sylvatic animal blood-sucking bugs, to meet men (Ferreira et al., 1996; Coura, 2007). Hence, the disease was characterized as a zoonosis, when men invaded the sylvatic habitat, deforesting and changing the ecological balance, and making triatomine bugs access to the residences.

---

* Danielle Misael[2], Cristina S. Silva[2], Denise Feder[1], Alice H. Ricardo-Silva[2], André L. S. Santos[3], Jacenir R. Santos-Mallet[2] and Teresa Cristina M. Gonçalves[2]
[1]*Laboratório de Biologia de Insetos, GBG, Universidade Federal Fluminense-UFF, Rio de Janeiro, RJ, Brasil*
[2]*Laboratório de Transmissores de Leishmanioses, Setor de Entomologia Médica e Forense, IOC-FIOCRUZ-Rio de Janeiro, RJ, Brasil*
[3]*Laboratório de Estudos Integrados em Bioquímica Microbiana, Instituto de Microbiologia Paulo de Góes (IMPG), Bloco E-subsolo, Universidade Federal do Rio de Janeiro (UFRJ), Rio de Janeiro, RJ, Brasil*

Therefore, the transmission cycle of *T. cruzi* is comprised by a sylvatic cycle, in which the parasite circulates among mammals and sylvatic vectors, and a domiciliary cycle, in which the infection is ensued by the contact of mammals, sylvatic vectors and sinantropic animals with domestic and domiciled animals, including men (Barretto, 1979).

Human Chagas disease, an antropozoonosis that evolved from a zoonosis, is strongly related with men's social class, type of work and habitation (Dias, 2000). During the 70's, the disease endemic area achieved at least 2,450 Brazilian cities, 771 of which were detected to have *Triatoma infestans*, the main disease vector in Brazil. At that time, there were over five million people affected by the disease in the country, with an incidence of approximately one hundred thousand new cases yearly and mortality above ten thousand deaths yearly. Less than five percent of blood banks used to control donors and over seven hundred cities had their homes infected by *T. infestans*. This situation led scientists to press the government to prioritize a national program against the disease. Homes from endemic areas were sprinkled with the appropriate insecticide and, in accordance with law; mandatory screening of blood donors was implemented throughout the country (Dias et al., 2002). The control program of the main vector in Brazil was recognized in 2006, with a certificate from the World Health Organization (WHO) for virtual elimination of *T. infestans* in Brazil (Dias, 2006). As the main vector was eliminated, currently there is a concern that other Triatominae species, formerly deemed secondary in the disease transmission, such as *Triatoma brasiliensis*, *Triatoma pseudomaculata* and *Panstrongylus megistus*, take the place of *T. infestans* in some locations, therefore becoming potential disease vectors in Brazil (Coura, 2009).

Despite the great progress in controlling vector and transfusion transmission in the countries from the Southern Cone, transmission is ongoing in other parts of the continent, and the issue of already infected people, most of whom are in the chronic phase of the disease, is still a challenge to public health (Urbina, 1999). Currently Chagas disease affects between twelve and fourteen million people in Latin America, and at least 60 million people live in areas with transmission risk (WHO, 2002). In Brazil, the disease notification became compulsory as per Ordinance V of Health Surveillance Secretary of Ministry of Health dated February 21, 2006.

2. Triatomines

The first report of triatomine existence was recorded by the Spanish Francisco López de Gomara, in 1514, when mentioning Darién region he said: "Hay muchas garrapatas y chinches com alas", apparently referring to *Rhodnius prolixus* (Stål, 1859) (León, 1962). *Cimex rubrofasciatus* (*Triatoma rubrofasciata*), was described in 1773 by De Geer, and later assigned by Laporte as the type species of *Triatoma* genus (Lent & Wygodzinsky, 1979). In Brazil, the first report of triatomine in domicile was possibly *Panstrongylus megistus* (Burmeister, 1835) (Gardner, 1942). However, the identification of *Trypanosoma cruzi* sylvatic isolates is contemporary to the discovery of this parasite and Chagas disease by Carlos Chagas in 1909. When they went to Lassance, Minas Gerais, Brazil, for malaria epidemics study, he identified flagellated forms in the intestine of triatomine of *Conorhinus megistus* (*Panstrongylus megistus*) in humans and cats, referring to them as *Schizotrypanum cruzi* (Chagas, 1909). Later Chagas (1912) isolated the parasite in armadillos (*Tatusia novemcincta*, now called *Daysipus novemcinctus*), identifying the *T. cruzi* sylvatic reservoirs, and in the same ecotope he found infected *Triatoma geniculata* (*Panstrongylus geniculatus*) specimens, establishing the disease sylvatic cycle (Coura & Dias, 2009).

Between 1913 and 1924 it became evident that the disease was not restricted to Brazil, being diagnosed in other countries in Central and South Americas, such as El Salvador, Venezuela, Peru and Argentina (Talice et al., 1940; Zeledón, 1981). In subsequent studies, Coura & Dias, 2009 mentions that Chagas (1924) demonstrated *T. cruzi* transmission cycle in the Amazon region with the identification of this parasite in monkeys of *Saimiri scirius* species.

In Rio de Janeiro state, the first Triatominae occurrence dated 1859, when Stal described *Conorhinus vitticeps* species, now called *Triatoma vitticeps*. At that time, Rio de Janeiro was assigned as type location, without defining whether it referred to the city or state.

Following this finding, Neiva (1914) recorded the occurrence of *T. vitticeps* in Conceição de Macabu, formerly Macaé city district, presently Conceição de Macabu city. Due to information accuracy, Lent (1942) suggested it would be considered as the type location of *T. vitticeps*.

Subsequently, Pinto (1931, as cited in Lent, 1942) pointed out its presence in Magé, and Lent (1942) in Nova Friburgo, at Secretario location in Petrópolis city and at Federal District, which was Rio de Janeiro at that time. In Minas Gerais state, it was observed by the first time by Martins et al (1940), and in Espírito Santo state, as mentioned by Lent (1942).

In Rio de Janeiro state other species were also found. Guimarães and Jansen (1943) collected *Panstrongylus megistus* specimens in a building by the hill, and identified *Trypanosoma cruzi* sylvatic reservoir (skunk), but did not find the sylvatic focus. Dias (1943) listed Chagas disease transmitters in Rio de Janeiro as being *Panstrongylus megistus, Panstrongylus geniculatus* (Latreille, 1811), *Triatoma vitticeps* (Stal, 1859), *Triatoma oswaldoi* (Neiva & Pinto, 1923), *Triatoma infestans* (Klug) and *Triatoma rubrofasciata* (De Geer, 1773), first recording the occurrence of *Schizotrypanum sp*-infected *P. megistus* in two districts in the capital of Republic (Santa Tereza and Botafogo). In 1953, in a survey performed at Araruama and Magé, Dias stated it was a relevant issue for the State, while Bustamante & Gusmão 1953 pointed out the presence of *T. infestans* at Resende and Itaverá cities. New findings have been identified, such as that of Coura et al. (1966), who found *P. megistus, Triatoma tibiamaculata* and *T. rubrofasciata* in three districts at Rio de Janeiro city, and that of Aragão & Souza (1971), who signalized the presence of *T. infestans* colonizing domiciles at two cities in Baixada Fluminense. In the same year, Coura et al. (1966) described some autochthonous instances of *T. infestans*-transmitted Chagas disease at Baixada Fluminense, and Becerra-Fuentes et al. (1971) recorded *T. rubrofasciata* occurrence at Morro do Telégrafo in the former Guanabara state. Silveira et al. (1982) performed an entomologic inquiry at Duque de Caxias and Nova Iguaçu cities (RJ), and only found *T. infestans* species. Ferreira et al. (1986) verified the occurrence of *T. vitticeps*, and positivity for *T. cruzi*-like forms, in 12 cities, of which the one with the highest incidence for both observations was Triunfo location at Santa Maria Madalena city. In 1989, a *P. geniculatus* specimen was found in a domicile at São Sebastião do Alto city (RJ) (personal communication with Teresa Cristina M. Gonçalves). The occurrence of *Rhodnius prolixus* (Stål, 1859) in Teresópolis was pointed out by Pinho et al. (1998), which caused questioning, once this species was restricted to the northern region of the country. Nowadays it is known this species does not occur in Brazil (Monteiro et al., 2000, 2003). *T. vitticeps* was found in Poço das Antas, Silva Jardim city, by Lisbôa et al. (1996), and in Santa Maria Madalena by Gonçalves et al. (1998). In both locations, biological and morphological characterization of *T. cruzi* isolates, obtained for both triatomine bugs and vertebrate hosts, confirmed the maintenance of enzootic disease form. In the period from 2008 to 2010 *T. vitticeps* was pointed out at Cantagalo, Tanguá, Trajano de Morais, and São Fidélis cities (Oliveira et al., 2010).

In Espírito Santo, where *T. vitticeps* incidence was also signalized, the rates of infection by *T. cruzi*-like forms were assessed in specimens collected in the domicile: 4% by Santos et al. (1969) at Alfredo Chaves (ES); 25.2% by Silveira et al. (1983) at Cachoeiro do Itapemirim and Guarapari (ES); 35.2% by Ferreira et al. (1986) in 12 cities from Rio de Janeiro state; 64.70% by Sessa & Carias (1986) in 19 cities from Espírito Santo state; and 70.2% and 51.8%, respectively, for females and males, by Dias et al. (1989).

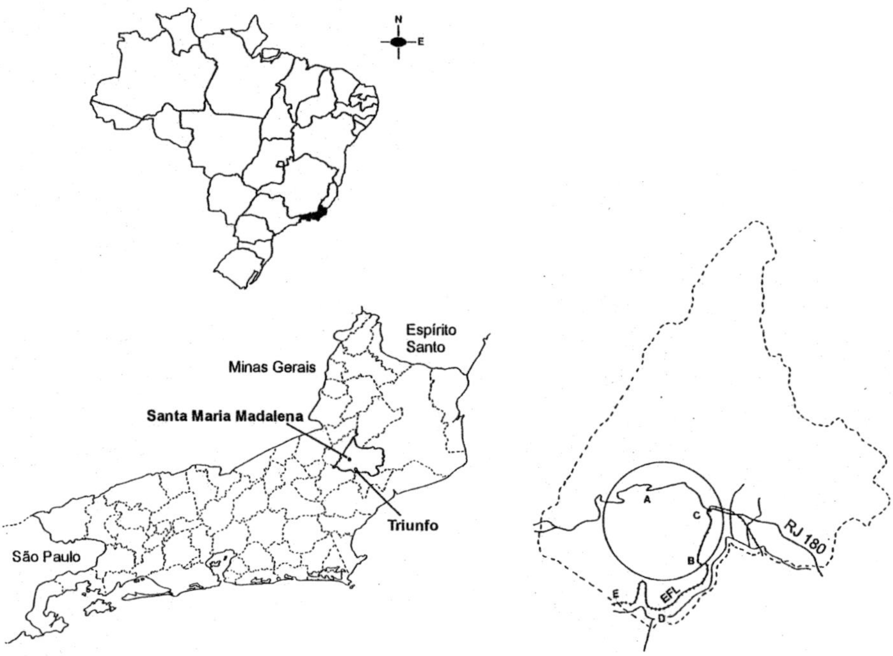

Fig. 1. Studied area and sites of capture of *Triatoma vitticeps* in Triunfo, Santa Maria Madalena, Municipal district, State of Rio de Janeiro, Brazil.

Data from National Health Foundation ("FUNASA") signalized *T. vitticeps* presence in the northern region of Rio de Janeiro state, and the number of notifications on adult form occurrence was increasing (Lopes et al., 2009; Dias et al., 2010). Although studies regarding *T. vitticeps* biology have suggested that this species would not represent a major concern from epidemiologic point of view (Dias, 1956; Heitzmann-Fontenelle, 1980; Silva, 1985; Diotaiuti et al., 1987; Gonçalves et al., 1988, 1989), reports of this species frequently invading the domicile with high *T. cruzi* infection rates (Gonçalves et al., 1998, Gonçalves, 2000) indicated its study was required. With sylvatic habit and unknown habitat, this species ecobiology was studied in further details at Triunfo district, Santa Maria Madalena city (RJ), in three areas (A, B and C) (Figure 1). Of the triatomine bugs collected, 68 *T. cruzi* samples were isolated, which showed heterogeneity in which refers to biology, histopathogenesis and differential expression of surface enzymes.

2.1 *Trypanosoma cruzi*

Trypanosoma cruzi (Figure 2) is a flagellated protozoan belonging to Trypanosomatidae family (Kent, 1880), Kinetoplastida order, *Trypanosoma* genus (Chagas, 1909a; Coura, 2006). Kinetoplastida order was established as a function of the presence of a single cytoplasmic structure, the kinetoplast (Wallace, 1966), where mitochondrial DNA or k-DNA is concentrated. Its form, size, and position are important for characterizing the different evolution forms of the parasite (Vickerman, 1985).

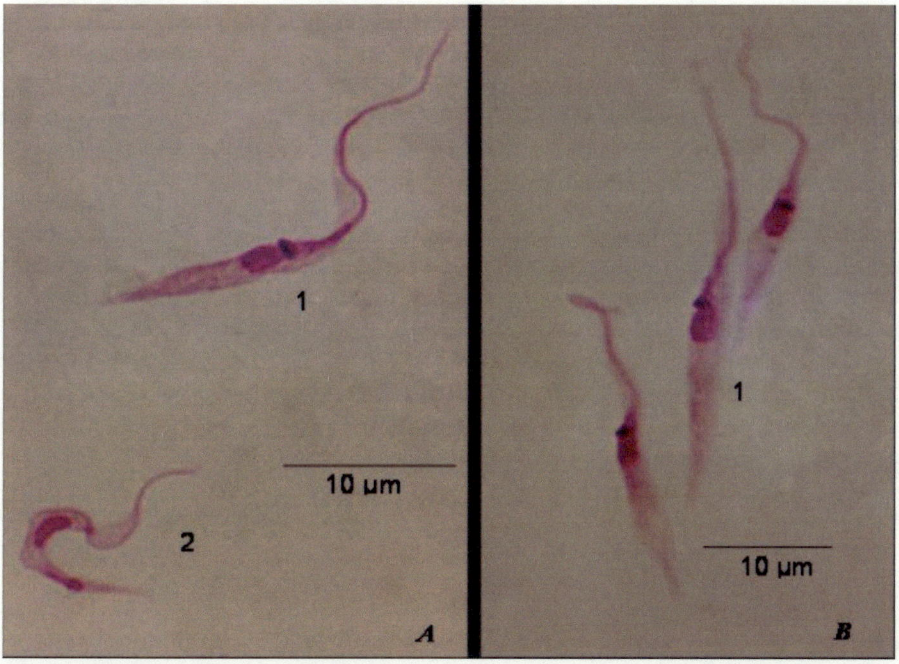

Fig. 2. Epimastigote (1) and tripomastigote (2) forms of *Trypanosoma cruzi* sylvatic isolates from Trinfo, Santa Maria Madalena municipal district, State of Rio de Janeiro - Brazil.

It is a euryxene and digenetic trypanosomatid, since part of its life cycle occurs inside a vertebrate or invertebrate host (Hoare, 1964). Vertebrate and invertebrate hosts are represented, respectively, by domiciled or domestic mammals and sylvatic triatomines.

The parasite cycle can be summarized as follows: the triatomine vector usually defecates during or at the end of blood sucking, eliminating metacyclic trypomastigote forms of *T. cruzi* on the vertebrate hosts. These forms found in dejections can penetrate the host through a continuity skin solution or skin mucosa. Inside the host cell, trypomastigotes transform into amastigotes and, approximately 35 hours later, the binary division begins. After five days, amastigotes transform into trypomastigotes, and as soon as they have long flagella, the cell disrupts releasing these forms into the bloodstream, so that they infect other cells or achieve different organs (Sousa, 2000). In triatomines, the blood-sucking trypomastigote forms ingested during hematophagy differentiate into epimastigotes in the digestive tract. Another differentiation occurs in the digestive tract, more specifically in its final portion and in rectus, when epimastigotes transform into metacyclic trypomastigotes, which is infectious for the vertebrate host and eliminated with the feces (Zeledón et al., 1977; Garcia & Azambuja, 2000).

T. cruzi is found as a parasite in a considerable number of mammals and in a wide range of tissues and niches in these hosts (Deane et al., 1984). Such eclecticism has characterized *T. cruzi* as one of the most successful microorganism in presenting parasitary life (Jansen et al., 1999). Therefore, this protozoan comprises a wide set of heterogeneous populations that circulate through very diverse vertebrate and invertebrate hosts, with a variation of different genotype predominance. The parasite has several morphological, physiological and ecological variations, and also in which refers to its infectivity and pathogenicity (Miles et al., 1978, 1980, 2009), which can warrant the various clinical manifestation forms of Chagas disease observed in different geographic regions (Miles et al., 1981a). Many studies have been performed seeking molecular markers that could correlate the parasite genotype with varying types of this infirmity clinical manifestation. Several works tried to clarify the multiple factors related with population epidemiology and genetics.

T. cruzi has a great phenotypic and genotypic variability in its strains, and therefore this protozoan has the ability to perform genetic exchanges through an unusual mechanism of nuclear fusion, forming a polyploidy progeny, which can suffer recombination among alleles, and after losing its chromosome, can return to diploid status. Some studies provided strong evidence that sexual reproduction is absent in *T. cruzi*, and that its population structure is clonal (Gaunt et al., 2003; Lewis et al., 2009).

3. Molecular profile of *T. cruzi* populations

Early investigations on the genetic of *T. cruzi* populations are based on electrophoretic profiling of isoenzymes (zimodeme analysis), a technique used to explore the genetic diversity of microorganisms. Enzymatic electrophoresis uses soluble raw-materials and extracts from an organism to assess the activity of a protein, and its product is revealed by means of a colorimetric reaction. Under controlled conditions, differences in isoenzymatic mobility imply genetic differences (Miles, 1985; Miles & Cibulkis, 1986). Toye (1974) was the first to use isoenzymes to classify trypanosomas from the New World, reporting differences among *T. cruzi* samples. By the end of the 70's and beginning of the 80's, several studies on isoenzymatic variability among *T. cruzi* populations were performed in Brazilian Northeast, and later in different regions within the country, by employing six enzymes: ALT (alanine aminotransferase), AST (aspartate aminotransferase), glucose phosphate isomerase (GPI), glucose-6-dehydrogenase phosphate (G6PDH), malic enzyme (ME) and phosphoglucomutase (PGM), characterizing three enzymatic profiles belonging to parasite groups called zymodemes I (Z1), II (Z2) and III (Z3). Z1 and Z3 are related with the sylvatic transmission cycle and Z2 with the domestic transmission cycle of the parasite (Miles et al., 1977, 1978, 1980, 1981a, b). As the number of analyzed isoenzymes has been amplified and sub-populations circulating among domestic and sylvatic vertebrates and invertebrates have been studied, an elevated degree of *T. cruzi* heterogeneity was verified (Miles et al., 1980; Bogliolo et al., 1986; Tibayrenc et al., 1986; Tibayrenc & Ayala, 1988; Barnabe et al., 2000).

With technologic advancement and the discovery of new molecular biology tools, it was possible to study the diversity of *T. cruzi* by means of DNA analysis, allowing for molecular characterization of this parasite strains (Devera et al., 2003). Therefore, the genetic diversity was corroborated by randomly amplified polymorphic DNA (RAPD) and restriction fragment length polymorphism (RFLP) analyses, DNA fingerprinting, microsatellites and molecular karyotyping (reviewed by Zingales et al., 1999). Analyses of gene sequences with lowest evaluative rates, such as ribosomal RNA genes, classic evolution markers and mini-exon genes, indicated dimorphism in *T. cruzi* isolates, rating them into two groups (Souto et al., 1996). Mini-exon gene that is present in Kinetoplastid nuclear genome at approximately 200 copies in a tandem type array is composed by three different regions: exon, intron and intergenic regions. Exon is a highly preserved sequence between de order compounds, added to nuclear messenger RNA post-transcription (Devera et al., 2003). Intron is moderately preserved between species of the same genus or sub-genus, and the intergenic region is particularly different among species. In *T. cruzi*, the amplification of mini-exon intergenic region by Polimerase Chain Reaction (PCR) allowed us to classify the different isolates into two main taxonomic groups: *T. cruzi* I and *T. cruzi* II (Fernandes, 1996; Souto et al., 1996; Fernandes et al., 1998). Thereafter, PCR amplification assay were standardized, allowing for rapid molecular typing, which started to be broadly used. Thereby the use of multiplex PCR based on intergenic region allowed us to classify the isolates as *T. cruzi* I, *T. cruzi* II, *T. cruzi* Z3 or *T. rangeli* with 200, 250, 150 pb and 100 pb, respectively (Fernandes et al., 2001a).

Aiming at standardizing double lines and hybrid isolates, a committee settled the lines were referred to as *T. cruzi I* and *T. cruzi II* "groups" (Zingales et al., 1999). Such denomination was not attributed to hybrid isolates, and additional studies are recommended to better characterize them (Zingales, 2011). From hybrid isolate gene sequence analysis, it has been shown that events of genetic exchanges with these parasites originated four distinct isolate groups (Sturm & Campbell, 2009). Thus, by using multilocus enzyme electrophoresis (MLEE) and RAPD markers, it was suggested that the group *T. cruzi* II was divided into five subgroups, including the four hybrid groups (Freitas et al., 2006; Brisse et al., 2000). *T. cruzi III*, a third ancestral group, was proposed from the analysis of microsatellites and mitochondrial DNA.

In 2009, the scientific community felt the need to standardize once again *T. cruzi* groups' nomenclature, aiming at clarifying questions on biology, eco-epidemiology and pathogenicity (Zingales et al., 2009). In this respect, it was recommended that *T. cruzi* was divided into six groups (*T. cruzi I–VI*), and that each group was called Discreet Taxonomic Units (DTUs) I, IIa, IIb, IIc, IId, IIe (Figure 3), defined as groups of isolates that are genetically similar and can be identified through molecular or immune markers (Tibayrenc, 1998), with DTU I corresponding to *T. cruzi* line I and DTU IIb corresponding to *T. cruzi* line II, and sub-lines IIa and IIc-e associated with hybrid strains and those belonging to zymodeme 3 (Brisse et al., 2000). The distribution of haplotypes from five nuclear genes and one satellite DNA was analyzed in isolates that were representative of the six DTUs by net genealogy and Bayesian phylogeny. Such data indicated that DTUs *T. cruzi I* and *T. cruzi II* are monophyletic and the other DTUs have different combinations of *T. cruzi I* and *T. cruzi II* haplotypes and DTU-specific haplotypes (Tomazi et al., 2009; Ienne et al., 2010). One of the possible interpretations for this observation is that *T. cruzi I* and *T. cruzi II* are two different species and that DTUs II-IV are hybrid resulting from independent hybridization/genomic combination events (Zingales, 2011).

In this setting, the characterization of these parasites extracted from different hosts aim at helping clarify the biological meaning and repercussion of this variability for clinics and for Chagas disease epidemiology (Lainson et al., 1979). However, the great majority of studies performed are related to parasite populations belonging to TCI and TCII groups, with scarce works performed with Z3 group.

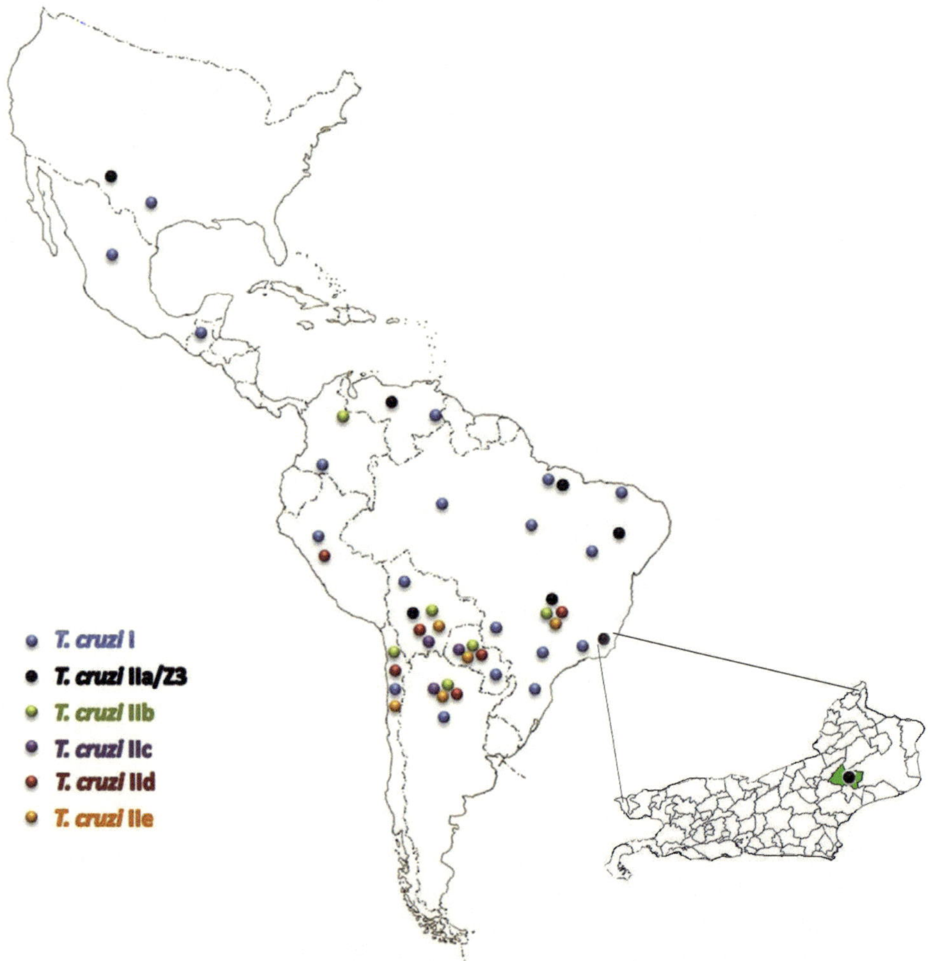

Fig. 3. General pattern of distribution of *T. cruzi* lineages and sublineages; the sylvatic isolates from Rio de Janeiro (extended map showing in green Triunfo, Santa Maria Madalena municipal district) were typed as *T. cruzi* IIa/Z3. (Adaptated map by Noireau F. Vet. Res. (2009)).

3.1 *T. cruzi* isolates from Rio de Janeiro

Therefore, this work was performed from *T. cruzi* samples isolated from *Triatoma vitticeps* (Figure 1) by Gonçalves in 2000, at Triunfo location, 2nd district of Santa Maria Madalena city, Rio de Janeiro state (Figure 2). Four hundred sixty five (465) *Triatoma vitticeps* specimens were collected: 294 females, 156 males, and 15 nymphs from five different areas:

area A, located at 250-meter altitude and 3.5 km distant from the district headquarters, very modified by deforestation for banana farming; area B, located at 130-meter altitude and 4 km distant from the headquarters, placed in a valley with preserved vegetation (secondary forest). These areas are 2-km distant to each other, separated by a mountain (Figure 3). Area C, the district headquarters, at 40-meter distance, was totally modified by pasture formation, and areas D and E were totally preserved and placed at 10 and 12-km distances from the headquarters, respectively. *T. cruzi* isolates used in this study were extracted from triatomines captured from areas A, B and F (Table 1). Area F was located in Vista Alegre, a city neighboring Conceição de Macabu, at Northern region of Rio de Janeiro State (Gonçalves, 2000).

Isolates (Samples)	Area	Host	Geographical origin
SMM10	A	Tv	Triunfo
SMM53	A	Tv	Triunfo
SMM88	A	Tv	Triunfo
SMM98	A	Tv	Triunfo
SMM36	B	Tv	Triunfo
SMM82	B	Tv	Triunfo
SMM1	F	HCD	Conceição de Macabu

SMM (Santa Maria Madalena)

Tv – *Triatoma vitticeps*; HCD (Haemoculture of the swiss mouse) – the parasites were inoculated in mice and was done haemoculture.

Table 1. *Trypanosoma cruzi* samples isolated from *Triatoma vitticeps* captured on the State of Rio de Janeiro, Brazil

Those *T. cruzi* samples isolated from *Triatoma vitticeps*, collected in Rio de Janeiro State, were classified by our group as Z3 based on mini-exon gene (Santos-Mallet et al., 2008) and showed great heterogeneity regarding growth curve and mouse virulence patterns (Silva, 2006), susceptibility to benznidazole (Sousa, 2009), total protein pattern and proteolytic activity profile (Gomes et al., 2006; Gomes et al., 2009). This heterogeneity observed in samples collected from the same region leads to questionings on how this diversity could influence the parasite-host cell interaction.

3.2 Molecular profile of *T. cruzi* isolates from Rio de Janeiro

The results obtained by means of molecular analysis revealed that the isolates have similar profiles, except for sample SMM1 (area F). Samples SMM10, SMM53, SMM88, SMM98 (area A), SMM36 and SMM82 (area B) revealed the presence of 150 bp, indicating that they belong to the zymodeme III group (Z3; Figure 4). Likewise, sample SMM1 from area F showed similarity to Z3 (150 bp), but also presented another band that may be related to the TcII profile (250 bp) and was very similar to the reference strain CL Brener (Figure 4). The phylogenetic position of Z3 has been much debated. According to some authors, the numerical taxonomy based on 24 isoenzymatic Z3 profiles is more closely associated with Z1 (TcII) than with Z2 (TcI) (Ready & Miles, 1980). However, other works place Z3 in an intermediate position between Z1 and Z2 (Stothard et al., 1998). Our study revealed one isolate (SMM1) with a hybrid profile associated with Z3 and TcII. This result may corroborate the hypothesis that this isolate is the product of a mixture of parasite populations, since the vector in wild environments may feed on several vertebrate hosts. This complexity was demonstrated in the State of Rio de Janeiro by Fernandes et al. (1999), who showed a preferential association of the two lineages of *T. cruzi* with different hosts. They suggest that the vector *T. vitticeps* is involved in the transmission cycle among mammals infected by lineage 2 in the municipality of Teresópolis, and in the transmission cycle of primates in municipality of Silva Jardim. The hybrid profile found in these samples may indicate a possibility that the vector *T. vitticeps* does not only participate in the wild cycle of the disease.

The main purpose of typing of isolates of *T. cruzi* is to identify strains with different epidemiological and/or clinical characteristics of Chagas disease. Our results corroborate other descriptions in the literature, and contribute to the knowledge and records of the profile of some additional wild isolates of *T. cruzi* in regions not yet affected by the disease. Added to the complexity observed between the isolates is the finding that the Z3 profile is divided into two groups, called Z3a and Z3b (Mendonça et al., 2002). Our laboratory is interested in investigating whether such a dichotomy occurs among the Z3 isolates obtained from *T. vitticeps* in this area of study.

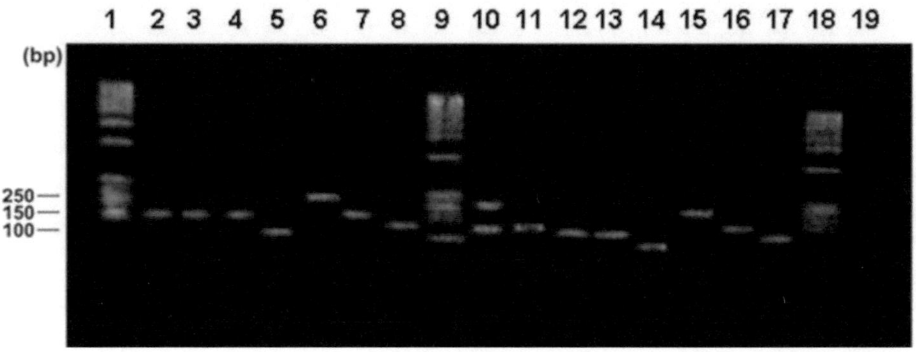

Fig. 4. PCR Multiplex - Mini-exon. The gel of agarose for electrophoresis was amplified using isolates of *Trypanosoma cruzi* of reference that possess approach bands of TCI, compared to TCII, Z3 and *Trypanosoma rangeli* and with *T. cruzi* sylvatics isolates from Rio de Janeiro. The isolates was performed using 25 ng of genomic DNA extracted using the phenol–chloroform method. Five primers were used: for Tc1 (5'-TTG CTC GCA CAC TCG GCT GCAT-3'), for Tc2 (5'-ACA CTT TCT GTG GCG CTG ATC G-3'), for Z3 (CCG CGW ACA ACC CCT MAT AAA AAT G-3'), for Tr (CCT ATT GTG ATC CCC ATC CCC ATC TTC G-3'), and for the mini-exon (5' TAC CAA TAT AGT ACAGAA ACT G-3'). Lane 1. Molecular weight marker (100bp DNA ladder), 2. SMM98, 3. SMM36, 4. SMM82, 5. *T. rangeli*, 6. CL Brener, 7. DM28c, 8. JJ, 9. Molecular weight marker (100bp DNA ladder), 10.SMM1, 11. SMM10, 12. SMM53, 13. SMM88, 14. *T. rangeli*, 15. CL Brener, 16. DM28c, 17. JJ, 18. Molecular weight marker (100bp DNA ladder), 19. negative control (no DNA added). bp = base pairs.

3.3 Proteolytic enzymes

Despite the existing knowledge of this flagellate genome and its main families of proteins, little is known about these parasites isolated from triatomines captured in the field, as well *T. cruzi* in mammals of wild origin. Proteolytic enzymes are reported to play an important role in determining the virulence of these microorganisms.

Proteases are essential for all life forms. They are involved in a multitude of physiological reactions, ranging from simple digestion of proteins for nutritional purposes, to highly-regulated metabolic cascades (e.g. proliferation and growth, differentiation, signaling and death pathways), and are essential for homeostatic control in both prokaryote and eukaryote cells (Rao et al., 1998). Proteases are also essential molecules in viruses, bacteria, fungi and protozoa, for their colonization, invasion, dissemination and evasion of host immune responses, mediating and sustaining the infectious disease process. Collectively, proteases participate in different steps of the multifaceted interaction events between microorganism and host structures, being considered as virulent attributes. Consequently, the biochemical characterization of these proteolytic enzymes is of interest not only for understanding proteases in general, but also for understanding their roles in microbial infections, and thus, their use as targets for rational chemotherapy of microbial diseases (Santos, 2010) (dos Santos, 2011).

Proteases are subdivided into two major groups, depending on their site of action: exopeptidases and endopeptidases. Exopeptidases cleave the peptide bond proximal to the amino (NH_2) or carboxyl (COOH) termini of the proteinaceous substrate, whereas endopeptidases cleave peptide bonds within a polypeptide chain. Based on their site of action at the NH_2 terminal, the exopeptidases are classified as aminopeptidases, dipeptidyl peptidases or tripeptidyl peptidases that act at a free NH_2 terminus of the polypeptide chain and liberate a single amino acid residue, a dipeptide or a tripeptide, respectively. Carboxypeptidases or peptidyl peptidases act at the COOH terminal of the polypeptide chain and liberate a single amino acid or a dipeptide (which can be hydrolyzed by the action of a dipeptidase). Carboxypeptidases can be further divided into three major groups: serine, metallo and cysteine carboxypeptidases, based on the functional group present at the active site of the enzymes. Similarly, endopeptidases are classified according to essential catalytic residues at their active sites in: serine, metallo, glutamic, threonine, cysteine and aspartic endopeptidases. Conversely, there are a few miscellaneous proteases that do not precisely fit into the standard classification (dos Santos, 2010, 2011).

Cysteine peptidases from parasitic protozoa have been characterized as factors of virulence and pathogenicity in several human and veterinary diseases. *T. cruzi* contains a major cysteine peptidase named cruzipain (also known as cruzain or GP57/51), which is present in different developmental forms of the parasite, although at variable levels (Dos Reis et al., 2006). Cruzipain is a papain-like peptidase that shares biochemical characteristics with both cathepsin L and cathepsin B (Cazzulo et al., 1990b). Cysteine peptidases have already been detected in many species of Trypanosomatidae, and are regarded as essential for the survival of several parasitic protozoa. The enzyme has been shown to be lysossomal, and is located in an epimastigote-specific pre-lysossomal organelle called the 'reservossome', which contains proteins that are digested during differentiation to metacyclic trypomastigotes (Soares et al., 1992). Some authors have suggested a second location of enzyme isoforms in the plasma membrane, associated with a glycosylphosphatidylinositol (GPI) anchor (Elias et al., 2008). These isoforms were present in epimastigotes, amastigotes and trypomastigotes, and reacted with polyclonal anti-cruzipain sera, thereby becoming an immunodominant antigen that is recognized by the sera of human patients with chronic Chagas disease (Martínez et al., 1991). Recently, the peptidase expression analysis of fresh field sylvatic isolated strains of *T. cruzi* showed a heterogeneous profile of cysteine proteolytic activities in the main phylogenetic groups TCI and TCII (Fampa et al., 2008).

Gomes et al (2009) investigated the production of peptidases, especially cruzipain, as well as the protein surface distribution in four newly sylvatic isolates of *T. cruzi* belonging to the Z3 genotype.

3.4 Proteolytic profile of *T. cruzi* isolates from Rio de Janeiro

The differences in peptidase expression between TCI and TCII phylogenetic groups have recently been investigated. Since *T. cruzi* isolates from sylvatic triatomines were included in the third phylogenetic group, named Z3, our investigation contributes to investigate the expression of surface polypeptides and the major cysteine peptidase from the Z3 parasite population, thereby furthering understanding on the genetic variability in the pathogenesis of Chagas disease. In this context, we carried out an identification of the protein profile and peptidase from epimastigotes (replicative forms of this parasite) of sylvatic isolates of *T. cruzi* (classified as Z3) from triatomines captured in Santa Maria Madalena (SMM) in the State of Rio de Janeiro. The separation of soluble whole proteins revealed a different protein profile, with approximately 35 polypeptides presenting apparent molecular masses from 118 to 25 kDa in all the samples. The proteolytic activity was determined by zymograms analysis of all the samples, using SDS-polyacrylamide gel electrophoresis containing gelatin as substrate. Our main results demonstrate a major band of 45 kDa sensible to E-64, a powerful cysteine peptidase inhibitor, in all the samples. In order to confirm this data, western blotting was performed using the anti-cruzipain polyclonal antibody. These findings showed a strong polypeptide band with an apparent molecular mass between 40 and 50 kDa in all the sylvatic isolates: SMM10; SMM53; SMM88 and SMM98 respectively and also Dm28c (Figure 5).

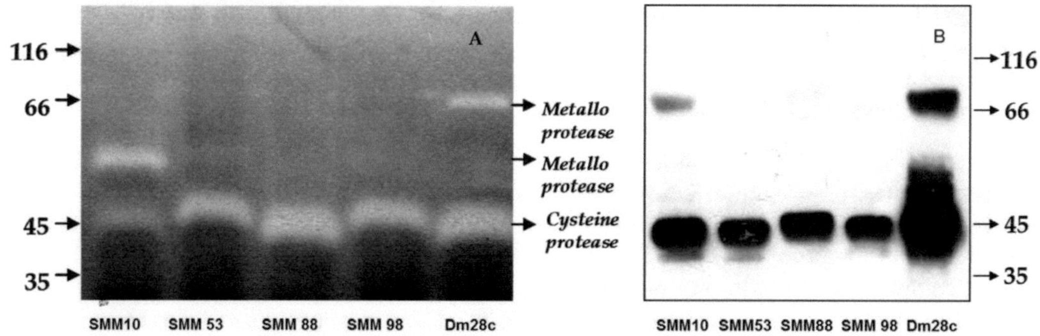

Fig. 5. A - Gelatin-SDS-PAGE showing the proteolytic activity profiles of *T. cruzi* sylvatic isolates. Parasites (SMM10, SMM53, SMM88, SMM98, and Dm28c) grown for 7 days were harvested and lysed by SDS. The gel was incubated in 50 mM sodium phosphate buffer, pH 5.5, supplemented with 2 mM DTT for 40 h at 37°C; B- Western blotting showing the reactivity of cellular polypeptides of *T. cruzi* sylvatic isolates with the anti-cruzipain polyclonal antibody. Numbers on the left indicate the relative molecular mass markers, expressed in kilodaltons.

These results show the presence of a main cysteine peptidase, cruzipain, in the sylvatic isolates of *T. cruzi* from Santa Maria Madalena, in the State of Rio de Janeiro (Gomes et al., 2009). We also observed another gelatinolyti activity of 66 kDa that was recognized by the anti-cruzipain antibody, probably a cruzipain isoform; since cruzipain is a high mannose-type glycoprotein containing about 10% carbohydrate, its molecular mass can be estimated from the sequence, considering two high-mannose oligosaccharide chains, as about 40 kDa. However, this enzyme can present anomalous behavior in SDS-PAGE, yielding apparent molecular mass values of 35 to 60 kDa depending on the experimental conditions. The cysteine peptidases from parasites, including *T. cruzi*, have proven to be valuable targets for chemotherapy. Due to the biological importance of cruzipain in the life cycle of *T. cruzi*, many studies have sought to build specific inhibitors against the active core of this enzyme, in order to obtain a new drug capable of providing protection against human infection by *T. cruzi*.

4. Conclusion

Trypanosoma cruzi shows considerable heterogeneity among populations isolated from sylvatic and domestic cycles. Despite of knowledge concerning the genome of these flagellated organisms and their main protein families, very little is known about these parasites isolated from triatomine bugs captured from field, as well as *T. cruzi* extracted from sylvatic mammals. In this context, we do hereby highlight the importance of molecular studies on *T. cruzi* sylvatic isolates collected by blood culture from vertebrate hosts and/or from triatomine vectors, *Triatoma vitticeps*, in Triunfo location, 2nd district of Santa Maria Madalena city, Northern region of Rio de Janeiro State, Brazil. The results of our investigations with *T. cruzi* samples isolated from sylvatic triatomine insects revealed that these parasites belong to a phylogenetic group called ZIII, and proteolytic analyzes evidenced the presence of a key peptidase cysteine, cruzipain, in all samples of sylvatic *T. cruzi* isolates from Santa Maria Madalena - Rio de Janeiro (Brazil), which was confirmed by anti-cruzipain antibody recognition. Taken together, our results can corroborate in understanding the role of proteolytic enzymes in determining the virulence of these microorganisms, as well as genetic variability of Z3 population in Chagas disease pathogenesis.

5. Acknowledgment

The authors would like to thank all the members of Setor de Entomologia Forense from Laboratório de Transmissores de Leishmanioses at Instituto Oswaldo Cruz- FIOCRUZ for the encouragement and help, especially to Prof. Catarina Macedo Lopes, who helped and made some figures of this chapter. The financial support CAPES, CNPq, FAPERJ and FIOCRUZ.

6. References

Aragão, M.B. & Souza, S.A.(1967). Triatoma infestans colonizado em domicílios da baixada fluminense, Estado do Rio de Janeiro, Brasil. *Revista da Sociedade Brasileira de Medicina Tropical*, Vol.5, No.1, (August 1971), ISSN 0037-8682.

Ávila-Pires, F.D. (1976). Ecology of small mammals in relation to sylvan and domestic transmission cycles. In new approaches in American tripanosomiasis research. *Pan American Health organization Scientific Publication*, Vol. 318, (March 1976), pp.301-306, ISSN 1020-4989.

Barnabé, C.; Brisse, S. & Tibayrenc, M. (1908). Population structure and genetic typing of *Trypanosoma cruzi*, the agent of Chagas disease: a multilocus enzyme electrophoresis approach. *Parasitology*, Vol.120, No.5, (May 2000), pp.513-526, ISSN 0031-1820.

Barretto, M.P. (1967). Estudos sobre reservatórios e vectores silvestres do *Trypanosoma cruzi*. XXII. Modificações dos focos naturais da tripanossomose americana e suas conseqüências. *Revista da Sociedade Brasileira de Medicina Tropical*, Vol.1, (November 1967), pp.167-173, ISSN 0037-8682.

Barretto, M.P. (1979). Epidemiologia. In: *Trypanosoma cruzi* e *doença de Chagas*, Brener, Z. and Andrade, Z.A. pp. 89-15, Guanabara Koogan ISBN 85-277-0563-X, Rio de Janeiro, Brasil.

Becerra-Fuentes, F.; Coura, J.R. & Ferreira, L.F. (1967). Observações sobre o *Triatoma rubrofasciata* (De Geer, 1773) no Estado da Guanabara. *Revista da Sociedade Brasileira de Medicina Tropical*, Vol. 5, (may 1971), pp. 47-54, ISSN 0037-8682.

Bogliolo, A.R.; Chiari, E.; Silva-Pereira, R.O. & Silva-Pereira, A.A. (1981). A comparative study of *Trypanosoma cruzi* enzyme polymorphism in South America. *Brazilian Journal of Medical and Biological Research*, Vol. 19, No.6, (jan 1986), pp.673-683, ISSN 1678-4510.

Brisse, S.; Barnabe, C. & Tibayrenc, M. (1971). Identification of six *Trypanosoma cruzi* phylogenetic lineages by random amplified polymorphic DNA and multilocus enzyme electrophoresis. *International Journal for Parasitology*, Vol.30, No.1, (January 2000), pp.35–44, ISSN 0020-7519.

Cazzulo, J.J.; Hellman, U., Couso, R., Parodi, A.J. (1980). Amino acid andcarbohydrate composition of a lysosomal cysteine proteinase from *Trypanosoma cruzi*. Absence of phosphorylated mannose residues. *Molecular and Biochemical Parasitology*, Vol.38, No.1, (January 1990b), pp.41–48, ISSN 0166-6851.

Chagas, C. (1909). Nova tripanossomíase humana: Estudos sobre a morfologia e o ciclo evolutivo do *Schizotrypanum cruzi* n. gen., n. sp., agente etiológico de nova entidade mórbida do homem. *Memórias do Instituto Oswaldo Cruz*, Vol.1, No.2, (August 1909), pp.159-218, ISSN 0074-0276.

Chagas, C. (1912). Sobre um *Trypanosoma* do tatu, *Tatusia novemcincta*, transmitido pelo *Triatoma geniculata* Latr (1811). Possibilidade de ser o tatu um depositário do *Trypanosoma cruzi* no mundo exterior. Nota prévia. *Braz Med*, Vol.26, (August 1912), pp.305-306.

Chagas, C. (1924). Infection naturelle dês singes du Pará (*Chrysotrix sciureus*) par *Trypanosoma cruzi*. *Comp Rend Séanc Soc Biol Ses Fin*, Vol.90, (February 1924), pp.873-876.

Coura J.R. (1967).Transmission of chagasic infection by oral route in the natural history of Chagas disease. *Revista da Sociedade Brasileira de Medicina Tropical*, Vol.39, No.Suppl 3, (November/December 2006), pp. 113-117, ISSN 0037-8682.

Coura, J.R. & Dias, J.C.P. (2009). Epidemiology, control and surveillance of Chagas disease - 100 years after its discovery. *Memórias do Instituto Oswaldo Cruz*, Vol.104, Suppl. 1, (July 2009), pp.31-40, ISSN 0074-0276.

Coura, J.R. (1909). Chagas disease: what is known and what is needed-a background article. *Memórias do Instituto Oswaldo Cruz*, Vol.102, Suppl. I, (October 2007), pp.113-122, ISSN 0074-0276.

Coura, J.R.; Ferreira, L.F.; Silva, F.R. (1959). Triatomíneos no Estado da Guanabara e suas relações com o domicílio humano. *Revista do Instituto de Medicina Tropical de São Paulo,* Vol.8, No.4, (July/August 1966), pp.162-166, ISSN 0036-4665.

Deane, M.P.; Lenzi, H.L. & Jansen, A.M. (1909). *Trypanosoma cruzi*: vertebrate and invertebrate cycles in the same mammals host, the opossum *Didelphis marsupialis*. *Memórias do Instituto Oswaldo Cruz,* Vol.79, No.4, (October/December 1984), pp.513-515, ISSN 0074-0276.

Devera, R.; Fernandes, O. & Coura, J.R. (1909). Should *Trypanosoma cruzi* be called 'cruzi' complex? A review of the parasite diversity and potential of selecting population after in vitro culturing and mice infection. *Memórias do Instituto Oswaldo Cruz,* Vol.98, No.1, (January 2003), pp.1-12, ISSN 0074-0276.

Dias, C.M.G.; Bezerra, I.O.; Moza, P.G.; Braga, J O.; Silva, J.L.; Santos, H.R.; Souza, M.S.B.; Fonseca Filho, M.; Pacheco, S R R.; Gonçalves, T.C.M.; Santos-Mallet, J.R.; Lopes, C.M. (2010). Investigação do risco de Doença de Chagas no município Trajano de Morais - Região Serrana do Estado do Rio de Janeiro. 26ª Reunião de Pesquisa Aplicada em Doença de Chagas/ 14ª Reunião de Pesquisa Aplicada em Leishmanioses. pp.17.

Dias, E. (1909). Observações sobre eliminação de dejeções e tempo de sucção em alguns triatomíneos sul-americanos. *Memórias do Instituto Oswaldo Cruz,* Vol. 54, No.1, (June 1956) pp.115-124, ISSN 0074-0276.

Dias, E. (1909). Presença do *Panstrongylus megistus* infectado por *Schizotrypanum* no Rio de Janeiro, DF. *Memórias do Instituto Oswaldo Cruz,* Vol.38, No.2, (April 1943), pp.177-180, ISSN 0074-0276.

Dias, J.C. (1967). Notes about of *Trypanosoma cruzi* and yours bio-ecology characteristics with agents of the transmission by meals. *Revista da Sociedade Brasileira de Medicina Tropical,* Vol.39, No.4, (Jul-Aug 2006), 370-375, ISSN 0037-8682.

Dias, J.C.P. (2000). Chagas disease control and the natural history of human Chagas disease: a possible interaction? *Memórias do Instituto Oswaldo Cruz,* Vol.95, Suppl. II, pp.14-20, ISSN 0074-0276.

Dias, J.C.P.; Feitosa, V.R.; Ferraz Filho, N.A.; Rodrigues, V.L.C.; Alencar, A.S. & Sessa, P.A. (1909). Fonte alimentar e potencial vetorial de *Triatoma vitticeps* (Stal, 1859), com relação à doença de Chagas humana no Estado do Espírito Santo, Brasil (Hemiptera, Reduviidae). *Memórias do Instituto Oswaldo Cruz,* Vol.84, Suppl IV, (November 1989) pp.165-173, ISSN 0074-0276.

Dias, J.C.P.; Silvera, A.C. & Schofield, C.J. (1909). The impact of Chagas disease control in Latin América. A Review. *Memórias do Instituto Oswaldo Cruz,* Vol.97, No.5, (July 2002), pp.603-612, ISSN 0074-0276.

Diotaiuti, L.; Bronfen, E.; Perilo, M.M.; Machado, G.B.N. & Loiola, C.F. (1967). Aspectos do comportamento do *Triatoma vitticeps* na transmissão da doença de Chagas. *Revista da Sociedade Brasileira de Medicina Tropical,* Vol.20, Suppl., (February 1987), pp.87, ISSN 0037-8682.

Dos Reis, F.C.G.; Júdice, W.A.F.; Juliano, M.A.; Juliano, J.; Scharfstein, J.; Lima, A.P.C.A. (1977). The substrate specificity of cruzipain 2, a cysteine protease isoform from *Trypanosoma cruzi*. *FEMS Microbiology Letters,* Vol.259, No.2, (June 2006), pp.215-220, ISSN 0378-1097.

Dos Santos A.L. (2009) HIV aspartyl protease inhibitors as promising compounds against Candida albicans. *World Journal of Biological Chemistry,* Vol.26, No.1(2), (February 2010), pp.21-30, ISSN 1949-8454.

Dos Santos A.L. (2010). Protease expression by microorganisms and its relevance to crucial physiological/pathological events. *World Journal of Biological Chemistry,* Vol.26, No.2, (March 2011), pp.48-58, ISSN 1949-8454.

Elias, C.G.; Pereira, F.M.; Dias, F.A.; Silva, T.L.; Lopes, A.H.; d'Avila-Levy, C.M.; Branquinha, M.H.; Santos, A.L. (1951/52). Cysteine peptidases in the tomato trypanosomatid *Phytomonas serpens*: influence of growth conditions, similarities with cruzipain and secretion to the extracellular environment. *Experimental Parasitology,* Vol.120, No.4, (December 2008), pp.343-52, ISSN 0014-4894.

Fampa, P.; Lisboa, C.V.; Jansen, A.M.; Santos, A.L.S.; Ramirez, M.I. (1908). Protease expression analysis in recently field-isolated strains of *Trypanosoma cruzi*: a heterogeneous profile of cysteine protease activities between TC I and TC II major phylogenetic groups. *Parasitology,* Vol.135, No.9, (August 2008), pp.1093–1100, ISSN 0031-1820.

Fernandes, O. Análise da estrutura primária do gene de mini-exon em diferentes tripamosomatídeos e sua utilização como marcador molecular. (PhD Thesis, Instituto Oswaldo Cruz, 1996), 268p.

Fernandes, O.; Mangia, R.H.; Lisboa, C.V.; Pinho, A.P.; Morel, C.M.; Zingales, B.; Campbell, D.A. & Jansen, A.M. (1908). The complexity of the sylvatic cycle of *Trypanosoma cruzi* in Rio de Janeiro state (Brazil) revealed by the non-transcribed spacer of the mini-exon gene. *Parasitology,* Vol.118, No. Pt2, (February 1999), pp.161–166, ISSN 0031-1820.

Fernandes, O.; Santos, S.S.; Cupolillo, E.; Mendonça, B.; Derre, R.; Junqueira, A.C.; Santos, L.C.; Sturm, N.R.; Naiff, R.D.; Barret,T.V.; Campbell, D.A. & Coura, J.R. (1908). A mini-exon multiplex polymerase chain reaction to distinguish the major groups of *Trypanosoma cruzi* and *T. rangeli* in the Brazilian Amazon. *Transactions of the Royal Society of Tropical Medicine and Hygiene,* Vol.95, No.1, (January/February 2001), pp.97-99, ISSN 0035-9203.

Fernandes, O.; Souto, R.P.; Castro, J.A.; Pereira, J.B.; Fernandes, N.C.; Junqueira, A.C.; Naiff, R.D.; Barrett, T.V.; Degrave, W.; Zingales, B.; Campbell, D.A. & Coura, J.R. (1952). Brazilian isolates of *Trypanosoma cruzi* from humans and triatomines classified into two lineages using mini-exon and ribosomal RNA sequences. *American Journal of Tropical Medicine and Hygiene,* Vol.58, No.6, (June 1998), pp.807-811, ISSN 0002-9637.

Ferreira, E.; Souza, P,S.; Fonseca Filho, M. & Rocha, I. (1986). Nota sobre a distribuição do *Triatoma vitticeps* (Stal, 1859), (Hemiptera, Reduviidae), no Estado do Rio de Janeiro, Brasil. *Revista Brasileira de Malariologia e Doenças Tropicais,* Vol.38, (1986), pp.11-14, ISSN 0034-7256.

Ferreira, M.S.; Lopes, E.R.; Chapadeiro, E.; Dias, J.C.P. & Luquetti, A.O. (1996). Doença de Chagas. In: *Tratado de Infectologia* (9ª ed.), Veronesi, R. & Foccacia, R., pp.1175-1213, Atheneu, ISBN 85737, São Paulo, Brasil.

Freitas, J.M.; Augusto-Pinto, L.; Pimenta. J.R.; Bastos-Rodrigues, L.; Goncalves, V.F.; Teixeira, S.M.; Chiari, E.; Junqueira, A.C.; Fernandes, O.; Macedo, A.M.; Machado, C.R. & Pena, S.D. (2005). Ancestral genomes, sex, and the population structure of

Trypanosoma cruzi. *PLoS Pathogens*, Vol.2, No.3, (March 2006), pp.e24, ISSN 1553-7366.

Garcia, E.S. & Azambuja, P. (2000). Fisiologia de triatomíneos: desenvolvimento, reprodução e interação com o *Trypanosoma cruzi*. In: *Trypanosoma cruzi e Doença de Chagas* (2 ed.), Brener, Z.; Andrade, Z.A. & Barral-Neto, M., pp.41-47, Guanabara Koogan, ISBN 85-277-0563-X, Rio de Janeiro, Brasil.

Gardner, G. (1942). *Viagens no Brasil, principalmente nas províncias do norte e nos distritos do ouro e do diamante durante os anos de 1836-1842* (2 ed.), EDITORA Nacional, ISBN 978-85-232-0587-4, São Paulo, Brasil.

Gaunt, M.W.; Yeo, M.; Frame, I.A.; Stothard, J.R.; Carrasco, H.J.; Taylor, M.C.; Solis Mena, S.; Veazy, P.; Miles, G.A.J.; Acosta, N.; Rojas de Arias, A. & Miles, M.A. (1869). Mechanism of genetic exchange in American trypanosomes. *Nature*, Vol.421, (February 2003), pp. 936–939, ISSN 0028-0836.

Gomes, S.A.O.; Fonseca de Souza, A.L.; Silva, B.A.; Kiffer-Moreira, T.; Santos-Mallet J.R.; Santos A.L.S.; Meyer-Fernandes J.R. (1951/52). *Trypanosoma rangeli*: differential expression of cell surface polypeptides and ecto-phosphatase activity in short and long epimastigote forms. *Experimental Parasitology*, Vol.112, No.4, (April 2006), pp. 253–262, ISSN 0014-4894.

Gomes, S.A.O.; Misael, D.; Silva, B.A., Feder, D.; Silva, C.S; Gonçalves, T.C.M.; Santos, A.L.S. and Santos-Mallet, J.R. (1987). Major cysteine protease (cruzipain) in Z3 sylvatic isolates of *Trypanosoma cruzi* from Rio de Janeiro, Brazil. *Parasitology Research*, Vol.105, No.3, (September 2009), ISSN 1432-1955.

Gonçalves, T.C.M. Aspectos ecológicos de *Triatoma vitticeps* (Stal, 1859) (Hemiptera, Reduviidae), com caracterização das amostras de *Trypanosoma cruzi* Chagas, 1909 (Kinetoplastida, Trypanosomatidae) isoladas desse triatomíneo, no município de Santa Maria Madalena, Estado do Rio de Janeiro. (PhD Thesis, Instituto Oswaldo Cruz, 2000), 125 p.

Gonçalves, T.C.M.; Oliveira, Edson.; Dias, L.S.; Almeida, M.D.; Nogueira, W.O. & Ávila-Pires, F.D. (1909). An Investigation on the Ecology of *Triatoma vitticeps* (Stal, 1859) and its Possible Role in the Transmission of *Trypanosoma cruzi*, in the Locality of Triunfo, Santa Maria Madalena Municipal District, State of Rio de Janeiro, Brazil. *Memórias do Instituto Oswaldo Cruz*, Vol.93, No.6, (November/December 1998), pp.711-717, ISSN 0074-0276.

Gonçalves, T.C.M.; Victorio, V.M.N.; Jurberg, J. & Cunha, V. (1909). Biologia do *Triatoma vitticeps* (Stal, 1859) em condições de laboratório (Hemiptera: Reduviidae: Triatominae). I. Ciclo evolutivo. *Memórias do Instituto Oswaldo Cruz*, Vol.83, No.4, (October/December 1988), pp.519:523, ISSN 0074-0276.

Gonçalves, T.C.M.; Victorio, V.M.N.; Jurberg, J. & Cunha, V. (1909). Biologia do *Triatoma vitticeps* (Stal, 1859) em condições de laboratório (Hemiptera: Reduviidae: Triatominae). II. Resistência ao jejum. *Memórias do Instituto Oswaldo Cruz*, Vol.84, No.1, (January/March 1989), pp.131-134, ISSN 0074-0276.

Guimarães, F.N. & Jansen, G. (1909). Um foco potencial de tripanosomiase americana na cidade do Rio de Janeiro (Distrito Federal). *Memórias do Instituto Oswaldo Cruz*, Vol.39, No.3, (December 1943), pp.405-420, ISSN 0074-0276.

Heitzmann-Fontenelle, T.J. (1976). Bionomia comparativa de triatomíneos. IV. *Triatoma vitticeps* (Stal, 1859) (Hemiptera, Reduviidae). *Ecossistema*, Vol.5, No.1, (August 1980), pp.39-46, ISSN: 0100-4107.

Hoare, C.A. (1964). Morphological and taxonomic studies on mammalian trypanosomas. X: Revision of the systematics. *Journal of Protozoology*, Vol.11, (May 1964), pp.200-207, ISSN 0022-3921.

Ienne, S.; Pedroso, A.; Carmona e Ferreira, R.; Briones, M.R.S. & Zingales, B. (2001). Network genealogy of 195-bp satellite DNA supports the superimposed hybridization hypothesis of *Trypanosoma cruzi* evolutionary pattern. *Infection, Genetics and Evolution*, Vol.10, No.5, (July 2010), pp.601–606, ISSN 1567-1348.

Jansen, A.M.; Pinho, A.P.S.; Lisboa, C.V.; Cupolillo, E.; Mangia, R.H. & Fernandes, O. (1909). The sylvatic cycle of *Trypanosoma cruzi*: a still unsolved puzzle. *Memórias do Instituto Oswaldo Cruz*, Vol.94, Suppl I, (September 1999), pp.203-204, ISSN 0074-0276.

Lainson, R.; Shaw, J.J.; Fraiha, H.; Miles, M.A. & Drapes, C.C. (1908). Chagas' disease in the Amazon basin: I. *Trypanosoma cruzi* infections in silvatic mammals, triatomine bugs and man in state of Pará, north Brazil. *Transactions of the Royal Society of Tropical Medicine and Hygiene*, Vol.73, No.2, (March 1979), pp.193-204, ISSN 0035-9203.

Lent, H. & Wygodzinsky, P. (1979). Revision of the Triatominae (Hemiptera: Reduviidae), and their significance as vectors of Chagas' disease. *Bulletin of the American Museum of Natural History*, Vol.163, No.3, (July 1979) pp.123-520, ISSN 0003-0090.

Lent, H. (1942). Transmissores da moléstia de Chagas no estado do Rio de Janeiro. *Rev F de Medicina*, Vol.6, pp.3-13 ISSN 0036-4665.

León, L.A. (1962). Contribución a la historia de los transmissores de la enfermedad de Chagas (Del siglo XVI a XIX).In: *Anais Congresso Internacional sobre doença de Chagas* 3, Eds. pp.761- 770, Editora, ISBN 00029637, Rio de Janeiro, Brasil.

Lewis, M.D.; Llewellyn, M.S.; Gaunt, M.W.; Yeo, M.; Carrasco, H.J., & Miles, M.A. (1971). Flow cytometric analysis and microsatellite genotyping reveal extensive DNA content variation in *Trypanosoma cruzi* populations and expose contrasts between natural and experimental hybrids. *International Journal for Parasitology*, Vol.39, No.12, (October 2009) pp.1305–1317, ISSN 0020-7519.

Lisbôa, C.V.; Mangia, R.H.; Menezes-Trajano, V.; Ivo, A.; Nehme, N.S.; Morel, C.M. & Jansen AM (1909). Ecological aspects of the circulation of *Trypanosoma cruzi* in the sylvan environment. *Memórias do Instituto Oswaldo Cruz*, Vol.91, Suppl, pp.279, (November 1996) ISSN 0074-0276.

Lopes, C.M.; Mallet, J.R.S.; Ramos, L.B.; Giordano, C.; Silva, J.L.; OLIVEIRA, M. L. R.; Misael, D.S.; Silva, C.S.; Filho, M.F.; Rodrigues, M.L.J. & Gonçalves, T.C.M. (2009). Visitação e Colonização do Ambiente Domiciliar, por populações de Triatomineos em Regiões do Estado do Rio de Janeiro. In: *XXV Reunião de Pesquisa Aplicada em Doença de Chagas*. pp. 09, Editora UFTM- Uberaba, MG, Brasil.

Martinez, J.; Campetella O.; Frasch, A.C.; Cazzulo J.J. (1970). The major cysteine proteinase (cruzipain) from *Trypanosoma cruzi* is antigenic in human infections. *Infection and immunity*, Vol.59, No.11, (November 1991), pp.4275-4277, ISSN 0019-9567.

Martins, A.; Versiani, V. & Tupinambá, A. (1909). Estudos sobre a tripanosomíase americana em Minas Gerais. *Memórias do Instituto Oswaldo Cruz*, Vol.35, No.2, (July 1940), pp.286-301, ISSN 0074-0276.

Mendonça, M.B.A.; Nehme, N.S.; Santos, S.S.; Cupolillo, E.; Vargas, N.; Junqueira, A.; Naiff, R.D.; Barrett, T.V.; Coura, J.R.; Zingales, B. & Fernandes, O. (1908). Two main clusters within *Trypanosoma cruzi* zymodeme III are defined by distinct regions of the ribosomal RNA cistron. *Parasitology*, Vol.124, No.Pt2, (February 2002), pp.177–184, ISSN 0031-1820.

Miles, M.A. & Cibulkis, R.E. (1985). Zymodeme characterization of *Trypanosoma cruzi*. *Parasitology Today*, Vol.2, No.4, (January 1986), pp.94-97, ISSN 0169-4758.

Miles, M.A. (1985). *Trypanosoma cruzi*: analysis of isozymes and antigenic expression. *Annales de la Societe Belge de Medecine Tropical*, Vol.65, Suppl I, (March-April 1985), pp.67-69, ISSN 0365-6527.

Miles, M.A., Souza, A.A., Póvoa, M., Shaw, J.J., Lainson, R., Toye, P.J. (1869). Isozymic heterogeneity of *Trypanosoma cruzi* in the first autochthonous patients with Chagas' disease in Amazonian Brazil. *Nature*, Vol.272, (April 1978), pp.819-821, ISSN 0028-0836.

Miles, M.A.; Cedillos, R.A.; Povoa, M.M.; de Souza, A.A.; Prata, A. & Macedo, V. (1823). Do Radically dissimilar *Trypanosoma cruzi* strains (zymodemes) cause Venezuelan and Brazilian forms of Chagas' disease? *Lancet*, Vol.20, No.1, (June 1981b), pp.1338-1340, ISSN 0140-6736.

Miles, M.A.; Lanham, S.M.; Souza, A.A. & Póvoa, M. (1980). Further enzymic characters of *Trypanosoma cruzi* and their evalution for strain identification. *Transactions of the Royal Society of Tropical Medicine and Hygiene*, Vol.74, No.2, pp.221-237, ISSN 0035-9203.

Miles, M.A.; Llewellyn, M.S.; Lewis, M.D.; Yeo, M.; Baleela, R.; Fitzpatrick, S.; Gaunt, M.W. & Mauricio, I.L. (1908). The molecular epidemiology and phylogeography of *Trypanosoma cruzi* and parallel research on *Leishmania*: looking back and to the future. *Parasitology*, Vol.136, No.12, (October 2009), pp.1509-1528, ISSN 0031-1820.

Miles, M.A.; Póvoa, M.; Souza, A.A.; Lainson, R.; Shaw, J.J. & Ketteridge, D.S. (1908). Chagas'disease in the Amazon Basin: II. The distribuition *of Trypanosoma cruzi* zymodemes 1 and 3 in Pará State, north Brazil. *Transactions of the Royal Society of Tropical Medicine and Hygiene*, Vol.75, No.5, (May 1981a), pp.667-674, ISSN 0035-9203.

Miles, M.A.; Toye, P.J.; Oswald, S.C. & Godfrey, D.G. (1908). The identification by isoenzyme patterns of two district strain-groups of *Trypanosoma cruzi,* circulating independently in a rural area of Brazil. *Transactions of the Royal Society of Tropical Medicine and Hygiene*, Vol.71, No.3, (July 1977), pp.217-225, ISSN 0035-9203.

Monteiro, F.A.; Barrett, T.V.; Fitzpatrick S.; Cordón-Rosales, C.; Feliciangeli, D. & Beard CB. (1952). Molecular phylogeography of the Amazonian Chagas disease vectors *Rhodnius prolixus* and *R. robustus*. *Molecular Ecology*, Vol.12, No.4, (April 2003), pp.997-1006.

Monteiro, F.A.; Wesson, D.M.; Dotson, E.M.; Schofield, C.J. & Beard CB. (2000). Phylogeny and molecular taxonomy of the Rhodniini derived from mitochondrial and nuclear DNA sequences. *American Journal of Tropical Medicine and Hygiene*, Vol.62, No.4, (April 2000), pp.460-465.

Neiva, A. (1914). Presença em uma localidade do Estado do Rio de um novo transmissor da moléstia de Chagas encontrado infectado em condições naturais (Nota prévia). *Brasil Med*, Vol.28, pp.333-335.

Noireau, F.; Diosque, P.; Jansen A.M. (2009). Trypanosome cruzi: adaptation to its vectors and its hosts. Vet. Res. Vol.40, No.26 (February 2009), pp. 1 to 23, ISSN 0928 4249

Oliveira, M.L.R.; Lopes, C.M.; Gonçalves, T.C.M.; Mallet, J.R.S.; Misael, D.S.; Silva, A.H.R. & Duarte, R. (2010). Determinação de Fontes Alimentares de *Triatoma vitticeps* nas regiões serrana, norte e noroeste do estado do Rio de Janeiro. In: *XXVI Reunião de Pesquisa Aplicada em Doença de Chagas*. pp. 12 , Editora UFTM, Uberaba, Brasil.

Pinho, A.P.; Gonçalves, T.C.M.; Mangia, R.H.; Russel, N.N. & Jansen, A.M. (1909). The occurrence of *Rhodnius prolixus* Stal, 1859, naturally infected by *Trypanosoma cruzi* in the State of Rio de Janeiro, Brazil (Hemiptera, Reduviidae, Triatominae). *Memórias do Instituto Oswaldo Cruz*, Vol.93, No.2, (March/April 1998), pp.141-143, ISSN 0074-0276.

Pinto, C. (1931). Valor do rostro e antenas na caracterização dos gêneros de Triatomídeos. Hemíptera, Reduvioidea. *Boletim de Biologia*, Vol.19, pp.45-136, ISSN 00063185.

Rao, M.B.; Tanksale, A.M.; Ghatge, M.S.; Deshpande, V.V. (1997). Molecular and biotechnological aspects of microbial proteases. Microbiology and Molecular Biology Reviews, Vol.62, No.3, (September 1998), pp. 597-635, ISSN 1098-5557.

Ready, P.D. & Miles, M.A. (1908). Delimitation of *Trypanosoma cruzi* zymodemes by numerical taxonomy. *Transactions of the Royal Society of Tropical Medicine and Hygiene*, Vol.74, No.2, (June 1980), pp.38–242, ISSN 0035-9203.

Santos, A. L. S. (2010). Aspartic peptidase inhibitors as potential bioctive pharmacological compounds against human fungal pathogens. In: Combating Fungal Infection: Problems and Remedy, pp.289, ed Springer Heidelberg Dordrechithondon, ISBN 978-3-642-12172-2, e-ISBN 978-3-642-12173-9, New York.

Santos, U.M.; Pinto, A.F.S.; Almeida, A.Z.; Zaganelli, F.L.; Carrancho, P.V.; Netto, N.A. (1967). Doença de Chagas no estado do Espírito Santo. III: Vetores do *Trypanosoma*. *Revista da Sociedade Brasileira de Medicina Tropical*, Vol.3, No.1, (December 1969), pp.51-52, ISSN 0037-8682.

Santos-Mallet, JR, Silva CS, Gomes SAO, Oliveira DL, Santos CL, Sousa DM, Oliveira LR, Pinheiro NL, Gonçalves TCM (2008). Molecular characterization of *Trypanosoma cruzi* sylvatic isolates from Rio de Janeiro, Brazil. Parasitology Research (June 2008) volume 103, No. 5, pp.1041-1045, ISSN: 1432-1955.

Sessa, P.A. & Carias, V.R.D. (1967). Infecção natural de triatomíneos do Espírito Santo por flagelados morfologicamente semelhantes ao *Trypanosoma cruzi*. *Revista da Sociedade Brasileira de Medicina Tropical*, Vol.19, No.2, (April/June 1986), pp.99-100, ISSN 0037-8682.

Silva C.S. Estudo morfobiológico e histopatológico de amostras silvestres de *Trypanosoma cruzi* isoladas de *Triatoma vitticeps* (Stal, 1959) no estado do Rio de Janeiro. (Master Thesis, Universidade Federal Rural do Rio de Janeiro, 2006), 85p.

Silva, I.G. Influência da temperatura na biologia de 18 espécies de triatomíneos (Hemiptera: Reduviidae) e no xenodiagnóstico. (MSC Thesis, Universidade Federal do Paraná, Curitiba, 1985), 169p.

Silveira, A.C.; Alencar, T.A. & Máximo, M.H.C. (1983). Sobre o *Triatoma vitticeps* (Stål, 1859), no estado do Espírito Santo, Brasil. In: *X Reunião Anual Pesquisa Básica doença de Chagas*, Caxambu, Brasil. p.58.

Silveira, A.C.; Sakamoto, T.; Faria Filho, O.F. & Gil, H.S.G. (1949). Sobre o foco de triatomíneos domiciliados na baixada fluminense. *Revista Brasileira de Malariologia e Doenças Tropicais*, Vol.34, (mar 1982), pp.50-58.

Soares, M.J.; Souto-Padrón, T.; De Souza W. (1966). Identification of a large pre-lysosomal compartment in the pathogenic protozoon *Trypanosoma cruzi*. *Journal of cell science*, Vol.102, No.Pt 1, (May 1992), pp.157–167, ISSN 0021-9533.

Sousa D.M. Aspectos morfobiológicos, bioquímicos e susceptibilidade ao benznidazol de isolados silvestres de *Trypanosoma cruzi*. (Master Thesis, Instituto Oswaldo Cruz, 2009), 85p.

Sousa, M.A. Tripanosomatídeos de insetos e plantas: análise do crescimento, diferenciação e divisão celular, biometria e fenômenos sugestivos de sexualidade. Valor taxonômico. (PhD Thesis, Instituto Oswaldo Cruz, 2000), 161 pp.

Souto, R. P.; Fernandes, O.; Macedo, A. M.; Campbell, D. A.; Zingales, B. (1980). DNA markers define two major phylogenetic lineages of *Trypanosoma cruzi*. *Molecular and Biochemical Parasitology*. Vol.83, No.2, (December 1996), pp.141-152, ISSN 0166-6851.

Stothard, J.R.; Frame, I.A.; Carrasco, H.J. & Miles, M.A. (1908). On the molecular taxonomy of *Trypanosoma cruzi* using riboprinting. *Parasitology*, Vol.117, Pt3, (September 1998), pp.243–247, ISSN 0031-1820.

Sturm, N.R., Campbell, D.A. (2009). Alternative lifestyles: The population structure of *Trypanosoma cruzi*. *Acta Tropica*;Vol.115, No.1-2, (July/August 2010), pp.35-43, ISSN 0001-706X.

Talice, R.V.; Costa, R.S.; Rial, B. & Osimani, J.J. (1940). Los 100 primeros casos agudos confirmados de Enfermedad de Chagas (Triapanosomiasis americana) en el Uruguay. A. Monteverde y Cia., ISBN 0066-4170, Montevideo.

Tibayrenc, M. & Ayala, F.J. (1947). Isoenzyme variability in *Trypanosoma cruzi*, the agent of Chagas disease: genetical, taxonomical and epidemiological significance. *Evolution*, Vol.42, No.2, (March 1988), pp.277-292, ISSN 0014-3820.

Tibayrenc, M. (1971). Genetic epidemiology of parasitic protozoa and other infectious agents: the need for an integrated approach. *International Journal for Parasitology*, Vol.28, No.1, (January 1998), pp.85-104, ISSN 0020-7519.

Tibayrenc, M.; Ward, P.; Moya, A. & Ayala, F.J. (1915). Natural populations of *Trypanosoma cruzi*, the agent of Chagas disease, have a complex multiclonal structure. *Proceedings of the national academy of sciences of the United States of America (PNAS)*, Vol.83, No.1, (January 1986), pp.115-119, ISSN 0027-8424.

Tomazi, L.; Kawashita, S. Y.; Pereira, P. M.; Zingales, B. & Briones, M.R.S. (2002). Haplotype distribution of five nuclear genes based on network genealogies and Bayesian inference indicates that *Trypanosoma cruzi* hybrid strains are polyphyletic. *Genetics and Molecular Research*, Vol.8, No.2, (April 2009), pp.458-476, ISSN 1676-5680.

Urbina, J.A. (1909). Parasitological cure of Chagas disease: is it possible? Is it relevant? *Memórias do Instituto Oswaldo Cruz*, Vol.94, Suppl I, (September 1999), pp.49-355, ISSN 0074-0276.

Vikerman, K. (1943/44). Development cycle and biology of pathogenic trypanosomes. *British Medical Bulletin*, Vol.41, No.2, (April 1985), pp.105-114, ISSN 0007-1420

Wallace, F.G. (1951/52). The trypanosomatid parasites of insects and arachnids. *Experimental Parasitology*, Vol.18, No.1, (February 1966), pp.124-193, ISSN 0014-4894.

WHO. Expert Commitee on the Control of Chagas disease (2000: Brasilia, Brazil). Control of Chagas disease: Second Report of the WHO expert committee (WHO technical report series; 905) (2002), pp. 1-120, ISBN 9241209054.

Zeledón, R.; Alvarenga, N.J. & Schosinsky. (1977). Ecology of *Trypanosoma cruzi* in the insect vector. In new approaches in American trypanosomiasis research. *Pan American Health Organization Scientific Publication*, Vol. 347, (july 1977)pp.59-70, ISSN 1020-4989.

Zingales, B (2011). *Trypanosoma cruzi*: um parasita, dois parasitas ou vários parasitas da doença de Chagas? *Trypanosoma cruzi*: one parasite, two parasites or several parasites of Chagas disease? Revision. *Revista da Biologia*, Vol.6b, (June 2011) pp.44-48, ISSN 1984-5154.

Zingales, B.; Andrade, S.G.; Briones, M.R.S.; Campbell, D.A.; Chiari, E.; Fernandes, O.; Guhl, F.; Lages-Silva, E.; Macedo, AM.; Machado, CR.; Miles, M.A.; Romanha, A.J.; Sturm, N.R.; Tibayrenc, M. & Schijman, A.G. (1909). A new consensus for *Trypanosoma cruzi* intraspecific nomenclature: second revision meeting recommends TcI to TcVI. *Memórias do Instituto Oswaldo Cruz*, Vol.104, No.7, (November 2009), pp.1051-1054, ISSN 0074-0276.

Zingales, B.; Stolf, B.S.; Souto, R.P.; Fernandes, O. & Briones, M.R. (1909) Epidemiology, biochemistry and evolution of *Trypanosoma cruzi* lineages based on ribosomal RNA sequences. *Memórias do Instituto Oswaldo Cruz*, Vol.94, Suppl I, (September 1999), pp.159-164, ISSN 0074-0276.

10

Pulsed Field Gel Electrophoresis in Molecular Typing and Epidemiological Detection of Methicillin Resistant *Staphylococcus aureus* (MRSA)

Velazquez-Meza Maria Elena[1]*, Vázquez-Larios Rosario[2], Hernández Dueñas Ana Maria[2] and Rivera Martínez Eduardo[2]
[1]*Instituto Nacional de Salud Pública, Cuernavaca Morelos*
[2]*Instituto Nacional de Cardiología "Dr. Ignacio Chávez"*
México D. F.

1. Introduction

Methicillin-resistant *Staphylococcus aureus* (MRSA) is an important threat to hospitalized patients worldwide and is responsible for a wide range of human diseases, including septicemia, endocarditis, pneumonia, osteomyelitis, toxic shock syndrome, and bacteremia (Tenover & Gaynes, 2000). This species nevertheless represents a serious public health burden, particularly the clones which are resistant to methicillin and other classes of antibiotics; the emergence of penicillin-methicillin-, and recently high-level vancomycin-resistant strains emphasize the importance and urgency of such rational prescribing policy for the treatment of MRSA infections (Appelbaum, 2007; Goldstein 2007). Multiple studies have shown clonal spreads of epidemic MRSA strains within hospitals, between hospitals within a country (Breurec et al., 2011; Nübel et al., 2010), and also between countries and continents (Breurec et al., 2011; Deurenberg et al., 2009; Diekema et al. 2001). There are only a limited number of nosocomial MRSA clones spread worldwide (the Iberian [ST247-SCC*mec* I], the Brazilian [ST239-IIIA], the Hungarian [ST239-III], the New York/Japan [ST5-II], the Pediatric [ST5-VI], the Berlin [ST45-IV], EMRSA-15 [ST22-IV], and the EMRSA-16 [ST36-II] clones) (Enright et al., 2002; Oliveira et al., 2001).

Molecular typing of MRSA is used to support infection control measures. Although Pulsed-field gel electrophoresis (PFGE) is well known and considered as golden standard, for establishing clonal relationships at the local level, its detection capacity seems to make it also too discriminative for global comparisons (McDougal et al.2003; Murchan et al., 2003). Recently multilocus sequence typing (MLST) has been proven to be the most adequate method both for long-term and global epidemiologic studies and for population genetic studies. Typing methods based on sequencing of more stable housekeeping genes (MLST) allow the creation of Internet-based curate databases and inter-laboratory data exchange

* Corresponding Author

(Enright et al., 2000) The combination of these methods allows the unambiguous assignment of collections of MRSA isolates or new MRSA clones (Enright et al., 2000).

The prevalence of MRSA in Mexico differs widely from one hospital to another and according to different studies performed; an increasing frequency of MRSA (7% in 1989, 14% in 2001 and 36% in 2004) are documented by reports of routine oxacillin disk diffusion tests only (Alpuche et al., 1989; Calderón et al., 2002; Chávez, 2004). This is of great concern, because it is a common experience that once MRSA is introduced in a hospital it is difficult to eradicate it (Creamer et al., 2011; Rebmann & Aureden, 2011). However, reports from Mexico documenting the clonality of MRSA isolates are very scarce, Aires de Sousa *at al.* in 2001 (Aires de Sousa et al., 2001) reported dominant and unique MRSA clone designated the Mexican clone (I::NH::M), identified by PFGE among isolates collected in 1997, 1998 from a pediatric hospital in Mexico, which had a rather limited resistance profile. In more recent studies which involve strains collected for the period 1997 to 2003 in two Mexican hospitals, PFGE distributed the MRSA isolates into two types M (clone EMRSA-16-U.K) and C (clone New York/Japan) these two clones were distinguished by antibiogram and other molecular properties (Echaniz et al., 2006; Velazquez et al., 2004).

The aim of this study was to identify MRSA clones circulating in a tertiary care hospital in Mexico City and their prevalence in the course of time 2002-2009. For this purpose, we used a phenotypic characterization and a combination of different molecular typing methods, including PFGE, hybridization with a Tn*554* and *mecA* probes, staphylococcal cassette chromosome *mec* (SCC*mec*) and MLST.

2. Material and methods

2.1 Hospital setting

The Instituto Nacional de Cardiologia "Dr. Ignacio Chavez" (CAR) is a tertiary-care cardiology hospital located, in Mexico City with 246 beds, distributed 10 wards: surgery, adults and pediatric cardiology, neumology, nephrology, coronary unit and others. In addition the hospital has 17 external services. The microbiology laboratory receives an average of 18,000 samples annually. The hospital has 5,800 admission and 5,700 discharges per year.

2.2 Bacterial isolates

We studied a total of ninety single-patient clinical MRSA isolates, between January 2002 and December 2009. The strains were collected from several clinical sources: bronchial secretions (n=34); wound secretions (n= 25), blood (n= 16); catheter (n=3); pleural liquid (n=3); peritoneal fluid (n=1) and others (n=13). MRSA strains were collected from different wards: pediatric surgery, adult surgery, coronary unit, nephrology, surgery and cardiology. Of the 90 MRSA isolates, 24 were from children and 66 were from adults.

2.3 Antimicrobial susceptibility testing

Antimicrobial susceptibility testing for MRSA isolates was performed using the automated method of MicroScan® (DADE-BEHRING, Sacramento, CA) for: penicillin, oxacillin, amoxicillin, cefotaxime, cephalothin, cefazolin, imipenem, trimethoprim-sulfamethoxazole, ciprofloxacin, chloramphenicol, clindamycin, erythromycin, clarithromycin, gentamicin, rifampin, tetracycline and vancomycin, following the Clinical Laboratory Standards Institute guidelines (Clinical Laboratory Standards Institute [CLSI], 2009).

2.4 Molecular typing

The whole genomic DNA was prepared as described previously (Chung et al., 2000). After digestion with *Sma*I endonuclease, DNA was separated in a CHEF-DRII apparatus (Bio-Rad, Birmingham, U.K) (Chung et al., 2000). Strains HU25, HPV107, HDE288, BK2464, JP27 and 96/32010, representing the Brazilian, Iberian, Pediatric, New York/Japan-USA, New York/Japan-Japan and EMRSA-16-U.K clones, were included in the PFGE gels as controls. The control strains were kindly provided by Prof. Herminia de Lencastre from the Molecular Genetics laboratory Institute de Tecnología Química e Biologica da Universidad Nova de Lisboa. Criteria of Tenover were used to compare different clones (Tenover et al., 1995). Strains BK2464 and HDE288 were used as SCC*mec* controls. Hybridization of *SmaI* digests with *mecA* and Tn554 probes (de Lencastre et al., 1994), SCC*mec* typing (Oliveira & de Lencastre, 2002) and MLST (Enright et al., 2000) were performed as previously described. Briefly, MLST is based in internal fragments of seven housekeeping genes (*arcC, aroE, glpF, gmk, pta, tpi, yqiL*) for each isolates, the alleles at the seven loci defined the allelic profiles, which corresponded to a sequence type (ST). ST designations were those assigned the MLST data base (http://www.mlst.net). The SCC*mec* typing system is defined by combining the class of the *mec* gene complex with the cassette chromosome recombinase gene (*ccr*) allotypes. The polymorphism in the vicinity of the *mecA* gene detected by probe *ClaI*-digested DNAs with a *mecA* probe and transposon Tn*554* insertion patterns detecting by probing *ClaI* digestion DNAs with a specific probe (de Lencastre et al., 1994, Enright et al., 2000; Oliveira & de Lencastre, 2002).

2.5 Computer-fingerprinting analysis

The computer analysis of the banding patterns obtained by PFGE was done using the NTSYSpc software version 2.0.2.11 (Applied Biostatistics Inc.) after visual inspection. Each gel included reference strain *S. aureus* NCTC 8325 to normalize the PFGE profiles. For clusters analyses, the Dice coefficients were calculated to compute the matrix similarity and were transformed into an agglomerative cluster by the unweighted pair group method with arithmetic average (UPGMA).

3. Results

3.1 Antimicrobial susceptibility

The 90 isolates showed resistance to penicillin (100%), oxacillin (99.3%), amoxicillin (100%) , cefotaxime (100%), cephalothin (100%), cefazolin (100%), chloramphenicol (100%), imipenem (99.3%), ciprofloxacin (87.7%); eleven strains (12.2%) showed low susceptibility for clindamycin, erythromycin, clarithromycin and were susceptible to ciprofloxacin; two strains (2.2%) showed low susceptibility for oxacillin (MIC 4μg/mL) and imipenem. All strains were susceptible to rifampin, tetracycline, gentamicin, trimethoprim-sulfamethoxazole and vancomycin.

3.2 Molecular typing

3.2.1 PFGE analysis

The PFGE analysis separated the MRSA strains into three types, A (5 subtypes), B (3 subtypes) and C (6 subtypes) (Figure 1). PFGE pattern C and subtypes were predominant in this isolates n=72 (80%), Clone A, n=11 (12.2%) and B, n=7 (7.8%) were only found in the isolates of 2002, and these two clones (A and B) were totally replaced by clone C in 2004 and continue until 2009. The results produced by a computer analysis of the banding patterns show clearly the division of the three clone groups (A, B and C); interestingly, the A and B clone isolates have very similar PFGE patterns (coefficient similarity 95%). Nevertheless, the three clones A, B and C could easily be distinguished by antibiograms and other molecular properties as well (Table 1). The three clones were multiresistant, however, each one of them showed a characteristic resistance pattern; clone A was resistant to ß-lactams and showed a low susceptibility to clarithromycin, clindamycin, erythromycin and was susceptible to ciprofloxacin; while clones B and C were resistant to ß-lactams, clarithromycin, clindamycin, erytromycin and ciprofloxacin; only the strains with subtypes B1 and B2 showed low susceptibility for oxacillin and imipenem.

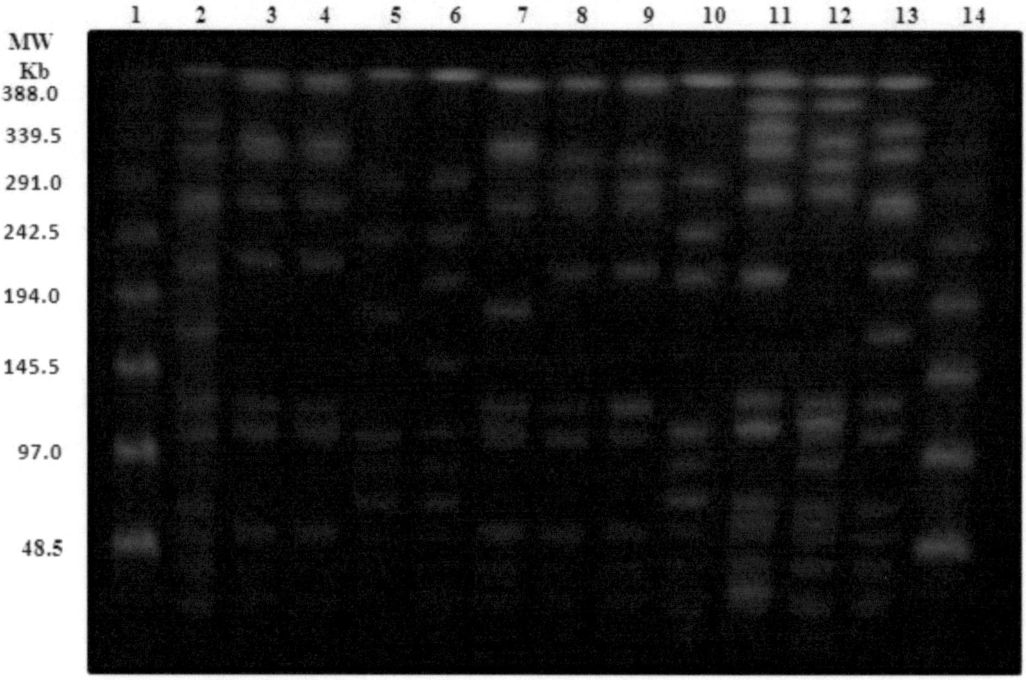

Fig. 1. Pulsed field gel electrophoresis profiles of MRSA clinical isolates from the Instituto Nacional de Cardiología "Dr. Ignacio Chávez", Mexico and representatives of international clones. Lanes: 1-14 lambda ladder used a molecular size (MW) markers; 2 and 13 reference strain NCTC8325; 3-4 (44CAR and 47CAR) pattern C; 5 (2CAR) pattern A; 6 (20CAR) pattern B; 7-12 (HDE288, BK2464, JP27, EMRSA16, HPV107 and HU25) control strains representative of Pediatric, New York/Japan-USA, New York/Japan-Japan, EMRSA16-U.K, Iberian and Brazilian clones.

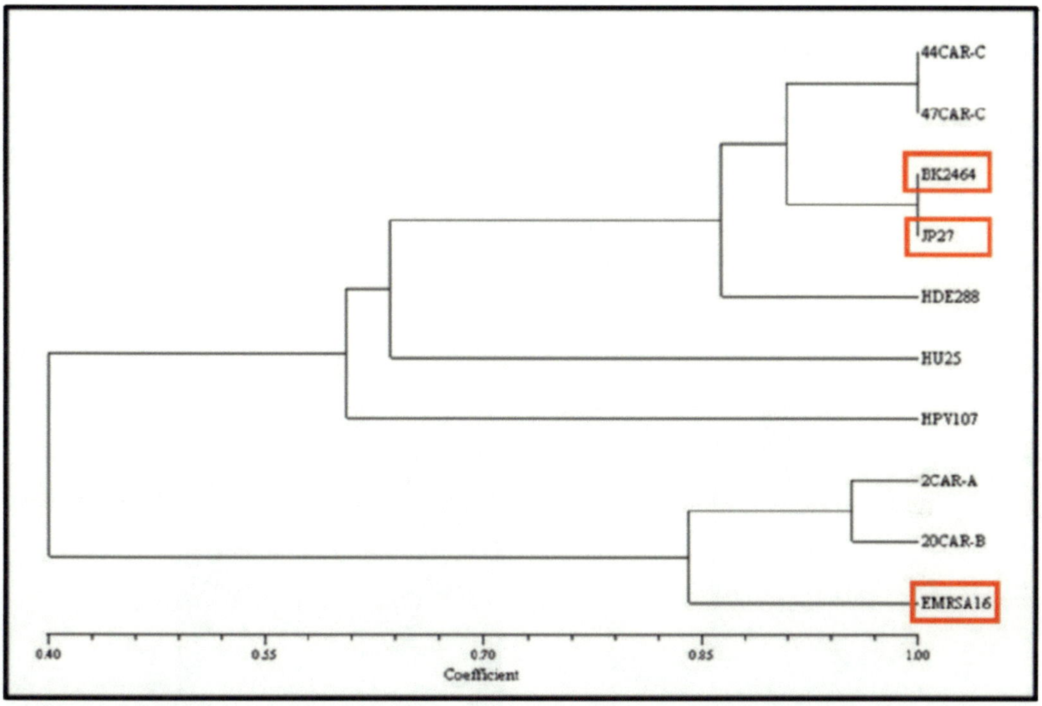

Fig. 2. Dendrogram comparing MRSA clones A, B and C from the Instituto Nacional de Cardiología "Dr. Ignacio Chávez", Mexico with different international MRSA clones: BK2464-New York/Japan-USA clone; JP27-New York/Japan-Japan clone; HDE288-Pediatric clone; HU25-Brazilian-clone; HPV107-Iberian clone; EMRSA-16-U.K. clone. For cluster analysis, Dice coefficients were calculated to compute matrix similarity a transformed into an agglomerative cluster with the unweighted pair group method with arithmetic average.

Property	Clone A	Clone B	Clone C
Antibiotype[1] (Resistance)	ß-lactams, (CLA,CD,ERY)[2]	ß-lactams, (CIP, CLA,CD, ERY)	ß-lactams, (CIP, CLA,CD, ERY)
Number of subtypes	5	3	6
SSC*mec* type[3]	IV	II	II
Hybridization bands(Kb) *SmaI-mecA*	~180	~211[4]	~211
Hybridization bands(Kb) *Sma-Tn554*	~180	~211-640	~211-640
ST	30	30[5]/36	5

[1]Antibiotic abbreviations: CLA- clarithromycin; CD - clindamycin; ERY - erythromycin GEN - gentamicin; CIP - ciprofloxacin. [2]Intermediate resistance pattern. [3]Staphylococcal cassette chromosome *mec*. [4]Except B1 and B2; [5]Sequence typing (MLST), only the patterns B1 and B2.

Table 1. Antibiotype and Genotypic characterization of the MRSA clones presented in the Instituto Nacional de Cardiología "Dr. Ignacio Chávez", Mexico (2002-2009).

3.2.2 Hybridization pattern

The hybridization patterns with *mecA* and Tn*554* probes indicated that the MRSA strains accompanying clone A carried the *mecA* gene on a *Sma*I fragment of approximately 180 Kb, while the *mecA* gene of the clones B and C were found on a fragment of approximately 211 Kb (Figure 3-A). One Tn*554* copy was identified usually on the fragment approximately 180 Kb between the isolates of clone A; while the MRSA strains accompanying clones B and C usually carried two identified Tn*554* copies between the *SmaI* fragments of approximately 211 and 640 Kb; only the strains 3CAR, (B1) and 8CAR, (B2), carried the transposon Tn*554* in a fragment of 640 Kb (Figure 3-B).

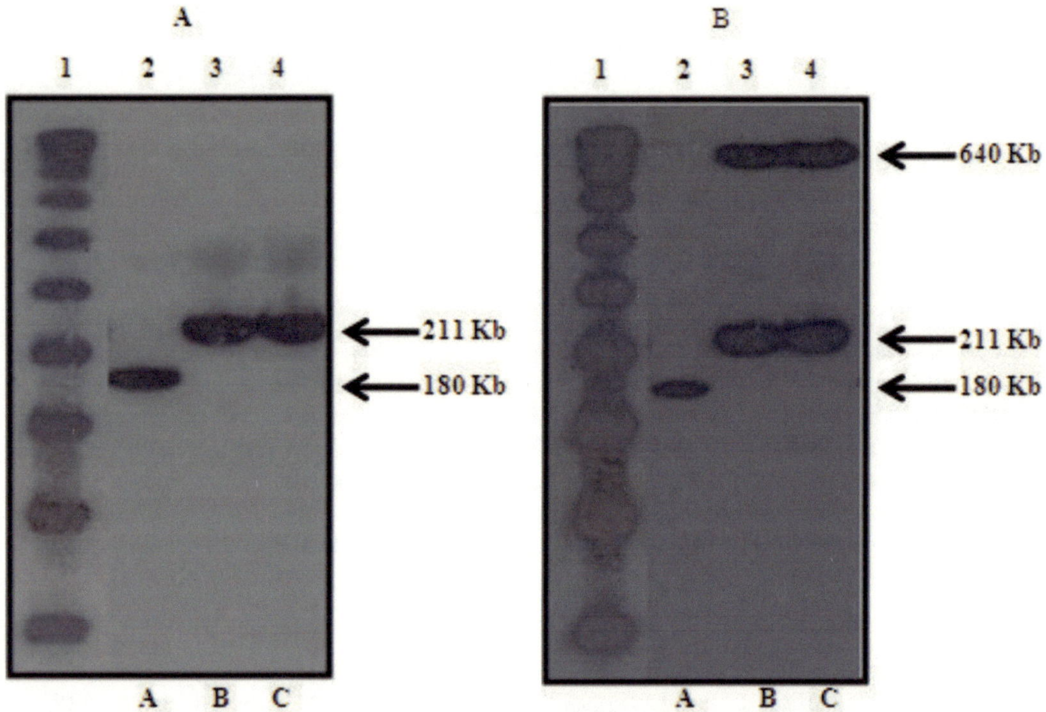

Fig. 3. (A) *SmaI-mecA* and (B) *SmaI* Tn*554* patterns identified among the clones A, B and C of the Instituto Nacional de Cardiología "Dr. Ignacio Chavéz, Mexico. Lane 1, molecular weight markers, lambda ladder; lane 2, 2CAR (pattern A); lane 3, 20CAR (pattern B) and lane 4, 44CAR (pattern C).

Two strains collected in this hospital 3CAR, (pattern B1) and 8CAR, (pattern B2) did not hybridize with the *mecA* DNA probe, interestingly both presented a low susceptibility to oxacillin and these isolates were subtypes of pattern B. The only difference was found in the *SmaI* hybridization fragment, which contains the *mecA* gene: in the two isolates, this fragment had a smaller molecular size (145 instead of 180 Kb) and did not react with the *mecA* probe, indicating a deletion of approximately 35 Kb, which must have included both the *mecA* gene and part of the *mec* element (Figure 4A and 4B). All the isolates accompanying clone A presented SCC*mec* type IV and sequence type 30 (ST30) whereas the

MRSA strains of clones B and C had SCC*mec* type II, sequences type 36 and 5 (ST36 and ST5) respectively, except B1 and B2 which did not amplify SCC*mec*, this isolates showed sequence type 30 (ST30) .

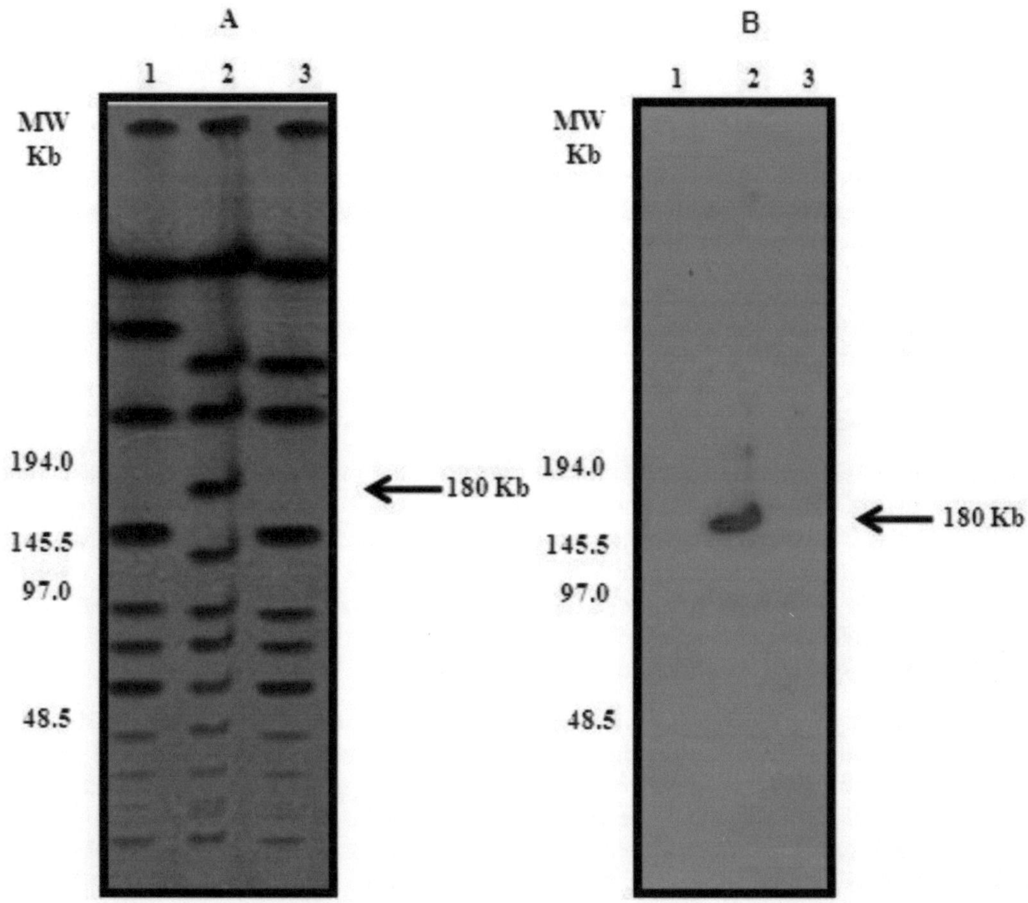

Fig. 4. (A) PFGE patterns of MRSA strains from the Instituto Nacional de Cardiología "Dr. Ignacio Chávez", Mexico. Lane 1, 3CAR, (pattern B1); lane 2, control strain and lane 3, 8CAR (pattern B2).

3.2.3 Homology pattern

One isolate belonging to each type of clones (A, B and C) were compared to strains belonging to previously characterized MRSA clones, i. e., representatives of the pediatric clone and isolates belonging to the New York-Japan clone and also to other international pandemic clones, namely, the Iberian, Brazilian and EMRSA-16 clones (Figure 1). Clone C showed a high degree of similarity to the pediatric (85.5%) and the New York-Japan (89.5%) clones. Clones A and B showed a high degree of similarity to the EMRSA-16 (80%) clone (Figure 2).

4. Discussion

The emergence of strains resistant to methicillin and other antibacterial agents has become a major concern especially in the hospital environment, because of the higher mortality due to systemic methicillin-resistant *S. aureus* infections (Cosgrove et al., 2003; Handberger et al., 2011). Seven major pandemic MRSA (the so-called Brazilian, Hungarian, Iberian, New York-Japan, pediatric, EMRSA 16 and Berlin clone (EMRSA15) have been identified as the cause for the majority of hospital-acquired *S. aureus* infections in the world (Oliveira et al., 2002), indicating that they represent successful clones in terms of their ability to cause infections, persist and spread from one geographic zone to another, including across continents.

The combination of different molecular typing methods used in the present study allowed us to register epidemiologically relevant features of MRSA populations in the Instituto Nacional de Cardiología "Dr. Ignacio Chávez" in Mexico and document the coexistence of MRSA clones of international distribution.

All 90 strains were resistant to at least eleven antibiotics (amoxicillin, cefotaxime, cephalothin, cefazolin, chloramphenicol, imipenem, clindamycin, erythromycin and clarithromycin) in addition to penicillin and oxacillin and 94.4% were resistant to ciprofloxacin as well. The phenotypes of resistance to the antimicrobial agents are shown in the Table 1.

As a response to the emergence and worldwide spread of antibiotic-resistant *S. aureus* there was an urgent need for the creation of international surveillance systems with methodologies that could help hospital infection prevention and control such organisms. MRSA causing nosocomial infections have been reported in other hospitals in Mexico showing a wide geographic spread of MRSA specific clones (Aires de Sousa et al., 2001; Echaniz et al., 2006; Velazquez et al. 2004) similar spread has been observed by other clones in USA and Europe (Da Silva, 2003; Johnson, 2011; Oliveira & de Lencastre 2002).

Interestingly, only three PFGE types were found during the period of the study, designed A, B and C. Previous studies had documented that MRSA clones may spread in and between hospitals, cities and countries and even intercontinental spread may occur (Auken et al., 2002; Nübel et al., 2010). The multiresistant clone C (New York/Japan clone) was present in more than 50% of MRSA that were recovered from a variety of infections sites and hospital wards. Previously, this clone had already been reported in two hospitals in Mexico: Hospital Civil de Guadalajara "Fray Antonio Alcalde" and Hospital de Pediatría del Centro Medico Nacional Siglo XX1-IMSS and it has been circulating in these hospitals since 1999, and 2001 respectively (Echaniz et al., 2006; Velazquez et al. 2004). The results of these studies showed that clone C (New York/Japan clone) had, sequence type 5 and SCC*mec* type II. In this study we found that pattern C was very similar (89.5%) to the multiresistant New York-Japan clone (Figure.2), which correspond to our last year´s results, proving with this the capacity of this clone to persists for long periods of time within the hospitals; as well as its capacity to spread to other hospitals (epidemic clone). whose evidence is the existence of this clone in other hospital of third level in Mexico Instituto Nacional de Cancerología (INCan). It is important to mention that the existence of clone C (New York/Japan clone) had not been present in the INCan before 2006 (Cornejo et al., 2010). All these results are of relevant importance if we consider that the first high-level VRSA (vancomycin-resistant *S. aureus*) (MIC = 1024 μg/mL vancomycin), belonged to the New York lineage (Weigel et al., 2003) and the fact that the descending MRSA strains of this clone are circulating in our population, together with the few means of antibiotic restriction it could represent a potential short term risk for the VRSA appearance in the hospitals of our country. Clone A and B were only found in the isolates in 2002, these clones showed a high degree of similarity to the EMRSA-16 clone, this clone is one of the dominant types of MRSA found in a UK hospital (Moore & Lindsay, 2002) and was widely disseminated in Canada (Simor et al., 2002), Greece (Aires de Sousa et al., 2003) and Mexico (Aires de Sousa et al., 2001). Interestingly, both clones (A and B) are very similar (95%) (Figure 2), nevertheless, clone A showed a reduced resistance profile as clone B, and this is because of the existence of the SCC*mec* IV in these isolates, this chromosomal cassette was found in relation to isolated MRSA strains in the community (CA-MRSA) (Coombs et al., 2011). Different reports of several infections caused by CA-MRSA in Latin America (Uruguay, Rio de Janeiro, Colombia, Argentina and Mexico) have been published (Alvaréz et al., 2006; Ma et al., 2005; Reyes et al., 2009; Ribeiro et al., 2005; Velazquez et al., 2011). All pattern of PFGE of the clones A, B and C showed subtypes. Probably the PFGE subtypes indicate the continued evolutionary divergence of these clones during its massive geographic expansion.

Relative genetic instability of the *mecA* element was observed in two strains and this was associated with an apparent deletion of the *mec* element, these isolates were very similar to profile B (B1 and B2) and presented a low susceptibility to oxacillin. In the literature there are reports of *S. aureus* strains with low-level methicillin resistance (MIC 2-4 µg/mL) which are not associated to the presence of *mecA* gene, Tomasz et al. reported one class of borderline methicillin-resistant strains having PBP1 and PBP2 with altered methicillin-binding affinities and overproduction of PBP4 (Tomasz et al., 1989). Another class of low-susceptibility has been reported and was attributed to overproduction of penicillinase (McDougal & Thornsberry, 1986). Hackbarth et al. studied the nucleotide sequence of the PBP2 gene and identified a point mutation near the penicillin-binding motive of transpeptidase (Hackbarth et al., 1995). An MRSA clinical strain with significant methicillin resistance (MIC 64µg/mL) despite absence of *mec A* was reported (Yoshida et al., 2003).

5. Conclusion

The combination of molecular typing methods (PFGE, *mecA*, Tn*554* probes, SCC*mec*, and MLST) with epidemiologic and clinical information allows the detection of MRSA clusters and outbreaks and therefore provides a rationale for appropriate infection control intervention. Our study emphasizes the need of national and international collaborations to monitor the spread of current epidemic strains as well as the emergence of new ones in our country. The mechanisms of spread in different areas are poorly understood and further studies are necessary to understand the dynamics involved in the predominance of unique MRSA clones.

6. Acknowledgment

We thank PhD. Lilia Chihu for style review of the paper

7. References

Aires de Sousa, M. Miragaia, M. Santos, I. Avila, S. Adamson, I. Casagrande, S. Brandileone, C. Palacio, R. Dell'Acqua, L. Hortal, M. Camou, T. Rossi, A. Velazquez, ME.

Echaniz, G. Solorzano, F. Heitmann, I. & de Lencastre, H. (2001). Three-year assessment of methicillin-resistant *Staphylococcus aureus* clones in Latin America from 1996 to 1998. *Journal Clinical Microbiology*, Vol.39, No.6, (June 2001), pp. 2197-3205, ISSN 0095-1137

Aires de Sousa, M. Bartzavali, C. Spiliopoulou, I. Santos, I. Crisostomo, I & de Lencastre, H. (2003). Two international methicillin-resistant *Staphylococcus aureus* clones endemic in a university hospital in Patras, Greece. *Journal of Clinical Microbiology,* Vol.41, No.5, (May 2003), pp. 2027-2031, ISSN 0095-1137

Alpuche, C. Avila, C. Espinoza, LE. Gómez, D. & Santos, I. (1989). Perfiles de sensibilidad antimicrobiana de *Staphylococcus aureus* en un hospital pediátrico: prevalencia de resistencia a meticilina. *Boletín Medico del Hospital Infantil de México*, Vol.11, pp. 697-699. ISSN 1665-1146

Alvaréz, C. Barrientes, O. Leal, A. Contreras, G. Barrero, L. Rincón, S. Díaz, L. Vanegas, N. & Arias, C. (2006). Community-associated methicillin-resistant *Staphylococcus aureus*, Colombia. *Emerging Infectious Diseases*, Vol.12, No.12, (December 2006), pp. 2000-2001, ISSN 1080-6040

Appelbaum, P. (2007). Reduced glycopepdide susceptibility in methicillin-resistant *Staphylococcus aureus* (MRSA). *International Journal of Antimicrobial agents*, Vol.30, No.4, (November 2007), pp. 398-408, ISSN 0924-8579

Auken, H. Ganner, M. Murchan, S. Cookson, D. & Johnson, P. (2002). A new UK strain of epidemic methicillin-resistant *Staphylococcus aureus* (EMRSA-17) resistant to multiple antibiotics. *Journal Antimicrobial Chemotheraphy*, Vol.50, No.2, (August 2002), pp.171-175, ISSN 0305-7453

Breurec, S. Zriouil, . Fall, C. Boisier, P. Brisse, S. Djibo, S. Etienne, J. Fonkoua, M Perrier-Gros-Claude, J. Pouillot, R. Ramarokoto, C. Randrianirina, F. Tall, A. Thiberge, J, Working Group on *Staphylococcus aureus* infections, Laurent, F. & Garin, B. (2011) Epidemiology of methicillin-resistant *Staphylococcus aureus* lineages in five major African towns: emergence and spread of atypical clones. *Clinical Microbiology and Infectious Diseases*, Vol.17, No.2, (February 2011), pp. 160-165, ISSN 1198-743X

Calderón, E. Espinoza, LE. & Avila, R. (2002). Epidemiology of drug resistance: The case of *Staphylococcus aureus* and coagulase negative staphylococci infections. *Salud Pública de México*, Vol.44, No.2, (October 2001), pp. 108-12. ISSN 0036-3634

Chávez, B. (2004). Infecciones intrahospitalarias ¿Que ha pasado en 23 años? *Enfermedades Infecciosas y Microbiología*, Vol.24, No.3, (July-September 2004), pp. 89-92.

Chung, M. de Lencastre, H. Matthews, P. Tomasz, A. Adamsson, I. Aires de Sousa, M. Camou, T. Cocuzza, C. Corso, A. Couto, I. Dominguez, A. Gniadkowski, M. Goering, R. Gomes, A. Kikuchi, K. Marchese, A. Mato, R. Melter, O. Oliveira, D. Palacio, R. Sá-Leão, R. Santos, I. Song, J. Tassios, P. & Villari, P. (2000). Molecular typing of methicillin-resistant *Staphylococcus aureus* by pulsed-field gel electrophoresis: comparison of results obtained in a multilaboratory effort using identical protocols and MRSA strains. *Microbial Drug Resistance*, Vol.6, No.3, (October 2000), pp. 189-98, ISSN 1076-6294

[CLSI] Clinical and Laboratory Standards Institute. (2009). Performance Standards for Antimicrobial Susceptibility Testing. Nineteenth Informational Supplement; M100-S19. *Clinical and Laboratory Standards Institute*, Vol.29, No.3, (January 2009), pp. 1-149, ISBN 1-56238-690-5, Wyne, Pennsylvania.

Coombs, G. Monecke, S. Pearson, J. Tan, H. Chew, Y. Wilson, L. Ehricht, R. O'Brien, F. & Christiansen, K. (2011). Evolution and diversity of community-associated methicillin-resistant *Staphylococcus aureus* in a geographical region. *BMC Microbiology*, Vol.11, (September 2011), pp. 1-212 ISSN 1471-2180

Cornejo, P. Volkow, P. Sifuentes, J. Echániz, G. Díaz, A. Velázquez, C. Bobadilla M. Gordillo, P. & Velazquez, ME. (2010). Tracing the source of an outbreak of methicillin-resistant *Staphylococcus aureus* in a tertiary-care oncology hospital by epidemiology and molecular methods. *Microbial Drug Resistance,* Vol.16, No.3, (September 2010), pp. 203-208, ISSN 1076-6294

Cosgrove, S. Sakoulas, G. Perencevich, E. Schwaber, M. Karchmer, A. & Carmeli, Y. (2003). Comparison of mortality associated with methicillin-resistant and methicillin-susceptible *Staphylococcus aureus* bacteremia: a meta analysis. *Clinical Infectious Diseases*, Vol.36, No.1, (January 2003), pp. 53-59, ISSN 1058-4838

Creamer, E. Galvin, S. Dolan, A. Sherlock, O. Dimitrov, B. Fitzgerald-Hughes. D. Thomas, T. Walsh, J. Moore, J. Smyth, E. Shore, A. Sullivan. D. Kinnevey, P. O'Lorcain, P. Cunney, R. Coleman, D. & Humphreys, H. (2011). Evaluation of screening risk and nonrisk patients for methicillin-resistant *Staphylococcus aureus* on admission in an acute care hospital. *American Journal of Infection Control*, Vol.30, [Epub ahead of print] ISSN 0196-6553

Da Silva, M. Silva, M. Wisplinghoff, H. Hall, G. Tallent. S. Wallance, S. Edmond, M. Figueiredo, A. & Wenzel. R. (2003). Clonal spread of methicillin-resistant *Staphylococcus aureus* in a large geographic area of the United States. *The Journal of Hospital Infection*, Vol.53, No.2, (February 2003), pp. 103-110, ISSN 0195-6701

de Lencastre, H. Couto, I. Santos, I. Melo, J. Torres, A. & Tomasz, A. (1994). Methicillin-resistant *Staphylococcus aureus* disease in a Portuguese hospital: characterization of clonal types by a combination of DNA typing methods. *European Journal of Clinical Microbiology and Infectious Diseases*, Vol.13, No.1, (January 1994), pp. 64-73, ISSN 0934-9723

Deurenberg, R. Nulens, E. Valvatne, H. Sebastian, S. Driessen, C. Craeghs, J. De Brauwer, E. Heising, B. Kraat, Y. Riebe, J. Stals, F. Trienekens, T. Scheres, J. Friedrich, A. van Tiel, F. Beisser, P. & Stobberingh, E. (2009). Cross-border dissemination of methicillin-resistant *Staphylococcus aureus*, Euregio Meuse-Rhin region. *Emerging infectious diseases*, Vol.15, No.5, (May 2009), pp. 727-734, ISSN 1080-604

Diekema, D. Pfaller, M. Schmitz, F. Smayevsky, J. Bell, J. Jones, R. & Beach, M. (2001). Survey of infections due to *Staphylococcus* species: frequency, occurrence and antimicrobial susceptibility of isolates collected in the United States, Canada, Latin America, Europe and the Western Pacific region for the SENTRY Antimicrobial Surveillance Program, 1997-1999. *Clinical of Infectious Diseases*, Vol.15, No.32, (May 2001), pp. S114-132, ISSN 1058-4838

Echaniz, G. Velazquez, ME. Aires de Sousa, M. Morfín, R. Rodríguez-Noriega, E. Carnalla, N. Esparza, S. & de Lencastre H. (2006). Molecular characterization of a dominant methicillin-resistant *Staphylococcus aureus* (MRSA) clone in a Mexican hospital (1999-2003). *Clinical Microbiology and Infectious Diseases*, Vol.12, No.1, (January 2006), pp. 12-28, ISSN 1198-743X

Enright, M. Day, N. Davies, C. Peacock, S. & Spratt, B. (2000). Multilocus sequence typing for characterization of methicillin-resistant and methicillin-susceptible clones of

Staphylococcus aureus. *Journal of Clinical Microbiology*, Vol.38, No.3, (March 2000), pp. 1008-1015, ISSN 0095-1137

Enright, M. Robinson, D. Randle, G. Feil, E. Grundmann, H. & Spratt, B. (2002). The evolutionary history of methicillin-resistant *Staphylococcus aureus* (MRSA). *Proceedings of the National Academy of Sciences of the United States of America,* Vol. 99, No.11, (May 2002), pp. 7687-7692, ISSN 0027-8424

Goldstein, F. (2007). The potential clinical impact of low-level antibiotic resistance in *Staphylococcus aureus*. *Journal Antimicrobial Chemotherapy*, Vol.59, No.1, (January 2007), pp. 1-4, ISSN 0305-7453.

Hackbarth, C. Kocagoz, T. Kocagoz, S. & Chambers. H. (1995). Point mutation in *Staphylococcus aureus* PBP2 gene affects penicillin-binding kinetics and is associated with resistance. *Antimicrobial Agents and Chemotherapy*, Vol.39, No.1, (January 1995), pp. 103-106, ISSN 0066-4804

Hanberger, H. Walther, S. Leone, M. Barie, P. Rello, J. Lipman J. Marshall, J. Anzueto A. Sakr, Y. Pickkers, P. Engoren, M. Vincent, J. & EPIC II Group of investigators. (2011). Increased mortality associated with meticillin-resistant *Staphylococcus aureus* (MRSA) infection in the intensive care united: results from the EPIC II study. *International Journal of Antimicrobial Agents*, Vol.38, No.4, (October 2011), pp. 331-335, ISSN 0924-8579

Johnson, A. (2011). Methicillin-resistant *Staphylococcus aureus*: the European landscape. *The Journal of antimicrobial chemotherapy*, Vol.66, No.4, (May 2011), pp. 43-48, ISSN 0305-7453

Ma, XX. Galiana, A. Pedreira, W. Mowszowicz, M. Christophersen, I. Machiavello, S. Lopez, L. Benaderet, S. Buela, F. Vincentino, W. Albini, M. Bertaux, O. Constenla, I. Bagnulo, H. Llosa, L. Ito, T. & Hiramatsu, K. (2005). Community-acquired methicillin-resistant *Staphylococcus aureus*, Uruguay. *Emerging Infectious Diseases*, Vol.11, No. 6, (June 2005), pp. 973-976, ISSN 1080-604

McDougal, L. & Thornsberry, C. (1986). The role of ß-lactamase in staphylococcal resistance to penicillinase-resistant penicillins and cephalosporins. *Journal of Clinical Microbiology*, Vol.23, No.5, (May 1986), pp. 832-839, ISSN 0095-1137

McDougal, L. Steward, C. Killgore, G. Chaitram, J. McAllister, S. & Tenover, F. (2003). Pulsed-field gel electrophoresis typing of oxacillin-resistant *Staphylococcus aureus* isolates from the United States: establishing a national database. *Journal of Clinical Microbiology*, Vol.41, No.11, (November 2003), pp. 5113-5120, ISSN 0095-1137

Moore, P. & Lindsay. J. (2002). Molecular characterization of the dominant UK methicillin-resistant *Staphylococcus aureus* strain, EMRSA-15 and EMRSA-16. *Journal of Medical Microbiology*, Vol.51, No.6, (June 2002), pp. 516-521, ISSN 0022-2615

Murchan, S. Kaufmann, M. Deplano, A. de Ryck, R. Struelens, M. Zinn, C. Fussing, V. Salmenlinna, S. Vuopio-Varkila, J. El Solh, N. Cuny, C. Witte, W. Tassios, P. Legakis, N. van Leeuwen, W. van belkum, A. Vindel, A. Laconcha, I. Garaizar, J. Haeggman, S. Olsson-Liljequist, B. Ransjo, U. Coombes, G. & Cookson, B. (2003). Harmonization of pulsed-field gel electrophoresis protocols for epidemiological typing of strains of methicillin-resistant *Staphylococcus aureus*: a single approach developed by consensus in 10 European laboratories and its application for tracing the spread of related strains. *Journal of Clinical Microbiology,* Vol.41, No.4, (April 2003), pp. 1574-85, ISSN 0095-1137

Nübel, U. Dordel, J. Kurt, K. Strommenger, B. Westh, H. Shukla, S. Zemlicková, H. Leblois, R. Wirth, T. Jombart, T. Balloux, F & Witte, W. (2010). A timescale for evolution, population expansion, and spatial spread of an emerging clone of methicillin-resistant *Staphylococcus aureus*. *PLoS pathogens*, Vol.6, No.4, (April 2010), pp. 1-12, ISSN 1553-7366

Oliveira, D. Tomasz, A. & de Lencastre, H. (2001). The evolution of pandemic clones of methicillin-resistant *Staphylococcus aureus*: identification of two ancestral genetic backgrounds and the associated *mec* elements. *Microbial Drug Resistance*, Vol.7, No.4, (December 2001), pp. 349-361, ISSN 1076-6294

Oliveira, D. & de Lencastre, H. (2002). Multiplex PCR strategy for rapid identification of structural types and variants of the *mec* element in methicillin-resistant *Staphylococcus aureus*. *Antimicrobial Agents and Chemotherapy*, Vol.46, No.7, (July 2002), pp. 2155-2161, ISSN 0066-4804

Oliveira, D. Tomasz, A. & de Lencastre, H. (2002). Secrets of success of a human pathogen: molecular evolution of pandemic clones of methicillin-resistant *Staphylococcus aureus*. *The Lancet infectious diseases*, Vol.2, No.3, (March 2002), pp. 180-189, ISSN 1473-3099

Rebmann, T. & Aureden, K. (2011). Preventing methicillin-resistant *Staphylococcus aureus* transmission in hospitals: an Executive Summary of the Association for Professionals in Infection Control and Epidemiology, Inc, Elimination Guide. *American Journal of Infection Control*, Vol.39, No.7, (September 2011), pp. 595-598, ISSN 0196-6553

Reyes, J. Rincón, S. Díaz, L. Panesso, D. Contreras, G. Zurita, J. Carrillo, C. Rizzi, A. Guzmán, M. Adachi, J. Chowdhury, S. Murray, B. & Arias, C. (2009). Dissemination of methicillin resistant *Staphylococcus aureus* USA 300 sequence type 8 linage in Latin America. *Clinical of Infectious Diseases*, Vol.49, No. 12, (December 2009), pp. 1861-1867, ISSN 1058-4838

Ribeiro, A. Dias, C. Silva-Carvalho, M. Berquó, L. Ferreira, F. Santos, R. Ferreira-Carvalho, B. & Figueiredo, A. (2005). First report of infection with community-acquired methicillin-resistant *Staphylococcus aureus* in South America. *Journal of Clinical Microbiology*, Vol.43, No.4, (April 2005), pp. 1985-1988, ISSN 0095-1137

Simor, A, Ofner-Agostini, M. Bryce, E. McGeer, A. Paton, S. & Mulvey, M. (2002). Laboratory characterization of methicillin-resistant *Staphylococcus aureus* in Canadian hospitals: results of 5 years of national surveillance, 1995-1999. *The Journal of Infectious Disiases*, Vol.186, No.5, (September 2002), pp. 652-660, ISSN 0022-1899

Tenover, F. Arbeit, R. Goering, R. Mickelsen, P. Murray, B. Persing, D. & Swaminathan, B. (1995). Interpreting chromosomal DNA restriction patterns produced by pulsed-field gel electrophoresis: criteria for bacterial strain typing. *Journal of Clinical Microbiology*, Vol.33, No.9, (September 1995), pp. 2233-2239, ISSN 0095-1137

Tenover, F. & Gaynes, R. The epidemiology of *Staphylococcus* infections, In: V. A. Fischetti, R. Novick, J. Ferretti, D. Portnoy, and J. Rood, (Ed.). (2000). *Gram-positive pathogens*. American Society for Microbiology, Washington, DC. pp. 414-421

Tomasz, A. Drugeon, H. de Lencastre, H. Jabes, D. McDougall, D. & Bille, J. (1989). New mechanism for methicillin resistance in *Staphylococcus aureus*: clinical isolates that lack the PBP2a gene and contain normal penicillin-binding proteins with modified penicillin-binding capacity. *Antimicrobial. Agents and Chemotherapy*, Vol.33, No.11, (November 1989), pp. 1869-1874, ISSN 0066-4804

Velazquez, ME. Aires de Sousa, M. Echaniz, G. Solórzano, F. Miranda, G. Silva, J, & de Lencastre, H. (2004). Surveillance of methicillin- resistant *S. aureus* in a pediatric hospital in Mexico City during a 7-year period (1997 to 2003): clonal evolution and impact of infection control. *Journal of Clinical Microbiology*, Vol.42, No.8, (August 2004), pp. 3877-3880, ISSN 0095-1137

Velazquez, ME. Ayala, J. Carnalla, N. Soto, A. Guajardo, C. & Echaniz, G. (2011). First report of community-associated methicillin-resistant *Staphylococcus aureus* (USA300) in Mexico. *Journal of Clinical Microbiology*, Vol.49, No.8, (August 2011), pp. 3099-3100, ISSN 0095-1137

Weigel, L. Clewell, D. Gill, S. Clark, N. McDougal, L. Flannaga, S. Kolonay, J. Shetty, J. Killgore, G. & Tenover, F. (2003). Genetic analysis of a high-level vancomycin-resistant isolate of *Staphylococcus aureus*. *Science*, Vol.302, No.5650, (November 2003), pp. 1569-1571, ISSN 0036-8075

Yoshida, R. Kuwahara-Arai, K. Baba, T. Cui, L. Richardson, J. & Hiramatsu, K. (2003). Physiological and molecular analysis of a *mec*A-negative *Staphylococcus aureus* clinical strain that expresses heterogeneous methicillin-resistance. *The Journal Antimicrobial Chemotherapy*, Vol.51, No.2, (February 2003), pp. 247-255, ISSN 0305-7453

11

Usefulness of Pulsed Field Gel Electrophoresis Assay in the Molecular Epidemiological Study of Extended Spectrum Beta Lactamase Producers

Patrick Eberechi Akpaka[1] and Padman Jayaratne[2]
[1]*Department of Para-Clinical Sciences, The University of the West Indies, St. Augustine*
[2]*Department of Pathology & Molecular Medicine*
McMaster University, Hamilton, Ontario
[1]*Trinidad & Tobago*
[2]*Canada*

1. Introduction

A major problem in several health institutions, countries and regions is to categorically define or delineate the source or index case of any microbial organism/s during an outbreak of an infection. Understanding bacterial distribution and their relatedness is essential for determining the epidemiology of nosocomial infections and aiding in the design of rational pathogen control methods. The role of bacterial typing is to determine if epidemiologically identical or related isolates are also genetically related **[Singh A et al, 2006].** Based on phenotypic and genotypic typing methods, multi-drug resistant bacteria organisms such as extended spectrum beta lactamase (ESBL) enzyme producing pathogens e.g. *Escherichia coli* and *Klebsiella pneumoniae* can be traced to have been transferred from one hospital to another, from one country or region to another. Such information and knowledge have greatly assisted clinicians and health care policy makers to determine the best approach of stopping or eliminating such spreads and transfers of the pathogenic organisms involved in the infection.

As noted in the reviews by Singh A et al, the use of molecular methods for typing of nosocomial pathogen has assisted in efforts to obtain a more fundamental assessment of strain interrelationship **[Singh A et al, 2006].** Establishing clonality of pathogens can aid in the identification of the source (environmental or personnel) of organisms, distinguish infectious from non infectious strains, and distinguish relapse from reinfection. Many of the species that are key hospital-acquired causes of infection are also common commensal organisms, and therefore it is important to be able to determine whether the isolate recovered from the patient is a pathogenic strain that caused the infection or a commensal contaminant that likely is not the source of the infection. Likewise, it is important to know whether a second infection in a patient is due to reinfection by a strain distinct from that causing the initial infection or whether the infection is likely a relapse of the original infection. If the infection is due to relapse, this may be an indication that the initial treatment regimen was not effective, and alternative therapy may be required **[Singh A et al, 2006].**

Gel electrophoresis and in particular Pulsed-Field-Gel-Electrophoresis (PFGE) is a tool that has made genotyping of bacterial isolates possible. The PFGE is a laboratory technique used for separation of large deoxyribonucleic acid (DNA) molecules if electric current that periodically changes direction is applied to it. The PFGE is the "gold standard "technique used in this discipline of molecular epidemiological studies and it is basically the comparison of large genomic DNA fragments after digestion with a restriction enzyme that cuts infrequently. Since the bacterial chromosome is typically a circular molecule, the digestion by the enzyme yields several linear large DNA molecules. Moving these large DNA molecules posed a problem but Schwartz and Cantor in 1984 introduced a voltage gradient that gave better resolution and the ability to move large molecules **[Schwartz DC & Cantor CR, 1984]**

Conventional agarose gel electrophoresis can only be used for the separation of DNA fragments that ranges between 20 - 25 base pair (kbp) by using specialized apparatus no matter how long it is run. The distance between DNA bands of a given length is determined by the percent agarose in the gel. The disadvantage of higher concentrations is the long run times (sometimes days). PFGE uses a special type of agarose that has a larger matrix pore sizes even at a higher percentages such as 1%. The most commonly utilized PFGE methods approaches include the contour-clamped homogenous electric field (CHEF) and field inversion gel electrophoresis (FIGE) **[Carle, G. F et al, 1986; Finney, M. 1993].** Field inversion gel electrophoresis utilizes a conventional electrophoresis chamber in which the orientation of the electric field is periodically inverted by 180° and has an upper limit of resolution about 200kbp. CHEF uses a more complex electrophoresis chamber with multiple electrodes to achieve highly efficient electric field conditions for separation; typically the electrophoresis apparatus reorients the DNA molecules by switching the electric fields at 120° angles. CHEF can separate even up to 2-3 Mbp.

Interpreting DNA fragment patterns generated by PFGE and relating or associating them into epidemiologically useful information for typing nosocomial pathogens, the clinical microbiologist or researcher must understand how to compare PFGE patterns and how random genetic events can alter these patterns. Ideally, the PFGE isolates representing an outbreak strain will be indistinguishable from each other and distinctly different from those of epidemiologically unrelated strains. If this occurs, the outbreak is relatively easy to identify. A random genetic activity such as mutation in a DNA can occur and when this happens, it will change the restriction fragment profile obtained during the course of the outbreak **[Hall LMC, 1994; Quintiliani R., Jr., & P. Courvalin, 1996; Thal LA et al, 1997]** These random variations in the fingerprints will depend on the organism and the time period of the outbreak

The aim of this study is to demonstrate the usefulness of PFGE techniques as a tool to be used in identifying outbreaks of bacterial infection and hence can be used as a tool for infection control measures in a hospital. Also to determine its importance in delineating the clonal relatedness or diversity of bacterial strains isolated from several regional hospitals in Trinidad and Tobago. The PFGE has been shown to be useful in the determination of the sources, clonal relatedness and spread of bacterial isolates in hospitals and countries where the isolates have been recovered or encountered.

2. Materials & methods

More than 230 strains of *Klebsiella pneumoniae* and *Escherichia coli* obtained routinely from three major regional hospitals in Trinidad and Tobago were used for this study. These non consecutive bacterial isolates were identified using standard microbiological methods as had been previously reported **[Akpaka PE & Swanston WH; 2008].** The initial screening for ESBL production by these pathogens using MIC values at concentrations and breakpoints recommended by the CLSI for ESBL screening **[CLSI 2010]** were performed with the automated micro dilution MicroScan WalkAway-96 System (Siemens, USA). Structured standardized questionnaire was used to extract epidemiological information from hospital records of the patients yielding these isolates. Such information included bio data, gender, hospital facilities where the patients were attended to, clinical signs and symptoms, diagnosis, other forms of investigations and treatments, treatment failures and complications.

2.1 Confirmation of ESBL phenotypes

In accordance with the protocols from the manufacturer to phenotypically determine the ESBL production by bacterial isolates, the E-test strips (AB Biodisk, Solna Sweden), a very sensitive and convenient assay to use was employed to confirm ESBL production in the isolates. The control strain for all the phenotypic testing were *E. coli* ATCC 25922 (negative control) and *K. pneumoniae* ATCC 700603 (ESBL positive).

2.2 Multiplex PCR amplification

The detection of gene sequences coding for the TEM, SHV, and CTX-M enzymes were carried out using multiplex PCR as previously described with some modifications **[Monstein HJ et al, 2007].** The cycling conditions used in the PCR assays were as previously described **[Paterson DL et al, 2003; Boyd DA et al, 2004].** The oligonucleotide primer sets specific for the SHV, TEM and CTX-M genes and the cycling conditions used in the PCR assays were as described previously and are depicted in the Table 1 below.

Gene	Primer	bp Sizes	Reference
*bla*SHV	5'-ATG CGT TAT ATT CGC CTG TG-3' 5'-TGC TTT GTT ATT CGG GCC AA-3'	747-bp	Paterson DL *et al*
*bla*TEM	5'-TCG CCG CAT ACA CTA TTC TCA GAA TGA-3' 5'-ACG CTC ACC GGC TCC AGA TTT AT-3'	445-bp	Boyd DA *et al*
*bla*CTX-M	5'-ATG TGC AGY ACC AGT AAR GTK ATG GC-3' 5'- TGG GTR AAR TAR GTS ACC AGA AYCAGC GG-3'	593-bp	Boyd DA *et al*

Table 1. Showing primers used for amplifications of the genes in ESBL producers

A Multiplex PCR method previously described **[Woodford et al, 2006]** for detection of *bla*CTX-M alleles was used to identify the CTX-M phylogenetic group of positive isolates. All PCR reactions were carried out using 2µl bacterial cell suspension (density of 70%T in Vitek Colorimeter) as the DNA template. Respective genes were detected by the size separation PCR amplicons by agarose gel electrophoresis.

2.3 DNA Electrophoresis

The molecular genotyping method employed to compare the DNA of the ESBL producing isolates was the PFGE. This was carried out as previously described **[Kaufmann ME, 1998]** with some modifications. Briefly, the bacterial isolate suspensions were embedded in agarose plugs. The cells were lysed and the proteins digested. The plugs were washed to remove cellular debris and they were sectioned. Restriction analysis of chromosomal DNA with *Xba*1 (New England BioLabs, Beverly, MA) was carried out, and separation of the DNA was performed using 1% pulsed-field gel agarose (Bio-Rad Laboratories, La Jolla, CA). The pulsed-field gel electrophoresis was performed using a contour-clamped homogeneous electric field apparatus set (CHEF DRIII, Bio-Rad Hercules, CA, USA) as in Figure 1 below.

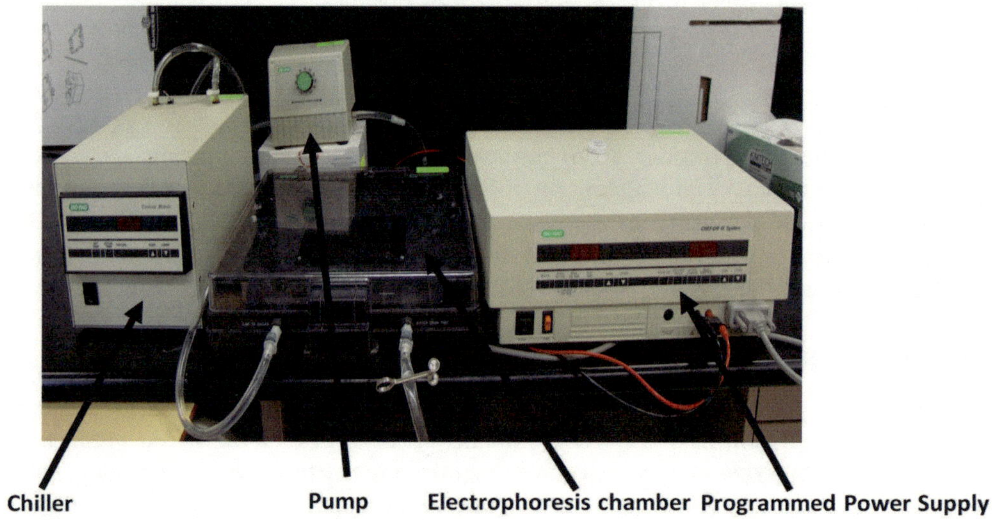

The different components of the CHEF system are indicated as electrophoresis chamber, chiller, pump, and the programmed power supply in the above figure. Alternating the electric field between spatially distinct pairs of electrodes causes large and small DNA fragments to re-orient and move at different speeds through the pores in an Agarose gel

Fig. 1. Picture of CHEF DRIII, Bio-Rad Hercules, CA, USA used for the Pulsed field gel Electrophoresis for the microbial agent DNA separation technique.

The gels were stained and images captured on the Gel Doc imaging system using Quantity One software version 4.4.1 (Bio-Rad Laboratories, Hercules CA, USA), Figure 2 below. After viewing the banding patterns, the results were compared and analyzed by manual visualization from a computer monitor following previously established criteria **[Tenover FC et al, 1995]** so as to determine potential outbreak patterns or spread of the infections from one patient to another or hospital facility to another.

The established criteria or guidelines proposed by Tenover *et al.* were used for the interpretation of PFGE **[Tenover FC et al, 1995].** With these guidelines, a banding pattern difference of up to three fragments could have occurred due to a single genetic event and thus these isolates are classified as highly related, differences of four to six restriction fragments are likely due to two genetic events, and differences of equal to or greater than seven restriction fragments are due to three or more genetic events. Isolates that differ by three fragments in PFGE analysis may represent epidemiologically related subtypes of the same strain. Conversely, isolates differing in the positions of more than three restriction fragments may represent a more tenuous epidemiologic relation. Some studies using PFGE and other typing methods indicate that single genetic events, such as those that may alter or create a new restriction endonuclease site or DNA insertions/deletions associated with plasmids, bacteriophages, or insertion sequences, can occur unpredictably even within the time span of a well-defined outbreak (1 to 3 months) **[Arbeit RD et al 1990; Sader HS et al 1993; Tenover FC et al 1995]**. With the detection of two genetic variation events by differences in fragment patterns compared to the outbreak strain, the determination of relatedness to an outbreak falls into a gray zone. The results may indicate that these isolates are related (especially if isolates were collected over a long period of time, such as 3 to 6 months), but there is also a possibility that strains are unrelated and not part of the outbreak, hence demonstrating the usefulness of PFGE techniques as a tool in infection control measures in a hospital. PFGE results should always be considered in conjunction with the epidemiologic information and data. The bacterial isolates may also show some degree of clonal relatedness or diversity, thus helping in the determination of the sources, clonal relatedness and spread of the bacterial isolates in hospitals and countries where the isolates have been encountered.

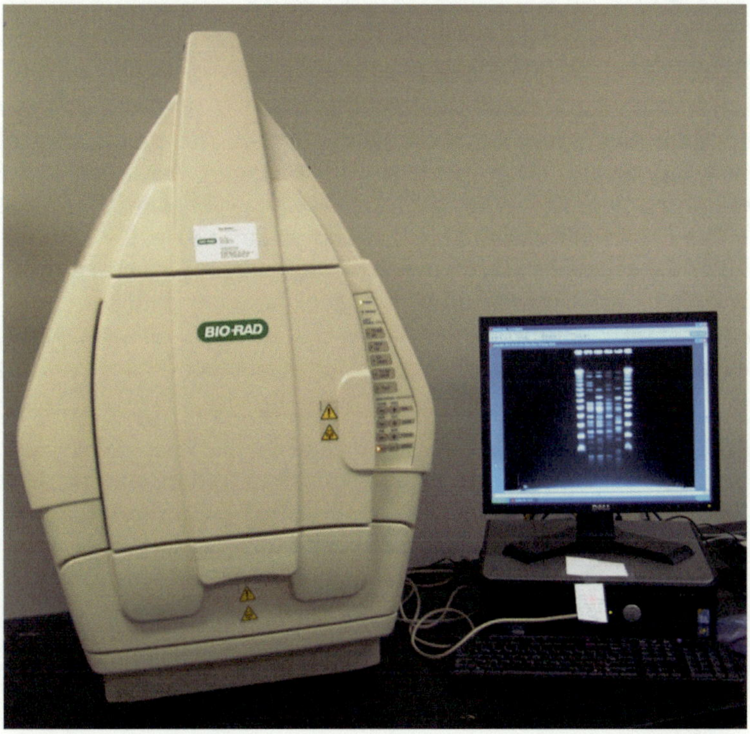

The Gel Doc image system captures picture of the stained gel with the bands and this is transmitted to a computer and monitor for better visualization and analysis

Fig. 2. Picture of Gel Doc (Bio-Rad Hercules, CA, USA) imaging system used to visually analyze images captured after staining the bands formed in the gels.

3. Results

3.1 Bacterial isolates

One hundred and ninety-eight bacterial isolates comprising 120 *K. pneumoniae* and 78 *E. coli* isolates from patients with ages ranging from 2 days old to 85 years had higher MIC values from the E-test assays and thus fulfilled the criteria for further molecular characterization. More than 70% of the isolates were recovered from female patients. The isolates involved in urinary tract infections were 60%. Skin and soft tissue isolates and infections contributed 30% of the isolates and the rest of the isolates were either from respiratory tract system (5%), blood streams (4%) or central nervous systems (1%) respectively. Most of these ESBL isolates were recovered from adult patients admitted into the medical (48%) and surgical (35%) facilities of the hospital. The rest were from patients seen in the paediatrics wards (9%), Obstetrics and gynaecology ward (5%), and intensive care units (3%).

3.2 Multiplex PCR gene detection

The multiplex PCR assay detected the 100% *bla*TEM genes, 25% *bla* SHV and 52% *bla* CTX-M genes among the *E. coli* isolates. Similarly, 94% *bla*TEM, 42% *bla*SHV and 70% *bla*CTX-M genes in the *K. pneumoniae* isolates were detected. All CTX-M genes were identified as alleles belonging to the phylogenetic group I.

3.3 Pulsed-field gel electrophoresis

The PFGE picture of all the isolates used for this study is partly represented in Figure 3 below. As depicted in the figure, the PFGE typing of the ESBL-producing isolates revealed various different and diverse DNA banding profiles among the *E. coli* and *K. pneumoniae* isolates. Bacteriophage lambda ladder PFGE marker (New England Biolabs) is all depicted on the lanes λ. The *E. coli* isolates are demonstrated on lanes 1 - 6, 7 - 12 while lanes 13 - 18 and 19 - 24 shows the *K. pneumonia* isolates. Note that except for lanes 4 and 7, all the lanes containing the isolates have significantly divergent banding patterns. From these results therefore, one would interpret the data as stating that all the *E. coli* or *K. pneumoniae* isolates are distinguishable by the PFGE and divergent from each other (>7 band difference).

There was no major clonal similarity or relatedness of either the *K. pneumoniae* or *E. coli* producing ESBL isolates regardless of which hospital facility the patient was admitted to or specimen the bacterial pathogen was recovered from. In addition, one could notice that from the Figure 2, the bands of the DNA were not separated (i.e. no resolution) for isolates in lanes 4 and 7. This phenomenon is called smearing and occurs when there is a contamination of the nucleases (agarose plug, buffer or reagents), or use of abnormal temperature and concentration of the buffers or wrong conditions which may all affect the enzyme. All these were the case with these isolates because when the tests were repeated with only the two isolates the bands were fully separated.

4. Discussions

The chromosomal DNA is the most fundamental component of identity of the cell and therefore represents a preferred method to assess the relatedness of the strains. And the PFGE method is the gold standard for now in assessing this property of the microbial agents

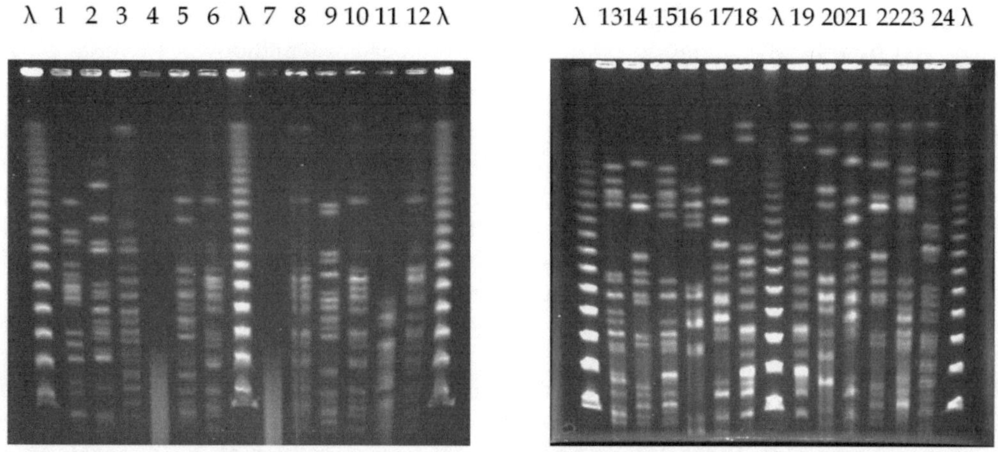

Figure 3 Picture depicting patterns generated by PFGE of *xba*1-digested chromosomal DNA obtained from *bla* TEM, SHV and CTX-M genes produced by *Escherichia coli* and *Klebsiella pneumoniae* isolates. Lane λ, bacteriophage lambda ladder PFGE marker (New England Biolabs), lanes 1 - 6, 7 - 12 *E. coli* isolates and lanes 13 - 18 and 19 - 24, *K. pneumonia* isolates. Smearing phenomenon occurred in lanes 4 and 7 hence the DNA particles were not completely separated or resolved.

Fig. 3. PFGE picture of *Escherichia coli* and *Klebsiella pneumoniae* ESBL producers.

including bacterial cells. All state public health laboratories in the USA as well as Centers for Disease Control and Prevention (CDC) perform molecular epidemiology testing using the PFGE. The PFGE assay can adequately be used to type several organisms including the ones involved in nosocomial infections or pathogens associated with food-borne diseases. PFGE is one of the most reproducible and highly discriminatory typing methods that is available and is a method of choice for many epidemiologic evaluations.

The PFGE typing method used in this study to characterize the ESBL-producing *Klebsiella pneumoniae* and *Escherichia coli* isolates showed various DNA banding profiles. These banding profiles were in no way similar or related to each other indicating their independent origin. This clonal diversity detected among these ESBL-producing isolates suggests that most of the strains have been unable to be maintained or spread in the different wards or facilities of the hospitals from where the bacterial isolates used in this present study were recovered from. This observation may challenge the many conventional thoughts about the nosocomial epidemiology of antibiotic resistance in the hospitals where the isolates were recovered in Trinidad and Tobago. But it is clearly obvious from the PFGE picture that the isolates were in no way closely related as there were different band patterns produced after restriction by the same enzymes under the same physical conditions.

The smearing phenomenon whereby the DNA particles were not completely separated or resolved that occurred and was observed in lanes 4 and 7 in Figure 2 highlights some of the drawbacks to using the PFGE method in studying molecular studies. Once the experimental or laboratory errors are eliminated, results obtained are perfect. Again, an arguement can be made or put forward that the procedures takes several days to be completed. The PFGE process can take less than 48 hours to complete. More time is expended in recovering the bacterial isolates in pure cultures from the clinical specimen because this is the time required for incubation and identification of the bacterial isolate. Thus the turnaround time tends to be long and this can also be a point against the use of PFGE. But despite the longer turnaround time PFGE method in performing molecular epidemiology of bacterial isolates still remains a gold standard for now.

Using the questionnaire to retrospectively review the laboratory and medical records, it was observed that these isolates did not significantly share the same patient demographics and occurrence periods. Despite being isolated mostly from urine of patients admitted in the medical and surgical facilities of the hospitals sharing significant patient demographics and isolate characteristics yet the ESBL enzymes differed. This clearly indicated that most ESBL-producing isolates were not sporadic but that multiple clones were widespread in the hospitals. The occurrence of these ESBL producing pathogens were definitely not from spread from one patient to another or from one ward to another. It must probably therefore be as a result of antibiotic use pressure stemming from the use and overuse of antibiotics such as third generation cephalosporins in these facilities and hospitals as already has been reported in the country **[Pinto Pereira LM et al, 2004; Akpaka PE et al, 2010].** This therefore calls for a need for continuous and active surveillance measures; and effective infection controls practices, most especially antibiotic stewardship which is nonexistent in these hospitals.

This is the first study to report bla*TEM*, bla*SHV* and bla*CTX-M* in the country that reveals that phylogenetic group 1 is the predominant CTX-M types prevalent in the hospitals. This study clearly indicated that CTX-M, mainly CTX-M-1 for ESBL-producing *E. coli* and *K. pneumoniae* was highly prevalent and probably endemic in Trinidad & Tobago. Most ESBL producers were resistant to oxyiminocephalosporins and other non-beta-lactam agents at high levels and exhibited a high rate of the MDR phenotype. The spread of ESBL-producing bacteria appeared to be polyclonal, and none of the major epidemic strains were identified.

5. Conclusions

In summary, this study reports the first extensive study regarding the prevalence and molecular characterization of ESBL genes and the epidemiology of ESBL-producing *E. coli* and ESBL-producing *K. pneumoniae* isolates causing infections in Trinidad and Tobago that was specifically and clearly delineated by the use of the PFGE method.

6. Acknowledgement

The authors wish to acknowledge the several laboratory technicians from the microbiology laboratory of the regional hospitals in Trinidad and Tobago who assisted in collecting the bacterial isolates used in this study.

Conflicts of Interests: None to declare

7. References

Akpaka PE, Legall B, Padman J. 2010. Molecular detection and epidemiology of extended-spectrum beta-lactamase genes prevalent in clinical isolates of *Klebsiella pneumoniae* and *E coli* from Trinidad and Tobago. West Indian Med J. 59(6):591-6.

Akpaka PE, Swanston WH. 2008. Phenotypic Detection and Occurrence of Extended-Spectrum Beta-Lactamases in Clinical Isolates of *Klebsiella pneumoniae* and *Escherichia coli* at a Tertiary Hospital in Trinidad & Tobago. The Brazilian Journal of Infectious Diseases;12(6):516-520.

Arbeit RD, Arthur M, Dunn R, Kim C, Selander RK, Goldstein R. 1990. Resolution of recent evolutionary divergence among *Escherichia coli* from related lineages: the application of pulsed-field electrophoresis to molecular epidemiology. J Infect Dis; 161:230-35

Boyd DA, Tyler S, Christianson S, McGeer A, Muller MP, Willey BM, Bryce E, Gardam M, Nordmann P, Mulvey MR, and the Canadian nosocomial infection surveillance program, Health Canada; 2004. Complete nucleotide sequence of a 92-kilobase plasmid harbouring the CTX-M-15 extended-spectrum beta-lactamase involved in an outbreak in long-term-care facilities in Toronto, Canada. Antimicrob Agents Chemother; 48:3758–64.

Carle, G. F., M. Frank, and M. V. Olson. 1986. Electrophoretic separations of large DNA molecules by periodic inversion of the electric field. Science 232: 65–68.

Clinical and Laboratory Standard Institute (CLSI) 2010. Performance standards for antimicrobial susceptibility testing: Twentieth Informational Supplement. CLSI document M100-S20; Vol 30 No. 1. CLSI, Wayne, Pa

Finney, M. 1993. Pulsed-field gel electrophoresis, p. 2.5.9–2.5.17. *In* F. M. Ausubel, R. Brent, R. E. Kingston, D. D. Moore, J. G. Seidman, J. A.Smith, and K. Struhl (ed.), Current protocols in molecular biology, vol. 1.Greene-Wiley, New York, N.Y.

Hall, L. M. C. 1994. Are point mutations or DNA rearrangements responsible for the restriction fragment length polymorphisms that are used to type bacteria? Microbiology 140:197–204.

Kaufmann, M E. 1998. Pulsed-field gel electrophoresis. Methods Mol. Med; 15:17-31

Monstein HJ, Östholm-Balkhed Å, Nilsson MV, Dornbusch, Nilsson LE. 2007. Multiplex PCR amplification assay for rapid detection of *bla* SHV, *bla* TEM and *bla* CTX-M genes in Enterobacteriaceae. APMIS; 115:1400-8

Paterson DL, Hujer KM, Hujer AM, Yeiser B, Bonomo MD, Rice LB, Bonomo RA, and the International Klebsiella study group. 2003. Extended spectrum b-lactamases in *Klebsiella pneumoniae* bloodstream isolates from seven countries: dominance and widespread prevalence of SHV-and CTX-M-type b-lactamases. Antimicrob Agents Chemother; 47:3553–60.

Pinto Pereira LM, Phillips M, Ramlal H, Teemu K, Prabhakar P. Third generation cephalosporin use in a tertiary hospital in Port of Spain, Trinidad: need for an antibiotic policy. BMC Infectious Diseases 2004; 4:59 doi: 10.1186/ 1471-2334-4

Quintiliani, R., Jr., and P. Courvalin. 1996. Conjugal transfer of the vancomycin resistance determinant *vanB* between enterococci involves the movement of large genetic elements from chromosome to chromosome. FEMS Microbiol. Lett. 119:359–364.

Sader HS, Pignatari AC, Lemme LL, Burattini MN, Taneresi R, Hollis RJ, Jones RN. 1993. Epidemiologic typing of multiple drug resistant *Pseudomonas aeruginosa* isolated from an outbreak in an intensive care unit. Diagn Microbiol Infect Dis; 17: 13-18

Schwartz DC, Cantor CR. 1984. Separation of yeast chromosome-sized DNAs by pulsed field gradient gel electrophoresis. Cell; 37(1):67–75

Singh A, Goering RV, Simjee S, Foley SL, Zervos MJ. 2006. Application of molecular techniques to the study of hospital infection. Clin Microbiol Reviews; 19 (3) 512 -3

Tenover FC, Abreit RD, Goering RV, Michelsen PA, Murray BE, Persing DH, et al. 1995. Interpreting chromosomal DNA restriction patterns produced by pulsed-field gel electrophoresis: criteria for bacterial strain typing. J Clin Microbiol; 33:2233—9.

Thal, L. A., J. Silverman, S. Donabedian, and M. J. Zervos. 1997. The effect of Tn*916* insertions on contour-clamped homogenous electrophoresis patterns of the *Enterococcus faecalis*. J. Clin. Microbiol. 35:969–972.

Woodford N, Fagan EJ, Ellington MJ. 2006. Multiplex PCR for rapid detection of genes encoding CTX-M extended-spectrum B-lactamases. J Antimicrobial Chemotherapy; 57(1):154-5

Part 3

Electrophoresis Application in the Analysis of Protein-Nucleic Acid Interaction and Chromosomal Replication

12

Electrophoretic Mobility Shift Assay: Analyzing Protein – Nucleic Acid Interactions

Carolina Alves and Celso Cunha
Center for Malaria and Tropical Diseases
Institute of Hygiene and Tropical Medicine, New University of Lisbon
Portugal

1. Introduction

Interactions between proteins and nucleic acids mediate a wide range of processes within a cell from its cycle to the maintenance of cellular metabolic and physiological balance. These specific interactions are crucial for control of DNA replication and DNA damage repair, regulation of transcription, RNA processing and maturation, nuclear transport, and translation.

The characterization of protein-nucleic acid interactions is essential not only for understanding the wide range of cellular processes they are involved in, but also the mechanisms underlying numerous diseases associated with the breakdown of regulatory systems. These include, but are far from being limited to, cell cycle disorders such as cancer and those caused by pathogenic agents that rely on or interfere with host cell machinery. More recently, it has been hypothesized that many neurological disorders such as Alzheimer's, Huntington's, Parkinson's, and polyglutamine tract expansion diseases are a consequence, at least in part, of aberrant protein-DNA interactions that may alter normal patterns of gene expression (Jiménez, 2010).

The electrophoretic mobility shift assay (EMSA), also known as gel retardation assay, is a regularly used system to detect protein-nucleic acid interactions. It was originally developed with the aim of quantifying interactions between DNA and proteins (Fried & Crothers, 1981; Garner & Revzin, 1981) and since then evolved to be suitable for different purposes including the detection and quantification of RNA-protein interactions. EMSA is most commonly used for qualitative assays including identification of nucleic acid-binding proteins and of the respective consensus DNA or RNA sequences. Under proper conditions, however, EMSA can also be used for quantitative purposes including the determination of binding affinities, kinetics, and stoichiometry.

EMSA is a commonly used method in the characterization of transcription factors, the most intensely studied DNA-binding proteins, and the largest group of proteins in humans, second only to metabolic enzymes. Their purification and identification is crucial in understanding gene regulatory mechanisms. Transcription factors are sequence specific DNA binding proteins that are usually assembled in complexes formed prior to transcription initiation. They bind discreet and specific DNA sequences in the promoter region functioning either as an activator or repressor of expression of the targeted gene through protein-protein interactions (reviewed by Simicevic & Deplancke, 2010). Transcription factors play essential roles during development and differentiation. It is well established that disruption of normal function of tissue-specific transcription factors, as a result of mutations, is often associated with a number of diseases including most forms of cancer, neurological, hematological, and inflammatory diseases. Additionally, transcription factors are often found differentially expressed in different pathologies suggesting an at least indirect involvement on the onset or progression of diseases. One of the most prominent examples of the involvement of transcription factors in development and progression of diseases is perhaps the p53 protein. p53 is a transcription factor involved in the modulation of expression of several genes that regulate essential cellular processes such as cell proliferation, apoptosis, and DNA damage repair (reviewed by Puzio-Kuter, 2011). Mutations in p53 that cause loss of function were reported in about 50% of all cancers. It is believed that this loss of function makes cancer cells more prone to the accumulation of mutations in other genes thus facilitating and accelerating the formation of neoplasias (reviewed by Goh et al., 2011).

In our laboratory, research is mainly directed to the study of host-pathogen interactions during hepatitis delta virus (HDV) replication and infection. HDV is the smallest human pathogen so far identified and infects human hepatocytes already infected with the hepatitis B virus (HBV). Both viruses have the same envelope proteins that are coded by the HBV DNA genome. HDV is, thus, considered a satellite virus of HBV. The HDV genome consists of a single-stranded, circular, RNA molecule of about 1700 nucleotides. This genome contains only one open reading frame from which two forms of the same protein, the so-called delta antigen, are derived by an editing mechanism catalyzed by cellular adenosine deaminase I. Both forms, small and large delta antigen, were shown to play crucial roles during virus replication: the small delta antigen is necessary for virus RNA accumulation and the large delta antigen plays an important role during envelope assembly (reviewed by Rizzetto, 2009). However, neither protein seems to display any known enzymatic activity. Accordingly, HDV is highly dependent on the host cell machinery for virus replication. It has been shown through EMSA that the small delta antigen binds *in vitro* to RNA and DNA without any specificity, which is in agreement with one of the roles attributed to the protein as a chaperone (Alves et al., 2010). Making use of different experimental approaches it was possible to identify a number of cellular proteins that interact with HDV antigens or RNA (reviewed by Greco-Stewart & Pelchat, 2010). However, the precise role played by most host factors during the virus life cycle remains elusive. Furthermore, it is highly consensual among HDV researchers that many other cellular factors that interact with delta antigens or HDV RNA remain to be identified and it is crucial to find those that interact with HDV RNA for a better insight on its replication and as possible targets for new therapies.

In this chapter we will review the principles of EMSA and its advantages and limitations for the quantitative and qualitative analysis of protein-nucleic acid interactions. The key parameters influencing the quality of protein samples, binding to nucleic acids, complex migration in gels, and sensitivity of detection will be discussed. Finally, an overview of the principles, advantages and disadvantages of methods that are an alternative to gel retardation assays will be provided.

2. Advantages and limitations

Since its first publication, in 1981, several improvements and variant techniques of EMSA were reported. Originally described as a method to qualitatively detect protein-DNA interactions, gel retardation assays rapidly became one of the most popular methods to map interaction sequences and domains not only in DNA but in RNA-protein interactions as well. EMSA was also adapted in order to allow the determination of quantitative parameters including complex stoichiometry, binding kinetics and affinity.

Several features made EMSA one of the most popular methods among researchers that study protein-nucleic acid interactions. Probably, the main advantages of EMSA when compared to other methods, as we will further discuss in the next sections, may be considered as follows: (1) EMSA is a basic, easy to perform, and robust method able to accommodate a wide range of conditions; (2) EMSA is a sensitive method, using radioisotopes to label nucleic acids and autoradiography, it is possible to use very low concentrations (0.1nM or less) and small sample volumes (20 μL or less; Hellman & Fried, 2007). Even though, less sensitive, non-radioactive labels are often used as well. These labels can further be detected using fluorescence, chemiluminescence or immunohistochemical approaches. Although less sensitive then radioisotopes, the wide variety of labels that can be used makes EMSA a very versatile method; (3) EMSA can also be used with a wide range of nucleic acid sizes and structures as well as a wide range of proteins, from small oligonucleotides to heavy transcription complexes; (4) Under the right conditions a gel retardation assay can separate the distribution of proteins between several nucleic acids within a single sample (Fried & Daugherty, 1998) or distinguish between complexes with different protein stoichiometry and/or binding site distribution (Fried & Crothers, 1981); (5) Finally, but not less important, it is possible to use both crude protein extracts and purified recombinant proteins enabling the identification of new nucleic acid-interacting proteins or characterization of specific proteins and its targets.

Despite its sensitivity, versatility and usually easy to perform protocols, EMSA is often considered to bear a number of limitations. Dissociation can occur during electrophoresis since samples are not at equilibrium during the run, thus preventing detection. Additionally, complexes that are not stable in solution may be stable in the gel requiring very short runs so that the observed pattern relates to what happens in solution. EMSA does not provide a straightforward measure of the weights or entities of the proteins as mobility in gels is influenced by several other factors. Also, EMSA does not directly provide information on the nucleic acid sequence the proteins are bound to. However, this problem may usually be overcome using footprinting approaches as described further ahead. Kinetic studies using EMSA are limited since the time resolution for a regular EMSA protocol consists of the time required to mix the binding reaction and for the electrophoretic migration to occur before the mix enters the gel. Only processes that have relaxation times larger than the interval required for solution handling are suitable for kinetic studies.

3. How complexes migrate in gels

In this section, we will start with a simple account of the characteristics of the electrophoretic mobility of nucleic acids alone, and afterwards we will discuss how the formation of protein-nucleic acid complexes alters these characteristics.

In a non-denaturing agarose or polyacrylamide gel and conventional buffer conditions the nucleic acids, being negatively charged, will migrate towards the anode when electric current is applied. The gel will then act as a sieve selectively impeding the migration in proportion to the nucleic acid molecular weight, which is generally proportional to its charge. Therefore, and as the weight is approximately related to chain length, the length of nucleic acid is estimated by its migration. There is though another property that affects gel migration that is the topology of the nucleic acid (conformation, circularity) making the molecules seem longer or shorter than they really are. Secondary and tertiary structures can be removed using denaturing agents (for example, formaldehyde, formamide and urea) allowing for the electrophoretic mobility to become a simple function of molecular weight. Obviously, this denaturing step cannot be applied in a gel retardation assay as it would impede the interaction between the protein and nucleic acid.

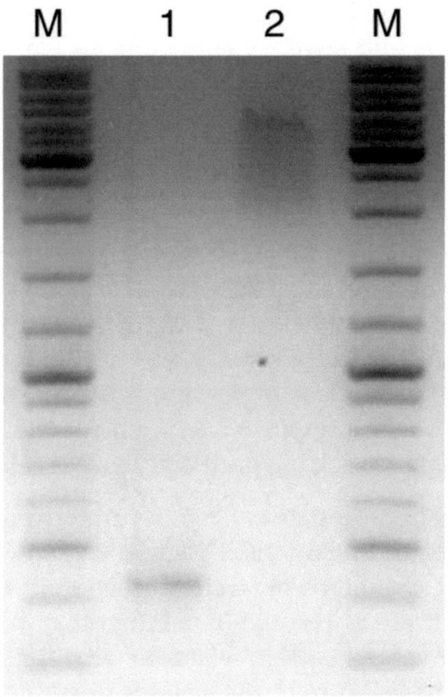

Fig. 1. Example of an electrophoretic mobility shift assay. An unlabeled DNA of 400 base pairs (bp) was incubated in a phosphate buffer (137mM NaCl, 2.7mM KCl, 4.3mM Na_2HPO_4, 1.5mM KH_2PO_4, pH 7.4) in the absence (1) or presence (2) of 2μM of small delta antigen. The samples were loaded onto a 1.5% agarose gel and after electrophoresis in TAE buffer (40mM Tris acetate, 1mM EDTA) the DNA was stained with ethidium bromide. (M) represents the molecular weight marker (GeneRuler DNA Ladder mix, Fermentas).

When a protein is added to the mix and interacts with the nucleic acid forming complexes it results in a change in gel migration relative to that of the free nucleic acid. This shift is mainly due to an obvious increase in the molecular weight, the adjustment of charge and eventual changes in the nucleic acid conformation. In figure 1 we give an example of an EMSA study where the small delta antigen was added to DNA. It is clear that the addition of the small delta antigen (Fig.1. well 2) to a 400bp DNA fragment results in the formation of a complex with decreased gel mobility when compared with the unbound DNA (Fig.1. well 1). We can conclude that under our *in vitro* binding conditions, the small delta antigen interacts with the given 400bp DNA fragment causing a clear mobility shift.

It is expected that when protein binds a nucleic acid fragment there will be a decrease in relative mobility and if the protein doesn't induce any appreciable bend on the nucleic acid then the conformational contribution to the decrease is small. Although an increase in the protein molecular weight results in reduction of gel migration it has been reported that the increase of the nucleic acid length can have the opposite effect. This was reported for the Lac repressor bound to DNA fragments of increasing sizes, which resulted in an increase of relative mobility (Fried, 1989). This observation indicates that the ratio of protein and nucleic acid weights is more important in the migration than the absolute weight of the complex. Another interesting study reports that the binding of protein to a nucleic acid can accelerate mobility. This was observed for relatively large linear DNA binding to a protein from the hyperthermophilic *Methanothermus fervidus* that was shown to induce nucleic acid condensation (Sandman et al., 1990). In this case the conformational change of the DNA is a stronger factor than the weight increase, causing acceleration rather than a decrease in relative mobility.

Overall, the conformational features that influence gel migration of protein-nucleic acid complexes are not thoroughly studied and questions are only raised when exceptions emerge such as the ones mentioned above. Nowadays, the EMSA method is almost exclusively used to analyze the interaction between proteins and nucleic acids and to a lesser extent its conformations that can influence gel migration. When exceptions arise and the retardation pattern is not exactly as predicted, it can still point out clearly whether the molecules are interacting or not. In the end, the exact location of the resulting gel bands cannot be predicted but the answer is usually unambiguous.

External factors can also influence the separation of the bound or unbound nucleic acid such as the nature of the gel matrix and temperature during electrophoresis. Generally, the best resolution is obtained with the smallest pore diameter that allows the migration of unbound nucleic acid. However, if large complexes are expected there should be a compromise in pore size so that they can enter the gel matrix. As will be discussed below, polyacrylamide gels offer the best conditions for small complexes and nucleic acid fragments. On the other hand, agarose gels are more suitable for larger aggregates.

The detection of a protein-nucleic acid complex within a gel depends critically on the resolution obtained between unbound nucleic acid and the formed complexes as well as its stability within the gel matrix. In most cases, the gel matrix is expected to stabilize the preformed complex as it impedes the diffusion of dissociating components maintaining the concentration of protein and nucleic acid (and complex) at levels as high or higher than those achieved in the equilibrium binding reaction. This of course is compromised if for instance the salt concentration in the binding reaction differs largely from that in the electrophoresis/gel buffer, resulting in an adjustment in salt concentration that could disrupt the complexes formed. As the gel retardation method is an *in vitro* assay, when extrapolating to the *in vivo* conditions one must be careful as the former may provide favorable binding conditions that are not achieved at physiological concentrations.

4. The method

There are five focal steps in a conventional EMSA protocol that involve different variables susceptible to optimization: (1) preparation of protein sample; (2) synthesis and labeling of nucleic acid; (3) binding reaction; (4) non-denaturing gel electrophoresis and (5) detection of the outcome. In this segment we will discuss each step separately mentioning the key variables in each one and the options available for any given situation. Figure 2 represents schematically the regular steps in a gel retardation assay that will be discussed below. Whenever possible we will also refer to examples in the literature.

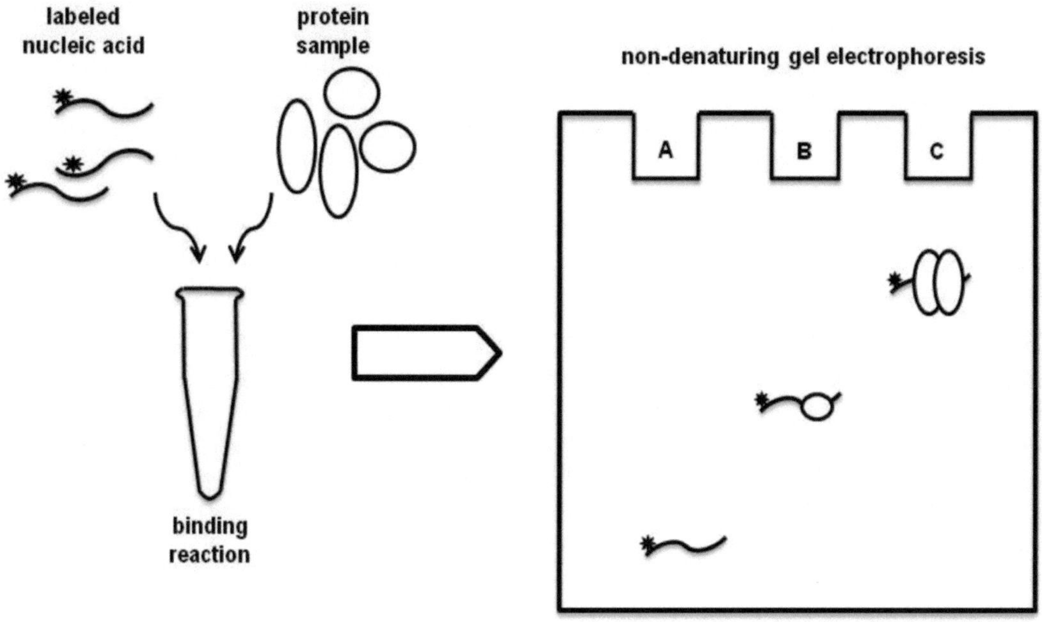

Fig. 2. A schematic representation of a conventional EMSA protocol. The labeled nucleic acid, simplified as lines with a star representing the label, is mixed with the protein sample, represented by the oval shapes, in a binding reaction and then loaded into a non-denaturing gel. After electrophoresis the result is detected according to the label in the nucleic acid. On the schematic gel (A) represents a well on which only the labeled nucleic acid was loaded. The free nucleic acid is expected to have more mobility than the bound molecules. In well (B) is symbolized a labeled nucleic acid binding to one small peptide and in well (C) is binding to two larger proteins. The heavier complex (in C) is expected to display the lowest mobility during electrophoresis and therefore is closer to the beginning of the gel.

4.1 Preparation of the protein sample

Regarding the protein sample, the EMSA can be divided into two categories based on whether the nucleic acid-interacting protein is known or not. Therefore, preparing the protein sample will depend on which category it falls, in order to obtain an optimal performance.

When faced with a putative nucleic acid-binding protein or complex of completely unknown subcellular origin, whole cell extracts must be used. If there is an educated guess on the nature of the protein, it is advisable to isolate nuclear and cytoplasmic proteins from crude extracts improving the results. Particularly, if the binding protein is thought to be nuclear and in low abundance, the isolation of nuclear extracts will prevent the dilution that would occur if whole cell extracts were used, which could render the concentration too low for the protein to be even detected.

Cell extracts are easy and relatively fast to obtain and the methods are commonly derived from the protocol described by Dignam and collaborators almost three decades ago (Dignam et al., 1983). This method isolates both nuclear and/or cytoplasmic proteins suitable for later analysis using EMSA. One disadvantage in preparing cell extracts is its crudeness; they generally degrade faster than purer preparations due to the presence of cellular proteases. To limit protein degradation or alteration the protocol should be performed on ice or at 4ºC and protease inhibitors should be added. A control test can easily be performed to assess the viability of the extract by using ubiquitous DNA probes (Kerr, 1995). If these fail than the cell extract might be "dead". Despite its disadvantages cell extracts are needed when the interest lies in identifying new nucleic acid-binding proteins or when a complex of different proteins is needed to interact with the target nucleic acid as sometimes one recombinant protein cannot bind by itself. Tissue samples can also be a source of protein sample for these assays. The same care should be taken as in whole cell extracts to minimize the activity of proteases.

If the nucleic acid-binding protein is known then recombinant proteins can be expressed and purified. Recombinant or heterologous proteins are commonly expressed in bacteria or an eukaryotic cell line of interest. Fusion proteins of the target are generally constructed with a tag to facilitate purification. Common tags, such as glutathione-S-transferase (GST), tandem affinity purification tag (TAP tag), maltose binding protein (MBP) or 6xHistidine, are cloned in frame with the protein. Sometimes it is possible to include a protease cleavage site between the protein of interest and the tag so the latter can be easily removed after purification. Even though a tag can be very helpful, it should be taken into account that it can alter the recombinant protein conformation and even disrupt its binding ability. On the other hand they can be helpful in stabilizing the protein terminus they are close to. A careful study is needed when choosing the tag and usually small peptides are preferred to minimize its impact on the recombinant protein of interest.

There are several systems available for the production of heterologous proteins of which bacterial extracts of *Escherichia coli* are one of the most widely used. This Gram-negative bacterium remains an attractive host due to its ability to grow rapidly and with high density using inexpensive substrates. Its genetics has been well characterized for quite some time and there is a wide range of cloning vectors as well as mutant host strains that make it such a versatile system. Typically, the heterologous complementary DNA is cloned into a compatible plasmid which is then transfected into the bacteria to achieve a high gene dosage. This doesn't necessarily guarantee the accumulation of high levels of a full-length active form of the recombinant protein but other efforts can be made to improve that. To achieve high-level production in *E. coli* strong promoters should be used such as the bacteriophage T7 late promoter, and usually the T7 polymerase is also present under IPTG (isopropyl-β-D-1-thiogalactopyranoside)-induction. In the past years several strains have been engineered to improve the recombinant protein yields through efforts to increase mRNA stability as well as improve transcription termination and translational efficiency (reviewed by Baneyx, 1999 and Makino et al., 2011). However, this extensively used system for protein overexpression has an important drawback when studying eukaryotic proteins. The bacterial systems are not able to perform post-translational modifications that would eventually happen *in vivo* in eukaryotic cells.

When working with recombinant nucleic acid-binding proteins it should be taken into account the importance of post-translational modifications on the protein's binding ability. A careful research of previous reports might hint if it is necessary to perform modifications prior to the binding reaction. In some cases post-translational modifications change the sequence-specificity of the binding. For example, genotoxic stress induces modifications on the C-terminus of the tumor suppressor protein p53 that modulate its DNA-binding specificity (Apella & Anderson, 2001). If the modifications are crucial, rather than using bacterial extracts a more biologically relevant host should be considered. Transient gene expression in mammalian cells has become a routine approach to express proteins in cell lines such as human embryonic kidney cells. The benefits are obvious for the production of eukaryotic proteins in mammalian cells as post-translational modifications will likely be native or near-native, solubility and correct folding are more likely to occur as well as expression of proteins in their proper intracellular compartments. These methods, however, tend to be more expensive as cells need a more complex growth media and there is a lower diversity in cloning vectors. To get out of the latter limitation an alternative approach uses baculovirus-infected insect cells. In this method a recombinant virus is produced either by site-specific transposition of an expression cassette into the shuttle vector or through homologous recombination (reviewed by Jarvis, 2009).

When expressing recombinant proteins, sometimes, the heterologous genes interfere severely with the survival of the host cell. For toxic proteins produced in *E. coli* strains there are some techniques available to get around this problem. A highly toxic gene can be defined as a gene that, when introduced into a cell, causes cell death or severe growth and maintenance defects even prior to expression induction. The best solution for expressing a highly toxic gene is to enable the host to tolerate it during the growth phase, so that after induction an efficient expression ensures a rapid and quantitative production of the toxic protein before the cell dies (reviewed by Saida et al., 2006). This can be achieved by different strategies such as manipulation of the gene's transcriptional and translational control elements, for example, by suppressing basal expression of the toxic protein from leaky inducible promoters. Managing the coding sequence to produce reversible inactive forms or controlling the plasmid copy number is also an option as well as selecting less susceptible *E. coli* strains or adding stabilizing sequences.

Cell-free systems are also available to express recombinant proteins including *in vitro* transcription\translation systems such as rabbit reticulocyte systems, wheat germ based systems or *E. coli* cell-free protein expression systems (reviewed by Endo & Sawasaki, 2006). Here, proteins can be expressed directly from cDNA templates obtained through PCR, avoiding subcloning which makes it a faster method by skipping this step, and eventually cheaper. It can also be used to express proteins that seriously interfere with the cell physiology such as the toxic proteins mentioned above. On the other hand these methods usually achieve smaller yields than for instance bacterial extracts approaches.

4.2 Synthesis and labeling of nucleic acids

One of the key advantages of EMSA is its versatility as it can be performed using a wide range of nucleic acid structures and sizes. This method can characterize both double- and single-stranded DNA as well as RNA, triplex and quadruplex nucleic acids or even circular fragments. The probe design and synthesis depends on the application or purpose of the study and is a significant aspect, as it will influence the detection and therefore the sensitivity of the results. There are two main aspects to consider in this step: the length of the nucleic acid and its labeling.

Unlabeled nucleic acids can be used in a gel retardation assay and be detected by post-electrophoretic staining with chromophores or fluorophores that bind nucleic acids or in the "classical way" using ethidium bromide. However the use of labeled nucleic acids is usually preferred as it can facilitate detection and add sensitivity to the method. The most common choice is radioisotope labeling as it offers the best sensitivity without interfering with the structure of the probe. A higher sensitivity makes it ideal for assays that have a limited amount of starting material. The radioisotope, usually ^{32}P, can be incorporated in the nucleic acid during its synthesis, by the use of labeled nucleotides, or afterwards via end labeling using a kinase or a terminal transferase. With a radioactive label the EMSA results can be easily detected by autoradiography. Even if radioisotope labeling confers high sensitivity to the method it implies handling hazardous radioactive material requiring extra safety measures that may not be available. Other labels can be used as alternatives that, even though are less sensitive, are a lot safer to manipulate and more stable such as fluorophores, biotin or digoxigenin (Holden & Tacon, 2011). When these molecules are used detection is achieved by chemiluminescence or immunohistochemistry. Although, in general radioisotope labeling achieves higher sensitivity there are some reports that similar results can be obtained with other labels such as Cyano dye Cy5 (Ruscher et al., 2000).

Although the most common approach is the labeling of the nucleic acid probe there are protocols available that employ protein labeling at the same time. For example, Adachi and co-workers suggest the use of an iodoacetamide derivative labeling of the thiol residue of cysteins (Adachi et al., 2005). Using radioisotope labeled DNA mixed with a nuclear protein extract they perform a conventional EMSA and after detection by autoradiography the complexes are eluted from excised gel bands and treated with 5-iodoacetamidofluorescein for protein labeling. The sample is then loaded onto a denaturing gel and after electrophoresis is transferred to a membrane and detected with anti-fluorescein antibody. This allows the characterization of the proteins in the complex giving information on how many proteins are present and their molecular weight. However it is not able to detect proteins without cystein residues.

Regarding the length of the nucleic acid probe, it depends on what is being studied. If one is looking for specific binding sites, small probes can be used to assess with each segment the protein will interact. The use of short nucleic acids has several advantages as they are easily synthesized and inexpensive to purchase; a small sequence has less non-specific binding sites (it should be particularly advantageous when a protein has low sequence-specificity); the electrophoretic resolution between complexes and free nucleic acid is higher so shorter electrophoresis times can be used. Nevertheless, in a short sequence the binding sites are closer to the molecular ends which can cause aberrant binding and it can be tricky to resolve the free nucleic acid from the complexes formed if these have a very high molecular weight.

On the other hand, the longer nucleic acid targets avoid these problems but will have more non-specific binding sites and the mobility shift is generally smaller requiring longer electrophoresis times as they run more slowly. A compromise needs to be reached depending on what the EMSA study is trying to achieve.

4.3 Binding reaction

The interaction between proteins and nucleic acid is sensitive to salt concentration and pH as it will influence the protein charge and conformation. However, the experimental conditions are very versatile in that different buffers can achieve good results. The most commonly used are Tris based buffers but other options include 4-(2-hydroxyethyl)-1-piperazineethanesulfonic acid (HEPES), 3-(N-morpholino)-propanesulfonic acid (MOPS), and glycine or phosphate buffers. Naturally, it is advisable to provide an environment as close as possible to physiological conditions so the data obtained *in vitro* can be related to what happens *in vivo*.

Additives can be included in the binding reaction either if the interactions require the presence of co-factors or stabilizing agents, or as helpful components to minimize non-specific binding. Glycerol or other small neutral solutes, for example sucrose, can be added to the binding mixture to stabilize labile proteins or enhance the stability of the interaction (Vossen et al., 1997). These solutes are used at final concentrations of 2M or less, as higher concentrations might interfere with the sample's viscosity and complicate handling. Other assays may require the presence of co-factors for a correct interaction such as the presence of cAMP for the *E. coli* CAP protein (Fried & Crothers, 1984) or ATP for human recombinase Rad51 (Chi et al., 2006). Non-ionic detergents are used to maximize protein solubility. In this case, the concentrations used depend on the detergent and system under study. Nuclease and phosphatase inhibitors can be useful as well as protease inhibitors, which as mentioned before, are particularly important when the protein sample comes from cell extracts. These inhibitors are commercially available and the concentration depends on the manufacturer's instructions. Some of the additives mentioned, particularly those involved in stabilizing the formed complexes can be included not only in the binding mixture but also in the gel buffers.

To minimize non-specific loss of protein the addition of a carrier protein (less than 0.1mg/mL) such as bovine serum albumin can be very helpful. The addition of unlabeled competing nucleic acids is suitable when there are secondary binding activities that mask the relevant one. Of course this only works if the protein interacts with the target nucleic acid with greater affinity then its competitor and the secondary binding does not discriminate between the sequences. Since the presence of a competing nucleic acid will always reduce the amount of specific binding, testing different competitors and concentrations is needed to optimize the assay. Another option to circumvent the problem of non-specific binding is the addition of salt at concentrations that will disrupt non-specific ionic bonds but leave the more specific interactions unimpaired.

4.4 Non-denaturing gel electrophoresis

After the binding reaction the free nucleic acid is separated from the formed complexes by non-denaturing gel electrophoresis. EMSA can be performed on polyacrylamide or agarose gels depending mainly on the size of the nucleic acid and desired resolution. The average pore size is estimated to be around 5 to 20nm in diameter for 10 and 4% acrylamide gels respectively (Lane et al., 1992). Typically the higher concentration gels are used for oligonucleotides and small RNAs and the lowest concentration for DNA fragments of around 100bp. A polyacrylamide gradient gel is sometimes preferred over linear gels as the gradient in pore size increases the range of molecular weight fractioned in a single run, which is particularly important when the complex has a much higher weight than the free nucleic acid (Walker, 1994). When complexes of different composition are formed, the gradient gels are also more likely to separate those with close molecular weight.

Agarose gels, on the other hand, have a pore size of around 70 to 700nm (Lane et al., 1992) in diameter and are therefore mostly used in assays with larger nucleic acid fragments or when large protein complexes are expected. Overall, polyacrylamide gels offer a better resolution for nucleic acid-protein complexes with a molecular weight of up to 500,000Da (Fried, 1989 as cited in Hellman & Fried, 2007).

Regarding the electrophoresis buffers, it should be taken into account the fact that the interaction between nucleic acids and proteins involves an ionic component. Therefore, the buffer's ionic strength and pH are important features that play a role in the complex stability. Although this is a very important factor there hasn't been, to our knowledge, any thorough study on the subject. The choice of electrophoresis buffers is varied and generally low ionic strength buffers are preferred and sometimes coincide with the buffer used in the binding reaction. Buffers with a medium salt concentration help stabilize the complexes, generate less heat during electrophoresis and also increase the speed of migration. High salt concentrations not only disrupt the complexes but also interfere with its movement into the gel matrix and lead to significant heating during the electrophoresis. Too low salt concentrations can also disrupt the stability of the preformed complexes as well as separate a double stranded DNA template (Kerr, 1995). The most common buffers are TBE (90mM Tris-Borate, 2mM EDTA, pH 8) and TAE (40mM Tris-Acetate, 1mM EDTA, pH 8). However, there are some complexes that cannot be detected with the classical buffers. For example the complexes formed between phage Mu repressor and its operators have an electrophoresis buffer-dependent stability and require Tris-glycine buffer at pH 9.4. (Alazard et al., 1992 as cited in Lane et al., 1992).

Particularly, in agarose gels it is important to monitor the temperature during electrophoresis to prevent the gel from heating up which could result in dissociation of the nucleic acid-protein complexes. Some cases may require that pre-cooling of the gel or even that the electrophoresis proceeds at lower than room temperatures, which can be achieved with special refrigeration devices.

4.5 Detection

The detection of an EMSA result will naturally depend on the labels used if any has been used. The results uncovered can involve the detection of the mobility shift between free nucleic acid and the complexed form or the detection of the mobility shift of free protein and the complexes.

Looking at the nucleic acid component without any label added the shift in mobility can be detected by staining with molecules that bind nucleic acids. Different products can be used ranging from the classic but hazardous ethidium bromide to other chromophores or fluorophores such as RedSafe DNA Stain (ChemBio) or SYBR® Safe DNA gel stain (Invitrogen). When the nucleic acid has been previously labeled the detection methods depend on the nature of the label. A ^{32}P radioisotope is one of the easiest and most sensitive methods to detect nucleic acids but it's a hazardous material to work with. Other very common labels are biotin, digoxigenin or fluorophores. These labels are innocuous but usually give less sensitive results and the detection procedure can involve extra steps such as transfer to a membrane and incubation with primary and secondary antibodies as well as intermediate washing steps. The results in these cases can be observed by immunohistochemistry or chemiluminescence approaches.

The detection of protein mobility shift involves less direct methods, meaning, extra steps such as a denaturing step and electrotransfer onto a membrane, may be necessary as they are usually immunodetected. If the protein of interest is known, and a specific antibody is available, it can be used in detection. If not, a method such as the one discussed above, proposed by Adachi and colleagues that involves labeling the thiol group of cysteins and using an antibody against the label. Stepwise, the easier way to detect protein in an EMSA is by labeling it with radioisotope, a method designated by reverse EMSA that will be discussed ahead. This procedure has the disadvantage of working with radioactive material but the mobility shift can be visualized by autoradiography.

5. EMSA applications

The gel retardation assay has been used under different conditions in order to achieve specific results. The method is useful in studying not only the interaction between proteins and nucleic acids but also in assessing nucleic acid conformational characteristics. It can be used to characterize bends in the DNA double helix with polyacrylamide gels and comparative measurements (for an example Crothers & Drak, 1992) or to detect complexes formed with super coiled DNA being sometimes designated as topoisomer gel retardation (for examples see Palecek, 1997; Nordheim & Meese, 1988). In this section we mention how a gel retardation assay can help characterize protein-nucleic acid interactions.

5.1 Binding constants

Although EMSA is most commonly used as a qualitative assay it can, under certain conditions, provide quantitative data for relatively stable complexes. One of its earliest applications was in the measurement of kinetic and thermodynamic parameters. The association rates are determined by mixing the complex components at known concentrations and loading them in a running gel at precise intervals (for an example Spinner et al., 2002). For dissociation rates, a time course experiment is done by addition of competing nucleic acid to the preformed complexes (Fried & Crothers, 1981). The binding constant can be determined by the amount of complex formed as a function of protein concentration at equilibrium or as a ratio of the association and dissociation constants (for an example Demarse et al., 2009). An alternative method to measure kinetic and thermodynamic constants is the nitrocellulose filter binding assay that will be mentioned below.

As an example we show in figure 3 the titration of a DNA with the small delta protein to assess binding constants. The binding reaction was done by incubating the samples in a phosphate buffer during the same period of time (10 minutes) and then loading them onto an agarose gel for electrophoresis. It is clear that when the protein is present at only 0.25μM it does not interfere with the DNA mobility (Fig.3. well 2) as the band covered the same distance as the first sample, in which the protein was not present (Fig.3. well 1). But when 1.5μM of the small delta antigen are present in the binding reaction there is almost no free DNA present and the majority of the molecules are bound in a complex (Fig3. well 5). In the intermediate concentrations it can be clearly observed the decreasing presence of free DNA and increasing DNA-protein complexes as the protein concentration raises. We can consider that the dissociation constant can be estimated by quantifying the disappearance of the free DNA band (Demarse et al., 2009). From figure 3 we can say that the apparent dissociation constant is between 1 and 1.5μM.

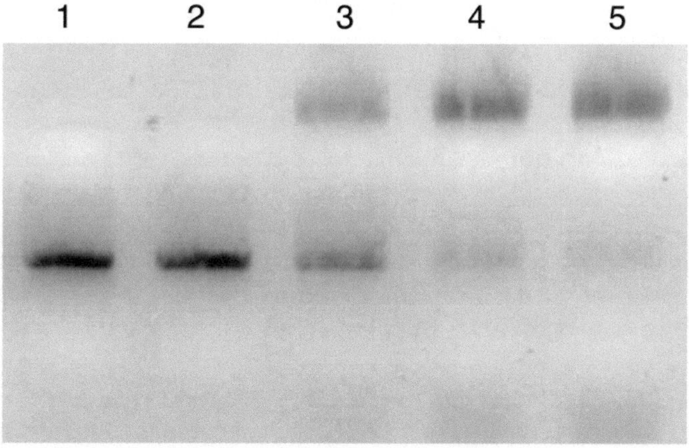

Fig. 3. Titration of a 500bp DNA fragment with the small delta antigen to estimate binding constants. An unlabeled 500bp DNA complementary to part of the HDV RNA was incubated, in a phosphate buffer (137mM NaCl, 2.7mM KCl, 4.3mM Na_2HPO_4, 1.5mM KH_2PO_4, pH 7.4), with increasing concentrations of small delta antigen of 0; 0.25; 0.5; 1; and 1.5μM and samples were loaded onto wells 1, 2, 3,4 and 5, respectively. Electrophoresis was in a 1.5% agarose gel in TAE buffer and the DNA was stained with ethidium bromide.

5.2 Cooperativity

Proteins can bind nucleic acids in a cooperative manner, that is, the complexes formed involve the binding of more than one protein to a specific nucleic acid segment. These multiprotein complexes may be a consequence of direct protein-protein interaction needed for nucleic acid binding, or a protein-induced deformation of the nucleic acid is a prerequisite to facilitate the binding of a second protein, or it may result from the bringing together of molecules bound at distinct sites in the nucleic acid sequence. The cooperativity can be inferred in a gel retardation assay from the underrepresentation of intermediate complexes between the unbound and saturated states. Multiprotein complexes can be comprised of a single protein species forming a homomultimer or of different proteins. The latter can be easily characterized by EMSA by the stability of the complexes formed with one protein in the presence or absence of the other(s).

5.3 Stoichiometry

Determining the important parameter that is stoichiometry is not as easy a task as it seems. The apparent weight changes estimated from the complexes' gel mobility are not applicable in determining the stoichiometry due to complications of charges and conformational effects on gel migration. A different approach is needed. The presence of truncated or extended protein derived from the wild-type but with the same binding and multimerization capacity will originate new bands that can reflect the monomers bound to the nucleic acid (Hope & Struhl, 1987). A similar method that will be discussed in the next segment is the supershift EMSA that uses an antibody specific for the binding protein recognizing an epitope that is accessible while the protein is bound to the nucleic acid. The addition of the antibody to the preformed complex can provide an estimate of the number of proteins bound by the extent of increments in retardation (Michael N & Roizman B, 1991 as cited in Lane & Prentki, 1992).

A more complex approach has been proposed in 1988 to determine a complex's stoichiometry (Granger-Schnarr et al., 1988). After the separation of the free and the complexed nucleic acid on a non-denaturing gel, the proteins are transferred to a membrane after sodium dodecyl sulfate (SDS) denaturation. This then allows the detection of proteins directly or indirectly using a specific antibody. The protein bands as well as the nucleic acids autoradiograph are then quantified by densiometry and the relative stoichiometry can be determined. The need for a specific antibody limits this method to complexes formed by well known proteins with available antibodies.

6. EMSA variants

Over the years variations or coupling of the EMSA protocol with other methods has been proposed to enhance its results or obtain more information from one experiment. Some examples of these EMSA-based approaches will be presented.

6.1 Reverse EMSA (rEMSA)

A reverse EMSA consists in labeling the protein sample rather than the nucleic acid (Filion et al., 2006). This method shows the difference in mobility between the free protein and nucleic acid-bound protein. It is an approach that can facilitate the determination of the protein binding affinity using different nucleic acids. Because the label used is ^{35}S instead of ^{32}P it is less sensitive than the conventional EMSA due to the isotope's energy.

6.2 Supershift EMSA

The supershift EMSA uses the same protocol as a regular EMSA except in that an antibody against the binding protein is added. As a result there is a more marked mobility shift during electrophoresis because the antibody will increase the overall complex molecular weight, hence the term supershift. This method can help identify if the proteins present in the complex have a specific epitope and is also used to validate previously identified proteins. It can also improve resolution when the difference between free nucleic acid and the complex is very small.

6.3 Multiplexed competitor EMSA (MC-EMSA)

The multiplexed EMSA was developed in 2008 by Smith and Humphries to characterize nuclear protein and DNA interactions, namely with transcription factors. In this method the nuclear extract is incubated with a pool of unlabeled DNA consensus competitors prior to adding the labeled DNA probe. An initial EMSA run will determine which cocktail competes with the probe binding to nuclear proteins which will then run individually in another EMSA to determine the precise competitor (Smith & Humphries, 2008). It is a competition-based method to identify uncertain DNA binding proteins requiring only a prior knowledge of transcription factor consensus sequences.

6.4 Two-dimensional EMSA (2D-EMSA)

The two-dimensional EMSA is a process that combines EMSA with proteomic or sequencing techniques to identify the proteins or the nucleic acid sequences that are present in the formed complexes. Two slightly different protocols have been developed to identify the interacting proteins and another method aims at the target nucleic acid sequence.

An initial approach was proposed by Woo and colleagues as they tried to identify and characterize transcription factors (Woo et al., 2002). A crude nuclear extract is partially purified by gel filtration and the resulting fractions are then bound to the nucleic acid probe and analyzed by EMSA. Meanwhile, in parallel, the pI and molecular weight of the putative interacting protein(s) is estimated as the fractions are analyzed by isoelectric focusing or SDS-Polyacrylamide Gel Electrophoresis (SDS-PAGE) in order to characterize possible candidates. Next, spots with the predetermined pI and molecular weight of the candidates are excised from a two-dimensional array of nuclear proteins and the proteins are eluted, renatured and tested for their binding ability through EMSA and the spots are afterwards analyzed by mass spectrometry for protein identification. This method is limited to proteins that can re-form into functional nucleic acid-binding conformations after the denaturing SDS-PAGE step, although EMSA can still show results even if renaturation efficiency is low. Because the final EMSA step that confirms the binding is performed with protein eluted from single spots it is only possible to identify proteins that interact with the nucleic acids as monomers or homomultimers. Proteins that only interact when complexed with other proteins will give a negative result on the validation EMSA.

A similar 2D-EMSA technique has since then been developed that incorporates EMSA into a two-dimensional proteomics approach by replacing the isoelectric focusing with EMSA as the first dimension of the 2D method (Stead et al., 2006). The protein sample, in the presence or absence of the nucleic acid, is separated by native PAGE as in a conventional EMSA. The protein bands from both conditions are then separated in a second dimension by denaturing SDS-PAGE. The proteins showing the nucleic acid dependent shift in mobility can be extracted from the gel for mass spectrometry identification. This approach does not require any previous knowledge of the chemical or physical properties of the binding protein and does not require protein renaturation after gel excision. It is also not limited to identify proteins that bind by themselves or as homomultimers and allows the characterization of complexes composed of different proteins.

These 2D approaches were developed by the two groups to study transcription factors, therefore, double stranded DNA is used as a nucleic acid probe but they can also adapted to other nucleic acid probes making them quite versatile methods to identify nucleic acid-interacting proteins.

Chernov and collaborators have developed a similar protocol with two dimensions but instead of aiming to identify the interacting protein(s) it characterizes and maps the specific protein target sites in regions of the human genome (Chernov et al., 2006). This approach is also based on first separating the complexes from the free nucleic acid in a non-denaturing gel and afterwards separating it under denaturing conditions (Vetchinova et al., 2006). The group used a pool of radioisotope-labeled short DNA sequences covering the genome region of interest and mixed it with a nuclear extract from a specific cell line. The formed complexes were separated in a non-denaturing one-dimensional standard EMSA. The complexes were localized by autoradiography and the gel strip containing them was excised and treated with a denaturing agent, SDS, to disrupt the preformed complexes. The strip is then loaded onto the second-dimension denaturing gel and another electrophoresis is performed. The gel is autoradiographed to determine the location of the freed DNAs, which are afterwards cut from the gel to be analyzed. By pairing this method with high-throughput sequencing the authors were able to identify a multitude of specific protein binding sites within a given genomic region.

6.5 EMSA-three-dimensional-electrophoresis (EMSA-3DE)

A three dimensional approach has very recently emerged to purify nucleic acid binding proteins from complexes separated by EMSA (Jiang et al., 2011). This method focuses on recovering the protein in high yield for subsequent analysis and has been developed to study low abundant transcription factors. In this EMSA-based purification procedure the complexes formed are extracted after a native PAGE retardation assay and applied to two-dimensional electrophoresis, isoelectric focusing and SDS-PAGE. The EMSA conditions are systematically optimized to reduce non-specific binding and increase protein yield. After the three electrophoreses the sample can then be electrotransfered onto a nitrocellulose or polyvinylidene difluoride membrane for southwestern and western blotting analysis to further characterize the complexes. Spots of interest can be cut from the gel or the membrane for protein identification by mass spectrometry.

7. Alternatives to EMSA

There are several alternatives to EMSA used in the analysis of nucleic acid-protein interactions with its own advantages and disadvantages when compared to EMSA.

7.1 Footprinting

Footprinting is essentially a protection assay used to characterize the binding site recognized by a given protein. It relies on the fact that a protein bound to the nucleic acid will protect it and interfere with the modification of the sequence it is bound to. The modification can be chemical or enzymatic and it is usually the endonuclease cleavage of radioisotope-labeled nucleic acid previously mixed with the protein(s) of interest. After cleavage the resulting ladder is analyzed on denaturing polyacrylamide gel and visualized by autoradiography. The gaps in the ladder are indicative of sites protected by the protein or proteins in the mixture (reviewed by Hampshire et al., 2007). This method was originally developed to characterize sequence selectivity but it is also helpful in estimating the binding strength through a footprinting reaction over a range of protein concentrations. For slow binding reactions footprinting can also be applied to assess the reaction kinetics estimating the association and dissociation rates. Although it is a widely used method, there are other approaches that provide higher throughput as the ones described ahead.

A variant on DNA footprinting is the *in vivo* approach, a technique that enables the detection of DNA-protein interactions as they occur in the cell. *In vivo* footprinting also relies on the fact that the bound protein protects the nucleic acid, at its binding site, from cleavage by endonucleases or modification by a chemical agent. The difference is that the cleavage of DNA is carried out within the nucleus following the *in vivo* binding of the proteins to chromatin. Footprints and endonuclease hypersensitive sites that are due to deformations of DNA in chromatin can be detected by this *in vivo* method. This method has been coupled with deep sequencing to identify DNaseI hypersensitive sites in the genome of different cell lines. It enabled the precise identification of a large number of specific cis-regulatory protein binding events with a single experiment (Boyle et al., 2011). Accordingly, the data obtained by this procedure may be more significant and representative of true events when compared with data obtained by the previously described *in vitro* footprinting.

7.2 Nitrocellulose filter binding

Nitrocellulose filter binding assays were developed in the 70s as a rapid enough method to allow kinetic as well as equilibrium studies of DNA-protein interactions (Riggs et al., 1968 and Riggs et al., 1970 as cited in Helwa & Hoheisel, 2010). The manipulation required is rapid enough to allow such measurements. The assay is based on the premise that proteins can bind to nitrocellulose without losing the ability to bind DNA. After the binding reaction the mixture is separated by electrophoresis and then blotted onto a nitrocellulose membrane. Only protein bound DNA remains on the membrane as the free double-stranded DNA will not be retained on nitrocellulose. The amount of DNA on the membrane can be quantified by measuring the label on the nucleic acid. However, this method has its limitations such as the fact that the proteins involved are not identified or the proportion in which they bind DNA. It also provides no information on the DNA sequence the protein interacts with unless well defined nucleic acid fragments are used and is limited to double stranded DNA as single stranded DNA can bind to nitrocellulose under certain conditions resulting in undesirable background.

7.3 Microfluidic mobility shift assay (MMSA)

The capillary microfluidic mobility shift assay (MMSA) is a method that uses fluorescence-based multi-well capillary electrophoresis to characterize protein-nucleic acid interactions. For example, it has been used effectively in characterizing RNA-protein binding in a study of the interaction between human immunodeficiency virus 1 transactivator of transcription and the transactivation-responsive RNA (Fourtounis et al., 2011). This technique requires only nanoliter amounts of sample that are introduced into microscopic channels and separated by pressure-driven flow and application of a potential difference. The free molecules or complexes are visualized by LED-induced fluorescence, discarding the need for hazardous radiolabeling. With the ability to perform 384-well screening this method has an increased capacity over regular EMSA to be compatible with high-throughput screenings.

7.4 Yeast hybrid systems

The yeast one-hybrid is an approach used to identify proteins that bind a given nucleic acid sequence as opposed to the methods that are suited to identify the nucleic acid sequences preferably recognized by a known protein. The protocol is based on a hybrid prey protein fused to a transcription activation domain that allows the expression of a reporter gene when the prey protein interacts with the DNA bait (reviewed by Deplancke et al., 2004). This method allows for a proteome-scale analysis depending on the prey protein library but only detects monomers that bind the target nucleic acid. Although it is an *in vivo* approach it is performed in yeast (*Saccharomyces cerevisiae*), which may not be the endogenous context, and is limited to DNA-protein interactions.

RNA-protein interactions can be studied with a yeast three-hybrid system that involves the expression in yeast cells of not one but three chimerical molecules, which assemble in order to activate two reporter genes (Kraemer et al., 2000). It represents a modification of the yeast two-hybrid system, widely used to identify protein-protein interactions, that was designed to allow high sensitivity *in vivo* detection of RNA-protein interactions. The yeast three-hybrid system includes: a fusion protein consisting of a DNA binding protein and a RNA-binding protein; a hybrid protein consisting of a transcription activating domain and a peptide thought to interact with a particular RNA; a RNA intermediate that promotes the interaction of the two hybrid proteins, this RNA includes the RNA that interacts with the system's RNA-binding protein and the RNA molecule to be investigated. The successful interaction of these 3 components allows the reconstitution of a transcription factor and subsequent activation of reporter genes (Hook et al., 2005 and Wurster & Maher, 2010)

7.5 ChiP assays

Chromatin immunoprecipitation (ChiP) is a commonly used method to study DNA-binding proteins *in vivo* and a standard method for the identification of transcription binding sites and histone modification locations (reviewed by Massie & Mills, 2008). In this method a cross-linking agent (e.g. formaldehyde) is added to cells to covalently bind proteins and chromatin that are in direct contact. Afterwards, the cells are lysed and chromosomal DNA is isolated and fragmented. Specific antibodies are used to immunoprecipitate the targeted proteins with the cross-linked DNA. The bound nucleic acid is released by reverting the cross-linking and then analyzed. Classically, the DNA was characterized by polymerase chain reaction (PCR) which required some previous knowledge of the candidate DNA regions. Nowadays, the DNA bound to protein is more commonly characterized through more powerful tools either coupled with microarrays that represent the genome (ChIP-chip) or state-of-the-art high-throughput sequencing (ChIP-seq). The improvements in DNA sequencing technology allow tens of millions of sequence reads, therefore ChIP-seq has a major advantage of increased sensitivity and resolution to add to the fact that it is not limited to predetermined probe sets as ChIP-chip. The major strength of the ChIP-based approaches is that they capture complexes *in vivo* and the binding reactions can be studied under different cellular conditions and at different time points. However it also has important limitations. The method requires high-quality antibodies that are available only for a limited number of proteins. To circumvent this, epitope-tagged proteins could be used although it usually implies the introduction of modified genes into the endogenous locus in order to obtain expression at physiological levels. This method does not distinguish between proteins that bind directly to the genomic DNA and those that only interact with other proteins that do bind.

7.6 SELEX

The Systematic Evolution of Ligands by Exponential Enrichment (SELEX) is a well established method that enables the selection of enriched sequences from a random library that bind recombinant proteins. This procedure starts with the synthesis of the oligonucleotide library and then incubating the generated sequences with the putative interacting protein(s). The sequences that bind are eluted, amplified by PCR and subjected to more rounds of selection with increasing stringency conditions. This allows the identification of the tightest-binding sequences. It is a widely used approach to obtain transcription factors binding motifs as it requires low amounts of purified proteins (Matys et al., 2006). This approach becomes very complicated to use when large numbers of nucleic acid-binding proteins are analyzed as it then requires multiple rounds of selection. Another limitation is the fact that it is aimed at the identification of the best binding DNA targets *in vitro* and does not allow the characterization of the exact *in vivo* selectivity.

7.7 Protein microarray

A protein microarray is a method that allows high-throughput analysis in which labeled nucleic acids are queried against proteins immobilized on a chip (reviewed by Hu et al., 2011). In a functional protein microarray, thousands of purified recombinant proteins can be immobilized in a glass slide in discrete locations forming a high-density protein matrix, providing a flexible platform to characterize different protein activities. It is a very versatile method as it can perform a semi-quantitative analysis of protein binding to a wide range of molecules (nucleic acids, other proteins, antibodies, lipids, glycans…). In theory, it is feasible to print arrays of all the annotated proteins of a given organism originating a whole proteome microarray. However, it implies the expression and purification of each individual protein and several conditions need to be optimized to render the proteins apt for this method. Since the protein is immobilized it is crucial to guarantee that its structural integrity remains intact especially the binding domains that are to be studied.

7.8 Nucleic acid microarrays

Nucleic acid microarrays can also be used for a direct analysis of protein-nucleic acid interactions. In this case it is the nucleic acid that is immobilized and not the protein. Nucleic acid chips are a powerful and versatile tool in biological research. They consist of high-density arrays of oligonucleotides or complementary DNA that can cover a whole genome (reviewed by Stoughton, 2005). For protein-interaction studies, the protein(s) of interest is expressed usually with an epitope tag, and purified. The tag serves two purposes; it helps to isolate the protein through affinity purification, and allows detection by an epitope-specific reporter antibody. After incubation of the protein with the nucleic acid chip the signal intensities at the several array spots can be measured.

7.9 Ribonucleoprotein Immunoprecipitation – Microarray (RIP-chip)

RNA immunoprecipitation and chip hybridization (RIP) is a protocol very similar to ChIP-chip except that it targets RNA-protein interactions rather than DNA-protein (Keene et al.,

2006). RIP-chip is an approach that consists on a microarray profiling of RNAs obtained from immunoprecipitated RNA-protein complexes. Genome-wide arrays are used to identify messenger RNAs (mRNAs) that are present in endogenous messenger ribonucleoprotein complexes making it a great tool to identify the physiological substrates of mRNAs. The endogenous complexes are immunoprecipitated from cell lysates which limits this study to kinetically stable interactions. Even though it can identify RNA-protein complexes with heteromultimers, at least one of the proteins has to be previously known to be the basis of immunoprecipitation and "fish out" the whole complex.

7.10 Crosslinking and Immunoprecicipation (CLIP) and Photoactivable-Ribonucleoside-Enhanced Crosslinking and Immunoprecipitation (PAR-CLIP)

The RIP-chip method that has just been described is limited to studies of very stable RNA-protein complexes; to remediate this problem another method is available to study RNA-binding proteins. The crosslinking and immunoprecipitation (CLIP) approach uses *in vivo* UV crosslinking prior to the complexes immunoprecipitation to identify less stable interactions (Ule et al., 2003). After immunoprecipitation RNA molecules are separated and cDNA sequencing is carried on. However, this method is not perfect as the commonly used UV 254nm RNA-protein crosslinking has low efficiency and it is difficult to distinguish between crosslinked RNAs from background non-crosslinked fragments that can be detected in the sample due to the presence of abundant cellular RNAs.

A more recent approach tries to further improve the CLIP method using photoreactive ribonucleoside analogs such as 4-thiouridine or 6-thioguanosine (Hafner et al., 2010). In this photoactivatable-ribonucleoside-enhanced crosslinking and immunoprecipitation (PAR-CLIP) protocol the photoreactive nucleosides are incorporated into nascent transcripts within living cells. The irradiation is performed with UV light of 365nm, which induces an efficient crosslink of the labeled cellular RNA to its interacting proteins. The labeled RNAs are isolated after co-immunoprecipitation, and converted into cDNA for deep sequencing. The precise crosslinking position can be identified by mutations in the sequenced cDNA making it possible to distinguish the crosslinked fragments from background.

7.11 High-Throughput Sequencing – Fluorescent Ligand Interaction Profiling (HisT-FLIP)

Very recently a new method was developed to characterize DNA-protein interactions using second-generation sequencing instruments (Nutiu et al., 2011). This method allows high throughput and quantitative measurement of DNA-protein binding affinity. This High-Throughput Sequencing - Fluorescent Ligand Interaction Profiling (HiTS-FLIP) uses the optics of a high-throughput sequencer to visualize *in vitro* binding of a protein to the sequenced DNA in a flow cell. The new method was initially used on a *Saccharomyces cerevisiae* transcription factor. The fluorescently tagged protein was added at different concentrations to a flow cell containing around 88 million DNA clusters, the equivalent of over 160 yeast genomes. The traditional EMSA was used as an independent validation of the dissociation constants obtained and found a high correlation with values obtained with the new method and those from EMSA as reported in literature. This high-throughput method has an obvious advantage in the fact that it can provide hundreds of millions of measurements but is limited to DNA-protein interactions and requires expensive equipment. Another advantage is that the sequencing instrument can measure multiple fluorescent wavelengths allowing hetero and homodimeric forms to be measured in the same run, using distinct tags on individual proteins.

8. Conclusion

Since the first report, 30 years ago, EMSA became one of the most popular methods for detection and characterization of protein-nucleic acid interactions. Hundreds of protocols have been published accommodating modifications in virtually every parameter influencing the experimental outcome. Improvements were made in all EMSA steps including the methods for preparation of protein samples and purification, synthesis and labeling of nucleic acids, and detection. This allowed enlarging and diversifying the applications of EMSA and resulting in a number of variants of the method.

However, despite the large amount of available literature and protocols trial and error will ultimately be the way to optimize the EMSA conditions for the nucleic acid-protein complex to be analyzed. The guidelines discussed above help to provide an initial protocol adjusted to each study but slight changes may be needed to improve binding and detection of the complexes.

In recent years, the use of highthrouhput approaches to detect biologically relevant interactions, including those between proteins and nucleic acids, was reported. Development of these approaches was made possible, at least in part, by the availability of more sensitive and specific equipment and tools. Although EMSA cannot achieve a high throughput level it remains a valuable tool to confirm the detected interactions.

9. Acknowledgements

We are grateful to Dr. Cristina Branco for constructive comments. Work in the authors' laboratory is supported by Fundação para a Ciência e Tecnologia (PTDC/SAU-MII/098314/2008). CA is a recipient of a FCT PhD grant.

10. References

Adachi, Y.; Chen, W.; Shang, W. & Kamata, T. (2005). Development of a direct and sensitive detection method for DNA-binding proteins based on electrophoretic mobility shift assay and iodoacetamide derivative labelling. *Analytical Biochemistry,* Vol.342, pp. 348-351

Alves, C.; Cheng, H.; Roder, H. & Taylor, J. (2010). Intrinsic disorder and oligomerization of the hepatitis delta virus antigen. *Virology,* Vol. 370, pp 12-21

Apella, E. & Anderson, C. (2001). Post-translational modifications and activation of p53 by genotoxic stresses. *European Journal of Biochesmistry,* Vol.268, No.10, pp. 2764-2772

Baneyx, F. (1999). Recombinant protein expression in Escherichia coli. *Current Opinion in Biotechnology,* Vol.10, pp. 411-421

Boyle, A.; Song, L.; Lee, B.; London, D.; Keefe, D.; Birney, E.; Iyer, V.; Crawford, G. & Furey, T. (2011). High-resolution genome-wide in vivo footprinting of diverse transcription factors in human cells. *Genome Research,* Vol.21, No.3, pp. 456-464

Chernov, I.; Akopov, S.; Nikolaev, L. & Sverdlov, E. (2006). Identification and mapping of DNA binding proteins target sequences in long genomic regions by two-dimensional EMSA. *BioTechniques*, Vol.41, pp. 90-96

Chi, P.; Van Komen, S.; Sehorn, M.; Sigurdsson, S. & Sung, P. (2006). Roles of ATP binding and ATP hydrolysis in human Rad51 recombinase function. *DNA Repair*, Vol.5, No.3, pp. 381–391

Crothers, D. & Drak, J. (1992). [3] Global features of DNA structure by comparative gel electrophoresis. *Methods in Enzymology*, Vol 212, pp. 46-71

Demarse, N.; Ponnusamy, S.; Spicer, E.; Apohan, E.; Baatz, J.; Ogretmen, B. & Davies, C. (2009). Direct binding of glutharaldehyde 3-phosphate dehydrogenase to telomeric DNA protects telomeres against chemotherapy-induced rapid degradation. *Journal of Molecular Biology*, Vol.394, No.4, pp. 789-803

Deplancke, B.; Dupuy, D.; Vidal, M. & Walhout, A. (2004). A gateway-compatible yeast one-hybrid system. *Genome Research*, Vol.1, 2093-2101

Dignam, J.; Lebovitz, R. & Roeder, R. (1983). Accurate transcription initiation by RNA polymerase II in a soluble extract from isolated mammalian nuclei. *Nucleic Acids Research*, Vol.11, No.5, pp. 1457-1489

Endo, Y. & Sawasaki, T. (2006). Cell-free expression systems for eukaryotic protein production. *Current Opinion in Biotechnology*, Vol.17, pp. 373-380

Filion, G.; Fouvry, L. & Defossez, P.-A. (2006). Using reverse electrophoretic mobility shift assay to measure an compare protein-DNA binding affinities. *Analytical Biochemistry*, Vol.357, pp. 156-158

Fourtounis, J.; Falgueyret, J.-P. & Sayegh, C. (2011). Assessing protein-RNA interactions using microfluidic capillary mobility shift assays. *Analytical Biochemistry*, Vol.411, pp. 161-163

Fried, M. (1989). Measurement of protein-DNA interaction parameters by electrophoresis mobility shift assay. *Electrophoresis*, Vol. 10, pp. 366-376

Fried, M. & Crothers, D. (1981). Equilibria and kinetics of lac repressor-operator interactions by polyacrylamide gel electrophoresis. *Nucleic Acid Research*, Vol.9, No.23, pp. 6505-6525

Fried, M. & Crothers, D. (1984). Equilibrium studies of the cyclic AMP receptor protein-DNA interaction. *Journal of Molecular Biology*, Vol.172, No.3, pp. 241-262

Fried, M. & Daugherty, M. (1998). Electrophoretic analysis of multiple protein-DNA interactions. *Electrophoreis*, Vol.19, pp. 1247-1253

Garner, M. & Revzin, A. (1981). A gel electrophoresis method for quantifying the binding of proteins to specific DNA regions: application to components of the Escherichia coli lactose operon regulatory system. *Nucleic Acids Research*, Vol.9, No.13, pp. 3047-3060

Goh, A.; Coffil, C. & Lane, D. (2011). The role of mutant p53 in human cancer. *The Journal of Pathology*, Vol.223, No.2, pp. 116-126

Granger-Schnarr, M.; Lloubes, R.; de Murcia, G. & Shnarr, M. (1988). Specific protein-DNA complexes: immunodetection of the protein component after gel electrophoreis and Western blotting. *Analytical Biochemistry*, Vol.174, No.1, pp. 235-238

Greco-Stewart, V. & Pelchat, M. (2010). Interaction of host cellular proteins with components of the hepatitis delta virus. *Viruses*, Vol.2, No.1, pp 189-212

Haffner, M.; Landthaler, M.; Burger, L.; Khorshid, M.; Hausser, J.; Berninger, P.; Rothballer, A.; Ascano, M.; Jungkamp, A.-C.; Munschauer, M.; Ulrich, A.; Wardle, G.; Dewell, S.; Zavola, M. & Tuschi, T. (2010). PAR-CliP – A method to identify transcriptome-wide binding sites of RNA binding proteins. *Journal of Visualized Experiments*, Vol.41, pp.2034

Hampshire, A.; Rusling, D.; Broughton-Head, V. & Fox, K. (2007). Footprinting: a method for determining the sequence selectivity, affinity and kinetics of DNA-binding ligands. *Methods,* Vol.42, pp. 128-140

Hellman, L. & Fried, M. (2007). Electrophoretic mobility shift assay (EMSA) for detecting protein-nucleic acid interactions. *Nature Protocols,* Vol.2, No.8, pp. 1849-1861

Helwa, R. & Hoheisel, J. (2010). Analysis of DNA-protein interactions: from nitrocellulose filter binding assays to microarray studies. *Analytical & Bioanalytical Chemistry,* Vol.398, pp. 2551-2561

Holden, N. & Tacon, C. (2011). Principles and problems of the electrophoretic mobility shift assay. *Journal of Pharmacological and Toxicological Methods,* Vol.63, pp. 7-14

Hope, I. & Struhl, K. (1987). GCN4, a eukaryotic transcriptional activator protein, binds as a dimer to target DNA. *The EMBO Journal,* Vol.6, No.9, pp. 2781-2784

Hook, B.; Bernstein, D.; Zhang, B. & Wickens, M. (2005). RNA-proteins interactions in the yeast three-hybrid system: affinity, sensitivity, and enhanced library screening. *RNA,* Vol.11, No.2, pp. 227-233

Hu, S.; Xie, Z.; Qian, J.; Blackshaws, S. & Zhu, H. (2011). Functional protein microarray technology. *WIREs Systems Biology and Medicine,* Vol.3, pp. 255-268

Jarvis, D. (2009). Baculovirus-insect cell expression systems. *Methods in Enzymology,* Vol.463, pp. 191-222

Jiang, D.; Jia, Y. & Jarrett, H.W. (2011). Transcription factor proteomics: Identification by a novel gel mobility shift-three-dimensional electrophoresis method coupled with southwestern blot and high-performance liquid chromatography-electrospray-mass spectrometry analysis. *Journal of Chromatography A,* Vol.1218, No.39, pp. 7003-7015

Jiang, S.; Macias, M. & Jarrett, H. (2010). Purification and identification of a transcription factor, USF2, binding to E-box elements in the promoter of human telomerase reverse transcriptase (hTERT). *Proteomics,* Vol.10, No.2, pp. 203-211

Jimenez, J. (2010). Protein-DNA interaction at the origin of neurological diseases: a hypothesis. *Journal of Alzheimer's Disease,* Vol.22, No.2, pp. 375-391

Keene, J.; Komisarow, J. & Friedersdorf, M. (2006). RIP-Chip: the isolation and identification of mRNAs, microRNAs and protein components of ribonucleoprotein complexes from cell extracts. *Nature Protocols,* Vol.1, pp. 302-307

Kerr, L. (1995). [42] Electrophoretic mobility shift assay. *Methods in Enzymology,* Vol.254, pp. 619-632

Kraemer, B.; Zhang, B.; SenGupta, D.; Fields, S. & Wickens, M. (2000). Using the yeast three-hybrid system to detect and analyze RNA-protein interactions. *Methods in Enzymology,* Vol.328, pp. 297-321

Lane, D.; Prentki, P. & Chandler, M. (1992). Use of gel retardation to analyze protein-nucleic acid interactions. *Microbiological Reviews,* Vol.56, No.4, pp. 509-528

Makino, T.; Skretas, G. & Georgiou, G. (2011). Strain engeneering for improved expression of recombinant proteins in bacteria. *Microbial Cell Factories,* Vol.10, No.1, pp. 32-42

Massie, C. & Mills, I. (2008). ChIPping away at gene regulation. *EMBO Reports,* Vol.9, No.4, pp. 337-343

Matys, V.; Kel-Margoulis, O.; Fricke, E.; Liebich, I.; Land, S.; Barre-Dirrie, A.; Reuter, I.; Checkmenev, D.; Krull, M.; Hornischer, K.; Voss, N.; Stegmaier, P.; Lewicki-Potapov, B.; Saxel, H.; Kel, A. & Wingender, E. (2006). TRANSFAC and its module TRANSComple: transcriptional gene regulation in eukaryotes. *Nucleic Acids Research,* 1 34 Database issue D108-110

Nordheim, A. & Meese, K. (1988). Topoisomer gel retardation: detection of anti-Z-DNA antibodies bound to Z-DNA within supercoiled DNA minicircles. *Nucleic Acids Research,* Vol.16, No.1, pp. 21-37

Nutiu, R.; Friedman, R.; Luo, S.; Khrebtukova, I.; Silva, D.; Li, R.; Zhang, L.; Schroth, G. & Burge, C. (2011). Direct measurement of DNA affinity landscapes on a high-throughput sequencing instrument. *Nature Biotechnology,* Vol.29, No.7, pp. 659-664

Palecek, E.; Vlk, D.; Stankova, V.; Brazda, V.; Vojtesek, B.; Hupp, T.; Schaper, A. & Jovin, T. (1997). Tumor suppressor protein p53 binds preferentially to supercoiled DNA. *Oncogene,* Vol.15, No.18, pp. 2201-2209

Puzio-Kuter, A. (2011). The role of p53 in metabolic regulation. *Genes & Cancer,* Vol.2, No.4, pp. 385-391

Rizzetto, M. (2009). Hepatitis D: Thirty years after. *Journal of Hepatology,* Vol.50, pp. 1043-1050

Ruscher, K.; Reuter, M.; Kupper, D.; Trendelenburg, G.; Dirnagl, U. & Meisel, A. (2000). A fluorescence based non-radioactive electrophoretic mobility shift assay. *Journal of Biotechnology,* Vol78, pp. 163-170

Saida, F.; Uzan, M.; Odaert, B. & Bontems, F. (2006). Expression of highly toxic genes in E. coli: special strategies and genetic tools. *Current Protein and Peptide Science,* Vol.7, pp. 47-56

Sandman, K.; Krzycki, J.; Dobrinski, B.; Lurz, R. & Reeve, J. (1990). HMf, a DNA-binding protein isolated from the hyperthermophilic archaeon Methanothermus fervidus, is most closely related to histone. *Proceedings of the National Academy of Sciences of the United States of America,* Vol.87, No.15, pp. 5788-5791

Simicevic, J. & Deplancke, B. (2010). DNA-centered approaches to characterize regulatory protein-DNA interaction complexes. *Molecular BioSystems,* Vol.6, No.3, pp. 462-468

Spinner, D.; Liu, S.; Wang, S-W. & Schmidt, J. (2002). Interaction of the myogenic determination factor myogenin with E12 and a DNA target: mechanism and kinetics. *Journal of Molecular Biology,* Vol.317, No.3, pp. 431-445

Stead, J.; Keen, J. & McDowall, K. (2006). The identification of nucleic acid-interacting proteins using a simple proteomics-based approach that directly incorporates the electrophoretic mobility shift assay. *Molecular & Cellular Proteomics,* Vol.5, No.9, pp. 1697-1702

Stoughton, R. (2005). Applications of DNA microarrays in biology. *Annual Review of Biochemistry,* Vol.74, pp. 53-82

Ule, J.; Jensen, K.; Ruggiu, M.; Mele, A.; Ule, A. & Darnell, R. (2003). CLIP identifies Nova-regulated RNA networks in the brain. *Science,* Vol.302, No.5648, pp. 1212-1215

Vetchinova, A.; Akopov, S.; Chernov, I.; Nikolaev, L. & Sverdlov, E. (2006). Two-dimensional electrophoretic mobility shift assay: identification and mapping of transcription factor CTCF target sequences within an FXYD5-COX7A1 region of human chromosome 19. *Analytical Biochemistry,* Vol.354, pp. 85-93

Vossen, K.; Wolz, R.; Daugherty, M. & Fried, M. (1997). Role of macromolecular hydration in the binding of the Escherichia coli cyclic AMP receptor to DNA. *Biochemistry,* Vol.36, No.39, pp. 11640-11647

Walker, J. (1994). Gradient SDS polyacrylamide gel electrophoresis of proteins. *Methods in Molecular Biology,* Vol.32, pp. 35-38

Woo, A.; Dods, J.; Susanto, E.; Ulgiati, D. & Abraham, L. (2002). A proteomics approach for the identification of DNA binding activities observed in the electrophoretic mobility shift assay. *Molecular & Cellular Proteomics,* Vol.1, No.6, pp. 472-478

Wurster, S. & Maher, L. (2010). Selections that optimize RNA display in the yeast three-hybrid system. *RNA,* Vol.16, No.2, pp. 253-258

13

Analysis of Chromosomal Replication Progression by Gel Electrophoresis

Elena C. Guzmán[1] and Enrique Viguera[2]
[1]*Departamento de Bioquímica, Biología Molecular y Genética*
Facultad de Ciencias, Universidad de Extremadura, Badajoz
[2]*Área de Genética, Facultad de Ciencias, Universidad de Málaga, Málaga*
Spain

1. Introduction

An absolute requirement for life is the preservation of genome integrity and the faithful duplication of chromosomes before segregation. A proliferating cell must duplicate its entire complement of DNA with exquisite precision facing a barrage of impediments of different nature. Single-strand breaks (SSBs) in the template DNA, either pre-existing or arising from abnormal DNA structures (folded DNA, cruciform structures or cross-links, etc.), collisions with DNA-bound proteins such as transcription complexes or DNA structural barriers restrain replication progression. For instance, during *Escherichia coli* DNA replication, the two forks initiated at the single origin of replication, *oriC*, move along the chromosome with high probability of pausing, stalling or even collapse (Maisnier-Patin et al., 2001). Replication arrest is a source of genetic instability in all types of living cells (Michel, 2000; Carr, 2002; Kolodner et al., 2002). As a consequence, cells have developed several effective strategies to tackle with replication fork arrest and/or repairing the double strand breaks (DSBs) generated at the stalled replication forks (Bierne et al., 1994; Kuzminov, 1995). Considerable evidence has been accumulated in the past decade demonstrating the involvement of recombination proteins in either direct or bypass repair of the lesions or structures blocking replication fork progression (reviewed in Courcelle et al., 2004; Kreuzer, 2005; Hanawalt, 2007; Michel et al., 2007 and references herein).

Conventional agarose DNA electrophoresis is one of the most frequently used techniques in molecular biology for the isolation or identification of DNA fragments. However, Pulse Field Gel Electrophoresis (PFGE) and Two-Dimensional (2D) Agarose Gel Electrophoresis techniques have been used to study biological processes such as the progression of the replication fork along a DNA fragment. In this work we introduce how these techniques has been used in bacteria to (i) verify and quantify the presence of stalled replication forks (ii) recognize DNA structure at the stalled replication fork, and (iii) understand how the replication fork could be restarted.

2. Pulse Field Gel Electrophoresis (PFGE)

Separation of DNA fragments by standard agarose gel electrophoresis is based on the capacity of the molecules to pass through the pore generated inside the matrix gel. Using this feature as the only separation mechanism, the large DNA molecules cannot be discriminated from each other. The practical range of resolution is up to approximately 50kb; making impossible the direct genomic analysis of large DNA molecules as those generated by the presence of complex DNA structures, or the DSBs involved in stalled replication forks in the *E. coli* chromosome.

In 1983, PFGE was developed as a method to circumvent this limitation, allowing fractionation of very large DNA molecules up to a million base pairs in size (Schwartz & Cantor, 1984; Herschleb et al., 2007). PFGE allows the separation of these large DNA molecules through abrupt electrical perturbations to the paths crawling molecules take trough the gel. In PFGE, the *direction of the electrical field* is periodically changed (usually 120°), requiring electrophoresing molecules to reorientate (Fig. 1). The *time* required to complete the orientation process scales with the size of the DNA, so that increasing the size of the DNA molecules, it takes more time between changes in the direction of the electric field (Fig. 1). These intervals vary depending on the size of the fragments that have to be resolved, a few seconds for small fragments to hours for fragments larger than 5Mb. The principle of PFGE is that large DNA fragments require more time to reverse the direction in an electric field than small DNA fragments. Alternating current direction during gel electrophoresis can resolve DNA fragments of 100 to 1,000 kb.

The *equipment* required to perform PFGE is also different from that used in traditional electrophoresis. The tank contains a set of *electrodes* (6-8), instead of a couple of them, being thicker and disposed to allow the different orientations of the electric field. Maintaining a constant the *temperature* (usually 14°C) during the process is important to avoid temperature variations through the gel, which could affect the resolution of DNA fragments. Accordingly, the system should include a cooling device. Agarose gel preparation does not differ from that reported for conventional electrophoresis.

Due to the fragility of the very large DNA fragments to be separated, preparing the *sample* is the most critical step for PFGE. To avoid breakage of genomic DNA during manipulation, the DNA is not extracted, but the cells are embebed in agarose plugs and then fixed into the wells.

2.1 Verifying replication fork reversal by detection of DNA breakage at the stalled replication forks

The progression of the replication fork can be halted by several causes, including deficiencies in replication enzymes and obstacles such as DNA-bound proteins, transcription complexes, nicks, gaps, DNA damage or topological constrictions. Replication arrest is a source of DNA breakage and rearrangement in all organisms (reviewed in Aguilera & Gómez-González, 2008); consequently stalled replication forks create the need for replication reactivation, and different ways of restarting replication have been proposed (Michel et al., 2004; Michel et al., 2007). In bacteria, the consequences of replication blockage have been studied mainly in *E. coli*. In several *E. coli* replication mutants, the stalled forks generated upon inactivation of the mutant enzyme are reversed and result in the formation of a Holliday junction (HJ) adjacent to a DNA double strand end, a reaction called 'replication fork reversal' (RFR) (Fig. 2A) (Michel et al., 2004; Seigneur et al., 1998; Seigneur et al., 2000). In a *rec* proficient background this intermediary could be processed without

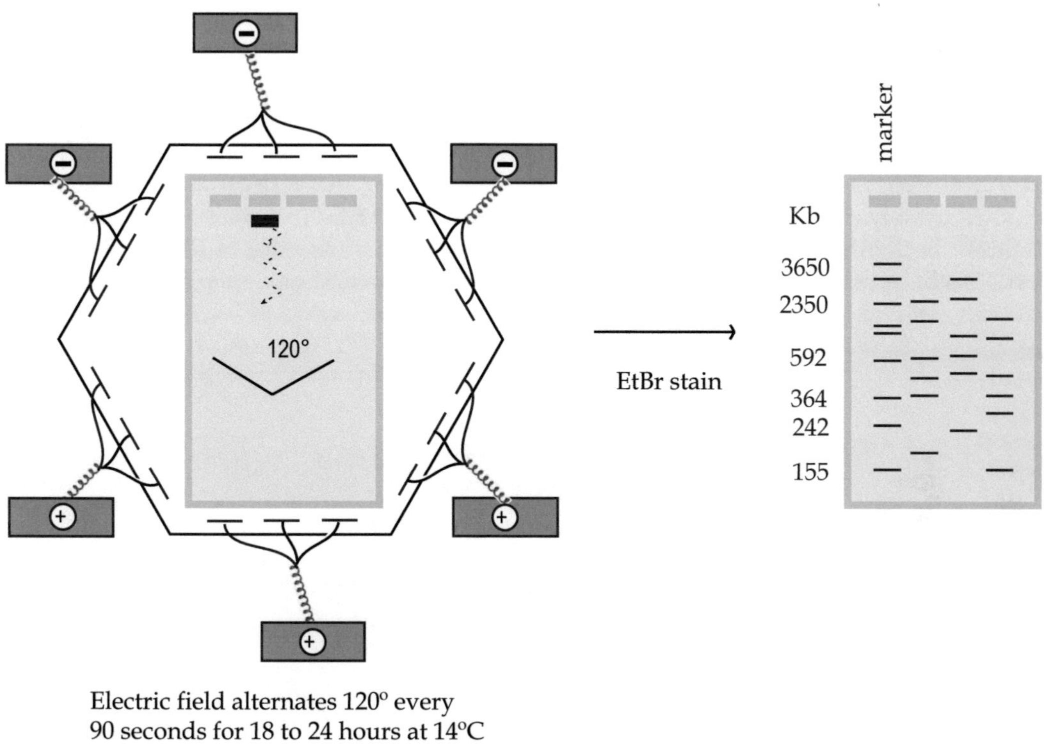

Fig. 1. Schematic diagram of PFGE instrumentation. Contoured clamped homogeneous electric field (CHEF) systems use a hexagonal gel box that alters the angle of the fields relative to the agarose gel. After running the gel by PFGE, DNA fragments are visualized by staining with ethidium bromide.

generating DSBs by using the recombination proteins RecBCD, RecA, and by the HJ-specific resolvase RuvABC (Fig. 2B) (Seigneur et al., 1998). This is a key aspect of the RFR model as it allows restarting of the blocked forks without generating chromosome instability. Nevertheless, in the absence of RecBCD activity (Fig. 2C), resolution of the RFR-produced HJ is done by RuvABC resolvase and leads to fork breakage. These particular DSBs are dependent on RuvABC activity in a *recB* deficient background.

If RFR does not take place at the stalled fork, at least two situations may arise. On the one hand, there would be an increase of DSBs independent of RuvABC activity and generated by another unknown endonuclease (Fig. 2E) as in the case of the thymine starvation (Guarino et al., 2007b). On the other hand, there would be no increase in the amount of DSBs probably because the stalled forks are not susceptible to the endonuclease action, and the restarting of the forks would take place without the generation of fork breakage. This situation has been described in *gyrB* mutants (Grompone et al., 2003), and when replication termination sequences *ter* were placed at ectopic positions on the bacterial chromosome (Bidnenko et al., 2002). Using the system described above, the fate of the stalled replication forks caused by any condition can be studied.

To verify the RFR process by PFGE, a *recB* deficient background should be used (i) to inhibit the degradation or the recombinational repair of the DNA tail created by the regression of the fork (Miranda & Kuzminov, 2003), allowing RuvABC resolvase to transform this tail in a DSB; (ii) to inhibit the repair of the DSBs generated by RuvABC resolvase (Fig. 2C). According to the RFR model, the occurrence of this process at the stalled forks generated under restrictive conditions can be verified by testing whether there is an increase of DSBs in a *recB* deficient background, and determining whether these DSBs are dependent on RuvABC resolvase activity by measuring the amount of DSBs in a *recB* and *recB ruvABC* deficient background (Fig. 2C) (Seigneur et al., 1998). The occurrence of RFR at the stalled forks has been detected by this system in several replication mutants, such as in the helicase mutants *rep* and *dnaBts* (Michel et al., 1997; Seigneur et al., 2000), in *holD*^{G10} (Flores et al., 2001, 2002), in *dnaEts* at 42°C and in the *dnaNts* mutant at 37°C (Grompone et al., 2002) and finally in *nrdA101ts* (Guarino et al., 2007a, 2007b).

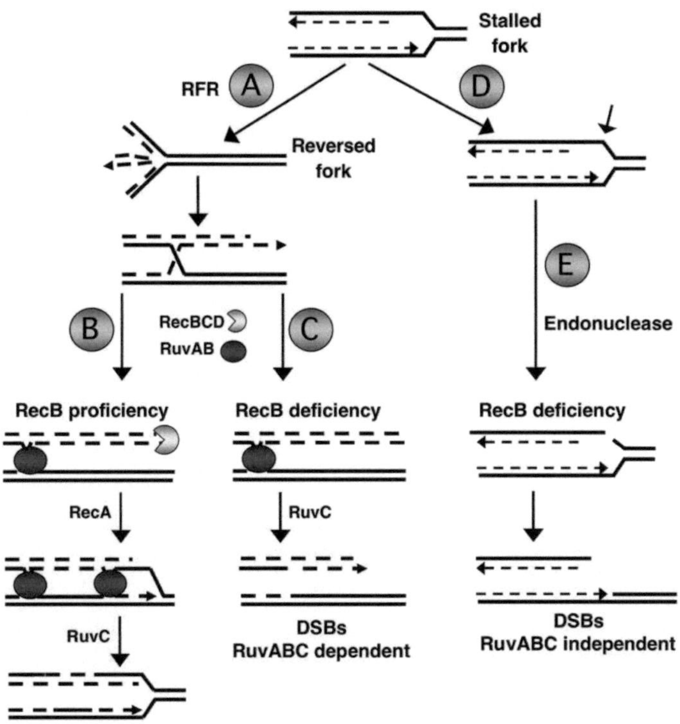

Fig. 2. The fate of the stalled forks. In the first step (A), the replication fork is arrested, causing fork reversal. The reversed fork forms a HJ (two alternative representations of this structure are shown - open X and parallel stacked X). In Rec$^+$ cells (B), RecBCD initiates RecA-dependent homologous recombination, and the resulting double HJ is resolved by RuvABC. In the absence of RecBCD (C), resolution of the HJ by RuvABC leads to DSBs at the stalled replication fork. Alternatively, the replication fork is arrested without being regressed (D) and it is susceptible to be cut by an endonuclease, generating DSBs at the stalled replication fork (E). Continuous line (parental chromosome); dashed lines (newly synthesized strands); disk (RuvAB); incised disk (RecBCD). Adapted from Guarino et al., 2007b.

The amount of linear DNA resulting from DSBs can be estimated by using PFGE combined with cell lysis in agarose plugs (Michel et al., 1997). Briefly, cultures of *recB* and *recB ruvABC* strains growing in M9 minimal medium are labelled by addition of 5 μCi/ml [*methyl*-^{3}H] thymidine (100 Ci/mmol). When cultures reached 0.2 OD_{450nm}, 1ml of cells were collected, washed in cold minimal medium and resuspended in 100μl of TEE buffer. Cells were incubated at 37°C for ten minutes, mixed with 100 μl of low melting agarose 2% in TEE at 55°C and poured into the mould. Once agarose had solidified, cell lysis was performed in the plugs. This ensures only linear chromosomes to enter the gels, while circular molecules remain in the wells (Michel et al., 1997; Seigneur et al., 1998). Plugs were incubated with lysozyme at 5 mg/ml and sarcosyl 0.05% in TEE for 2 h with gently shaking. Then, plugs were retrieved and incubated with lysis solution (1 mg/ml Proteinase K , 1% SDS in TEE) at 56°C overnight. PFGE were run for 48 h at 4°C as described (Seigneur et al., 1998); initial run 500 sec, final run 500 sec, 3 volts/cm and 106° reorientation angle. DNA was visualized by ethidium bromide staining. Lanes were cut into slices and the proportion of migrating DNA was determined by calculating the amount of tritium present in each slice with respect to the total amount of tritium present in the corresponding lane plus the well (Fig. 3A). All the PFGE linear DNA data were analyzed by the least-squares statistical approach, considering measures as highly significantly different if $p < 0.01$.

A typical profile of gel migration for the different strains analyzed is shown in figure 3B. In this case, results indicate that the amount of DSBs in the strain *nrdA101 recB* was greater than in the strain *nrdA+ recB*, suggesting an increase in the number of the stalled forks induced by the presence of a defective ribonucleotide reductase (RNR) at the permissive temperature. To establish the possible origin of the DSBs induced by the *nrdA101 recB* background we investigated whether the formation of DSBs resulted from the action of the RuvABC resolvase (Fig. 2). The DSBs levels estimated in *nrdA101 recB ruvABC* and *nrdA+ recB ruvABC* strains were markedly lower than in the respective Ruv+ counterpart strains. As RuvABC is a specific resolvase for HJ, according the RFR model (Fig. 2), it generates DSBs at arrested replication forks in *recB* deficient background (Seigneur et al., 1998); these results indicated the occurrence of replication fork reversal in *nrdA101* mutant. As RFR is one of the mechanisms to restart the stalled replication forks, we could infer that the *nrdA101* strain growing at 30°C increases the number of stalled replication forks that would proceed with the help of RFR process in a Rec+ proficient context (Guarino et al., 2007a).

2.2 Replication fork collapse at natural arrest sites

In *E. coli*, replication termination occurs by the encounter of two opposite migrating forks at the terminus region or at specific arrest sites named *Ter*, when one of the forks reaches the terminus before the other. Tus protein binds the *Ter* sites forming a complex that acts as a polar replication fork barrier by preventing the action of the DnaB helicase (Neylon et al., 2005).

PFGE was used to determine the analysis of replication forks blocked at terminator sequences *Ter* inserted at ectopic positions on the bacterial chromosome (Bidnenko et al., 2002). This strain requires the RecBCD pathway of homologous recombination for viability, although replication forks blocked at *Ter* are not broken nor reversed (Fig. 2A, D). The analysis of the structure of the chromosomes by PFGE showed linear fragments of about 2 Mb, corresponding to the distance between the origin and the ectopic *Ter* sites. A model of a collapse of replication forks at terminator sequences was proposed in which the blocked replication forks at Ter/Tus are stable, but they are re-replicated in a new replication round, generating 2 Mb linear fragments. These results suggest that natural and accidental replication arrest sites are processed differentially in the cell (Bidnenko et al., 2002).

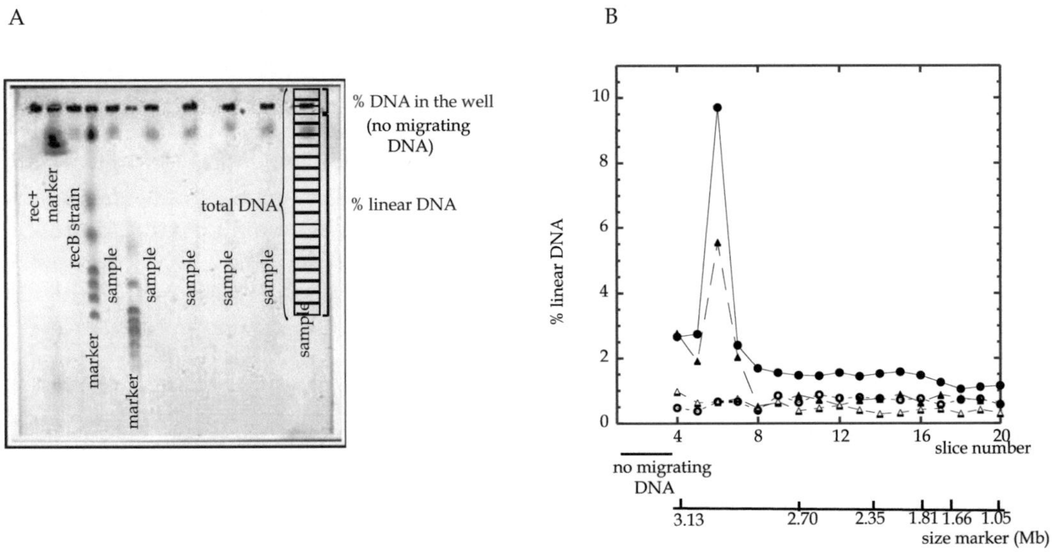

Fig. 3. (A) Visualization of a representative agarose gel stained with ethidium bromide (B) Typical profile of a PFGE experiment. *nrdA+ recB* (▲), *nrdA101, recB* (●), *nrdA+ recB ruvABC* (Δ), and *nrdA101 recB ruvABC* (O). Agarose plugs were prepared as described in the text. Gels were cut in 3 mm slices and the amount of [*methyl*-^{3}H]thymidine present in each slice was measured, and the ratio of the total amount of [*methyl*-^{3}H]thymidine in the lane was calculated for each slice. The gel origin is not shown; only the migrating DNA is shown. The position of size marker is shown (*Hansenula wingei*, Bio-Rad). The amount of linear DNA was calculated from slice 4 to 12. Total proportions of migrating DNA in this experiment from slice 4 to 12 were 16.65% *nrdA+ recB*, 25.21% *nrdA101, recB*, 5.59% *nrdA+ recB ruvABC*, and 5.73% *nrdA101 recB ruvABC*. Adapted from Guarino et al., 2007a.

2.3 Detection of branched DNA structures by PFGE

In PFGE, linear DNA migrates according to its size; however, circular chromosomes or branched DNA structures do not enter in the gel and remain trapped in the wells (Bidnenko et al., 2002). Moreover, replication or recombination intermediates, which are Y or X structures, prevent migration of linear DNA fragments in PFGE. These molecules remain also trapped in the wells after PFGE so that the measurement of the amount of such DNA fragments allows quantification of the formation of abnormal structures in a DNA region.

2.3.1 Reinitiation events under thymine starvation

Thymine starvation results in cellular death in thymine requiring strains. This is a phenomena know as *thymineless death* (TLD), first described in the 50's (Cohen & Barner 1954). Some proposals postulated the formation of branched DNA as the source of the toxic effect of thymine starvation (Nakayama, 2005). Nakayama and co-workers demonstrated the presence of complex DNA structures by digesting DNA from cultures under thymine starvation with the restriction enzyme *XbaI* and separating it by PFGE (Nakayama et al., 1994). They called these structures "non-migrating DNA" (nmDNA), defined as the DNA that is unable to enter the gel and gets stuck in the well. The nmDNA was characterized as having single-stranded tails or gaps and branching with single-stranded arms. TLD has been related to DNA replication (Maaloe & Hanawalt, 1961; Hanawalt & Maaloe, 1961); nevertheless, ongoing replication does not appear to be required for TLD as same lethality is observed under thymine starvation in the presence of hidroxyurea (Morganroth & Hanawalt, 2006), an inhibitor of the DNA synthesis. By contrast, TLD is suppressed by the addition of rifampicin or cloramphenicol, both inhibitors of the new initiation events of the *E. coli* chromosomal DNA (Hanawalt, 1963).

To study whether the formation of nmDNA correlates with TLD under the above described replication conditions, we analyzed the generation of nmDNA after thymine starvation in the presence or absence of rifampicin, cloramphenicol and hidroxyurea (Mata & Guzmán, 2011). Mid-exponentially growing culture of *thyA* mutant MG1693 was starved for thymine in the presence or absence of rifampicin, cloramphenicol, or hydroxyurea. Two hours after the treatment, cells were collected, washed, embedded in agarose plugs, gently lysed and plugs treated with *XbaI* (50 U/100 μl) for two hours before being used for PFGE (Matushek et al., 1996; Gautom, 1997). The visualization of DNA bands was achieved by ethidium bromide staining (Fig. 4A). The amount of nmDNA was quantified by densitometry of the PFGE by using the *Imagen J* program. The nmDNA values were expressed as the ratio (%) between the arbitrary densitometric units of the gel well and those of the gel line plus the well (Table 1, Figure 4B). By using this experimental approach, we showed that nmDNA was generated under thymine starvation, and it was absent in the presence of rifampicin or chloramphenicol, as previously reported (Nakayama et al., 1994). This might suggest that TLD correlates with the generation of nmDNA. However, we found no nmDNA under thymine starvation in the presence of hydroxyurea, indicating that the generation of nmDNA is not a requirement for TLD (Mata & Guzmán, 2011).

Treatment	% nmDNA[1]	Treatment effect[2]	Lethality
None -Exponential culture	8.6 ± 3.6	1	no
-Thymine	21.4 ± 4.7	2.48	yes
-Thy+ 75 mM hydroxyurea	11.2 ± 3.8	1.30	yes
-Thy+ 150 μg/ml rifampicin	9.0 ± 3.9	1.04	no
-Thy+ 200 μg/ml chloramphenicol	9.5 ± 4.4	1.10	no

Table 1. Percentage of nmDNA in MG1693 cells after 2 h of thymine starvation in the presence or absence of hydroxyurea, rifampicin or chloramphenicol. [1] The percentage of nmDNA is expressed as the mean ± standard deviation. [2] The percentage of nmDNA relative to the exponential culture.

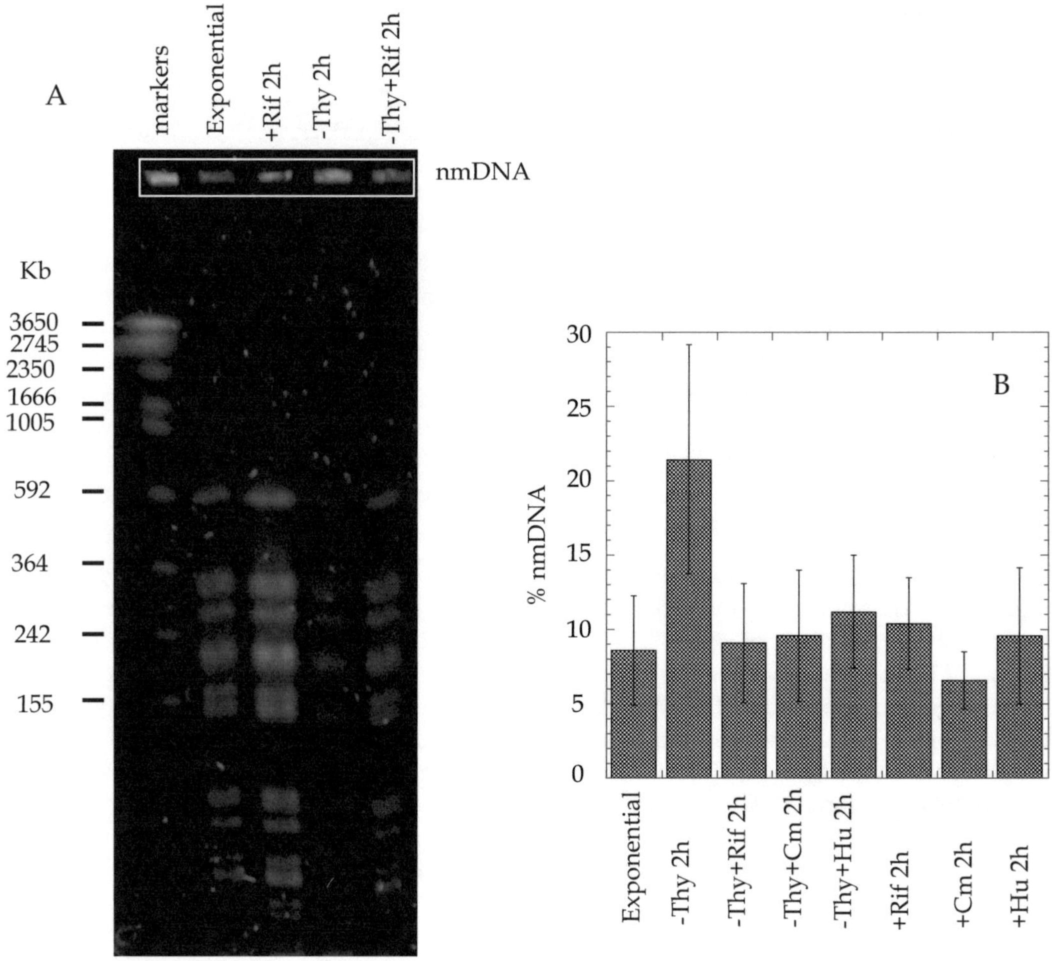

Fig. 4. Visualization and quantification of nmDNA by PFGE after treatment with *XbaI*. (A) Visualization of a representative agarose gel stained with ethidium bromide. (B) Quantification of the amount of nmDNA after different treatments of thymine starvation in the absence or the presence of the drugs, by using the data obtained by densitometry analysis. The standard deviation from 2-3 independent experiments is shown.

2.3.2 Collision between replication and transcription machines

PFGE has been also used for the characterization of collisions of the replication and transcription complexes. Genetic instability following head-on collisions of replication and transcription has been described in bacteria (Vilette et al., 1995) and yeast (Torres et al., 2004; Prado & Aguilera, 2005). Replication fork barriers are present on the rDNA of all eucaryotic cells described until now, for example in yeast (Brewer & Fangman, 1988; Kobayashi et al., 1992; López-Estraño et al., 1998; Sánchez-Gorostiaga et al., 2004), *Xenopus* (Wiesendanger et al., 1994) and plants (Hernández et al., 1993; López-Estraño et al., 1998) and they bound specific proteins. In yeast, these barriers are polar since they block replication forks facing transcription from the pre-rRNA 35S (Brewer & Fangman, 1988; Linkens & Huberman,

1988). It has been proposed that the function of those barriers could be to prevent head-on collisions between replication forks and the highly expressed rRNA genes (Takeuchi et al., 2003).

In order to avoid head-on collisions, ribosomal operons (*rrn*) are transcribed in the direction of replication in bacteria. By using genetic approaches together with PFGE analysis, the laboratory of B. Michel (Boubakri et al., 2010) identified that the three *E. coli* DNA helicases DinG, Rep and UvrD are recruited to the replication fork to allow replication across oppositely oriented highly transcribed ribosomal operons. Strains containing an inversion of an *rrn* operon were used in such a way that a region of increased head-on collisions between replication and transcription were created. Increased level of DNA trapping at wells in PFGE experiments was correlated with a high level of *rrn* transcription, suggesting the formation of abnormal structures in certain genetic backgrounds (Boubakri et al., 2010).

3. Two-dimensional agarose gel electrophoresis

The movement of a DNA molecule through an agarose gel is determined either by factors intrinsic to the electrophoretic conditions (agarose concentration, the strength of the electric field, the presence of intercalating agents, etc.) as well as the size and shape of the molecule.

The most evident example of the influence of the shape of a DNA molecule on the electrophoretic mobility in an agarose gel is observed when circular DNA molecules are analyzed: supercoiled DNA molecules and the corresponding relaxed-nicked DNA forms do not migrate necessarily at the same position than a linear DNA molecule of the same mass.

Taking into account this property, neutral/neutral 2D agarose gel electroforesis technique was developed to study the shape of recombination intermediates (Bell & Byers, 1983). Later on, it was adapted to study the DNA replication intermediates (RIs) (Brewer & Fangman, 1987). Since then, 2D agarose gel electrophoresis was used to map and characterize replication origins (Brewer & Fangman, 1988; Gahn & Schildkraut, 1989; Liu & Botchan, 1990; Schvartzman et al., 1990; Linskens & Huberman, 1990 b; Friedman & Brewer, 1995; Bach et al.; 2003), to analyze the progression of DNA replication along a DNA fragment (Azvolinsky et al., 2006), to characterize replication fork barriers (Brewer & Fangman, 1988; Linskens & Huberman, 1988; Hernandez et al., 1993; Wiesendanger et al., 1994; Samadashwily et al., 1997, López-Estraño et al., 1998, Possoz et al., 2006; Mirkin et al., 2006, Boubakri et al., 2010), replication termination (Zhu et al., 1992; Santamaría et al., 2000a,b), origin replication interference (Viguera et al., 1996), RIs knotting (Viguera et al., 1996; Sogo et al., 1999), fork reversal (Viguera et al.; 2000; Fierro-Fernandez et al., 2007a) or the topology of partially replicated plasmids (Martín-Parras et al., 1998; Lucas et al., 2001). See (Schvartzman et al., 2010) for an excellent review in plasmid DNA replication analyzed by 2D-gel.

2D agarose gel electrophoresis consists of two successive electrophoreses in which the second dimension occurs perpendicular to the first. Two different migration conditions are used so that the first dimension conditions (low voltage, low agarose concentration) minimize the effect of molecular shape on electrophoretic mobility, whereas this effect is maximized during the second dimension (high voltage and high agarose concentration, in the presence of an intercalating agent) (Friedman & Brewer, 1995). As a consequence, a branched DNA molecule like a recombination or a replication intermediate is separated from a linear molecule of the same mass during the second dimension.

As DNA replication is a continuous process, a sample of DNA isolated from an exponentially growing culture should contain all the replication intermediates (RIs), ranging from the linear non-replicative forms (named 1.0X) to molecules almost completely replicated (2.0X) (Fig. 5A). See (Krasilnikova & Mirkin, 2004), for a detailed protocol of isolation of RIs in *E. coli* and *S. cerevisiae*.

The different migration patterns of a RI digested with a specific restriction enzyme are revealed after southern blotting hybridization with a specific probe and it indicates the mode it has been replicated (Fig. 5). Electrophoresis conditions must be adapted to the fragment size in order to obtain a good separation of the different patterns (Friedman & Brewer, 1995). Different situations can be discerned by using 2D gels. (i) A single fork that moves from one end to the other end of the fragment generates a simple-Y pattern indicating that the DNA fragment is replicated passively and does not contain neither a replication origin nor a replication terminus (Fig. 5A). (ii) Two forks that move convergently generate a double-Y pattern, indicating that replication termination occurs within the analyzed fragment (Fig. 5C); and (iii) two forks that have initiated at some specific point in the analyzed fragment and progress divergently, generate a bubble pattern, indicating that DNA replication has been initiated inside this fragment (Fig. 5B).

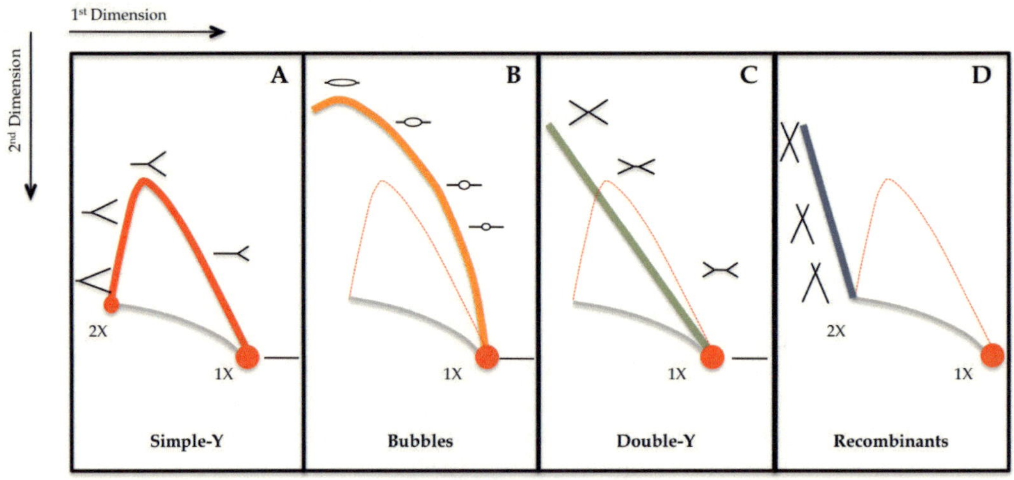

Fig. 5. 2-D gel hybridization patterns generated by replication and recombination intermediates after two-dimensional agarose gel electrophoresis. Replication and recombination intermediates of the restriction fragment are shown above the different 2D-gel pattern. In panels B, C and D, the simple-Y arc is presented as a reference. See text for details.

Moreover, the relative proportion of a particular RI in the population increases as a consequence of the stalling of the replication forks at a specific site. This accumulation is detected as a discrete signal on top of the corresponding arc produced by the RIs. In order to map the region where replication is paused, different restriction fragments must be analyzed to confirm that the signal corresponding to the accumulated molecules move along the arc.

Recently, this technique was used to get insight into the nature of the elements that causes the trapping of the DNA in PFGE experiments in *E. coli* mutant strains containing an inverted *rrn* operon (Boubakri et al., 2010). No RIs were detected in the non-inverted strains or the inversion mutants that express all helicases. However, a simple-Y arc that corresponds to the accumulated Y-shape restriction intermediates was detected in all *dinG*, *rep* and *uvrD* helicase mutants in which the Inv-fragment was trapped in PFGE wells. Moreover, an intense elongated spot was observed over the simple-Y arc. These results indicate that a specific accumulation of RIs occur at the 3′ end of the *rrn* operon (Figure 6).

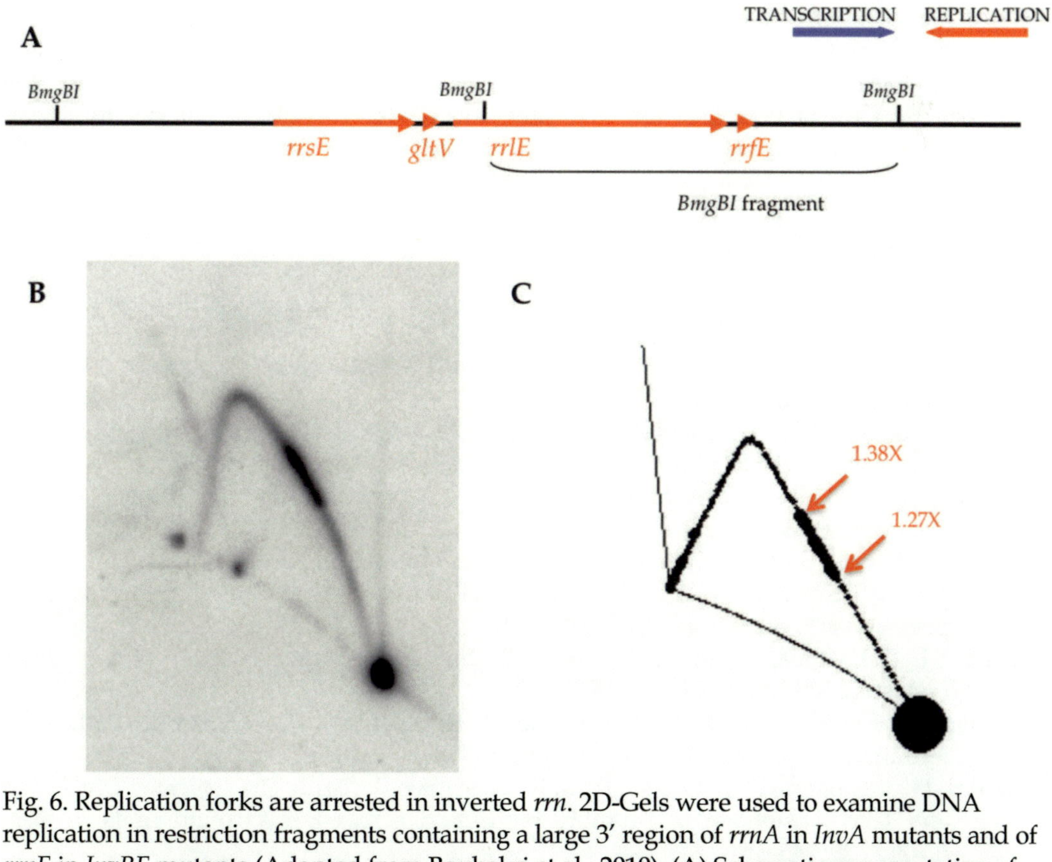

Fig. 6. Replication forks are arrested in inverted *rrn*. 2D-Gels were used to examine DNA replication in restriction fragments containing a large 3′ region of *rrnA* in *InvA* mutants and of *rrnE* in *InvBE* mutants (Adapted from Boubakri et al., 2010). (A) Schematic representation of the restriction fragment used for 2D gels (only *InvBE* is presented in this figure). The position of rrn and restriction sites is shown. (B) DNA from *InvBE dinG rep* mutant was digested with *BmgBI* , analyzed by 2D gels and probed for the sequence just downstream of *rrnE*. A simple-Y arc is clearly detected. On top of this arc, an enlarged signal corresponding to arrested forks is detected. (C) A simulation of replication arrest in this fragment of about 500 pb around the *rrn* transcription terminator sequence was obtained by using the 2D-Gel computer program (Viguera et al., 1998).

4. Concluding remarks

Replication arrest is a source of genetic instability in all types of living cells. As a consequence, cells have developed several effective strategies to tackle with replication fork arrest and/or repairing the double DSBs generated at the stalled replication forks. We have reviewed how PFGE and 2D gels can be used to elucidate some features related to the progression of the replication forks. By using these two non-conventional electrophoresis it can be verified the presence of stalled replication forks, understanding how they have been generated and how they could be restarted.

5. Acknowledgements

We are very grateful to Bénédicte Michel for bacterial strains and continuous support and advice. We especially thank Estrella Guarino, Israel Salguero, Carmen Mata, Encarna Ferrera and Hasna Boubakri for their works and technical help. This work was supported by grants BFU2007-63942 to EG and BFU2007-64153, and P09-CVI-5428 to EV from the Ministerio de Ciencia e Innovación and Junta de Andalucía. EV is grateful to Dr. JB Schvartzman for training in 2D electrophoresis and helpful discussions along the years. EV is grateful to Dr. JB Schvartzman for training in 2D electrophoresis and helpful discussions.

6. References

Aguilera, A. & Gómez-González, B. (2008). Genome instability: a mechanistic view of its causes and consequences. *Nat Rev Genet* 9: 204-217.

Azvolinsky, A., Dunaway, S., Torres, J., Bessler, J.B., & Zakian, V. (2006). *Genes & Development* 20: 3104–3116.

Bach, T., Krekling, M.A., & Skarstad, K. (2003). Excess SeqA prolongs sequestration of *oriC* and delays nucleoid segregation and cell division. *EMBO J* 22: 315-323.

Bell, L. & Byers, B. (1983). Separation of branched from linear DNA by two-dimensional gel electrophoresis. *Analytical Biochemistry* 130: 527-535.

Bierne, H. & Michel, B. (1994). When replication forks stop. *Mol. Microbiol.* 13: 17-23.

Brewer, B.J. & Fangman, W.L. (1987). The localization of replication origins on ARS plasmids in S. cerevisiae. *Cell* 51: 463-471.

Brewer, B.J. & Fangman, W.L. (1988). A replication fork barrier at the 3' end of yeast ribosomal RNA genes. *Cell* 55: 637-643.

Bidnenko, V., Ehrlich, S.D., & Michel, B. (2002). Replication fork collapse at replication terminator sequences. *EMBO J.* 21:3898-3907.

Boubakri, H., de Septenville, A.L., Viguera, E., & Michel, B. (2010) The helicases DinG, Rep and UvrD cooperate to promote replication across transcription units *in vivo*. *EMBO J* 29: 145–157.

Cohen, S.S. & Barner, H.D. (1954). Studies on unbalanced growth in *Escherichia coli*. *Proc Natl Acad Sci USA* 40: 885-893.

Courcelle, J., Belle, J.J., & Courcelle, C.T. (2004). When replication travels on damaged templates: bumps and blocks in the road. *Research in Microbiol.* 155: 231-237.

Fierro-Fernandez, M., Hernández, P., Krimer, D.B., & Schvartzman, J.B. 2007a. Replication fork reversal occurs spontaneously after digestion but is constrained in supercoiled domains. *Journal of Biological Chemistry* 282: 18190–18196.

Flores, M.J., Bierne, H., Ehrlich, S.D., & Michel. B. (2001). Impairment of lagging strand synthesis triggers the formation of a RuvABC substrate at replication forks. *EMBO J.* 20: 619-629

Flores, M.J., Ehrlich, S.D., & Michel, B. (2002). Primosome assembly requirement for replication restart in the *Escherichia coli* holDG10 replication mutant. *Mol. Microbiol.* 44: 783-792.

Friedman, K.L. & Brewer, B.J. (1995). Analysis of replication intermediates by two-dimensional agarose gel electrophoresis. In *DNA Replication*, Campbell, J.L., (Ed.), 613-627, Academic Press Inc. (San Diego, CA).

Gahn, T.A. & Schildkraut, C.L. (1989). The Epstein-Barr virus origin of plasmid replication, *oriP*, contains both the initiation and termination sites of DNA replication. *Cell* 58: 527-535.

Gautom, R.K. (1997). Rapid pulsed-field gel electrophoresis protocol for typing of *Escherichia coli* O157:H7 and other gram-negative organisms in 1 day, *J. Clin. Microbiol.* 35: 2977-2980.

Grompone, G., Seigneur, M., Ehrlich, S.D., & Michel, B. (2002). Replication fork reversal in DNA polymerase III mutants of *Escherichia coli*: a role for the β clamp. *Mol. Microbiol.* 44:1331-9.

Grompone, G., Ehrlich, S.D., & Michel, B. (2003). Replication restart in *gyrB Escherichia coli* mutants. *Mol. Microbiol.* 48:845-54.

Guarino, E., Jiménez-Sánchez, A., & Guzmán, E.C. (2007a). Defective ribonucleoside diphosphate reductase impairs replication fork progression in *Escherichia coli. J Bacteriol* 189: 3496-501.

Guarino, E., Salguero, I., Jiménez-Sánchez, A., & Guzmán, E.C. (2007b). Double-strand break generation under deoxyribonucleotide starvation in *Escherichia coli. J Bacteriol* 189: 5782-6.

Hanawalt, P.C., Maaloe, O., Cummings, D.J., & Schaechter, M. (1961). The normal DNA replication cycle. II. *J. Mol. Biol.* 3: 156-165.

Hanawalt, P.C. (1963). Involvement of synthesis of RNA in thymineless death. *Nature* 198: 286.

Hanawalt, P.C. (2007) Paradigms for the three Rs: DNA Replication, Recombination and repair. *Mol. Cell* 28: 702-707.

Hernández, P., Martín-Parras, L., Martínez-Robles, M.L., & Schvartzman, J.B. (1993). Conserved features in the mode of replication of eukaryotic ribosomal RNA genes. *EMBO J* 12: 1475-1485.

Herschleb, J., Ananiev, G., & Schwartz D.C. (2007). Pulsed-field gel electrophoresis. *Nature Protocols* 2: 677-684.

Horiuchi, T., Nishitani, H., & Kobayashi, T. (1995). A new type of *E. coli* recombinational hotspot which requires for activity both DNA replication termination events and the Chi sequence. *Adv. Biophys.* 31:133-47.

Kobayashi, T., Hidaka, M., Nishizawa, M., & Horiuchi, T. (1992). Identification of a site required for DNA replication fork blocking activity in the rRNA gene cluster in *Saccharomyces cerevisiae*. *Mol Gen Genet.* 233: 355-62.

Krasilnikova, M. & Mirkin, S. (2004). Analysis of Triplet Repeat Replication by Two-Dimensional Gel Electrophoresis. *Methods in Molecular Biology.* 277: 19-28

Kreuzer, K.N. (2005). Interplay between DNA Replication and Recombination n prokaryotes. *Ann. Rev. Microbiol.* 59: 43-67.

Kuzminov, A. (1995). Instability of inhibited replication forks in *E. coli. Bioessays* 17:733-41.

Linskens, M.H.K. & Huberman, J.A. 1988. Organization of replication of ribosomal DNA in *Saccharomyces cerevisiae. Molecular and Cellular Biology* 8: 4927–4935.

Linskens, M.H.K. & Huberman, J.A. (1990b). The two faces of eukaryotic DNA replication origins. *Cell* 62: 845–847.

Liu, Y. & Botchan, M. (1990). Replication of bovine papillomavirus type-1 DNA initiates within an E2-responsive enhancer element. *Journal of Virology* 64: 5903–5911.

López-Estraño, C., Schvartzman, J.B., Krimer, D.B., & Hernández, P. (1998). Colocalization of polar replication fork barriers and rRNA transcription terminators in mouse rDNA. *J Mol Biol* 277: 249.

Lucas, I., Germe, T., Chevrier-Miller, M., & Hyrien, O. (2001). Topoisomerase II can unlink replicating DNA by precatenane removal. *EMBO J* 20: 6509-6519.

Maaloe, O. & Hanawalt, P.C. (1961). Thymine deficiency and the normal DNA replication cycle. I, *J. Mol. Biol.* 3: 144-155.

Maisnier-Patin, S., Nordstrom, K., & Dasgupta, S. (2001). Replication arrests during a single round of replication of the *Escherichia coli* chromosome in the absence of DnaC activity. *Mol. Microbiol.* 42: 1371-1382.

Martin-Parras, L., Lucas, I., Martínez-Robles, M.L., Hernandez, P., Krimer, D.B., Hyrien, O., & Schvartzman, J.B. (1998). Topological complexity of different populations of pBR322 as visualized by two-dimensional agarose gel electrophoresis. *Nucleic Acids Res* 26: 3424-3432.

Mata Martín, C. & Guzmán, E.C. (2011). DNA replication initiations as a key element in thymineless death. *DNA Repair* 10: 94-101.

Matushek, M.G., Bonten, M.J. & Hayden, M.K. (1996). Rapid preparation of bacterial DNA for pulsed-field gel electrophoresis. *J Clin Microbiol.* 34: 2598-2600.

Michel, B., Ehrlich, S.D. & Uzest, M. (1997). DNA double-strand breaks caused by replication arrest. *EMBO J.* 16: 430-8.

Michel, B., Grompone, G., Flores, M.J. & Bidnenko, V. (2004). Multiple pathways process stalled replication forks. *Proc. Natl. Acad. Sci. U S A* 101: 12783-8.

Michel, B., Boubakri, H., Baharoglu, Z., Lemasson, M. & Lestini, R. (2007). Recombination proteins and rescue of arrested replication forks. *DNA Repair (Amst).* 6: 967-80.

Miranda, A. & Kuzminov, A. (2003). Chromosomal lesion suppression and removal in *Escherichia coli* via linear DNA degradation. *Genetics* 163:1255-71.

Mirkin, E., Castro, D., Nudler, E. & Mirkin S. (2006). Transcription regulatory elements are punctuation marks for DNA replication. *Proc. Natl. Acad. Sci. USA* 103: 7276-7281

Morganroth, P.A. & Hanawalt, P.C. (2006). Role of DNA replication and repair in thymineless death in *Escherichia coli. J. Bacteriol.* 188: 5286-5288.

Nakayama, K., Kusano, K., Irino, N. & Nakayama, H. (1994). Thymine starvation-induced structural changes in Escherichia coli DNA. Detection by pulsed field gel electrophoresis and evidence for involvement of homologous recombination. *J. Mol. Biol.* 243: 611-620.

Nakayama, H. (2005). *Escherichia coli* RecQ helicase: a player in thymineless death. *Mutat. Res.* 577: 228-236.

Neylon, C., Kralicek, A.V., Hill, T.M. & Dixon, N.E. (2005). Replication termination in *Escherichia coli*: structure and antihelicase activity of the Tus-Ter complex. *Microbiol Mol Biol Rev.* 69:501-26.

Prado, F. & Aguilera, A. (2005). Impairment of replication fork progression mediates RNA polII transcription-associated recombination. *EMBO J* 24: 1267–1276

Possoz, C., Filipe, S., Grainge, I. & Sherratt D. (2006) Tracking of controlled Escherichia coli replication fork stalling and restart at repressor-bound DNA *in vivo. EMBO J.* 25: 2596-2604

Samadashwily, G., Raca, G. & Mirkin, S. (1997). Trinucleotide repeats afect DNA replication *in vivo. Nature Genetics.* 17: 298-304

Sánchez-Gorostiaga, A., López-Estraño, C., Krimer DB, Schvartzman J.B. & Hernández P. (2004). Transcription terminator factor reb1p causes two replication fork barriers at its cognate sites in fission yeast rDNA *in vivo. Molecular and Cellular Biology* 24: 398–406

Santamaría, D., Hernández, P., Martínez-Robles, M.L., Krimer, D.B., & Schvartzman, J.B. (2000a). Premature termination of DNA replication in plasmids carrying two inversely oriented ColE1 origins. *J Mol Biol* 300: 75-82.

Santamaría, D., Viguera, E., Martínez-Robles, M.L., Hyrien, O., Hernandez, P., Krimer, D.B., & Schvartzman, J.B. (2000b). Bi-directional replication and random termination. *Nucleic Acids Res* 28: 2099-2107.

Schwartz, D.C. & Cantor, C.R. (1984). Separation of yeast chromosome-sized DNAs by pulsed field gradient gel electrophoresis. *Cell.* 37: 67-75.

Schvartzman, J.B., Adolph, S., Martín-Parras, L., & Schildkraut, C.L. (1990). Evidence that replication initiates at only some of the potential origins in each oligomeric form of Bovine Papillomavirus Type 1 DNA. *Mol Cell Biol* 10: 3078-3086.

Schvartzman J.B., Martínez-Robles, M.L., Hernandez, P. & Krimer, D.B. (2010). Plasmid DNA replication and topology as visualized by two-dimensional agarose gel electrophoresis. *Plasmid* 63: 1–10.

Seigneur, M., Bidnenko, V., Ehrlich, S.D., & Michel, B. (1998). RuvAB acts at arrested replication forks. *Cell* 95:419-30.

Seigneur, M., Ehrlich, S.D. & Michel, B. (2000). RuvABC-dependent double-strand breaks in *dnaBts* mutants require RecA. *Mol. Microbiol.* 38:565-74.

Sogo, J.M., Stasiak, A., Martínez-Robles, M.L., Krimer, D.B., Hernández, P. & Schvartzman, J.B. (1999). Formation of knots in partially replicated DNA molecules. J Mol Biol 286, 637-643.

Takeuchi, Y., Horiuchi, T. & Kobayashi, T. (2003) Transcription-dependent recombination and the role of fork collision in yeast rDNA. *Genes Dev* 17: 1497–1506.

Torres, J.Z., Schnakenberg, S.L. & Zakian, V.A. (2004) *Saccharomyces cerevisiae* Rrm3p DNA helicase promotes genome integrity by preventing replication fork stalling: viability of rrm3 cells requires the intra-S-phase checkpoint and fork restart activities. *Mol Cell Biol* 24: 3198–3212

Viguera, E., Hernández, P., Krimer, D.B., Boistov, A.S., Lurz, R., Alonso, J.C., & Schvartzman, J.B. (1996). The ColE1 unidirectional origin acts as a polar replication fork pausing site. *J Biol Chem* 271: 22414-22421.

Viguera, E., Rodríguez, A., Hernandez, P., Krimer, D.B., Trelles, O., & Schvartzman J.B. (1998) A computer model for the analysis of DNA replication intermediates by two-dimensional agarose gel electrophoresis. *Gene* 217: 41–49.

Viguera, E., Hernandez, P., Krimer, D.B., & Schvartzman J.B. (2000). Visualisation of plasmid replciation intermediates containing reversed forks. *Nuc. Acids Res.* 28: 498-503.

Vilette, D., Ehrlich, S.D., & Michel, B. (1995). Transcription-induced deletions in *Escherichia coli* plasmids. *Mol Microbiol* 17: 493–504.

Voineagu, I., Narayanan, V., Lobachev, K., & Mirkin, S. (2008). Replication stalling at unstable inverted repeats: Interplay between DNA hairpins and fork stabilizing proteins. *Proc. Natl. Acad. Sci. USA* 105: 9936-9941.

Wiesendanger, B., Lucchini, R., Koller, T., & Sogo, J.M. (1994). Replication fork barriers in the *xenopus* rDNA. *Nucleic Acids Res* 22: 5038-5046.

Zhu, J., Newlon, C.S., & Huberman, J.A. (1992). Localization of a DNA replication origin and termination zone on chromosome III of *Saccharomyces cerevisiae*. *Mol Cell Biol* 12: 4733-4741.

Part 4

Electrophoresis Application in Enzymology

14

Polyacrylamide Gel Electrophoresis an Important Tool for the Detection and Analysis of Enzymatic Activities by Electrophoretic Zymograms

Reyna Lucero Camacho Morales, Vanesa Zazueta-Novoa, Carlos A. Leal-Morales, Alberto Flores Martínez, Patricia Ponce Noyola and Roberto Zazueta-Sandoval
University of Guanajuato
México

1. Introduction

Gel electrophoresis of enzymes is a very useful and powerful analytical method, which is at present widely used in many distinct fields of both biological and medical sciences and successfully applied in many different fields of human activity. The tremendous expansion of this methodology is mainly due to its simplicity and its high ability to separate both isoenzymes and alloenzymes, which have proven to be very useful genetic markers. The most important step of enzyme electrophoresis is the detection of native enzymes on electrophoretic gels; it means the procedure of obtaining electropherograms, or zymograms. Detection of enzymes on electrophoresis gels means the visualization of gel areas occupied by specific enzyme molecules after their electrophoretic separation. From this point of view, sometimes the testing of an enzyme-coding DNA sequence for expression of catalytically active enzyme is performed by zymograms, where the use of this technique for this purpose is very effective, cheap, and time saving. The number of applications of zymogram techniques for testing cloning enzyme-coding genes for their expression at the protein level is growing (Pfeiffeer-Guglielmi et al., 2000; Lim et al., 2001; Okwumabua et al., 2001). An absolute prerequisite for this is the specific and sensitive zymogram technique suitable for detection of the enzyme inside the gel and the use of the appropriate substrates.

The zymograms has been used to detect a variety of oxidoreductases (Bergmeyer, 1983), including isoenzymes (Jeng & Wayman, 1987) and to classify various genera of yeast based upon the relative mobility of the activity bands produced by selected enzymes (Goto & Takami, 1986; Yamasaki & Komagata, 1983). Electrophorezed gels are placed in a staining solution containing a reduced substrate such as an alcohol, oxidized cofactor such as NAD^+ or $NADP^+$, a dye such as nitroblue tetrazolium, and an electron acceptor-donor such as phenazine methosulphate. At the location of the appropriate enzyme catalyzing oxidation of substrate and reduction of cofactor, a dark-purple band appears as a result of the precipitate that forms upon reduction of the dye.

Previously we have investigated the use of zymogram staining of native electrophoretic gels as an initial approach to the identification of carbonyl reductase activities against both aliphatic (Silva et al., 2009; Zazueta et al., 2008) and aromatic hydrocarbons (Durón et al., 2005; Zazueta et al., 2003) in *Mucor circinelloides* YR1, an indigenous fungus isolated from petroleum contaminated soil.

Oil spills sometimes occur during routine operations associated with the exploration and production of crude oil. Crude oils vary widening in composition depending on factors such as source bed type and generation temperatures (Hunt, 1979). Biodegradation rates for crude oils will vary due to differences in composition, as reflected by hydrocarbon class distribution: saturates, aromatics, and polars, and the amount *n*-alkanes *versus* branched and cyclic alkanes within the saturated hydrocarbon class (Cook et al., 1974). In nature exist many types of microorganisms useful in the biodegradation processes of contaminant compounds (Atlas, 1995), such as the polycyclic aromatic hydrocarbons (PAH's) that are persistent soil contaminants and many of which have toxic and carcinogenic properties (Hyötyläinen and Oikari, 1999; Cerniglia, 1997).

In bacterial aerobic degradation of aromatic compounds, reactions of metabolic pathways generally lead to the formation of aromatic intermediates containing two hydroxyl constituents, which are subsequently ring-cleaved by excision dioxygenases (Neidle et al., 1992). In many catabolic pathways the formation of such intermediates is carried out by two successive enzymatic steps namely dihydroxylation of the polyaromatic substrate to produce *cis*-diols followed by dehydrogenation (Harayama & Timmis, 1989). The ring hydroxylation is catalyzed by multi-component dioxygenases, while the dehydrogenation is catalyzed by *cis*-diol-dehydrogenases. In mammalian tissues the enzyme dihydro-diol dehydrogenase (DD, EC 1.3.1.20) exists in multiple forms (Hara et al., 1990; Higaki et al., 2002) and catalyses the $NADP^+$-linked oxidation of *trans*-dihydro-diols of aromatic hydrocarbons to the corresponding catechols (Penning et al., 1999). Studies on the metabolism of aromatic hydrocarbons by fungi are limited, nevertheless have been shown to posses the ability to metabolize aromatic compounds (Auret et al., 1971; Ferris et al., 1976) and the aryl oxidative enzymes of fungi appear to be similar to monooxygenases of hepatic microsomes (Cerniglia & Gibson, 1977; Ferris et al., 1976). Smith & Rosazza (1974) have also presented evidence that naphthalene is metabolized to 1-naphtol by six different genera of fungi.

In this work we analyze the cytosolic fraction of YR-1 strain by electrophoretic zymograms, methodology that there is not described in the literature for the $NADP^+$-dependent dihydrodiol dehydrogenase (DD) activities. We analyze all the activity bands corresponding to proteins with DD activity present in an enzymatic extract in only one lane of the electrophoretic gel. Our results show eleven different DD activity bands, five of them are constitutive, DD1-5, since they appears when the strain is growth on glucose, and the others six are induced by different compound added to the culture media as a sole carbon source. Some biochemical-enzyme characteristics as pH, optimal temperature, cofactor dependence, substrate specificity and the effect of cations, EDTA and pyrazole were investigated for DD activities when YR-1 strain was grown in naphthalene as sole carbon source.

2. Materials and methods

2.1 Organisms used and culture conditions

Mucor circinelloides strain YR-1 originally isolated from petroleum-contaminated soil in Salamanca, Guanajuato, Mexico was used as enzymatic source. A defined media containing yeast-peptone-glucose-agar (YPGA) (Bartnicki-García & Nickerson, 1962) was used for strain maintenance, spore collection and mycelium growth. Aerobic mycelium growth was also carried out on salt minimal medium (Alvarado et al. 2002) added with 0.1% (w/v) peptone (sMMP). As a carbon source we added D-glucose (1% w/v) or glycerol (1.0% v/v) or ethanol (2.0% v/v) or *n*-decanol (1.0% v/v) or *n*-pentane (1.0% v/v) or *n*-decane, (1.0% v/v) or *n*-hexadecane (1.0% w/v) or naphthalene (0.5% w/v) or anthracene (0.5% v/v) or phenanthrene (0.5% w/v) or pyrene (0.5% w/v). Liquid cultures (600 ml) inoculated with spores at a final cell density of 5 x 10^5/ml were propagated in 2-l Erlenmeyer flasks and incubated in a reciprocating water bath at 28°C for 22 h at 125 rpm for all substrates except glucose that was incubated for 12 h at same other conditions.

2.2 Preparation of purified fractions

Mycelium of 22 h of incubation was harvested by filtration and exhaustively washed with cold sterile-distilled water; mycelial mass was suspended in 15 ml of 20 mM Tris-HCl pH 8.5 buffer containing 1 mM phenylmethanesulphonyl fluoride (previously dissolved in ethanol). Approximately 20 ml of cells was mixed with an equal volume of glass beads (0.45-0.50 mm diameter) and disrupted in a Braun model MSK cell homogenizer (Braun, Melsungen, Germany) for four periods of 30 sec each under a CO_2 stream. The homogenate (crude extract) was centrifuged at 4,300*g* for 10 min in a J2-21 Beckman rotor in a Beckman JA-20 centrifuge to remove cell walls and unbroken cells, a 1 ml sample of the supernatant was saved. The rest of the supernatant (low speed supernatant) was centrifuged at 31,000*g* for 20 min in a 70Ti Beckman rotor in a Beckman L8-80 ultracentrifuge and samples of 1 ml of the supernatant was saved; the resulting pellet (mitochondrion rich sample) was resuspended in 2 ml of buffer and saved. The rest of the supernatant was high-speed centrifuged at 164,500*g* for 45 min in a 70Ti Beckman rotor at 4°C in a Beckman L8-80 ultracentrifuge; the supernatant (cytosolic fraction) was put aside, and the pellet, the mixed membrane fraction (MMF), was resuspended in 2 ml and saved. In all cases samples of different fractions were kept at -70 °C for further studies.

2.3 Gel electrophoresis

The slab gels were 1.5 mm-thick contained 6% (w/v) acrylamide/4% (w/v) bisacrilamide, loaded with the cytosolic fraction of each culture and run in the mini-gel system manufactured by Bio-Rad. The continuous buffer system described by Laemmli (1970) without SDS (native conditions) was used to run for 2.5 h at 80 V. The *Rm* values were calculated as the ratio of the distance migrated by the stained band divided by the distance migrated by tracking dye; standard deviation was calculated with Excel from three independent experiments and each experiment was made by triplicate on each substrate.

2.4 Enzymatic assays

All enzyme assays were carried out in a final volume of 1 ml and incubated for different times at 25 °C. NAD^+-dependent ADH activity was assayed in the oxidative direction according to Bergmeyer (1983). The enzymatic assays contained 25 mM Tris-HCl (pH 8.5), 2 mM NAD^+ or $NADP^+$, cell-free extract (100-200 μg protein), and 100 mM of the substrate (1R, 2S)-*cis*-1,2-di-hydro-1,2-naphthalene-diol. The reaction was started by dihydrodiol addition, and reduction of NAD^+ or $NADP^+$ was monitored by the increase in absorbance at

340 nm in a Beckman DU-650 spectrophotometer. One unit of enzyme activity was defined as the amount required reducing 1 µmol of NAD^+ or $NADP^+$ per minute at 25°C. Specific dihydrodiol dehydrogenase (DD) activity was expressed as units per milligram of protein.

For DD activity in gels we developed an appropriate methodology because there is not any report in the literature about the detection of these enzymes by means of electrophoretic zymograms, so for we modified the method described for Nikolova & Ward (1991) for alcohol dehydrogenase. Briefly, after non-denaturing 6% (w/v) PAGE, described above, the activity was revealed as follows. The gel was submerged for 120 min in 4 ml of 0.5 M Tris-HCl buffer pH 8.5 containing 0.5 mg phenazine methosulphate (PMS), 7.5 mg *p*-nitro-blue tetrazolium (PNBT), 14.34 mg $NADP^+$ or NAD^+, 1 mM EDTA, 1 mM DTT and 100 mM of (1R, 2S)-*cis*-1, 2-dihydro-1, 2-naphthalene-diol as substrate. After incubating at 25 °C for 30 min (in dark) with gentle shaking at 80 rpm, the dihydrodiol dehydrogenases or ADH electro-morphs were observed as blue-dark bands.

When substrate specificity of DD was tested, different single alcohols were added to the mixture reaction at a final concentration of 100 mM. The following substrates were tested: *N*-decanol, *n*-hexadecanol, *n*-octadecanol, hexane-1,2,3,4,5,6-hexaol, benzyl alcohol, cholesterol, *cis*-naphthalene-diol, ethylene-glycol, poly-ethylene-glycol 3350, and sorbitol, were previously dissolved in dioxan and others were prepared in water: methanol, ethanol, propane-1-ol, propane-2-ol, butane-1-ol, pentane-1-ol, propane-1,2,3-triol and methyl propane-1-ol.

The pH, optimal temperature, substrate specificity, and effect of cations, EDTA and pyrazole were performed after a non-denaturing gel, 6% acrylamide, loading 300 µg of protein. The pH determination was performed from 3 to 9 with citrate buffer for 3 to 5, phosphate buffer for 5 to 7 and Tris/HCl buffer for 7 to 9. The temperature effect was tested in a range of 4 to 45 °C, using a freezer or metabolic bath at the desired temperature. The cation effect was tested using 1 mM of $CaCl_2$, $MgSO_4$, $ZnSO_4$ and $FeSO_4$, and for the EDTA, 1mM was also used. The assays were performed in the presence of *cis*-naphthalene-diol as substrate and $NADP^+$ as electron acceptor; the enzymatic activity was measured over a range of pH values in the forward reaction dihydrodiol → diol.

2.5 Miscellaneous

Protein concentration was measured according Lowry et al. (1951), using bovine serum albumin as standard. Phenylmethanesulphonyl fluoride and *cis*-naphthalene-diol were purchased from Sigma (St. Louis, MO, USA), the alcohol used as substrates were from J.T. Baker (Phillipsburg, NJ, USA). All reagents were analytical grade.

Densitometric analysis was performed in a Gene Genius Bio-Imaging System V. 6.05.01, SYNGENE, Synoptics Systems. Software used was Gene Tools V. 3.06.02, Syn. Ltd.

3. Results

3.1 Sub-cellular distribution of dihydrodiol dehydrogenase activity

The first approach was to know the sub-cellular distribution of the dihydrodiol dehydrogenase (DD) activity by means of a differential-centrifugation procedure and the spectrophotometer detection of the DD activity from *M. circinelloides* YR-1 grown in different carbon sources, using a variation of the method described by Bergmeyer (1983). For this purpose we use the commercial substrate *cis*-naphthalene-diol. If the low speed supernatant is compared, the enzymatic activity was almost 8 times higher when naphthalene rather than glucose was the carbon source and $NADP^+$ was used as electron acceptor (Table 1).

	DD activity (x 10^{-2})					
	$NADP^+$			NAD^+		
Sample	Carbon source					
	Glucose	Ethanol	Naphthalene	Glucose	Ethanol	Naphthalene
4,300 x *g*						
Supernatant	42	4.0	270	131	12	39
31,000 x *g*						
Pellet	1.7	ND[a]	ND	23	ND	12
Supernatant	4.5	1.3	91	57	23	61
164,500 x *g*						
Pellet (MMF)	0.7	0.1	0.1	5.2	1.2	ND
Supernatant (Cytosol)	21	0.5	178	59	2.8	1.4

Table 1. $NADP^+$ or NAD^+-dependent dihydrodiol dehydrogenase activities present in sub-cellular fractions of *Mucor circinelloides* YR-1 grown on different carbon sources. Mycelial cells, grown in the indicated carbon sources, were broken (Braun) and fractions obtained by differential-centrifugation. DD activity of the different fractions was measured with *cis*-naphthalene-diol as substrate and $NADP^+$ or NAD^+ as electron acceptor. The values are the means of three independent experiments with triplicate determinations.
[a] ND, no detected.

This suggests that at least some of the detected activity could be inducible, and as can be seen, the major enzymatic activity is present in the soluble fractions. When NAD+ was used as electron acceptor, the activity found in the low speed supernatant when the fungus was grown in glucose as a carbon source is more than 3 times higher than the one present when naphthalene was used, and more than ten times higher if compared with the activity obtained with ethanol as a carbon source.

These results enhance the interest to investigate how many different activities will be revealed by electrophoretic zymograms in the cytosolic fraction of the fungus when it grown on different carbon sources.

3.2 Use of zymograms to reveal the presence of several dihydrodiol dehydrogenase activities in cytosolic fraction of *M. circinelloides* YR-1 grown on different carbon sources

Aerobically mycelium grown in different carbon sources (see Materials and Methods) was used to obtain the corresponding cytosolic fraction and each one was run on no-denaturing polyacrylamide gels and stained for $NADP^+$-dependent dihydrodiol dehydrogenase activity with *cis*-naphtalen-diol as substrate.

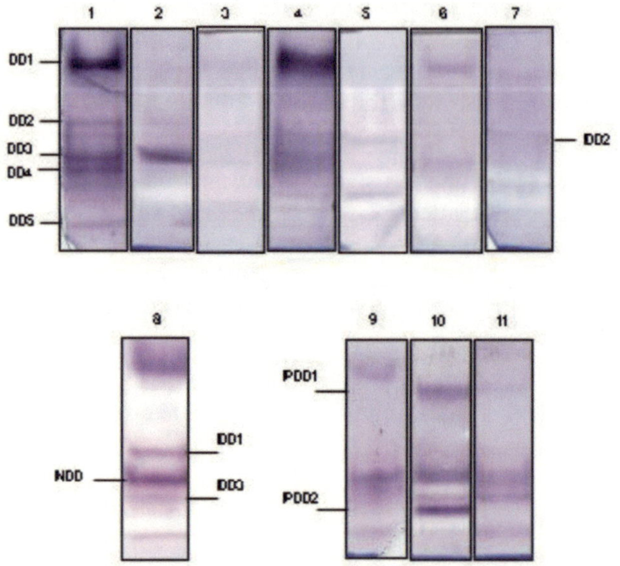

Fig. 1. Dihydrodiol dehydrogenase enzymatic activity present in cytosolic fraction of *M. circinelloides* YR-1 grown in different carbon sources. Mycelia in lane 1 was grown for 12 h and lanes 2-11 were grown for 22 h at 28°C on sMMP medium added with the carbon source indicated. *Lanes*: *1,* D-glucose; *2,* glycerol; *3,* ethanol; *4, n*-decanol; *5, n*-pentane; *6, n*-decane; *7, n*-hexadecane; *8,* naphthalene; *9,* anthracene; *10,* phenanthrene; *11,* pyrene, at the concentrations described in Material and Methods. The extracts were electrophoresed and stained as described in Materials and Methods. In all cases 300 µg of protein were loaded in each lane. In this gel 100 mM *cis*-naphthalene-diol was the substrate and $NADP^+$ the electron acceptor. These results are representative gels and mycelia were grown up and run on the gels at least three times. The induction patterns were always reproducible.

Under the conditions tested, five bands were seen in glucose as carbon source (Fig. 1, lane 1) which were considered as constitutive dihydrodiol dehydrogenases and identified with a number (1-5) considering their decreasing *Rm* (DD1-5) (Table 2), and six inducible bands of activity were detected, depending of the carbon source in the culture media used for growth (Fig. 1; Table 3). One of the inducible bands (iDD1) was seen when *n*-decanol or *n*-pentane or *n*-hexadecane or naphthalene was the carbon source (Fig. 1, lanes 4, 5, 7 and 8 respectively, Table 3).

A second inducible $NADP^+$-dependent dihydrodiol dehydrogenase activity was seen when *n*-decanol or *n*-pentane was the carbon source (iDD2) (Fig. 1, lanes 4 and 5; Table 3). The third inducible enzymes (iNDD) was seen only when aromatic hydrocarbons were used as sole carbon source in the growth media (Fig. 1, lanes 8 to 11, Table 3). A fourth inducible naphthalene-diol dehydrogenase (iDD3) was induced by some of the alcohols, alkanes and aromatic polycyclic compounds tested (Fig. 1, lanes 3, 4, 6, 8, 10 and 11; Table 3). When phenanthrene was used as the carbon source, two new bands with different relative motilities were revealed, iPDD1 and iPDD2, (Fig. 1, lane 10; Table 3). The iPDD1 was also observed when pyrene was the carbon source (Fig. 1, lane 10 and 11; Table 3).

Carbon source	*Rm*[a] of DD constitutives				
	1	2	3	4	5
D-Glucose	0.22±0.04	0.4±0.03	0.62±0.01	0.69±0.02	0.90±0.01
Glycerol	-	-	-	-	-
Ethanol	-	-	-	-	-
n-Decanol	0.21±0.01	-	0.61±0.02	0.69±0.03	0.89±0.03
n-Pentane	-	-	-	0.7±0.03	-
n-Decane	0.21±0.01	-	-	-	-
n-Hexadecane	-	-	-	-	-
Naphthalene	0.21±0.01	-	-	-	0.89±0.006
Anthracene	0.22±0.006	-	-	-	0.89±0.02
Phenanthrene	-	-	-	-	0.89±0.01
Pyrene	-	-	-	-	0.89±0.01

Table 2. Constitutive $NADP^+$-dependent dihydrodiol dehydrogenase activities in cytosolic fraction of *Mucor circinelloides* YR-1 grown in different carbon sources.
[a]*Rm*; was calculated as described in Materials and Methods section as its standard deviation.

Carbon source	*Rm*[a] of DD inducibles					
	iDD1	iDD2	iDD3	iPDD1	iPDD2	iNDD
D-Glucose	-	-	-	-	-	-
Glycerol	-	-	-	-	-	-
Ethanol	-	-	-	-	-	-
n-Decanol	0.55±0.002	0.66±0.005		-	-	-
n-Pentane	0.55±0.001	0.66±0.003	0.74±0.002	-	-	-
n-Decane	-	-	-	-	-	-
n-Hexadecane	-	-	0.73±0.010	-	-	0.66±0.005
Naphthalene	0.56±0.003	-	0.73±0.020	-	-	0.67±0.008
Anthracene	-	-	-	-	-	0.67±0.008
Phenanthrene	-	-	0.74±0.010	0.23±0.006	0.78±0.003	0.67±0.008
Pyrene	-	-	0.74±0.004	0.22±0.020	-	0.67±0.008

Table 3. Inducible $NADP^+$-dependent dihydrodiol dehydrogenase activities in cytosolic fraction of *Mucor circinelloides* YR-1 grown in different carbon sources.
[a]*Rm*; was calculated as described in Materials and Methods section as its standard deviation.

3.3 Bands intensity of dihydrodiol dehydrogenase activities in cytosolic fraction of *M. circinelloides* YR-1

With the comparing purpose the activity showed for the denominated iNDD enzyme that is induced when naphthalene was used as a carbon source was taking as a 100% and the others enzymes where referred to this value (Table 4).

In all cases *cis*-naphthalene-diol and $NADP^+$ were used in the enzymatic assay. When phenanthrene was used as a carbon source there are four different inducible enzymes, iDD3, iPDD1, iPDD2 and iNDD, being iPDD1 and iPDD2 best induced by this carbon source, and the latest is the one that showed the highest induction value of all inducible enzymes (103.5%) (Table 4). The iPDD2 enzyme is the only one that it is induced by only one carbon source (Table 3). In the case of the iDD1, the highest induction value obtained was when naphthalene was used as a carbon source as iDD3 and iNDD enzymes (Table 4). The iDD2 showed its best induction value when n-pentane was used as a carbon source (Table 4). It is noticeable that glycerol, ethanol and *n*-decane do not induce any of the DD activities.

Carbon source	Band intensity[a] (relative units)					
	iDD1	iDD2	iDD3	iPDD1	iPDD2	iNDD
D-Glucose	0.0	0.0	0.0	0.0	0.0	0.0
Glycerol	0.0	0.0	0.0	0.0	0.0	0.0
Ethanol	0.0	0.0	0.0	0.0	0.0	0.0
n-Decanol	10.2	5.3	0.0	0.0	0.0	0.0
n-Pentane	8.7	18.2	14.6	0.0	0.0	0.0
n-Decane	0.0	0.0	0.0	0.0	0.0	0.0
n-Hexadecane	0.0	0.0	6.3	0.0	0.0	4.5
Naphthalene	56.8	0.0	16.3	0.0	0.0	100.0
Anthracene	0.0	0.0	0.0	0.0	0.0	90.5
Phenanthrene	0.0	0.0	8.1	82.9	103.5	84.5
Pyrene	0.0	0.0	10.2	8.0	0.0	38.4

Table 4. Relative inducibility of $NADP^+$-dependent dihydrodiol dehydrogenase activities in cytosolic fraction of *Mucor circinelloides* YR-1 grown in different carbon sources.
[a]Relative units were obtained by densitometry, using the value from iNDD as 100% when the fungus was growth on naphthalene.

3.4 Effect of ethanol as substrate and NAD^+ as electron acceptor on induced dihydrodiol dehydrogenase activities in cytosolic fraction of *M. circinelloides* YR-1 grown on different carbon sources

It is interesting to compare if the inducible DD enzymes are able to use NAD^+ as electron acceptor and/or ethanol as substrate because some dehydrogenases are able to use both of them. In the presence of NAD^+ as electron acceptor and *cis*-naphthalene-diol as substrate, there was not any staining in the region of either constitutive or inducible dihydrodiol dehydrogenase activities (Fig. 2A, lanes 1-5; Table 5). In the presence of ethanol as substrate and $NADP^+$ as electron acceptor, two ADH activity in cytosolic fraction from mycelium grown on glucose were revealed (Fig. 2B, lane 1) one with a *Rm* of 0.42±0.008 (denominated ADH1) and the other with a *Rm* of 0.84±0.003 (denominated ADH2). Under these conditions we also observed the inducible dihydrodiol dehydrogenase enzymes denominated DD3, suggesting that this enzyme also possesses an ADH activity $NADP^+$-dependent (Fig. 2B, lane 1). Also, under these assay conditions, two bands were observed when phenanthrene was used as a carbon source to growth the mycelia (Fig. 2B, lane 4). The bands correspond to the DD5 (*Rm* of 0.90±0.010) and a new ADH, denominated ADH3 with a *Rm* of 0.94±0.010 (Fig. 2B, lane 4).

When we used NAD^+ as electron acceptor and ethanol as substrate the denominated ADH1, ADH3 and DD5 activity bands were revealed when the cytosolic fraction from mycelium grown on glucose (Fig. 2C, lane 1).

As a control, a sample of the culture media lacking carbon source (Fig. 2A to C, lane 6) and an assay lacking substrate in the reaction mixture for the activity in zymograms (Fig. 2A to C, lane 7) did not showed any enzymatic activity.

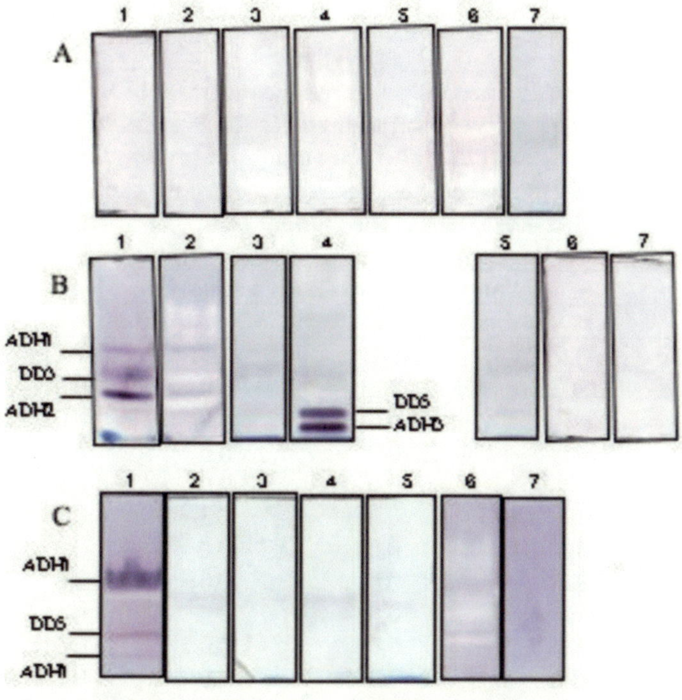

Fig. 2. Detection of dihydrodiol dehydrogenase activities in cytosolic fraction of *M. circinelloides* YR-1 by activity-stained gels. All mycelia were obtained after grown for 22 h at 28°C on the carbon source indicated. A. *Lanes 1,* 1.0% glucose; *2,* 0.5% naphthalene; *3,* 0.5% anthracene; *4,* 0.5% phenanthrene; *5,* 0.5% pyrene; *6,* a sample of the culture media without carbon source; *7,* without substrate in the activity-reaction mixture. The reaction was performed with NAD^+ and 100 mM cis-naphthalene-diol as substrate, described in Materials and Methods section. B. identical samples as A, but the activity was developed with $NADP^+$ and 100 mM ethanol as substrate. C. identical samples as A, but the activity was developed with 100 mM ethanol as substrate. The amount of protein loaded per track was equalized to 300 µg. These results are representative gels and mycelia were grown up and run on the gels at least three times. The activity patterns were always reproducible.

In order to compare the observed activities we performed a densitometric analysis to the bands intensity and the Table 5 shows the obtained values under the condition where the activity have its highest value, taking the iNDD activity as a 100%. As show, the denominated DD1 enzymes, a constitute one, has the highest activity of all, in contrast the activity denominated iDD3 and DD5 measured with ethanol and NAD^+ are the lowest

(Table 5). It is interesting that only the denominated ADH 1 to 3 have activity with ethanol as a substrate, being the ADH3 the enzyme with the highest activity (Table 5). Surprisingly not a single one activity was revealed when *cis*-naphthalene-diol and NAD^+ were used as a substrate (Table 5).

3.5 General properties of NDD activities

DD activities were assayed only in crude cell-free extracts of aerobically-naphtahalene grown mycelia cells because at these moments, we were strongly interested in the NDD activity.

In all cases, the DD activities were tested in cell-free extracts of *M. circinelloides* YR-1.grown in 0.5% of naphthalene as sole carbon source for 22 h at 28 °C. Cytosolic fraction was separated in a native electrophoresis and the amount of protein loaded per track equalized and was equivalent to 300 μg, $NADP^+$ as electron acceptor and 100 mM *cis*-naphthalene-diol as enzyme substrate were used to reveal the zymograms.

	Band intensity (relative units)[a]				
	cis-naphthalene-diol		Ethanol		
Enzyme	$NADP^+$	NAD^+	$NADP^+$	NAD^+	*Rm*[b]
DD 1	104.2	0.0	0.0	0.0	0.21±0.006
DD 2	1.2	0.0	0.0	0.0	0.40±0.010
DD 3	3.4	0.0	0.0	0.0	0.61±0.020
DD 4	3.2	0.0	0.0	0.0	0.69±0.030
DD 5	1.2	0.0	67.6	0.4	0.90±0.010
iDD1	56.8	0.0	0.0	0.0	0.56±0.010
iDD2	18.2	0.0	0.0	0.0	0.66±0.050
iDD3	16.3	0.0	3.6	0.4	0.73±0.010
iPDD1	82.9	0.0	0.0	0.0	0.23±0.006
iPDD2	103.5	0.0	0.0	0.0	0.78±0.003
iNDD	100.0	0.0	0.0	0.0	0.67±0.008
ADH1	0.0	0.0	1.2	3.1	0.42±0.008
ADH2	0.0	0.0	45.2	0.0	0.84±0.003
ADH3	0.0	0.0	74.8	0.0	0.94±0.010

Table 5. Activities of *cis*-naphthalene-diol and alcohol dehydrogenase of cytosolic fraction of *M. circinelloides* YR-1 grown in the best inducer for each one. Densitometric analysis was carried out as described in Materials and Methods. The enzymatic determination was on the gel with *cis*-naphthalene-diol or ethanol as the substrate and $NADP^+$ or NAD^+ as the electron acceptor.
[a] Relative units were obtained by densitometry, using the value from iNDD as 100% when naphthalene was the carbon source.

3.5.1 pH dependence

The Fig. 3 shows that the optima pH value for all five activities expressed with n-naphthalene as carbon source and $NADP^+$ and naphthalene-diol in the enzymatic reaction was 8.5. It is noticeable that only the iNDD show activity at pH 3 and little DD activities were showed at pH values below 8.5. It is important to say that the background in the lane for activity revealed at pH 9, was darken because of pH.

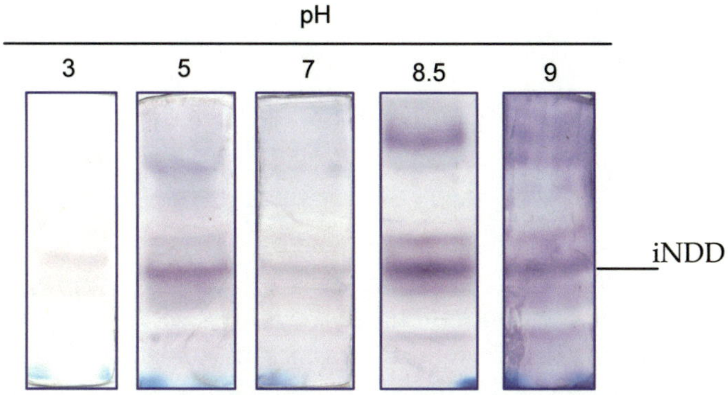

Fig. 3. Effect of pH on dihydrodiol dehydrogenase activities present in cytosolic fraction of *M. circinelloides* YR-1grown in naphthalene. Each track was cut and stained at the indicated pH value.

3.5.2 Temperature

The effect of the temperature on DD activities was tested on cytosolic fraction in a range of temperatures oscillating between 4 and 45 °C. The optimum temperature was 37 °C, notice that even at 45 °C the activity band corresponding to the iNDD can be seen in the zymogram (Fig. 4). It is important to specify that the background in the lanes for activity revealed at 37 or 45 °C were darker because of the incubation temperature.

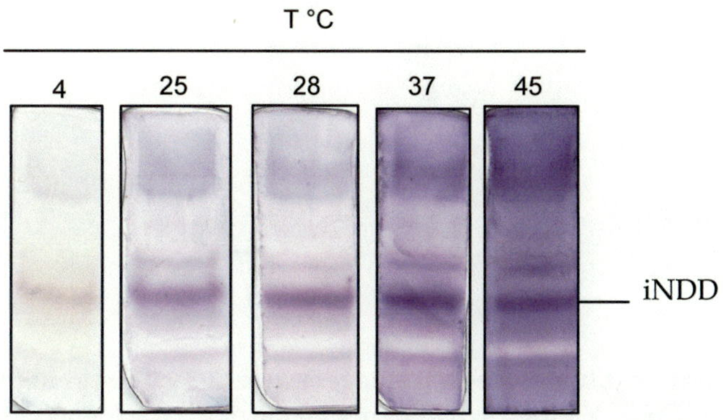

Fig. 4. Temperature effect on dihydrodiol dehydrogenase activity present in cytosolic fraction of *M. circinelloides* YR-1. Each track was cut and activity developed at the indicated temperature.

3.5.3 Requirement of different divalent ions

Different divalent ions were used to prove if some of them were required for DD activities. The Fig. 5 shows that only Ca^{2+} had an enhancing effect on DD activities meanwhile the other divalent metals tested and also EDTA were inhibitory $Fe^{2+} > Zn^{2+} >$ EDTA> Mg^{2+}.

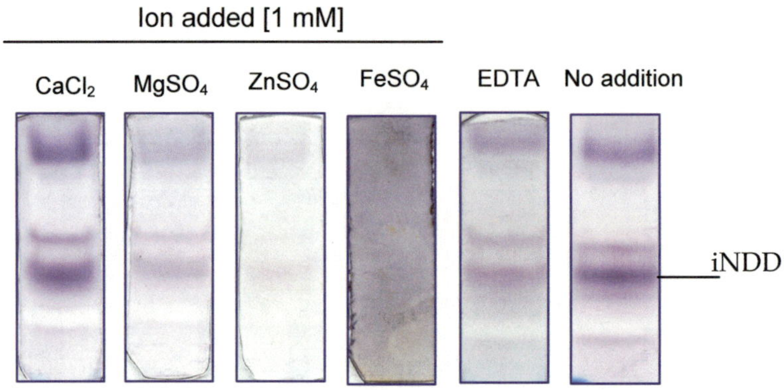

Fig. 5. Divalent ions effect on dihydrodiol dehydrogenase activity present in cytosolic fraction of *M. circinelloides* YR-1. Each track was cut and stained adding to the reaction mixture the divalent ion indicated at the concentrations described in Materials and Methods.

3.5.4 Pyrazole effect

Pyrazole is a well known ADH competitive inhibitor (Pereira et al., 1992) and this is the principal reason we decide to prove its effect on the different DD activities present in crude cell-free extracts obtained from *M. circinelloides* YR-1 mycelia grown in naphthalene as the sole carbon source. As can be seen in Fig. 6, pyrazole has a little inhibitory effect on the different DD's. In addition, iNDD showed a mild decrease in the level of its activity when measured by staining for activity in gels in presence of pyrazole (Fig. 6)

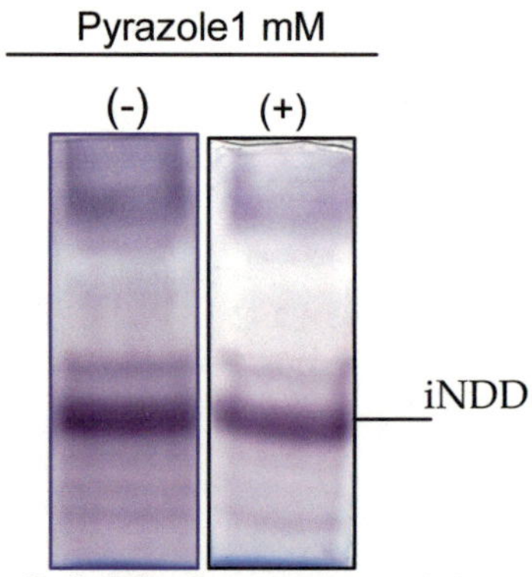

Fig. 6. Pyrazole effect on dihydrodiol dehydrogenase activity present in the cytosolic fraction of *M. circinelloides* YR-1 grown in naphthalene. Each track was cut and stained adding or not to the reaction mixture 1.0 mM pyrazole.

3.5.5 Substrate specificity of the different inducible and constitutive dihydrodiol dehydrogenase activities

To test the substrate specificity we chose naphthalene as the carbon source in the culture media since we can observe two constitutive band of activity (DD1 and DD5) and three inducible bands (iDD1, iDD3 and iNDD) (Fig. 1, Table 2 and 3). A variety of substrates were tested in the gel making the assay with $NADP^+$ as electron acceptor. The constitutive DD1 enzyme it is the one that shows in a majority of substrates, 14 out of 18, being ethanol, propane-1-ol, benzyl alcohol and sorbitol the substrates where the activity did not show, but when show its intensity is really low (Fig. 7). In contrast, the band with the highest intensity is the DD2 enzyme when propane-1,2,3-triol was the substrate (Fig. 7, lane 10), even when this enzyme cannot be seeing when cis-naphtalene-diol is used as a substrate (Fig. 7, lane 18). As this DD2 enzyme there are others enzymes that did not show with cis-naphthalene-diol and can be seen with others substrates, as it is DD3 and iPDD2, that shows with three and four different substrates respectively (as an example see Fig. 7, lane 10). It is surprising that the iPDD2 enzyme that only showed when phenanthrene was the carbon source to growth the fungus, it is present here depending on the substrate used, propane-1,2,3-triol, 2-methyl propane-1-ol, ethylene-glycol and benzyl alcohol (Fig. 7, lanes 10, 11, 13 and 15).

In the particular case of NDD1, it was present only when naphthalene was the carbon source (Fig. 7) but it was absent in all other aromatic hydrocarbons used as carbon source (Fig. 1) and this enzyme practically only uses *cis*-naphthalene-diol as substrate (Fig. 7, lane 18). In the case of iDD3 and DD5, both present broad substrate specificity, showing a special preference for short-chain alcohols, including 1-decanol (Fig. 7). There are five bands that show with different substrates, but its intensity is really low and they were not taken in account (Fig. 7).

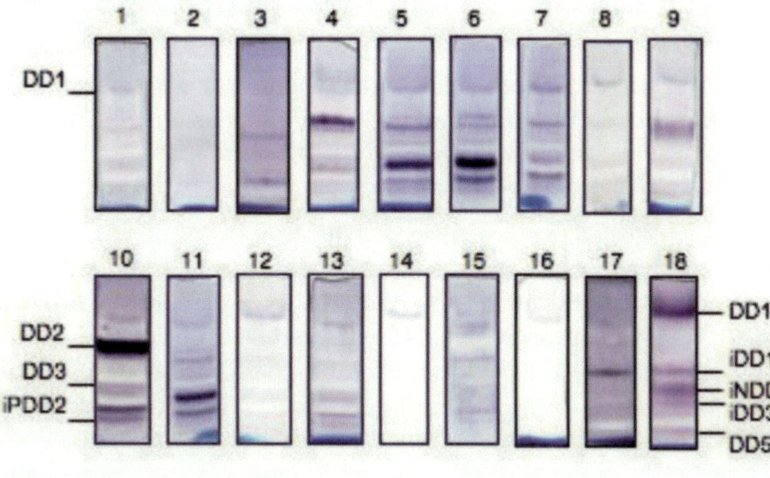

Fig. 7. Substrate specificity of constitutive and inducible DD activities of cytosolic fraction of *M. circinelloides* YR-1 grown on naphthalene, revealed by activity-stained gels. A variety of substrates were used with $NADP^+$ as electron acceptor. All substrates were tested at 100 mM of final concentration. Lane *1* methanol; *2* ethanol; *3* propane-1-ol; *4* propane-2-ol; *5* butane-1-ol; *6* pentane-1-ol; *7* 1-decanol; *8* 1-hexadecanol; *9* hexane-1,2,3,4,5,6- hexaol; *10* propane-1,2,3-triol; *11* 2-methyl propane-1-ol; *12* 1-octadecanol; *13* ethylene-glycol; *14* poly-ethylene-glycol 3350; *15* benzyl alcohol; *16* cholesterol; *17* sorbitol; *18* *cis*-naphthalen-diol.

4. Discussion and conclusion

Many studies have been done on NAD^+-dependent *cis*-dihydrodiol dehydrogenases (DD) in bacteria (Jouanneau & Meyer, 2006; Van Herwijnen et al., 2003). In the case of $NADP^+$-

dependent *trans*-dihydrodiol dehydrogenases, almost all investigations have been done in mammalian tissues (Carbone et al., 2008; Chang et al, 2009; Chen et al., 2008) but only a few reports about these important enzymes have been done in fungi (Bezalel et al., 1997; Hammel, 1995; Sutherland et al., 1993) particularly in *Phanerochaete chrisosporium* (Bogan & Lamar, 1996; Muheim et al., 1991). At date, there is no any report about the detection of dihydrodiol dehydrogenase activities by means of electrophoretic zymograms in any organism. This methodology represents an interesting approach because in this way it is possible to detect, study and compare the different isoenzymes present in the cell-free extracts of the organism used as enzymatic source. In our own work, YR-1 strain possesses extraordinary metabolic machinery that premises it to survive in a very dangerous place how is a petroleum-contaminated soil.

The results about the localization of DD activities in a differential centrifugation procedure from YR-1 grown in different carbon sources (Table 1), revealed that the activity measured with *cis*-naphthalene-diol as substrate and $NADP^+$ as electron acceptor was only present in the supernatant fractions of each centrifugation speed, suggesting that all DD activity observed must be a soluble enzyme. At date, we cannot discard the possibility that the DD activity could be located in the lumen of some kind of microsomal bodies, because of the drastic ballistic treatment used to homogenize the cells. Actually, we are conducting different experiments employing density gradients and electron transmission microscopy to resolve this question.

Complementary analysis of DD activities by electrophoretic zymograms led us to detect eleven different activities and all of them were $NADP^+$-dependent (Fig. 2) this represents the first report about the detection of DD activities by electrophoretic zymograms, a non-denaturing gel electrophoresis stained with a colored product of the enzymatic reaction.

Of the eleven bands detected, we described five different constitutive DD activities, DD 1-5, since them were observed when D-glucose was the carbon source and only DD-2 was solely induced by this sugar, since the others are induced at least for another carbon source. When *n*-decanol was used as a carbon source, we observed four out of five of the constitutive bands, lacking only the DD-2 band. Its noticeable that only when glycerol, ethanol, n-pentane and n-hexadecane were the carbon source to grow the fungus, not a single constitutive band was observed, may be due to the fact that these compounds only can be metabolized specifically by the induced enzymes. In glucose grown mycelium, all inducible dihydrodiol dehydrogenase activities were absent suggesting that they could be subject to carbon-catabolite repression (Fig. 2).

Surprisingly all the activities described here as DD are able to use *cis*-naphthalene-diol, since this substrate has been describes as bacterial specific (Cerniglia & Gibson 1977). The substrate reported for eukaryotic cells is the *trans*-naphthalene-diol (Cerniglia 1977).

Phenanthrene was the best inducer since when used as a carbon source four out of six inducible bands were observed, *n*-decanol and naphthalene were the second best inducers since each one led the induction of three different enzymes, sharing the bands denominated iDD1 and iDD2. Also in the case of the inducible enzymes glycerol, n-pentane and n-hexadecane were unable to induce any activity. The specific induction of an activity must be due to substrate specificity.

We have shown that on naphthalene, anthracene, phenanthrene or pyrene used as sole carbon source, there exist three different inducible $NADP^+$-dependent dihydrodiol dehydrogenase activities. One of them iNDD, was the only isoenzyme inducible by all aromatic hydrocarbons which presumably is involved in the aromatic hydrocarbon biodegradation pathway in YR-1 strain (Fig. 1). In particular iNDD was capable to use only *cis*-naphthalene-diol as substrate (Fig. 7) suggesting that this enzyme is specific part of the metabolic pathway of the naphthalene; all other activity bands are ADH substrate unspecific

It is interesting that iPDD2 has broad substrate specificity when $NADP^+$ was the electron acceptor, suggesting that it could be one of the different dehydrogenases belonging to the microsomal system of alcohol (ethanol) oxidation [MEOS (Krauzova et al., 1985)]. No one of all inducible DD's activities showed to be able to use NAD^+ as electron acceptor.

For iNDD, naphthalene was the best inducer and pyrene the worst. In the case of both iPDD1 and 2, phenanthrene was the only inducer of these enzymes, however pyrene shows a very low inducer effect on iPDD1 (Table 3). The finding of two inducible (iNDD and iPDD1) and one constitutive (DD1) enzymes that uses specifically a dihydrodiol as substrate is in agreement with the number of three possible DD's of predicted function, reported in database of *Mucor circinelloides* (Torres-Martínez et al., 2009).

With regard to the constitutive dihydrodiol dehydrogenase activities present in YR-1 strain, four of them use only *cis*-naphthalene-diol as substrate: DD1, 2, 3 and 7; DD2, use both *cis*-naphthalene-diol and with high efficiency propane-1,2,3-triol, indicating that it can be the glycerol dehydrogenase-1 (iGlcDH1 inducible by 1-decanol) described previously by ourselves in YR-1 (Camacho et al., 2010).

Our above-mentioned findings with *M. circinelloides* YR-1 dihydrodiol dehydrogenase activities are indicative of developmental regulation of the different DD's enzymes; this interpretation is supported by following observations: in zymograms for DD's activities when YR-1 was grown in different carbon sources is showed a differential pattern of the activity bands depending of the carbon source used in the culture media. The present results suggest the existence of eleven enzymes with dihydrodiol dehydrogenase activity. Particularly important the DD1 that could be the constitutive DD, and iNDD iPDD1, that could be part of the aromatic hydrocarbon biodegradation pathway in YR-1 strain for naphthalene or the others aromatic hydrocarbon, respectively. Future genomic analysis after isolation of the respective genes should prove the existence of one gene for each constitutive or inducible activity in agreement with the *M. circinelloides* data base prediction. The details of the possible interaction between alcohols or hydrocarbons metabolism remain to be determined.

5. Acknowledgments

Support for this research by Universidad de Guanajuato (México), is gratefully acknowledged.

6. References

Alvarado, C. Y., Gutiérrez-Corona, F. & Zazueta-Sandoval, R. (2002). Presence and physiologic regulation of alcohol oxydase activity in an indigenous fungus isolated from petroleum-contaminated soils. *Applied Biochemistry and Biotechnology*. Vol. 98-100, No. 1-9 (March 2002), pp. 243-255. ISSN: 0273-2289.

Atlas, R. M. (1995). Bioremediation. *Chemical and Engineering News*. Vol. 73, No. 14 (April 1995), pp. 32-42. ISSN 0009-2347. DOI: 10.1021/cen-v073n014.p032

Auret, B. J., Boyd, D. R., Robinson, P. M. & Watson, C.G. (1971). The NIH Shift During the Hydroxylation of Aromatic Substrates by Fungi. *Journal of the Chemical Society D: Chemical Communications* No. 24, (December 1971), pp. 1585-1587. ISSN 1364-548X.

Bartnicki-García, S. & Nickerson, W. J. (1962). Nutrition, Growth and Morphogenesis of *Mucor rouxii*. *Journal of Bacteriology*. Vol. 84, No. 4 (October, 1962), pp. 841-858. ISSN 0021-9193.

Bergmeyer, H. U. (1983). Alkoholdehydrogenase. *In*: Bergmeyer, H.U. (ed) *Methods of enzymatic analysis*, Verlag Chemie, Weinheim, Vol. 11, 3rd edn. pp-139-145.

Bezalel, L., Hadar, Y. & Cerniglia, C.E. (1997). Enzymatic Mechanisms Involved in Phenanthrene Degradation by the White-Rot Fungus *Pleurotus ostreatus*. *Applied and Environmental Microbiology*. Vol. 63, No. 7, (July 1997) pp. 2495-2501. ISSN 0099-2240.

Bogan, B. W. & Lamar, R. T. (1996). Polycyclic Aromatic Hydrocarbon-Degrading Capabilities of *Phanerochaete laevis* HHB-1625 and Its Extracellular Ligninolytic enzymes. *Applied and Environmental Microbiology*. Vol. 62, No. 5, (May 1996), pp. 1597-1603. ISSN 0099-2240.

Camacho, M. R. L., Durón, C. A. & Zazueta-Sandoval, R. (2010). Analysis of Glycerol Dehydrogenase Activities Present in *Mucor circinelloides* YR-1. *Antonie Van Leeuwenhoek*. Vol. 98, No. 4, (November 2010), pp. 437-445, ISSN 0003-6072.

Carbone, V., Hara, A. & El-Kabbani, O. (2008). Structural and Functional Features of Dimeric Dihydrodiol Dehydrogenases. *Cellular and Molecular Life Science*. Vol. 65, No. 10, (May 2008), pp. 1464-1474, ISSN 1420-682X.

Cerniglia, C. E. (1997). Fungal Metabolism of Polycyclic Aromatic Hydrocarbons: Past, Present and Future Applications in Bioremediation. *Journal of Industrial Microbiology and Biotechnology*. Vol. 19, No. 5-6 (November 1997), pp. 324-333. ISSN 1367-5435.

Cerniglia, C. E. & Gibson, D. T. (1977). Metabolism of Naphthalene by *Cunninghamella elegans*. *Applied and Environmental Microbiology*. Vol. 34, No. 4, (October 1977), pp. 363-370. ISSN 0099-2240.

Chang, C. H., Chen, L. Y., Chan, P. C., Yeh, T. K., Kuo, J. S., Ko, J. C. & Fang, Y. H. (2009). Overexpression of Dihydrodiol Dehydrogenase as a Prognostic Marker in Resected Gastric Cancer Patients. *Digestive Diseases and Sciences*. Vol. 54, No. 2 (February 2009), pp. 342-347. ISSN 0163-2116.

Chen, J., Adikari, M., Pallai, R., Parckh, K. H. & Simpkins, H. (2008). Dihydrodiol Dehydrogenases Regulate the Generation of Reactive Oxygen Species and the Development of Cisplatin Resistance in Human Ovarian Carcinoma Cells. *Cancer Chemotherapy and Pharmacology*. Vol. 61, No. 6 (May 2008), pp. 979-987. ISSN 0344-5704.

Cook, F. D., Jobson, A., Phillippe, R. & Westlake, D. W. S. (1974). Biodegradability and crude oil composition. *Canadian Journal of Microbiology*. Vol. 20, No. 7 (July 1974), pp. 915-928. ISSN 0008-4166.

Durón-Castellanos, A., Zazueta-Novoa, V., Silva-Jiménez, H., Alvarado-Caudillo, Y., Peña Cabrera, E. & Zazueta-Sandoval, R. (2005). Detection of NAD^+-Dependent Alcohol Dehydrogenase Activities in YR-1 Strain of *Mucor circinelloides*, a Potential Bioremediator of Petroleum-Contaminated Soils. *Applied Biochemistry and Biotechnology*. Vol. 121-124, No. 1-3 (March 2005), pp. 279-288. ISSN 0273-2289.

Ferris, J. P., MacDonald, L. H., Patrie, M. A. & Martin, M. A. (1976). Aryl Hydrocarbon Hydroxylase Activity in the Fungus *Cunninghamella bainieri*: Evidence for the Presence of Cytochrome P-450. *Archives of Biochemistry and Biophysics*. Vol. 175, No. 2 (August 1976), pp. 443-452. ISSN 0003-9861.

Goto, S. & Takami, H. (1986). Classification of Ascoideaceous Yeasts Based on the Electrophoretic Comparison of Enzymes and Coenzymes Q Systems. *The Journal of General and Applied Microbiology*, Vol. 32, No. 4 (June 1986), pp. 271-282. ISSN 1349-8037.

Hammel, K. E. (1995). Mechanisms for Polycyclic Aromatic Hydrocarbon Degradation by Ligninolytic Fungi. *Environmental Health Perspectives*, Vol. 103. Suppl. 5 (June 1995), pp. 41-43.

Hara, A., Taniguchi, H., Nakayama, T. & Sawada, H. (1990). Purification and Properties of Multiple Forms of Dihydrodiol Dehydrogenases from Human Liver. *Journal of Biochemistry*. Vol. 108, No. 2 (August 1990), pp. 250-254. ISSN 0021-924X.

Harayama, S. & Timmis, K. N. (1989). *In: Genetics of bacterial diversity*. Hopwood, D.A., and Chater K.E.. eds. Academic Press, pp. 151-174. New York, USA.

Higaki, Y., Kamiya, T., Usami, N., Shintani, S., Shiraishi, H., Ishikura, S., Yamamoto, I. & Hara, A. (2002). Molecular Characterization of Two Monkey Dihydrodiol Dehydrogenases. *Drug Metabolism and Pharmacokinetics*. Vol. 17, No. 4 (September 2002), pp. 348-356, ISSN 1347-4367.

Hunt, J. M. (1979). *Petroleum geochemistryand geology*. W. M. Freeman, San Francisco, CA. pp. 221. ISBN 0 7167 1005 6.

Hyötyläinen, T. & Olkari, A. (1999). The Toxicity and Concentrations of PAHs in Creosote-Contaminated Lake Sediment. *Chemosphere*. Vol. 38, No. 5 (February 1999), pp. 1135-1144. ISSN 0045-6535.

Jeng, R. S., Shiyuan, Y. & Wayman M. (1987). Isoenzyme and Protein Patterns of Pentose-Fermenting Yeasts. *Canadian Journal of Microbiology*. Vol. 33, No. 11 (November 1987), pp. 1017-1023. ISSN 0008-4166.

Jouanneau, Y. & Meyer, C. (2006). Purification and Characterization of an Arene *Cis*-Dihydrodiol Dehydrogenase Endowed with Broad Substrate Specificity Toward Polycyclic Hydrocarbon Hydrodiols. *Applied and Environmental Microbiology*. Vol. 72, No. 7 (July 2006), pp. 426-434. ISSN 0099-2240.

Krauzova, V. I., Il'chenko, A. P., Sharyshev, A. A. & Lozinov, A. B. (1985). Possible Pathways of the Oxidation of Higher Alcohols by Membrane Fractions of Yeasts Cultured on Hexadecane and Hexadecanol. *Biochemistry*. Vol. 50, No. 5 (May 1985), pp. 609-615. ISSN 0006-2960.

Laemmli, U. K. (1970). Cleavage of Structural Proteins During the Assembly of Head of Bacteriophage T6. *Nature*. Vol. 227, No. 5259 (August 1970), pp. 680-685. ISSN 0028-0836.

Lim, W. J., Park, S. R., Cho, S. J. Kim, M. K., Ryu, S. K., Hong, S. Y., Seo, W. T., Kim, H. & Yun, H. D. (2001). Cloning and characterization of an intracellular isoamylase gene from *Pectobcterium chrysantemi* PY35. *Biochemical and Biophysical Chemical Communications*. Vol. 287, No. 2 (September 2001), pp.348-354. ISSN 0006-921X.

Lowry, O. H., Rosebrough, N. J., Farr, A. L. & Randall, R. J. (1951). Protein Measurement with the Folin Phenol Reagent. *Journal of Biological Chemistry*, Vol. 193, No. 1 (November 1951), pp. 265-275. ISSN 0021-9258.

Muheim, A., Waldner, R., Sanglard, D., Reiser, J., Schoemaker, H. E. & Leisola M. S. (1991). Purification and Properties of an Aryl-Alcohol Dehydrogenase from the White-Rot Fungus *Phanerochaete chrysosporium*. *European Journal of Biochemistry*. Vol. 195, No. 2 (January 1991), pp. 369-375. ISSN 0014-2956.

Neidle, E., Hartnett, C., Ornston, L. N., Bairoch, A., Rekik, M. & Harayama, S. (1992). *Cis*-Diol Dehydrogenases Encoded by the TOL pWWO Plasmid *xylL* Gene and the *Acinetobacter calcoaceticus* Chromosomal *benD* Gene Are Members of the Short –

Chain Alcohol Dehydrogenase Superfamily. *European Journal of Biochemistry*. Vol. 204, No. 1, (February 1992), pp. 113-120. ISSN 0014-2956.

Nikolova, P. & Ward, O. P. (1991). Production of L-phenylacetyl Carbinol by Biotransformation: Product and By-Product Formation and Activities of the Key Enzymes in Wild Type and ADH Isoenzyme Mutants of *Saccharomyces cerevisiae. Biotechnology and Bioengineering*. Vol. 38, No. 5 (August 1991), pp. 493-498. ISSN 1097-0290.

Okwumabua, O., Persaud, J. S., & Reddy, P. G. (2001). Cloning and characterization of the gene encoding the glutamato dehydrogenase of *Streptococcus suis* serotype 2. *Clinical and Diagnostic Laboratory Immunology*. Vol. 8, No. 2 (March 2001), pp. 251-257. ISSN 556-6811.

Penning, T. M., Burczynski, M. E., Hung, C. F., McCoull, K. D. Palackal, N. T. & Tsuruda, L. S. (1999). Dihydrodiol dehydrogenases and polycyclic aromatic hydricarbon activation: Generation of reactive and redox *o*-quinones. *Chemical Research in Toxicology*. Vol. 12, No. 1 (December 1998), pp. 1-18. ISSN 0893-228X.

Pereira, E. F., Aracava, Y., Aronstam, R. S., Barreiro, E. J. & Alburquerque, E. S. (1992). Pyrazole, an Alcohol Dehydrogenase Inhibitor, Has a Dual Effects on N-methyl-D-asp. *Journal of Pharmacology and Experimental Theory*. Vol. 261, No.1 (April 1991), pp. 331-340. ISSN 0022-365.

Pfeiffeer-Guglielmi, B., Broer, S., Broer, A. & Hamprecht, B. (2000). Isoenzyme pattern of glycogen phosphorylase in the rat nervous system and rat astroglia-rich primary cultures: electrophoretic and polymerase chain reaction studies. *Neurochemical Research*, Vol. 25, No. 11 (November 2000), pp. 1485-1491. ISSN 0364-3190.

Silva, J. H., Zazueta-Novoa, V., Durón, C. A., Rodríguez, R. C., Leal-Morales, C. A. & Zazueta-Sandoval, R. (2009). Intracellular Distribution of Fatty Alcohol Oxidase Activity in *Mucor circinelloides* YR-1 Isolated from Petroleum Contaminated Soils. *Antonie Van Leeuwenhoewk*. Vol. 96, No. 4 (November 2009), pp. .527-535. ISSN 0003-6072.

Smith, R. V. & Rosazza, J. P. (1974). Microbial Models of Mammalian Metabolism. Aromatic Hydroxylation. *Archives of Biochemistry and Biophysics*. Vol. 161, No. 2 (April 1974), pp. 551-558. ISSN 003-9861

Sutherland, J. B., Fu, P. P., Yang, S. K., Von Tungeln, L. S., Casillas, R. P., Crow, S. A. & Cerniglia, C. E. (1993). Enantiomeric Composition of the *Trans*-Dihydrodiols Produced by Phenanthrene by Fungi. *Applied and Environmental Microbiology*. Vol. 59, No. 7 (July 1993), pp. 2145-2149. ISSN 0099-2240.

Torres-Martínez, S., Garre, V., Corrochano, L., Eslava, A. P., Baker, S.E. & others. (2009). Project CBS277.49, v1.0 http//genome.jgi-PSF.org.

Van Herwijnen, R., Springael, D., Slot, P., Govers A. J. H. & Parsons, R. J. (2003). Degradation of Anthracene by *Mycobacterium* sp. Strain LB501T Proceeds Via a Novel Pathway Throught o-Phtalic Acid. *Applied and Environmental Microbiology*. Vol. 69, No. 1 (January 2003), pp. 186-190. ISSN 0099-2240.

Yamazaki, M. & Komagata, K. (1983). An Electrophoretic Comparison of Enzymes of Ballistosporogenous Yeasts. *The Journal of General Applied Microbiolgy*. Vol. 29, No. 2 (February 1983), pp. 115-143. ISSN 0022-1260.

Zazueta-Sandoval, R., Durón, C. A. & Silva, J. H. (2008). Peroxidases in YR-1 strain of *Mucor circinelloides* a Potential Bioremediator of Petroleum-Contaminated Soils. *Annals of Microbiology*. Vol. 58, No. 3 (September 2008), pp. 421-426. ISSN 1590-4261.

Zazueta-Sandoval, R., Zazueta-Novoa, V., Silva-Jiménez, H. & Ortíz, R. C. (2003). A Different Method of Measuring and Detecting Mono and Di-Oxygenase Activities: Key Enzymes in Hidrocarbon Biodegradation. *Applied Biochememistry and Biotechnology*. Vol. 108, No. 1-3 (March 2003), pp. 725-736. ISSN 0273-2289.

15

Applications of Zymography (Substrate-SDS-PAGE) for Peptidase Screening in a Post-Genomic Era

Claudia M. d'Avila-Levy[1], André L. S. Santos[2], Patrícia Cuervo[1], José Batista de Jesus[1,3] and Marta H. Branquinha[2]

[1]*Instituto Oswaldo Cruz, Fundação Oswaldo Cruz, Rio de Janeiro*

[2]*Departamento de Microbiologia Geral, Instituto de Microbiologia Paulo de Góes, Universidade Federal do Rio de Janeiro, Rio de Janeiro*

[3]*Universidade Federal de São João Del Rei, São João Del Rei*

Brazil

1. Introduction

Peptidases are enzymes that catalyze the hydrolysis of peptide bonds in proteins or peptides. The hydrolysis can be specific or unspecific, leading to highly regulated cleavage of specific peptide bonds, or to complete degradation of proteins to oligopeptides and/or amino acids. Peptidases can be classified as endo- or exopeptidases, the latter only act near the ends of the polypeptide chain. Endopeptidases are divided into six major families by virtue of the specific chemistry of their active site: aspartic, serine, metallo-, cysteine, glutamic and threonine peptidases (Rawlings et al. 2010).

Zymography is an electrophoretic technique, based on sodium dodecyl sulfate polyacrylamide gel electrophoresis (SDS-PAGE) and a substrate (e.g. gelatin, casein, albumin, hemoglobin, etc.) co-polymerized with the polyacrylamide matrix. Proteins are prepared by the standard SDS-PAGE buffer under non-reducing conditions (no boiling and no reducing agent), and are separated by molecular mass in the standard denaturing SDS-PAGE co-polymerized with a protein substrate. After electrophoresis, peptidases are re-natured by the removal of the denaturing SDS by a non-ionic detergent, such as Triton X-100, followed by incubation in conditions specific for each peptidase activity (time, temperature, ions, ionic strength), when the enzymes hydrolyze the embedded substrate, then proteolytic activity can be visualized as cleared bands on a Coomassie stained background (Heussen and Dowdle, 1980). Therefore, only endopeptidases can be detected by substrate-SDS-PAGE, which requires a considerable degradation of the substrate for visualization of the degradation haloes. Alternatively, an overlay with specific chromogenic or fluorogenic peptide substrate can be done after SDS-PAGE separation of the proteins and renaturation with Triton X-100, which allows the detection of specific peptidases in complex biological samples.

This technique has many benefits: (1) it is relatively inexpensive, requires short assaying times, and peptidases with distinct molecular masses can be detected on a single gel; (2)

separation of proteins by molecular mass through non-reducing electrophoretic migration allows a presumptive correlation with known peptidases; (3) incubation with proteolytic inhibitors provides powerful information about enzyme classification; (4) pH and temperature changes help to assess peptidase characteristics; (5) several substrates can be co-polymerized to assess peptidase degradation capacity; (6) densitometry can be used for quantitative analysis. Ultimately, in organisms with complete genome sequences, bioinformatic analysis provides rich information on putative peptidases, such as: peptidase classification, approximated molecular mass, possible cellular localization through classical motifs, evolutionary and functional relationships, and so on. However, it cannot be ascertained if the described ORFs are indeed expressed and active. Therefore, a zymographic assay coupled with bioinformatic analysis may allow the detection of functionally active enzymes.

The advantages of this technique are exemplified by its application nowadays to unveil peptidases in biological systems, which possesses genome information, but still zymography is the method of choice for peptidase screening, identification and characterization. Wilder and colleagues, for instance, report that zymography can selectively distinguish cathepsins K, L, S and V in cells and tissues by its electrophoretic mobility and by simply manipulating substrate and pH. The sequence homology among these cathepsins leads to a substrate promiscuity, which precludes desired specificity for in solution assays with specific chromogenic or fluorogenic peptide substrate (Wilder et al. 2011). Zymography allows the detection of a 37 kDa (cathepsin K), 35 kDa (cathepsin V), 25 kDa (cathepsin S) and 20 kDa (cathepsin L). Cathepsin K activity disappeared and V remained when incubated at pH 4.0 instead of 6.0, allowing the visualization of each enzyme (Wilder et al. 2011). Kupai and colleagues also highlighted that substrate zymography is the method of choice, among several analyzed, to detect the activity of the different matrix metallopeptidase (MMP) isoenzymes from a wide range of biological samples (Kupai et al. 2010). Also, it allows high throughput screening of specific MMP inhibitors, especially because the nature of the residues in the enzyme's active site is highly conserved among the different MMPs, therefore, once again, in solution enzymatic assays are not applicable (Devel et al. 2006, Kupai et al. 2010). Also, for the screening of tissue inhibitors of metallopeptidases (TIMPs), reverse zymography is a powerful approach. This technique is based on the ability of the inhibitors to block gelatinase activity of a MMP, usually MMP-2. A calibrated solution of gelatinase-A (MMP-2) is co-polymerized with gelatin in the polyacrylamide gel. The samples possibly containing TIMPs are then separated by electrophoresis, SDS is removed and the gel is incubated in a buffer that allows the gelatinase to digest the gelatin, except where it is inhibited by TIMP proteins. After staining with Coomassie blue, the result is a gel with a pale blue background (where gelatin was degraded by the gelatinase) with blue bands showing the positions and relative amounts of TIMPs (Snoek-van Beurden and Von den Hoff, 2005).

In view of this, below we will present comments on peptidase screening through zymography discussing possible protocol variations and its implications, and then we present and discuss practical examples of the application of zymography to generate critical data in organisms that still do not possess genome information. Finally, we will discuss the possibility of direct peptidase identification through two-dimensional zymography coupled to mass spectrometry.

2. Comments on peptidase screening through substrate-SDS-PAGE

Several research groups perform substrate-SDS-PAGE to assess, screen and characterize peptidases in complex or purified biological preparations. After the original publication from Heussen and Dowdle (1980), several adaptations have been implemented to improve the detection of a specific peptidase class. Below we will present a generalized protocol indicating possible variations.

2.1 Sample homogenization

The preparation of the biological sample is critical for the success of the zymography, all the procedure must be performed at 4ºC, the addition of detergents such as Triton X-100, SDS or CHAPS (3-[(3-cholamidopropyl)dimethylammonio]-1-propanesulfonate) for the solubilization and recovery of hydrophobic enzymes is necessary, if one is interested in such enzymes, also addition of proteolytic inhibitors to undesired peptidase classes is also an interesting strategy. Alternatively, the separation of hydrophobic from hydrophilic proteins can be achieved during phase partition in solutions of Triton X-114, which occurs at 37ºC preserving enzyme integrity (Figure 1) (Bouvier et al. 1987). After sample preparation, SDS-PAGE sample buffer is added to the biological sample (62.5 mM Tris HCl, pH 6.8, 2% SDS, 10% (v/v) glycerol, and 0.002% bromophenol blue). A concentrated sample buffer can be used to avoid sample dilution, which is critical for the detection of low abundant enzymes. The proteins are not denatured since sample is kept at 4ºC, there is no sample boiling, nor the addition of reducing agents such as dithiothreitol (DTT) or 2-mercaptoethanol, as usual in sample preparation for standard SDS-PAGE analysis. The sample must maintain its native form due to the the further step of substrate degradation.

2.2 Polyacrylamide gels containing sodium dodecyl sulfate and co-polymerized substrates

Here, to the standard Laemmli protocol (Laemmli, 1970), a substrate can be co-polymerized to the gel (Heussen and Dowdle, 1980). Alternatively, an overlay with fluorogenic or chromogenic peptide substrates can be done (Cadavid-Restrepo et al. 2011). The acrylamide concentration in gels varies more commonly from 7 to 15%, which impact on protein separation; low molecular mass proteins usually require higher acrylamide concentration for better protein resolution. The co-polymerized substrate can be virtually any protein. Gelatin is commonly used as a protein substrate because it is easily hydrolyzed by several peptidases and does not tend to migrate out of the resolving gel in electrophoretic tests performed at 4°C, and is inexpensive (Michaud et al. 1996). In addition to gelatin, several other proteins have been used, such as: casein, bovine serum albumin, human serum albumin, hemoglobin, mucin, immunoglobulin, and collagen (d'Avila-Levy et al. 2005; Pereira et al. 2010a). Also, complex mixtures of proteins can be used, which may reflect a functional role of the enzyme. For instance, our research group employed gut proteins from an insect to co-polymerize in acrylamide gels. Then, extracts from a protozoan believed to interact with the insect gut were assayed, revealing the peptidases capable of degrading the insect gut proteins (Pereira et al. 2010a). An example of zymographies performed with a set of eight distinct proteinaceous substrates, as well as, a densitometric measure of the degradation halos can be seen in Figure 2.

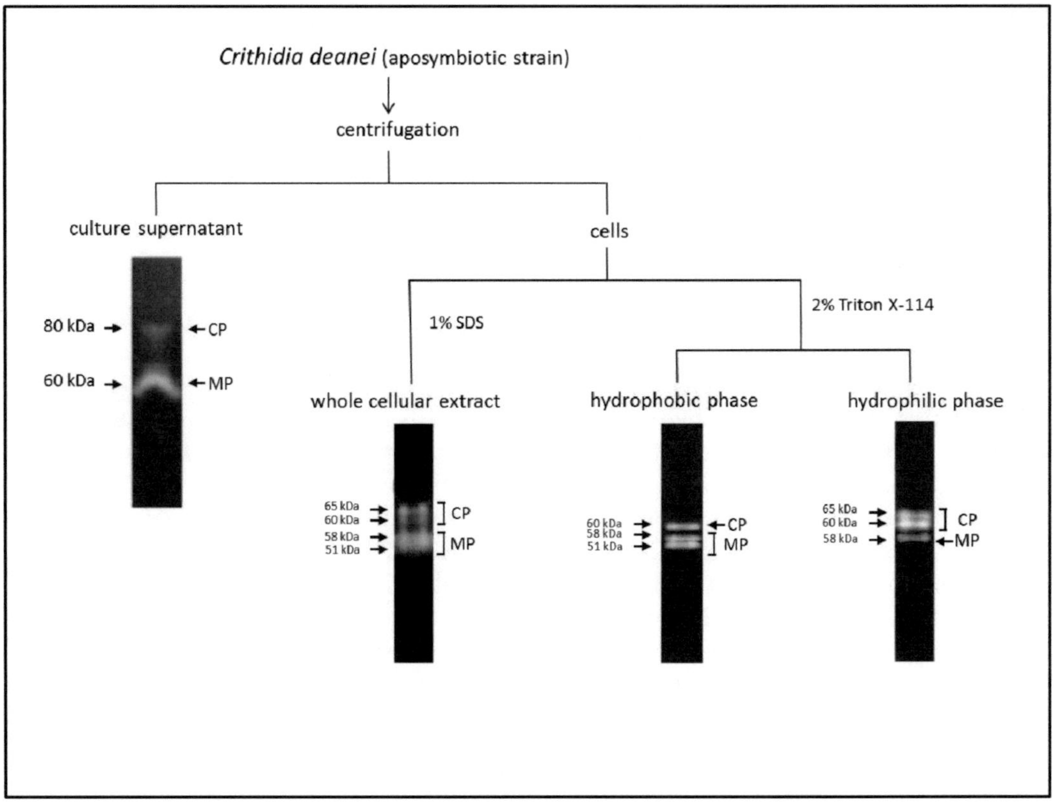

Fig. 1. Extracellular and cell-associated proteolytic enzymes of *Crithidia deanei* cells, an insect trypanosomatid. Parasites were cultured in a complex medium (brain heart infusion) for 48 h at 28°C. Then, cells were harvested by centrifugation, the culture supernatant was filtered in Millipore membrane 0.22 µm and concentrated 50-fold by dialysis (cut-off 9000 Da) against polyethylene glycol 4000 overnight at 4°C. The cells were lysed by: the addition of SDS, generating whole cellular extract, or by Triton X-114 to obtain the hydrophilic (cytoplasmatic and intravesicular fraction) and hydrophobic (membrane fraction) phases. The extracellular and cellular extracts were applied on gelatin-SDS-PAGE to evidence the proteolytic enzymes. The gels were incubated in 50 mM sodium phosphate buffer pH 5.5 supplemented with DTT 2 mM at 37°C for 24 h. MP, metallopeptidase and CP, cysteine peptidase. For experimental details see d'Avila-Levy et al. 2001, 2003.

2.3 Enzyme renaturation and proteolysis

After electrophoresis, the enzymes are renaturated by replacement of the anionic detergent SDS by the non-ionic detergent Triton X-100, through gel washing. Then, gels are incubated under conditions ideal for detection of the desired peptidase. For instance, metallopeptidases are known to require neutral to basic pH for activity, while cysteine peptidases require an acidic pH and a reducing agent, usually DTT (Branquinha et al. 1996). However, it is common to screen biological samples, where there is no previous clue on what peptidase class shall be present, nor the best conditions for proteolysis. Therefore, it is necessary to assess several parameters, such as incubation time, pH, temperature, influence of ions or reducing agents and finally assess the inhibition profile. A general flowchart for establishing such conditions is shown in Figure 3, and a general view of Gelatin-SDS-PAGE screening in an uncharacterized organism can be seen in Figure 4.

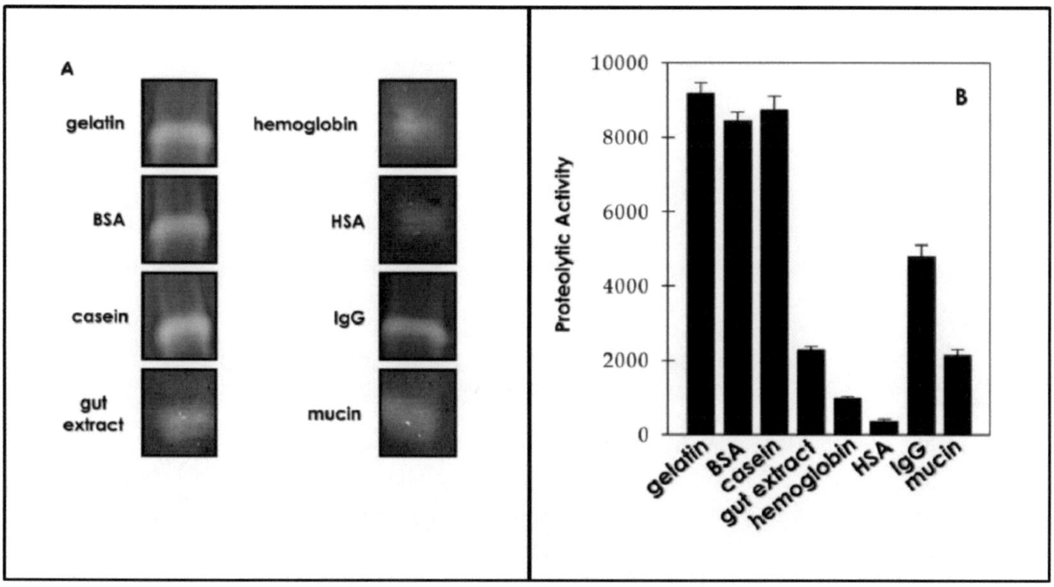

Fig. 2. Degradation of different proteinaceous substrates co-polymerized to SDS-PAGE by a surface metallopeptidase from *Herpetomonas samuelpessoai*, an insect trypanosomatid. The following substrates were individually incorporated into SDS-PAGE to evidence the proteolytic activity: gelatin, bovine serum albumin, human serum albumin, casein, immunoglobulin G (IgG), hemoglobin, mucin and gut extract from *Aedes aegypti*. The gels were incubated for 20 h at 37°C in 50 mM sodium phosphate buffer pH 6.0 supplemented with DTT 2 mM (A). The degradation halos, which correlate with degradation capability, were densitometric measured and expressed as arbitrary units of proteolytic activity (B). For experimental details see Pereira et al. 2010b. Reprinted with permission of *Protist*.

3. Practical examples of peptidases screening through SDS-PAGE-substrate

3.1 A first glance on *Bodo* sp. peptidases

Bodo sp. is a free-living flagellate that belongs to the family Bodonidae, order Kinetoplastida. This bodonid isolate still has its taxonomic position unsolved, but it is phylogenetically related to *Bodo caudatus* and *Bodo curvifilus*, which are considered ancestral to the trypanosomatids. The Trypanosomatidae family comprises parasites that are of particular interest due to their medical importance, such as the etiologic agent of Chagas' disease (*Trypanosoma cruzi*), African trypanosomiasis (*Trypanosoma brucei* complex) and the various forms of leishmaniasis caused by *Leishmania* spp.. Due to their medical relevance, this family has been the focus of extensive research (Wallace, 1966; Vickerman, 1994). Peptidase characterization in *Bodo* sp. and comparison to peptidases from closely related pathogenic protozoa may help to understand peptidase function and evolution in general. The gold standard approach for such comparison would be a bioinformatic analysis of the *Bodo*

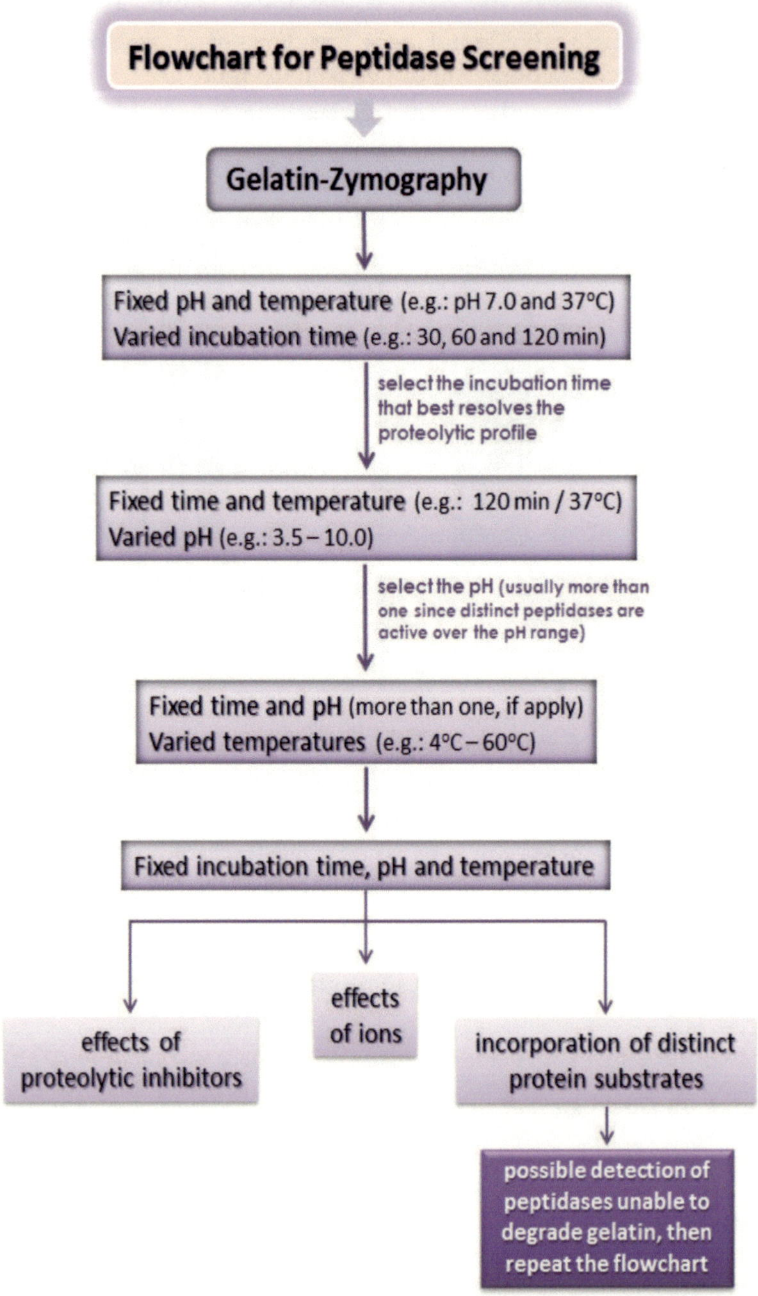

Fig. 3. Flowchart for peptidase screening. Several parameters must be assessed to resolve the proteolytic profile in an unknown biological sample. This scheme represents a suggestion of a step-by-step analysis of these parameters, which are: amount of sample, incubation time, pH, temperature, effect of ions, effect of reducing agents, effect of peptidase inhibitors, and ability to degrade distinct proteinaceous substrates. Usually, gelatin is the proteinaceous substrate of choice for initial screening because it is easily hydrolyzed by several peptidases,

does not tend to migrate out of the resolving gel and is inexpensive, then with an arbitrary pH and temperature (usually neutral pH at 37ºC), the incubation time is varied from minutes to even 72 h, depending on the sample. After selecting the incubation time that allows the detection of the higher number of enzymes without band overlapping, variations on the pH allows the determination of this biochemical characteristic of each band. After this assay, the peptidase(s) of interest can be tested over a range of temperatures, ions, reducing agents or proteolytic inhibitors. Finally, distinct proteinaceous substrates can be co-polymerized to the gels, revealing either the ability of the detected peptidases to degrade other substrates, which ultimately gives a glance on peptidase function, or even revealing enzymes not capable of degrading gelatin (see figure 2). The proper combination of these parameters may reveal interesting enzymes, such as peptidases strictly dependable on metal ions, stimulated by reducing agents, active only at acidic or alkaline conditions and so on.

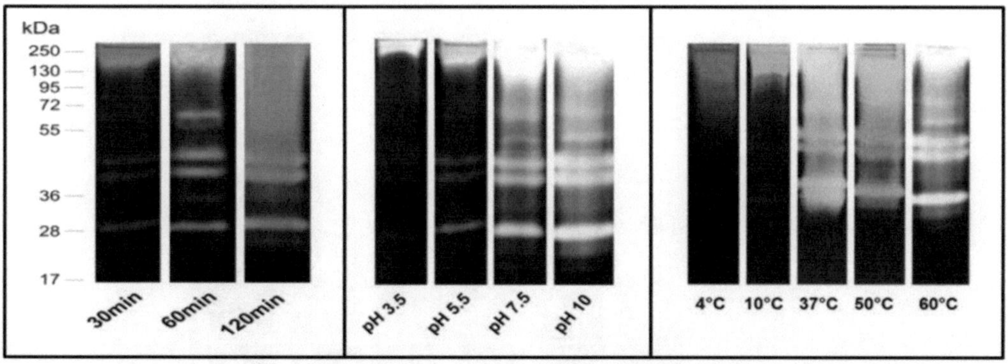

Fig. 4. Gelatin-SDS-PAGE screening of peptidases in homogenates from *Aedes albopictus* pupa. The following parameters were assessed: incubation time, pH and temperature. Peptidase activities were detected after incubation of the gels for 30, 60 or 120 min at 37ºC in 100 mM Tris-HCl buffer pH 7.5. The numbers on the left indicate apparent molecular masses of the active bands expressed on kiloDaltons (kDa). Afterwards, 60 min incubation was selected and the gels were incubated in reaction buffer containing 100 mM sodium acetate at pH 3.5 or 5.5 or 100 mM Tris-HCl at pH 7.5 or 10.0. Finally, the effect of temperature on the proteolytic activities was assayed by incubating of the gels for 60 min at 4, 10, 37, 50 or 60ºC in reaction buffer containing 100 mM Tris-HCl at pH 7.5. Saboia-Vahia et al. unpublished data.

genome coupled to more defined biochemical characterization of individual peptidases. However, in the absence of a *Bodo* genome, we have employed substrate-SDS-PAGE to assess peptidases in this bodonid, which presents serine peptidases ranging from 250 to 75 kDa, with a slight preference for acidic pH. This finding is dissimilar to what has been described in related pathogenic protozoa (Figure 5) (d'Avila-Levy et al., 2009). Curiously, all the analyzed closely related parasitic trypanosomatids, as well, as *Cryptobia salmonistica,* reveal through gelatin-SDS-PAGE only metallo- and cysteine peptidases, which are prototypal peptidases and virulence factors. In trypanosomatids, for instance, serine peptidases can only be detected by in-solution assays or after enrichment processes (Grellier et al. 2001). It is somewhat intriguing that cysteine and metallopeptidases are either not resistant to the denaturation/refold process and/or are not abundantly expressed by this free living bodonid, because they were not detected by zymography. This may reflect substantial differences among the peptidases from these organisms. The raw data revealed by susbtrate-zymography provided the first observation on possible differences in peptidase profile among the families Bodonidae, Trypanosomatidae, and Cryptobiidae, forming the basis for future research.

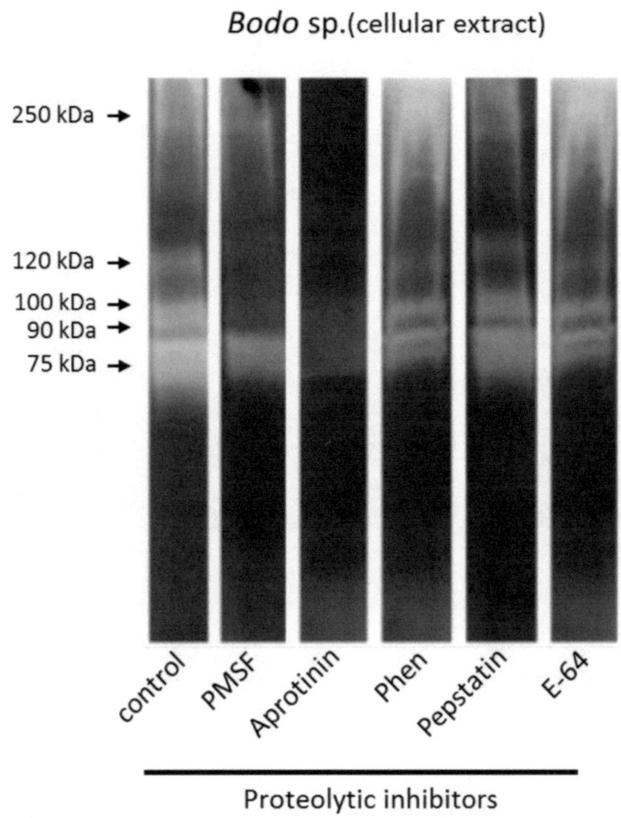

Fig. 5. Inhibition profile of cellular peptidases of *Bodo* sp. in gelatin–SDS-PAGE. In order to determine the enzymatic class, after electrophoresis, the gels were incubated for 48 h at 28°C in 50mM phosphate buffer pH 5.5 in the absence (control) or in the presence of the following proteolytic inhibitors: 1mM phenylmethylsulfonyl fluoride (PMSF), 1 mg/ml aprotinin, 10mM 1,10-phenanthroline (Phen), 1 mM pepstatin A, or 10 μM *trans*-epoxysuccinyl L-leucylamido-(4-guanidino) butane (E-64). Numbers on the left indicate relative molecular mass of the peptidases. For experimental details see d'Avila-Levy et al. 2009. Reprinted with permission of *The Journal of Eukaryotic Microbiology.*

3.2 Identification of peptidases in *Herpetomonas* spp. and possible biological functions proposed by overlay gel approaches

In addition to the heteroxenic parasites that are of particular interest in the Trypanosomatidae family due to their medical importance, several genera are composed of monoxenic parasites of the gut of a wide range of insects. The *Herpetomonas* genus is composed of insect trypanosomatids that display promastigote, paramastigote and opisthomastigote developmental stages during its life cycle (McGhee and Cosgrove, 1980), being used as a model to study the complex events of cell differentiation process. Also, these traditionally "non-mammalian and non-pathogenic" microorganisms have been used as experimental models of the Trypanosomatidae family for exploring their basic mechanisms at the genetic, physiological, ultrastructural and biochemical levels. In the same way, several research groups have described common structures/molecules produced by monoxenous and heteroxenous parasites belonging to the Trypanosomatidae family (Lopes et al. 1981; Breganó et al. 2003; Santos et al. 2006, 2007; Elias et al. 2008). Interestingly, *Herpetomonas* species have been detected not only in insects, but repeatedly in plants and mammals, including immunosuppressed humans, mainly in HIV-infected individuals, in whom the parasites caused either visceral or cutaneous lesions (reviewed by Chicharro and Alvar, 2003; Morio et al. 2008), showing the ability to develop digenetic life style under certain conditions. Collectively, these studies emphasize the need for further investigation in the biochemical machinery of these intriguing insect trypanosomatids.

Whole cellular extracts of *H. samuelpessoai* promastigotes when analyzed by gelatin-SDS-PAGE revealed two major peptidase classes: a prominent metallopeptidase of 66 kDa (actually a broad hydrolytic activity ranging from 60 to 80 kDa), inhibited by 10 mM 1,10-phenanthroline, and a minor cysteine peptidase activity of 45 kDa, restrained by 1 μM E-64 and leupeptin (Santos et al. 2003).

The 66 kDa metallopeptidase activity was detected in the parasite membrane fraction after Triton X-114 partition (Etges, 1992; Schneider and Glaser, 1993; Santos et al. 2003) or after treatment of living cells with phospholipase C (Santos et al. 2002; Santos et al. 2006) as well as in the extracellular environment as the major secreted peptidase component (Santos et al. 2001, 2003, 2006; Elias et al. 2006). This metallopeptidase produced by *H. samuelpessoai* cells shares common biochemical and immunological properties (Elias et al. 2006; Santos et al. 2006) with the major metallopeptidase expressed by *Leishmania* species, called leishmanolysin or gp63, a virulence factor that participates in different stages of the parasite life cycle such as adhesion and escape from host immune response (Yao, 2010). The incorporation of different proteinaceous substrates into SDS-PAGE demonstrated that leishmanolysin-like molecule from *H. samuelpessoai* was able to degrade hemoglobin, casein, immunoglobulin G, mucin, human and bovine albumins as well as the gut protein extract from *Aedes aegypti* (Figure 2) (Pereira et al. 2010a), an experimental model to study the trypanosomatids-insect interplay (reviewed by Santos et al. 2006), culminating in the generation of peptides and amino acids required for parasite growth and development, as well as it might cleave structural barriers in order to improve its dissemination. Also, the pH dependence of the 66 kDa metallopeptidase of *H. samuelpessoai* was also determined by overlay gels, presenting a broad spectrum of pH (ranging from 5 to 10) and temperature (26 to 50°C), showing maximum hydrolytic activity at pH 6.0 at 37°C (Pereira et al. 2010a). These large spectra of pH and temperature retain maximum flexibility for the trypanosomatid to survive under different environmental conditions. In this sense, the surface leishmanolysin-like molecules of *H. samuelpessoai* cells participate in adhesive properties during the interaction with invertebrate gut (Pereira et al. 2010a) and mammalian macrophages (Pereira et al. 2010b). Other *Herpetomonas* species, including *H. megaseliae* and *H. anglusteri*, produce at least one metallopeptidase similar to the leishmanolysin, which is a conserved molecule with ancestral functions during the insect colonization.

The 45 kDa cysteine peptidase synthesized by *H. samuelpessoai* cells had its activity reduced during the parasite growth at 37ºC in comparison to 26ºC, and when cultured up to 72 h in the presence of the differentiation-eliciting agent, dimethylsulfoxide. The modulation in the 45 kDa cysteine peptidase expression is connected to the differentiation process, since both temperature and dimethylsulfoxide are able to trigger the promastigote into paramastigote transformation in *H. samuelpessoai* (Santos et al. 2003; Pereira et al. 2009). In contrast, the expression of leishmanolysin-like molecules was not modulated during the differentiation in *H. samuelpessoai* (Pereira et al. 2009, 2010b).

The cultivation of *H. megaseliae* and *H. samuelpessoai* in different growth media induced the production of distinct profiles of both cellular and extracellular peptidases as revealed by a simple inspection using substrate-SDS-PAGE (Branquinha et al. 1996; Santos et al. 2002, 2003; Nogueira de Melo et al. 2006). In addition, the incorporation of different proteinaceous substrates into SDS-PAGE allowed the identification of substrate specific proteolytic activity in a complex cellular extract. For example, cellular cysteine peptidase (115–100, 40 and 35 kDa) and metallopeptidase (70 and 60 kDa) activities of *H. megaseliae* were detected in both casein and gelatin zymograms (Nogueira de Melo et al. 2006). Additionally, the use of casein in the gel revealed a distinct acidic metallopeptidase of 50 kDa when the parasite was cultured in the modified Roitman's complex medium. However, no proteolytic activity was detected when hemoglobin was used as co-polymerized substrate (Nogueira de Melo et al. 2006).

3.3 Proteases produced by *Herpetomonas* species: Taxonomic marker

Insect trypanosomatids have been traditionally allocated to a number of genera that were described based on morphological features, host and geographical origin (Wallace et al. 1983; Momen, 2001). However, for identification purposes, these criteria proved to be impractical and insufficient, because the same trypanosomatid species may be recovered from diverse species of insects and the same insect species may harbor various species of trypanosomatids. In addition, the morphology of trypanosomatid cells can be modified by environmental factors (Podlipaev, 2001; Momen, 2001, 2002). Therefore, there is a need to develop more effective means of trypanosomatid identification. With this task in mind, the expression of proteolytic activities in the Trypanosomatidae family was explored as a potential marker to discriminate between the morphologically indistinguishable flagellates isolated from insects and plants (Branquinha et al. 1996; Santos et al. 1999, 2005, 2008). For instance, many trypanosomatids have been erroneously placed in the genus *Herpetomonas* or, conversely, many *Herpetomonas* spp. may remain hidden in other genera. Santos and co-workers (2005) proposed an additional tool for trypanosomatid identification, including species belonging to the *Herpetomonas* genus by using *in situ* detection of proteolytic activities on gelatin-SDS-PAGE, in association with specific peptidase inhibitors. The results showed that nine distinct *Herpetomonas* species (*H. anglusteri*, *H. samuelpessoai*, *H. mariadeanei*, *H. roitmani*, *H. muscarum ingenoplastis*, *H. muscarum muscarum*, *H. megaseliae*, *H. dendoderi* and *Herpetomonas* sp. isolated from the salivary gland of a phytophagous insect) produced species specific cellular peptidase profiles (Table 1), which can be useful in the correct identification of these parasites. The exception for this observation was seen in *H. samuelpessoai* and *H. anglusteri*, which presented a similar cell-associated proteolytic pattern. However, these two *Herpetomonas* species excreted distinct proteolytic activities, which may be a reflection of changes in the nutritional requirements during the life-cycle of the flagellates. Therefore, the authors infer that profiles of both cellular and extracellular peptidases represent an additional criterion to be used in the identification of trypanosomatids (Santos et al. 2005).

Herpetomonas species	Host		Predominant evolutive stage in culture	Number of cell-associated peptidases	Molecular masses of peptidases in kDa	
	Family	Species			Metallo-peptidases[a]	Cysteine peptidases[b]
Herpetomonas sp.	Hemiptera: Coreidae	*Phthia picta*	Promastigote	4	72, 60	45, 40
H. anglusteri	Diptera: Sarcophagidae	*Liopygia ruficornis*	Promastigote	2	60	45
H. dendoderi	Diptera: Culicidae	*Haemagogus janthinomys*	Promastigote	5	130, 110, 95	60, 45
H. mariadeanei	Diptera: Muscidae	*Muscina stabulans*	Promastigote	2	nd[c]	42, 38
H. megaseliae	Diptera: Phoridae	*Megaselia scalaris*	Promastigote	8	100, 80, 67, 60	95, 45, 40, 35
H. muscarum ingenoplastis	Diptera: Calliphoridae	*Phormia regina*	Promastigote	2	80, 67	nd
H. muscarum muscarum	Diptera: Muscidae	*Musca domestica*	Promastigote	6	100, 80	95, 50, 45, 40
H. roitmani	Diptera: Syrphidae	*Ornidia obesa*	Opisthomastigote	1	50	nd
H. samuelpessoai	Hemiptera: Reduviidae	*Zelus leucogrammus*	Promastigote	2	60	45

[a] The metallopeptidase activities were completely blocked by 10 mM 1,10-phenanthroline.
[b] The cysteine peptidases were inhibited by 10 μM E-64.
[c] Non detected (nd).

Table 1. Peptidase profiles in different *Herpetomonas* species detected in gelatin-SDS-PAGE.

3.4 Peptidase screening in *Crithidia*

Among the insect trypanosomatids, the genus *Crithidia* comprises monoxenic trypanosomatids of insects that were originally characterized by the presence of choanomastigote forms in their life cycles (Hoare and Wallace, 1966). The first studies employing zymograms in order to detect proteolytic activity in *Crithidia* spp. were performed by Frank and Ashall in 1990. In the two studies published in that year, the activity in *Crithidia fasciculata* extracts was compared to *T. cruzi*. In this sense, it is worth mentioning that *C. fasciculata*, among all non-pathogenic trypanosomatid species, has been considered an excellent model organism for many studies concerning trypanosomatids, because it can be cultivated in high yields and do not require specific bio-safety precautions (Vickerman, 1994).

In a first approach (Ashall, 1990), parasite extracts were made by the use of 0.5% Nonidet P-40 and mixed with SDS-PAGE sample buffer in non-denaturing conditions. After electrophoresis in SDS-PAGE, gels were overlaid with 0.75% agarose containing the chromogenic substrate Bz-Arg-pNA at 0.5 mM and at pH 8.0 and incubated for 4-6 h at 37°C, and then photographed using a blue filter to reveal yellow bands containing *p*-nitroaniline. A single component with molecular mass >200 kDa that hydrolyzed this substrate was detected in *C. fasciculata* as well as in *T. cruzi* crude extracts. A modified procedure was also employed (Ashall et al. 1990), in which electrophoresis was followed by shaking the gels with 2% Triton X-100 and then the incubation of gels for 10 min in the presence of a range of amidomethylcoumarin substrates containing arginine adjacent to the amidomethylcoumarin moiety, at pH 8.0. Fluorescent bands were visualized in gels by ultraviolet light by the hydrolysis of each substrate. A single band of substrate hydrolysis occurred with all six substrates tested, in both *T. cruzi* and *C. fasciculata,* with the same electrophoretic mobility (150-200 kDa). Incubation of gel strips with various peptidase inhibitors showed that the enzyme was strongly inhibited by diisopropyl phosphorofluoridate (DFP), N-α-tosyl-L-lysinyl-chloromethylketone (TLCK), leupeptin and a peptidyldiazomethane containing lysine at P_1, but not by E-64, PMSF, pepstatin A, 1,10-phenanthroline and a peptidyldiazomethane containing methionine at P_1. This enzyme was characterized as an alkaline peptidase, probably from the serine-type, that cleaves peptide bonds on the carboxyl side of arginine residues at pH 8.0 (Ashall, 1990).

Following this set of experiments, Etges (1992) employed surface radioiodination of living cells, fractionation by Triton X-114 extraction and phase separation, and zymogram analysis by fibrinogen-SDS-PAGE in order to describe the presence of a surface metallopeptidase in *C. fasciculata* with biochemical similarities to the gp63 from *Leishmania* spp. This peptidase is one of major surface molecules in all *Leishmania* species and play vital roles in the different stages of *Leishmania* life cycle, being suggested its participation in many aspects of the infection inside the mammalian host (Yao, 2010). The presence of a similar neutral-to-alkaline metallopeptidase at the surface of *C. fasciculata* led to the suggestion that gp63 should not be involved in the infection of the mammalian host by *Leishmania,* but rather contributes to the survival of the trypanosomatid inside the digestive tract of the insect (Santos et al. 2006).

The work of Etges (1992) opened the possibility to use the same technique in order to analyze the proteolytic profiles in different members of distinct trypanosomatid genera. With this task in mind, our group has analyzed the proteolytic profiles of a great number of species from 8 different genera of trypanosomatids by the use of SDS-PAGE containing 0.1% co-polymerized gelatin as substrate (Branquinha et al. 1996; Santos et al. 2005; 2008). In those studies, it became clear that two distinct proteolytic activities can be detected in total cell lysates: cysteine- and metallopeptidases. For detection of cysteine peptidase activity, the optimal conditions were established to be an acidic pH value (5.0-6.0) and the presence of a reducing agent, such as DTT, which was essential for detection of this activity. The use of specific inhibitors, such as E-64, prevented the development of cysteine peptidase activity bands. Metallopeptidases were consistently observed in a broad pH range (5.0-10.0), and the zinc-chelator 1,10-phenanthroline completely inhibited their activity.

In our first work (Branquinha et al. 1996), three *Crithidia* species were studied: *C. fasciculata, C. guilhermei* and *C. luciliae.* Cells were lysed by the addition of SDS-PAGE sample buffer in non-denaturing conditions, and peptidases were characterized by electrophoresis on 7-15% gradient SDS-PAGE with 0.1% gelatin co-polymerized as substrate. Cell lysates of the three species produced similar patterns of proteolysis at 28°C: two cysteine peptidase bands in the 80-110 kDa range and a minor cysteine peptidase activity detected at 45 kDa; and a metallopeptidase band detected in the 55-66 kDa range. When cells were lysed with the non-ionic detergent Triton X-114, cysteine peptidases were detected in the aqueous phase, whereas the metallopeptidase partitioned into the detergent-rich phase, which suggested that the latter is membrane-associated. Interestingly, cell lysates of these species were also employed by our group in another comparative study (Santos et al. 2005), in which a single cysteine peptidase was found at 50 kDa and two metallopeptidases were detected at 70 kDa and at 90 kDa. In the latter study, peptidases were analyzed in 10% linear polyacrylamide gels containing gelatin, and the temperature of incubation after electrophoresis was 37°C. As explained by Martinez and Cazzulo (1992), the apparent molecular mass of each band varies depending on the experimental conditions, including acrylamide concentration and temperature of incubation, which may explain this discrepancy. Despite this fact, both studies highlighted the common proteolytic profile between these species, possibly reflecting their phylogenetic proximity.

Unlike the similarities detected in the three *Crithidia* spp. described above, heterogeneous proteolytic profiles were observed in different members of this genus. For instance, *Crithidia acantocephali* produced 4 cysteine peptidases of 80, 75, 70 and 50 kDa, while *Crithidia harmosa* presented 3 metallopeptidases at 63, 50 and 45 kDa (Santos et al. 2005). These data suggested the value of proteolytic enzymes in distinguishing between trypanosomatid species that cannot be differentiated on structural grounds (Santos et al. 2005). In order to study the distribution of metallopeptidases in trypanosomatids, our group also investigated cell-associated proteolytic activities in distinct species by gelatin-SDS-PAGE in conditions that favor the detection of this subgroup, specifically alkaline conditions (pH 9.0) and proteolytic inhibitors that putatively identified these enzymes, such as 1,10-phenanthroline (Santos et al. 2008). The analysis confirmed the previous results in all the species cited above, showing a great heterogeneity of expression of metallopeptidases not only in *Crithidia* spp. but in a wide range of trypanosomatids as well.

In a similar approach, our group described the differential expression of peptidases in endosymbiont-harboring *Crithidia* species in comparison to members of this genus that naturally lacks a bacterium in the cytoplasm (d'Avila-Levy et al. 2001). In this genus, the trypanosomatids *Crithidia deanei, Crithidia desouzai* and *Crithidia oncopelti* have been described to contain a bacterium symbiont in the cytoplasm, known as endosymbiont, which can be eliminated by the use of antibiotics, leading to the generation of cured strains (reviewed by De Souza and Motta, 1999). Gelatin-SDS-PAGE analysis was used to characterize the cell-associated and extracellular peptidases in these organisms, and our survey showed that a similar proteolytic profile was observed in cells of *C. desouzai* and in wild and cured strains of *C. deanei*: two cysteine peptidases migrating at 60-65 kDa and two metallopeptidases at 51-58 kDa. An additional cysteine peptidase was detected in wild strains at 100 kDa. A subsequent study from our group showed that, after Triton X-114 extraction preformed in *C. deanei* cells, a 65-kDa cysteine peptidase partitioned exclusively in the aqueous phase, possibly present in intracellular compartments, and a 51-kDa metallopeptidase was only detected in the detergent-rich phase (d'Avila-Levy et al. 2003). The remaining enzymes, at 60 kDa and at 58 kDa, which corresponds to a cysteine-type and to a metallo-type peptidase, respectively, were found in both aqueous and detergent-rich phases. In cells of *C. oncopelti*, two metallopeptidases were detected in 59-63 kDa range (d'Avila-Levy et al. 2001) (Figure 1).

The analysis of the spent culture medium showed a similar profile among the above-mentioned species: *C. desouzai* and both strains of *C. deanei* displayed an 80-kDa cysteine peptidase and a 60-kDa metallopeptidase, and *C. oncopelti* showed four bands of protein degradation migrating at 101 kDa, 92 kDa, 76 kDa and 59 kDa, all belonging to the metallopeptidase class. For comparison, *C. fasciculata* displayed a more complex extracellular profile, comprising five metallopeptidases migrating at 101 kDa, 92 kDa, 76 kDa, 60 kDa and 43 kDa (d'Avila-Levy et al. 2001). In summary, the proteolytic profiles of *C. deanei* and *C. desouzai* are identical, and distinct from *C. oncopelti*, which is in accordance to a revision in *Crithidia* taxonomy proposed previously by Brandão et al. (2000) and d'Avila-Levy et al. (2004) and recently confirmed by molecular phylogenetic analyses (Teixeira et al. 2011). In this sense, this genus must be subdivided into three groups: the first one (*Angomonas*) must include *C. deanei* and *C. desouzai*, the second one (designated as *Strigomonas*) must include *C. oncopelti* and the remaining *Crithidia* spp. would remain in the originally described genus.

In the same work (d'Avila-Levy et al. 2001), the availability of *C. deanei* wild and cured strains allowed us to study whether the presence of the endosymbiont induces any alteration in the proteolytic profile. The absence of the cell-associated 100-kDa cysteine peptidase in the cured strain was the only qualitative difference found, and may possibly be related to the absence of the endosymbiont. In addition, the activity of extracellular peptidases was enhanced in the cured strain, which provides evidence that the presence of the endosymbiont diminishes the secretion of proteolytic enzymes, mainly the metallopeptidase (d'Avila-Levy et al. 2001).

Extracellular peptidases were also the focus of studies in some species belonging to the genus *Crithidia*. Unlike cell-associated enzymes, qualitative differences were observed when extracellular proteolytic enzymes were analyzed. In all the species tested, only metallopeptidases were detected, and 3 bands in the 60-80 kDa range were common to *C. fasciculata, C. guilhermei* and *C. luciliae*. Nevertheless, bands with lower molecular mass (30-40 kDa) were found exclusively in *C. fasciculata*, while higher molecular mass bands (90-100 kDa) were only detected in *C. fasciculata* and *C. guilhermei* (d'Avila-Levy et al. 2001; Santos et al. 2005). Interestingly, the extracellular proteolytic profile of *C. luciliae* was also analyzed by Jaffe and Dwyer (2003), but only two metallopeptidases were detected at 97 kDa and at 50 kDa, which could be explained by the smallest amount of spent culture medium employed as well as by the reduced incubation period for proteolysis development.

Melo et al. (2002) characterized the extracellular peptidases from *C. guilhermei* through the incorporation of different protein substrates into SDS-PAGE. When cells were grown in yeast extract-peptone-sucrose medium, the extracellular proteolytic zymogram comprised four bands with gelatinolytic activity migrating at 80 kDa, 67 kDa, 60 kDa and 55 kDa. All bands were inhibited by 1,10-phenanthroline, which classified these enzymes as metallopeptidases, and these gelatinases remained active over a broad pH range, being the maximum activity reached at pH 5.0, which is in accordance to their proper activity in the insect gut. Interestingly, these enzymes were mainly detected at 37°C; when gels were incubated at 28°C, which corresponds to the room temperature and to the expected value in the insect gut, the proteolytic activity was reduced and the 55-kDa band was not detected, possibly reflecting the adaptations of the parasite to the different environments it might confront during its life cycle. Besides the gelatinolytic activity, the 60-, 67- and 80-kDa bands were also able to degrade casein incorporated into SDS-PAGE, but with minor activity, and no proteolytic activity was detected when bovine serum albumin incorporated into the gel. A distinct pattern of degradation was observed when hemoglobin was used as substrate: a 43-kDa metallopeptidase was exclusively detected in these conditions. These hemoglobinases are possibly involved in supplying exogenous iron and heme for the parasite.

Besides the characterization of these peptidases when *C. guilhermei* cells were grown in yeast extract-peptone-sucrose medium, log-phase cells grown in different culture medium composition were obtained and analyzed. The proteolytic zymograms displayed no qualitative difference, only quantitative variations. In this sense, the replacement of sucrose by glucose enhanced the proteolytic activity of the four bands, while either the replacement of sucrose by glycerol or the cultivation of cells in BHI decreased the proteolytic detection. These results pointed out to the influence of the culture medium composition in the production of extracellular peptidases in this microorganism (Melo et al. 2002).

4. Two-dimensional zymography coupled to peptidase identification through mass spectrometry: Possibilities and technical difficulties

For decades, one-dimensional (1D) zymographic gel systems have been broadly used for the analysis and characterization of proteolytic activities in several organisms. Especially in protozoa parasites, this technique has been extensively useful to detect and identify peptidases involved in virulence of pathogenic protozoa (North and Coombs 1981; Coombs and North 1983; Lockwood et al. 1987; Williams and Coombs 1995; Cuervo et al. 2006; De Jesus et al, 2009). Also, through this technique, crucial roles of these enzymes during the cell cycle of parasites have been revealed (Brooks et al. 2001; De Jesus et al. 2007). In the post-genomic era, this methodology is shedding light on the biochemical traits of organisms of unknown genomes (Santos et al. 2005; Pereira et al. 2009; d'Avila-Levy et al. 2001), and has the potential of increasing the functional annotation of the genome for those organisms yet sequenced. However, information regarding on isoforms of proteolytic enzymes, isoelectric point of peptidases, and even a higher resolution of complex proteolytic profiles cannot be obtained by 1D zymographic systems. In superior eukaryotes, a broader analysis of functional peptidases has been achieved by combining zymographic techniques with proteomic technologies, specifically two-dimensional electrophoresis (2D) and mass spectrometry that enable a better resolution of peptidase arranges and the direct identification of peptidase species (Ong and Chang 1997; Park et al. 2002; Zhao and Russell 2003; Wilkesman and Schröder 2007; Lee et al. 2011). Nevertheless, this combined approach has been little used in the study of protozoan parasites (De Jesus et al. 2009).

Proteomic approaches intend to produce the widest possible resolution of individual proteins from a protein mixture, followed by protein identification by mass spectrometry (MS). The fractionation of complex cellular extracts by 2D is attained by combining two independent electrophoretic separations, the isoelectric focusing (IEF) in the first dimension and SDS-PAGE in the second dimension (MacGillivray and Rickwood 1974; O'Farrell 1975). After, protein spots are excised from the gel, submitted to enzymatic digestion and the resulting peptides are analyzed by MS. The developments of soft ionization sources for protein MS analysis, such as matrix-assisted laser desorption/ionization (MALDI) and electrospray ionization (ESI) enabled the reliable identification of proteins (Karas and Hillenkamp, 1988; Tanaka et al. 1988; Fenn et al. 1989). In this way, the combination of MALDI or ESI with several different mass analyzers and increasingly powerful bioinformatics tools allows the identification of thousands protein components from a complex biological sample. Although protein identification relies on genome sequences data, several algorithms based on homology analyses yet permit to identify proteins of organisms with unknown genomes (Shevchenko et al. 2001; Waridel et al. 2007). The expressive contribution of 2D and MS approaches to the understanding of several aspects of the biology of protozoan parasites such as pathogenic trypanosomatids has been recently reviewed (Cuervo et al. 2010). In these parasites, proteomics studies have contributed to catalogue global protein profiles, provide experimental evidence for gene expression, reveal changes in protein expression during development, assign potential functions to the hypothetical proteins, elucidate the subcellular localization, and determine potential drug and vaccine targets (Cuervo et al. 2010, 2011). Despite all the advantages of 2D, the determination of enzymatic activity in this technique is hampered due to the use of chaotropic agents and additional denaturant components present in the sample buffer used for IEF.

The potentialities of both approaches, i. e., the capability to resolve complex protein mixtures by 2D and the capability to reveal functional (active) peptidases by zymography are merged in the two-dimensional zymography (2DZ) methodology (Figure 6). This technique, coupled with mass spectrometry for protein identification make possible the broader mapping of active proteolytic enzymes present in a protein extract (Zhao et al. 2004; De Jesus et al. 2009; Saitoh et al. 2007; Paes-Leme et al. 2009; Larocca et al. 2010; Lee et al. 2011). Two main strategies are used for 2DZ analysis: the first one consist on the separation of protein sample by 2DZ or 2D reverse zymography in parallel with separation by denaturing or non-denaturing 2D followed by staining with MS compatible stain. After migration, the comparison and overlapping of both gel images, using appropriated gel image analysis software, allow the assigning of proteolytic spots to protein spots which are carefully excised from the gel and further identified by MS (Métayer et al. 2002; Park et al. 2002; Choi et al. 2004; Taiyoji et al. 2009; Lee et al. 2011). Using this strategy our group identified active cysteine peptidases in whole extracts of the two *Trichomonas vaginalis* isolates exhibiting high and low virulence phenotypes (De Jesus et al. 2007) (Figure 6). Whole extracts analyzed by 2DZ gels showed both qualitative and quantitative differences in the cysteine peptidase spots between the isolates. According to the pH distribution across the gel strip, proteolytic spots displayed p*I* values between 4.2 and 6.5, a biochemical characteristic that cannot be obtained from 1DZ. It was also observed that the qualitative and quantitative differences in the cysteine peptidases (CP) expression revealed by 2DZ may be related to the virulence pattern of the *T. vaginalis* isolate (Figure 6). After identification of the active "cysteine peptidase fingerprint" expressed by each *T. vaginalis* isolate by tandem MS analysis (MS/MS) it was corroborated that distinct isoforms of CP4 are expressed between the isolates, specifically differentiated by a change in one amino acid of a main peptide. Whereas low-virulence parasites expressed NSWGTAWGEK-containing CP4 isoforms, the virulent isolate expressed a NSWGTTWGEK-containing CP4 isoform (De Jesus et al. 2007). The NSWGTTWGEK-containing CP4 isoform is present in several virulent isolates, is secreted and can induce apoptosis in the epithelial cells (Sommer et al. 2005; De

Jesus et al. 2007). Another important contribution of 2DZ analysis in this work was to reveal that only a limited number of active gelatinase-CPs are expressed *in vitro*, which contrast to a high number of CP genes present in the parasite genome. Alternatively, the employment of different substrates may reveal other peptidase activities. Additionally, in this work, 2DZ allowed preliminary mapping of active forms of low-abundance CPs, which are not easily visualized in 2D Coomassie-stained gels (De Jesus et al. 2007).

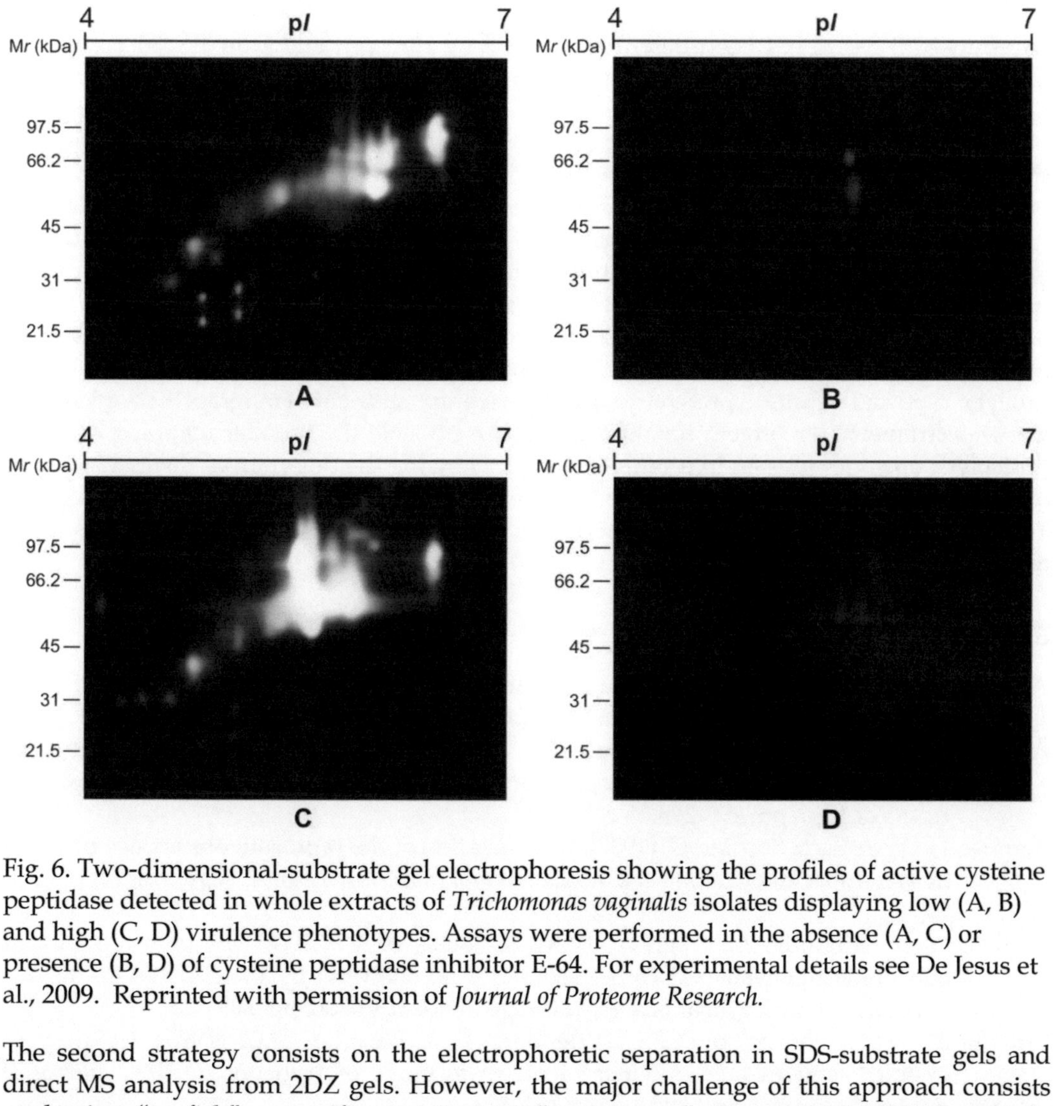

Fig. 6. Two-dimensional-substrate gel electrophoresis showing the profiles of active cysteine peptidase detected in whole extracts of *Trichomonas vaginalis* isolates displaying low (A, B) and high (C, D) virulence phenotypes. Assays were performed in the absence (A, C) or presence (B, D) of cysteine peptidase inhibitor E-64. For experimental details see De Jesus et al., 2009. Reprinted with permission of *Journal of Proteome Research.*

The second strategy consists on the electrophoretic separation in SDS-substrate gels and direct MS analysis from 2DZ gels. However, the major challenge of this approach consists on having "to fish" a specific protein in a "protein sea". To overcome this drawback, fluorescent substrates are used (Zhao et al 2004; Thimon et al. 2008). Proteins are separated by 2D, gels are further incubated with fluorescent peptide substrate and the emitted fluorescence is observed under an UV transilluminator. As the substrate is not embedded in the gel, it can be easily washed, and the protein spot can be excised from the gel for MS/MS analysis. It is clear that the 2DZ-MS techniques should be preceded by broad biochemical characterization of the proteolytic profile of the organism as suggested in the flowchart (Figure 1). The use of 2DZ approaches combined with MS/MS analysis might be a shortcut in the identification of the active degradome and, associated to conventional 2D mapping, might allow the identification of active and inactive peptidases without the use of specific antibodies or laborious purification methods.

5. Conclusion

Substrate-SDS-PAGE has been described in 1980 by Heussen and Dowdle, 30 years after description of this technique, several and important advances in methodological approaches to unveil biological systems have been achieved, such as automated DNA sequencing and protein identification through mass spectrometry just to cite a few. In spite of this, this simple and inexpensive methodology still provides powerful and unique information about peptidases. In organisms with complete genome sequences, bioinformatic analysis provides rich information on putative peptidases, however, it cannot be ascertained if the ORFs are indeed expressed and active. It is common to observe an elevated number of putative peptidases in the organisms' genome, with a limited number of active peptidases. While in organisms without genome information, this technique allows the detection and assessment of several biochemical characteristics of the enzymes, such as preferable pH, temperature, catalytic type and substrate preference. Finally, two-dimensional zymography coupled with mass spectrometry for protein identification make possible the broader mapping of active proteolytic enzymes present in a protein extract, allowing the detection of distinct isoforms of peptidases differentiated by a single change in one amino acid of a main peptide. Therefore, thirty years after its first description, zymography still is a powerful approach to unveil peptidases.

6. References

Ashall, F. (1990). Characterisation of an alkaline peptidase of *Trypanoma cruzi* and other trypanosomatids. *Molecular and Biochemical Parasitology* Vol.38, No.1, pp.77-87.

Ashall, F., Harris, D., Roberts, H., Healy, N. & Shaw, E. (1990). Substrate specificity and inhibitor sensitivity of a trypanosomatid alkaline peptidase. *Biochimica et Biophysica Acta* Vol.1035, pp. 293-299.

Bouvier, J., Etges, R., & Bordier, C. (1987). Identification of the promastigote surface protease in seven species of *Leishmania*. *Molecular and Biochemical Parasitology* Vol. 24, pp. 73–79.

Brandão, A.A., Miranda, A., Degrave, W.M. & Sousa, M.A. (2000). The heterogeneity of choanomastigote-shaped trypanosomatids as analyzed by their kDNA minicircle size: taxonomic implications. *Parasitology Research* Vol.86, pp. 809-812.

Branquinha, M.H., Vermelho, A.B., Goldenberg, S. & Bonaldo, M.C. (1996). Ubiquity of cysteine- and metalloproteinase in a wide range of trypanosomatids. *Journal of Eukaryotic Microbiology* Vol.43, pp. 131–135.

Breganó, J.W., Picão, R.C., Graça, V.K., Menolli, R.A., Jankevicius, S.I., Filho, P.P. & Jankevicius, J.V. (2003). *Phytomonas serpens*, a tomato parasite, shares antigens with *Trypanosoma cruzi* that are recognized by human sera and induce protective immunity in mice. *FEMS Immunology and Medical Microbiology* Vol.39, pp. 257–264.

Brooks, D.R., Denise, H., Westrop, G.D., Coombs, G.H. & Mottram, J.C. (2001). The stage-regulated expression of *Leishmania mexicana* CPB cysteine proteases is mediated by an intercistronic sequence element. *The Journal of Biological Chemistry* Vol. 276, pp. 47061-47069.

Cadavid-Restrepo, G., Gastardelo, T.S., Faudry, E., de Almeida, H., Bastos, I.M., Negreiros, R.S., Lima, M.M., Assumpção, T.C., Almeida, K.C., Ragno, M., Ebel, C., Ribeiro, B.M., Felix, C.R. & Santana, J.M. (2011). The major leucyl aminopeptidase of *Trypanosoma cruzi* (LAPTc) assembles into a homohexamer and belongs to the M17 family of metallopeptidases. *BMC Biochemistry* Vol. 12, pp. 46.

Chicharro, C. & Alvar, J. (2003). Lower trypanosomatids in HIV/AIDS patients. *Annual Tropical Medical Parasitology* Vol.97, pp. 75–80.

Choi, N.S., Yoo, K.H., Yoon, K.S., Maeng, P.J. & Kim, S.H. (2004). Nano-scale proteomics approach using two-dimensional fibrin zymography combined with fluorescent SYPRO ruby dye. *Journal of Biochemistry and Molecular Biology* Vol.37, pp. 298-303.

Coombs, G.H. & North, M.J. (1983). An analysis of the proteinases of *Trichomonas vaginalis* by polyacrylamide gel electrophoresis. *Parasitology* Vol.86, pp. 1-6.

Cuervo, P., Sabóia-Vahia, L., Costa Silva-Filho, F., Fernandes, O., Cupolillo, E., D & Jesus, J.B. (2006). A zymographic study of metalloprotease activities in extracts and extracellular secretions of *Leishmania (Viannia) braziliensis* strains. *Parasitology* Vol. 132, pp. 177-185.

Cuervo, P., Domont, G.B. & De Jesus, J.B. (2010). Proteomics of trypanosomatids of human medical importance. *Journal of Proteomics* Vol.10, pp. 845-867.

Cuervo, P., Fernandes, N. & de Jesus, J.B. (2011). A proteomics view of programmed cell death mechanisms during host-parasite interactions. *Journal of Proteomics* Vol. 75, pp. 246-256.

d'Avila-Levy, C.M., Melo, A.C.N., Vermelho, A.B. & Branquinha, M.H. (2001). Differential expresson of proteolytic enzymes in endosymbiont-harboring *Crithidia* species. *FEMS Microbiology Letters*, Vol.202, pp. 73-77.

d'Avila-Levy, C.M., Souza, R.F., Gomes, R.C., Vermelho, A.B. & Branquinha, M.H. (2003). A metalloproteinase extracellularly released by *Crithidia deanei*. *Canadian Journal of Microbiology* Vol.49, pp. 625-632.

d'Avila-Levy, C.M., Araújo, F.M., Vermelho, A.B., Branquinha, M.H., Alviano, C.S., Soares, R.M. & Santos, A.L.S. (2004). Differential lectin recognition of glycoproteins in choanomastigote-shaped trypanosomatids: taxonomic implications. *FEMS Microbiol Lett.* Vol. 231, pp. 171-176.

d'Avila-Levy, C.M., Araújo, F.M., Vermelho, A.B., Soares, R.M., Santos, A.L.S. & Branquinha, M.H. (2005). Proteolytic expression in *Blastocrithidia culicis*: influence of the endosymbiont and similarities with virulence factors of pathogenic trypanosomatids. *Parasitology* Vol. 130, pp. 413-420.

d'Avila-Levy, C.M., Volotão, A.C.C., Araújo, F.M., De Jesus, J.B., Motta, M.C.M., Vermelho, A.B., Santos, A.L.S. & Branquinha, M.H. (2009). *Bodo* sp., a free-living flagellate, expresses divergent proteolytic activities from the closely related parasitic trypanosomatids. *The Journal of Eukaryotic Microbiology* Vol. 56, pp. 454–458.

De Jesus, J.B., Cuervo, P., Junqueira, M., Britto, C., Silva-Filho, F.C., Soares, M.J., Cupolillo, E., Fernandes, O. & Domont, G.B. (2007). A further proteomic study on the effect of iron in the human pathogen *Trichomonas vaginalis*. *Proteomics* Vol. 7, pp. 1961-1972.

De Jesus, J.B., Cuervo, P., Britto, C., Sabóia-Vahia, L., Costa E Silva-Filho, F., Borges-Veloso, A., Barreiros Petrópolis, D., Cupolillo, E. & Barbosa Domont, G. (2009). Cysteine peptidase expression in *Trichomonas vaginalis* isolates displaying high- and low-virulence phenotypes. *Journal of Proteome Research* Vol. 8, pp. 1555-1164.

De Souza, W. & Motta, M.C.M. (1999). Endosymbiosis in protozoa of the Trypanosomatidae family. *FEMS Microbiology Letters* Vol.173, No.1, pp. 1-8.

Devel, L., Rogakos, V., David, A., Makaritis, A., Beau, F., Cuniasse, P., Yiotakis, A.& Dive, V. (2006). Development of selective inhibitors and substrate of matrix metallo proteinase-12. *Journal of Biological Chemistry* Vol. 281, pp. 11152-11160.

Elias, C.G.R., Pereira, F.M., Silva, B.A., Alviano, C.S., Soares, R.M.A. & Santos, A.L.S. (2006). Leishmanolysin (gp63 metallopeptidase)-like activity extracellularly released by *Herpetomonas samuelpessoai*. *Parasitology* Vol.132, pp. 37–47.

Elias, C.G.R., Pereira, F.M., Dias, F.A., Silva, T.L., Lopes, A.H., d'Avila-Levy, C.M., Branquinha, M.H. & Santos, A.L.S. (2008). Cysteine peptidases in the tomato trypanosomatid *Phytomonas serpens*: influence of growth conditions, similarities with cruzipain and secretion to the extracellular environment. *Experimental Parasitology* Vol. 120, pp. 343–352.

Etges, R. (1992). Identification of a surface metalloproteinase on 13 species of *Leishmania* isolated from humans, *Crithidia fasciculata* and *Herpetomonas samuelpessoai*. *Acta Tropica* Vol.50, pp. 205–217.

Fenn, J., Mann, M., Meng, C.K., Wong, S.F. & Whitehouse, C.M. (1989). Electrospray ionization for mass spectrometry of large biomolecules. *Science* Vol. 246, pp. 64-71.

Grellier, P., Vendeville, S., Joyeau, R., Bastos, I.M., Drobecq, H., Frappier, F., Teixeira, A.R., Schrével J, Davioud-Charvet E, Sergheraert C & Santana JM. (2001). *Trypanosoma cruzi* prolyl oligopeptidase Tc80 is involved in nonphagocytic mammalian cell invasion by trypomastigotes. *Journal of Biological Chemistry* Vol. 276, pp. 47078-47086.

Heussen, C. & Dowdle, E.B. (1980). Electrophoretic analysis of plasminogen activators in polyacrylamide gels containing sodium dodecyl sulphate and copolymerized substrates. *Analytical Biochemistry* Vol. 102, pp. 196–202.

Hoare, C.A. & Wallace, F.G. (1966). Developmental stages of trypanosomatid flagellates: a new terminology. *Nature*, Vol.212, pp. 1385-1386.

Jaffe, C.L. & Dwyer, D.M. (2003). Extracellular release of the surface metalloprotease, gp63, from Leishmania and insect trypanosomatids. *Parasitology Research* Vol.91, pp. 229-37.

Karas, M. & Hillemkamp, F. (1988). Laser desorption ionization of proteins with molecular masses exceeding 10000 kDa. *Analytical Chemistry* Vol. 60, pp. 2299-2301.

Kupai, K., Szucs, G., Cseh, S., Hajdu, I., Csonka, C., Csont, T. & Ferdinandy, P. (2010). Matrix metalloproteinase activity assays: Importance of zymography. *Journal of Pharmacological and Toxicological Methods* Vol. 61, pp. 205-209.

Laemmli, U.K. (1970). Cleavage of structural proteins during the assembly of the head of bacteriophage T4. *Nature* Vol. 227, pp. 680-685.

Larocca, M., Rossano, R. & Riccio, P. (2010). Analysis of green kiwi fruit (Actinidia deliciosa cv. Hayward) proteinases by two-dimensional zymography and direct identification of zymographic spots by mass spectrometry. *Journal of the Science of Food and Agriculture* Vol. 90, pp. 2411-248.

Lee, K.J., Kim, J.B., Ha, B.K., Kim, S.H., Kang, S.Y., Lee, B.M. & Kim, D.S. (2011). Proteomic characterization of Kunitz trypsin inhibitor variants, Tia and Tib, in soybean [Glycine max (L.) Merrill]. *Amino Acids*, in press.

Lockwood, B.C., North, M.J., Scott, K.I., Bremner, A.F. & Coombs, G.H. (1987). The use of a highly sensitive electrophoretic method to compare the proteinases of trichomonads. *Molecular and Biochemical Parasitology* Vol. 24, pp. 89-95.

Lopes, J.D., Caulada, Z., Barbieri, C.L. & Camargo, E.P. (1981). Cross-reactivity between *Trypanosoma cruzi* and insect trypanosomatids as a basis for the diagnosos of Chagas' disease. *American Journal of Tropical Medicine and Hygine* Vol.30, pp. 1183-1188.

MacGillivray, A.J. & Rickwood, D. (1974). The heterogeneity of mouse-chromatin nonhistone proteins as evidenced by two-dimensional polyacrylamide-gel electrophoresis and ion-exchange chromatography. *European Journal of Biochemistry* Vol. 41, pp. 181-190.

Martinez, J. & Cazzulo, J.J. (1992). Anomalous electrophoretic behaviour of the major cysteine proteinase (cruzipain) from *Trypanosoma cruzi* in relation to its apparent molecular mass. *FEMS Microbiology Letters* Vol.74, pp. 225-229.

McGhee, R.B. & Cosgrove, W.B. (1980). Biology and physiology of the lower Trypanosomatidae. *Microbiology Reviews* Vol.44, pp. 140–173.

Melo, A.C.N., d'Avila-Levy, C.M., Branquinha, M.H. & Vermelho, A.B. (2002). *Crithidia guilhermei*: gelatin- and haemoglobin-degrading extracellular metalloproteinases. *Experimental Parasitology* Vol.102, pp. 150-156.

Métayer, S., Dacheux, F., Dacheux, J.L. & Gatti, J.L. (2002). Comparison, characterization, and identification of proteases and protease inhibitors in epididymal fluids of domestic mammals. Matrix metalloproteinases are major fluid gelatinases. *Biology of Reproduction* Vol. 66, pp. 1219-1229.

Michaud, D., Cantin, L., Raworth, D.A. & Vrain, T.C. (1996). Assessing the stability of cystatin/cysteine proteinase complexes using mildly-denaturing gelatin-polyacrylamide gel electrophoresis. *Electrophoresis* Vol.17, pp. 74-79.

Momen, H. (2001). Some current problems in the systematics of trypanosomatids. *International Journal for Parasitology* Vol.31, pp. 640-642.

Momen, H. (2002). Molecular taxonomy of trypanosomatids: some problems and pitfalls. *Archives of Medical Research* Vol.33, pp. 413–415.

Morio, F., Reynes, J., Dollet, M., Pratlong, F., Dedet, J.P. & Ravel, C. (2008). Isolation of a protozoan parasite genetically related to the insect trypanosomatid *Herpetomonas samuelpessoai* from a human immunodeficiency virus positive patient. *Jounal of Clinical Microbiology* Vol.46, pp. 3845-3847.

Nogueira de Melo, A.C., d'Avila-Levy, C.M., Dias, F.A., Armada, J.L.A., Silva, H.D., Lopes, A.H.C.S., Santos, A.L.S., Branquinha, M.H. & Vermelho, A.B. (2006). Peptidases and gp63-like proteins in *Herpetomonas megaseliae*: possible involvement in the adhesion to the invertebrate host. *International Journal for Parasitology*, Vol.36, pp. 415–422.

North, M.J. & Coombs, G.H. (1975). Proteinases of *Leishmania mexicana* amastigotes and promastigotes: analysis by gel electrophoresis. *Molecular and Biochemical Parasitology* Vol. 3, pp. 293-300.

O'Farrell, P.H. (1975). High resolution two-dimensional electrophoresis of proteins. *The Journal of Biological Chemistry* Vol. 250, pp. 4007-4021.

Ong, K.L. & Chang, F.N. (1997). Analysis of proteins from different phase variants of the entomopathogenic bacteria *Photorhabdus luminescens* by two-dimensional zymography. *Electrophoresis* Vol. 18, pp. 834-839.

Paes Leme, A.F., Kitano, E.S., Furtado, M.F., Valente, R.H., Camargo, A.C., Ho, P.L., Fox, J.W. & Serrano, S.M. (2009). Analysis of the subproteomes of proteinases and heparin-binding toxins of eight Bothrops venoms. *Proteomics* Vol. 9, pp. 733-745.

Park, S.G., Kho, C.W., Cho, S., Lee, D.H., Kim, S.H. & Park, B.C. (2002). A functional proteomic analysis of secreted fibrinolytic enzymes from *Bacillus subtilis* 168 using a combined method of two-dimensional gel electrophoresis and zymography. *Proteomics.* Vol. 2, pp. 206-211.

Pereira, F.M., Elias, C.G., d'Avila-Levy, C.M., Branquinha, M.H. & Santos, A.L.S. (2009). Cysteine peptidases in *Herpetomonas samuelpessoai* are modulated by temperature and dimethylsulfoxide-triggered differentiation. *Parasitology*, Vol.136, pp. 45-54.

Pereira, F.M., Dias, F.A., Elias, C.G., d'Avila-Levy, C.M., Silva, C.S., Santos-Mallet, J.R., Branquinha, M.H. & Santos, A.L.S. (2010a). Leishmanolysin-like molecules in *Herpetomonas samuelpessoai* mediate hydrolysis of protein substrates and interaction with insect. *Protist* Vol.161, pp. 589-602.

Pereira, F.M., Santos-Mallet, J.R., Branquinha, M.H., d'Avila-Levy, C.M. & Santos, A.L.S. (2010b). Influence of leishmanolysin-like molecules of *Herpetomonas samuelpessoai* on the interaction with macrophages. *Microbes and Infection* Vol.12, pp. 1061-170.

Podlipaev, S. (2001). The more insect trypanosomatids under study-the more diverse Trypanosomatidae appears. *International Journal for Parasitology* Vol.31, pp. 648–652.

Rawlings, N.D., Barrett, A.J. & Bateman, A. (2010). MEROPS: the peptidase database. *Nucleic Acids Research.* Vol. 38, pp. D227-D233.

Saitoh, E., Yamamoto, S., Okamoto, E., Hayakawa, Y., Hoshino, T., Sato, R., Isemura, S., Ohtsubo, S. & Taniguchi, M. (2007). Identification of cysteine proteases and screening of cysteine protease inhibitors in biological samples by a two-dimensional gel system of zymography and reverse zymography. *Analytical Chemistry Insights* Vol. 18, pp. 51-59.

Santos, A.L.S., Ferreira, A., Franco, V.A., Alviano, C.S. & Soares, R.M.A. (1999). Characterization of proteinases in *Herpetomonas anglusteri* and *Herpetomonas roitmani*. *Current Microbiology* Vol.39, pp. 61–64.

Santos, A.L.S., Abreu, C.M., Alviano, C.S. & Soares, R.M.A. (2002). Activation of the glycosylphosphatidylinositol anchored membrane proteinase upon release from *Herpetomonas samuelpessoai* by phospholipase C. *Current Microbiology* Vol.45, pp. 293–298.

Santos, A.L.S., Batista, L.M., Abreu, C.M., Alviano, C.S., Angluster, J. & Soares, R.M.A. (2001). Developmentally regulated protein expression mediated by dimethylsulfoxide in *Herpetomonas samuelpessoai*. *Current Microbiology* Vol.42, pp. 111–116.

Santos, A.L.S., Rodrigues, M.L., Alviano, C.S., Angluster, J. & Soares, R.M.A. (2003). *Herpetomonas samuelpessoai*: dimethylsulfoxide-induced differentiation is influenced by proteinase expression. *Current Microbiology* Vol.46, pp. 11–17.

Santos, A.L.S., Abreu, C.M., Alviano, C.S. & Soares, R.M.A. (2005). Use of proteolytic enzymes as an additional tool for trypanosomatid identification. *Parasitology* Vol.130, pp. 79–88.

Santos, A.L.S., Branquinha, M.H. & d'Avila-Levy, C.M. (2006). The ubiquitous gp63-like metalloprotease from lower trypanosomatids: in the search for a function. *Anais da Academia Brasileira de Ciências* Vol.78, pp. 687–714.

Santos, A.L.S., d'Avila-Levy, C.M., Elias, C.G.R., Vermelho, A.B. & Branquinha, M.H. (2007). *Phytomonas serpens*: immunological similarities with the human trypanosomatid pathogens. *Microbes and Infection* Vol.9, pp. 915–921.

Santos, A.L.S., Soares, R.M.A., Alviano, C.S. & Kneipp, L.F. (2008). Heterogeneous production of metallo-type peptidases in parasites belonging to the family Trypanosomatidae. *European Journal of Protistology* Vol.44, pp.103–113.

Schneider, P. & Glaser, T. A. (1993). Characterization of a surface metalloprotease from *Herpetomonas samuelpessoai* and comparison with *Leishmania major* promastigote surface protease. *Molecular and Biochemical Parasitology* Vol.58, pp. 277–282.

Shevchenko, A., Sunyaev, S., Loboda, A., Shevchenko, A., Bork, P., Ens, W. & Standing, K.G. (2001). Charting the proteomes of organisms with unsequenced genomes by MALDI-quadrupole time-of-flight mass spectrometry and BLAST homology searching. *Analytical Chemistry* Vol.73, pp. 1917-1926.

Sommer, U., Costello, C.E., Hayes, G.R., Beach, D.H., Gilbert, R.O., Lucas, J.J. & Singh, B.N. (2005). Identification of *Trichomonas vaginalis* cysteine proteases that induce apoptosis in human vaginal epithelial cells. *Journal of Biological Chemistry* Vol. 280, pp. 23853-23860.

Snoek-van Beurden, P.A. & Von den Hoff, J.W. (2005). Zymographic techniques for the analysis of matrix metalloproteinases and their inhibitors. *Biotechniques*. Vol. 38, pp. 73-83.

Taiyoji, M., Shitomi, Y., Taniguchi, M., Saitoh, E. & Ohtsubo, S. (2009). Identification of proteinaceous inhibitors of a cysteine proteinase (an Arg-specific gingipain) from *Porphyromonas gingivalis* in rice grain, using targeted-proteomics approaches. *Journal of Proteome Research* Vol. 8, pp. 5165-5174.

Tanaka, K., Waki, H., Ido, Y., Akita, S., Yoshida, Y. & Yoshida, T. (1988). Protein and polymer analyses up to m/z 100000 by laser ionization time-of-flight mass spectrometry. *Rapid Communications in Mass Spectrometry* Vol. 2, pp.151-153.

Teixeira, M.M., Borghesan, T.C., Ferreira, R.C., Santos, M.A., Takata, C.S., Campaner, M., Nunes, V.L., Milder, R.V., de Souza, W. & Camargo, E.P. (2011). Phylogenetic validation of the genera *Angomonas* and *Strigomonas* of trypanosomatids harboring bacterial endosymbionts with the description of new species of trypanosomatids and of proteobacterial symbionts. *Protist* Vol.162, pp. 503-524.

Thimon, V., Belghazi, M., Labas, V., Dacheux, J.L. & Gatti, J.L.(2008). One- and two-dimensional SDS-PAGE zymography with quenched fluorogenic substrates provides identification of biological fluid proteases by direct mass spectrometry. *Analytical Biochemistry* Vol. 375, pp. 382-384.

Vickerman, K. (1994). The evolutionary expansion of the trypanosomatid flagellates. *The International Journal for Parasitology,* Vol.24, pp. 1317-1331.

Wallace, F.G. (1966). The trypanosomatid parasites of insects and arachnids. *Experimental Parasitology* Vol. 18, pp. 124-193.

Wallace, F.G., Camargo, E.P., McGhee, R.B. & Roitman, I. (1983). Guidelines for the description of new species of lower trypanosomatids. *Journal of Protozoology* Vol.30, pp. 308–313.

Waridel, P., Frank, A., Thomas, H., Surendranath, V., Sunyaev, S., Pevzner, P. & Shevchenko, A. (2007). Sequence similarity-driven proteomics in organisms with unknown genomes by LC-MS/MS and automated de novo sequencing. *Proteomics.* Vol. 7, pp. 2318-2329.

Wilder, C.L., Park, K., Keegan, P.M. & Platt, M.O. (2011). Manipulating substrate and pH in zymography protocols selectively distinguishes cathepsins K, L, S, and V activity in cells and tissues. *Archives of Biochemistry and Biophysics* Vol. 516, pp. 52-77.

Williams, A.G. & Coombs, G.H. (1995). Multiple protease activities in *Giardia intestinalis* trophozoites. *The International Journal of Parasitology* Vol. 25, pp. 771-778.

Wilkesman, J.G. & Schröder, H.C. (2007). Analysis of serine proteases from marine sponges by 2-D zymography. *Electrophoresis* Vol. 28, pp. 429-436.

Yao, C. (2010). Major surface protease of trypanosomatids: one size fits all? *Infection and Immunity* Vol.78, pp. 22-31.

Zhao, Z. & Russell, P.J. (2003). Trypsin activity assay in substrate-specific one- and two-dimensional gels: a powerful method to separate and characterize novel proteases in active form in biological samples. *Electrophoresis* Vol. 24, pp. 3284-3248.

Zhao, Z., Raftery, M.J., Niu, X.M., Daja, M.M. & Russell, P.J. (2004). Application of in-gel protease assay in a biological sample: characterization and identification of urokinase-type plasminogen activator (uPA) in secreted proteins from a prostate cancer cell line PC-3. *Electrophoresis* Vol. 25, pp. 1142-1148.

Part 5

Temporal Temperature Gel Electrophoresis

16

Temporal Temperature Gel Electrophoresis to Survey Pathogenic Bacterial Communities: The Case of Surgical Site Infections

Romano-Bertrand Sara[1,2], Parer Sylvie[1,2], Lotthé Anne[1,2], Colson Pascal[3]
Albat Bernard[4] and Jumas-Bilak Estelle[1,2]
[1]*University Montpellier 1, Equipe pathogènes et environnements, UMR 5119 ECOSYM*
[2]*University Hospital of Montpellier, hospital hygiene and infection control team*
[3]*University Hospital of Montpellier, Cardio-thoracic intensive care unit*
[4]*University Hospital of Montpellier, Cardio-thoracic surgery unit*
France

1. Introduction

The main objective of this chapter is to review 16S rRNA gene-based PCR-Temporal Temperature Gel Electrophoresis (TTGE) methods with emphasis on its use in medical microbiology and infectious diseases. As an example of application, we describe optimization, validation and results of an original approach for exploring the microbiology of surgical site infections with a focus on particular constraints related to low microbial load found in this setting. Finally, PCR-TTGE will be situated in the evolution of medical microbiology and infectious disease medicine toward the analysis of complex microbial communities.

2. State of the art

2.1 From human microbiome to pathogenic bacterial communities

At the early 21th century, studies on human-associated bacteria showed that there are at least 10 times as many bacterial cells as human cells, i.e. 10^{14} bacterial cells in the human gut (Turnbaugh et al., 2007). The current estimation of the number of genes in the human genome is about 23000 (Wei & Brent, 2006). Based on the diversity of gut microbes and the average number of genes contained in one bacterial genome, the diversity of bacterial genes in human gut was guessed to be 100 times greater than that of our human genome (Bäckhed et al., 2005). This number seems to be underestimated, since a more recent publication estimates to more than 9.000.000 the number of unique genes in human gut bacterial community (Yang et al., 2009). This huge community is named the human microbiome, a term coined by Joshua Lederberg in 2001 "to signify the ecological community of commensal, symbiotic, and pathogenic microorganisms that literally share our body space" (Lederberg & McCray, 2001). A new concept is to consider human organism as an assemblage of human and bacterial cells organized into organs, tissues, and cellular communities amounting to a super- or a hetero-organism. Metagenomics show clear differences between microbiomes in various body sites and together with metatranscriptomics and metaproteomics reveal how microbiomes contribute to organ and tissue functions. Consequently, human biology can no longer concern itself only with human cells: microbiomes at different body sites and functional metagenomics must be considered part of systems biology (Pflughoeft & Versalovic, 2011). The biological concept of hetero-organism further evolves into new medical conception on health and disease, involving not only human genes and cells but also related genomes from environment around and inside human body. For instance, the change in gut microbiome is associated with local pathology such as Crohn's disease but also with systemic diseases such as obesity (Turnbaugh et al., 2009), diabetes (Cani et al., 2007) and hypertension (Holmes et al., 2008), among other chronic diseases (Proal et al., 2009; Lampe, 2008). The microbiome of various human body sites has already been described for the digestive tract, the mouth, the skin, the genital tract and it will continue with the « Human Microbiome Project » (Turnbaugh et al., 2007).

Microbiome and its disturbances are now a major area of microbiology research, as attested by the leading publications produced in this field. However, these advances in 'microbiomology' have not yet change the patterns of medical reflection. In the field of infectious diseases, bacteriologists still strive to obtain a pure culture preferentially isolating known pathogens and focus on the morphology, physiology, and genetics of the small subset of the microorganisms in the body that are known to cause disease. Hence, most bacteria present in a community are neglected and so are the complex relationships among members of the community. It is well-known that a large range of endogenous bacteria can cause opportunistic infections such healthcare associated infections (HAI), consecutive to microbiome disequilibrium, and that strict pathogens must cross the cutaneous or mucosal barriers covered with local microbiota before developing their virulence. Therefore, considering the whole human microbiota is required to better understand the physiopathology of infections and improve their prevention and treatment of bacterial infectious diseases that remain major challenges in public health. Confronting microbiomes analyses to clinical issues will warrant these analyses to be good predictors and markers in infectious disease. This implies that microbiomes analyses become routine, which is not yet the case.

2.2 How to explore bacterial communities?

Bacterial communities are classically assessed through culture-dependent methods based on colony isolation on solid medium, sometimes after enrichment by growth in liquid medium. This way, the description of complex communities inevitably involved time-consuming steps of growth on multiple media under varying conditions, while the full description of the community is not insured. Indeed, it is now obvious that the real microbial diversity is poorly represented by the cultured fraction, and conventional culture techniques have been shown to explore less than 1% of the whole bacterial diversity in environment samples, such as soil samples (Riesenfeld et al., 2004). It is also estimated that as much as 20% to 60% of the human-associated microbiome, depending on body site, is unculturable (NIH HMP Working Group et al., 2009). A renewal of studies on complex bacterial communities has been made possible by molecular methods that allow direct analysis of bacteria present in a sample, while avoiding the bias of cultivability. The emergence of molecular ecology and metagenomics offers the potential of determining microbial diversity in an ecosystem without prior laboratory enrichment, isolation and growth on artificial media thanks to sophisticated methodological and computational tools. Generally, culture-independent approaches allow a precise description of an ecosystem by assessing its genetic diversity. As an example of molecular ecology efficiency, Dekio et al (2005) show that the total skin microflora explored by culture-independent molecular profiling is greater than previously believed, finding 22 potentially novel members, comprising 9 species and 13 phylotypes not yet described as members of skin microbiota (Dekio et al., 2005).

Metagenomics examine the complexity of the community by sequencing genomic libraries made from DNA extracted directly from the sample containing a complex mix of different bacteria. The complete metagenomic approach will give the total gene content of a community, thus providing data about biodiversity but also function and interactions (Tyson et al., 2004). For the purpose of biodiversity studies, metagenomics can focus on one common gene shared by all members of the community. The most commonly used culture-independent method relies on amplification and analysis of the 16S rRNA genes in a microbiome (Nossa et al., 2010).

16S rRNA genes are widely used for documentation of the evolutionary history and taxonomic assignment of individual organisms because they have highly conserved regions for construction of universal primers and highly variable regions for identification of individual species (Woese, 1987). The 16S rRNA gene, in spite of some recognized pitfalls (von Wintzingerode et al., 1997), remains today the most popular marker for studying the specific diversity in a bacterial community. The different 16S rRNA genes representative of the community are amplified by PCR and then separated and identified either by cloning and Sanger sequencing or by direct pyro-sequencing (Nossa et al., 2010). Tools for sequence-specific separation after bulk PCR amplification, such as T-RFLP (Terminal-Restriction Fragment Lenght Polymorphism) (Kitts, 2001), D-HPLC (Denaturing High Performance Liquid Chromatography) (Penny et al., 2010), CDCE (Constant Denaturing Capillary Electrophoresis) (Thompson et al., 2004), SSCP (Single Stand Conformation Polymorphism) (Ege et al., 2011), DGGE (Denaturing Gradient Gel Electrophoresis) (Muyzer et al. 1993), TGGE (Temperature Gradient Gel Electrophoresis) (Zoetendal et al., 1998) and TTGE (Temporal Temperature Gradient Gel Electrophoresis) (Ogier et al., 2002), can also be used. Methods based upon separation in denaturing electrophoresis appear particularly suitable for the routine follow-up of microbiomes with low or medium diversity (Roudière et al., 2007). They provide a "fingerprint" of the community diversity.

2.3 16S rRNA gene PCR-TTGE: Advantages and limitations

PCR-TTGE is a PCR-denaturing gradient gel electrophoresis that allows separation of DNA fragments in a temporal gradient of temperature (Yoshino et al., 1991; Ogier et al., 2002). PCR amplicons of the same size but with different sequences are separated in the gel. In a denaturing acrylamide gel, DNA denatures in discrete regions called melting domains, each of them displaying a sequence specific melting temperature. When the melting temperature (*Tm*) of the whole amplicon is reached, the DNA is denatured creating branched molecules. This branching reduces DNA mobility in the gel. Therefore, amplicons of the same size but with different nucleotide compositions can be separated based on differences in the behavior of their melting domains. When DNA is extracted and amplified from a complex community, TTGE lead to the separation of the different amplicons and produce a banding pattern characteristic of the community. The gradient obtained by varying the temperature over time in TTGE generally produces more clear and reproducible profiles than does the chemical gradient in DGGE.

Direct observation and counting bands on the TTGE profile provides a diversity score that roughly corresponds to the number of molecular species in the sample. However, it must be remembered that two amplicons with different sequences can give identical migration distances when their *Tm* are identical. The banding profile can be further analyzed by the affiliation of each band to a species or other taxon. Affiliation can be realized by comparing the migration distance of each band to a molecular ladder, named diversity ladder (Roudière et al., 2009; Ogier et al., 2002), constructed by using amplicons corresponding to species known to be representative of the community under study. A more accurate analysis of each band can be achieved by cutting bands from the gel, extracting DNA from bands and sequencing. This way the diversity ladder will be completed and updated. A method associating the use of diversity ladder with sequencing has shown its efficiency in describing bacterial communities of low complexity such as the gut microflora of neonates (Roudière et al., 2009). Such an approach is simple enough to survey dynamics of bacterial communities on a wide range of samples, particularly in health and disease (Jacquot et al., 2011).

PCR-TTGE appears not suitable for investigating highly diverse communities. This limit is due to number of bands that can be separated within the length of the gel. Optimization of TTGE conditions allows separation of bands by a minimum of 0.1 mm over all the gel length. We showed that the number of specific bands separable by TTGE could not exceed about 50 for the migration of an artificial diversity ladder. For instance, a ladder contained 53 different bacterial species found in stool of neonates can be efficiently separated in the optimized TTGE conditions (Roudière et al., 2009). For samples obtained from natural ecosystem, TTGE would be difficult to interprete if the diversity exceeds 25 to 30 bands.

Prior to TTGE migration, other technical steps are limitative and should be carefully considered and optimized. Particularly, DNA should be recovered and amplified from all the genotypes in the community, i.e. extraction and PCR should be as universal as possible. Special attention should be given to *Firmicutes* and *Actinobacteria* because they display thick and resistant cell wall. The extraction efficiency should be tested on a wide panel of bacteria to scan a large range of bacterial types. Extraction is generally improved by the use of large-spectrum lytic enzymes and/or by a mechanical grinding (Roudière et al., 2009; Le Bourhis et al., 2007).

The PCR itself is a cause of limitations in the PCR-TTGE approach. Molecular methods are often praised for their sensitivity. However, this detection sensitivity can fail when complex samples are analyzed. For example, detection thresholds of 10^3-10^4 CFU/mL are currently described for universal PCR-TTGE or PCR-DGGE (Le Bourhis et al., 2007; Temmerman et al., 2003; Roudière at al., 2009). The detection limit in PCR-TTGE cannot be easily assessed as it depends on both CFU/g count of each species and the relative representation of species in the community. Minor populations of less than 1% of total population are undetectable by PCR-TTGE. This breakpoint is commonly reported for denaturing-gel-based methods used in microbial ecology (Ogier et al., 2002; Zoetendal et al., 1998; Roudière et al, 2009).

PCR can also be affected by preferential PCR amplification that may hinder the detection of some genotypes when a complex mix of DNA molecules is used as template. Preferential PCR amplification can be caused by primer mismatches at the annealing sites for some genotypes or by a lower rate of primer hybridization to certain templates due to a low local denaturation (Kanagawa, 2003). PCR carried out on complex bulk DNA can produce heteroduplexes particularly in later cycles when primer concentration decreases and the concentration of PCR products is high (Kanagawa, 2003). Chimeric amplicons can also be formed in later PCR cycles when template concentration is high enough to allow the re-annealing of templates before primer extension (von Wintzergerode et al., 1997; Kanagawa, 2003). All these artifacts can generate additional signals that do not correspond to genotypes in the sample. Heteroduplexes and chimera produce additional bands in the TTGE pattern that lead to an overestimation of the diversity. These artifactual bands can be detected either on the basis of their very short migration distance or by sequencing. Consequently, the crude diversity index determined by simple band count should be optimized after exclusion of heteroduplexes and chimeric bands.

Last but not least, a major parameter in community studies is the choice of a molecular marker allowing the genotyping of the whole community. The notion developed by Woese that rRNA genes could identify living organisms by reconstructing phylogenies resulted in the adoption of 16S rRNA gene in microbiology (Woese, 1987). Its universality and the huge number of sequences stored in databases have established 16S rRNA gene as the "gold standard" not only in microbial phylogeny, systematics, and identification but also microbial ecology (Case et al., 2007).

For the purpose of identifying an isolated bacterial strain, the complete 16S rRNA gene (1500 bp) is generally used, giving accurate affiliation to a species in most cases. In PCR-TTGE experiments, the amplified fragments are short (200 to 400 bp) to allow migration in polyacrylamide gel. The fragment amplified should contain hypervariable regions of the 16S rRNA gene in order to compensate for the lack of information due to the small sequence size by a high rate of mutation. Bacterial 16S rRNA genes comprise nine hypervariable regions, V1-V9, exhibiting sequence diversity among species (Van de Peer et al., 1996). In most studies, the V3 region located in the 5′ part of the gene is chosen (Jany & Barbier, 2008). However, the phylogenetic information is sometimes insufficient to achieve species identification. Depending on the bacterium, sequences provide identification to the genus or family level only. Consequently, the diversity of the community is not described by a list of bacterial species but by a list of operational taxonomic units (OTUs) corresponding to the lower taxonomic level being accurately identified.

At the genomic level, rRNA genes are generally organized in multigene families (Acinas et al., 2004). The members of a rRNA multigene family are subject to a homogenization process allowing the multiple gene copies to evolve in concert. In a concerted evolution mode, mutation occurring in one copy will be fixed in all of them or lost from all. Thus, rRNA sequences show low variability within species, subspecies or genome (Liao, 2000). However, intra-genomic heterogeneity in the form of nucleotide differences between 16S rRNA gene copies are often described. For examples micro-heterogeneity has been identified in *Escherichia coli, Mycobacterium terrae, Paenibacillus polymyxa,* members of the classes *Mollicutes,* and *Actinomycetales* (Teyssier et al., 2003). Analyzing of complete genome sequences has recently assessed intra-genomic heterogeneity. For genomes with more than one rRNA operon, 62% display some degree of sequence divergence between 16S rRNA loci in a same genome (Case et al., 2007). In PCR-TTGE, the intra-genomic 16S rDNA heterogeneity can lead to multiple bands for a single OTU and then to an overestimation of OTU diversity. This pitfall inherent to the 16S rRNA gene marker will be avoided by band sequencing. However, as a pre-requisite of diversity analysis by PCR-TTGE, the major known species expected in a particular ecosystem should be individually studied by TTGE in order to explore heterogeneity in 16S rRNA gene copies (Roudière et al, 2007; Michon et al., 2011).

Alternative markers can also be proposed such as *rpoB* (Case et al., 2007) but universal *rpoB* PCR primers allowing the exploration of the whole bacterial diversity can not be designed (personal data) and the databases remain poor in *rpo* sequences.

Several authors remarked that culture-independent methods regularly fail to identify species obtained using culture-dependent methods (Jany & Barbier, 2008). By contrast, culture-dependent methods have yielded information on the structure of microbial populations but they are limited by the in vitro growth capacity of most bacteria in a community. Such a discrepancy is not observed for all the communities studied. For instance, culture dependant and independent approaches of the premature neonate gut microbiome give globally congruent results (Roudière et al., 2009). However, it is accepted that culture-independent methods remain the only approach for monitoring the rapid dynamics of microbial communities. Nevertheless, the two types of methods reveal different images of the same community and combining culture-dependent and culture-independent methods may be worthwhile to obtain a more accurate view of the structure of the microbial community (Case et al., 2007).

In spite of the limited growth capacity of most bacteria, culture-dependent methods remain the sole approach available for monitoring sub-populations selected on the basis of phenotypic traits such as dependence to metabolites or resistance to antimicrobial agents. In this context, PCR-TTGE can be used after culture in specific conditions in order to describe the diversity of cultivable population. The colonies growing in diverse conditions can be bulked and further analyzed by PCR-TTGE as described before. This culture- and genetic-based mixed approach is particularly suitable to describe dynamics of populations according to their level of resistance to antimicrobial drugs in natural environments (Vanhove et al., 2011).

Culture-independent approaches have previously shown their interest in cardiology to detected new or atypical infectious agents (Marchandin et al., 2009; Daïen et al., 2010). Considering the interest and limitations of 16S rRNA PCR-TTGE, we will proposed a protocol for describing the diversity and following the dynamics of the bacterial community that colonize surgical wound of the patient during hospitalization for cardiac surgery. We will show how 16S rRNA PCR-TTGE is particularly suited to the low bacterial diversity encountered in aseptic surgical settings, where antibiotic prophylaxis and cutaneous antisepsis effectively reduce the bacterial load of patients.

In addition, one example of the use of 16S PCR-TTGE in a culture-dependent analysis will be detailed. This approach associates determination of Minimal Inhibitory Concentration (MIC) at the community level and determination of the diversity by 16S PCR-TTGE in the resistant sub-population at each concentration of antimicrobial agent.

3. An original method of PCR-TTGE to learn more about the physiopathology of surgical site infection in cardiac surgery

3.1 Surgical site infections: State of the art

3.1.1 Epidemiology of surgical site infection

Surgical site infections (SSIs) are among the most frequent healthcare associated infections (HAIs), along with urinary tract and pulmonary infections, and remain an unresolved problem for modern medicine, their occurrence having significant impact on patient morbidity, length of stay and cost of care. Data from longitudinal surveillance studies show SSI rates of 1 to 5% (Klevens et al., 2007; Astagneau et al., 2009; de Lissovoy et al., 2009), whereas higher rates are reported from interventional studies, where control groups can have up to 8.5% infection rates (Bode et al., 2010; Perl et al., 2002), owing to different population case mixes. Large-scale epidemiological studies have identified risk factors for SSI that can be grossly classified as related to patient condition, surgical procedure and environment. The American National Nosocomial Infection Surveillance system (NNIS) developed an easily calculated risk index that combines the patient-related risk assessment of the American Society of Anesthesiologists (ASA) score, and 2 surgical procedure-related factors: type of surgery as defined by pre- or per-operative microbial contamination (Altemeier classification, from I - clean surgery- to IV - septic surgery), and duration of operation exceeding 75th percentile for a given procedure. For all categories of surgery, there is a linear increase in the incidence of SSI when the NNIS risk index increases (Coello et al., 2005). Beside these surveillance-derived risk assessment scores, prospective controlled studies identified many more factors associated with a higher risk of SSI. Most important are poorly controlled diabetes mellitus, malignant diseases, smoking, advanced age, per operative hypothermia, emergency surgery (Coello et al., 2005). Specific risk factors have been identified for cardiac surgery: obesity, pre operative myocardial infarction, chronic obstructive broncho-pulmonary disease, duration of extra corporeal circulation, early post operative bleeding, combined valve and coronary bypass procedures… (Filsoufi et al., 2009).

In the mid 1980s, 30 to 35% of nosocomial (i.e. hospital-acquired) infections were deemed evitable (Haley et al., 1985). A recent analysis of infection control interventional studies estimates that as many as 26 to 54% of SSIs could be avoided by comprehensive implementation of evidence-based prevention strategies, foremost of which are pre operative cutaneous antisepsis, no pre operative shaving of surgical site, timely antibiotic prophylaxis and strict per operative glycemic control (Umsheid et al., 2011).

Still, even when all known preventive measures are implemented, even in low-risk (i.e. clean, non urgent) surgery for low risk patients, SSI can occur. This seemingly irreducible rate of "inevitable" infections raises the problem of how surgical site infections develop. Success of preventive measures based on antisepsis and optimization of patient status compounds the hypothesis that infection results from disequilibrium between host defense mechanisms and microbial infectiveness. In a 10-to-1 inequity between human cells and colonizing microbes, the balance of power in ensured by integrity of skin and mucous membranes, both obviously disrupted by surgery (Wenzel, 2010). However, the intimate mechanisms of infection are not known, starting with the origin of germs involved.

3.1.2 Are surgical site infections related to endogenous or exogenous bacteria?

SSIs are 4.5 times more frequent in patients with nasal carriage of *Staphylococcus aureus* compared to non-carriers. In carrier patients who developed SSI, *S. aureus* isolated from infection site was identical to the one isolated from anterior nares in 84% of cases (Perl et al., 2002). The case for an endogenous bacterial origin is indirectly made by the 60% reduction rate in *S. aureus* SSIs obtained by thorough pre operative decontamination (Bode et al., 2010). Nasal carriage of *S. aureus* was found to be associated with a higher rate of SSIs in orthopedic surgery in a French multicentric study (Berthelot et al., 2010). However, in this study, only 27% of *S. aureus* infections were molecularly linked to an endogenous strain. This can be due to insufficient sensitivity of carriage detection, or to the fact that infections don't necessarily have an endogenous origin.

Indeed, bacteria involved in SSIs can also originate from an exogenous source in the per operative environment. Several studies report cases of cross contamination of surgical sites with bacteria molecularly linked to health care professionals or other patients (Perl et al., 2002). Contamination of surgical instruments and devices increases over time inside the operating room (Dalstrom et al., 2008). Under different air treatment devices, air contamination can vary from 8 to 34% (Knobben et al., 2006). In the air above the surgical site, *S. aureus* and coagulase negative staphylococci can be found, mostly molecularly linked to nasal and pharyngeal carriage by operating team personnel (Edminston et al., 2005).

3.1.3 From contamination to infection

The origin of germs notwithstanding, what induces ordinarily commensal germs to become pathogenic is mostly unexplained. Indeed, in spite of the frequent presence of germs around and in the surgical site (up to 4.1% deep tissue samples found positive in clean orthopedic surgery (Byrne et al., 2007)), only very few infections actually develop.

Antibiotic prophylaxis certainly thwarts the development of most SSIs, at least insofar as bacterial load remains low and antibiotic spectrum and pharmacokinetics are adapted to the germs and tissues involved (Classen et al., 1992). However, the occurrence of SSIs in spite of adequate prophylaxis leads to consider the mechanism of infection from a bacterial point of view. Infections are often the consequences of disequilibrium in the relationships inside the microbiota or between the microbiota and the host. SSIs are typically infections related to microbiota disequilibrium, their major etiologic agents being members of the skin microbiota such as *S. aureus,* coagulase-negative Staphylococci and *Propionibacterium acnes.* The mechanism of selection of these bacteria from the skin community remains unknown and could be highlighted by dynamic survey of the microbial re colonization by a suitable method. Exploring the microbiota disequilibrium and its dynamics should provide insights into the mechanisms involved in SWI infections.

We hereafter present the use of TTGE in culture-independent and culture-based approaches to study the dynamics of bacterial communities involved in SSIs. The study was undertaken in the setting of clean cardiothoracic surgery, because SSI rates remain unacceptably high in coronary artery bypass grafts (CABG), in spite of ongoing efforts to minimize patient- and procedure-related risk factors (Filsoufi et al., 2009).

3.2 Culture-independent approach: Use of TTGE to survey the dynamics of bacterial communities involved in surgical wound colonization and infection

3.2.1 Patients and samples

Forty cotton swabs were collected from 5 patients who underwent CABG surgery at the Service of Thoracic and Cardiovascular Surgery of the Montpellier University Hospital (France). For each patient, swabs were sampled during intervention at the surgical site, superficially and deeply. The first sample was taken on the skin after cutaneous antisepsis just before incision and the second one in sub-cutaneous tissue once incision made. Thirdly, sternum edges were sampled after sawing. The fourth sample was mediastinal tissue after positioning of sternal retractors. At the end of the operation, mediastinum, sternum edges, sub-cutaneous tissue and skin were sampled again.

3.2.2 Cell lysis, extraction of bacterial DNA and PCR amplification

Bacterial genomic DNA was directly extracted from bulk cells present on cotton swabs using an enzymatic method (MasterPure Gram positive DNA purification kit, EPICENTRE Biotechnologies ®) according to the manufacturer's recommendations with modifications as described by (Roudière et al., 2009). This method has been previously described as efficient on a wide range of bacteria including Gram-positive bacteria (Roudière et al., 2009; Jacquot et al., 2011). A fragment about 1465 bp of the 16SrRNA gene was amplified using the primers 27f (5'-GTGCTGCAGAGAGTTTGATCCTGGCTCAG-3') and 1492r (5'-CACGGATCCTACGGGTACCTTGTTACGACTT-3'). The PCRs were carried out in 50 µL of reaction mixture containing 200 nM of each primer (Sigma Genosys), 200 nM each dNTP (Fermentas), 1U of *Taq* polymerase (Promega) in the appropriate reaction buffer, and 1 µL of crude DNA extract as the template. PCR conditions were 30 cycles of 1 min at 94°C, 1 min at 65°C, and 2 min at 72°C.

The 199-bp fragment (from position 338 to position 536, *Escherichia coli* numbering) overlapping the 16S rDNA V2-V3 variable region (Neefs et al., 1993; Sundquist et al., 2007) was amplified using the primers HDA1-GC (primer HDA1 with a fragment rich in GC - the 'GC clamp' - added to the 59 extremity) and HDA2 (Ogier et al., 2002). The reaction mixture (50 µl) consisted of 200 nM of each primer (Sigma Genosys), 200 mM each dNTP (Fermentas), 2.5 U FastStart Taq DNA polymerase (Roche, France) in the appropriate reaction buffer, with 1.8 mM $MgCl_2$. One µL of DNA previously amplified was added to the reaction buffer and the thermal cycling was as follows: 95°C for 2 min; 35 cycles of 95°C for 1 min, 62°C for 30 s, 72°C for 1 min ; and 72°C for 7 min. PCR products were checked by electrophoresis in a 1.5 % agarose gel before TTGE migration.

3.2.3 TTGE migration

TTGE migration was performed in the DCode Universal Mutation Detection System (Bio-Rad Laboratories). Gels were composed of 8 % (w/v) bisacrylamide (37.5: 1), 7 M urea, 40 ml N,N,N9,N9-tetramethylethylenediamine, and 0.1 % (w/v) ammonium persulfate, and were run in 16 Tris/acetate/EDTA buffer at pH 8.3. DNA was loaded on the gel with in-house dye marker (50% sucrose, 0.1% bromophenol blue) using capillary tips. The electrophoresis conditions were 46 V for 16 h with an initial temperature of 63°C and a final temperature of 70°C corresponding to an increase of 0.4°C h^{-1}. In order to obtain thin discrete bands, a pre-migration for 15 min at 63°C and 20 V was done after loading. Pre-migration and migration were performed with additional magnetic shaking in the electrophoresis chamber. Gels were stained for 15 min with 0.5 mg ethidium bromide ml^{-1} in 1x TAE buffer, washed for 45 min in 1x TAE buffer, and photographed under UV illumination.

3.2.4 TTGE band sequencing

Each TTGE band for further analysis was cut out of the gel with a disposable sterile scalpel to avoid contamination between bands. Gel slices were washed twice in molecular biology grade water and incubated overnight at 37 °C in 10 mM Tris buffer (pH 8.5) to allow DNA diffusion. Amplification of a single 16S rRNA gene V3 region copy was performed using 1 µl band eluate and the primers HDA1 without a GC-clamp and HDA2. The PCR was carried out in 50 ml reaction mixture containing 200 nM of each primer, 200 mM each dNTP, 2.5 mM $MgCl_2$ and 2.5 U Taq DNA polymerase (Promega) in the appropriate buffer. PCR conditions were 94 °C for 2 min; 35 cycles of 45 s at 95 °C, 30 s at 62 °C, 1min at 72 °C; and 10 min at 72 °C. PCR products were checked by electrophoresis in a 1.5 % agarose gel and sequenced on an ABI 3730xl sequencer (Cogenics). Each sequencing chromatograph was visually inspected and corrected. The sequences were analyzed by comparison with Genbank (http://www.ncbi.nlm.nih.gov/) and RDPII databases (http://rdp.cme.msu.edu/) using Basic Local Alignment Search Tool (BLAST) and Seqmatch programs, respectively and affiliated to an OTU as recommended (Drancourt et al., 2000; Stackebrandt & Goebel, 1994).

3.2.5 Optimization of DNA amplification in samples with low load of template

In most assays, the bacterial load is low because samples originated from quasi-sterile sites, i.e. the mediastinum during the surgical intervention. When applied on DNA samples extracted from a low number of bacteria (<10^3/tube) (Roudière et al., 2009), a single PCR produced a faint or no signal in agarose gel electrophoresis (data not shown). We showed that the signal was equivalent to the positive control for DNA extracted from skin, a site with a resident microbiota but faint for the samples taken from sub-cutaneous and mediastinal tissues, sites expected to be almost sterile (data not shown). To gain in sensitivity, we performed a nested-PCR approach with pre-amplification of the almost complete 16S rRNA gene, further used as template for a second PCR focused on the V2-V3 hypervariable regions of the gene. As expected, nested-PCR resulted in an improved signal in agarose gel for all samples including samples with low bacterial load (Figure 1A and 1B). However, nested-PCR could induce biases such as amplification of contaminant DNA, preferential amplification of a sub-population of template and heteroduplex formation, among others (Park & Crowley, 2010. Consequently, the enhanced quantity of amplified DNA in agarose gel had to be confirmed in TTGE in order to assess the quality of amplified DNA and qualitatively compare single and nested PCR. Figure 2 shows the TTGE profiles after single- and nested-PCR obtained for 4 samples. We observed that nested-PCR led to an enrichment of the profiles while all the bands detected by single PCR-TTGE were also observed in nested PCR-TTGE. The PCR-TTGE profiles presented in Figure 2 displayed neither obvious heteroduplex bands nor smears due to amplification of miscellaneous contaminant DNA. However, preferential amplification was observed particularly for sample AD6 In the corresponding TTGE profile, the faint bands seen in single PCR-TTGE profile were very faint in the nested PCR-TTGE profile while original and intense bands appeared in the latter profile. The reproducibility is also satisfactory (data not shown).

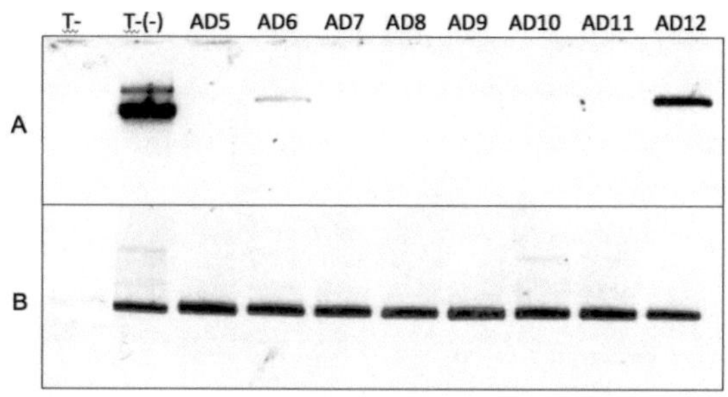

Fig. 1. Migration in agarose gel after single (A) and nested PCR (B) for samples from the patient AD. AD5, skin before incision; AD6, sub-cutaneous tissue after incision; AD7, sternum edges; AD8, mediastinum at the beginning of the operation; AD9, mediastinum at the end of intervention; AD10, edges of sternum before closing; AD11, sub-cutaneous tissue before suture; AD12, skin at the end of the procedure.

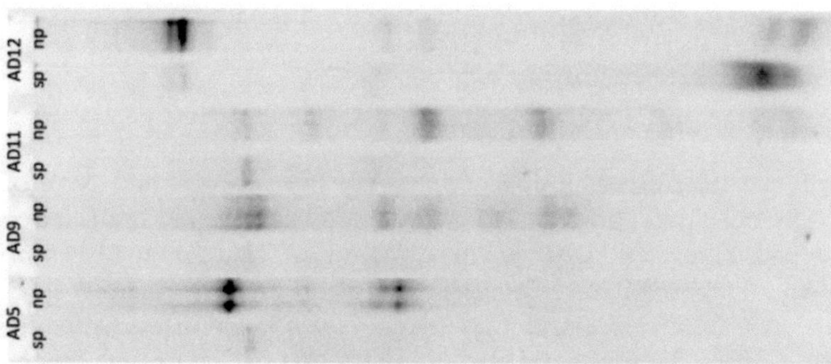

Fig. 2. TTGE migration after single PCR (sp) and nested PCR (np) on samples from the patient AD. AD6 sub-cutaneous tissue after incision; AD9, mediastinum before suture; AD11, sub-cutaneous tissue before suture; AD12, skin at the end of the surgery.

3.2.6 PCR-TTGE profiles and diversity index

About fifteen samples from the sternal region were recovered during hospital stay of each patient undergoing programmed CABG surgery. TTGE profiles were obtained for each sample after nested-PCR and showed clear bands easily separated from each other (data not shown). This result confirmed that 16S rRNA gene PCR-TTGE was a suitable method to survey the bacterial community from skin and wound tissues in cardiac surgery. By numbering the bands in each profile, crude diversities index (DI) were determined. Crude DI varied from 2 to 12 (mean value = 6). The bacterial communities presenting the lower mean value of crude DI (DI=4.2) were sampled on the skin just after preoperative antisepsis. Subcutaneous tissues sampled just before the closure of the wound at the end of the operation displayed the higher crude DI (DI=8.2). The dynamics of evolution of mean crude DI over the surgical operation showed that the bacterial diversity globally increased from the start to the end of the surgery, from 4.2 to 7.6. The increase of crude DI was also observed at each site of sampling: 4.2 to 7.6 for skin, 7.4 to 8.2 for subcutaneous tissues, 4.4 to 5.2 for sternum banks and 5.4 to 6.8 for mediastin. Figure 3 showed a representative DI dynamics for a patient.

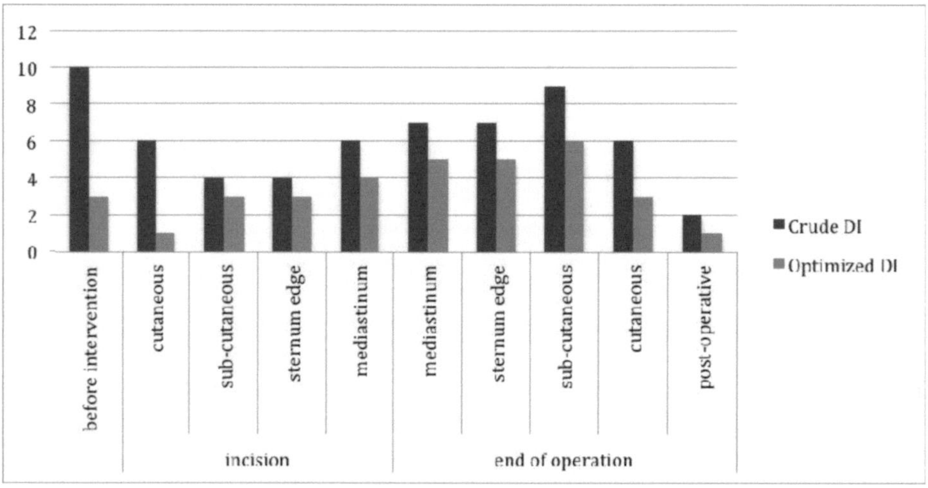

Fig. 3. Representative evolution of crude and optimized DI during the course of the surgical operation of the patient AD.

3.2.7 Operational Taxonomic Units diversity

Each band with specific migration distance was cut from the gel and the eluted DNA was reamplified, sequenced and compared to genetic databases for OTU affiliation. For the five patients presented here, we detected 16 different OTUs (Table 1). Four major OTU, *Staphylococcus* sp., *Flavobacteriaceae*, *Comamonadaceae* and *Propionibacterium* sp., were detected in all patients. *Acinetobacter* sp. and *Corynebacterium* sp. are also frequently detected. In contrast, OTUs such as *Bosea* sp. and *Bacillus* sp. were specific of one single patient. Most major OTUs were present in all patient at all step of the surgery operation and in all sampling sites. *Flavobacteriaceae* and *Comamonadaceae* were atypical Gram-negative bacilli recently described in the skin metagenome (Grice et al., 2009). These atypical Gram-negative bacilli were particularly represented in subcutaneous and in deep tissues (Figure 4). Band identification by sequencing allowed refining the crude DI by the elimination of multiple bands corresponding to a single OTU, as observed for bacteria that displayed 16S rRNA genes heterogeneity. In some cases, the DI dramatically decreased as shown in Figure 3.

3.3 Culture-dependent approach: Use of PCR-TTGE to assess the impact of antimicrobial prophylaxis on the whole skin microbiome

3.3.1 Methods

Sternal skin microbiome was sampled with cotton swabs, before hospitalization, in four patients requiring CABG surgery. This sample reflects the normal resident skin microbiome of the patients. Sternal skin samples were streaked on 5% blood and nutrient agar plates supplemented with different concentrations of cefamandole (0, 0.125, 0.25, 0.5, 2, 4, 8 and 32 µg/mL) and incubated for 24h at 37°C in anaerobiosis and aerobiosis. The resulting colonies

OTU	patient				
	AB	AD	AE	AF	AG
Staphylococcus sp.	X	X	X	X	X
Clostridiaceae				X	
Enterococcus sp.		X			
Acinetobacter sp.		X		X	X
Lactococcus sp.	X				
Streptococcus sp.				X	
Flavobacteriaceae	X	X	X	X	X
Comamonadaceae	X	X	X	X	X
Bosea sp.	X				
Corynebacterium sp.	X	X		X	X
Anaerococcus sp.				X	
Bacillus sp.			X		
Micrococcaceae	X	X	X		
Rhizobiaceae		X			X
Parococcus sp.	X	X		X	X
Propionobacterium sp.	X	X	X	X	X

Table 1. Repartition of the OTUs detected in the skin microbiota of five patients.

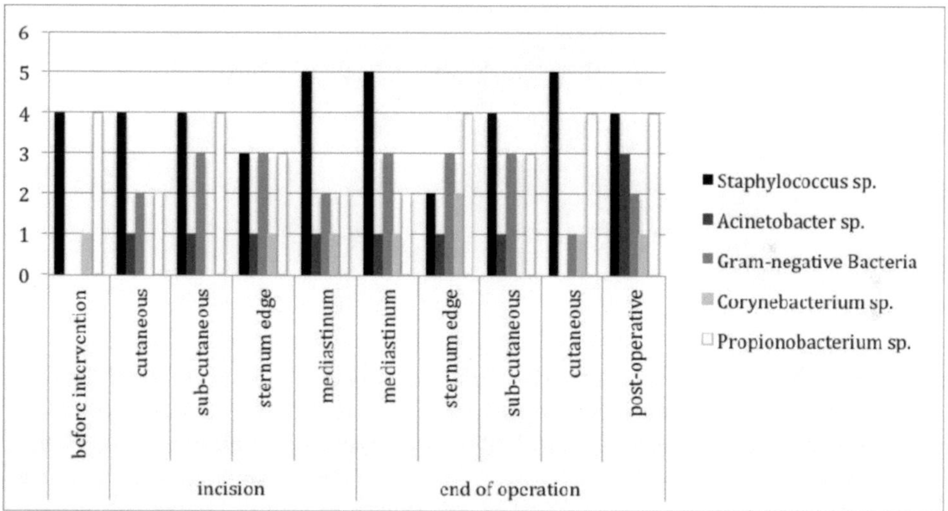

Fig. 4. Number of patients carrying the major OTUs according to the step of the surgery procedure.

were counted and the whole bacterial cultures were then harvested for DNA extraction. DNA extraction, single PCR HDA, TTGE migration and band affiliation to an OTU were performed as described in previous section.

3.3.2 Results: Impact of antimicrobial prophylaxis on skin microbiota

Most samples displayed cultivable microbiota susceptible to cefamandole, i.e. no culture was observed on agar plates with 8mg/L, which was the cefamandole critical concentration used to define susceptibility versus resistance. At lower concentrations (from 0.125 mg/L to 4 mg/L), we observed a decrease of colonies in number and in diversity, comparatively to the control plate. The mean colony counts went from 80 CFU/plate without antibiotics to 30 CFU/plate with 0.125 mg/L cefamandole and to 6 CFU/plate with 1mg/L cefamandole. All the colonies were bulked and analyzed by 16S rRNA gene PCR-TTGE. Profiles comparison allowed to determine the concentration of antimicrobial agents that inhibited 25%, 50%, 75% and 90% of the microbial diversity and to identify the resistant bacteria to the species level. These data will be related to dynamics of wound re-colonization after surgery.

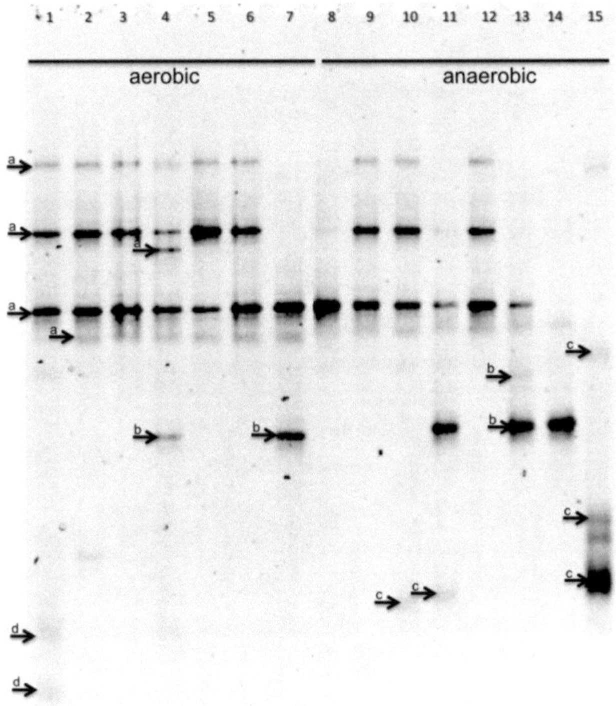

Fig. 5. TTGE gel illustrating the culture-dependent approach on skin sample of the patient CB3. The gel shows differences in banding patterns according to the concentration of cefamandole and the culture atmosphere. Lanes 1 and 8, culture without cefamandole; Lanes 2 and 9, cefamandole 0.125mg/L; Lanes 3 and 10, cefamandole 0.25mg/L; Lanes 4 and 11, cefamandole 0.5mg/L; Lanes 5 and 12, cefamandole 1 mg/L; Lanes 6 and 13, cefamandole 2mg/L; Lanes 7 and 14, cefamandole 4mg/L; Lane 15, cefamandole 8mg/L. Bands corresponding to OTUs are shown with arrow: a for *Staphylococcus* sp., b for *Bacillus* sp., c for *Corynebacterium* sp. and d for *Propionibacterium* sp.

The TTGE profiles from colonies harvested on control plates without antibiotics showed the predominant cultivable population. For the four patients studied herein, the corresponding bacteria belonged to the genera *Staphylococcus* (4 to 6 species) and *Propionibacterium*. For two patients, the cultivable skin microbiota was susceptible to cefamandole because no colonies were observed on plates with more than 8 mg/L, which was the cefamandole critical concentration used to define susceptibility versus resistance. For the two other patients, colonies were observed for 8 mg/L plates incubated in aerobiosis and even for 32 mg/L plates incubated in anaerobiosis. Resistant bacteria belonged to the genera *Bacillus* and *Corynebacterium*. For lower concentrations, which could correspond to the tissue concentrations in the surgical wound (0.125mg/L to 4mg/L), TTGE profiles showed a modification of the microbiota with loss of bands affiliated to the genus *Staphylococcus* and appearance of other bands affiliated to the genera *Corynebacterium* and *Micrococcus* or to *Bacillus cereus* group. These bands corresponding to OTUs with a decreased susceptibility to cefamandole revealed minor OTUs that were undetected in the total cultivable microbiota, i.e. on plates without antibiotics. For instance in the patient CB (Figure 5), a band corresponding to *Bacillus* sp. was observed from 0,5 à 4mg/L cefamandole and a band corresponding to *Corynebacterium* sp. was observed in at least 8mg/L cefamandole.

4. Conclusion: Why should 'microbiomology' be on the benchtop of the medical microbiologist?

In this experiment we applied PCR TTGE to the study of microbiome dynamics in skin and tissues during a cardiac surgery procedure. We present the results obtained for 5 patients in a culture-independent approach and in 4 patients in a culture-dependent approach. Our results enable us to describe the bacteria involved in surgical site colonization during the procedure. The bacteria detected belonged to OTUs previously described in the skin microbiome (Grice et al., 2009). Gram-positive bacteria belonged to *Bacilli* and *Clostridia* in the phylum *Firmicutes*, and to *Corynebacterium*, *Propionibacterium* and *Micrococcaceae* in the phylum *Actinobacteria*. Gram-negative bacteria mainly belonged to *Proteobacteria* but also to *Flavobacteria*. This result confirmed that the nested-PCR TTGE approach gave a good representation of the skin microbiome even after antisepsis when the bacterial load is very low.

We have showed for the first time that the skin microbiota contaminated the surgical wound at all steps of the surgery in quantities that can be detected by molecular methods. The PCR-TTGE appeared as a suitable method to follow-up the microbiology of a clean wound such as a surgical wound. The approach could also give insights about the activity of antibiotics, e.g. agents used for antibioprophylaxis, at the community level and not only on isolated strains as performed currently.

In order to find relationships between the dynamics of microbiome and the clinical evolution of patients, a cohort of 120 patients will be studied for their surgical wound microbiota and their clinical outcome. Such clinical studies are necessary to understand to physio-pathological mechanism of wound re-colonization and SSI. PCR-TTGE appeared as easy and cost-effective enough to provide microbiological arguments at the community level in large clinical studies. Other very efficient methods such as pyrosequencing are until now not convenient for clinical studies on large cohorts of patients.

More generally the study presented here underlines that is now time for microbiome studies to exit from research labs and to take place on the benchtop of the medical microbiologist.

Current medical microbiology still rests on the shoulders of Leeuwenhoek, Koch and Pasteur. Although every microbiologist knows that a bacterium does not live alone but in complex communities, routine practice still consists in watching a small subset of isolated microbes under microscope and on Petri dishes. It is now clear that the human microbiome has profound effects on health and disease. Ambitious project are in progress such as the international Human Microbiome Consortium and the Human Microbiome Project. High-throughput sequencing and metagenomics allow us to explore new fields of the relationships between human and bacteria, considering these relationships at the community level. Facing the huge volume of data generated by metagenomics, microbiologists and even more so physicians are perplexed. For instance, Grice et al. admit that "...hairy, moist underarms lie a short distance from smooth dry forearms, but these two niches are likely as ecologically dissimilar as rainforests are to deserts" (Grice et al., 2009). A conceptual revolution is occurring while in medical microbiology practice, skin is still shown as a sheet covered by *S. epidermidis*.

How can we deal with this new world we discover within and around ourselves with our PCR and sequencers? To change our understanding of health and disease, it is now time to develop ways of seeing patterns and interpret them among the staggering biodiversity of microbes. Extending microbial ecology to healthy and diseased microbiota as previously done for mouth (Bik et al., 2010), stomach (Bik et al., 2006), gut (Yang et al., 2009) and skin (Grice et al., 2009) could elucidate the physio-pathology of diseases. For instance, description of the baseline skin microbiome is a step toward testing the therapeutic potential of manipulating the microbiome in skin disorders (Grice et al., 2009). Studies on psoriasis (Gao et al., 2008) describe selective microbial shifts associated with diseases and suggest that therapies might require not only inhibiting the growth of pathogenic bacteria, but also promoting the growth of mutualistic bacteria. Moreover, antibiotic exposure and hygienic practices such as antisepsis modify the skin microbiome. Methods that help understand naturally occurring mutualistic microbial communities will provide insights into the conditions that favor the emergence of antibiotic-resistant organisms.

It is now urgent to confront 'microbiomology' data to clinical data in order to define new indicators for re-evaluating the risk of disease and for adapting therapeutics. Before the generalization of high-throughput metagenomics on the microbiologist's benchtop, alternative methods such as PCR-TTGE or other methods based on denaturant electrophoresis should be considered. Although they are less exhaustive and powerful than pyrosequencing, these approaches are easy to handle in all laboratories and are cost-effective. They can be used to study large cohorts of patients, thus deriving clinical benefits from the conceptual revolution brought by contemporary researches on human microbial ecology.

5. References

Acinas Sylvia G, Marcelino LA, Klepac-Ceraj V & Polz MF. (2004). Divergence and Redundancy of 16S rRNA Sequences in Genomes with Multiple rrn Operons. *Journal of Bacteriology*, Vol.186, No.9 (May 2004), pp 2629–2635, ISSN 1098-5530

Astagneau Pascal, L'Hériteau F, Daniel F, Parneix P, Venier AG, Malavaud S, Jarno P, Lejeune B, Savey A, Metzger HM, Bernet C, Fabry J, Rabaud C, Tronel H, Thiolet JM, Coignard B, for the ISO-RAISIN Steering Group. (2009). Reducing surgical site infection incidence through a network: results from the French ISO-RAISIN surveillance system. *Journal of Hospital Infection*, Vol.72, No.2 (June 2009), pp 127-134, ISSN 0195-6701

Bäckhed Fredrik, Ley RE, Sonnenburg JL, Peterson DA, Gordon JI. (2005). Host-bacterial mutualism in the human intestine. *Science*, Vol.307, No.5717 (March 2005), pp. 1915–1920, ISSN 1095-9203

Berthelot Philippe, Grattard F, Cazorla C, Passot JP, Fayard JP, Meley R, Bejuy J, Farizon F, Pozzetto B & Lucht F. (2010). Is nasal carriage of Staphylococcus aureus the main acquisition pathway for surgical-site infection in orthopaedic surgery? *European Journal of Clinical Microbiology and Infectious Diseases*, Vol.29, No.4 (January 2010), pp 373–382, ISSN 1435-4373

Bik Elisabeth M, Eckburg PB, Gill SR, Nelson KE, Purdom EA, Francois F, Perez-Perez G, Blaser MJ & Relman DA. (2006). Molecular analysis of the bacterial microbiota in the human stomach. *Proceedings of the National Academy of Sciences of the United States of America*, Vol.103, No.3 (January 2006), pp 732–737, ISSN 1091-6490

Bik Elisabeth M, Long CD, Armitage GC, Loomer P, Emerson J, Mongodin EF, Nelson KE, Gill SR, Fraser-Liggett CM & Relman DA. (2010). Bacterial diversity in the oral cavity of 10 healthy individuals. *The ISME Journal*, Vol.4, No.8 (August 2010), pp 962–974, ISSN 1751-7370

Bode Lonneke GM, Kluytmans JAJW, Wertheim HFL, Bogaers C, Vandenbroucke-Grauls CMJE, Roosendaal R, Troelstra A, Box ATA, Voss A, van der Tweel I, van Belkum A, Verbrugh HA & Vos MC. (2010). Preventing surgical-site infections in nasal carriers of Staphylococcus aureus. (2010). *The New England Journal of Medicine*, Vol.362, No.1 (January 2010), pp 9-17, ISSN 1533-4406

Byrne Ann Maria, Morris S, McCarthy T, Quinlan W & O'Byrne JM. (2007). Outcome following deep wound contamination in cemented arthroplasty. *International Orthopaedics*, Vol.31, No.1 (February 2007), pp 27-31, ISSN 1432-5195

Cani Patrice D, Amar J, Iglesias MA, Poggi M, Knauf C, Bastelica D, Neyrinck AM, Fava F, Tuohy KM, Chabo C, Waget A, Delmée E, Cousin B, Sulpice T, Chamontin B, Ferrières J, Tanti J-F, Gibson GR, Casteilla L, Delzenne NM, Alessi MC & Burcelin R. (2007). Metabolic endotoxemia initiates obesity and insulin resistance. *Diabetes*, Vol.56, No.7 (July 2007), pp 1761–1772, ISSN 1939-327X

Case Rebecca J, Boucher Y, Dahllöf I, Holmström C, Doolittle WF & Kjelleberc S. (2007). Use of 16S rRNA and rpoB genes as molecular markers for microbial ecology studies. *Applied and Environmental Microbiology*, Vol.73, No.1 (January 2007), pp 278-288, ISSN 1098-5336

Classen David C, Evans RS, Pestotnik SL, Horn SD, Menlove RL & Burke JP. (1992). The timing of prophylactic administration of antibiotics and the risk of surgical-wound infection. *The New Englang Journal of Medecine*, Vol.326, No.5 (January 1992), pp 281-286, ISSN 1533-4406

Coello Rosa, Charlett A, Wilson J, Ward V, Pearson A & Borriello P. (2005). Adverse impact of surgical site infections in English hospitals. *Journal of Hospital Infection*, Vol.60, No.2 (June 2005), pp 93-103, ISSN 0195-6701

Daïen Claire I, Cohen JD, Makinson A, Battistella P, Jumas-Bilak E, Jorgensen C, Reynes J & Raoult D. (2010). Whipple's endocarditis as a complication of tumour necrosis factor-alpha antagonist treatment in a man with ankylosing spondylitis. *Rheumatology*, Vol.49, No.8 (February 2010), pp 1600-1602, ISSN 1462-0332

Dalstrom Daivd J, Venkatarayappa I, Manternach AL, Palcic MS, Heyse BA, & Prayson MJ. (2008). Time-dependent contamination of opened sterile operating-room trays. *The Journal of Bone and Joint Surgery*, Vol.90, No.5 (May 2008), pp 1022-1025, ISSN 1535-1386

Dekio Itaru, Hayashi H, Sakamoto M, Kitahara M, Nishikawa T, Suematsu M & Benno M. (2005). Detection of potentially novel bacterial components of the human skin microbiota using culture- independent molecular profiling. *Journal of Medical Microbiology*, Vol.54, No.12 (December 2005), pp 1231–1238, ISSN 0022-2615

Drancourt Michel, Bollet C, Carlioz A, Martelin R, Gayral JP & Raoult D. (2000). 16S ribosomal DNA sequence analysis of a large collection of environmental and clinical unidentifiable bacterial isolates. *Journal of Clinical Microbiology*, Vol.38, No.10 (October 2000), pp 3623–4630, ISSN 0095-1137

Edminston Charles E, Seabrook GR, Cambria RA, Brown KR, Lewis BD, Sommers JR, Krepel CJ, Wilson PJ, Sinski S & Towne JB. (2005). Molecular epidemiology of microbial contamination in the operating room environment: is there a risk for infection? *Surgery*, Vol.138, No.4 (October 2005), pp 573-582, ISSN 1532-950X

Ege Markus J, Mayer M, Normand AC, Genuneit J, Cookson WO, Braun-Fahrländer C, Heederik D, Piarroux R, von Mutius E & GABRIELA Transregio 22 Study Group. (2011). Exposure to environmental microorganisms and childhood asthma. *The New England Journal of Medicine*, Vol.364, No.8 (February 2011), pp 701-709, ISSN 1533-4406

Filsoufi Farzan, Castillo JG, Rahmanian PB, Broumand SR, Silvay G, Carpentier A & Adams DH. (2009). Epidemiology of deep sternal wound infection in cardiac surgery. *Journal of Cardiothoracic and Vascular Anesthesia*, Vol.23, No.4 (August 2009), pp 488-494, ISSN 1053-0770

Gao Zhan, Tseng CH, Strober BE, Pei Z & Blaser MJ. (2008). Substantial alterations of the cutaneous bacterial biota in psoriatic lesions. *Public Library of Science one*, Vol.3, No.7 (July 2008), e2719, ISSN 1932-6203

Grice Elisabeth A, Kong HH, Conlan S, Deming CB, Davis J, Young AC; NISC Comparative Sequencing Program, Bouffard GG, Blakesley RW, Murray PR, Green ED, Turner ML, Segre JA. (2009). Topographical and temporal diversity of the human skin microbiome. *Science*, Vol.324, No 5931 (May 2009), pp 1190-1192, ISSN 1095-9203

Haley Robert W, Culver DH, White JW, Morgan WM, Emori TG Munn VP & Hooton TM. (1985). The efficacy of infection surveillance and control programs in preventing nosocomial infections in US hospitals. *American Journal of Epidemiology*, Vol.121, No.2 (February 1985), pp 182-205 ISSN 1476-6256

Holmes Elaine, Loo RL, Stamler J, Bictash M, Yap IK, Chan Q, Ebbels T, De Iorio M, Brown IJ, Veselkov KA, Daviglus ML, Kesteloot H, Ueshima H, Zhao L, Nicholson JK & Elliott P. (2008). Human metabolic phenotype diversity and its association with diet and blood pressure. *Nature*, Vol.453, No.7193 (May 2008), pp 396–400, ISSN 0028-0836

Jacquot Aurelien, Neveu D, Aujoulat F, Mercier G, Marchandin H, Jumas-Bilak E & Picaud JC. (2011). Dynamics and clinical evolution of bacterial gut microflora in extremely premature patients. *The Journal of Pediatrics*, Vol.158, No.3 (March 2011), pp 390-396, ISSN 1432-1076

Jany Jean-Luc & Barbier G. (2008). Culture-independent methods for identifying microbial communities in cheese. *Food Microbiology*, Vol.25, No.7 (October 2008), pp 839-848, ISSN 0740-0020

Kanagawa Takahiro. (2003). Bias and artifacts in multitemplate polymerase chain reaction (PCR). *Journal of Bioscience and Bioengineering,* Vol.96, No.4 (December 2003), pp 317–323, ISSN 1347-4421

Kitts Christopher L. (2001). Terminal restriction fragment patterns: a tool for comparing microbial communities and assessing community dynamics. *Current Issues in Intestinal Microbiology,* Vol.2, No.1 (March 2001), pp 17-25, ISSN 1466-531X

Klevens Monia R, Edwards JR, Richards CL, Horan TC, Gaynes RP, Pollock DA & Cardo DM. (2007). Estimating health care-associated infections and deaths in U.S. hospitals, 2002. *Public Health Reports,* Vol.122, No.2 (March-April 2007), pp 160-166, ISSN 0094-6214

Knobben Bas AS, van Horn JR, van der Mei HC & Busscher HJ. (2006). Evaluation of measures to decrease intra-operative bacterial contamination in orthopaedic implant surgery. *Journal of Hospital Infection,* Vol.62, No.2 (February 2006), pp 174-180, ISSN 1532-2939

Lampe Johanna W. (2008). The Human Microbiome Project: getting to the guts of the matter in cancer epidemiology. *Cancer Epidemiology, Biomarkers & Prevention,* Vol.17, No.10 (October 2008), pp 2523-2524, ISSN 1538-7755

Le Bourhis Anne-Gaëlle, Doré J, Carlier J, Chamba J, Popoff M & Tholozan J. (2007). Contribution of C. beijerinckii and C. sporogenes in association to C. tyrobutyricum to the butyric fermentation in Emmental type cheese. *International Journal of Food Microbiology,* Vol.113, No.2 (2007), pp 154–163, ISSN 1879-3460

Lederberg Joshua & McCray AT. (2001). 'Ome Sweet 'Omics - a genealogical treasury of words. *The Scientist,* Vol.15, No.7 (April 2001), pp 8-8, ISSN 1547-0806

Liao Daiqing. (2000). Gene conversion drives within genic sequences: concerted evolution of ribosomal RNA genes in bacteria and archaea. *Journal of Molecular Evolution,* Vol.51, No.4 (June 2000), pp 305-317, ISSN 1432-1432

de Lissovoy Gregory, Fraeman K, Hutchins V, Murphy D, Song D & Vaughn BB. (2009). Surgical site infection: incidence and impact on hospital utilization and treatment costs. *American Journal of Infection Control,* Vol.37, No.5, (June 2009), pp 387-397, ISSN 1527-3296

Marchandin Hélène, Battistella P, Calvet B, Darbas H, Frapier JM, Jean-Pierre H, Parer S, Jumas-Bilak E, Van de Perre P & Godreuil S. (2009). Pacemaker surgical site infection caused by Mycobacterium goodii. *Journal of Medical Microbiology,* Vol.58, No.4 (April 2009), pp 517-520, ISSN 1473-5644

Michon Anne-Laure, Aujoulat F, Roudière L, Soulier O, Zorgniotti I, Jumas-Bilak E & Marchandin H. (2010). Intragenomic and intraspecific heterogeneity in rrs may surpass interspecific variability in a natural population of Veillonella. *Microbiology,* Vol.156, No.7 (July 2010), pp 2080–2091, ISSN 1465-2080

Michon Anne-Laure, Jumas-Bilak E, Imbert A, Aleyrangues L, Counil F, Chiron R & Marchandin H. (2011). Intragenomic and intraspecific heterogeneity of the 16S rRNA gene in seven bacterial species from the respiratory tract of cystic fibrosis patients assessed by PCR-Temporal Temperature Gel Electrophoresis. *Pathologie Biologie,* Vol. 2956, (May 2011), pp 1-6, ISSN 1768-3114

Muyzer Gerard, De Waal E.C & Uittierlinden A.G. (1993). Profiling of Complex Microbial Populations by Denaturing Gradient Gel Electrophoresis Analysis of Polymerase

Chain Reaction-Amplified Genes Coding for 16S rRNA. *Applied And Environmental Microbiology*. Vol.59, No.3 (March 1993), pp 695-700, ISSN 0099-2240

Neefs Jean-Marc, Van de Peer Y, De Rijk P, Chapelle S & De Wachter R. (1993). Compilation of small ribosomal subunit RNA structures. *Nucleic Acids Research*, Vol.21, No.13 (July 1993), pp 3025–3049, ISSN 1362-4962

NIH HMP Working Group, Peterson J, Garges S, Giovanni M, McInnes P, Wang L, Schloss JA, Bonazzi V, McEwen JE, Wetterstrand KA, Deal C, Baker CC, Di FrancescoV, Howcroft TK, Karp RW, Lunsford RD, Wellington CR, Belachew T, Wright M, Giblin C, David H, Mills M, Salomon R, Mullins C, Akolkar B, Begg L, Davis C, Grandison L, Humble M, Khalsa J, Little AR, Peavy H, Pontzer C, Portnoy M, Sayre MH, Starke-Reed P, Zakhari S, Read J, Watson B & Guyer M. (2009). The NIH Human Microbiome Project. *Genome Research*, Vol.19, No.12 (December 2009), pp 2317-2323, ISSN 1424-859X

Nossa Carlos W, Oberdorf WE, Yang L, Aas JA, Paster BJ, Desantis TZ, Brodie EL, Malamud D, Poles MA & Pei Z. Design of 16S rRNA gene primers for 454 pyrosequencing of the human foregut microbiome. *World Journal og Gastroenterology*, Vol.16, No.33 (September 2010), pp 4135-4144, ISSN 1007-9327

Ogier Jean-Claude, Son O, Gruss A, Tailliez P & Delacroix-Buchet A. (2002). Identification of the bacterial microflora in dairy products by temporal temperature gradient gel electrophoresis. *Applied and Environmental Microbiology*, Vol.68, No. 8 (August 2002), pp 3691-701, ISSN 1098-5336

Park Joong-Wook & Crowley DE. (2010). Nested PCR bias: a case study of Pseudomonas spp. in soil microcosms. *Journal of environmental monitoring*, Vol.12, No.4 (April 2010), pp 985-988, ISSN 1464-0333

Penny Christian, Nadalig T, Alioua M, Gruffaz C, Vuilleumier S & Bringel F. (2010). Coupling of denaturing high-performance liquid chromatography and terminal restriction fragment length polymorphism with precise fragment sizing for microbial community profiling and characterization. *Applied and Environmental Microbiology*, Vol.76, No.3 (February 2010), pp 648-651, ISSN 1098-5336

Perl Trish M, Cullen JJ, Wenzel RP, Zimmerman B, Pfaller MA, Sheppard D, Twombley J, French PP, Herwaldt LA, & The Mupirocin and the Risk of Staphylococcus aureus Study Tean. (2002). Intranasal mupirocin to prevent postoperative Staphylococcus aureus infections. *The New England Journal of Medicine*, Vol.346, No.24 (June 2002), pp 1871-1877, ISSN 1533-4406

Pflughoeft Kathryn J & Versalovic J. (2011). Human Microbiome in Health and Disease. *Annual Review of Pathology: Mechanisms of Disease*, Vol.7, No.1 (September 2011) ISSN 1553-4014 [Epub ahead of print]

Proal Amy D, Albert PJ & Marshall T. (2009). Autoimmune disease in the era of the metagenome. *Autoimmunity Reviews*, Vol.8, No.8 (July 2009), pp 677-681, ISSN 1568-9972

Riesenfeld Christian S, Schloss PD & Handelsman J. (2004). METAGENOMICS: Genomic Analysis of Microbial Communities. *Annual Review of Genetics*, Vol.38 (July 2004), pp 525–552, ISSN 0066-4197

Roudière Laurent, Lorto S, Tallagrand E, Marchandin H, Jeannot JL & Jumas-Bilak E. (2007). Molecular fingerprint of bacterial communities and 16S rDNA intra-species heterogeneity: a pitfall that should be considered. *Pathologie Biologie*, Vol.55, No.8-9 (November 2007), pp 434-440, ISSN 1768-3114

Roudière Laurent, Jacquot A, Marchandin H, Aujoulat F, Devine R, Zorgniotti I, Jean-Pierre H, Picaud JC & Jumas-Bilak E. (2009). Optimized PCR-Temporal Temperature Gel Electrophoresis compared to cultivation to assess diversity of gut microbiota in neonates. *Journal of Microbiological Methods*, Vol.79, No.2 (November 2009), pp 156-165, ISSN 0167-7012

Stackebrandt Erko & Goebel BM. (1994). Taxonomic note: a place for DNA-DNA reassociation and 16S rRNA sequence analysis in the present species definition in bacteriology. *International Journal of Systematic Bacteriolgy*, Vol.44, No.4 (October 1994), pp 846–849, ISSN 0020-7713

Sundquist Andreas, Bigdeli S, Jalili R, Druzin ML, Waller S, Pullen KM, El-Sayed YY, Taslimi MM, Batzoglou S & Ronaghi M. (2007). Bacterial flora-typing with targeted, chip-based Pyrosequencing. *BMC Microbiology*, Vol.7, No.108 (November 2007) pp 1-11, ISSN 1471-2180

Temmerman Robin, Masco L, Vanhoutte T, Huys G & Swings J. Development and Validation of a Nested-PCR-Denaturing Gradient Gel Electrophoresis Method for Taxonomic Characterization of Bifidobacterial Communities. *Applied and Environmental Microbiology*, Vol.69, No.11 (November 2003), pp 6380–6385, ISSN 1098-5336

Teyssier Corine, Marchandin H Siméon De Buochberg, Ramuz M & Jumas-Bilak E. (2003). Atypical 16S rRNA gene copies in Ochrobactrum intermedium strains reveal a large genomic rearrangement by recombination between rrn copies. *Journal of Bacteriology*, Vol.185, No.9 (May 2003), pp 2901–2909, ISSN 1098-5530

Thompson Janelle R, Randa MA, Marcelino LA, Tomita-Mitchell A, Lim E & Polz MF. (2004). Diversity and dynamics of a north atlantic coastal Vibrio community. *Applied and Environmental Microbiology*, Vol70, No.7 (July 2004), pp 4103-4110, ISSN 1098-5336

Turnbaugh Peter J, Ley RE, Hamady M, Fraser-Liggett CM, Knight R & Gordon JI. (2007). The human microbiome project. *Nature*, Vol.449, No.7164 (October 2007), pp 804–810, ISSN 1476-4687

Turnbaugh Peter J, Hamady M, Yatsunenko T, Cantarel BL, Duncan A, Ley RE, Sogin ML, Jones WJ, Roe BA, Affourtit JP, Egholm M, Henrissat B, Heath AC, Knight R & Gordon JI. (2009). A core gut microbiome in obese and lean twins. *Nature*, Vol.457, No.7228 (January 2009), pp 480–484, ISSN 1476-4687

Tyson Gene W, Chapman J, Hugenholtz P, Allen EE, Ram RJ, Richardson PM, Solovyev VV, Rubin EM, Rokhsar DS & Banfield JF. (2004). Community structure and metabolism through reconstruction of microbial genomes from the environment. *Nature*, Vol.428, No.6978 (March 2004) pp 37–43, ISSN 1476-4687

Umsheid Craig A, Mitchell MD, Doshi JA, Agarwal R, Williams K & Brennan PJ. (2011). Estimating the proportion of healthcare-associated infections that are reasonably preventable and the related mortality and costs. *Infection Control and Hospital Epidemiology*, Vol.32, No.2 (February 2011), pp 101-114, ISSN 1559-6834

Van de Peer Yves, Chapelle S & De Wachter R. (1996). A quantitative map of nucleotide substitution rates in bacterial rRNA. *Nucleic Acids Research*, Vol.24, No.17 (September 1996), pp 3381–3391, ISSN 1362-4962

Vanhove Audrey, Jumas-Bilak E, Aujoulat F, Roger F, Vallaeys T, Marchandin H & Monfort P. An original approach for studying the bacterial resistance to antimicrobial agents at the community level: application to natural brackish lagoon water. FEMS, Geneva, Switzerland, June 2011

Wei Chaochun & Brent MR. (2006). Using ESTs to Improve the Accuracy of de novo Gene Prediction. *BMC Bioinformatics*, Vol.7, No.327 (July 2006), pp 1-10, ISSN 1471-2105

Wenzel Richard P. (2010). Minimizing surgical-site infections. The New England Journal of Medicine, Vol.362, No.1 (January 2010), pp 75-77, ISSN 1533-4406

von Wintzingerode Friedrich, Gobel UB & Stackebrandt E. (1997). Determination of microbial diversity in environmental samples: pitfalls of PCR-based rRNA analysis. *FEMS Microbiology Reviews*, Vol.21, No.3 (November 1997), pp 213-229, ISSN 1574-6976

Woese Carl R. (1987). Bacterial evolution. *Microbiological Reviews*, Vol.51, No.2 (June 1987), pp 221-271, ISSN 1740-1534

Yang Xing, Xie L, Li Y & Wei C. (2009). More than 9,000,000 unique genes in human gut bacterial community: estimating gene numbers inside a human body. *Public Library of Science One*, Vol.4, No.6 (June 2009), e6074, ISSN 1932-6203

Yoshino Kenji, Nishigaki K & Husimi Y. (1991). Temperature sweep gel electrophoresis: a simple method to detect point mutations. *Nucleic Acids Research*, Vol.19, No.11 (June 1991), pp 3153, ISSN 1362-4962

Zoetendal Erwin G, Akkermans AD & de Vos WM. (1998). Temperature gradient gel electrophoresis analysis of 16S rRNA from human faecal samples reveals stable and host-specific communities of active bacteria. *Applied and Environmental Microbiology*, Vol.64, No.10 (October 1998), pp 3854-3859, ISSN 0099-2240

Part 6

Two-Dimensional Gel Electrophoresis (2-DE)

17

Two-Dimensional Gel Electrophoresis Reveals Differential Protein Expression Between Individual *Daphnia*

Darren J. Bauer, Gary B. Smejkal and W. Kelley Thomas
Hubbard Center for Genome Studies, University of New Hampshire
USA

1. Introduction

Analysis of individual genetic variation is paramount to understanding how organisms and communities respond to changes in the environment and requires a model system with well-developed molecular resources and a solid foundation of ecological knowledge. Traditional genetic model systems (*E. coli*, yeast, fly, worm, and mouse) have served as workhorses in elucidating virtually all of the knowledge in modern molecular biology. While these systems were chosen for their robustness in laboratory studies, virtually nothing is known about their life histories in their native environment. By contrast, new-model systems, which have typically been studied in depth, from an ecological perspective are severely limited in regards to their molecular resources.

The model organism *Daphnia* has been utilized as an ecological model for centuries, and now with the sequencing of the genome complete and the development of the associated molecular resources, it is poised as one of the few model systems with the necessary molecular and ecological tools to answer the questions of response to the environment (Colborne et al., 2011). Long recognized as a model for ecological research, the freshwater crustacean *Daphnia* is rapidly maturing into a powerful model for understanding basic biological processes, within an ecological context. A common resident of lakes and ponds, *Daphnia* has been the subject of over a century of study in the areas of rapid environmental response, physiology, nutrition, predation, parasitology, toxicology and behavior (Edmondson, 1987). The reproductive cycle of *Daphnia* is ideal for experimental genetics. Generation time in the laboratory rivals that of almost all other model eukaryotic systems, reaching maturity within 5-10 days. Under favorable environmental conditions, *Daphnia* reproduce through parthenogenesis, allowing the conservation of genetic lines. Sexual reproduction is induced by environmental changes allowing the production of inbred or outbred lineages. The sexually produced diapausing eggs, termed ephippia, can be stored viably for considerable periods. Moreover, they have been hatched from lake sediments up to a century old (Hairston et al., 2001; Limburg & Weider, 2002) allowing tracking of genetic changes over ecological and evolutionary time scales. *Daphnia* are transparent throughout life, allowing for studies of tissue-specific gene expression at any life stage and direct observation of parasites and pathogens. As a result, there is a growing body of work in

Daphnia related to regulation of developmental genes, the genetic basis of evolutionary ecology, and parasite resistance and immunity.

Understanding and predicting how individual organisms respond at the molecular level to environmental change will provide new insight into the evolution of complex biological systems. This insight will lead to the development of new predictive models of host-pathogen interactions, environmental stress and community dynamics as a function of environment and genotype/phenotype (National Science Board, 2000) advancing the field of individualized molecular medicine. As the number of organisms with complete genome sequence increases and technological improvements allow more sequence to be generated at a lower cost, the ability to look at genetic variation in a variety of organisms is greater than ever. However, to understand the role of genetic variation in the context of the natural environment, a model system with two critical components, (1) well-developed molecular resources and (2) a well-understood ecological knowledge base, is essential. Until recently, model systems typically possessed one of these components in depth while the other was nominal or lacking altogether. The recent sequencing of the entire *Daphnia pulex* genome and the establishment of the still growing molecular toolbox (ESTs, genetic map, arrays, etc.) represents the first model system with both components in place.

Organisms respond to environmental change through relatively quick changes in gene expression or through evolutionary response over multiple generations. To better comprehend the effect of gene expression on phenotype, an understanding of genetic variation for gene expression is necessary. A comprehensive understanding of genetic variation is obtained by sampling between and within populations, including individual organisms, directly from their natural environment. The well-documented ecological understanding of *Daphnia* makes the system uniquely suited to this and allows researchers to collect and sample individuals of wild populations directly from their native environment.

Our goal was to demonstrate that it is possible to detect biologically relevant variation in protein expression from an individual *Daphnia*. Using pressure cycling technology (PCT) for sample preparation and two-dimensional gel electrophoresis (2-DE), we have demonstrated that differences in protein expression between individual *Daphnia* with distinct genotypes and exhibiting biologically relevant phenotypic differences are detectable. The ability to detect and analyze individual differences for a large number of proteins represents an important step towards understanding the connection between genotype/phenotype and the environment.

2. Materials and methods

2.1 *Daphnia* and algae cultures

2.1.1 Algae cultures

Starter cultures of the green algae *Ankistrodesmus falcatus* were obtained from UTEX, The Culture Collection of Algae at The University of Texas (Austin, TX, USA). *A. falcatus* was grown in 2 L aerated, air-filtered culture vessels containing GTk media at 25°C under continuous illumination. GTk contains the following macronutrients, 0.2 mM $CaCl_2$, 2.5 mM KNO_3, 0.3 mM $MgSO_4$, 0.4 mM Na_2HPO_4 and the following micronutrients, 150 μM EDTA,

Na_2, 20 μM $FeSO_4$, 2 μM $ZnSO_4$, 1 μM $NaMoO_3$, 0.6 μM $CuSO_4$, $CoCl_2$ and 14 μM $MnCl_2$ (Leland Jahnke, Personal Communication).

2.1.2 *D. magna* with unique phenotype

D. magna starter cultures were obtained from Sachs Systems Aquaculture (St. Augustine, FL, USA). Stabilized cultures were maintained in 8 L of 25% mineralized water (Vermont Spring Water Company, Brattleboro, VT, USA) at a density of 60-120 individuals/L. *D. magna* were cultured at 22° ± 1°C under constant illumination with standard fluorescent bulbs. Cultures were maintained at pH 7.0-7.4 by the addition of 100 g/L crushed coral (Tideline Aquatics, Hanahan, SC, USA) supplied in nylon bags. Starter cultures were fed daily with 1 mL/L of *Nanochloropsis* microalgae liquid concentrate (Reed Maricultures, Campbell, CA, USA) for the first four weeks, followed by 0.1 mL/L thereafter.

2.1.3 *D. magna* with unique genotype and *D. pulex*

D. magna clones Iinb1 and Xinb3 were isolated from Munich, Germany and Tvärminne, Finland, respectively (Rottu et al., 2010). *D. pulex* clone Log50 was obtained from the *Daphnia* Genomics Consortium stock (www.wfleabase.org/stocks). Xinb3 and Log50 are the clones for the respective, *D. magna* and *D. pulex* genome projects. Cultures were maintained in 8 L of COMBO media (Killham et al., 1998) at a density of 30 individuals/L at 20° ± 1°C under a 16:8 hours, light:dark, low intensity photoperiod, and fed 1mg Carbon/L of *A. falcatus*.

2.2 Harvesting of *Daphnia*

Daphnia gut contents were minimized by allowing the microcrustaceans to feed on copolymer microspheres of 4.3-micron mean diameter (Duke Scientific, Fremont, CA, USA) for one hour prior to harvesting. Microspheres were fed at a concentration equal to the number of algal cells previously supplied. *Daphnia* were harvested by filtration through 250 um Nitex mesh (Sefar America, Depew, NY, USA) and flash frozen. Average mass of adult *Daphnia pulex* was 0.1158 ± 8.3 mg fully hydrated and 0.05285 ± 10.60 mg dehydrated (n = 50). Average mass of adult *D. magna* was 1.37± 0.46 mg fully hydrated and 0.23 ± 0.06 mg when dehydrated (n = 64).

2.3 Pressure Cycling Technology (PCT)

PCT has been shown to be an effective means for isolating proteins from a variety of microorganisms, as well as many difficult-to-lyse samples such as *Caenorhabditis elegans* (Geiser et al., 2002; Smejkal et al., 2006b; Smejkal et al., 2007). PCT, which subjects samples to rapid cycles of pressure, facilitated the extraction of proteins from single *Daphnia magna* with and without ephippia and from single *Daphnia pulex*.

Daphnia were transferred to tared PULSE Tubes (Pressure BioSciences, Inc., South Easton, MA, USA) and suspended in 500 uL of 7M urea, 2M thiourea, and 4% CHAPS supplemented with 100 mM dithiothreitol (DTT) and protease inhibitor cocktail P-2714 (Sigma Aldrich Chemicals, St. Louis, MO). An additional 900 uL of mineral oil was added to accommodate the necessary volume for the PULSE Tubes. The tubes were placed in the Barocycler NEP-3229 (Pressure BioSciences, Inc., South Easton, MA, USA) for 60 pressure cycles, each cycle consisting of 10 seconds at 35,000 psi followed by rapid depressurization and held for 2 seconds at atmospheric pressure. Following PCT, each PULSE Tube was coupled to an Ultrafree-CL centrifugal filtration device with a 5-micron pore size (Millipore Corporation, Danvers, MA, USA) and evacuated by centrifugation for 1 minute at 1000 RCF. The PULSE Tube was removed and centrifugation continued for 4 minutes at 4000 RCF, followed by the removal of the mineral oil.

2.4 Reduction, alkylation, and ultrafiltration

Samples were transferred to ULTRA-4 ultrafiltration devices with 10 kDa molecular weight cut-off (Millipore Corporation, Danvers, MA, USA). Proteins larger than 10 kDa are retained in the ultrafiltration device. Centrifugation assisted the ultrafiltration, and the samples were exchanged with fresh UTC until the final DTT concentration was 10 mM. Reduction and alkylation of the samples were performed directly in the ultrafiltration devices using 5 mM tributylphosphine and 50 mM acrylamide as described (Smejkal et al., 2006a). The alkylation reaction was terminated by resuming centrifugation and ultrafiltrative exchange. Bradford Reagent (Sigma-Aldrich Chemicals, St. Louis, MO, USA) was used to measure the protein concentration in each sample.

2.5 IEF and 2-DE

Two-hundred uL of each sample was placed onto individual wells in IPG rehydration trays from Proteome Systems (Woburn, MA, USA). Bio-Rad ReadyStrip IPG strips with a pH range of 3-10, 4-7, or 7-10 (Hercules, CA, USA) were placed onto each sample, and the tray was placed into a humidifying Ziploc bag. Rehydration occurred over six hours until all the sample was visibly absorbed by the strip. At the termination of rehydration, strips were placed into isoelectric focusing trays and ran at 10,000 volts (maximum voltage) for 110,000 accumulative volt-hours. Strips were equilibrated twice for 10 minutes in 375 mM Tris-HCl containing 2.5% SDS, 3M urea, 0.01% and phenol red, then placed onto Criterion Tris-HCl 8-16% IPG+1 gels (Bio-Rad Laboratories, Hercules, CA) and ran at 120 V and 60 mA/gel. 2-DE gels were ran for all IPGs, only 4-7 gels are shown.

2.6 Digital image analyses

The 24-bit images were analyzed using PDQuest™ software (Bio-Rad, v.7.1). Background was subtracted, and protein spot density peaks were detected and counted. A reference pattern was constructed from one of the individual gels to which each of the gels in the matchset was matched. Numerous proteins that were uniformly expressed in all patterns were used as landmarks to facilitate rapid gel matching. After matching, the total spot count was determined for each gel.

3. Results and discussion

3.1 Protein variation between individual *D. magna* of distinct genotypes

Our first goal was to demonstrate that differences in protein expression could be detected between individual *Daphnia* of distinct genotypes. Individual *D. magna* from Iinb1 and Xinb3 genotypes were isolated, proteins extracted and analyzed in quadruplicate by 2-DE as describe above. Silver staining detected an average of 687 ± 11 protein spots from the Xinb3 gels and 692 ± 14 protein spots from the Iinb1 gels (figure 1). After normalization of the gel images based on total intensity, 679 spots were matched between the two gel images. One Hundred thirty six spots showed a two-fold or greater difference in spot intensity. Seventy-nine of these were more abundant in Xinb3 and 57 were more abundant in the Iinb1. To illustrate these differences, the silver stained gels were digitally colored. The Iinb1 gels were red and the Xinb3 gels were green. The gels were then superimposed. (Figure 2).

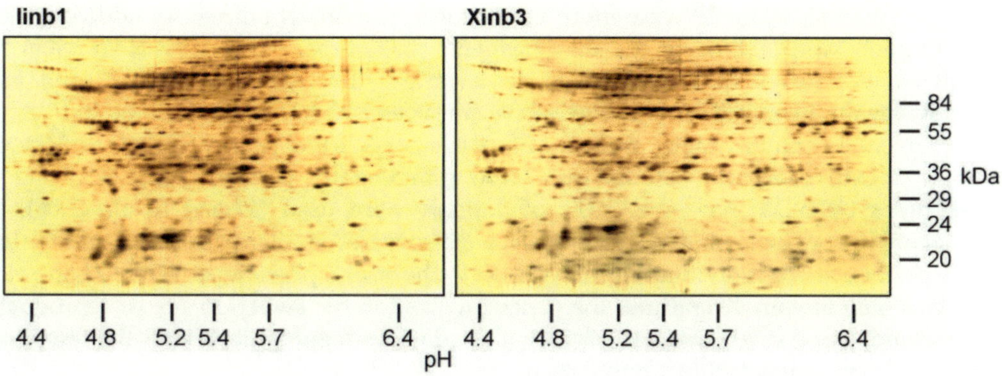

Fig. 1. Representative silver-stained 2D gels of individual *Daphnia magna* with distinct genotypes. Quadruplicate gels revealed 692 ± 14 spots in Iinb1 clone (left) and 687 ± 11 spots in Xinb3 clone (right).

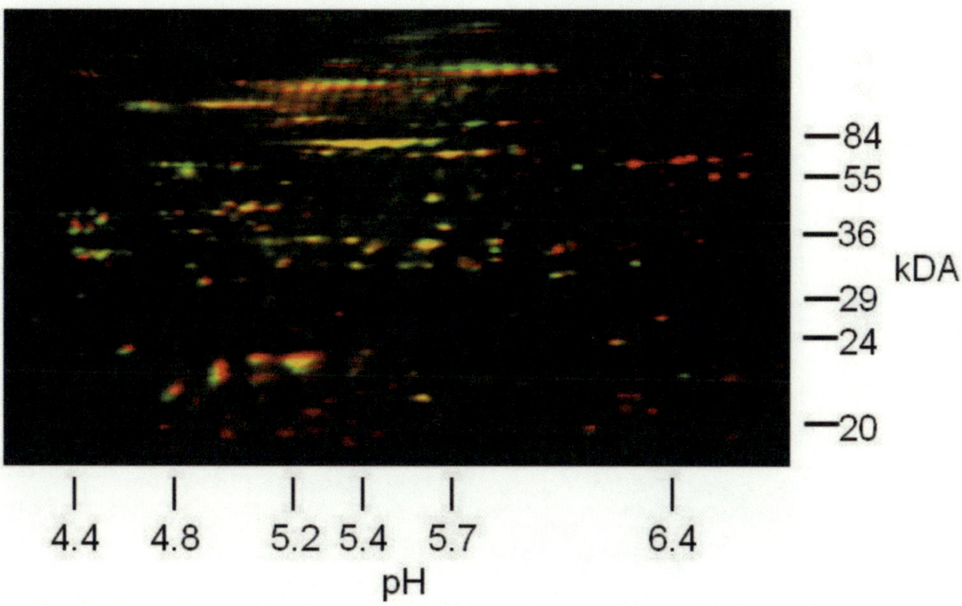

Fig. 2. Digitally enhanced, superimposed, silver-stained 2D gels of individual *Daphnia magna* with distinct genotypes. Red indicates spots unique to Iinb1, green indicates spots unique to Xinb3, yellow indicate spots shared by both genotypes.

3.2 Protein variation between individual *D. magna* of distinct phenotypes

We were also able to detect differences between individual *Daphnia* with distinct phenotypic differences. Individual *D. magna,* with and without ephippia, were isolated, proteins extracted and analyzed by 2-DE as described above. Silver staining detected an average of 524.5 ± 7.8 protein spots in 2D gels produced from single *D. magna,* with and without ephippia (figure 3). After normalization of the gel images based on total intensity, 386 spots were matched between the two gel images. Eighty-four spots showed a three-fold or greater difference in spot intensity. Fifty-five of these were more abundant in the parthenogenic (no ephippia) animal, while 29 were more abundant in the sexual animal. In addition, eleven protein spots were unique to the parthenogenic phenotype, while 49 protein spots were unique to the sexual phenotype. This demonstrates the feasibility of 2-DE and image analysis for the differentiation of *Daphnia* phenotypes isolated in the field as indicators of environmental variables. Other studies with parthenogenic and sexual *Daphnia carinata* were able to identify several proteins that were differentially expressed between the two phenotypes by 2-DE; however, 100's of animals were used (Zhang et al., 2006). It is interesting to note that using single animals, we discovered similar patterns of up-regulation in the parthenogenic individual in comparison to the sexual phenotype. While Zhang et al.'s goal was only to gain insight into the genes involved in the switch to sexual reproduction, our method, using single *Daphnia magna* with and without ephippia, allows the sampling of these candidate genes within a population.

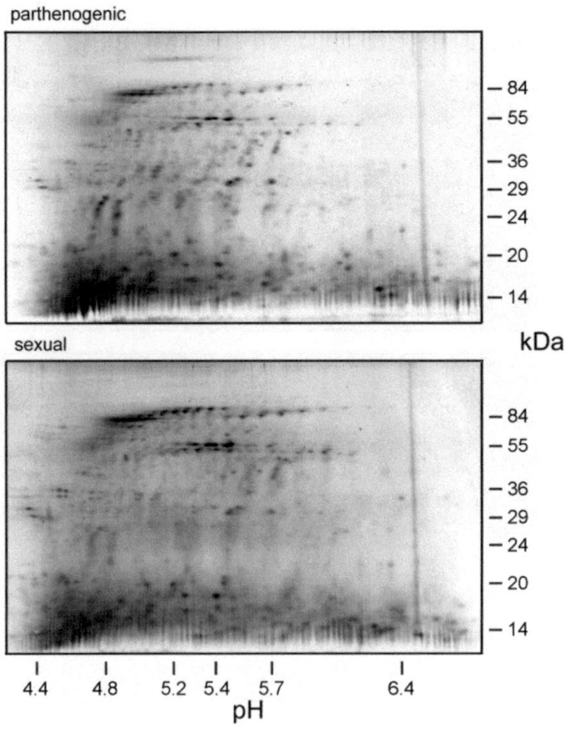

Fig. 3. Silver stained 2D gels of individual *Daphnia magna* with unique phenotypes.

3.3 Extending approach to single *D. pulex*

As *D. magna* is the largest of the *Daphnia* genus, we used PCT to extract protein from single *D. pulex* to demonstrate that our technique would be feasible with smaller individuals. Even when using the much smaller *Daphnia pulex,* we were able to detect approximately 900 spots from a single individual. It is reasonable to expect that differences in 2-DE between single *D. pulex* with phenotypic differences would also be detectable. As an indication of reproducibility of our method, we ran duplicate 2-DE of PCT- extracted proteins from 1, 2, 3, 4 and 5 *Daphnia pulex*. Figure 4 shows representative gels of 1, 2 or 3 *Daphnia pulex*. Figure 5 shows the number of protein spots detected and the standard deviation for *Daphnia pulex* gels of 1, 2, 3, 4 and 5 individuals. The low standard deviation indicates that PCT and 2-DE is an efficient and highly reproducible method of sample preparation and protein comparison.

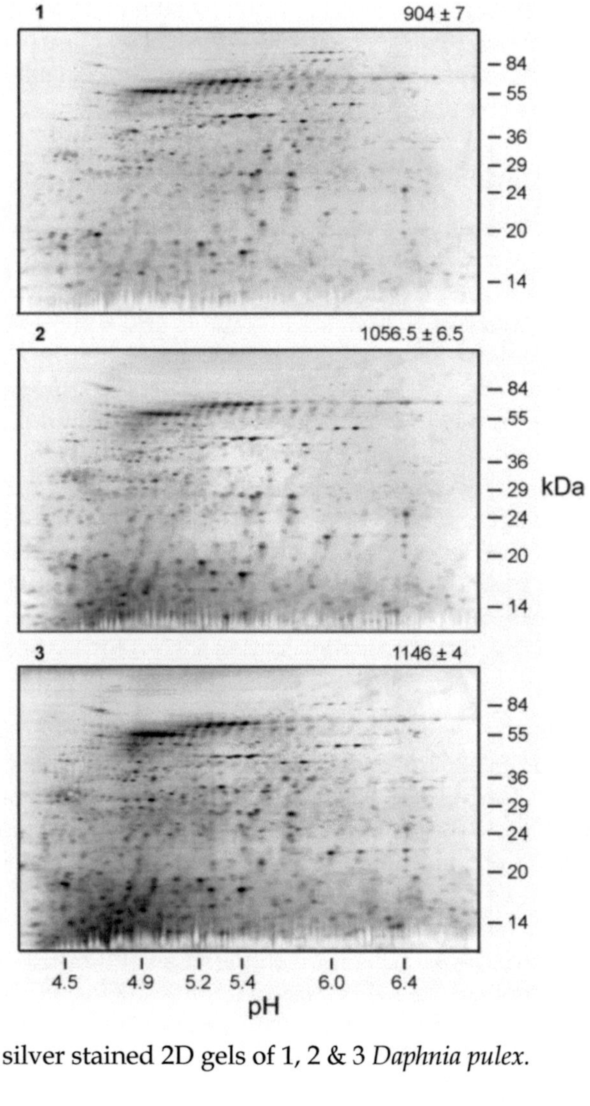

Fig. 4. Representative, silver stained 2D gels of 1, 2 & 3 *Daphnia pulex*.

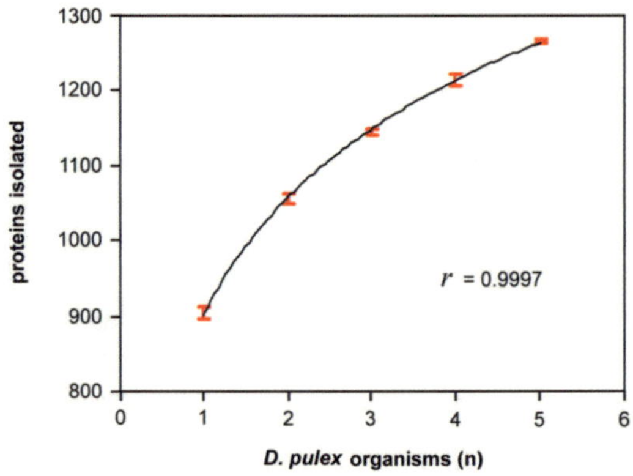

Fig. 5. Graph of detected protein spots from duplicate silver stained 2D gels of proteins extracted 1, 2, 3, 4 & 5 *Daphnia pulex*. The shape of the curve suggests that our method is an efficient and reproducible approach to protein extraction and the majority of the unique proteins are recovered from a single animal.

3.4 Protein functional diversity

To properly evaluate our method, it is necessary to understand the functional diversity of the proteins sampled. The proteins detected on any 2D gel will be biased towards the most abundant and the goal is to sample gene expression for proteins of diverse functions. To evaluate the likely diversity of proteins detectable and comparable by this method, we first generated a theoretical 2D gel for the 2000 most highly expressed genes in the *Daphnia pulex* genome. The correlation between mRNA expression levels and protein abundance is a debated topic; (for a brief review, see Greenbaum et al., 2003) however, recent studies have found a high correlation (Lu et al., 2007; Tuller et al., 2007). Using the recently completed draft of the *Daphnia pulex* genome, we found the top 2000 gene prediction models with the most Expressed Sequence Tag (EST) evidence using BLAST (Altschul et al., 1990). The theoretical pI and MW of these genes were calculated using the Compute pI/MW tool from ExPASy (Gasteiger et al., 2003) and the results graphed using Excel to create the theoretical 2D gel (Figure 6). Importantly, the pI range used on the actual gels (indicated by the rectangle) covers a significant portion of the predicted proteome.

To assay the functional diversity of these theoretical proteins, we utilized the 25 eukaryotic orthologous groups (KOGs) (Totusov et al., 2003) that were assigned to the *Daphnia pulex* genes as part of the genome sequence annotation project. Many *D. pulex* genes have no homology to any entries at NCBI; therefore, it is not surprising that of the approximately 30,000 predicted genes only 18,371 have been assigned to a KOG class (Colbourne et al., 2011) Of the 2000 most highly expressed genes, 298 had been assigned to a KOG class. Twenty four of the 25 KOG classes were represented, and only "coenzyme transport and metabolism" was absent. To understand the functional diversity that may be present on a single animal 2D gel of approximately 1000 spots, we looked at the KOG assignments of the 1000 most highly expressed genes. Ninety-seven have been assigned to a KOG class, with 21 of the 25 KOG classes being represented. Not represented were "coenzyme transport and metabolism", "secondary metabolites, biosynthesis, transport and catabolism", "nuclear structure" and "chromatin structure and dynamics" (Figure 7). In general, KOG classes that are well represented in the *Daphnia* genome are also well represented in the top 2000 of most highly expressed genes. As 84% of the KOG classes are represented in the top 1000 most highly expressed genes, we feel that a single animal 2D gel of approximately 1000 spots should represent a diverse sampling of biologically relevant proteins.

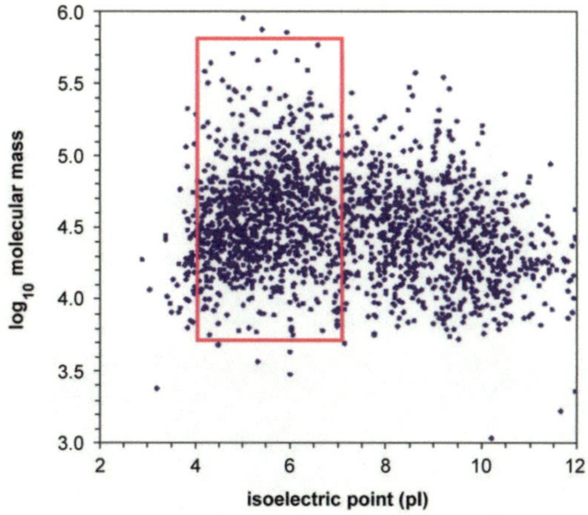

Fig. 6. Theoretical 2D gel of the top 2000 highly expressed *D. pulex* genes. The rectangle represents the separation range (MW and pI) of actual 2D gels (pH range 4 - 7).

pI range	number of theoretical proteins [a]	percent of total predicted	number of observed proteins	coefficient of variation
3-4	17	0.9	-	-
4-7	1027	51.4	904.0 ± 7.0 [b]	0.008
7-10	749	37.5	381.5 ± 26.5 [c]	0.069
> 10	207	10.4	-	-
total	2000	-	1285.5 ± 33.5	0.026

[a] Proteins predicted from sequence does not account for charge
[b] Spots detected in 2D gels from single D. pulex organisms.
[c] Spots detected in 2D gels of a multiple D. magna organisms.
(184.0 ± 2.6 mg or approximately 130 daphnids)

Table 1. Distribution of 2000 most abundant *Daphnia* proteins predicted from genome sequence

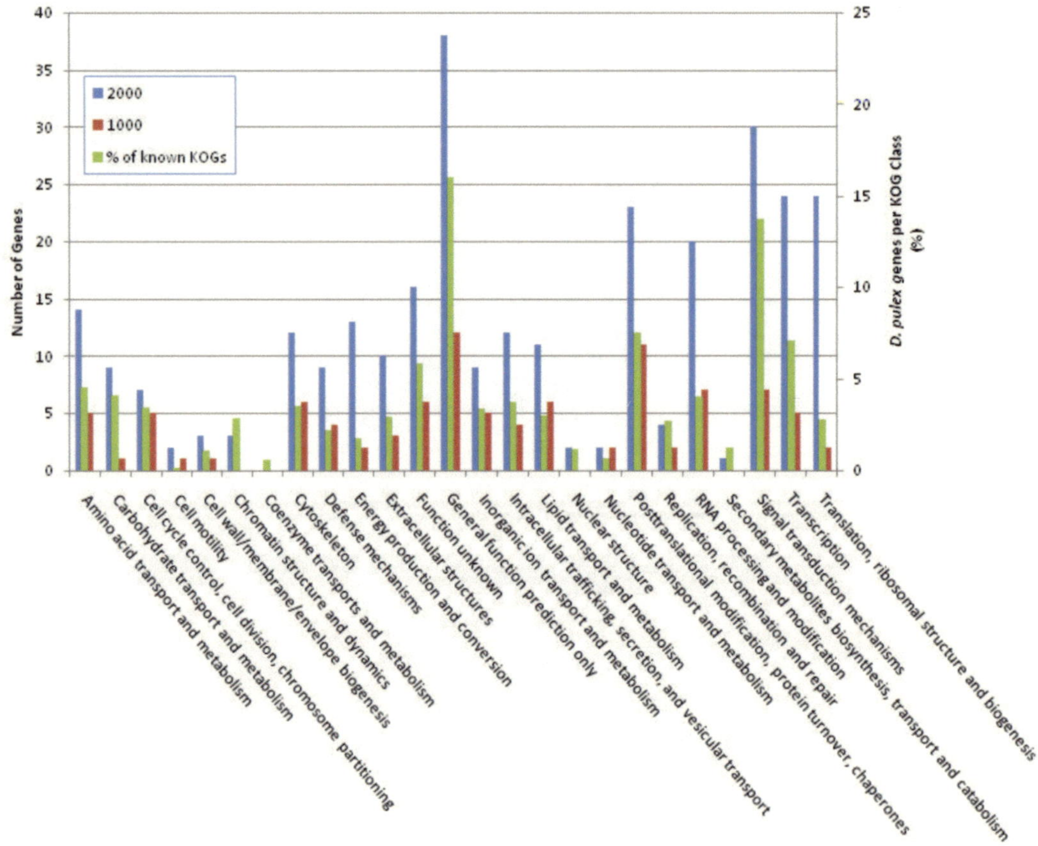

Fig. 7. Distribution of highly expressed *D. pulex* genes across 25 KOG classes. Blue and red bars indicate the number of genes in each class from the top 2000 and 1000, respectively, most highly expressed genes (Left-hand axis). Green bars indicate the total genes assigned to each KOG class as a percentage of the total number of genes (18,371) assigned to a KOG class (Right-hand axis). To further characterize the predicted protein gel, we compared it to our observed spot counts. Table 1 summarizes the number of predicted proteins and the number of actual proteins within specific pI ranges. Through the generation of several *D. magna* (both single animal and multiple-animal) 2D gels (not shown), we were able to visualize a total of 1285 protein spots. Seventy percent of these were observed in the 4-7 pI range, while the theoretical gel predicted 51% in this same range. It is important to note that the theoretical 2D does not account for post-translational modifications and was generated from *D. pulex* genes. Both of these factors will contribute to differences between the predicted (*D. pulex*) and observed (*D. magna*) number of proteins.

3.5 Basic proteins constituents of the *D. magna* proteome

Initially, single Daphnid extracts were analyzed on IPG strips with a pH range of 3-10. Since more than 80% of the proteins visualized by silver staining were in the pH 4-7 range, subsequent analyses were performed using IPGs pH 4-7. However, the theoretical 2D does predict a significant number of proteins (47%) in the basic range (7-10). Due to their relative low abundance, the visualization of very basic proteins (pH 7-10) in Daphnids required many more organisms. For this, 184 ± 3 mg of *D. magna* (approximately 135 organisms) were cultured under either normal or hypoxic conditions and processed in 1.3 mL of ProteoSOLVE IEF Reagent. The samples were concentrated two-fold by ultrafiltration, and IEF was performed on IPGs pH 7-10, followed by silver staining and image analysis. Silver staining detected 355 and 408 spots (gels not shown) in pH range 7-10 from *D. magna* cultured under normal or hypoxic conditions respectively, further illustrating the utility of 2-DE for detecting phenotypic differences influenced by specific environments.

4. Conclusion

Organisms live in ever changing environments. Understanding how individuals respond and adapt to these environments on a molecular level forms the basis for advances in personalized medicine and requires model systems with both well-developed ecological knowledge and molecular resources. The freshwater crustacean *Daphnia* now has these two requirements in place. We have demonstrated the ability of 2-DE to identify protein differences between single *Daphnia magna* with distinct genotypes (Iinb1 and Xinb3), distinct phenotypic differences (presence or absence of ephippia) and when cultured in different environments (normal or hypoxic conditions).

We predict that the detectable proteins on a single animal 2D gel, while biased towards the most abundant proteins, represent a functionally diverse set of proteins. This technique represents an important step to a greater understanding of individual variation of gene expression and is critical to advancing the field of EEFG. However, as the use of silver stain convolutes downstream mass spectrometry, the number of protein spots that can be confidently identified is significantly curtailed. The full potential of single animal gels will be realized with the development of comprehensive 2D maps of the *Daphnia* proteome.

5. References

Altschul SF, Gish W, Miller W, Myers EW, & Lipman DJ. 1990. Basic local alignment search tool. *Journal of Molecular Biology*, 215(3):403-410.

Colbourne J.K., *et. al.*, 2011. The Ecoresponsive Genome of *Daphnia pulex*. *Science*, Feb 4; 331(6017): 555-61.

Edmondson WT. 1987. *Daphnia* in Experimental Ecology: Notes on Historical Perspectives. In: Peters RH, R. dB, editors. *Daphnia. Pallanza: Memorie dell'Instituto Italiano di Idrobiologia*, 45:11-30.

Gasteiger E, Gattiker A, Hoogland C, Ivanyi I, Appel RD & Bairoch A. 2003. ExPASy: The proteomics server for in-depth protein knowledge and analysis. *Nucleic Acids Research*, 31(13):3784-3788.

Geiser HA, Hanneman AJ, & Reinhold V. 2002. HTP proteome-glycome analysis in Caenorhabditis elegans. *Glycobiology*, 12(10):650.

Greenbaum D, Colangelo C, Williams K, & Gerstein M. 2003. Comparing protein abundance and mRNA expression levels on a genomic scale. *Genome Biology*, 4(9):117.

Hairston NG, Jr., Holtmeier CL, Lampert W, Weider LJ, Post DM, Fischer JM, Caceres CE, Fox JA, & Gaedke U. 2001. Natural selection for grazer resistance to toxic cyanobacteria: Evolution of phenotypic plasticity?, *Evolution*, 55(11): 2203-2214.

Kilham SS, Kreeger DA, Lynn SG, Goulden CE, & Herrera L. 1998. COMBO: a defined freshwater culture medium for algae and zooplankton. *Hydrobiologia*, 377:147-159.

Limburg PA, & Weider LJ. 2002. `Ancient' DNA in the resting egg bank of a microcrustacean can serve as a palaeolimnological database. *Proceedings of the Royal Society of London:Series B*, 269:281-287

Lu P, Vogel C, Wang R, Yao X, & Marcotte EM. 2007. Absolute protein expression profiling estimates the relative contributions of transcriptional and translational regulation. *Nature Biotechnology*, 25(1):117-124.

National Science Board. 2000. Environmental Science and Engineering for the 21st Century: The role of the National Science Foundation. Report number: NSB 00-22.

Rottu J, Jansen B, Colson I, De Meester L, & Ebert D. 2010. The first-generation *Daphnia magna* linkage map. *BMC Genomics*, 11:508.

Smejkal GB, Li C, Robinson MH, Lazarev AV, Lawrence NP, & Chernokalskaya E. 2006a. Simultaneous reduction and alkylation of protein disulfides in a centrifugal ultrafiltration device prior to two-dimensional gel electrophoresis. *Journal of Proteome Research*, 5(4):983-987.

Smejkal GB, Robinson MH, Lawrence NP, Tao F, Saravis CA, & Schumacher RT. 2006b. Increased protein yields from Escherichia coli using pressure-cycling technology. *Journal of Biomolecular Techniques*, 17(2):173-175.

Smejkal GB, Witzmann FA, Ringham H, Small D, Chase SF, Behnke J, & Ting E. 2007. Sample preparation for two-dimensional gel electrophoresis using pressure cycling technology. *Analytical Biochemistry*, 363(2):309-311.

Tatusov RL, Fedorova ND, Jackson JD, Jacobs AR, Kiryutin B, Koonin EV, Krylov DM, Mazumder R, Mekhedov SL, Nikolskaya AN, Rao BS, Smirnov S, Sverdlov AV, Vasudevan S, Wolf YI, Yin JJ, & Natale DA. 2003. The COG database: an updated version includes eukaryotes. *BMC Bioinformatics*, 4:41.

Tuller T, Kupiec M, Ruppin E. 2007. Determinants of Protein Abundance and Translation Efficiency in S. cerevisiae. *PLoS Computational Biology*, 3(12):e248.

Zhang M-F, Zhao Y-L, & Zeng C. 2006. Differential protein expression between parthenogenetic and sexual female of *Daphnia* (*Ctenodaphnia*) *carinata*. *Acta Zoologica Sinica*, 52(5):916 -923.

18

Two-Dimensional Gel Electrophoresis and Mass Spectrometry in Studies of Nanoparticle-Protein Interactions

Helen Karlsson, Stefan Ljunggren, Maria Ahrén, Bijar Ghafouri,
Kajsa Uvdal, Mats Lindahl and Anders Ljungman
Linköping University; County Council of Östergötland
Sweden

1. Introduction

1.1 Nanoparticles

Adverse health effects have been associated with the exposure to particulate matter (PM) ever since the London smog in the winter of 1952. Recent estimates attribute about 12,000 excess deaths to have occurred because of acute and persisting effects of the London smog (Bell & Davis, 2001). Over the years a number of epidemiological studies have shown that PM from combustion sources such as motor vehicles contributes to respiratory and cardiovascular morbidity and mortality (Kreyling et al., 2002, 2006, Wick et al., 2010). Especially so do the ultra-fine particles (UFPs) with a diameter less than 0.1 micrometer. UFPs from combustion engines are capable to translocate over the alveolar–capillary barrier (Rothen-Rutishauser et al., 2007). When nano-sized PM (nanoparticles, NP), which are small enough to enter the blood stream, do so they are likely to interact with plasma proteins and this protein-NP interaction will probably affect the fate of and the effects caused by the NPs in the human body. Herein we present results showing that several proteins indeed are associated to NPs that have *in vitro* been introduced to human blood plasma.

NPs are atoms and molecules defined as particles less than 100 nanometers in at least one dimension (Elsaesser & Howard, 2011). Due to the plethora of NPs being produced in various forms (e.g. spherical, fibers, rods, clusters) or by different processes (e.g. flame-spray synthesis, chemical vapor deposition), defining the characteristics of a NP is not an easy task even when it comes to manufactured NPs and when considering those formed unintentionally during processes such as combustion in motor vehicles it becomes an even harder task. This variation in properties according to the respective composition of NPs is also the basis of the wide range of potential applications, from medicine to consumer products. Due to the unique physicochemical properties of nanomaterials, there are plenty of possibilities for NPs to enter the human body, either deliberately as medicines or unintentionally as environmental contaminants and thus potentially cause adverse human health effects (Elsaesser & Howard, 2011; Stern & McNeil, 2008). Although many characteristics have been highlighted as driving the potential adverse health effects associated with NP exposure, it has been specifically the size and increased surface area of NPs that has been concluded as elucidating any such adverse effects observed (Elsaesser & Howard, 2011; Stern & McNeil, 2008).

UFPs from combustion sources, such as motor vehicles, are capable to promote atherosclerosis, thrombogenesis and other cardiovascular events mainly via the ability to induce inflammatory and protrombotic responses (McAuliffe & Perry 2007). Thus, NP induced effects within the lung has been studied over the past twenty years (Mühlfeld et al., 2008; Rothen-Rutishauser et al., 2007). Indeed, exposure to most forms of NPs will initially be via inhalation, especially when considering occupational exposure, and thus will affect the respiratory system. The respiratory system is by far the main port of entry even though the gut and skin are also possible ways of entry. However, NP localization and fate is not only restricted to their portal of entry. NPs can be distributed to organs distal to their site of exposure, so that potential NP toxicity can occur in any secondary site. Research into the possible secondary toxicity of NPs is quite limited. Even so studies investigating the effects of NP translocation to secondary organs have shown that NPs can elicit negative effects to the liver, brain, GI tract (following inhalation), spleen, reproductive systems and the placenta (Kreyling et al., 2002, 2006; McAuliffe & Perry 2007; Wick et al., 2010). NPs toxic mechanisms at the cellular level includes protein misfolding and protein fibrillation causing major problems in the brain and chronic inflammation as a result of nanoparticle exposure, for example in the lung and other organs, via frustrated phagocytosis or production of reactive oxygen species. A vulnerable target for possible toxicological effects of nanoparticles is the fetus. Gold nanoparticles have been shown to cross the maternal-fetal barrier and fullerenes were found to have a fatal effect on mouse embryos (Elsaesser & Howard, 2011; Stern & McNeil, 2008).

When nanoparticles enter the blood vessels, or any biological fluid, e.g. saliva or mucus they are most likely surrounded by a layer of proteins. This dynamic protein "corona" depends on the concentration in the biological fluid and hence the composition of the layer varies in different parts of the body (Cedervall et al., 2007; Lundqvist et al., 2008; Lynch et al., 2007; Walczyk et al., 2010,). Thus, the reactions in the body to such a NP-protein complex is most likely different from that induced by the bare NP and possibly affecting bio-distribution and thereby causing unwanted side effects (Adiseshaiah et al., 2010; Leszczynski 2010,). The function of protein coating is not fully known but since most nanoparticles show strong affinity for proteins, it is of importance to investigate this interaction in different fluids. In blood, plasma proteins constituting the NP corona is possibly affecting a wide range of effects such as phagocytosis via immunoglobulins and complement (Dobrovolskaia & McNeil 2007), coagulation via prothrombin (Dobrovolskaia 2009) and the distribution of lipoproteins (Benderly et al., 2009; Hellstrand et al., 2009; Zensi et al., 2010).

The availability and toxicity of any substance to a biological organism is determined by both the concentration/dose that the organism is exposed to, as well as the "toxicokinetics" of the substance. These include the uptake, transport, metabolism and sequestration to different compartments by the organism, as well as the elimination of the substance from the biological organism. These parameters are essential since the potential toxicity of substances is dependent upon the specific organs or cell types exposed, which form the substance is in (e.g. bound to serum protein, aggregated, dissolved, oxidized), as well as the period of time the substance interacts/remains at the site of primary and secondary exposure. These parameters are influenced by the physical-chemical characteristics of the substance, therefore a detailed characterization of the substance is pivotal in order to allow generalizable conclusions and should therefore be given ample attention. Furthermore, most of these toxicological parameters involve proteins and/or actions carried out by proteins. Thus it is pivotal to the understanding of NP toxicology, and thereby the possibility to predict health effects caused by NPs, to understand NPs interactions with proteins. One of the best, if not the best, technique to separate proteins is two-dimensional gel electrophoresis (2-DE). Preferably, this is combined with peptide mass fingerprinting and MALDI-TOF MS analyses for fast identification of the separated proteins, which then may be followed by tandem MS analyses for sequence information. Here we present our results regarding serum protein interactions with metal-nanoparticles (Al_2O_3, ZnO/Al6%, SiO_2) and Carbon Nanotubes, obtained using 2-DE and MS. Furthermore, we elaborate on and exemplify different aspects of the 2-DE/peptide mass fingerprinting-technique to further improve the approach.

1.2 General introduction to the 2-DE technique

Two-dimensional gel electrophoresis (2-DE) is an excellent method for separation of proteins from most kinds of tissues and complex mixtures of proteins (O`Farrel-1975). Both qualitative characterization of the protein expression, including post-translational modifications and quantitative characterization comparing the protein expression in different individuals or groups, are possible by this technique. Two steps are included, the isoelectric focusing (IEF) step, where the proteins are separated according to their isoelectric point (pI) in a pH-gradient, and the sodium dodecyl sulphate polyacrylamide gel electrophoresis (SDS-PAGE) step, where the proteins are separated according to their molecular weight. Since it is less common that two proteins have the same isoelectric point and molecular weight, this will result in each protein migrating to its own unique position. The 2-DE technique allows, depending on the nature of the sample, the separation of 500-3000 protein spots and the resolution can be improved, e.g. by removal of abundant proteins or by composite gels from overlapping pH-gradients. Proteins separated by gel electrophoresis can be visualized by a number of methods using different types of stains. Various stains interact differently with the proteins and some of the stains used are not even specific for proteins. The degree of sensitivity is also different. Processing data from stained protein gels by computers includes the gel images being digitized by an imaging system and then analyzed using computer software allowing a number of different measurements such as number, size, and intensity of the stained protein spots. Separated proteins are then identified by mass spectrometry (MS). The proteins are in-gel digested and extracted peptides analyzed by peptide mass fingerprinting or peptide sequencing. Two widely used MS instruments used for these respective analyses are matrix assisted laser desorption/ionization-time of flight-mass spectrometry (MALDI-TOF MS) where peptides are transferred from solid phase to gas phase, and electrospray ionization tandem mass spectrometry (ESI MS/MS) where peptides are transferred from liquid phase to gas phase. The key advantage with 2-DE is the ability to separate protein isoforms. On the other hand, very large and hydrophobic proteins are underrepresented in 2-DE and the need of peptide extraction from the in-gel digests may influence to amounts of analytes available for MS protein identification. Nevertheless, combining the separation and analytical ability of the 2-DE technique with the identification power of MS provides a powerful tool in human toxicology.

2. Methods

2.1 Characterization of nanoparticles

Commercial $SiO_{2\ size\ 0.007\mu}$ (S-3051), ZnO/6% Al doped $_{<50nm}$(677450), and $Al_2O_{3\ <50nm}$ (544833) were purchased from Sigma Aldrich. As comparison, single walled Carbon Nanotubes (CNO, 704121) from Sigma Aldrich was used. Nanoparticles, characterized by manufacturer, were dispersed in Milli-Q water and/or PBS (137 mM NaCl, 2.7 mM KCl, 8.45 mM Na_2HPO_4, 1.47 mM KH_2PO_4 pH 7.3) and sonicated on ice for 10 min. The hydrodynamic sizes of the particles were analyzed by Dynamic Light Scattering (DLS).

DLS measurements were performed using an ALV/DLS/SLS-5022F system (ALV-GmbH, Langen Germany) and a HeNe laser at 632.8 nm with 22 mW output power. The scattering angle was 90° and the temperature 22°C. For temperature stabilizing purposes, samples were placed in a thermostat bath (22°C) for at least 10 minutes prior to the measurements. Samples were diluted in Milli-Q water or PBS. Ultrasonication of two different kinds was performed to decrease the degree of agglomeration; either an ultrasonic bar homogeniser (Sonoplus HD 2200, Bandelin electronics, Germany) was used or samples were placed in an ultrasonic bath (USC300T, VWR, Sweden). The viscosity of PBS was set to 0.9782 mPa's by linear interpolation between tabulated values for 20°C (0.911) and 25 °C (1.023) (Hackley & Clogston, 2007).

Data analysis was performed using a nonlinear fit model via ALV-Regularized Fit (ALV-Correlator Software Version 3.0. using ALV-Regularized in nonlinear fit model http://www.alvgmbh.de/).

2.2 Preparation of human plasma

Human plasma, collected in sodium citrate tubes, was prepared from three healthy volunteers. After cooling, the blood was centrifuged in 800g for 10 min and plasma, free from red blood cells, was drawn from the top of the tube. SiO_2, ZnO/6% Al doped, Al_2O_3 and CNO were then exposed to three plasma samples respectively. In all nanoparticle exposures fresh plasma was used.

2.3 Nanoparticle/plasma incubation

Nanoparticles were dissolved (final concentration 2 mg/ml) in PBS and incubated with 1% plasma at 37°C for 1h. As control, to ensure there was no protein precipitation, one sample was prepared without nanoparticles. Unbound proteins were separated from nanoparticles by centrifugation for 40 min at 50 000g and 4°C. The supernatant was discarded and the particle pellet was washed in Dithiothreitol (DTT, Sigma-Aldrich) 20mM/-acetone buffer followed by a second centrifugation step for 10 min at 50 000g and 4°C. The supernatant was discarded and the pellet was air dried. The pellets containing nanoparticles and attached proteins were then dissolved in denaturing solution containing 9M Urea (Sigma-Aldrich), 65mM DTT and 4% (3-(3-cholamidopropyl)-dimethylamino)-1-propanesulfonate (CHAPS). The solution was incubated at room temperature for 30 min before denatured proteins were separated from the nanoparticles by centrifugation for 30 min at 50 000g and 4°C. The supernatant was collected and the samples were analyzed in triplicates. 20 µl (40µg protein) was applied on the IEF strip in each 2-DE analysis.

2.4 Isolation of lipoproteins

2.4.1 HDL isolation

Preparation of high density lipoprotein (HDL) was performed by a method described by Sattler et al.1994, with slight modifications (Karlsson et al., 2005). Blood samples in EDTA-containing tubes were obtained from healthy volunteers after an overnight fast. After centrifugation (10 min, 700*g*) at room temperature, plasma was collected. EDTA (1 mg/mL) and sucrose (final concentration 0.5%) were added to prevent HDL oxidation and aggregation, respectively. Five milliliters of EDTA-plasma adjusted to a density of 1.24 g/mL with solid KBr (0.3816 g/mL) was layered in the bottom of a centrifuge tube (Beckman, Ultraclear tube). The EDTA plasma fraction was gently overlayered with KBr/PBS solution (0.0834 g KBr/mL, total density 1.063 g/mL). In one centrifuge tube, proteins were stained with Coomassie Brilliant Blue to be used as a reference while collecting the HDL fraction. Ultracentrifugation was performed in a Beckman XL-90 equipped with a Ti 70 rotor (fixed angle; Beckman Instruments, Fullerton, CA, USA) for 4 h at 290 000*g* and 15°C. By this procedure the lipoprotein fractions with a density lower than 1.063 g/mL (low density and very low density lipoprotein) are located at the top of the tube and HDL is located in the middle of the tube. HDL was collected by penetrating the tube with a syringe. To avoid contamination by serum proteins, HDL were then further purified by a second centrifugation step. KBr/PBS solution (0.3816 g KBr/mL) was added to the HDL (total density 1.24 g/mL) and the centrifugation was performed under the same conditions as described above, but for 2 h. HDL was collected from the top of the tube and desalted using desalting buffer (NH_4HCO_3, 12mM, pH 7.1) and PD 10 columns (Sephadex™ G-25 M, GE Healthcare, Buckinghamshire, United Kingdom). Protein concentration in the HDL solution was determined with Bio-Rad protein assay (Bio-Rad, Richmond, CA, USA). Sample (3.5 mL) was lyophilized and dissolved in 0.25 mL sample solution (9 M Urea, 4% CHAPS, 2% Pharmalyte , 65 mM DTT, 1% bromophenol blue) according to Görg et al.1988.

2.4.2 Immunoaffinity chromatography

Anti-ApoA-I antibodies were attached to a 5 mL HiTrap NHS-activated HP column (GE Healthcare) according to manufacturer's instructions. 2.5 mL plasma were desalted by the use of PD-10 columns and the eluted sample were diluted to 4 mL with 50 mM Tris-HCl, 0.15M NaCl, pH 7.5. The ApoA-I coupled immunoaffinity column were equilibrated by allowing 10 column volumes flow through it. Desalted sample were applied into the column and allowed to recirculate for 40 minutes. Loop were disconnected and washed with 10 column volumes of 50 mM Tris-HCl, 0.15M NaCl, pH 7.5 followed by ten column volumes of 50 mM Tris-HCl, 0.5M NaCl, pH 7.5. ApoA-I adsorbed to the column were eluted with 20 mL of 0.1M Glycin-HCl, pH 2.2. Sample was collected in fractions of 0.4 mL in tubes which each contained 20 µL 1M Tris, pH 9.0 for pH-neutralization of the sample. Fractions containing proteins were pooled and desalted using PD-10 columns.

2.5 Albumin and IgG removal

The removal of the high abundance proteins albumin and IgG from plasma was performed using an Albumin and IgG removal kit (GE Healthcare). Briefly, the column was equilibrated with binding buffer (20mM $Na_2H_2PO_4$, 0.15M NaCl, pH 7.4), 50 µl of plasma was diluted to a volume of 100 µl and applied to the column. After 5 min incubation, the depleted sample was collected in an eppendorf tube by centrifugation at 784g. The total protein concentration before/after depletion was determined with Bio-Rad protein assay. After depletion of high abundant proteins the samples were desalted using PD-10 columns.

2.6 2-DE analysis

2-DE was performed using IPGphor and Multiphor (GE healthcare). Briefly, the proteins were resuspended in 150 µL of a 2-DE sample buffer containing 9 M urea, 65 mM DTT, 2% Pharmalyte (GE Healthcare), 4% CHAPS, and 1% bromophenol blue and then centrifuged at 4°C and 23000g for 30 min to remove debris. The supernatant was then mixed with a rehydration buffer consisting of 8 M urea, 4% CHAPS, 0.5% IPG buffer 3-10 NL (GE Healthcare), 19 mM DTT, and 5.5 mM Orange G to a final volume of 350 µL. The first dimension was performed by in-gel rehydration for 12 h in 30 V on 18 cm pH 4-7 linear or pH 3-10 nonlinear IPG strips (Immobiline DryStrips, GE Healthcare). The proteins were then focused at 53000 Vh at a maximum voltage of 8000 V (Görg et al., 2000). The second dimension (SDS-PAGE) was performed by transferring the pI focused proteins (IPG strips) to homogeneous or gradient home-cast gels on gel bonds. The electrophoresis was performed at 40-800 V, 10°C, 20-40 mA, overnight.

2.7 Staining and image analysis

Sypro Ruby (Bio-Rad) staining were done according to manufacturer's instructions. In short, gels to be stained with Sypro Ruby were directly placed in a fixing solution containing 10% methanol and 7% acetic acid for at least 20 minutes after 2-DE. Gels were then washed 3x10 minutes under agitation with Milli-Q water before approximately 400 mL of Sypro Ruby stain were added and incubated in room temperature over night.

Silver staining of gels were done according to Shevchenko et al. 1996, with some few modifications. Proteins were fixed by incubating the gel in 50% methanol and 5% acetic acid for at least 20 minutes directly after 2-DE and then incubated with 50 % Methanol for 5 minutes, followed by Milli-Q water for 10 minutes. In the sensitizing step the gel was incubated with 0.02 % sodium thiosulphate for 1 minute, followed by 2x1 minutes washing with Milli-Q water. The gel was then immersed in 0.1 % silver nitrate solution for 20 minutes before excess of silver was washed away by 2x1 minute in Milli-Q water. Next, the gel was developed in 0.04 % formaldehyde in 2 % sodium bicarbonate solution for 2x1 minute. The exact developing time was optimized depending of the protein amount in the gel. Finally, the reaction was stopped by incubation 1x5 min in 0.5 % glycine and the gel washed with Milli-Q water for 2x20 min.

The images of the protein patterns were analyzed by a CCD (Charge-Coupled Device) camera digitizing at 1340*1040 pixel resolution in a UV scanning illumination mode for Sypro Ruby stained gels or at 1024*1024 pixel resolution in white light mode for silver stained gels using a Flour-S-Multi Imager in combination with a computerized imaging 12-bit system (PDQuest 2-D gel analysis software, version 7.1.1). The unit of the UV light source is expressed in counts while the unit of the white light source is expressed as optical density (OD). Gel images were evaluated by spot detection, spot intensities and geometric properties.

2.8 Isolation of protein spots

Protein spots were excised from the gels using a syringe and transferred to eppendorf tubes. For silver destaining, 25 µL of 100 mM sodium thiosulphate and 25 µL of potassium ferricyanide were added to the gel pieces (Gharahdaghi et al., 1999). When the pieces were completely destained, the chemicals were removed by washing (6×5 min with Milli-Q water) before addition of 50 µL of 200 mM ammonium bicarbonate and incubation for 20 min at room temperature. The gel pieces were washed (3x5 min with Milli-Q water) and dehydrated with 100% acetonitrile (ACN) for 5 min or until the gel pieces were opaque white. After removal of the ACN, the gel pieces were dried in a SpeedVac vacuum concentration system (Savant, Farmingdale, NY). Protein spots excised from Sypro Ruby stained gels were washed in 50% ACN/25 mM ammonium bicarbonate 2x30 min prior to dehydration with 100% ACN.

2.9 Digestion

2.9.1 Tryptic digestion

The protein spots were excised from the gel with a syringe and transferred to small eppendorf tubes (0.5 mL). Proteins from fluorescently stained gels were visualized and excised on a blue light transluminator (DR-180 B from Clara Chemical Research, Denver, CO, USA) wearing darkened amber glasses. After destaining and dehydration, about 25 µL trypsin (20 mg/mL in 25 mM ammonium bicarbonate, Promega, Madison, WI, USA) was added to each gel piece. To minimize autocatalytic activity, the samples were kept on ice for 30 min, prior to incubation in 37° C over night. The supernatant was transferred to a separate tube and the peptides were further extracted from the gel piece by incubation in 50% ACN/5% trifluoroacetic acid (TFA, Sigma-Aldrich) for 5 h at room temperature. The supernatant from the two steps was then pooled and dried in SpeedVac until complete dryness. If not dissolved in 5 µL 0.1% TFA for further MS preparation, the proteins were stored at -70°C.

2.9.2 Glu-C digestion

Glu-C (Roche, Basel, Switzerland) was diluted with 25 mM NH_4HCO_3, pH 7.8, to a concentration of 20 µg/mL. 25 µL was added to each gel piece and incubated at room temperature over night. The supernatant was then dried in SpeedVac until complete dryness. If not dissolved in 5 µL 0.1% TFA for further MS preparation the proteins were stored at -70°C.

2.9.3 Cyanobromide digestion

One Cynobromide (CNBr, Sigma-Aldrich) crystal was dissolved in 250 µL 70% TFA (in dH_2O). 25 µL CNBr in 70% TFA was added to the dried gel piece and incubated in darkness in room temperature overnight. The supernatant was then dried in SpeedVac until complete dryness. If not dissolved in 5 µL 0.1% TFA for further MS preparation the proteins were stored at -70°C.

2.9.4 Endoproteinase Asp-N digestion

Asp-N (P3303, Sigma-Aldrich) was diluted with 100 mM NH_4HCO_3 pH 8.5 to an enzyme concentration of 8 µg/mL. 25 µL was added to one tube containing one gel piece. To minimize autocatalytic activity, the samples were kept on ice for 30 min, prior to incubation in 37° C over night. The supernatant was then dried in SpeedVac until complete dryness. If not dissolved in 5 µL 0.1% TFA for further MS preparation the proteins were stored at -70°C.

2.9.5 Enzymatic deglycosylation

HDL (500 µg) was lyophilized and dissolved in 100 µL 2-DE sample buffer (described earlier).The sample was then incubated with 20 U of PNGase F (Sigma-Aldrich, P7367) in 37° C over night. After incubation, sample were stored in -20°C until 2-DE.

HDL was desalted in a PD-10 column and eluted in a 50 mM Na_2HPO_4, pH 6.0 buffer. A volume corresponding to 1200 µg of HDL proteins were then incubated with 10 U of Neuraminidase (Sigma-Aldrich, N3786) in 37° C for 4 hours. After incubation, sample was desalted in a PD-10 column with desalting buffer (NH_4HCO_3, 12mM, pH 7.1) and subsequently frozen in -70°C. Sample was lyophilized before 2-DE.

2.10 ZipTip

After digestion and drying of peptide samples, some were desalted and purified by the use of $ZipTip_{C18}$® pipette tips (Millipore, Billerica, MA, USA). Samples were diluted up to 10 µL with 0.1% TFA. A ZipTip was wet by loading 3x10 µL with 50% ACN and discarding the liquid. Ziptip was equilibrated by loading 3x10 µL of 0.1% TFA and discarding the fluid before peptide sample was carefully loaded into the ZipTip by pipetting. The ZipTip was washed with 5x10 µL of 0.1% TFA before peptides were eluted with 10 µL 50% ACN.

2.11 Mass spectrometry

Peptides obtained after digestion were mixed 1:1 with matrix, α-Cyano-4-hydroxycinnamic acid (CHCA, 0.02 mg/mL) or 2,5-dihydroxybenzoic acid (DHB, 0.02 mg/mL) in 70% ACN/0.3% TFA, and then spotted onto a stainless steel target plate. Analyses of peptide masses were performed using MALDI TOF-MS (Voyager DE PRO; Applied Biosystems) equipped with a 337 nm N2 laser operated in reflector mode with delayed extraction. Positive ionization, a delay time of 200 ns, and an accelerating voltage of 20 kV were used to collect spectra in the mass range of 600–3600 Da. Data processing of the spectra was performed in a Data Explorer TM Version 4.0 (Applied Biosystems). External mass calibration with a standard peptide mixture and internal calibration using known trypsin autolysis peaks (*m/z* 842.5100, 1045.5642, 2211.1046) were also performed prior to the database search. For tandem MS analysis, the digested peptides were dried and dissolved in 10 µL 0.1% TFA. The peptides were desalted and purified by using $ZipTip_{C18}$® columns. Elution was acidified by the addition of 1% formic acid. About 2 µL of the sample was applied into a silver-coated glass capillary and analyzed by a hybrid (triple quadrupole-TOF) mass spectrometer (API Q-STAR Pulzer; Applied Biosystems) equipped with a nanoelectrospray ion source (MDS-Protana, Odense, Danmark) operated in the nanopositive mode. Data processing was performed with Analyst QS software (Applied Biosystems). Fragmentation spectra were interpreted manually.

2.12 Database search

Peptide masses (major peaks) in the spectra were submitted to database search. NCBI and Swiss-Prot were used with Aldente or MS-Fit as search engines. Restrictions were human species, mass tolerance >75 ppm in most of the searches, maximum one missed cleavage, and cysteine modification by carbamidomethylation. MS-Digest, MS product, and BLAST search was used for protein identification of the derived tags resulting from amino acid sequencing with MS/MS. In peptide mass fingerprinting, protein matches with p-values below 0.05 are used and with LC-MS/MS analyses an FDR ≤ 1 % is considered significant.

2.13 Western blot

Plasma proteins were separated on 2-DE. Proteins were then transferred to a 0.2 µm PVDF membrane. After blocking with 5% non-fat dry milk in Tris buffered saline (TBS) overnight, the membrane was washed two times with Tween-Tris buffered saline (TTBS, pH 7.5) and then incubated overnight with primary rabbit anti human C-III antibodies (Abcam, 21032, 1:5000) in 2% non-fat dry milk in TTBS (pH 7.5) at room temperature. After washing four times with TTBS, the membrane was further incubated for 1h with secondary goat anti rabbit antibodies conjugated with horse radish peroxidase (HRP, 170-6515, BioRad,1:40 000) in 2% non-fat dry milk in TTBS (pH 7.5) at room temperature. In order to visualize the proteins the PVDF membrane was treated with ECL Plus Western Blotting Detection System (GE Healthcare) and then exposed to X-ray film (AGFA Medical, Mortsel, Belgium).

3. Results and discussion

3.1 Nanoparticle characterization

DLS, which is also known as Photon Correlation Spectroscopy or Quasi-Electron Light Scattering, is a technique used to study the size and size distribution of particles suspended in a liquid. The technique is based on the scattering of light of particles in diffusive random (brownian) motion. The average displacement for the Brownian motion is defined by the translational diffusion coefficient (D). The particle diffusive motion in liquid is size dependent, and a larger particle has a slower motion as compared to a smaller particle. This brownian motion can be investigated by irradiating the sample with a coherent laser and studying the intensity fluctuations of the scattered light (Finsy, 1994).

Particle sizing can be done in several ways. Typically the information retrieved from different techniques is to some extent diverse, as each technique is sensitive to it´s specific properties of the particles. That means that a technique which is based upon the scattering intensity does not deliver the same size or size distribution as a technique that is based upon the projected area or the density of a nanoparticle. For nano sized particles, transmission electron microscopy (TEM) is frequently used in purpose to study the size and the shape of particles. In TEM, the sample preparation together with the measurement is relatively time consuming and furthermore the measurement is limited only to a very small fraction of the sample. This means that a lot of replicates must be studied in order to achieve good statistics. A dynamic light scattering measurement on the other hand, is fast and convenient as it usually takes only a few minutes to perform. Data recording procedure is thus short but the analysis and interpretation requires knowledge and care. DLS measurements must be performed with highly diluted solutions, to avoid multiple scattering phenomenons and misleading artifacts are frequently present in DLS studies.

Particle sizes obtained when measuring with DLS are by default larger than those obtained when analyzing the material with TEM. The size calculated from the translational diffusion coefficient in DLS generally is referred to as the hydrodynamic diameter, e.g the diameter of a sphere having the same diffusion coefficient as the particle, while the size of the nanoparticles obtained from TEM is the core size of the nanoparticle investigated. It should be noted that in many cases when the sample consists of a mixture of nanoparticles with a range of sizes and/or mixture of particle shapes, results should be taken as an estimation only but clearly trends can be observed.

The autocorrelation function of the scattered intensity results in an average value of the product of the intensity at time t and the intensity at a time delay later t+dt. The value obtained from the correlation function is large for short delays, since the intensities are highly correlated. The value will be low for longer delays; i.e. the autocorrelation can be described as a decaying function of time delay. From the autocorrelation function, the diffusion constant D can be determined. Furthermore, by using Stokes-Einstein equation the corresponding size distribution is calculated according to:

$$d = k_B T / 3\pi\eta D \qquad (1)$$

where k_B is the Boltzman constant, T the temperature, η the viscosity of the solvent and d the hydrodynamic diameter of the particles. This implies that the temperature must be constant during the measurement and the viscosity of the sample solvent must be known. It should be noted that the formula shown above (Equation 1) is valid only for non-interacting spherically shaped particles, i.e experimental data are fitted to a model assuming spherical particles (Finsy, 1994).

Figure 1 shows five different functions representing the typical information retrieved in DLS in the actual measurement and by using algorithms. The measured data in DLS is the correlation function (Figure 1A). This function holds information about the diffusion of particles in the sample and can be transformed to a graph showing the decay time of light scattering fluctuations (Figure 1B). From the decay time distribution function, the values of the particle radius (Figure 1C) can be calculated using the Stokes-Einstein equation.

The amount of light that is scattered from a particle is dependent on the particle size. According to the Rayleigh theory (Sorensen, 2008), the scattering factor roughly is proportional to the sixth power of the particle size. This means that a small particle scatter light less than a larger particle and thus different weights have to be applied to transform the intensity weighted data to a useful size distribution. These weights can be mass based (Figure 1D) or number based (Figure 1E). Powder based samples that are dispersed in a liquid often are severely aggregated. Ultrasonic baths can be used to decrease the aggregation in the solution. Choice of solvent is also important and it clearly affects the capability of dispersing nanoparticles.

A number of examples of DLS results for commercial particles are shown in Figure 2, 3 and 4. The size distributions of Al_2O_3 nanoparticles , based upon number weighted fits of the data are shown in Figure 2. A set of samples were dispersed both in Milli-Q water and PBS. The size and size distribution were measured as a function of concentration. According to the supplier these particles are < 50 nm in size as measured by TEM, which should be taken as the core size of the nanocrystals in the Al_2O_3 material. A lot larger hydrodynamic diameters are achieved in our measurements, which show that the water based sample is composed of at least two populations with hydrodynamic radius of about 100 nm and 200-300 nm respectively. The smallest radius from this DLS study was achieved for Al_2O_3 dispersed in Milli-Q water and ultrasonicated with a bar homogenizer (radius approx. 30 nm). Also for samples dispersed in PBS, ultrasonication with a bar homogenizer indicates decrease of the aggregation and introduction of populations with smaller radius.

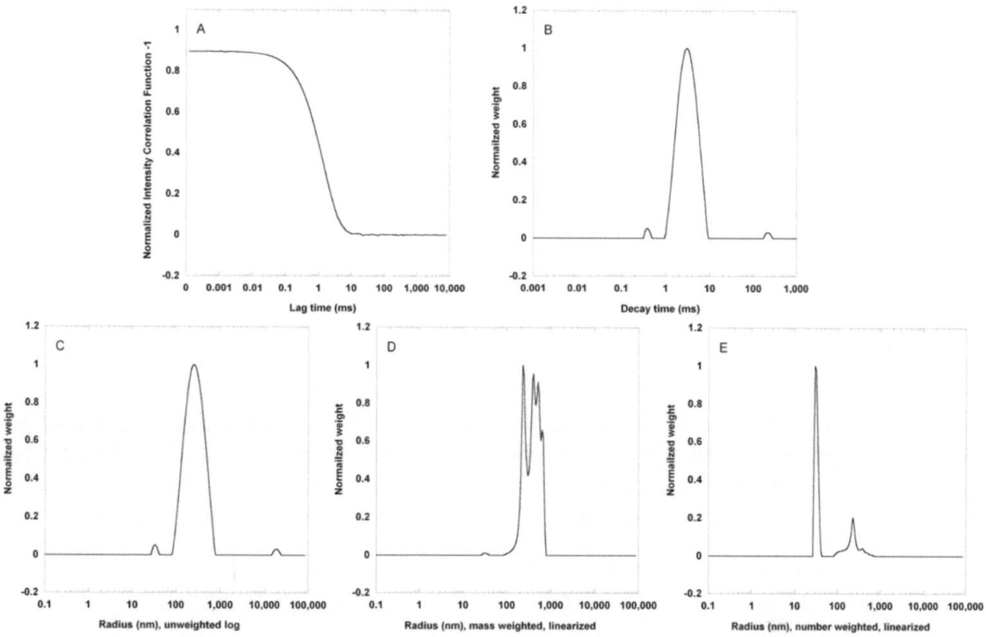

Fig. 1. Example of the typical information retrieved in a DLS measurement.
The measurement is performed on Al doped ZnO nanoparticles dissolved in MilliQ water. A) shows the correlation function which has a high correlation (close to 1) for really short lag times but decays to zero with a rate dependent on the particle size distribution in the sample. B) shows the normalized distribution function of the Decay time. C) shows the normalized distribution of the unweighted radius and the corresponding normalized distributions of the mass weighted radius and the number weighted radius are shown in D) and E) respectively.

Figure 3 shows the size distribution of Aluminium doped ZnO nanoparticles dispersed in water and PBS. Bar ultrasonication introduces smaller sized populations, in consistence with the results in Figure 2. The hydrodynamic radii of these particles in MilliQ water are around 30 nm as largest according to Figure 3. This could be compared with the information from the supplier that these particles are < 50 nm as measured by TEM. Again hydrodynamic diameter is always larger that the core size of nanocrystals obtained from TEM. In this case the sample is also aggregated in solution, which produces even larger sizes and size distributions.

The SiO_2 particles are 7 nm sized as primary particles according to the supplier based on calculations using the surface area as measured by the nitrogen adsorption method of Brunauer (Brunauer et al., 1938). The supplier also states that these particles commonly form

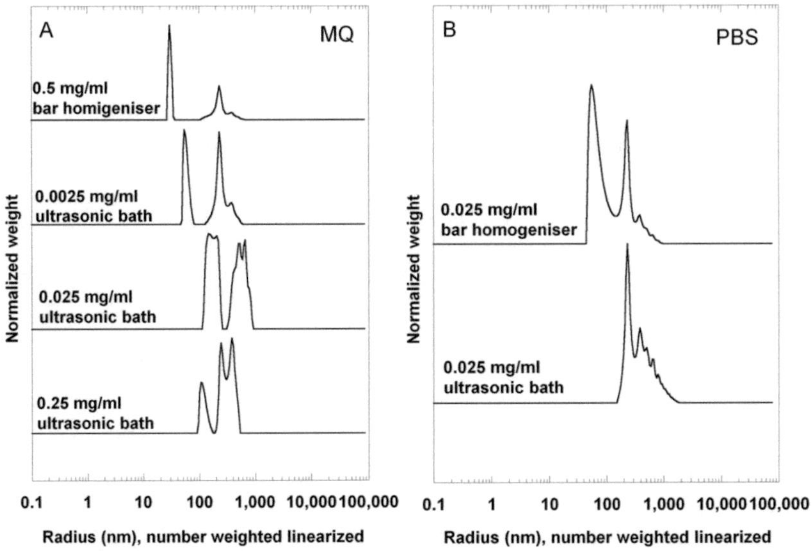

Fig. 2. DLS measurements of Al_2O_3 nanoparticles.
The normalized distribution of the number weighted radii of particles in Al_2O_3 nanopowder diluted to different concentrations in A) MilliQ and B) PBS as measured by DLS. As marked in the figure, two different kinds of ultrasonic treatment were used to decrease the aggregation before performing the measurement; either an ultrasonic bath or a bar homogenizer.

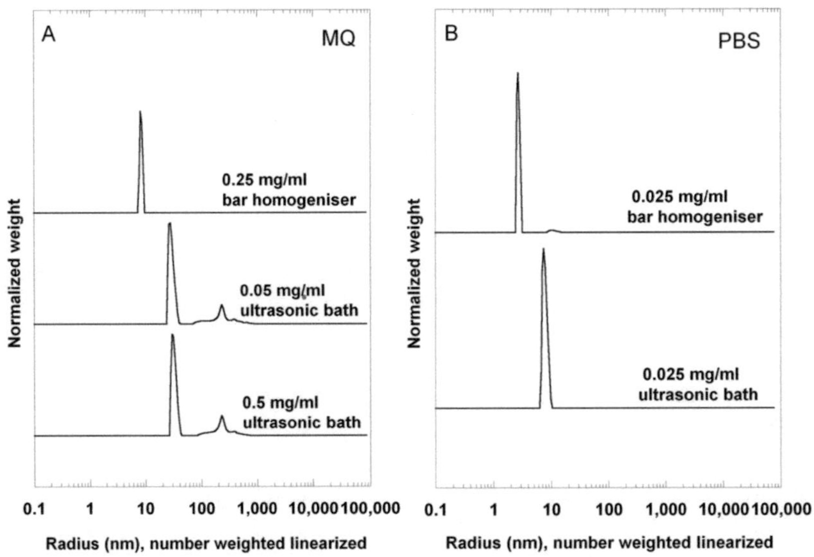

Fig. 3. DLS measurements of ZnO (Aluminium 6% doped) nanoparticles.
The normalized distribution of the number weighted radii of particles in ZnO (Aluminium 6% doped) nanopowder diluted to different concentrations in A) MilliQ and B) PBS as measured by DLS. As marked in the figure, two different kinds of ultrasonic treatment were used to decrease the aggregation before performing the measurement; either an ultrasonic bath or a bar homogenizer.

some hundreds of nanometer long chainlike branched aggregates. The size distributions obtained in our DLS measurements are presented in Figure 4. For all three concentrations in Milli-Q water the particle size is below 100 nm in radius. When SiO_2 particles are dispersed in PBS (Figure 4B), large aggregates are definitely present, which are partly removed when ultrasonicated with a bar homogenizer. Multiple scattering, surface charge on the nanoparticles and water solubility should be considered when further evaluating these data. Furthermore, it is known that these specific samples (SiO_2) are inhomogenous and the sample is thus far from ideal, i.e does not contain spherical shaped particles. It is shown in previous studies that long chainlike branched aggregates are present.

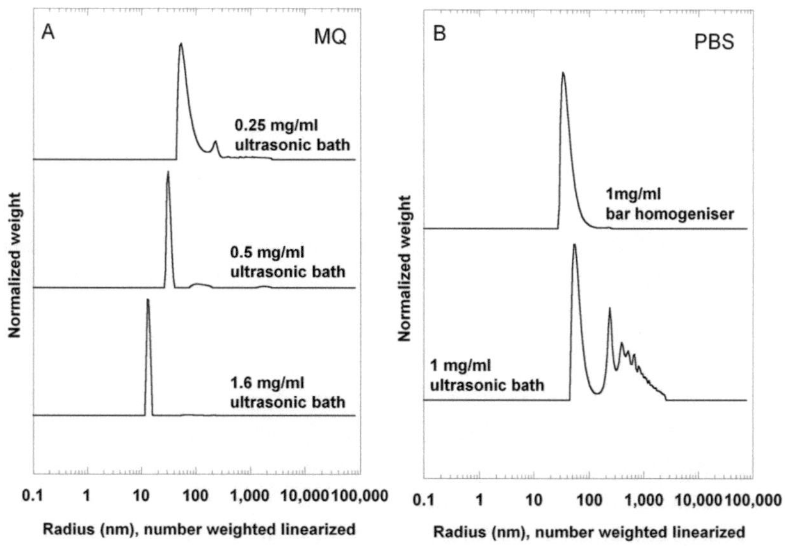

Fig. 4. DLS measurements of SiO_2 nanoparticles.
The normalized distribution of the number weighted radii of particles in SiO_2 nanopowder diluted to different concentrations in A) MilliQ and B) PBS as measured by DLS. As marked in the figure, two different kinds of ultrasonic treatment were used to decrease the aggregation before performing the measurement; either an ultrasonic bath or a bar homogenizer.

In conclusion, it could be said that size and size distribution of nanoparticle samples could be estimated by DLS. Valuable information as the trends in size distribution connected to sample preparation methods and choice of solvent can be obtained. Sample preparation methods are indeed very important as well as choice of solvent. Care should be taken when choosing fitting model and the model-inbuilt parameters. Inhomogeneous samples are less straight forward to analyze. Presence of aggregates is easily detected. In summary information obtained from DLS is important for everybody that is doing research on nanoparticles in liquids. The numbers given as product information i.e. the size and size distribution, are often relevant for the core-size of the nanocrystals within the material. However the nanoparticles are most often not soluble to that extent. Nanoparticles obtained in dry state and then dispersed in liquid usually form aggregates as shown in this study.

3.2 Nanoparticle-plasmaprotein interactions

In a previous study performed by us, the inflammatory response in human monocyte derived macrophages after exposure to wear particles generated from the interface of studded tires and granite containing pavement (Karlsson et al. 2011) was investigated. Particle characterization showed that dominating peaks in the EDX spectra were Silica and Aluminium. Particles of nanosize were also present (SMPS), but it was not possible to characterize them due to low abundance. As a result of their very small diameter (< 0.1 µm), inhaled nanoparticles are believed to be predominantly agglomerated and deposited in the periphery of the lungs, where they interact with cells such as macrophages and epithelial cells (Beck-Speier et al. 2005) but they may also translocate into the circulation, which is a critical step, since their fate *in vivo* in not known. Investigating plasma protein-nanoparticle interactions with a toxico-proteomic approach is a useful tool to improve our knowledge about the effects of nanoparticles of different origin, size and surface properties in biological systems.

In purpose to mimic a potential exposure to airborn nanoparticles translocated into the circulation, commercial SiO_2 and Al_2O_3 were mixed with plasma proteins. As comparison, commercial ZnO (Al-doped 6%) and a non-metal oxide; single walled Carbon Nanotubes was used. All preparations were performed in triplicates with three different subjects exposed to each type of particles. The protein patterns resulting from the three different exposures of commercial SiO_2, Al_2O_3, ZnO and CNO respectively were identical.

Particle characterization and estimation of particle agglomeration prior to exposure is crucial. In a recent study of nanoparticle-plasma protein interactions (Deng et al. 2009), the DLS spectra indicated large agglomerates prior to plasma protein exposure. Most likely, and in line with the authors suggestions (Deng et al. 2009), complexes with hydrodynamic size of 10000-100 000 nm do not result in the same protein patterns as the interactions of smaller particles/agglomerates and plasma proteins. In our optimized protocol, with different particle origin, less gentle bar sonication instead of in water bath and thereby reduced hydro dynamic sizes of the agglomerates (Fig 2-4) - an altered pattern of interacting proteins was indeed found (Figure 5, Table 1).

Interestingly, the interaction of SiO_2 and CNO with plasma proteins resulted in very similar protein patterns despite their different properties (Fig 5). It has to be stated though, that the fate of CNO in the lung may not be translocation into the circulation due to the tube like structure, but CNO is also of interest for medical applications (Wu et al. 2011).

The transport proteins Albumin and Alpha-2-HS-glycoprotein interacted with SiO_2, CNO and ZnO but not Al_2O_3 while Transferrin interacted with SiO_2, Al_2O_3 and notably also CNO but not ZnO. Supporting our results, the binding of albumin to single walled CNO has previously been described to promote uptake by the scavenger receptor in RAW cells (Dutta et al., 2007). In line, intravenous administration of CNO has in a different study resulted in high localization in the liver (Cherukuri et al., 2006). Another possible way for nanoparticles into the cells, are as Transferrin/particle complexes that are able to enter the cells via the Transferrin receptor. The Transferrin receptor is an interesting and relevant target in cancer research since its expression is increased in tumor cells. In a recent study, Transferrin covalently attached to silica nanoparticles carrying a hydrophobic drug caused an increase in mortality of the targeted cancer cells compared to cells exposed to nontargeted particles and free drug (Ferris et al. 2011). What demand further studies though, is the fate of the silica particles in a longer perspective, taken up by tumor cells as well as other cells. A third transport protein, the thyroxin transporting protein Transthyretin, also known to bind toxic components in the blood stream (Hamers et al., 2011), was found to interact with CNO and SiO_2. This finding confirms a previous study that pointed out that silica interaction (inhaled) with plasma Transthyretin is contributing to the stabilization of fibroids in rat lungs (Kim et al., 2005). The possible effects of CNO interaction with Transthyretin remains to be investigated.

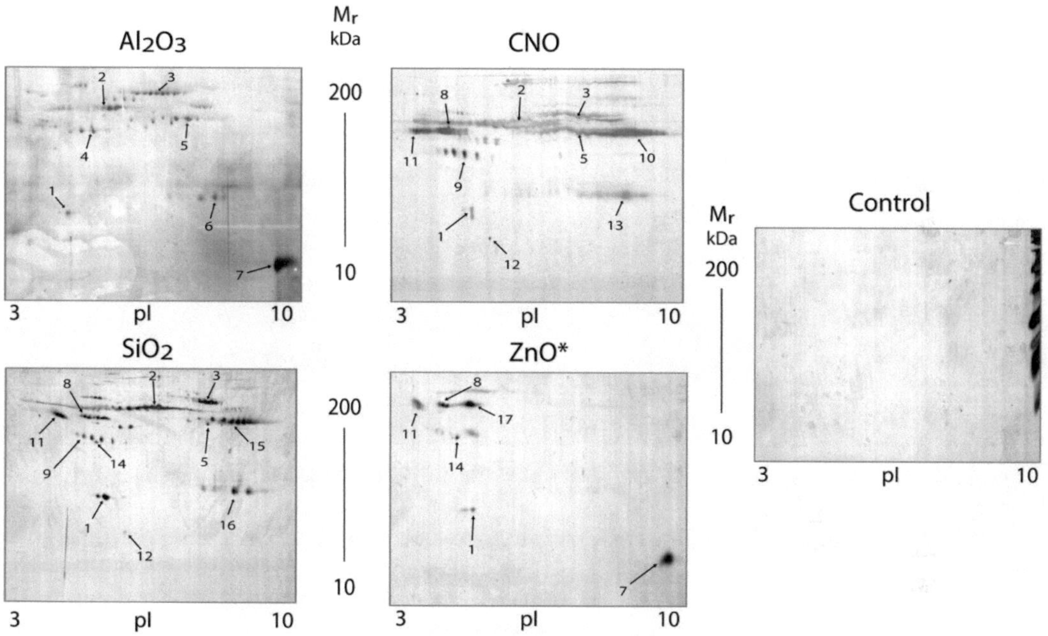

Fig. 5. Plasma protein binding profiles of different nanoparticles.
Four different nanoparticles; Al_2O_3, CNO, ZnO (* Aluminium-doped 6%) and SiO_2, were mixed with human plasma and isolated through ultracentrifugation. The protein contents were then separated by 2-DE and silver stained. Bound proteins were identified by MS as shown in Table 1.

The inflammatory marker Fibrinogen was found to interact with SiO_2, Al_2O_3 and CNO but interestingly not with ZnO (Fig 5). Fibrinogen binds foreign substances in the circulation and promotes macrophage activation - a mechanism that may result in retention of particle/protein complexes in the intima with accompanying cardiovascular complications (Shulz et al., 2005). On the other hand, Lysozyme C, a well known anti-bacterial agent, was only found on Al_2O_3 and ZnO. Under the present condition, it was unfortunately not determined if the binding to ZnO occurred due to the 6% Al doping of the ZnO. Lysozyme has previously been found to interact with nano-TiO_2 particles (Xu et al., 2010). They reported that the coexistence of nano-TiO_2 particles and Lysozyme resulted in the transition of Lysozyme conformation from α-helix into β-sheet secondary structure and a substantial inactivation of Lysozyme. Moreover the β-sheets are able to induce the formation of amyloid fibrils, a process which plays a major role in pathology.

Number in Figure 2	Protein	Found in Nano particle	Uniprot AccessionNumber	pI^a	Mw^a (Da)	Matched Peaks[b]	Sequence Coverage (%)	MOWSE Score
1	ApoA-I	Al_2O_3, CNO, SiO_2, ZNO	P02647	5.2	24500	28	74	9.65e+10
2	Albumin	Al_2O_3, CNO, SiO_2	P02768	6.0	68000	30	47	4.20e+14
3	Transferrin	Al_2O_3, CNO, SiO_2	Q53H26	6.7	77000	43	65	1.23e+22
4	Fibrinogen γ	Al_2O_3	P02679	5.4	50000	15	32	1.39e+6
5	Fibrinogen β	Al_2O_3, CNO	P02675	7.0	55000	17	37	5.83e+8
6	Ig Light chain	Al_2O_3	Q0KKI6	7.4	30000	5	32	2779
7	Lysozyme C	Al_2O_3, ZNO	P61626	9.4	15000	7	33	12888
8	α1-AT	CNO, SiO_2, ZNO	P01009	5.0	55000	31	63	5.36e+18
9	Haptoglobin	CNO, SiO_2	P00738	5.1	46000	8	20	2115
10	DKFZ	CNO	Q6N096	8.3	55000	14	34	1.10e+7
11	Alpha-2-HS	CNO, SiO_2, ZNO	P02765	4.8	55000	6	14	681
12	Transthyretin	CNO, SiO_2	P02766	5.4	16000	7	51	69267
13	Ig KC	CNO	Q6PJF2	8.0	30000	10	55	409936
14	ApoA-IV	SiO_2, ZNO	P06727	5.1	45000	10	23	72729
15	Igγ	SiO_2	P01859	7.7	55000	7	23	79647
16	Igγ-1 chain	SiO_2	P01857	8.5	35000	12	40	3.74e+7
17	Amyloid βA4	ZNO	B4DJT9	5.2	60000	20	15	53.1

Table 1. Identification of nanoparticle bound plasma proteins by peptide mass fingerprinting after 2-DE.
The table shows identified proteins with Uniprot accession number, isoelectric point (pI), molecule weight (Mw), number of peptide masses matched, sequence coverage and MOWSE score. [a] Isoelectric point (pI) and molecular weight (Mw) as estimated on gels. [b] Matched peak masses with a mass error tolerance of 75 ppm.

Furthermore, some antigen binding proteins; IgKC, DKFZ and IgLC, were also found to interact with Al_2O_3, CNO and SiO_2 but were not detectable on ZnO (Fig 5). Immunoglobulins are able to activate the complement system but they also often represent unspecific binding during protein purification caused by insufficient washing. Our results compared to previous findings indicates that increased hydrodynamic size of nanoparticle agglomerates seems to correlate to increased amounts of immunoglobulins. Overall ZnO was not binding as many proteins as the other particles and may even bind less without being Al-doped but an interesting finding in the ZnO preparation was a protein only described on transcript level and highly similar to the protein Amyloid β A4 precursor. This family of proteins acts as chelators of metal ions such as iron and zinc. They are also able to induce histidine-bridging between beta-amyloid molecules resulting in beta-amyloid-metal aggregates and it has been reported that extracellular zinc-binding increases binding of heparin to Amyloid β A4 (Uniprot 2011).

The protease inhibitor Alpha-1-antitrypsin, was found in the CNO, SiO_2 and ZnO preparations but not in Al_2O_3 while another antioxidant, Haptoglobin was found only on CNO and SiO_2. To our knowledge, the binding of Alpha-1-antitrypsin to the nanoparticles in this study has not been described previously but an increase of Haptoglobin has been reported in a study investigating acute phase proteins as biomarkers for predicting the exposure and toxicity of nanomaterials (Higashisaka et al., 2011).

At last, the HDL associated Apo A-I, with well known anti-endotoxin activity (Henning et al., 2006) and receptor interaction properties was found in all preparations but was most abundant after SiO_2 exposure. Apo A-I may be acting as a scavenger clearing the particles from the blood stream via the scavenger class B-I receptor (SRBI). The SRBI receptor is mainly located on the liver and plays an important role in cholesterol efflux (Verger et al., 2011) but is also present on other cells (Mooberry et al. 2010). Apolipoproteins in general are of interest for the pharmaceutical industry as carriers of nanoparticle bound drugs for brain uptake (Kreuter et al., 2005) since apo E and apo B-100 are taken up by the cells via receptor mediated endocytosis. The hypothesis is that the nano-particle/apolipoprotein complex mimics the natural lipoprotein particle. The identity of Apo A-I was, as the other proteins, confirmed by peptide mass fingerprinting using MALDI TOF MS (Table 1) and one dominating Apo A-I peptide in the MS spectra (Figure 6A) was in addition sequenced by MS/MS (Figure 6B) to further confirm the identity.

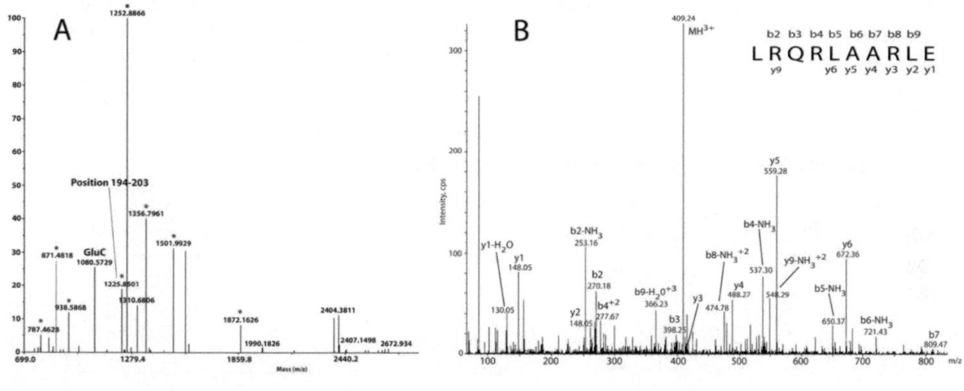

Fig. 6. Identification of nanoparticle bound apoliprotein A-I with mass spectrometry after 2-DE.
A: Peptide mass fingerprint spectrum obtained by MALDI-TOF mass spectrometry after endoproteinase GluC digestion. Asterisks represent peaks corresponding to peptide masses of apo A-I. B: Sequencing of a triply charged peptide (m/z 409.9) corresponding to position 194-203 of apo A-I by collision induced disassociation (CID) in a tandem mass spectrometer. The amino acid sequence with ions corresponding to the different fragments is shown in the upper right corner.

Overall the binding of plasma proteins to nanoparticles, based on previous and our findings, seems to vary with origin, surface properties, size and thereby also diameter of agglomerates. Particle characterization prior to exposure for plasma proteins or cells is therefore extremely important to receive reliable results that are possible to interpret. Interacting proteins under the present conditions are dominated by proteins involved in the immune defense and reverse transport to the liver but notably also proteins mediating brain uptake.

4. Methodological considerations and improvements of 2-DE and MALDI-TOF MS

4.1 Plasma sample preparations

Sample preparation is an important step that influences the separation of proteins with 2-DE. A common problem in most biological samples is the presence of salt ions. In 2-DE, salt concentrations >10 mM affects the isoelectric focusing step and markedly reduces the effectiveness of the charge separation. There are several easy ways to remove salts, e.g. by precipitation of the proteins or by gelfiltration in small desalting columns. Plasma samples are possible to analyze directly with 2-DE since the high protein concentration allows a simple dilution of the sample to lower the salt concentration. However, as illustrated in figure 7, a desalting step besides the dilution still improves the protein pattern. Another well-known problem with biological fluids is the presence of a few highly abundant proteins that may prevail over the low abundant proteins. In plasma, albumin and immunoglobulin G (IgG) constitutes a very large proportion (about 75%) of the total protein loaded on the gels. This may lead to the proportion of low abundant proteins being below the detection limit. Also, the staining of the abundant proteins may interfere with proteins with similar molecular mass and pI. It is therefore advisable to remove albumin and IgG and there are several commercial removal kits available, usually based on antibodies directed towards albumin and IgG. As shown in figure 7, such sample preparation step removes a large fraction of these proteins and increases the proportion of the other proteins in the sample. However, it is important to realize that this step also introduces unspecific removal of proteins and it is our experience that this unwanted loss of proteins varies considerable between the different available removal kits.

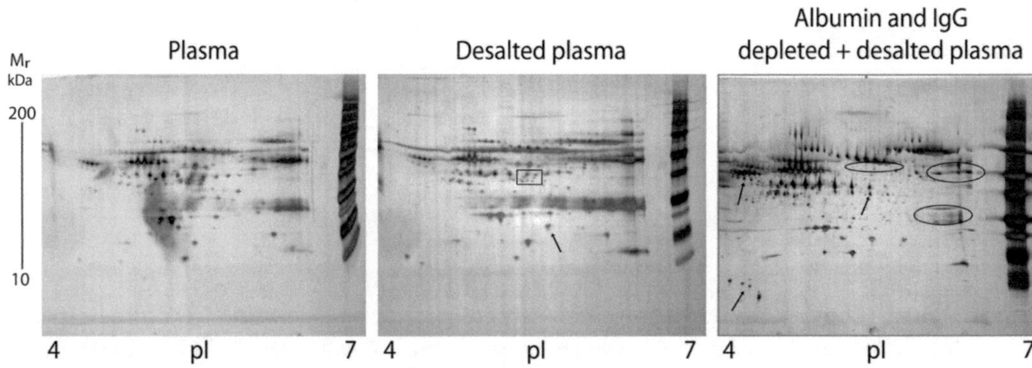

Fig. 7. Preparation of plasma for 2-DE
Untreated plasma (left), desalted plasma (middle) and desalted plasma after removal of albumin and immunoglobulin G (IgG), (right), were separated with 2-DE and silver stained. Arrows indicate protein or protein clusters with increased abundance after treatment compared to untreated plasma. The rectangle indicate an area with improved resolution after desalting. Positions for albumin and IgG chains are indicated with rings.

Plasma contains a wide variety of proteins, many of which are not detectable with 2-DE without further fractionation. In view of the lipid metabolism, important sub-fractions of plasma to study are the lipoproteins. As shown in table 1, one protein that interacts with nanoparticles is apo A-I, the major constituent of HDL. This implicates the need of more investigations of HDL as a possible target of nanoparticles that may influence the cholesterol metabolism and increase the risk of cardiovascular disease. HDL can be isolated based on density, size or protein content (e.g. apo A-I) using ultracentrifugation, size-exclusion chromatography or immune-affinity chromatography, respectively, each technique with its own merits and drawbacks. Thus, the rather harsh conditions during ultracentrifugation in high salt gradients may remove weakly associated proteins while the rather mild conditions during chromatography may favor unspecific co-purification of proteins with the cholesterol particle. We have previously mapped the protein content of HDL isolated by two-step density gradient ultracentrifugation (Karlsson et al., 2005). In this study we have compared ultracentrifugation and anti-apo A-I affinity chromatography to isolate HDL from the same plasma sample. As shown in figure 8, more proteins were obviously identified in immune-affinity purified HDL. However, some of the proteins must be considered as possible plasma contaminants as they were not, as e.g. the apolipoproteins, enriched in the HDL fraction.

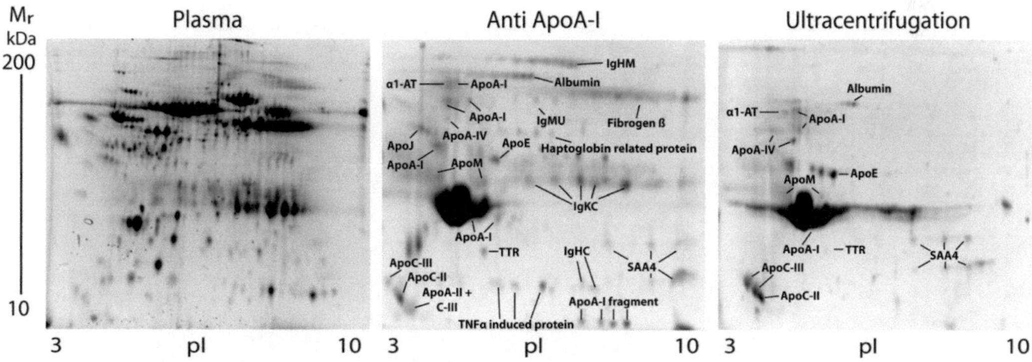

Fig. 8. Subfractionation of plasma with regard to HDL
Plasma (left), HDL purified according to apo A-I with immunoaffinity chromatography (middle) and HDL isolated according to density with ultracentrifugation (right) were separated with 2-DE and silver stained. Proteins were identified with mass spectrometry.

4.2 Protein detection

Gel-separated proteins can be visualized by several commonly used staining methods, including dyes (e.g. Coomassie Brilliant Blue and colloidal Coomassie), metals (e.g. silver staining) and fluorescent probes (e.g. Sypro staining and Cy-dyes) (Rabilloud, 2000). The stains interact differently with the proteins and have different limitations with regard to sensitivity, linear range, compatibility with mass spectrometry and type of proteins that stain best. In general, for staining of complex protein samples, silver staining can be considered the most sensitive technique (1-5 ng protein) and Coomassie Brilliant blue the least (50-100 ng) while the sensitivity of colloidal coomassie and the fluorescent dyes are in between. However, it is important to bear in mind that the different stains interact differently with the proteins and therefore one protein may stain very well with one staining method but not with the other (Fig. 9). For example, silver ions react with negatively charged groups and therefore stain glycoproteins containing negative sialic acid very well.

One the contrary, Sypro ruby that binds to proteins through hydrophobic interactions stains hydrophilic glycoproteins quite poorly. Being most sensitive, silver staining is obviously very useful for proteomic approaches. However, the advantage with silver is hampered by its rather low linear range, making it less suitable for quantification than the other staining techniques. For plasma analyses we have therefore adopted a double staining strategy. As illustrated in figure 9, the 2-DE gel is first stained with Sypro Ruby and the proteins are quantified within a high dynamic range. The gel then can be destained and restained by silver to detect additional proteins. As these additional proteins are less abundant their intensities usually are within the limited linear range of the silver staining technique. The proteins may then be selected for MS analyses. As the sample preparation protocol is more time-consuming for silver stained gels than Sypro stained gels it is convenient to pick as many proteins as possible after the first staining step. In this step it is also important to check the optimal destaining time before MS analyses, with plasma samples at least 90 minutes (fig 9). If the samples are not fully destained the signal to noise in the spectra are reduced.

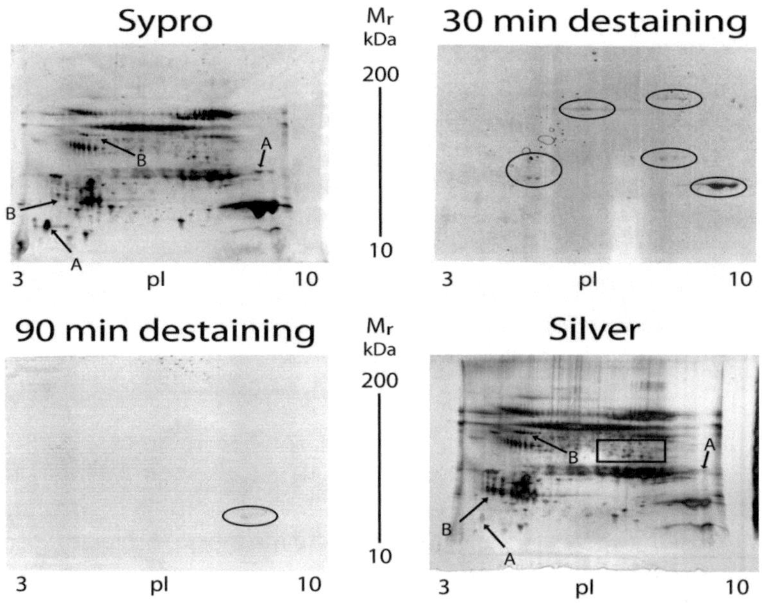

Fig. 9. Double staining of proteins
Plasma proteins were separated with 2-DE and first stained with Sypro Ruby. The gel was then destained with 25 mM ammonium bicarbonate/50 % acetonitrile buffer and the efficiency of the process checked after different time intervals. Finally, the proteins were re-stained with silver. A: Proteins more stained by Sypro compared to silver. B: Glycoproteins more stained by silver. The rectangle shows an area containing additional proteins detected by silver. Residual Sypro staining of proteins after destaining are indicated with rings.

4.3 Protein identification with peptide mass fingerprinting using MALDI-TOF MS

Peptide mass fingerprinting with MALDI-TOF MS is an excellent and robust technique for fast identification of plasma proteins after 2-DE (Lahm & Langen, 2000). However, several peptide peaks needs to be detected in the spectra with a high mass accuracy to avoid false positive results. There are several approaches to consider for improving the data from the analyses, such as digestion protocol, purification of peptides and choice of matrix.

4.3.1 Alternative digestion of samples

One of the most widely used ways to digest proteins before MS analyses is by trypsin, which cleaves C-terminally of lysine and arginine (not followed by proline). Although lysine and arginine often are distributed in the protein sequences in a way that provides sufficient number of peptides for identification after trypsin cleavage, this is not always so. Furthermore, it is sometimes necessary to use alternative digestion protocols in order to find specific peptides to e.g. characterize differences between protein isoforms. In these cases alternative enzymes, e.g. Asp-N (cleaves N-terminally of aspartic acid and cystein), Glu-C (C-terminally of glutamic acid) or chemical induced cleavage by CNBr that hydrolyzes C-terminally of methionine, is needed. In this study we used Asp-N as a complement to trypsin when identifying the plasma protein serum amyloid A4 (SAA4). This combined digestion approach generated almost 95 % sequence coverage of the protein (figure 10). SAA4 is a constitutively expressed protein which can be found in HDL as differently charged isoforms (fig 14, Karlsson et al., 2005). One explanation to these isoforms could be a small truncation of SAA4 in which one lysine and one tyrosine is removed C-terminally and thereby making the protein more acidic (Farwig et al., 2005). However, by using Asp-N we were able to detect both the intact C-terminal and the intact N-terminal peptide that were not possible after trypsin digestion, ruling out the presence of truncated SAA4 in our sample (figure 10). Besides the use of Asp-N to study SAA4, we have also used Glu-C to analyze apolipoprotein A-I (figure 6) and CNBr to study serum amyloid A-1/2 isoforms (figure 13).

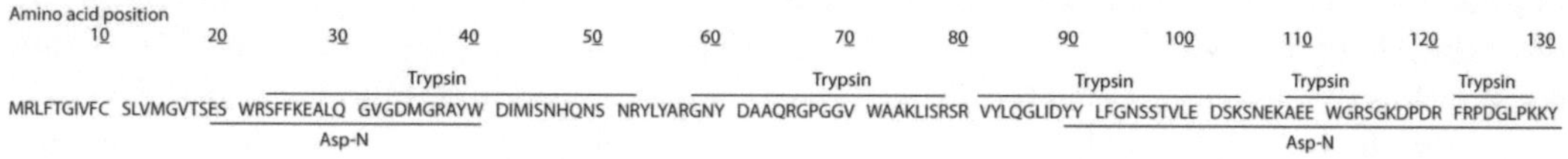

Fig. 10. Alternative digestion to improve sequence coverage of Serum amyloid A4 (SAA4).

Sequence coverage obtained through peptide mass fingerprinting with MALDI-TOF mass spectrometry by the use of trypsin and endoproteinase Asp-N as indicated by the lines. Sequence coverage (without the signal peptide in position 1-18) was 79.5 % with trypsin and 57.1 % with Asp-N. The combined sequence coverage was 93.8%.

4.3.2 Peptide sample cleaning

Peptide samples after in-gel digestion can be cleaned by adsorption to C-18 containing pipette tips (ZipTip®). This clean-up procedure done manually is rather time-consuming but is absolutely necessary before electrospray-quadrupole MS. On the other hand, with a MALDI-TOF instrument, being more insensitive to salts and other contaminants, it is not that obvious. To investigate the possible advantage with ziptip cleaning before MALDI-TOF MS we picked 17 sypro stained proteins after 2-DE. The proteins were digested with trypsin and an aliquot of the obtained peptide solution was purified by ZipTip® (50% ACN elution solution according to the protocol recommended by the supplier), mixed with the matrix CHCA and spotted on the MALDI-target plate and another aliquot of the peptide solution was mixed directly with the matrix and spotted on the same plate. All samples were then analyzed with MALDI-TOF MS with the same settings and with the laser induced collection of spectra in an automatic mode. The spectra were then used for NCBI database search with MS-fit using the same settings for all samples. All 17 proteins were identified with peptide mass fingerprinting in both ziptip cleaned and untreated samples. As shown in Table 2, ziptip cleaning significantly improved the peak intensities, signal to noise ratio and the mass accuracy. This illustrates that removal of salts and other contaminants that will compete with the peptides in the spectra increases the intensities of the peptide peaks and thereby increases the accuracy of the mass determinations and, as a consequence, also increases the reliability of the identifications. On the other hand, the number of peptides and sequence coverage found in ziptip cleaned samples were about the same as in the untreated samples (Table 2). In general, there was a clear tendency that in ziptip cleaned samples more peptides were detected in the lower mass region (<1000 m/z) while fewer peptides were detected in the higher mass region (>2000 m/z). This suggests that the removal of salt ions and low molecular chemicals with subsequent improved signal to noise increases the possibility to detect low molecular mass peptides but that this beneficial effect is counteracted by adsorption of larger peptides to the solid phase of the ziptip. To test this, 10 protein samples were sequential eluted from the ziptip with increasing ACN concentration, up to 90 %. Indeed, this procedure increased the number of peptides found and the sequence coverage increased from 52 +/- 16 % in the untreated samples to 58 +/- 16 in the ziptip cleaned samples ($p<0.05$). The effect varied among the different proteins but was in some samples quite profound, almost 2 times higher sequence coverage. It can be concluded from these experiments that purification of peptide samples with ziptip improves the results with MALDI-TOF MS. However, when it comes to the identification of proteins with peptide mass fingerprinting the beneficial effect of the cleaning procedure is quite limited as the number of peptides found, using the standard protocol, is not increased. Therefore, considering the work-load needed for the ziptip procedure, it is doubtful if it is practical to routinely clean samples with ziptip before MALDI-TOF analyses. However, for selected, low abundant, samples it can most likely make a significant difference for the identification. In these cases, sequential elution of the peptides from the ziptip with increasing acetonitrile concentrations is recommended.

	Average error (ppm)	**Sequence coverage (%)**	**Number of peptides**	**Signal/noise ratio**	**Peak Intensity**
Without Ziptip (n=17)	11.2 +/- 3.6	24.9 +/- 12.0	8 +/- 4	100 +/- 130	3500 +/- 2000
Ziptip (n=17)	7.9 +/- 4.4	24.1 +/- 11.4	8 +/- 3	200 +/- 480	6300 +/- 4000
Statistical significance	0.01	No	No	<0.001	<0.001

Table 2. Influence of sample cleanup of in-gel digested proteins on peptide mass fingerprinting data obtained with MALDI-TOF MS.

Proteins were separated by 2-DE and in-gel digested by trypsin. The same peptide samples were then purified by ZipTip, mixed with the matrix and spotted on the MALDI plate or directly mixed with the matrix and spotted on the plate. Statistical interpretations were done by Wilcoxons signed rank sum test.

4.3.3 Choice of matrix

To enhance the quality of the MALDI-TOF mass spectra and the number of desorbed peptides there are several matrices that could be considered to use. Different matrix compounds, both acidic and basic, have proved to work in sample preparation for MALDI mass analyzers. The far most commonly used matrix for peptide mass fingerprinting is CHCA, which is recommended for peptides with mass ions below 2500 Da (Beavis et al., 1992). Alternative matrices also used are sinapinic acid, mostly for masses higher than 25 kDa (Lewis et al., 2000) and DHB, originally suggested for glyco- or phosphopepides that are difficult to ionize (Strupat et al., 1991), but later also proven useful for silver stained proteins (Ghafouri et al., 2007). As illustrated in figure 11, CHCA and DHB have very different crystal structures on the target plate. Whereas CHCA usually has a homogeneously distributed spot appearance, making it ideal for automatic laser induced peptide desorption, the DHB crystals are needle shaped and often aggregated into fan-like structures directed from the outside towards the centre of the sample spot. Interestingly, we have found that peptides appear to be enriched in the base of the DHB structures (figure 11), significantly increasing the signal to noise ratio in spectra obtained from these areas. This is illustrated by the identification of transthyretin, one of the proteins that interact with silica nanoparticles and carbon nanotubes (figure 5). With DHB, the improved signal to noise in the spectrum displayed twice the number of peptides than with CHCA (figure 12). This dramatically increased the sequence coverage from 39 % obtained with CHCA to 72 % obtained with DHB. Thus, the use of DHB for low-abundant silver stained proteins from 2-DE is recommended.

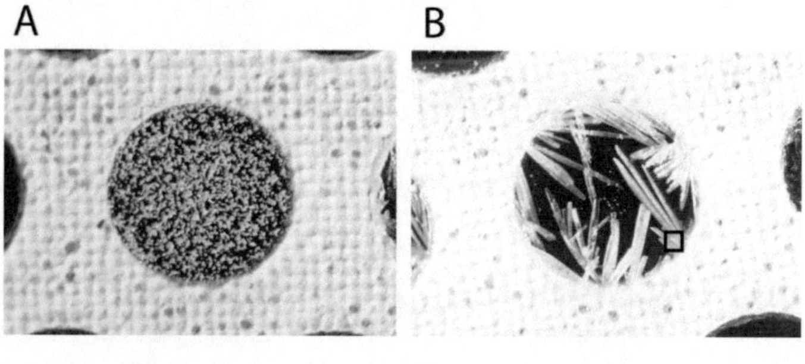

Fig. 11. Crystals of α-cyano-4-hydroxycinnamic acid (CHCA) and 2,5-dihydroxybenzoic acid (DHB).
A; CHCA and B; DHB as matrix on a MALDI plate. The marked area indicates the position of the laser where the best signal to noise was obtained with DHB.

4.4 Separation and characterization of isoforms

One of the main challenges for human proteomics is to identify and characterize co- and post-translational modifications to be able to study their relevance and place in systems biology. Most human proteins are expressed as different isoforms often depending on post-translational modifications. Two common modifications in plasma are truncations and glycosylations and here we have used the ability of 2-DE to separate such isoforms based on differences in isoelectric point and molecular mass.

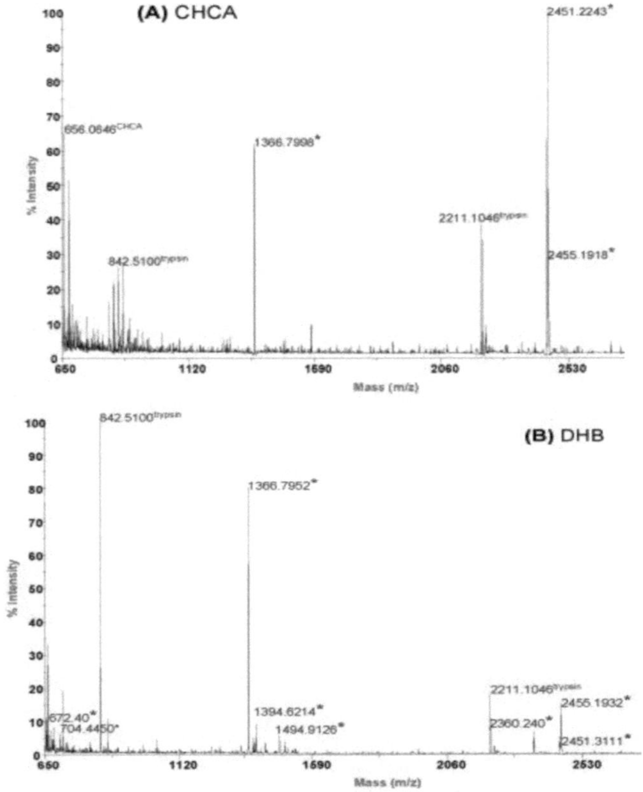

Fig. 12. Comparison of α-cyano-4-hydroxycinnamic acid (CHCA) and 2,5-dihydroxybenzoic acid (DHB) as matrices in MALDI-TOF MS of a silver stained protein.
Transthyretin was identified with sequence coverage of 39 % with CHCA (A) versus 72 % with DHB (B). Peptide peaks marked with an asterisk were matched to the theoretical masses with an accuracy <50 ppm.

Serum amyloid A1 and A2 (SAA1 and SAA2, respectively) are two acute phase proteins, which share more than 95 % sequence identity. Both SAA1 and SAA2 can also be expressed as an alpha- and a beta-form, which are discriminated from the others only in one amino acid position (Strachan et al., 1989). SAA1 and SAA2 are associated to HDL (Karlsson et al., 2005) and are heavily induced by endotoxins (Levels et al., 2011), which is highly relevant in particle toxicology to discriminate between different environmental agents (Karlsson et al. 2011). Based on differences in isoelectric points we were able to separate four isoforms of SAA1/2 in HDL (figure 13A). By peptide mass fingerprinting after trypsin digestion we identified two of the isoforms as SAA1α with pI 5.5 and 6 and two as SAA2α with pI 7 and 8 (figures 13A and 13B). The theoretical pI of SAA1α and SAA2α is 5.9 and 8.3, respectively. Thus, the pI of one of the isoforms of SAA1 and of SAA2 corresponded to the theoretical values while the other two had an acidic shift (pI 6→5.5 in SAA1α and pI 8→7 in SAA2α). N-terminal truncations of SAA1 and SAA2 that would produce such acidic shifts have previously been described (Ducret et al., 1996) and we therefore focused the MS analyses on the N-terminal peptide. As SAA1 and SAA2 contain arginine at the N-terminus we used CNBr, which cleaves before methionines, as an alternative digestion agent to detect the full length N-terminal peptide. These analyses showed that the more acidic isoforms of SAA1 and SAA2 comprised a mixture of truncations with the loss of one, two or four amino acids N-terminally; des-Arg, des-Arg-Ser and des-Arg-Ser-Phe-Phe, respectively, with the loss of arginine being the main explanation to the acidic shifts (figure 13D). On the other hand, the native peptide was only found in the more basic isoforms of SAA1 and SAA2. In total, four forms of SAA1α and four forms of SAA2α was identified. Interestingly, studies of these small molecular mass variants of SAA1 and SAA2 with SELDI-TOF MS indicates population cluster differences in HDL related to the truncations in response to endotoxin (Levels et al., 2011).

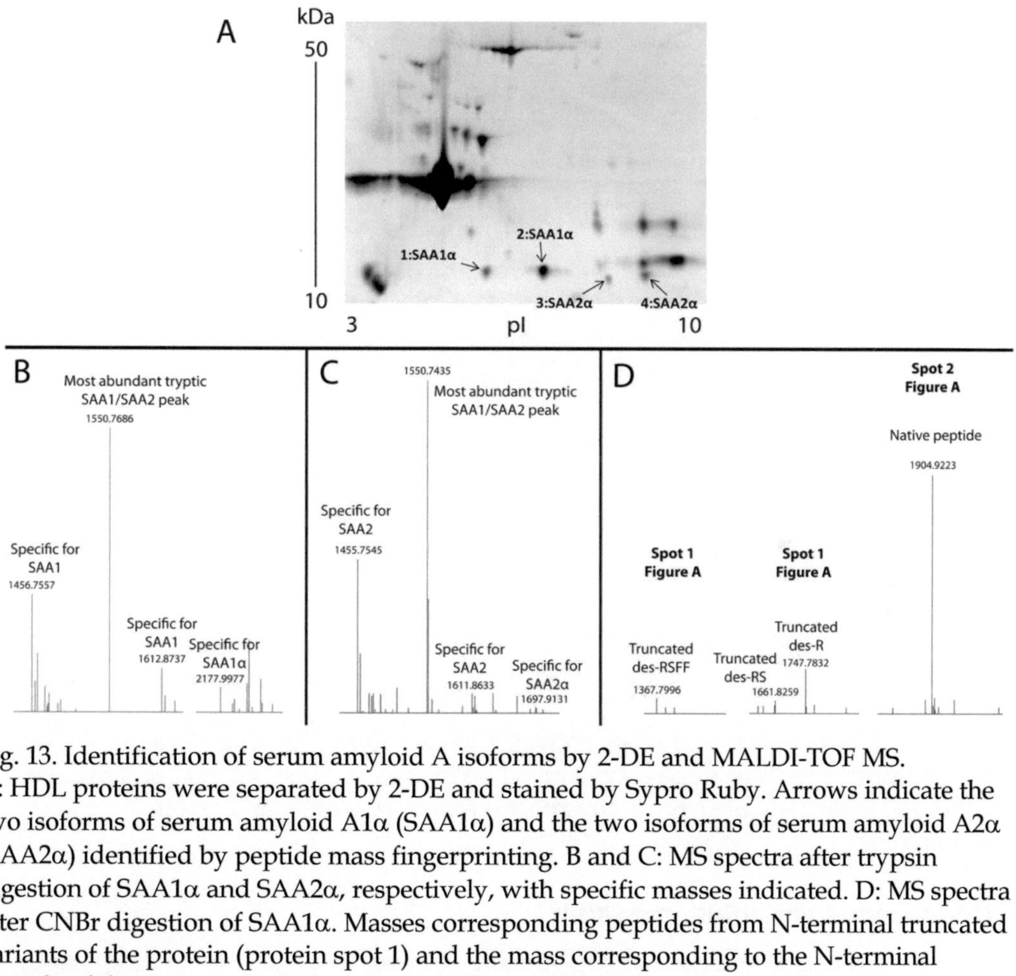

Fig. 13. Identification of serum amyloid A isoforms by 2-DE and MALDI-TOF MS.
A: HDL proteins were separated by 2-DE and stained by Sypro Ruby. Arrows indicate the two isoforms of serum amyloid A1α (SAA1α) and the two isoforms of serum amyloid A2α (SAA2α) identified by peptide mass fingerprinting. B and C: MS spectra after trypsin digestion of SAA1α and SAA2α, respectively, with specific masses indicated. D: MS spectra after CNBr digestion of SAA1α. Masses corresponding peptides from N-terminal truncated variants of the protein (protein spot 1) and the mass corresponding to the N-terminal peptide of the native protein (protein spot 2) are indicated.

2-DE makes it possible to separate proteins based on the degree of glycosylation. Hydrophilic sugars affect the binding of SDS and usually render the proteins an apparent higher molecular mass in the second dimension and the presence of negative sialyl-groups makes the proteins more acidic in the first dimension. We have therefore adapted two simple 2-DE mobility shift assays to demonstrate glycosylation of proteins and applied these to study glycosylated isoforms of plasma proteins in HDL. In the first we use endoglycosidase PNGase to cleave N-linked oligosaccharides from the protein backbone and the second is based on enzymatic removal of sialic acid with neuraminidase. As shown in figure 14A, SAA4 is usually expressed in HDL as 6 isoforms, three with molecular masses about 18k and three with molecular masses about 11k. After PNGase treatment it was clearly shown that the 18k isoforms are depending on N-linked glycosylation (Fig 14A). Another glycosylated protein in HDL is apo C-III that can be found as three isoforms; one di-sialylated, one mono-sialylated and one minor non-sialylated form (Bruneel et al., 2008). This was demonstrated by treatment with neuraminidase, which induced a mobility shift with the loss of the two sialylated isoforms and a substantial increase of the non-sialyated apo C-III form (figure 14B).

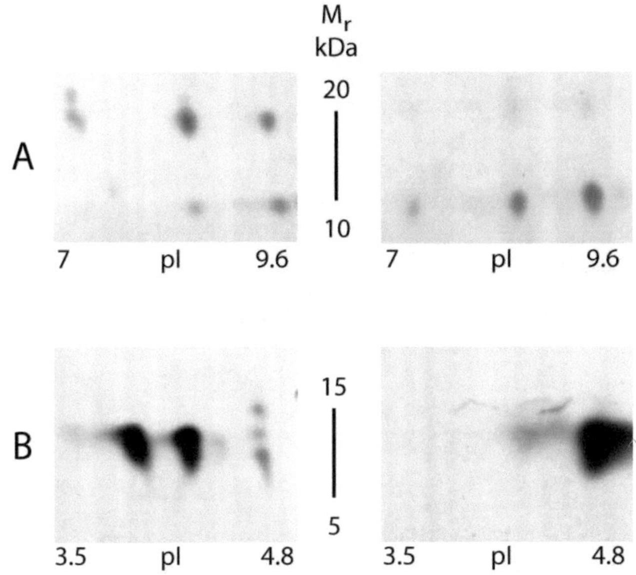

Fig. 14. 2-DE mass and charge mobility shift assays to demonstrate glycosylated protein isoforms.
A: SAA4 analyzed by 2-DE and silver stained. N-linked glycosylated serum amyloid A4 (SAA4) isoforms shown by deglycosylation with PNGase. B: Apo C-III analyzed by 2-DE and Western blots. Sialylated apo C-III isoforms shown by desialylation with neuraminidase.

5. Conclusions and future research

Overall the binding of plasma proteins to nanoparticles, based on our findings, seems to vary with origin, surface properties and size of the particles. A large portion of the interacting proteins we identified by 2-DE/MS are proteins involved in the immune defense and reverse cholesterol transport to the liver, but we also identified proteins mediating brain uptake. Most likely these protein patterns of the nanoparticles represent a mixture of particle-protein and protein-protein interactions. Extensive research in this field is therefore needed before conclusions could be drawn regarding potential health effects of nanoparticles and their associated protein "corona". One major difficulty to overcome is how to characterize the particles used in different studies in such a way that comparisons and generalized conclusions are allowed. Most types of nanoparticles seem to form aggregates, especially so in water suspensions and the "corona" seen might be heavily influenced by the size/diameter of these aggregates rather than by other particle characteristics. Thus, characterization of particles and the aggregates they form prior to exposure of plasma proteins, cells or other biological systems is therefore extremely important. One way of doing that, as we have showed herein, is by DLS. These analyses gives valuable information about the trends in size distribution connected to sample preparation methods and choice of solvent. Sample preparation methods are indeed very important as well as choice of solvent and care should be taken when choosing fitting model and the model-inbuilt parameters. Thus, information obtained from DLS is important for everybody that is doing research on nanoparticles in liquids. The numbers given as product information i.e. the size and size distribution are often relevant for the core-size of the nanocrystals within the material. However the nanoparticles are most often not soluble to that extent. Consequently, nanoparticles obtained in dry state and then dispersed in liquid usually form aggregates as shown in this study.

Given ample attention to the characterization of the NPs used, future studies of the NP–protein complex behavior in different biological systems are needed. Questions that need to be addressed are which properties of the NPs that govern the protein "corona" formed around the NPs in biological fluids and how these complexes interact with endothelial cells, platelets, cells of the immune system etc. One interesting finding in this study is Amyloid β A4, not previously identified in plasma, which was only associated to ZnO particles. This protein may act as a chelator forming metal-amyloid aggregates and needs further attention in toxicological studies.

In summary, the improved 2-DE/MS protocols shown herein underline this proteomic approach as a powerful tool in human nano-particle toxicology. Furthermore, thorough characterisation of the particles studied, e.g. with DLS, is crucial to evaluate the results.

6. References

Adiseshaiah, P.; Hall, J. & McNeil, S. E. (2010). Nanomaterial Standards for Efficacy and Toxicity Assessment. *Wiley Interdisciplinary Reviews – Nanomedicine and Nanobiotechnology*, Vol.2, No.1, pp. 99–112, ISSN 1939-0041

Beavis R.; Chaudhary T. & Chait B. (1992). α-Cyano-4-hydroxycinnamic acid as a matrix for matrixassisted laser desorption mass spectrometry. *Organic Mass Spectrometry*, Vol.27, No.2, pp. 156-158

Beck-Speier, I.; Dayal, N.; Karg, E.; Maier, K.; Schumann, G.; Schulz, H.; Semmler, M.; Takenaka, S.; Stettmaier, K.; Bors, W.; Ghio, A.; Samet, J. & Heyder, J. (2005). Oxidative stress and lipid mediators induced in alveolar macrophages by ultrafine particles. *Free Radical Biology and Medicine*, Vol.38, No.8, pp. 1080-1092, ISSN 0891-5849

Bell M. & Davis D. (2001). Reassessment of the lethal London Fog of 1952: Novel Indicators of Acute and Chronic Consequences of Acute Exposure to Air Pollution. *Environmental Health Perspectives*, Vol.109, pp. 389-394, ISSN 0091-6765

Benderly, M.; Boyko, V. & Goldbourt, U. (2009). Apolipoproteins and Long-Term Prognosis in Coronary Heart Disease Patients. *American Heart Journal*, Vol.157, No.1, pp. 103–110, ISSN 1097-6744

Brunauer, S.; Emmett, P. & Teller, J. (1938). Adsorption of Gases in Multimolecular Layers. *Journal of American Chemical Society*, Vol.60, pp.309-319.

Cedervall, T.; Lynch, I.; Lindman, S.; Berggard, T.; Thulin, E.; Nilsson, H.; Dawson, K. & Linse, S. (2007).

Understanding the Nanoparticle-Protein Corona Using Methods to Quantify Exchange Rates and Affinities of Proteins for Nanoparticles. *Proceedings of the National Academy of Sciences of the U.S.A*. Vol.104, No.7, pp. 2050–2055, ISSN 0027-8424

Cherukuri, P.; Gannon, C.; Leeuw, T.; Schmidt, H.; Smalley, R.; Curley, S. & Weisman R. (2006). Mammalian pharmacokinetics of carbon nanotubes using intrinsic near-infrared fluorescence. *Proceedings of the National Academy of Sciences of the U.S.A*, Vol.103, No. 50, pp. 18882-18886, ISSN 0027-8424

Da Silva, E.; Tsushida, T. & Terao, J. (1998). Inhibition of mammalian 15-lipoxygenase-dependent lipid peroxidation in low-density lipoprotein by quercetin and quercetin monoglucosides. *Archives of Biochemistry and Biophysics*, Vol.349, No.2, pp. 313-320, ISSN 0003-9861

Deng, Z.; Mortimer, G.; Schiller, T.; Musumeci, A.; Martin, D. & Minchin R. (2009). Differential plasma protein binding to metal oxide nanoparticles, *Nanotechnology*, Vol.20, No.45, ISSN 1361-6528

Dobrovolskaia, M. & McNeil, S. (2007). Immunological Properties of Engineered Nanomaterials. *Nature Nanotechnolology*, Vol.2, No.8, pp. 469–478, ISSN 1748-3395

Dobrovolskaia, M.; Patri, A.; Zheng, J.; Clogston, J.; Ayub, N.; Aggarwal, P.; Neun, B.; Hall, J. & McNeil, S. (2009). Interaction of Colloidal Gold Nanoparticles with Human Blood: Effects on Particle Size and Analysis of Plasma Protein Binding Profiles. *Nanomedicine*, Vol.5, No.2, pp. 106–117, ISSN 1549-9642

Ducret A.; Bruun C.; Bures E.; Marhaug G.; Husby G. & Aebersold R. (1996). Characterization of human serum amyloid A protein isoforms separated by two-dimensional electrophoresis by liquid chromatography/electrospray ionization tandem mass spectrometry. *Electrophoresis*, Vol.17, No.5, pp. 866-876, ISSN 0173-0835

Dutta, D.; Sundaram, S.; Teeguarden, J.; Riley, B.; Fifield, L.; Jacobs, J.; Addleman, S.; Kaysen, G.; Moudgil, B. & Weber, T. (2007). Adsorbed proteins influence the biological activity and molecular targeting of nanomaterials. *Toxicological Sciences*, Vol.100, No.1, pp. 303-315, ISSN 1096-6080

Elsaesser, A. & Howard, C. (2011). Toxicology of nanoparticles. *Advanced Drug Delivery Reviews*, ISSN 1872-8294

Farwig Z.; McNeal C.; Little D.; Baisden C. & Macfarlane R. (2005). Novel truncated isoforms of constitutive serum amyloid A detected by MALDI mass spectrometry. *Biochemical and Biophysical Research Communications* Vol.332, No.2, pp. 352–356

Ferris, D.; Lu, J.; Gothard, C.; Yanes, R.; Thomas, C.; Olsen, J.; Stoddart, J.; Tamanoi, F. & Zink, J. (2011). Synthesis of biomolecule-modified mesoporous silica nanoparticles for targeted hydrophobic drug delivery to cancer cells. *Small*, Vol.7, No.13, pp. 1816-1826, ISSN 1613-6829

Finsy R. (1994). Particle Sizing by Quasi-elastic light scattering, *Advances in colloid and interface science*, Vol.52, pp. 79-143

Ghafouri, B.; Karlsson, H.; Mortstedt, H.; Lewander, A.; Tagesson, C. & Lindahl, M. (2007). 2,5-Dihydroxybenzoic acid instead of alpha-cyano-4-hydroxycinnamic acid as matrix in matrix-assisted laser desorption/ionization time-of-flight mass spectrometry for analyses of in-gel digests of silver-stained proteins. *Analytical Biochemistry,* Vol.371, No.1, pp. 121–123, ISSN 0003-2697

Gharahdaghi F.; Weinberg C.; Meagher D.; Imai B. & Mische S. (1999). Mass spectrometric identification of proteins from silver-stained polyacrylamide gel: a method for the removal of silver ions to enhance sensitivity. *Electrophoresis,* Vol.20, No.3, pp. 601-605, ISSN 0173-0835

Görg, A.; Postel, W. & Gunther, S. (1988). The current state of two-dimensional electrophoresis with immobilized pH gradients. *Electrophoresis,* Vol.9, No.9, pp. 531-546, ISSN 0173-0835

Görg ,A.; Obermaier, C.; Boguth, G.; Harder, A.; Scheibe, B.; Wildgruber, R.; and Weiss, W. (2000). The current state of 2-dimensional electrophoresis with immobilized pH gradients. *Electrophoresis.* Vol. 21, No. 12 pp.1037-53

Hackley V. & Clogston J. (2007). Measuring the Size of Nanoparticles in Aqueous Media Using Batch-Mode Dynamic Light Scattering, *NIST-NCL Joint Assay Protocol PCC-1 Version 1.0,* Available from http://ncl.cancer.gov/NCL_Method_NIST-NCL_PCC-1.pdf

Hamers, T.; Kamstra, J.; Cenijn, P.; Pencikova, K.; Palkova, L.; Simeckova, P.; Vondracek, J.; Andersson, P.; Stenberg, M. & Machala, M. (2011) In vitro toxicity profiling of ultrapure non-dioxin-like polychlorinated biphenyl congeners and their relative toxic contribution to PCB mixtures in humans. *Toxicological Sciences,* Vol121, No.1, pp. 88-100, ISSN 1096-0929

Hellstrand, E.; Lynch, I.; Andersson, A.; Drakenberg, T.; Dahlback, B.; Dawson, K.; Linse, S. & Cedervall, T. (2009). Complete High-Density Lipoproteins in Nanoparticle Corona. *FEBS Journal,* Vol.276, No.12, pp. 3372–3381, ISSN 1742-4658

Henning, M.; Garda, H. & Bakas, L. (2006). Biophysical characterization of interaction between apolipoprotein A-I and bacterial lipopolysaccharide. *Cell Biochemistry and Biophysics,* Vol.44, No.3, pp. 490-496, ISSN 1085-9195

Higashisaka, K..; Yoshioka, Y.; Yamashita, K.; Morishita, Y.; Fujimura, M.; Nabeshi, H.; Nagano, K.; Abe, Y.; Kamada, H.; Tsunoda, S.; Yoshikawa, T.; Itoh, N. & Tsutsumi, Y. (2011). Acute phase proteins as biomarkers for predicting the exposure and toxicity of nanomaterials. *Biomaterials,* Vol.32, No.1, pp. 3-9, ISSN 1878-5905

Karlsson H.; Leanderson P.; Tagesson C, & Lindahl M. (2005). Lipoproteomics II: mapping of proteins in high-Density lipoprotein using two-dimensional gel electrophoresis and mass spectrometry. *Proteomics,* Vol.5, No.5, pp. 1431–1445, ISSN 1615-9853

Karlsson, H.; Lindbom, J.; Ghafouri, B.; Lindahl, M.; Tagesson, C.; Gustafsson, M. & Ljungman, A. (2011). Wear particles from studded tires and granite pavement induce pro-inflammatory alterations in human monocyte-derived macrophages: a proteomic study. *Chemical Research in Toxicology,* Vol.24, No.1, pp. 45-53, ISSN 1520-5010

Kim, Y.; Chung, S. & Lee, S. (2005). Roles of plasma proteins in the formation of silicotic nodules in rats. *Toxicology Letters,* Vol.158, No1, pp. 1-9, ISSN 0378-4274

Kreuter, J.; Michaelis, K.; Dries, S. & Langer, K. (2005). The role of apolipoproteins on brain uptake of nanoparticle-bound drugs, *Proceedings of 15th International symposium on Microencapsulation,* Parma, Italy, September 2005

Kreyling, W.; Semmler, M.; Erbe, F.; Mayer, P.; Takenaka, S. & Schulz, H. (2002). Translocation of ultrafine insoluble iridium particles from lung epithelium to extra pulmonary organs is size dependent but very low. *Journal of Toxicology and Environmental Health.* Vol.65, No.20, pp. 1513-1530, ISSN 1528-7394

Kreyling, W.; Semmler-Behnke, M. & Moller, W. (2006). Ultrafine particle-lung interactions: does size matter? *Journal of Aerosol Medicine,* Vol.19, No.1, pp. 74–83, ISSN 0894-2684

Lahm H. & Langen H. (2000). Mass spectrometry: a tool for the identification of proteins separated by gels. *Electrophoresis,* Vol.21, No.11, pp. 2105–2114, ISSN 0173-0835

Leszczynski, J. (2010). Bionanoscience: Nano Meets Bio at the Interface. *Nature Nanotechnology,* Vol.5, No.9, pp. 633–634, ISSN 1748-3395

Levels, J.; Geurts, P.; Karlsson, H.; Maree, R.; Ljunggren, S.; Fornander, L.; Wehenkel, L.; Lindahl, M.; Stroes, E.; Kuivenhoven, J. & Meijers, J. (2011). High-density lipoprotein proteome dynamics in human endotoxemia. *Proteome Science,* Vol.9, No.1, ISSN 1477-5956

Lewis, J.; Wei, J. & Siuzdak G. (2000). Matrix-assisted Laser Desorption/Ionization Mass Spectrometry in Peptide and Protein Analysis. In: *Encyclopedia of Analytical Chemistry,* Meyers, R., pp. 5880–5894, John Wiley & Sons Ltd, ISBN 978-0-471-97670-7, Chichester

Lundqvist, M.; Stigler, J.; Elia, G.; Lynch, I.; Cedervall, T. & Dawson, K. (2008). Nanoparticle Size and Surface Properties Determine the Protein Corona with Possible Implications for Biological Impacts. *Proceedings of the National Academy of Sciences of the U.S.A,* Vol.105, No.38, pp. 14265–14270, ISSN 1091-6490

Lynch, I.; Cedervall, T.; Lundqvist, M.; Cabaleiro-Lago, C.; Linse, S. & Dawson, K. (2007). The Nanoparticle-Protein Complexas a Biological Entity; a Complex Fluids and Surface Science Challenge for the 21st Century. *Advances in Colloid and Interface Science,* Vol.134-135, ISSN 0001-8686

McAuliffe, M. & Perry, M. (2007). Are nanoparticles potential male reproductive toxicants? A literature review. *Nanotoxicology.* Vol.1, No.3, pp. 204-210

Mooberry, L.; Nair, M.; Paranjape, S.; Mc Conathy, W. & Lacko, A. (2010). Receptor mediated uptake of pacilitaxel from synthetic high density lipoprotein nancarrier. *Journal of drug targeting,* Vol.18, No.1, pp. 53-58, ISSN 1029-2330

Mühlfeld, C.; Rothen-Rutishauser, B.; Blank, F.; Vanhecke, D.; Ochs, M. & Gehr, P. (2008). Interactions of nanoparticles with pulmonary structures and cellular responses. *American Journal of Physiology - Lung Cellular and Molecular Physiology.*Vol. 294, No.5, pp. L817-L829, ISSN 1040-0605

O'Farrell, P.H. (1975). High-resolution two-dimensional electrophoresis of proteins. *J Biol Chem.* 250, pp. 4007–4021

Rabilloud, T. (2000. Detecting proteins separated by 2-D gel electrophoresis. *Analytical Chemistry.* Vol.72, No.1, pp. 48A-55A, ISSN 0003-2700

Rothen-Rutishauser, B.; Muhlfeld, C.; Blank, F.; Musso, C. & Gehr, P. (2007). Translocation of particles and inflammatory responses after exposure to fine particles and nanoparticles in an epithelial airway model. *Particle and Fibre Toxicology,* Vol.4, No.9, ISSN 1743-8977

Sattler, W.; Mohr, D. & Stocker, R. (1994). Rapid isolation of lipoproteins and assessment of their peroxidation by high-performance liquid chromatography postcolumn chemiluminescence. *Methods in Enzymology,* Vol.233, pp. 469-489, ISSN 0076-6879

Schulz, H.; Harder, V.; Ibald-Mulli, A.; Khandoga, A.; Koenig, W.; Krombach, F.; Radykewicz, R.; Stampfl, A.; Thorand, B. & Peters, A. (2005). Cardiovascular effects of fine and ultrafine particles. *Journal of Aerosol Medicine,* Vol.18, No.1, pp. 1-22, ISSN 0894-2684

Shevchenko, A.; Wilm, M.; Vorm, O. & Mann, M. (1996). Mass spectrometric sequencing of proteins silver-stained polyacrylamide gels. *Analytical Chemistry,* Vol.68, No.5, pp. 850-858, ISSN 0003-2700

Shevchenko, A.; Chernushevich, I.; Shevchenko, A.; Wilm, M. & Mann, M. (2002). De novo sequencing of peptides recovered from in-gel digested proteins by nanoelectrospray tandem mass spectrometry. *Molecular Biotechnology,* Vol.20, **No.1,** pp. 107-118, ISSN 1073-6085

Sorensen, C. (2008). Scattering and Absorption of Light by Particles and Aggregates, *Handbook of Surface and Colloid Chemistry Third Edition,* Birdi K., CRC Press, ISBN 978-0-8493-7327-5, Boca Raton, Florida, US

Stern, S. & McNeil, S. (2008). Nanotechnology safety concerns revisited. *Toxicological Sciences,* Vol.101, No.1, pp. 4-21, ISSN 1096-6080

Strachan A.; Brandt W.; Woo P.; van der Westhuyzen D.; Coetzee G. & de Beer M. (1989). Human serum amyloid A protein. The assignment of the six major isoforms to three published gene sequences and evidence for two genetic loci. *Journal of Biological Chemistry,* Vol.264, No.31, pp. 18368-18373, ISSN 0021-9258

Strupat K.;,Karas M. & Hillenkamp F. (1991). 2,5-Dihydroxybenzoic acid: a new matrix for laser desorption—ionization mass spectrometry. *International Journal of Mass Spectrometry and Ion Processes,* Vol.111, pp. 86-102, ISSN 0168-1176

Uniprot. (2011). Amyloid beta A4 protein, 20th October 2011, Available from http://www.uniprot.org/uniprot/P05067

Vergeer, M.; Korporaal, S.; Franssen, R.; Meurs, I.; Out, R.; Hovingh, G.; Hoekstra, M.; Sierts, J.; Dallinga-Thie, G.; Motazacker, M.; Holleboom, A.; Van Berkel, T.; Kastelein, J.; Van Eck, M. & Kuivenhoven, J. (2011). Genetic variant of the scavenger receptor class B-I in humans. *The New England Journal of Medicine,* Vol.364, No.2, pp. 136-45, ISSN 1533-4406

Walczyk, D.; Bombelli, F.; Monopoli, M.; Lynch, I. & Dawson, K. (2010). What the Cell "Sees" in Bionanoscience. *Journal of the American Chemical Society,* Vol.132, No.16, pp. 5761–5768, ISSN 1520-5126

Wick, P.; Malek, A.; Manser, P.; Meili, D.; Maeder-Althaus, X.; Diener, L.; Diener, P-A.; Zisch, A.; Krug, H.F. & von Mandach, U. (2010). Barrier Capacity of Human Placenta for Nanosized Materials. *Environmental Health Perspectives.* Vol.118, No.3, pp. 432-436, ISSN 1552-9924

Wu, W. & Jiang, X. (2011). A practical strategy for constructing nanodrugs using carbon nanotubes as carriers. *Methods in Molecular Biology,* Vol.751, pp. 565-582, ISSN 1940-6029

Xu, Z.; Liu, X.; Ma, Y. & Gao, H. (2010). Interaction of nano-TiO2 with lysozyme: insights into the enzyme toxicity of nanosized particles. *Environmental Science and Pollutional Research International,* Vol.17, No.3, pp. 798-806, ISSN 1614-7499

Zensi, A.; Begley, D.; Pontikis, C.; Legros, C.; Mihoreanu, L.; Buchel, C. & Kreuter, J. (2010). Human Serum Albumin Nanoparticles Modified with Apolipoprotein a-I Cross the Blood-Brain Barrier and Enter the Rodent Brain. *Journal of Drug Targeting*, Vol.18, No.10, pp. 842–848, ISSN 1029-2330

19

Two Dimensional Gel Electrophoresis in Cancer Proteomics

Soundarapandian Kannan[1], Mohanan V. Sujitha[1], Shenbagamoorthy Sundarraj[1] and Ramasamy Thirumurugan[2]
[1]Proteomics and Molecular Cell Physiology Lab
Department of Zoology, Bharathiar University, Coimbatore
[2]Department of Animal Science, Bharathidasan University, Tiruchirappalli
India

1. Introduction

Two-dimensional electrophoresis (2-DE) is a powerful and widely used method for the analysis of complex protein mixtures extracted from cells, tissues, or other biological samples. This technique sort's protein according to two independent properties in two discrete steps: the first-dimension step, isoelectric focusing (IEF), separates proteins according to their isoelectric points (pI); the second-dimension step, SDS-polyacrylamide gel electrophoresis (SDS-PAGE), separates proteins according to their molecular weights (Mr, relative molecular weight). Each spot on the resulting two-dimensional array corresponds to a single protein species in the sample. Thousands of proteins can thus be separated, and information such as the protein pI, the apparent molecular weight, and the amount of each protein obtained. The separation of proteins by 2-DE dates back to the 1950s. The first 2-DE technique was developed by Smithies and Poulik in 1956 and O'Farrell, 1975 and Klose, 1975 significantly modified this method to elucidate protein profile. In the original technique, the first-dimension separation was performed in carrier ampholyte-containing polyacrylamide gels cast in narrow tubes.

The power of 2-DE as a biochemical separation technique has been recognized virtually since its introduction. Its application, however, has become significant only in the last few years because of a number of developments. The introduction of immobilized pH gradients and Immobiline™ reagents brought superior resolution and reproducibility to first-dimension IEF. Based on this concept, Görg *et al.*, 1989 and Gorg, 1991 developed the currently employed 2-D technique, where carrier ampholyte-generated pH gradients have been replaced with immobilized pH gradients and tube gels replaced with gels supported by a plastic backing. New mass spectrometry techniques have been developed that allow rapid identification and characterization of very small quantities of peptides and proteins extracted from single 2-D spots. More powerful, less expensive computers and software are now available, rendering thorough computerized evaluations of the highly complex 2-D patterns economically feasible. Data about entire genomes (or substantial fractions thereof) for a number of organisms are now available, allowing rapid identification of the gene encoding a protein separated by 2-DE. The World Wide Web provides simple, direct access to spot pattern databases for the comparison of electrophoresis results and genome sequence databases for assignment of sequence information.

A large and growing application of 2-DE in "proteome analysis." Proteome analysis is "the analysis of the entire Protein complement expressed by a genome". The analysis involves the systematic separation, identification, and quantification of many proteins simultaneously from a single sample. Two-dimensional electrophoresis is used e due to its unparalleled ability to separate thousands of proteins simultaneously. Two-dimensional electrophoresis is also unique in its ability to detect post- and co-translational modifications, which cannot be predicted from the genome sequence. Applications of 2-DEinclude proteome analysis, cell differentiation, and detection of disease markers, monitoring therapies, drug discovery, cancer research, purity checks, and microscale protein purification.

"Proteomics" is the large-scale screening of the proteins of a cell, organism or biological fluid, a process, which requires stringently controlled steps of sample preparation, 2-DE, image detection and analysis, spot identification, and database searches. Moreover, Proteomics studies lead to the molecular characterization of cellular events associated with cancer progression, cellular signaling, developmental stages etc. Proteomics studies of clinical tumor samples have led to the identification of cancer-specific protein markers, which provide a basis for developing new methods for early diagnosis and early detection and clues to understand the molecular characterization of cancer progression. A keystone of conventional proteomics is high-resolution 2D gel electrophoresis followed by protein identification using mass spectrometry.

As a technique with high-flux and high resolution, proteomics has been widely applied in proteome analysis of tumors. The onset and development of the tissues and cells can be detected at the entire protein level through analyzing the differential expression of proteins. The combination of 2-DE and mass spectrometry can be used to identify differential proteins between tumor cells and normal original cells, and these differential proteins imply a large quantity of biological information. Some of the special proteins are special markers of tumors. The most consistently successful proteomic method is the combination of two-dimensional gel electrophoresis (2DE) for protein separation, visualization, and mass spectrometric (MS) identification of proteins using peptide mass fingerprints and tandem MS peptide sequencing.

The experiments form the basis of proteomics, and present significant challenges in data analysis, storage and querying. The core technology of proteomics is 2-DE. At present, there is no other technique that is capable of simultaneously resolving thousands of proteins in one separation procedure. The replacement of classical first-dimension carrier ampholyte pH gradients with well-defined immobilized pH gradients has resulted in higher resolution, improved inter-laboratory reproducibility, higher protein loading capacity, and an extended basic pH limit for 2-DE. With the increased protein capacity, micropreparative 2-DE has accelerated spot identification by mass spectrometry and Edman sequencing. The remarkable improvements in 2-DE resulting from immobilized pH gradient gels, together with convenient new instruments for IPG-IEF, will make critical contributions to advances in proteome analysis.

A comprehensive understanding of protein–protein interactions is an important step in our quest to understand how the information contained in a genome is put into action. Although a number of experimental techniques can report on the existence of a protein–

protein interaction, very few can provide detailed structural information. NMR spectroscopy is one of these, and in recent years several complementary NMR approaches, including residual dipolar couplings and the use of paramagnetic effects, have been developed that can provide insight into the structure of protein–protein complexes.

Two-dimensional gel electrophoresis for separation of complex protein samples coupled with mass spectrometry for protein identification has been used to analyze protein expression patterns for many sample types. Inherent in the use of this technique is information on not only full-length protein expression, but expression of modified, splice variant, cleavage product, and processed proteins. Any protein modification that leads to a change in overall protein charge and/or molecular weight (MW) will generate a different spot on the 2-DE. Modification specific staining can identify whether a specific post-translational modification is responsible for the shift, and mass spectrometry can potentially identify the source of isoelectric point (pI) and/or MW differences. Due to the lack of complete coverage for a protein's amino acid sequence using either matrix-assisted laser desorption/ionization mass spectrometry (MALDI-MS) or high-performance liquid chromatography (HPLC) tandem mass spectrometry (LC-MS/MS), there has been limited success in using MS to identify isoforms and post-translational modifications. While the theoretical MW is often slightly higher than the MW of the fully processed protein due to cleavage of signal and pro-peptides, there can also be post-translational modifications that increase the protein's gel MW. Thus, an exploration into the causes of the difference in the theoretical MW and the MW as seen in the gel can yield information about the state of the protein. When the gel MW of a given protein is significantly lower than the calculated weight, the gel spot represents a protein fragment. The extent to which proteins are present as fragments or variants in tissues and fluids has not been determined, but the combination of 2-DE, Western blotting, and mass spectrometry–based protein identification makes such analyses possible. Two-dimensional gel electrophoresis of human mammary tissue, followed by immune blotting, resulted in multiple spots at significantly differing molecular weights. The function of protein fragments is dependent on activation processes and localization properties. This Chapter will be critically analyzed as per the contents given in the synopsis with up-to-date informations.

2. Overview of experimental design

2.1 Experimental design

2.1.1 Sample preparation

Efficient and reproducible sample preparation methods are a key to successful 2-DE (Rabilloud 1999, Macri *et al.* 2000, Molloy 2000). Sample preparation methods range from extraction with simple solubilization solutions to complex mixtures of chaotropic agents, detergents, and reducing agents. Sample preparation can include enrichment strategies for separating protein mixtures into reproducible fractions.

An effective sample preparation procedure will:

1. Reproducibly solubilize proteins of all classes, including hydrophobic proteins
2. Prevent protein aggregation and loss of solubility during focusing
3. Prevent postextraction chemical modification, including enzymatic or chemical degradation of the protein sample

4. Remove or thoroughly digest nucleic acids and other interfering molecules
5. Yield proteins of interest at detectable levels, which may require the removal of interfering abundant proteins or nonrelevant classes of proteins

Most protein mixtures will require some experimentation to determine optimum conditions for 2-D PAGE. Variations in the concentrations of chaotropic agents, detergents, ampholytes, and reducing agents can dramatically affect the 2-D pattern.

2.1.2 Solubilization

Solubilization of proteins is achieved by the use of chaotropic agents, detergents, reducing agents, buffers, and ampholytes. These are chosen from a small list of compounds that meet the requirements, both chemically and electrically, for compatibility with the technique of IEF in IPG strips. The compounds chosen must not increase the ionic strength of the solution, to allow high voltages to be applied during focusing without producing high currents. Thorough discussion of solubilization methods, including new variations, can be found in several books (Pennington and Dunn 2001, Rabilloud 2000).

2.1.3 Chaotropic agents

Urea is the most commonly used chaotropic agent in sample preparation for 2-D PAGE. Thiourea can be used to help solubilize many otherwise intractable proteins. Urea and thiourea disrupt hydrogen bonds and are used when hydrogen bonding causes unwanted aggregation or formation of secondary structures that affect protein mobility. Urea is typically used at 8 M. Thiourea is weakly soluble in water, but is more soluble in high concentrations of urea, so a mixture of 2 M thiourea and 5–8 M urea is used when strongly chaotropic conditions are required (Rabilloud 1998).

2.1.4 Detergents

Detergents are added to disrupt hydrophobic interactions and increase solubility of proteins at their pI. Detergents must be nonionic or zwitterionic to allow proteins to migrate according to their own charges. Some proteins, especially membrane proteins, require detergents for solubilization during isolation and to maintain solubility during focusing. Ionic detergents such as SDS are not compatible with IEF, but can be used with concentrated samples in situations where the SDS can be unbound from the proteins by IEF-compatible detergents that compete for binding sites. Nonionic detergents such as octylglucoside, and zwitterionic detergents such as CHAPS and its hydroxyl analog, CHAPSO, can be used. CHAPS, CHAPSO, or octylglucoside concentrations of 1–2% are recommended (Rabilloud 1999). New detergents are emerging that have great potential in proteomics, including SB 3-10 and ASB-14 (Chevallet *et al.* 1998). Some proteins may require detergent concentrations as high as 4% for solubility (Hermann *et al.* 2000).

2.1.5 Carrier ampholytes

A fundamental challenge with IEF is that some proteins tend to precipitate at their pI. Even in the presence of detergents, certain samples may have stringent salt requirements to maintain the solubility of some proteins. Salt should be present in a sample only if it is an absolute requirement, and then only at a total concentration less than 40 mM. This is problematic since any salt included will be removed during the initial high-current stage of focusing. Salt limits the voltage that can be achieved without producing high current, increasing the time required for focusing. Proteins that require salt for solubility are subject to precipitation once the salt is removed. Carrier ampholytes sometimes help to counteract insufficient salt in a sample. They are usually included at a concentration of ≤0.2% (w/v) in sample solutions for IPG strips. High concentrations of carrier ampholytes will slow down IEF until they are focused at their pI, since they carry current and hence limit voltage. Some researchers have increased resolution by varying the ampholyte composition.

2.1.6 Reducing agents

Reducing agents such as dithiothreitol (DTT) or tributylphosphine (TBP) are used to disrupt disulfide bonds. Bond disruption is important for analyzing proteins as single subunits. DTT is a thiol reducing agent added in excess to force equilibrium toward reduced cysteines. At 50 mM it is effective in reducing most cystines, but some proteins are not completely reduced by this treatment. If the concentration of DTT is too high it can affect the pH gradient since its pKa is around 8. TBP is a much more effective reducing agent than DTT. It reacts to reduce cystines stoichiometrically at low millimolar concentrations (Herbert *et al.* 1998). It is chemically more difficult to handle than DTT, but Bio-Rad has solved this problem by supplying it in a form safe for shipping and lab use.

2.1.7 Prefractionation

Reducing the complexity of the sample loaded on a 2-D gel can increase the visibility of minor proteins. Techniques such as differential extraction (Molloy *et al.* 1998), subcellular fractionation (Taylor *et al.* 2000, Morel *et al.* 2000), chromatography (Fountoulakis *et al.* 1999), or prefocusing in a preparative IEF device such as the Rotofor® system (Masuoka *et al.* 1998, Nilsson *et al.* 2000) have been used to reduce the complexity of samples.

2.1.8 Removal of albumin and IgG

The isolation of lower-abundance proteins from serum or plasma is often complicated by the presence of albumin and immunoglobulin G (IgG). Albumin is the most abundant protein (~60–70%) in serum and IgG is the second most abundant protein (10–20%). These two proteins effectively act as major contaminants, masking the presence of many co-migrating proteins, as well as limiting the amount of total serum protein that can be resolved on a 2-D gel. In the past, removal of albumin and IgG usually required separate chromatography methodologies for each of the two species. Now, Bio-Rad's Aurum™ serum protein kit allows selective binding and simultaneous removal of both albumin and IgG from serum or plasma samples prior to 2-DE.

2.1.9 Sequential extraction

One method for reducing sample complexity is the basis of the ReadyPrep™ sequential extraction kit. This protocol takes advantage of solubility as a third independent means of protein separation. Proteins are sequentially extracted in increasingly powerful solubilizing solutions. More protein spots are resolved by applying each solubility class to a separate gel, thereby enriching for particular proteins while simplifying the 2-D patterns in each gel. An increase in the total number of proteins is detected using this approach (Molloy *et al.* 1998).

2.1.10 Removal of DNA

The presence of nucleic acids, especially DNA, interferes with separation of proteins by IEF. Under denaturing conditions, DNA complexes are dissociated and markedly increase the viscosity of the solution, which inhibits protein entry and slows migration in the IPG. In addition, DNA binds to proteins in the sample and causes artifactual migration and streaking. The simplest method for removal of DNA is enzymatic digestion. Adding endonuclease to the sample after solubilization at high pH (40 mM Tris) allows efficient digestion of nucleic acids while minimizing the action of contaminating proteases. The advantage of the endonuclease method is that sample preparation can be achieved in a single step, by the addition of the enzyme prior to loading the first-dimension IPG.

2.1.11 Protein load

The amount of protein applied to an IPG strip can range from several micrograms to 1 mg or more (Bjellqvist *et al.* 1993a). Some of the factors affecting the decision of how much protein to load are:

a. Subsequent analysis. Enough of the protein of interest must be loaded for it to be analyzed. With the Ready Gel® mini system (7 cm IPG), detection of moderately abundant proteins in complex mixtures with Coomassie Brilliant Blue R-250 dye requires on the order of 100 µg total protein. With the same load, many low-abundance proteins can be detected with more sensitive stains such as silver or SYPRO Ruby protein gel stain.
b. The purpose of the gel. If the gel is being run solely for the sake of getting a good image of well-resolved proteins for comparative studies or for publication, the protein load would be the minimum amount that is stainable.
c. The abundance of the proteins of interest. If the purpose is to study low-copy-number proteins, a large mass of a protein mixture might be loaded (Wilkins *et al.* 1998).
d. The complexity of the sample. A highly complex sample containing many proteins of widely varying concentrations might require a compromise load so that high-abundance proteins don't obscure low-abundance proteins. By enriching a sample for specific types of proteins using prefractionation techniques, each individual protein will be at a higher relative concentration, which means that enough material can be loaded for detection of low-abundance constitutents.
e. pH range of IPG strip. In general, larger amounts of total protein can be loaded on a narrow-range IPG strip. Only the proteins with a pI within the strip pH range will be represented within the second-dimension gel.

2.2 The first dimension: Isoelectric Focusing (IEF)

2.2.1 Isoelectric point (pI)

Differences in proteins' pI are the basis of separation by IEF. The pI is defined as the pH at which a protein will not migrate in an electric field and is determined by the number and types of charged groups in a protein. Proteins are amphoteric molecules. As such, they can carry positive, negative, or zero net charge depending on the pH of their local environment. For every protein there is a specific pH at which its net charge is zero; this is its pI. Proteins show considerable variation in pI, although pI values usually fall in the range of pH 3-12, with the majority falling between pH 4 and pH 7. A protein is positively charged in solution at pH values below its pI and negatively charged at pH values above its pI.

2.2.2 IEF

When a protein is placed in a medium with a pH gradient and subjected to an electric field, it will initially move toward the electrode with the opposite charge. During migration through the pH gradient, the protein will either pick up or lose protons. As it migrates, its net charge and mobility will decrease and the protein will slow down. Eventually, the protein will arrive at the point in the pH gradient equal to its pI. There, being uncharged, it will stop migrating. If this protein should happen to diffuse to a region of lower pH, it will become protonated and be forced back toward the cathode by the electric field. If, on the other hand, it diffuses into a region of pH greater than its pI, the protein will become negatively charged and will be driven toward the anode. In this way, proteins condense, or are focused, into sharp bands in the pH gradient at their individual characteristic pI values. Focusing is a steady-state mechanism with regard to pH. Proteins approach their respective pI values at differing rates but remain relatively fixed at those pH values for extended periods. By contrast, proteins in conventional electrophoresis continue to move through the medium until the electric field is removed. Moreover, in IEF, proteins migrate to their steady state positions from anywhere in the system.

2.2.3 IPG strips

A stable, linear, and reproducible pH gradient is crucial to successful IEF. IPG strips offer the advantage of gradient stability over extended focusing runs (Bjellqvist *et al.* 1982). IPG strips are much more difficult to cast than carrier ampholyte gels (Righetti 1983); however, IPG strips are commercially available, for example as ReadyStrip™ IPG strips. pH gradients for IPG strips are created with sets of acrylamido buffers, which are derivatives of acrylamide containing both reactive double bonds and buffering groups. The general structure is CH2=CH-CO-NH-R, where R contains either a carboxyl [-COOH] or a tertiary amino group (e.g., N (CH3)2). These acrylamide derivatives are covalently incorporated into polyacrylamide gels at the time of casting and can form almost any conceivable pH gradient (Righetti 1990).

2.2.4 Choice of pH gradient ranges

Use of broad-range strips (pH 3-10) allows the display of most proteins in a single gel. With narrow-range and micro-range overlapping gradient strips, resolution is increased by expanding a small pH range across the entire width of a gel. Since many proteins are focused in the middle of the pH range 3-10, some researchers use nonlinear (NL) gradients to better resolve proteins in the middle of the pH range and to compress the width of the extreme pH ranges at the ends of the gradients. However, overlapping narrow-range and micro-range linear IPG strips can outperform a nonlinear gradient and display more spots per sample. This result is due to the extra resolving power from use of a narrower pI range per gel. Use of overlapping gradients also allows the ability to create "cyber" or composite gels by matching spots from the overlapping regions using imaging software.

2.2.5 IPG strip (2-D array) size

The 17 cm IPG strips and large-format gels have a large area to resolve protein spots; however, they take a long time to run. Using a mini system instead of, or as a complement to, a large gel format can provide significant time savings. A mini system is perfect for rapid optimization of sample preparation methods. Switching to a large format then allows thorough assessment of a complex sample and identification of proteins of interest. In many cases, a mini system consisting of narrow-range IPG strips can then be used to focus in on the proteins of interest. Throughput of the 2-D process is a consideration in choosing gel size. The ability to cast or run 12 gels at a time in any of 3 size formats is very useful in gathering proteomic results. In some cases, mini systems (7 cm ReadyStrip IPG strips with Mini-PROTEAN® 3 format gels, or 11 cm ReadyStrip IPG strips with Criterion™ precast gels) can completely replace large 2-D systems, providing speed, convenience, and ease in handling. The availability of narrow and micro overlapping pH-range ReadyStrip IPG strips can increase the effective width of pI resolution more than 5-fold after accounting for overlapping regions. When 3 narrow-range overlapping ReadyStrip IPG strips are used with the Criterion system, the resolution in the first dimension is increased from 11 to 26 cm. When micro-range strips are used, the resolution in the first dimension is expanded from 11 to 44 cm.

2.2.6 Estimation of pI

The pI of a protein can be estimated by comparing the position of the protein spot of interest to the position of known proteins or standards separated across the same pH gradient (Bjellqvist *et al.* 1993b, Garfin 2000). ReadyStrip IPG strips contain linear gradients, so the pI of an unknown protein can be estimated by linear interpolation relative to proteins of known pI.

2.2.7 Sample application

Commercial IPG strips are dehydrated and must be rehydrated to their original gel thickness (0.5 mm) before use. This allows flexibility in applying sample to the strips. There are 3 methods for sample loading: passive in-gel rehydration with sample, active in-gel rehydration with sample, or cup loading of sample after IPG rehydration. Introducing the sample while the strips are rehydrating is the easiest method. In some specific instances, it is best to rehydrate the strips and then apply sample through sample cups while current is applied.

2.2.8 Sample application during rehydration

For both active and passive rehydration methods, the sample is introduced to the IPG strip at the time of rehydration. As the strips hydrate, proteins in the sample are absorbed and distributed over the entire length of the strip (Sanchez *et al.* 1997). In the case of active rehydration, a very low voltage is applied during rehydration of the strips. Proteins enter the gel matrix under current as well as by absorption. The PROTEAN IEF cell has preprogrammed methods designed to accommodate active rehydration. Active rehydration is thought to help large proteins enter the strip by applying electrical "pull". Because the voltage is applied before all the solution and proteins are absorbed into the gel, the pH of a protein's environment will be the pH of the rehydration buffer, and the protein will move according to its mass-to charge ratio in that environment. Thus, small proteins with a higher mobility have a higher risk of being lost from the strip. With passive rehydration, proteins enter the gel by absorption only. This method allows efficient use of equipment since strips can be rehydrated in sample rehydration trays while other samples are being focused in the IEF cell.

Whether the strips are hydrated actively or passively, it is very important that they be incubated with sample for at least 11 hr prior to focusing. This allows the high molecular weight proteins time to enter the gel after the gel has become fully hydrated and the pores have attained full size. These sample application methods work because IEF is a steady-state technique, so proteins migrate to their pI independent of their initial positions.

The advantages of this approach are:

a. Sample application is simple (Görg *et al.* 1999)
b. Sample application during rehydration avoids the problem of sample precipitation, which often occurs with cup loading (Rabilloud 1999)
c. Shorter focusing times can be used because the sample proteins are in the IPG strip prior to IEF
d. Very large amounts of protein can be loaded using this method

2.2.9 Sample application by cup loading

Cup loading can be beneficial in the following cases (Cordwell *et al.* 1997, Görg *et al.* 2000):

- When samples contain high levels of DNA, RNA, or other large molecules, such as cellulose
- For analytical serum samples that have not been treated to remove albumin
- When running basic IPG strips; e.g., pH 7–10
- For samples that contain high concentrations of glycoproteins

Because of its relative difficulty and tendency toward artifacts, cup loading should be avoided if possible. When loading the protein sample from a cup, the IPG strips must be rehydrated prior to sample application. The IPG strips can be rehydrated in a variety of ways.

The rehydration tray is recommended although IPG strips are often rehydrated in 1 or 2 ml pipettes that have been sealed at both ends with Para film. Sample volumes of up to 100 µl can be loaded later onto each gel strip using a sample cup.

2.2.10 Power conditions and resolution in IEF

During an IEF run, the electrical conductivity of the gel changes with time, especially during the early phase. When an electrical field is applied to an IPG at the beginning of an IEF run, the current will be relatively high because of the large number of charge carriers present.

As the proteins and ampholytes move toward their pIs, the current will gradually decrease due to the decrease in the charge on individual proteins and carrier ampholytes. The pH gradient, strip length, and the applied electrical field determine the resolution of an IEF run. According to both theory and experiment, the difference in pI between two adjacent IEF-resolved protein bands is directly proportional to the square root of the pH gradient and inversely proportional to the square root of the voltage gradient at the position of the bands (Garfin 2000). Thus, narrow pH ranges and high-applied voltages yield high resolution in IEF. The highest resolution can be achieved using micro-range IPG strips and an electrophoretic cell, such as the PROTEAN IEF cell, capable of applying high voltages. IEF runs should always be carried out at the highest voltage compatible with the IPG strips and electrophoretic cell. However, high voltages in electrophoresis are accompanied by large amounts of generated heat. The magnitude of the electric field that can be applied and the ionic strength of the solutions that can be used in IEF are limited. Thin gels are better able to dissipate heat than thick ones and are therefore capable of withstanding the high voltage that leads to higher resolution. Also, at the completion of focusing, the current drops to nearly zero since the carriers of the current have stopped moving. The PROTEAN IEF cell is designed to provide precise cooling, allowing the highest possible voltages to be applied. (A default current limit of 50 µA per strip is intended to minimize protein carbamylation reactions in urea sample buffers. This limit can be increased to 99 µA per strip.)

Consistent and reproducible focusing requires that the time integral of voltage (volt-hours) be kept consistent. It is usually necessary to program IEF runs to reach final focusing voltages in stages. This approach clears ionic constituents in the sample from the IPG strips while limiting electrical heating of the strips. The PROTEAN IEF cell allows for multistep runs at durations set by time or volt-hours. The number of volt-hours required to complete a run must be determined empirically. A more complex sample in terms of number of proteins or even a different sample buffer might require increased volt-hours. The time needed to achieve the programmed volt-hours depends on the pH range of the IPG strip used as well as sample and buffer characteristics. If different strips are run at the same time, the electrical conditions experienced by individual strips will be different, perhaps exposing some strips to more current than desired, since the total current limit is averaged over all strips in a tray.

2.3 The second dimension: SDS-PAGE

2.3.1 Protein separation by molecular weight (MW)

Second-dimension separation is by protein mass, or MW, using SDS-PAGE. The proteins resolved in IPG strips in the first dimension are applied to second-dimension gels and separated by MW perpendicularly to the first dimension. The pores of the second-dimension gel sieve proteins according to size because dodecyl sulfate coats all proteins essentially in proportion to their mass. The net effect is that proteins migrate as ellipsoids with a uniform negative chargeto- mass ratio, with mobility related logarithmically to mass (Garfin 1995).

2.3.2 Gel composition

Homogeneous (single-percentage acrylamide) gels generally give excellent resolution of sample proteins that fall within a narrow MW range. Gradient gels have two advantages:

they allow proteins with a wide range of MW to be analyzed simultaneously, and the decreasing pore size along the gradient functions to sharpen the spots.

2.3.3 Single-percentage gels

The percentage of acrylamide, often referred to as %T (total percentage of acrylamide plus crosslinker) determines the pore size of a gel. Most protein separations use 37.5 parts acrylamide to 1 part bis-acrylamide (bis). Some researchers substitute piperazine bis-acrylamide (PDA), which can reduce silver staining background and give higher gel strength. If the total percentage of acrylamide plus crosslinker is higher, the smaller is the pore size. A suitable %T can be estimated from charts of mobility for proteins of different MW.

2.3.4 Gradient gels

Gradient gels are cast with acrylamide concentrations that increase from top to bottom so that the pore size decreases as proteins migrate further into the gels. As proteins move through gradient gels from regions of relatively large pores to regions of relatively small pores, their migration rates slow. Small proteins remain in gradient gels much longer than they do in single-percentage gels that have the same average %T, so both large and small molecules may be resolved in the same gel. This makes gradient gels popular for analysis of complex mixtures that span wide MW ranges. A gradient gel, however, cannot match the resolution obtainable with a properly chosen single concentration of acrylamide. A good approach is to use gradient gels for estimates of the complexities of mixtures. A proteomics experiment might start out with an 8–16%T gradient for global comparison. After interesting regions of the 2-D array have been identified, a new set of single-percentage gels may be run to study a particular size range of proteins. It is simplest and often most cost and labor effective to purchase commercially available precast gradient gels.

2.3.5 Precast gels

High-quality precast gels are preferred for high-throughput applications. They provide savings in time and labor, and the precision-poured gradients result in reproducibility among runs. Precast gels differ from handcast gels in that they are cast with a single buffer throughout and without SDS. During storage, different buffers in the stacking and resolving gels would mingle without elaborate means to keep them separate, and thus have no practical value. In addition, because the sample contains SDS, and the dodecyl sulfate ion in the cathode buffer moves faster than the proteins in the gel, keeping them saturated with the detergent, precast gels are made without SDS.

2.3.6 Transition from first to second dimension

The transition from first-dimension to second-dimension gel electrophoresis involves two steps: equilibration of the resolved IPG strips in SDS reducing buffer, and embedding of the strip on the top of the second-dimension gel. Proper equilibration simultaneously ensures that proteins are coated with dodecyl sulfate and that cysteines are reduced and alkylated. The equilibrated IPG strips are placed on top of the gel and fixed with molten agarose solution to ensure good contact between the gel and the strip.

2.3.7 Second dimension and high throughput

Since the first dimension can be run in batches of 12–24 strips at a time, it is desirable to run the same number of samples in the second dimension. Precast gels ensure high reproducibility among samples and help reduce the work involved in running large numbers of samples. Alternatively, gels can be hand cast 12 at a time under identical conditions with multi-casting chambers. The Dodeca cells save time, space, and effort, and help to ensure that gels are run under the same electrical conditions for highest throughput and reproducibility.

2.3.8 MW estimation

The migration rate of a polypeptide in SDS-PAGE is inversely proportional to the logarithm of its MW. The larger the polypeptide, the more slowly it migrates in a gel. MW is determined in SDS-PAGE by comparing the migration of protein spots to the migration of standards. Plots of log MW versus the migration distance are reasonably linear. Gradient SDS-PAGE gels can also be used to estimate MW. In this case, log MW is proportional to log (%T). With linear gradients, %T is proportional to distance migrated, so the data can be plotted as log MW vs. log (migration distance). Standard curves are actually sigmoid. The apparent linearit of a standard curve may not cover the full MW range for a given protein mixture in a particular gel. However, log MW varies sufficiently slowly to allow accurate MW estimates to be made by interpolation, and even extrapolation, over relatively wide ranges (Garfin 1995). Mixtures of standard proteins with known MW are available from Bio-Rad in several formats for calibrating the migration of proteins in electrophoretic gels. Standards are available unstained, prestained, or with tags for development with various secondary reagents (useful when blotting). Standards can be run in a reference well, attached to the end of a focused IPG strip by filter paper, or directly embedded in agarose onto the second-dimension gel

2.4 Detection of proteins in gels

2.4.1 Guidelines for detection of proteins in gels

Gels are run for either analytical or preparative purposes. The intended use of the gel determines the amount of protein to load and the means of detection. It is most common to make proteins in gels visible by staining them with dyes or metals. Each type of protein stain has its own characteristics and limitations with regard to the sensitivity of detection. Sometimes proteins are transferred to membranes by western blotting to be detected by immunoblotting, glycoprotein analysis, or total protein stain. If the purpose of gel electrophoresis is to identify low-abundance proteins (e.g., low-copy-number proteins in a cell extract, or contaminants in a purification scheme), then a high protein load (0.1–1 mg/ml) and a high-sensitivity stain, such as silver or a fluorescent stain, should be used (Corthals *et al.* 2000). When the intention is to obtain enough protein for use as an antigen or for sequence analysis, then a high protein load should be applied to the gel and the proteins visualized with a staining procedure that does not fix proteins in the gel. Quantitative comparisons require the use of stains with broad linear ranges of detection. The sensitivity that is achievable in staining is determined by: 1) the amount of stain that binds to the proteins; 2) the intensity of the coloration; 3) the difference in coloration between stained proteins and the residual background in the body of the gel (the signal-to-noise ratio). Unbound stain molecules can be washed out of the gels without removing much stain from the proteins. All stains interact differently with different proteins (Carroll *et al.* 2000). No stain will universally stain all proteins in a gel in proportion to their mass. The only observation that seems to hold for most stains is that they interact best with basic amino acids. For critical analysis, replicate gels should be stained with two or more different stains. Of all stains available, colloidal Coomassie Blue (Bio-Safe™ Coomassie) appears to stain the broadest spectrum of proteins. It is instructive, especially with 2-D PAGE gels, to stain a colloidal Coomassie Blue-stained gel with silver or to stain a fluorescently stained gel with colloidal Coomassie Blue or silver. Very often, this double staining procedure will show a few differences between the protein patterns. It is most common to stain gels first with Coomassie Blue or a fluorescent stain, and then restain with silver. However, the order in which the stains are used does not seem to be important, as long as the gels are washed well with high-purity water between stains.

2.4.2 Coomassie blue staining

Coomassie Brilliant Blue R-250 is the most common stain for protein detection in polyacrylamide gels. Coomassie Brilliant Blue R-250 and G-250 are wool dyes that have been adapted to stain proteins in gels. The "R" and "G" designations indicate red and green hues, respectively. Coomassie R-250 requires on the order of 40 ng of protein per spot for detection. Absolute sensitivity and staining linearity depend on the proteins being stained. The staining solution also fixes most proteins in gels. Bio-Safe Coomassie stain is made with Coomassie Brilliant Blue G-250. Bio-Safe Coomassie stain is a ready-to-use, single-reagent protein stain. Sensitivity can be down to 10 ng, and greater contrast is achieved by washing the gel in water after staining. Used stain can be disposed of as nonhazardous waste and the procedure does not fix proteins in the gel.

2.4.3 SYPRO ruby fluorescent staining

SYPRO Ruby protein gel stain has desirable features that make it popular in high-throughput laboratories. It is an endpoint stain with little background staining (high signal-to noise characteristics) and it is sensitive and easy to use. SYPRO Ruby protein stain does not detect nucleic acids. SYPRO Ruby protein stain is sensitive to 1–10 ng and can be linear over 3 orders of magnitude. It is compatible with high through put protocols and downstream analysis, including mass spectrometry and Edman sequencing (Patton 2000). It also allows detection of glycoproteins, lipoproteins, low MW proteins, and metalloproteins that are not stained well by other stains. This fluorescent stain is easily visualized with simple UV or blue-light transilluminators, as well as by the Molecular Imager FX™ Pro Plus multiimager and VersaDoc™ imaging systems.

2.4.4 Silver staining

Two popular methods for silver staining are recommended for 2-D analysis. They are based on slightly different chemistries but have similar sensitivities for protein. Bio-Rad's silver stain kit, based on the method of Merril *et al.* (1981), can be as much as 100 times more sensitive than Coomassie Blue R-250 dye staining and allows visualization of heavily glycosylated proteins in gels. Protein spots containing 10–100 ng of protein can be easily seen. Proteins in gels are fixed with alcohol and acetic acid, then oxidized in a solution of potassium dichromate in dilute nitric acid, washed with water, and treated with silver nitrate solution. Silver ions bind to the oxidized proteins and are subsequently reduced to metallic silver by treatment with alkaline formaldehyde. Color development is stopped with acetic acid when the desired staining intensity has been achieved. This method is not compatible with mass spectroscopic analysis since the oxidative step changes protein mass. The Silver Stain Plus stain from Bio-Rad requires only one simultaneous staining and development step and is based on the method developed by Gottlieb and Chavko (1987). Proteins are fixed with a solution containing methanol, acetic acid, and glycerol, and washed extensively with water. The gels are then soaked in a solution containing a silverammine complex bound to colloidal tungstosilicic acid. Silver ions transfer from the tungstosilicic acid to the proteins in the gel by means of an ion exchange or electrophilic process. Formaldehyde in the alkaline solution reduces the silver ions to metallic silver to produce the images of protein spots. The reaction is stopped with acetic acid when the desired intensity has been achieved. Because silver ions do not accumulate in the bodies of gels, background staining is light. Since this method lacks an oxidizing step, visualization of heavily glycosylated proteins and lipoproteins can be less sensitive than with the Merril stain. This method is better for use in proteomics when the end goal is identification by mass spectrometric analysis.

2.5 Image acquisition and analysis

2.5.1 Image acquisition instruments

Before 2-D gels can be analyzed with an image evaluation system, they must be digitized. The most commonly used devices are camera systems, densitometers, phosphor imagers, and fluorescence scanners. All of Bio-Rad's imaging systems are seamlessly integrated with PDQuest™ software, and they can export and import images to and from other software via TIFF files.

2.5.2 Densitometry

Densitometers compare the intensity of a light beam before and after attenuation by a sample. The GS-800™ calibrated imaging densitometer has been customized for analysis of gels, autoradiograms, and blots. The transmittance and true reflectance capabilities allow accurate scans of samples that are either transparent (gels and film) or opaque (blots). It provides high-quality imaging to resolve close spots and a variable resolution feature to preview and crop images. Wet 2-D gels may be scanned with red, green, and blue color CCD technology on the watertight platen.

2.5.3 Storage phosphor and fluorescence scanners

Digitization of 2-D gels stained with fluorescent dyes or radioactive compounds requires specific imaging systems (Patton 2000). The Molecular Imager FX™ Pro Plus system is flexible and expandable. 2-D gels of radiolabled proteins can be imaged using a Kodak phosphor screen more rapidly and accurately than with film. Popular proteomic fluorescent stains, including SYPRO Ruby protein gel and blot stains and SYPRO Orange protein gel stain can be imaged with single-color and multicolor fluorescence via direct laser excitation. This system permits detection of almost any fluorophore that is excited in the visible spectrum. The internal laser and external laser options allow optimal excitation of single-color or multicolor fluorescent samples. Computer-controlled, user-accessible filter wheels have eight filter slots, allowing detection of many multicolor combinations of dyes (Gingrich *et al.* 2000).

2.5.4 Computer-assisted image analysis of 2-D electrophoretic gels

Computer-assisted image analysis software is an indispensable tool for the evaluation of complex 2-D gels. It allows:

a. Storage and structuring of large amounts of collected experimental image data
b. Rapid and sophisticated analysis of experimental information
c. Supplementation and distribution of data among labs
d. Establishment of 2-D-protein data banks

Image analysis systems deliver error-fee comprehensive qualitative and quantitative data from a large number of 2-D gels (Miller 1989). PDQuest software from Bio-Rad is a popular analysis tool. Gel analysis of digitized gel images includes spot detection, spot quantitation, gel comparison, and statistical analysis. PDQuest software has the further advantage of seamless integration with any of Bio-Rad's image acquisition instruments, as well as the ability to control the ProteomeWorks™ spot cutter. The advanced annotation feature can be used to label spots with text, URL links, document links, or mass spectrometry data.

2.5.5 Spot detection and spot quantitation

Before the software automatically detects the protein spots of a 2-D gel, the raw image data are corrected and the gel background is subtracted. The process is executed with simple menus and "wizards." PDQuest software models protein spots mathematically as 3-D Gaussian distributions and uses the models to determine absorption maxima. This enables automatic detection a resolution of merged spots. Following this procedure, spot intensities are obtained by integration of the Gaussian function. The mathematical description of the spots is used both for data reduction and for increasing evaluation speed, since reevaluation of data after an image change takes only fractions of a second. The hit rate of automatic spot detection is highly dependent on the quality of the 2-D gels. Correction capabilities of PDQuest software can be used to add undetected spots to the list of spots or to delete spots that arise from gel artifacts.

2.5.6 Gel comparison

The next step in 2-D gel evaluation is the identification of proteins that are present in all gels of a series. This task is made difficult primarily because of inherent irreproducibility in gels, which affects the positions of spots within a gel series. Gel analysis software must detect minor shifts in individual spot position within the gel series. Many software packages for automatic gel comparison are created with the assumption that the relative positions of spots are altered only slightly relative to each other, and allocate the spots on this basis. Prior to automatic gel comparison, PDQuest software selects the best 2-D gel of a gel series as a reference or standard gel and compares all other 2-D gels to this gel. Proteins in a gel series that are not present in the reference gel are added manually so that the reference gel will include all proteins of a gel series.

Before the software can detect and document matching of different spots, a number of landmarks, or identical spots in the gel series, must be manually identified. The landmarking tool speeds the process by making "best guess" assignments of landmark spots to images in the gel series. With PDQuest software, it is possible to simultaneously display up to 100 enlarged details of 2-D gels on the screen. This simultaneous display of all 2-D gels of a test series enables rapid and error-free determination of the fixed points. Using the landmarks, the image analysis software first attempts to compare all spots lying very near these fixed points and then uses the matched spots as starting points for further comparisons. Thus, the entire gel surface is systematically investigated for the presence or absence of matching spots in a gel series. The results of the automatic gel comparison require verification, as does automatic spot detection.

Two tools assist this verification process in PDQuest: Either identical protein spots are labeled with matching letters and allocated section by section, or the deviations in the spot positions of a particular 2-D gel can be displayed as lines that show spot shifts in comparison to the reference gel.

2.5.7 Data analysis

With PDQuest software, all gels of an experiment are viewed as a unit. To compare gels from different experiments, the reference images are compared. In such comparisons, each spot is automatically assigned a number so that identical spots have identical numbers. Experimental data can also be analyzed statistically both parametric and nonparametric tests are available.

2.6 Identification and characterization of 2-D protein spots

2.6.1 Sequence data from 2-D gels

2-DE has the virtually unique capability of simultaneously displaying several hundred gene products. 2-D gels are an ideal starting point for protein chemical identification and characterization. Peptide mass fingerprint or sequence data can be derived following 2-DE with mass spectrometry or amino acid sequence analysis (Eckerskorn *et al.* 1988, Ducret *et al.* 1996). The sensitivity of currently available instruments makes 2-DE an efficient "preparative" analytical method. Most current protein identification depends on mass spectrometry of proteins excised from gels or blots.

2.6.2 Integration of image analysis with automated spot cutting

Image analysis software obtains quantitative and qualitative information about the proteins in a sample, and stores the information in files, which may also contain additional annotations. The ProteomeWorks™ spot cutter expands the capabilities of proteome labs by integrating PDQuest™ image analysis software. The image analysis files acquired by PDQuest direct automated spot cutting. Excised protein spots are deposited into microtiter plates ready for further automated processing. PDQuest software tracks the protein spots through spot cutting and protein identification. Downstream protein spot identifications are generally obtained from peptide mass fingerprint analysis using mass spectrometry. The ProteomeWorks spot cutter is a precision instrument with a small benchtop footprint. It is fully automatic to increase throughput and minimize the amount of hands-on time spent excising protein spots. The spot cutter individually excises even overlapping spots for unique identification.

2.6.3 Automated protein digestion

The ProteomeWorks spot cutter eliminates the first of two bottlenecks for excision and enzymatic digestion of protein spots. Driven by PDQuest software, it enables automated spot excision and deposition of cut gel spots into microtiter plate wells. Isolated proteins from the gel pieces are then digested to release peptides for detailed sequence analysis by mass spectrometry, leading to protein identification. Excised gel spots can be robotically destained, chemically modified (reduced and alkylated), and digested in preparation for either MALDI-TOF-MS or ESI-MS with the Micromass Mass*PREP* station. Each process is executed under fully automated software control with a range of standard protocols enabling high throughput and flexibility. Manual protein digestion is a tedious, time-consuming process that is subject to variability and keratin contamination. Automation of this process with the Mass*PREP* station eliminates a significant bottleneck for high-throughput protein identification.

Operational features of the Mass*PREP* station include a variable temperature control for optimized reduction, alkylation, and digestion of proteins, and onboard cooling capabilities for reagents and peptide digests to ensure reproducible digestion results. The station employs a variety of sample cleaning technologies (Mass*PREP* targets and Millipore ZipTip pipet tips) to prepare peptide digests prior to automated deposition of samples onto a M@LDI or Mass*PREP* target plate. Contamination of peptide samples is also minimized with the Mass*PREP* clean air enclosure.

2.6.4 Rapid, high-throughput protein identification by MALDI-TOF-MS

Peptide mass fingerprinting of protein digest products using matrix assisted laser desorption ionization time of flight mass spectrometry (MALDI-TOF-MS) provides an ideal method for protein identification when samples have been separated by 2-D PAGE. The M@LDI HT is one of a new generation of networked "2-D gel-MS" analyzers for high-throughput protein identification. M@LDI HT is the primary MS data acquisition device of the ProteomeWorks system, and features a fully automated target plate auto-changer for increased throughput. Networking enables distribution of data capture, protein assignment, and result presentation functions of ProteinLynx Global *SERVER* software within a secure clientserver architecture, maximizing computing power to quickly identify proteins. The M@LDI HT enables automated acquisition of optimized mass spectra and the derivation of monoisotopic peptide mass fingerprint information. Interrogation of multiple FASTA databases using Global *SERVER* software following capture of MS results provides rapid identification of proteins that fit the samples' peptide mass fingerprint, along with a confidence score indicating the validity of the identifications. Following MS identification, peptide mass fingerprint spectra and all of the identification results are available through electronic reports. In addition, protein identification results are seamlessly integrated with the gel image in PDQuest software. Using this system, the working time to process data from spot cutting to protein digestion to MS analysis and image annotation is reduced by over 50% compared to manual processing of gel samples, with a corresponding reduction in error. All of the instrumentation and software in this process is part of the integrated ProteomeWorks system, a set of powerful tools for proteomic analysis.

2.6.5 Advanced protein characterization with ESI-LC-MS and MALDI-TOF

MALDI-TOF MS provides an ideal high-throughput solution for protein identification; however, where protein identity is ambiguous, known databases must be searched with a higher degree of sequence information. The Micromass Q-Tof family of MS-MS instruments incorporates quadrupole/ orthogonal acceleration time-of-flight (Q/oa-TOF) technology, enabling exact mass measurement, and acquisition of the highest-level peptide sequence information for de novo sequencing and BLAST analyses. Protein digest samples in microtiter plates, prepared with the Mass*PREP* station, can be transferred directly to the Micromass CapLC (capillary HPLC) system for automated injection into the Q-Tof *micro* for integrated LC-MS-MS under MassLynx software control. The capability for MS to MS-MS switching "on the fly" with the Q-Tof family of instruments maximizes the amount of amino acid sequence information that can be generated with these instruments. MassSeq software also provides the capacity for automated de novo amino acid sequencing based on the MS results.

2.6.6 2DE in identification of bladder carcinoma protein marker (Calreticulin)

Susumu Kageyama et al., 2004 screened proteins as tumor markers for bladder cancer by proteomic analysis (2DE) of cancerous and healthy tissues and investigated the diagnostic accuracy of one such marker, *Calreticulin* (CRT) in urine.

They have produced two important findings in their experiments. The first is that increased production of CRT in bladder cancer tissue which was confirmed by proteome profiling by 2DE. Furthermore, we detected two isoforms of CRT, and full-length CRT which was more useful than cleaved CRT for distinguishing bladder cancer from healthy tissue.

In their study, although visual comparison of 2DE gels of TCC (transitional cell carcinoma) and noncancerous urothelium showed similar expression profiles, 15 protein spots (U-1 to U-15) were more intense in TCC samples (Fig. 1). They identified 10 of the proteins by use of a peptide mass fingerprinting method. One spot among them, with an apparent mass of 55 kDa and pI of 4.3, was identified as CRT (spot U-2 in Fig. 1 A). From NH2-terminal amino acid sequencing, 10 amino acids were sequenced (EPAVYFKEQF), and they were identical to residues 1–10 of mature human CRT according to the sequence homology search.

Further, to validate the 2DE finding of increased production of full-length CRT in bladder cancer tissue, they performed quantitative Western blot analysis in cancerous and healthy tissue using anti-COOH-terminus antibody. They compared CRT band intensities for 22 cancerous with 10 noncancerous tissues. For band quantification, they defined the CRT band derived from a total of 1 µg of heat-shocked HeLa cell extract as 1.0 unit/µg of protein. The mean (SD) concentrations in cancerous and healthy tissue were 1.0 (0.4) and 0.4 (0.3) units/µg of protein, respectively (Mann–Whitney *U*-test, $P = 0.0003$; Fig. 2). Among these tissue samples, six pairs of cancerous and noncancerous specimens were obtained from the bladders of patients who had undergone radical cystectomy. CRT concentrations were higher in all cancer tissues compared with the corresponding healthy urothelium.

To confirm the presence of isoforms of CRT, they performed two-dimensional Western blotting with two different antibodies: monoclonal antibody FMC75, which was produced against recombinant human CRT; and a polyclonal antibody that was produced using synthesized peptides of the human CRT COOH terminus (amino acids 388–400) as an immunogen. On Western blots with anti- COOH-terminus antibody, only one of the two spots was visualized, whereas both spots became visible on blots incubated with FMC75 (Fig. 3). One was the same as the 55-kDa (pI 4.3) spot, and the other had an apparent molecular mass of 40 kDa and pI of 4.5. This lower molecular- mass spot had the same NH2-terminal amino acid sequence as amino acids 1–10 of mature human CRT as shown by amino acid sequencing. Therefore, they suggested that the higher-molecular-mass spot was the full-length form and the other spot was a cleaved form that is truncated elsewhere in the COOH domain. Production of the full-length CRT in cancer tissue was increased compared with in healthy tissue, but the spots for cleaved CRT in cancerous and healthy urothelium had intensities that were similar and were reproducible on all silverstained 2DE gels.

Subsequently they tried to confirm whether anti-COOH terminus antibody binds to molecules other than full lengt1h CRT and performed immuno-precipitations (Fig. 4). The Western blot of the immuno-precipitate extracted from cancer tissue revealed only one band, and they concluded that anti-COOH-terminus antibody binds specifically to full-length CRT of ~55 kDa. They therefore judged that full-length CRT recognized by anti-COOH terminus antibody is appropriate as a tumor marker.

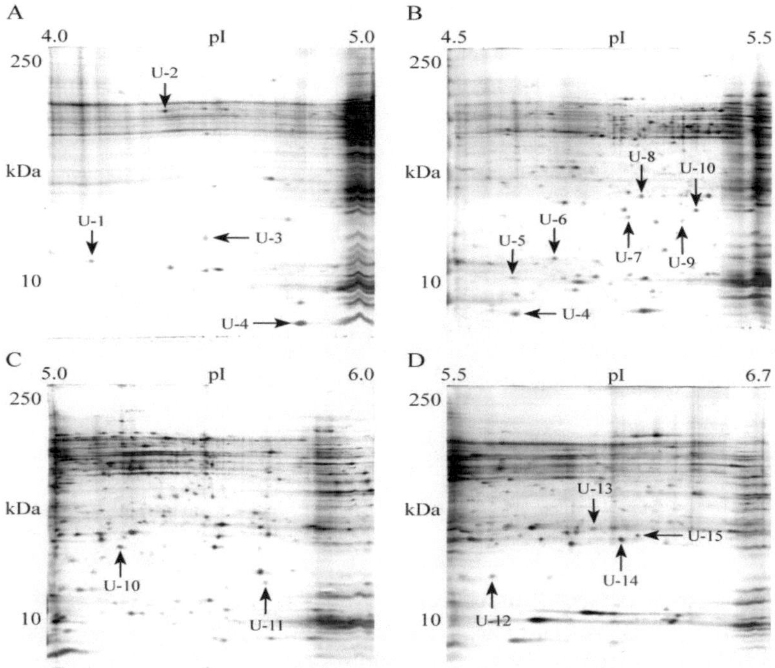

Fig. 1. Silver-stained images of analytical narrow-pH-range 2DE gels of proteins from bladder cancer. (A), pH 4.0–5.0; (B), pH 4.5–5.5; (C), pH 5.0–6.0; (D), pH 5.5–6.7. Arrows indicate spots (U-1 to U-15) containing higher amounts of protein. Spot U-2, ~55 kDa and pI 4.3, was confirmed to be CRT by a peptide mass fingerprinting method and NH2-terminal amino acid sequencing. *(Source: Susumu Kageyama et al., 2004 Clinical Chemistry, 50: 857–866)*

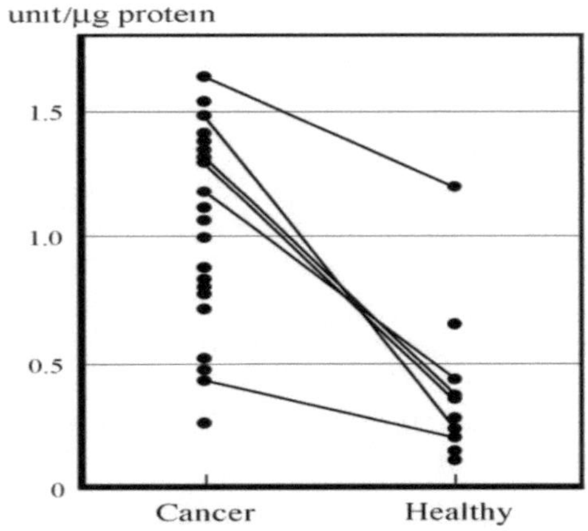

Fig. 2. Quantitative Western blot analysis of cancerous and healthy tissues using anti-COOH-terminus antibody. *Lines* show six pairs of cancerous and healthy specimens obtained from the bladders of patients who had undergone radical cystectomy.*(Source:* Susumu Kageyama *et al., 2004 Clinical Chemistry, 50: 857–866)*

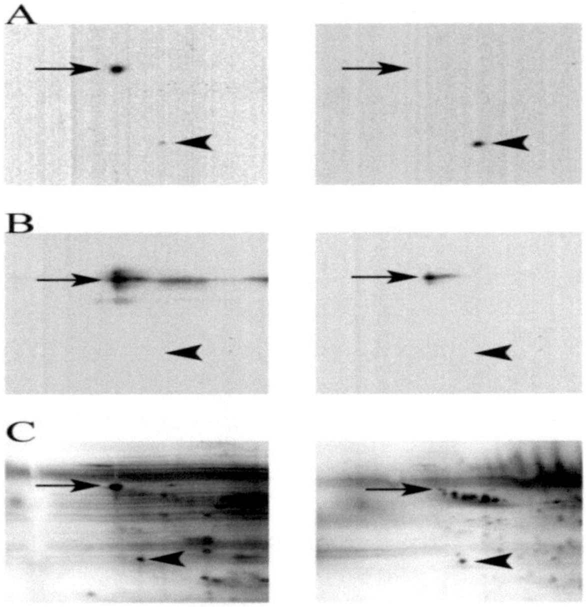

Fig. 3. Close-up sections of two-dimensional Western blot images obtained with two different antibodies, FMC75 (*A*) and anti-COOH terminus antibody (*B*), and proteins in silver-stained pH 4–7 gels (*C*). *Left panels* are bladder cancer and *right panels* are healthy urothelium. *Arrows* indicate full-length CRT (55 kDa; pI 4.3), and *arrowheads* indicate cleaved CRT (40 kDa; pI 4.5). These two CRT forms have the same NH2-terminal amino acid sequence (EPAVYFKEQF), but cleaved CRT is considered to lack the COOH terminus because of no immunoreactivity for anti-COOH-terminus antibody. *(Source: Susumu Kageyama et al., 2004 Clinical Chemistry, 50: 857–866)*

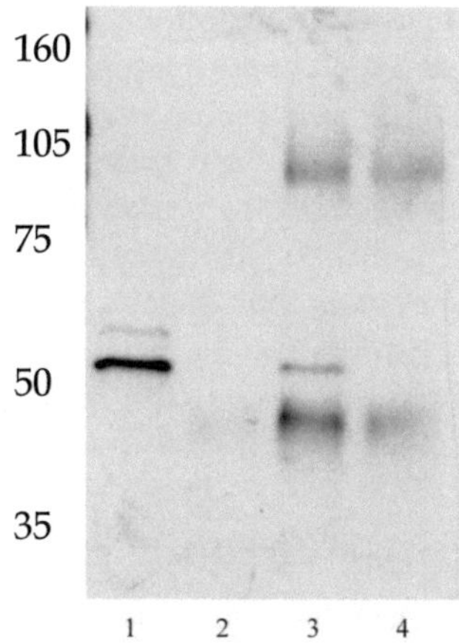

Fig. 4. Western blot with FMC75 antibody. *Arrow* indicates 55-kDa full-length CRT. *Lane 1,* total cell lysate; *lane 2,* extraction from protein A Sepharose beads that did not bind antibody; *lane 3,* immunoprecipitant eluted from beads binding anti-COOH-terminus antibody; *lane 4,* immunoprecipitant extracted from beads binding normal rabbit IgG indicates approximate molecular masses. *(Source: Susumu Kageyama et al., 2004 Clinical Chemistry, 50: 857–866)*

2.6.7 2-DE in identification of pancreatic carcinoma protein marker

Leucine-rich alpha-2-glycoprotein is characterized by its unusually high content of leucine, about 17% by weight. The primary structure of LRG suggests that it may be a membrane associated or membrane-derived protein. Aberrant regulation of LRG has been observed in patients with malignant disease and with virus infection.

Tatsuhiko Kakisaka et al., 2007 examined the plasma LRG expression levels of cancer patients who were not in the acute inflammatory phase. First, they selected cancer patients with normal C-reactive protein (CRP) concentration (Table 1, P6–10) to explore whether the increase in LRG levels paralleled the dynamics of the commonacute phase proteins. Second, they tested plasma from cancer patients with a normal level of CA19-9 a tumor marker commonly used for the diagnosis of pancreatic cancer, to examine the possibility of plasma LRG levels being using in a way complementary to existing tumor markers.

SDS-PAGE/Western blotting using an anti-LRG antibody showed consistent up-regulation of LRG in these patients, who were negative for CRP and/or CA19-9, compared with the noncancer bearing healthy donors (Fig. 5). Therefore, increased amounts of LRG may be independent of the regulation of other acute phase proteins and tumor markers. They also examined plasma samples from chronic pancreatitis patients and found that they tended to express lower LRG levels compared with the samples from pancreatic cancer patients. By correlating the expression levels of LRG with clinical information from a large sample set, they hope to validate the utility of LRG as a biomarker to monitor the status of patients. Some plasma samples from pancreatic cancer patients did not express high LRG levels, leading us to suggest that the examination of plasma LRG levels in combination with the existing biomarkers would increase the specificity and sensitivity of the diagnosis.

Some protein spots on 2D-PAGE gels overlapped across fractions in the anion-exchange chromatography even when they used the step-wise gradient method with system wash between intervals. They considered these overlapping spots to correspond to different isoforms of the same protein, and have therefore counted all protein spots on the 2D-PAGE gels. However, not every differentially expressed protein was considered to be a suitable tumor marker; for example, spots 8 and 11 (transthyretin) were differentially expressed between cancer patients and healthy donors, but they were very minute amounts of the total abundant transthyretin, and it was difficult to extract theseportion of the protein. On the contrary, LRG, which was also differentially expressed between cancer patients and healthy donors, was only expressed in one fraction and was therefore selected as a candidate for a tumor marker of pancreatic cancer.

The use of high-resolution 2D-PAGE with narrow-range IPG gels and large-format second dimension gels could solve this problem to some extent.

Patient informations of validation set 1

Case[a]	Age	Sex	Tumor location	Stage[b]	CA19-9 (U/ml)	CRP (mg/dl)
P6	56	Male	Head	IV	1	0.1
P7	45	Female	Body~tail	IV	3698	0.1
P8	55	Female	Body	III	<1	<0.1
P9	58	Male	Body	IV	25600	0.1
P10	65	Female	Body	III	804	<0.1

Case[a]	Age	Sex
N6	53	Male
N7	51	Female
N8	60	Female
N9	59	Male
N10	64	Female

[a] P: pancreatic cancer patients, N: non-cancer bearing healthy donors.
[b] The Union Internationale Contre le Cancer (UICC) classification [41].

Table 1. Patient informations of validation set 1

2.6.8 2-DE in identification of human gastric carcinoma protein marker

Gastric cancer is the second most common cause of cancer deaths worldwide and due to its poor prognosis, it is important that specific biomarkers are identified to enable its early detection. Through 2-D gel electrophoresis and MALDI-TOF-TOF-based proteomics approaches, Chien-Wei Tseng et al., 2011 found that 14-3-3β, which was one of the proteins that were differentially expressed by 5-fluorouracil-treated gastric cancer SC-M1 cells, was up regulated in gastric cancer cells. 14-3-3β levels in tissues and serum were further validated in gastric cancer patients and controls. The results showed that 14-3-3β levels were elevated in tumor tissues in comparison to normal tissues, and serum levels in cancer patients were also significantly higher than those in controls (Fig. 6). Elevated serum 14-3-3β levels highly correlated with the number of lymph node metastases, tumor size, and a reduced survival rate. Moreover, over-expression of 14-3-3β enhanced the growth, invasiveness, and migratory activities of tumor cells. Twenty-eight proteins involved in anti-apoptosis and tumor progression were also found to be differentially expressed in 14-3-3β-overexpressing gastric cancer cells. Overall, these results highlight the significance of 14-3-3β in gastric cancer cell progression and suggest that it has the potential to be used as a diagnostic and prognostic biomarker in gastric cancer.

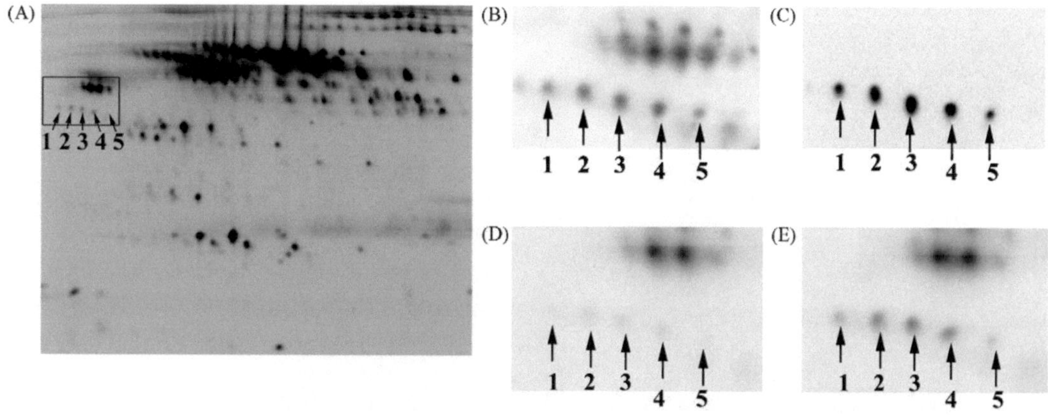

Fig. 5. Elevated level of plasma LRG in pancreatic cancer. The localizations of the five LRG spots are indicated by arrows 1–5 on the 2D image of the 150mM NaCl sample (A). The boxed area was transferred to a nitrocellulose membrane and scanned with a laser scanner to obtain the LRG spots on the membrane (B). The scanned membrane was reacted with an anti-LRG antibody and the antibody–antigen complexes were detected with an ECL system (C). The fluorescent signals of the LRG spots on the 2D-PAGE gels were compared between non-cancer bearing healthy donors (D) and pancreatic cancer patients (E). *(Source: Tatsuhiko Kakisaka et al., 2007 Journal of Chromatography B, 852: 257–267)*

2.6.9 2DE in identification of squamous cervical carcinoma protein markers

Recently, proteomic and genomic approaches to identify tumor markers are undergoing. Hellman and coworkers reported the protein expression patterns in primary carcinoma of the vagina. In relation to HPV, C33A cell line transfected with HPV *E7* gene and proteomic and genomic analyses were performed. But, until now, there was no report of SCC in cervix tissues. Prof. W.S. Ahn and his colleagues at Cancer Research Center of The Catholic University of Korea, South Korea contributed much more to understand the significance of 2DGE in diagnosis of cervical cancer. In general, screening in cervical cancer is progressing to find out candidate genes and proteins, which may work as biological markers and play a role in tumor progression. They examined the protein expression patterns of squamous cell carcinoma (SCC) tissues from Korean women using two-dimensional polyacrylamide gel electrophoresis (2-DE) and matrix-assisted laser desorption/ionization-time of fight (MALDI-TOF) mass spectrometer. A total of 35 proteins are detected in SCC. 17 proteins are up regulated and 18 proteins are down-regulated. Among the proteins identified, 12 proteins (pigment epithelium derived factor, annexin A2 and A5, keratin 19 and 20, heat shock protein 27, smooth muscle protein 22 alpha, alpha-enolase, squamous cell carcinoma antigen 1 and 2, glutathione S-transferase, apolipoprotein a1) are previously known proteins involved in tumor and 21 proteins were newly identified in this study. They concluded that the 2-DE offers total protein expression profiles of SCC tissues and further characterization of proteins that are differentially expressed will give a chance to identify tumor-specific diagnostic markers for SCC (Fig. 7 & 8).

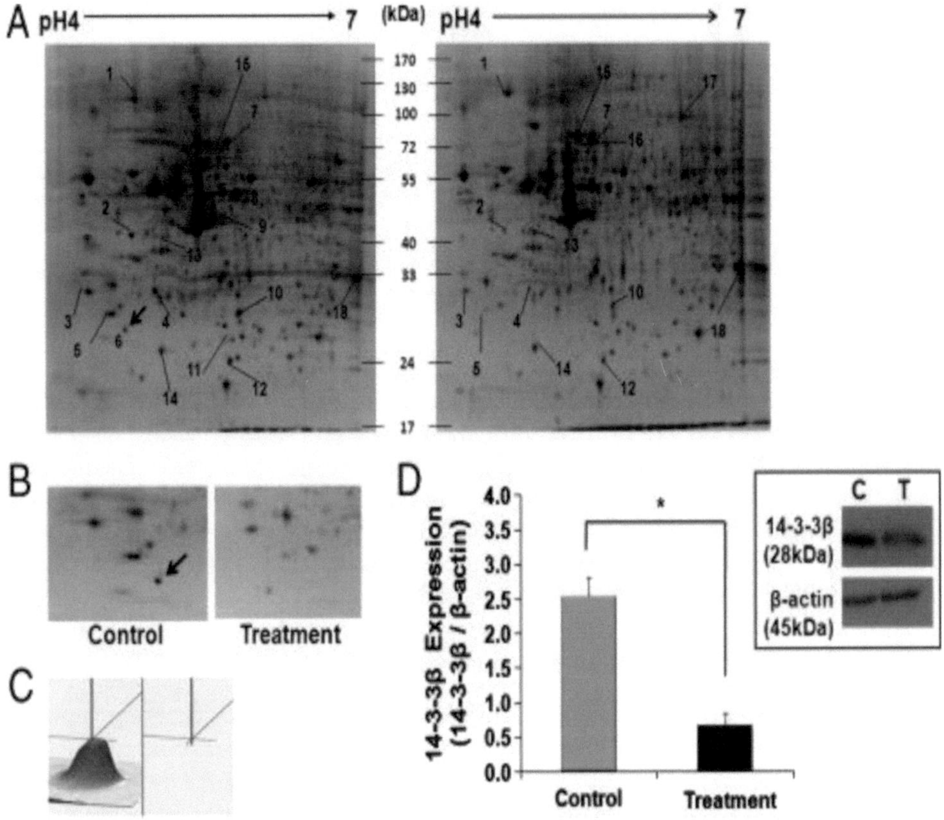

Fig. 6. 14-3-3β is differentially expressed after 5-FU treatment of SC-M1 cells. (A) Proteins from the 5-FU-treated (right) SC-M1 cells and the untreated control (left) were compared using 2-DE. Enlarged images and 3-D profiles of 14-3-3b on the gels are shown in (B) and (C). (D) 14-3-3β expression was significantly reduced after 5-FU treatment as confirmed by Western blot. *(Source: Chien-Wei Tseng et al., 2011 Proteomics, 11:2423–2439)*

In addition to cervical cancer, the 2-dimensional polyacrylamide gel electrophoresis (2-DE) has also been used to examine heterogeneity of protein expression in tissues from different tumors such as bladder, breast, colon–rectum, lung, and ovary. The advantage of 2-DE is that the complex protein expression is analyzed qualitatively and quantitatively. 2-DE combined with MALDI-TOF-MS has been applied to identify cancer-specific protein markers. These markers can provide a basis for developing new methods for early diagnosis and treatment.

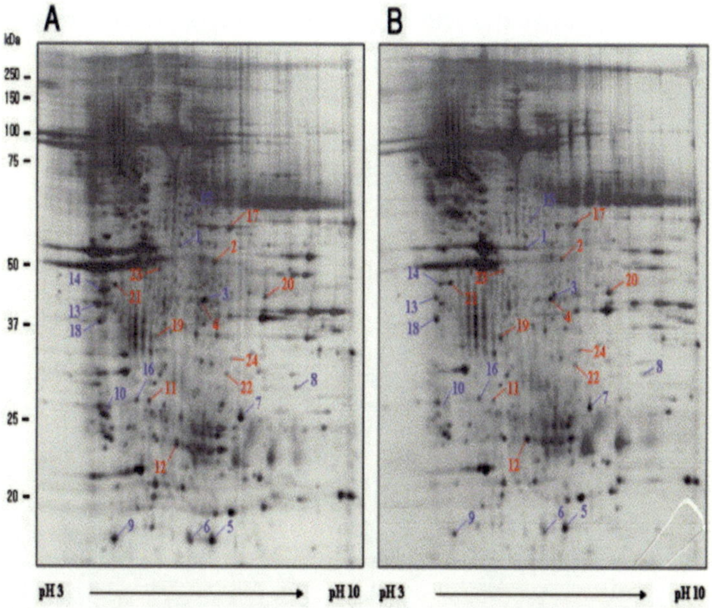

Fig. 7. Comparison of proteome by two-dimensional gel electrophoresis on normal tissues and cervical SCC tissues. Representative examples of 2-DE gels derived from a normal cervix tissue and cervical SCC tissue. Normal cervix (A) and cervical SCC (B) total proteins were separated by 2-DE using IPG strips pH 3–10 in the first and 12% SDS-PAGE in the second dimension. Identified protein spots are indicated by numbers. Proteins down- (A) or up-regulated (B) in cervical SCC are indicated. (*Source: Bae et al., 2005 Gynecologic Oncol., 99: 26-35)*

2.6.10 Breast cancer protein profiling

Franzen *et al.,* 1997 well documented that the two-dimensional electrophoresis (2-DE) analyses of human breast carcinoma reveals the following observations: (i) Analysis of samples from different areas of the same tumor showed a high degree of similarity in the pattern of polypeptide expression. Similarly, analysis of two tumors and their metastases revealed similar 2-DE profiles. (ii) In contrast, large variations have been observed between different lesions with comparable histological characteristics. Larger differences in polypeptide expression are pointed out in between potentially highly malignant carcinomas and comparisons of less malignant lesions. These differences are in the same order of magnitude as those observed comparing a breast carcinoma to a lung carcinoma. (iii) The levels of all cytokeratin forms resolved (CK7, CK8, CK15, and CK18) were significantly lower in carcinomas compared to fibro adenomas. (iv) The levels of high molecular weight tropo-myosins (1–3) were lower in carcinomas compared to fibro adenomas. The expression of tropomyosin-1 is 1.7-fold higher in primary tumors with metastatic spread to axillar lymph nodes compared to primary tumors with no evidence of metastasis ($p < 0.05$). (v) The expression of proliferating cell nuclear antigen (PCNA) and some members of the stress protein family (pHSP60, HSP90, and calreticulin) are higher in carcinomas. We conclude that malignant progression of breast carcinomas results in large heterogeneity in polypeptide expression between different tumors, but that some common themes such as decreased expression of cytokeratin and tropomyosin polypeptides can be discerned (Fig. 9 & 10). (Franzen, *et al.* 1997 *Electrophoresis,* 18 : 582–587.).

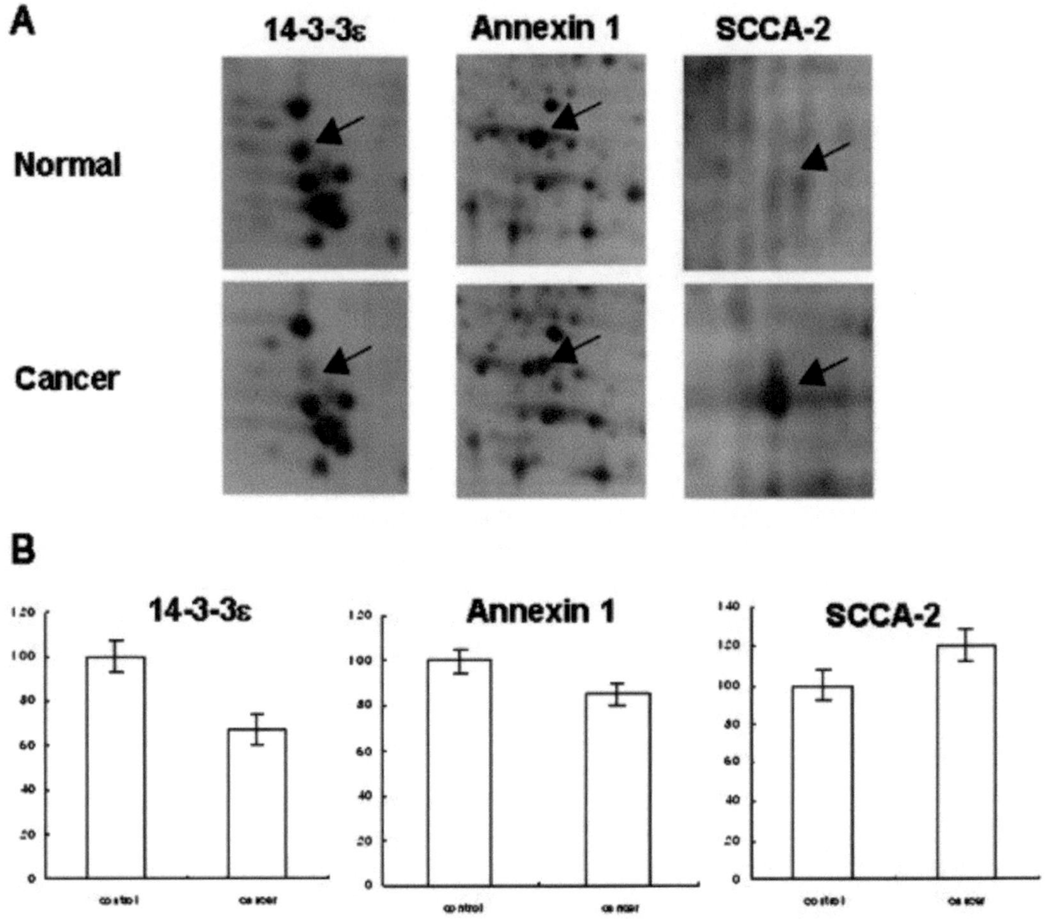

Fig. 8. Protein expression comparisons of normal samples and SCC samples. Up- and down-regulated proteins (14-3-3ε, annexin A1, and SCCA-2) were selected and magnified gel images were presented (A). From the PDQuest 2-D software quantification, the expression difference was statistically meaningful (*P* value < 0.05) (B). (*Source: Bae et al., 2005 Gynecologic Oncol., 99: 26-35)*

The same gel was post-stained with SyproRuby dye (*right panel*). 100 µg of each lysate from serum-starved cells were analyzed on a 9–16% gradient gel. *Circles* represent differentially expressed proteins detectable by both methods. *Arrows* represent spots detected by SyproRuby but not Cy dye labeling. *B*, the shift in molecular weight between the modified and unmodified proteins was visualized by image overlaying. The DIGE image (*Blue*) was overlaid with the SYPRO image (*Red*).

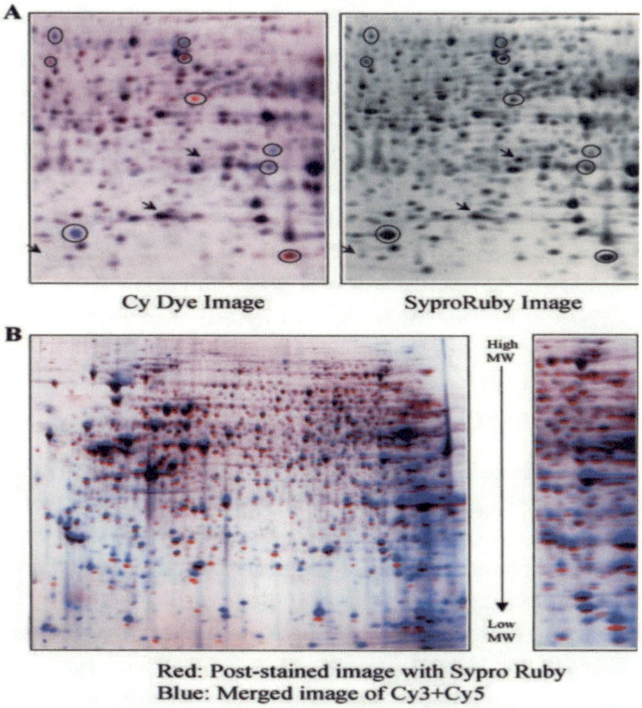

Fig. 9. Sensitivity of 2D-DIGE and compatibility with SYPRO gel staining. *A*, comparison of 2D-DIGE imaging and SyproRuby poststaining. Merged Cy dye image of HB4a lysate labeled with Cy3 (*red*) and HBc3.6 lysate labeled with Cy5 (*blue*) (*left panel*).

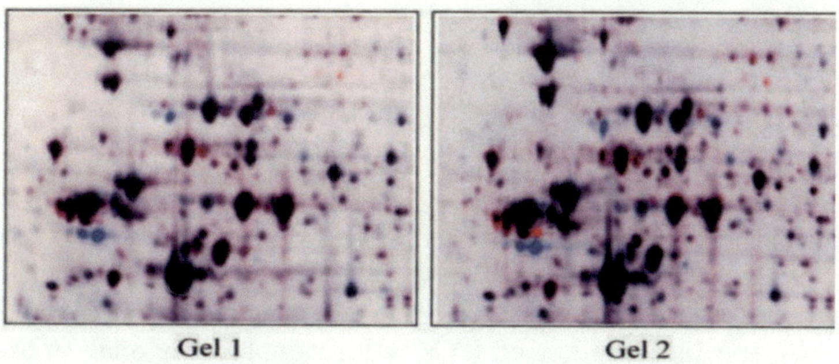

Fig. 10. 2D-DIGE is a reproducible detection method. Duplicate samples of HB4a and HBc3.6 were labeled separately with Cy3 (*red*) and Cy5 (*blue*), respectively.

2.6.11 2-DE protein pattern in classification of carcinoma cells

High-resolution two-dimensional polyacrylamide gel electrophoresis (2-D PAGE) is a powerful research tool for the analytical separation of cellular proteins. The qualitative and quantitative pattern of polypeptides synthesized by a cell represents its phenotype and thus defines characteristics such as the morphology and the biological behavior of the cell. By analyzing and comparing the protein patterns of different cells, it is possible to recognize the cell type and to identify the most typical features of these cells. In applied pathology it is often difficult to identify the tissue of origin and the stage or grade of a neoplasia by cellular morphology analyzed by classical or immunostaining procedures. The protein pattern itself is the most characteristic feature of a cell and should therefore contribute to the identification of the cell type. For this reason, we separated protein fractions originating from different lung tumor cell lines using 2-D PAGE and we compared the resulting patterns on a multivariate statistical level using correspondence analysis (CA) and ascendant hierarchical clustering (AHC). The results indicate that (i) protein patterns are highly typical for cells and that (ii) the comparison of the protein patterns of a set of interesting cell types allows the identification of potentially new marker proteins. 2-D PAGE is thus a unique and powerful tool for molecular cytology or histopathology, unveiling the protein expression level of tissues or cells.

2.6.12 2-DE in understanding Ovarian intratumoral heterogeneity

The process of tumor progression leads to the emergence of multiple clones, and to the development of tumor heterogeneity. One approach to the study of the extent of such heterogeneity is to examine the expression of marker proteins in different tumor areas. Two-dimensional gel electrophoresis (2-DE) is a powerful tool for such studies, since the expression of a large number of polypeptide markers can be evaluated. The tumor cells have been prepared from human ovarian tumors and analyzed by 2-DE and PDQUEST. As judged from the analysis of two different areas in each of nine ovarian tumors, the intratumoral variation in protein expression is low. In contrast, large differences have been observed when the protein profiles of different tumors are compared. The differences in gene expression between pairs of malignant carcinomas are slightly larger than the differences observed between pairs of benign tumors. Hence, the 2-DE analysis of intratumoral heterogeneity in ovarian cancer tissue indicates a low degree of heterogeneity.

2.6.13 Strengths and weaknesses of 2D-PAGE

Electrophoresis is an established technique that has undergone several advances that have enhanced resolution, detection, quantization, and reproducibility. The 2-D SDS-PAGE and 2DDIGE approaches to protein profiling are accessible and economical methods that possess high resolving power and enable the detection of hundreds of proteins on a single gel plate. Although reproducibility has been an issue with 2D-PAGE, especially when profiling two protein mixtures, it has been greatly improved with the use of 2D-DIGE. Resolution has been enhanced by the introduction of IPGs, which enable the analyst to tailor the pH gradient for maximum resolution using ultrazoom gels with a narrow pH gradient range. With modern 2D-PAGE, it is not unusual to resolve two proteins that differ in pI by 0.001 U. Although 2D-PAGE has been limited by its inability to resolve proteins that are too basic or too acidic, too large or too small, this limitation is continuously diminishing. For example, the separation of basic proteins can be analyzed using IPGs in the pH range of 4–12. Separation science is always evolving, and it will not be long before the remaining issues of gel electrophoresis are adequately resolved. The introduction of 2D-DIGE contributed immensely to solving problems of reproducibility and quantitation. The use of imagers and computers allows not only fast data mining, acquisition, and analysis but also spot detection, normalization, protein profiling, background correction, and reporting and exporting of data. As a separation, detection, and quantitation technique, 2D-DIGE is an important tool, especially for clinical laboratories involved in the determination of protein expression levels and disease biomarker discovery. When absolute biological variation between samples is the main objective, as in biomarker discovery, 2D-DIGE is the method of choice. While there has been significant progress in nongel (or solution-based) methods for coupling fractionation methods directly online with MS analysis, 2DPAGE has remained a popular technique for conducting proteomic studies. Though 2D-PAGE, like any fractionation scheme, has its advantages and disadvantages, there is no doubt that it will remain an essential technique for the characterization of proteomes for many years to come.

2.6.14 Two-dimensional electrophoresis for cancer proteomics

Proteome analysis is a direct measurement of proteins in terms of their presence and relative abundance (Wilkins *et al.*, 1996). The overall aim of a proteomic study is characterization of the complex network of cell regulation. Neither the genomic DNA code of an organism nor the amount of mRNA that is expressed for each gene product (protein) yields an accurate picture of the state of a living cell (Lubec *et al.*, 1999), which can be altered by many conditions. Proteome analysis is required to determine which proteins have been conditionally expressed, how strongly, and whether any posttranslational modifications are affected. Two or more different states of a cell or an organism (e.g., healthy and diseased tissue) can be compared and an attempt made to identify specific qualitative and quantitative protein changes. One of the greatest challenges of proteome analysis is the reproducible fractionation of these complex protein mixtures while retaining the qualitative and quantitative relationships. Currently, two-dimensional polyacrylamide gel electrophoresis (2-D PAGE) is the only method that can handle this task (Cutler et al., 1999, Fegatella et al., 1999, Görg et al., 2000), and hence has gained special importance. Since 2-D PAGE is capable of resolving over 1,800 proteins in a single gel (Choe & Lee, 2000), it is important as the primary tool of proteomics research where multiple proteins must be separated for parallel analysis. It allows hundreds to thousands of gene products to be analyzed simultaneously. In combination with computer assisted image evaluation systems for comprehensive qualitative and quantitative examination of proteomes, this electrophoresis technique allows cataloging and comparison of data among groups of researchers.

3. Conclusion

Two-dimensional gel electrophoresis and MALDI-MS are an effective strategy for determining the protein domains present in those gel spots that are observed at significantly lower MW values than are given in the database. While average sequence coverage is only 30%, the peptides detected are confined to a specific region of the protein, such as the protein N- or C-terminal. This information could easily be incorporated into protein identification tables. Regional coverage information is not readily available from either LC-MS/MS analysis of digests of cellular lysates or from epitope-specific antibodies. Some of the protein fragments correspond to chains produced by known cellular processing and activation pathways. Others have been detected as functional and structural domains during in vitro experiments or noted in other in vivo studies, indicating they function intra- or extra-cellularly. By using tools that allow both protein identification and measurement of MW, we can assess the abundance and distribution of protein fragments. Correlation of these results with targeted functional studies on specific proteins will elucidate the biological function of protein fragments.

4. References

Bae, SM., Lee, CH., Cho,YL., Nam, KH., Kim, YW., Kim, CK., Han, BD., Lee, YJ., Chun, HJ. & Ahn, WS. (2005). Two-dimensional gel analysis of protein expression profile in squamous cervical cancer patients. *Gynecologic Oncology* 99: 26-35.

Bjellqvist, B., Ek, K., Righetti, PG., Gianazza, E., Gorg, A., Westermeier, R. & Postel, W. (1982). Isoelectric focusing in immobilized pH gradients: principle, methodology and some applications, *J Biochem Biophys Methods* 6: 317–339.

Bjellqvist, B., Hughes, GJ., Pasquali, C., Paquet, N., Ravier, F., Sanchez, JC., Frutiger, S. & Hochstrasser, D. (1993b). The focusing positions of polypeptides in immobilized pH gradients can be predicted from their amino acid sequences, *Electrophoresis* 14: 1023–1031.

Bjellqvist, B., Sanchez, JC., Pasquali, C., Ravier, F., Paquet, N., Frutiger, S., Hughes, GJ. & Hochstrasser, D. (1993a). Micropreparative two-dimensional electrophoresis allowing the separation of samples containing milligram amounts of proteins, *Electrophoresis* 14: 1375–1378.

Carroll, K., Ray, K., Helm, B. & Carey, E. (2000). Two-dimensional electrophoresis reveals differential protein expression in high- and low-secreting variants of the rat basophilic leukaemia cell line, *Electrophoresis* 21: 2476–2486.

Chevallet, M., Santoni, V., Poinas, A., Rouquie, D., Fuchs, A., Kieffer, S., Rossignol, M., Lunardi, J., Garin, J. & Rabilloud, T. (1998). New zwitterionic detergents improve the analysis of membrane proteins by two-dimensional electrophoresis, *Electrophoresis* 19: 1901–1909.

Choe, LH. & Lee, KH. (2000). A comparison of three commercially available isoelectric focusing units for proteome analysis: The multiphor, the IPGphor and the protean IEF cell, *Electrophoresis* 21: 993–1000.

Cordwell, SJ., Basseal, DJ., Bjellqvist, B., Shaw, DC. & Humphery-Smith, I. (1997). Characterisation of basic proteins from *Spiroplasma melliferum* using novel immobilised pH gradients, Electrophoresis 18, 1393–1398.

Corthals,, GL., Wasinger, VC., Hochstrasser, DF. & Sanchez, JC. (2000). The dynamic range of protein expression: a challenge for proteomic research, *Electrophoresis* 21: 1104–1115.

Cutler, P., Bell, DJ., Birrell, HC., Connelly, JC., Connor, SC., Holmes, E., Mitchell, BC., Monte, SY., Neville, BA., Pickford, R., Polley, S., Schneider, K. & Skehel, JM. (1999). An integrated proteomic approach to studying glomerular nephrotoxicity, *Electrophoresis* 20: 3647–3658.

Ducret, A., Bruun, CF., Bures, EJ., Marhaug, G., Husby, G. & Aebersold, R. (1996). Characterization of human serum amyloid A protein isoforms separated by two-dimensional electrophoresis by liquid chromatography/electrospray ionization tandem mass spectrometry, *Electrophoresis* 17: 866–876.

Eckerskorn, C., Jungblut, P., Mewes, W., Klose, J. & Lottspeich, F. (1988). Identification of mouse brain proteins after two-dimensional electrophoresis and electroblotting by microsequence analysis and amino acid composition analysis, *Electrophoresis* 9: 830–838.

Fegatella, F., Ostrowski, M. & Cavicchioli, R. (1999). An assessment of protein profiles from the marine oligotrophic ultramicrobacterium, *Sphingomonas sp.* Strain RB2256, *Electrophoresis* 20: 2094–2098.

Fountoulakis, M., Takacs, MF., Berndt, P., Langen, H. & Takacs, B. (1999). Enrichment of low abundance proteins of *Escherichia coli* by hydroxyapatite chromatography, *Electrophoresis* 20: 2181–2195.

Franzén, B., Linder, S., Alaiya, AA., Eriksson, E., Fujioka, K., Bergman, AC., Jörnvall, H. & Auer, G. (1997). Analysis of polypeptide expression in benign and malignant human breast lesions. *Electrophoresis* 18: 582-587.

Garfin, DE. (1995). Electrophoretic methods, pp 53–109 in Glasel JA and Duetscher MP (eds) *Introduction to Biophysical Methods for Protein and Nucleic Acid Research*, Academic Press, San Diego.

Garfin, DE. (2000). Isoelectric focusing, pp 263–298 in Ahuja S (ed) *Separation Science and Technology*, Vol 2, Academic Press, San Diego.

Gingrich, JC. Davis, DR. & Nguyen, Q. (2000). Multiplex detection of quantification of proteins on western blots using fluorescent probes, *Biotechniques* 29: 636-642.

Görg, A. (1991). Two-dimensional electrophoresis, *Nature* 349: 545–546.

Görg, A., Obermaier, C., Boguth, G. & Weiss, W. (1999). Recent developments in twodimensional gel electrophoresis with immobilized pH gradients: wide pH gradients up to pH 12, longer separation distances and simplified procedures, *Electrophoresis* 20: 712–717.

Görg, A., Obermaier, C., Boguth, G., Harder, A., Scheibe, B., Wildgruber, R. & Weiss, W. (2000). The current state of two-dimensional electrophoresis with immobilized pH gradients, *Electrophoresis* 21: 1037–1053.

Görg, A., Postel, W., Domscheit, A. & Günther, S. (1989) Methodology of two dimensional electrophoresis with immobilized pH gradients for the analysis of cell lysates and tissue proteins, in Endler AT and Hanash S (eds) Two- Dimensional Electrophoresis. *Proceedings of the International Two-Dimensional Electrophoresis Conference*, Vienna, Nov. 1988, VCH, Weinheim FRG.

Gottlieb, M. & Chavko, M. (1987). Silver staining of native and denatured eukaryotic DNA in agarose gels, *Anal Biochem* 165: 33–37.

Herbert, BR., Molloy, MP., Gooley, AA., Walsh, BJ., Bryson, WG. & Williams, KL. (1998). Improved protein solubility in two-dimensional electrophoresis using tributyl phosphine as reducing agent, *Electrophoresis* 19: 845–851.

Hermann, T., Finkemeier, M., Pfefferle, W., Wersch, G., Kramer, R. & Burkovski, A. (2000). Two-dimensional electrophoretic analysis of *Corynebacterium glutamicu* membrane fraction and surface proteins, *Electrophoresis* 21: 654–659.

Klose, J. (1975). Protein mapping by combined isoelectric focusing and electrophoresis of mouse tissues. A novel approach to testing for induced point mutation in mammals. *Humangenetik* 26, 231–243.

Lubec, G., Nonaka, M., Krapfenbauer, K., Gratzer, M., Cairns, N. & Fountoulakis, M. (1999). Expression of the dihydropyrimidinase related protein 2 (DRP-2) in Down syndrome and Alzheimer's disease brain is downregulated at the mRNA and dysregulated at the protein level, *J Neural Transm Suppl* 57: 161–177.

Macri, J., McGee, B., Thomas, JN., Du, P., Stevenson, TI., Kilby, GW. & Rapundalo, ST. (2000). Cardiac sarcoplasmic reticulum and sarcolemmal proteins separated by two-dimensional electrophoresis: surfactant effects on membrane solubilization, *Electrophoresis* 21: 1685-1693.

Masuoka, J., Glee, PM. & Hazen, KC. (1998). Preparative isoelectric focusing and preparative electrophoresis of hydrophobic *Candida albicans* cell wall proteins with in-line transfer to polyvinylidene difluoride membranes for sequencing, *Electrophoresis* 19: 675–678.

Merril, CR., Goldman, D., Sedman, SA. & Ebert, MH. (1981). Ultrasensitive stain for proteins in polyacrylamide gels shows regional variation in cerebrospinal fluid proteins, *Science* 211: 1437–1438.

Miller, MJ. (1989). Computer-assisted analysis of two-dimensional gel electropherograms, pp 182–217 in Chrambach A, Dunn MJ and Radola BJ (eds) *Advances in Electrophoresis*, Vol. 3, Verlag Chemie, Weinheim.

Molloy, MP. (2000). Two-dimensional electrophoresis of membrane proteins using immobilized pH gradients, *Anal Biochem* 280: 1–10.

Molloy, MP., Herbert, BR., Walsh, BJ., Tyler, MI., Traini, M., Sanchez, JC., Hochstrasser, DF., Williams, KL. & Gooley, AA. (1998). Extraction of membrane proteins by differential solubilization for separation using two-dimensional gel electrophoresis, *Electrophoresis* 19: 837–844.

Morel, V., Poschet, R., Traverso, V. & Deretic, D. (2000). Towards the proteome of the rhodopsin-bearing post-Golgi compartment of retinal photoreceptor cells, *Electrophoresis* 21: 3460–3469.

Nilsson, CL., Larsson, T., Gustafsson, E., Karlsson, KA. & Davidsson, P. (2000). Identification of protein vaccine candidates from *Helicobacter pylori* using a preparative two-dimensional electrophoretic procedure and mass spectrometry, *Anal Chem* 72: 2148–2153.

O'Farrell, PH. (1975). High resolution two-dimensional electrophoresis of proteins. *J Biol Chem* 250, 4007–4021.

Patton, WF. (2000). A thousand points of light: the application of fluorescence detection technologies to two-dimensional gel electrophoresis and proteomics, *Electrophoresis* 21: 1123–1144.

Pennington, SR. & Dunn, MJ. (2001). *Proteomics. From Protein Sequence to Function*, Springer/Bios, New York.

Rabilloud, T. (1998). Use of thiourea to increase the solubility of membrane proteins in two-dimensional electrophoresis, *Electrophoresis* 19: 758–760.

Rabilloud, T. (1999). Solubilization of proteins in 2-DE. An outline, *Methods Mol Biol* 112: 9–19.

Rabilloud, T. (2000). *Proteome Research: Two Dimensional Gel Electrophoresis and Identification Tools*, Springer, Berlin.

Righetti, PG. (1983). *Isoelectric Focusing: Theory, Methodology and Applications*, Elsevier, Amsterdam.

Righetti, PG. (1990). Recent developments in electrophoretic methods, *J Chromatogr* 516: 3–22.

Sanchez, JC., Rouge, V., Pisteur, M., Ravier, F., Tonella, L., Moosmayer, M., Wilkins, MR. & Hochstrasser, DF. (1997). Improved and simplified in-gel sample application using reswelling of dry immobilized pH gradients, *Electrophoresis* 18: 324–327.

Smithies, O. & Poulik, MD. (1956). Two-Dimensional Electrophoresis of Serum Proteins. *Nature* 177, 1033.

Taylor, RS., Wu, CC., Hays, LG., Eng, JK., Yates, JR. & Howell, KE. (2000). Proteomics of rat liver Golgi complex: minor proteins are identified through sequential fractionation, *Electrophoresis* 21: 3441–3459.

Wilkins, MR., Gasteiger, E., Sanchez, JC., Bairoch, A. & Hochstrasser, DF. (1998). Two-dimensional gel electrophoresis for proteome projects: the effects of protein hydrophobicity and copy number, *Electrophoresis* 19: 1501–1505.

Wilkins, MR., Sanchez, JC., Gooley, AA., Appel, RD., Humphery-Smith, I., Hochstrasser, DF. & Williams, KL. (1996). Progress with proteome projects: Why all proteins expressed by a genome should be identified and how to do it. *Biotechnol Genet Eng Rev* 13: 19–50.

Wilkins, MR., Williams, KL; Appel, RD. & Hochstrasser, DF. (1998). Proteome Research: *New Frontiers in Functional Genomics*, Springer, Berlin.

Part 7

Other Applications of Gel Electrophoresis Technique

20

Enzymatic Staining for Detection of Phenol-Oxidizing Isozymes Involved in Lignin-Degradation by *Lentinula edodes* on Native-PAGE

Eiji Tanesaka, Naomi Saeki, Akinori Kochi and Motonobu Yoshida
Kinki University
Japan

1. Introduction

Lignocellulose is the most abundant organic compound in the terrestrial environment. Nonetheless, with the exception of basidiomycetous fungi, most organisms are either unable to degrade lignocellulose, or if they can, they do so with difficulty (Kirk & Fenn, 1982). Wood-decomposing basidiomycetes can be grouped into two categories: white-rot and brown-rot fungi. White-rot fungi have cellulases and lignin-degrading enzymes that decompose most cell wall components, whereas brown-rot fungi have enzymatic systems that selectively degrade cellulose and hemicelluloses, leaving brown shrunken lumps of tissue composed mainly of a loose lignin matrix (Enoki et al., 1988; Highley et al., 1985; Highley & Murmanis, 1987; Kirk & Highley, 1973). The name 'white-rot' is derived from the bleaching effect that this fungus has when degrading wood; the lignin-degrading enzymes that they secrete have the effect of promoting lignin loss and exposing the white cellulose fibrils. White-rot fungi are known to produce polyphenol oxidases (phenoloxidases), which, when the fungi are plated on agar media containing gallic or tannic acids, change the color of the agar to a dark reddish-brown in what is referred to as Bavendamm's polyphenol oxidase test or Bavendamm reaction (Bavendamm, 1928, as cited in Jørgensen & Vejlby, 1953). Based on this reaction, phenoloxidases are considered to be one of putative lignin-degrading enzymes (Higuchi 1990). Laccase (Lcc, EC 1.10.3.2), catechol oxidase (EC 1.10.3.1) and tyrosinase (monophenol monooxygenase, EC 1.14.18.1) are phenoloxidases with considerable overlap in their substrate affinities (Burke & Cairney, 2002). Lcc catalyze the reduction of O_2 to H_2O using a range of phenolics, aromatic amines, and other electron-rich substances as hydrogen donors (Thurston, 1994). Similar phenol-oxidizing activities are also observed in peroxidases (EC 1.11.1.x), which use H_2O_2 as an electron donor. Lignin peroxidase (ligninase, LiP, EC 1.11.1.14) was first discovered in *Phanerochaete chrysosporium* in which the H_2O_2-dependent C_α-C_β cleavage of non-phenolic lignin model compounds was first described (Tien & Kirk, 1983, 1984). Manganese peroxidase (MnP, EC 1.11.1.13) also strongly degrades lignin model compounds and the reaction is mediated by H_2O_2 and Mn^{2+} (Glenn et al., 1983; Glenn & Gold, 1985; Kuwahara et al., 1984). Whereas lignin can effectively be oxidized by LiP directly, as reviewed previously (Cullen & Kersten, 2004;

Gold & Alic, 1993), Mn^{2+} is considered to be an important physiological substrate for MnP. Further, while LiP expression has been observed in certain white-rot fungi (e.g. *Phanerochaete chrysosporium* and *Phlebia radiata*) under specific culture conditions (e.g., temperature, agitation, and nutritional constraints), MnP expression has been observed in a wide range of white-rot fungi (Gold & Alic, 1993), including cultivated edible fungi, such as *Agaricus bisporus* (Bonnen et al., 1994), *Ganoderma lucidum, Lentinula edodes,* and *Pleurotus* spp. (Orth et al., 1993).

The shiitake mushroom, *Lentinula edodes* (Berk.) Pegler, a white-rot basidiomycete, is one of the most valuable, cultured, edible mushrooms in the world (Chang & Miles 1989). Shiitake mushrooms were traditionally cultivated on Fagaceae logs, but they are now grown on sawdust-based media. The ability of white-rot basidiomycetes to degrade wood components, especially lignin, therefore affects both culture-time to harvesting and yields (Kinugawa & Tanesaka, 1990; Ohga & Kitamoto, 1997; Smith et al., 1988; Tanesaka et al., 1993). Although *L. edodes* secretes the lignin-degrading enzymes laccase (Lcc) and MnP when cultivated on sawdust-based media (Buswell et al., 1995; Leatham, 1985; Makker et al., 2001), it does not usually secrete these enzymes in liquid media. It was previously reported that the main isozyme produced by *L. edodes* cultured on sawdust was the manganese peroxidase, LeMnP2 (Sakamoto et al., 2009). In addition, we previously reported that a β-*O*-4 lignin model compound, 4-ethoxy-3-methoxyphenylglycerol-β-guaiacyl ether (Umezawa & Higuchi, 1985) was effectively degraded by *L. edodes* under MnP-induced conditions, but not under Lcc-induced conditions (Kochi et al., 2009). These observations supported the hypothesis that these enzymes, particularly MnP, play an important role in degrading sawdust during cultivation, and corroborating reports that the expression and properties of these enzymes is likely to influence mycelial growth and fruit body development (Smith et al., 1988; Wood et al., 1988). Several reports have been published on the purification and characterization of the lignin-degrading enzymes secreted by *L. edodes* using sophisticated biochemical procedures (Forrester et al., 1990; Nagai et al., 2002, 2003, 2007; Sakamoto et al., 2008, 2009). However, these methods are impracticable for routine isozyme analysis during breeding trials. Methods for isozyme detection by electrophoresis using enzyme catalytic properties - referred to as "protein activity staining" or "enzymatic staining" - are well established in histochemical studies and genetics (Pasteur et al., 1988). It was expected that Lcc, peroxidases (Per, EC 1.11.1.7), and MnP bands could be distinguished on the same gel by subtraction of newly appeared bands produced by sequential enzymatic staining. In practice, however, unexpected bands frequently appeared on gels exposed to conventional Lcc staining solutions. Indeed, in samples exhibiting strong MnP activity without Lcc activity, no additional bands appeared in subsequent staining procedures for either Per or MnP. We recently reported improved methods for enzymatic staining using native-PAGE to distinguish between Lcc and MnP isozymes induced in liquid cultures of *L. edodes* (Saeki et al., 2011).

In this chapter, we describe an assay system for the induction and identification of phenol-oxidizing enzymes produced by *L. edodes* grown under liquid culture conditions. In addition, the assay system was used to compare the glycosylation characteristics of these extra- and intracellular isozymes, as well as their modes of inheritance within monokaryotic progenies and β-*O*-4 lignin model compound degradation characteristics under Lcc- and MnP-induced conditions. Based these findings, the potential application of this assay system to elucidate the ligninolytic mechanisms employed by this fungus is also discussed.

2. Experimental procedures

2.1 Terminology

We use the term "phenol-oxidizing enzymes" to describe all phenoloxidases and peroxidases. We do so because of the ability of these enzymes to utilize the same substrates and produce the same catalytic products as described in the Introduction.

2.2 Fungi and culture conditions

The Hokken 600 variety of *L. edodes* (Hokken Co., Ltd., Tochigi, Japan; hereafter referred to as H600) and monokaryotic progenies derived from basidiospores were used in this study. To induce the phenol-oxidizing enzymes, mycelia were cultured in MYPG liquid medium (2.5 g malt extract; 1.0 g yeast extract; 1.0 g peptone; 5.0 g glucose in 1,000 ml of distilled water) supplemented with sawdust extract (MYPG-S). This sawdust extract was produced by adding 1 g of *Castanopsis cuspidata* (Thunb. ex Murray) Schottky sawdust to 30 ml of distilled water and then autoclaving the mixture for 15 min before filtering through filter paper (No. 1, Advantec, Tokyo) and collecting the extract. MYPG-S liquid media samples were then prepared from MYPG liquid medium by adding half a volume of sawdust extract instead of distilled water, which gave an MYPG-S extract that contained 500 mg sawdust in 30 ml media. Mycelia were sub-cultured at 25°C on MYPG 1% agar plates. After 14 days, three mycelial disks measuring 3 mm in diameter were harvested from the plates and used to inoculate 30 ml MYPG-S liquid in a 100 ml flask which was then statically cultured at 25°C. MnP activity was induced during culture on MYPG-S. Lcc activity was induced by adding 2 mM $CuSO_4 \cdot 5H_2O$ to the same media seven days after initial inoculation.

2.3 Enzyme assay

A schematic representation of the strategies employed to distinguish between individual phenol-oxidizing enzymes by subtractive activity assays (Szklarz et al., 1989) and sequential enzymatic staining of gels using native-PAGE is shown in Fig. 1.

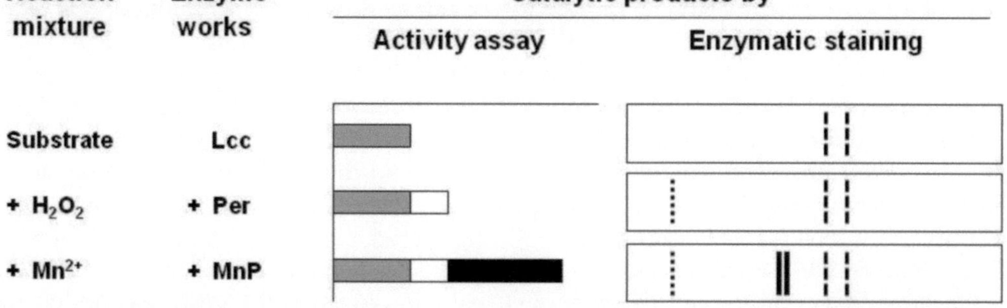

Fig. 1. Strategy underlying the subtractive activity assay and sequential enzymatic staining of a gel to distinguish between individual phenol-oxidizing enzymes. Grey, open, and solid squares represent the activities of Lcc, Per and MnP, respectively. Broken, dotted, and solid lines represent the sequential enzymatic staining of Lcc, Per and MnP in a gel, respectively.

To assay the activities of extracellular enzymes, 100 μl of culture liquid was sampled every two to three days during culture and centrifuged at 13,000 rpm for 10 min; this supernatant was used as a crude enzyme solution. The crude enzyme solution was assayed for Lcc, Per and MnP in identical 5 ml test tubes containing the following reaction mixtures: Lcc assay mixture consisted of 0.1 mM *o*-dianisidine in 0.1 M sodium tartrate buffer (pH 5.0), with additional H_2O_2 (final concentration 0.1 mM) added to the Lcc assay mixture to assess Per activity. Additional $MnSO_4 \cdot 5H_2O$ (final concentration 0.1 mM) was added to the Per assay mixture to assess MnP activity. Aliquots (20 μl) of crude enzyme solution were added to test tubes containing 980 μl of each reaction mixture, which were then incubated at 37°C for 10 min. The reactions were stopped by the addition of 50 μl of 40 mM NaN_3. To inactivate the enzymes in the control tubes, sodium azide was added to the control tubes containing the Lcc assay mixture before incubation. Catalytic products of the reaction were spectrophotometrically assayed using *o*-dianisidine as a substrate, and the activity of enzyme products was estimated by subtracting the respective absorbance values at 460 nm: i.e., Lcc activity = Lcc assay minus the control assay; Per activity = Per assay minus the Lcc assay; and MnP activity = MnP assay minus the Per assay, respectively. One unit (U) of enzyme activity was defined as the amount of enzyme required to catalyze 1 μmol of *o*-dianisidine in 1 min (ε_{460} = 29,400 $M^{-1}cm^{-1}$: Paszczynski et al., 1988).

2.4 Native PAGE and enzymatic staining

Each of the phenol-oxidizing isozymes was detected by native PAGE as described previously (Saeki et al., 2011). Briefly, whole cultures were filtered through a nylon stocking to separate the mycelia from the culture liquid. The collected mycelia were then ground with a ceramic mortar and pestle in two volumes (v/w) of crushing buffer (0.05 M Tris-HCl, pH 7.2, 0.1% β-mercaptoethanol) before being centrifuged at 12,000 rpm for 10 min. The resulting supernatant was considered to represent the intracellular enzyme sample. To prepare the extracellular enzyme sample, the culture liquid was centrifuged at 13,000 rpm for 10 min, and the supernatant was filtered (No.2 filter paper, Advantec) and then concentrated 15-fold by ultrafiltration using the centrifugal filter unit, Centriprep YM-10 (10-kDa cut-off membrane, Millipore, MA). Aliquots containing 15 μl enzyme sample, 1.5 μl glycerol and 1.5 mg bromophenol blue (BPB) as a dye marker were then loaded into the wells of 12.5% (for intracellular) or 17.5% (for extracellular) polyacrylamide gels. Native-PAGE gels were run at 15 mA for 15 min followed by 25 mA for 3–4 h. After electrophoresis, the gel was sequentially incubated at 37°C for 30 min in three different staining solutions. The first staining solution was an improved enzymatic staining solution containing additional ethylenediaminetetraacetic acid (EDTA) (for Lcc and Per) to remove the Mn^{2+} typically used in conventional staining solutions. Staining for enzymes was performed as follows. To the Lcc staining solution (LccS+EDTA), 1.8 mM *o*-dianisidine, 0.1 mM acetate buffer (pH 4.0) containing 130 mM EDTA (LccS+EDTA), an additional H_2O_2 (final concentration 1.0 mM) was added to produce the Per staining solution (PerS+EDTA). In the same way, additional $MnSO_4 \cdot 5H_2O$ (final concentration 0.1 mM) was added to the PerS without EDTA to produce the MnP staining solution (MnPS). The gels were rinsed with distilled water between each staining procedure to remove the previous staining solutions, particularly the EDTA from PerS+EDTA used for MnP staining. Isozyme nomenclature employed an (-e) or (-i) in Lcc-e, MnP-e or MnP-i to indicate whether the Lcc and MnP enzymes were extra- or intracellular. Numerals in parenthesis, e.g., MnP-e (52),

MnP-e (57) etc. indicated the relative mobility of each isozyme relative to the mobility of the bromophenol blue used as a dye.

2.5 Glycosidase treatment

To purify the enzymes in the crude enzyme solutions prior to electrophoresis, 1 ml acetone was added to a 100 μl aliquot of the enzyme solution and kept at −20°C for 3 h to precipitate the proteins. The proteins were then resuspended in 100 μl of 10 mM phosphate buffer (pH 6.0). To determine whether the enzymes were glycosylated, the protein suspension in phosphate buffer was incubated with glycosidase (Glycosidases 'Mixed', Seikagaku Biobusiness Corp., Japan) at final concentrations of 0.25−2.0% (w/v) with a protease inhibitor (Complete, Mini, EDTA-free; Roche Diagnostics, Germany) at 37°C overnight. Effects of the glycosidase treatment on activities of each of the isozymes were then examined by enzymatic staining after native-PAGE as described in section 2.4 above.

2.6 Identification of isozymes by mass spectrometry

Distinguishing between isozymes was performed as described previously (Saeki et al. 2011). Briefly, after native-PAGE had been conducted on the same sample solution in adjacent lanes, each gel was then subjected to enzymatic staining and Coomassie brilliant blue (CBB) staining. Bands of interest, such as those exhibiting the same mobility as bands in the enzymatic staining experiments, were then excised from the CBB-stained gel using a sterile surgical blade and placed in 1.5 ml microcentrifuge tubes. To remove the CBB dye, each polyacrylamide gel section was then repeatedly washed with 50, 30 and 50% v/v acetonitrile containing 25 mM NH_4HCO_3 under sonication for 20 min with a micromixer (Taitec, Tokyo), before finally being washed with 100% acetonitrile without NH_4HCO_3 for 5 min. The sections of polyacrylamide gel were then vacuum-dried for 5 min and recovered in 100 μl of 50 mM NH_4HCO_3 (pH 7.8) containing 10 ng/μl trypsin (Trypsin Gold, Mass Spectrometry Grade, Promega, WI) on ice for 30 min. Any extra trypsin solution was then removed and the sections of gel were incubated at 37°C for 16 h. The tryptic fragments in a gel were then extracted by immersing the gel sections in 50 μl of extraction buffer consisting 50% acetonitrile and 5% trifluoroacetic acid under sonication (Ultrasonic cleaner, SU-3T, Shibata, Japan) for 20 min. The extraction buffer was placed into new tube and replaced with 25 μl of fresh extraction buffer. The extraction process was repeated a further three times and the collected buffer containing the tryptic fragments was finally concentrated to approximately 5 μl by drying under vacuum. Analysis of the tryptic peptides by tandem mass spectrometry was performed on a nanoelectrospray ionization quadrupole time-of-flight (Q-TOF) hybrid mass spectrometer (Q-TOF Premier, Waters Micromass, MA) coupled with a nano-HPLC (Cap-LC; Waters Micromass). The peptides were separated on a BEH 130-C18 column (1.7 μm, 100 μm × 100 mm, Nano Ease, Waters, MA) at flow rate of 0.2 μl/min according to the manufacturer's instructions. The peptide sequences thus obtained were then either matched automatically to proteins in a non-redundant database (National Center for Biotechnology Information, NCBI, www.ncbi.nlm.nih.gov) using the Mascot MS/MS ions search algorithm (Mascot Server version 2.2, Matrix Science), or BLAST searches were manually performed against the DNA Data Bank of Japan database (DDBJ, www.ddbj.nig.ac.jp). Mascot search was also performed to calculate the false discovery rate (FDR) on acquired MS/MS data against decoy database.

2.7 RNA isolation and Northern blot analysis

Total RNA was extracted from mycelia after varying incubation periods under MnP- and Lcc- induced conditions using TRIzol Reagent (Invitrogen, CA). cDNA was synthesized from total RNA using an RNA PCR Kit Ver.3.0 (Takara Bio, Japan), and amplified using the primer set LeMnP2En5f (5′-TCCGACAGTGTCAATGACCTCGCTC) and LeMnP2En13r (5′-GTCAGTGGTGAGATTTGGGAAGGGC), which were designed based on the highly conserved *lemnp2* region (DDBJ Acc. No. AB306944; Sakamoto et al., 2009). A fragment measuring approximately 700 bp was then extracted from the 1% agarose–formaldehyde gel, purified, and sequenced using an ABI Prism 3100-*Avant* Genetic Analyzer (Applied Biosystems, CA) according to the manufacturer's instructions. Fragments with sequences matching *lemnp2* were then labeled using a PCR-based digoxigenin (DIG)-dUTP labeling kit (PCR DIG Probe Synthesis Kit, Roche Diagnostics) according to the manufacturer's instructions. The resulting DIG-labeled probe, *lemnp2*N, was then used for Northern hybridization following blotting of 10 µg of total RNA onto a Hybond-N$^+$ membrane (GE Healthcare, Switzerland) using an established protocol (Sambrook et al., 1989).

2.8 DNA isolation and Southern blot analysis

Genomic DNA was extracted from mycelia using cetyltrimethylammonium bromide (CTAB) isolation buffer (J. J. Doyle & J. L. Doyle, 1987, as cited in Milligan, 1998). To prepare the DIG-labeled probe, genomic DNA was used as a template for PCR with the primer set, LeMnP2En5f2 (5′-TCAGGAAAATTCCCGACTAT) and LeMnP2En12r (5′-GAACCTCGATG CCATCAA); this primer set was designed to amplify the region from exon 5 to exon 12 of *lemnp2*, including introns (DDBJ Acc. No. AB306944; Sakamoto et al., 2009), and the resulting probe was named *lemnp2*S. In addition, to examine cross-hybridization between *lemnp2* and a relative of the manganese peroxidases LeMnP1 coded by *lemnp1*, a probe for *lemnp1* was also prepared as described above using the primer set LeMnP1f1 (5′-GATTCCTGAGCCTTTCG) and LeMnP1r (5′-TTCGGGACGGGAATAAC); this primer set was designed to amplify the regions from exon 7 to exon 15 of *lenmp1* including introns (DDBJ Acc. No. AB241061; Nagai et al., 2007) and the resulting probe was named *lemnp1*S. These two probes were then used for Southern blot analysis (Sambrook et al., 1989).

2.9 Degradation assay of β-*O*-4 lignin model compound

2.9.1 Culture experiment

To assay the abilities of MnP and Lcc to degrade the β-*O*-4 lignin model compound, 4-ethoxy-3-methoxyphenylglycerol- β-guaiacyl ether (Fig. 2; β-*O*-4 compound, hereafter) was synthesized according to the method previously described (Umezawa & Higuchi 1985).

To prepare media containing the β-*O*-4 compound to media, 300 µg of the β-*O*-4 compound was diluted in 50 µl acetone and added to 30 ml MYPG-S. Whole liquid culture media was collected 14, 21 and 42 days after inoculation and assayed for phenol-oxidizing enzyme activity. The β-*O*-4 compound was then recovered from the liquid culture media by the addition of two volumes of ethyl acetate to separate the aqueous phase, before evaporating the ethyl acetate off and then precipitating the compound. To improve subsequent chromatographic analysis, the recovered β-*O*-4 compound was silylated using N-*O*-bis-(trimethylsilyl) trifluoroacetamide (BSTFA) to form a trimethylsilyl (TMS) derivative. This

TMS derivative was then subjected to gas chromatography-mass spectrometry (GC-MS) analysis (6890N, Agilent Technologies, CA), which was fitted with an capillary column (HP-5 MS, 30 m × 0.25 mm i.d., 0.25-μm; J&W Scientific, CA) coupled to an MS (JMS-K9, JEOL, Japan) according to the manufacturer's instructions. Helium was used as the carrier gas at 1.5 ml/min. GC oven conditions consisted of 150°C for 1 min, initially ramped at 10°C/min to 200°C and then at 5°C/min to 250°C. The electron impact mass spectra were obtained at an acceleration energy of 70 eV. The degradation rate (%) of the β-*O*-4 compound was calculated using the rate of quantities of the TMS derivative before and after culture, compensating for the recovery of the β-*O*-4 compound with 4,4′-dimethoxybenzoin (anisoin), which was used as an internal standard.

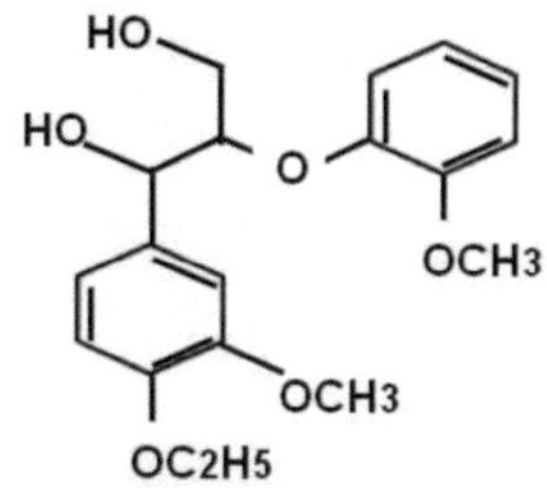

Fig. 2. β-*O*-4 lignin model compound used in this study; 4-ethoxy-3-methoxyphenylglycerol-β-guaiacyl ether (from Umezawa & Higuchi, 1985).

2.9.2 Incubation experiment

Degradation of the β-*O*-4 compound was also examined by incubation with the extracellular enzyme solution, which was prepared using the same procedures used for electrophoresis (section 2.4) with the following slight modifications. The extracellular enzyme solutions were diluted with 0.2 M sodium tartrate buffer (pH 5.0, final concentration of 0.1 M) and distilled water to bring the volume to 10 ml and keep the activity of MnP, Lcc and the mixed solution (MnP+Lcc) at 17 U/ml. For the MnP and MnP+Lcc reactions, additional H_2O_2 (final concentration 0.1 mM) and $MnSO_4 \cdot 5H_2O$ (final concentration 0.1 mM) were added to the reaction for Lcc. Then, 500 μg of the β-*O*-4 compound in 50 μl of acetone was added to the 10 ml enzyme solution and incubated at 37°C with agitation at 100 rpm for up to 10 days. The rate of degradation of the β-*O*-4 compound was evaluated using GC-MS as above.

3. Results and discussion

3.1 Selective induction of phenol-oxidizing enzyme

Phenol-oxidizing enzyme activities under different culture conditions are shown in Fig. 3. Neither MnP nor Lcc was induced when mycelia were cultured on MYPG liquid medium without sawdust extract (data not shown). Under MnP-induced conditions (i.e. when mycelia were cultured on MYPG+S), MnP activity increased suddenly on day 21, before reaching a maximum activity (95 U/ml) on day 35 and then decreasing thereafter (Fig. 3a).

We previously found that supplementing the MYPG liquid medium with wood chips or sawdust from members of the Fagacae, *C. cuspidata* or *Fagus crenata* Blume, induced MnP activity (Yoshikawa et al., 2004). The results of the present study show that sawdust extracts produced by autoclaving sawdust in hot water (section 2.2) induce MnP activity (Fig. 3a). Compared with mycelial growth on the MYPG (without extract) medium, the sawdust extract also had a marked effect on the promotion of mycelial growth (MYPG-S). Although less marked than that observed on MYPG-S (100 mg/30 ml), extracts produced using 100 mg sawdust in 30 ml media also promoted mycelial growth; however, MnP activity was not induced in cultures grown in MYPG-S media with lower extract concentrations for up to 35 days (data not shown). These observations suggested that MnP was induced by specific functional compounds in the sawdust extract, and not only due to mycelial growth.

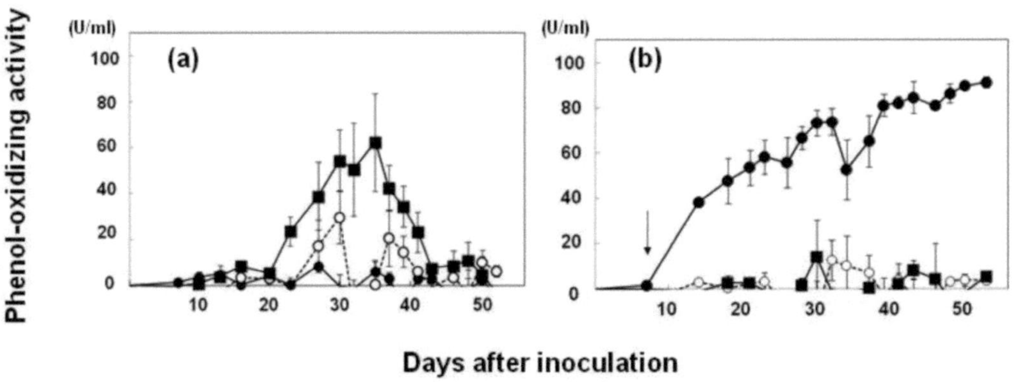

Fig. 3. Changes of phenol-oxidizing activities of (•) Lcc, (○) Per and (■) MnP in a liquid culture medium of *L. edodes* under (**a**) MnP-induced (MYPG-S) or (**b**) Lcc-induced conditions (MYPG-S with 2mM $CuSO_4 \cdot 5H_2O$) (reprinted from Saeki et al., 2011). The arrow in panel (**b**) indicates the day of 2 mM $CuSO_4 \cdot 5H_2O$ addition. Values are means with standard errors (vertical bars) for three replicate cultures.

Fourteen days after inoculation in MYPG-S containing Cu^{2+}, Lcc activity was detected (7 days after the addition of 2 mM $CuSO_4 \cdot 5H_2O$) (Fig. 3b). This Lcc activity increased gradually after day 52 while MnP activity was completely suppressed. Lcc has been shown to be induced by aromatic compounds and metallic ions such as copper (Collins & Dobson, 1997; Saparrat et al., 2002; Scheel et al., 2000; Shutova et al., 2008; Soden & Dobson, 2001). Indeed, copper has been reported to be a strong laccase inducer in the white-rot fungi *Pleurotus ostreatus* (Palmier et al., 2000), *Trametes pubescens* (Galhaup & Haltrich, 2001; Galhaup et al., 2002), and *T. versicolor* (Collins & Dobson, 1997). In *Trametes pubescens*, the transcription of the laccase gene is induced within 10 h after the addition of 2 mM $CuSO_4$ (Galhaup et al., 2002). Under the two culture conditions employed in this study, either Lcc or MnP activity were detected, but not both (Fig. 3). This finding suggests that the induction of MnP and Lcc are controlled by a negative feedback system, i.e., Lcc-induction suppresses MnP production, or more specifically, the addition of $CuSO_4 \cdot 5H_2O$ suppresses MnP production. Although the addition of several of the aromatic compounds that were tested did not induce Lcc - 2-methoxyphenol (guaiacol), 2,6-dimethoxyphenol (DMP), 4-anisidine, hydroquinone, or 1,2-benzenediol (catechol) - these substances except for DMP were observed to suppress MnP activity (data not shown).

3.2 Isozyme detection and identification

Of CBB staining bands subjected to protein identification, proteins with FDR ($q \leq 0.05$) were described below. Number of entry (MS/MS data) was ranged from 68 to 89. In the extracellular enzyme sample (culture liquid) prepared under MnP-induced conditions, two MnP-e isozyme bands, MnP-e (52) and MnP-e (57) in Fig. 4a, were detected. These two MnP isozymes were identified as the manganese peroxidase, LeMnP2, a major MnP isozyme that is secreted into sawdust medium by *L. edodes* (Sakamoto et al., 2009). Other enzyme, exo-β-1,3-glucanase, was also detected under MnP-induced conditions (Fig. 4b). In the extra-cellular enzyme samples prepared under Lcc-induced conditions, two major Lcc isozyme bands, Lcc-e (61) and Lcc-e (67), were detected together with broad tailing smears (Fig. 4c). These two isozymes were identified as being laccases (Lcc1; Fig. 4d), and are known to be an extracellular laccase produced by *L. edodes* (Nagai et al., 2002; Sakamoto et al., 2008). These results, combined with enzyme assay data and results of isozyme detection using PAGE, indicate that MnP and Lcc isozyme detection using the improved LccS+EDTA, PerS+EDTA and MnPS enzymatic staining methods can be used to successfully distinguished between each of the phenol-oxidizing enzymes (Saeki et al., 2011).

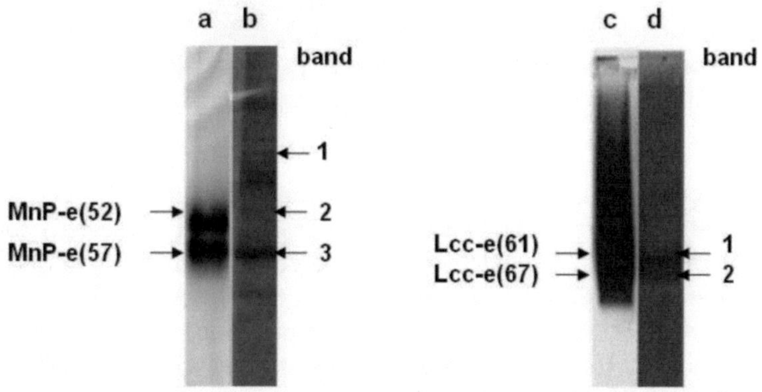

Fig. 4. Protein bands detected by (**a, c**) enzymatic staining and corresponding (**b, d**) CBB staining on native-PAGE (reprinted from Saeki et al., 2011). Lanes (**a**) and (**b**) show bands detected under MnP-induced conditions at 22 days; lanes (**c**) and (**d**) show bands detected under Lcc-induced conditions at 30 days. Protein bands identified by Q-TOF mass spectrometry in (**b**): band 1, exo-β-1,3-glucanase; band 2, manganese peroxidase (LeMnP2); and band 3, manganese peroxidase (LeMnP2) and in (**d**): band 1, laccase (Lcc1); band 2, laccase (Lcc1).

3.3 Comparisons of intracellular and extracellular Lcc isozymes

Expression patterns of extra- and intracellular Lcc isozymes during culture under Lcc-induced conditions are shown in Fig. 5.

Three major extracellular Lcc isozymes were detected: Lcc-e (61), which was expressed from day 12 (5 days after the addition of $CuSO_4 \cdot 5H_2O$), had a constant intensity with a broad smear tails, an Lcc-e (67) from day 17, and another Lcc-e (74) from day 22. All of these enzymes were expressed until the end of culture on day 47. The observed changes in the total band intensity of the three Lcc extracellular isozymes was generally associated with changes in Lcc activity in the culture liquid (refer Fig. 3). Three major intracellular Lcc isozymes, Lcc-i (61), Lcc-i (67) and Lcc-i (74), were also detected, and all exhibited the same mobilities as their respective extracellular Lcc isozyme counterparts. The intracellular Lcc were coincidentally expressed with the extracellular Lcc isozymes.

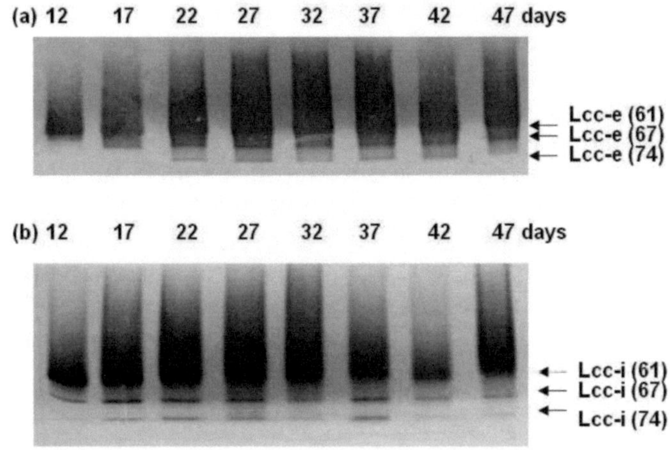

Fig. 5. Laccase isozyme banding patterns detected as (**a**) extracellular and (**b**) intracellular isozymes during culture under Lcc-induced conditions. Days after inoculation are shown above each lane.

Although we successfully extracted total RNA from mycelia under Lcc-induced conditions to examine the transcription of Lcc1, extraction of native (undigested) total RNA was unsuccessful. Native total RNA, which was prepared from mycelia under MnP-induced conditions (described in section 3.4 below), was degraded considerably quicker and to a greater extent after the addition of small amounts of cell lysate obtained under Lcc-induced conditions compared to when cell lysate obtained under MnP-induced conditions was added (data not shown). This relatively quicker degradation of native total RNA suggests a relatively high internal RNase activity in the cell lysate of the Lcc-induced condition, which may be attributed to the decrease observed in mycelial growth after the addition of $CuSO_4 \cdot 5H_2O$ and subsequent induction of Lcc, as well as the antagonistic expression of Lcc in the mycelial contact zone of any adjacent and competing basidiomycetes or other fungi (Iakovlev & Stenlid, 2000; Mercer, 1982; White & Boddy, 1992).

3.4 Manganese peroxidase gene transcription

The sequence of the fragment amplified from cDNA, which was prepared from a mixed pool of total RNAs obtained from the mycelia of 10-, 15- and 18-day-old cultures under MnP-induced conditions, was identical to that of *lemnp2a* but slightly different from *lemnp2b* (data not shown), corroborating the results obtained from the protein identification deduced by Q-TOF mass spectrometry (section 3.2). The finding that these sequences were similar also indicated that hot-water sawdust extracts induced the secretion of the same isozyme, LeMnP2, which is a major MnP isozyme that is secreted into sawdust media (Sakamoto et al. 2009). The results of the Northern blotting experiments with *lemnp2* are shown in Fig. 6. Under MnP-induced conditions, a detectable amount of *lemnp2* mRNA was present at

10–19-days during the initial stage of culture, with transcription increasing from day 22 to day 25 and then decreasing at day 28. These changes in *lemnp2* transcription occurred several days prior to the changes observed in MnP activities in the liquid culture medium.

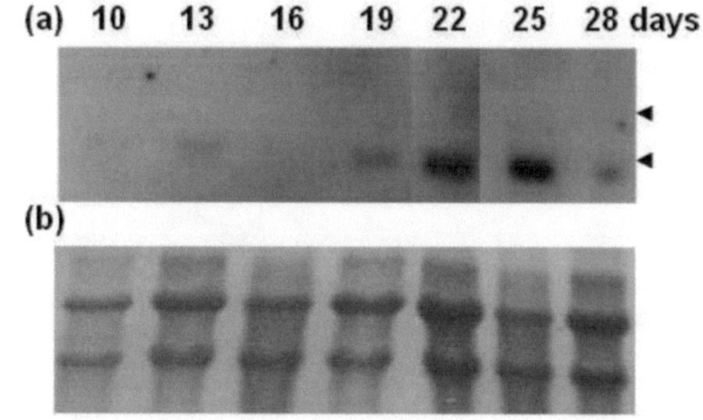

Fig. 6. Northern blot analysis of *lemnp2* gene transcript under (**a**) MnP-induced conditions and (**b**) ribosomal RNA used as a loading control. Days after inoculation are shown above each lane. Arrowheads indicate position of 26S and 18S rRNA.

3.5 Comparisons between intracellular and extracellular MnP isozymes

Expression patterns of extra- and intracellular MnP isozymes during culture under MnP-induced conditions are shown in Fig. 7. The extracellular MnP isozymes, MnP-e (52) and MnP-e (57), were strongly expressed during the initial stage of culture on days 11 to 19, before gradually decreasing until day 43. Four major bands were considered to be intracellular MnP isozymes, and of these, two bands, MnP-i (52) and MnP-i (57), exhibited the same mobility as extracellular MnP isozymes, while the other two bands, MnP-i (63) and MnP-i (66), were strictly intracellular. The intracellular MnP isozymes were expressed during the initial stage of culture, either several days before, or coincident with, the expression of the extracellular MnP isozymes. Compared to the intracellular MnP isozymes, the extracellular MnP isozymes maintained relatively high activities for up to 43 days of culture. However, changes in the intensities of bands that were neither extracellular nor intracellular MnP isozymes coincided with changes in MnP activity in the liquid culture medium during culture. Although we have no experimental data to explain why this may have occurred, it is worth noting that the intracellular enzyme solution (cell lysate) did not exhibit any phenol-oxidizing activities when assayed spectrophotometrically. In addition, the addition of intracellular enzyme solution to extracellular enzyme solutions caused marked inactivation of the latter (data not shown). Taken together, these observations either imply that the cell lysate contained a specific inhibitor of phenol-oxidizing enzymes when these enzymes and the inhibitor in cell lysate were not separated on a gel, or that there was an error in the manner in which the different experimental culture lots were processed, including the replicate flasks.

While treatment with glycosidase completely inactivated the two strictly intracellular MnP isozymes, glycosidase treatment had no effect on the activities of the extracellular MnP isozymes (Fig. 8). This finding indicates that the intracellular isozymes were active as glycosylated proteins, and implies that a relationship exists between the secretion of MnP and the simultaneous expression of β-glucanase detected by Q-TOF mass spectrometry.

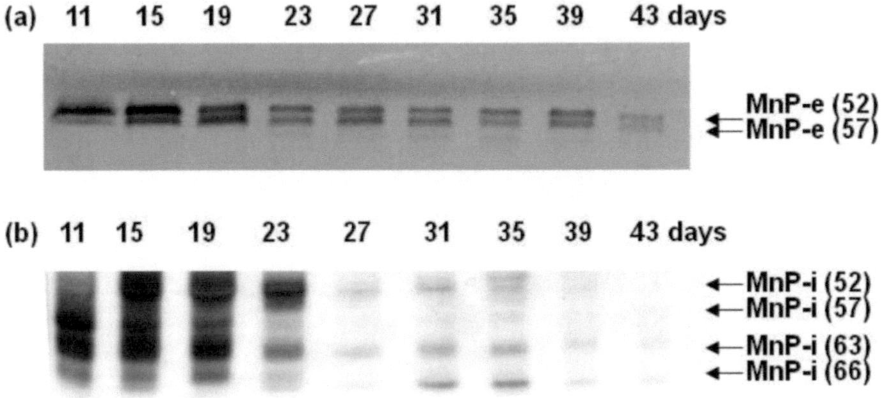

Fig. 7. Manganese peroxidase isozyme banding patterns detected as (**a**) extracellular and (**b**) intracellular isozymes during culture under MnP-induced conditions. Days after inoculation are shown above each lane.

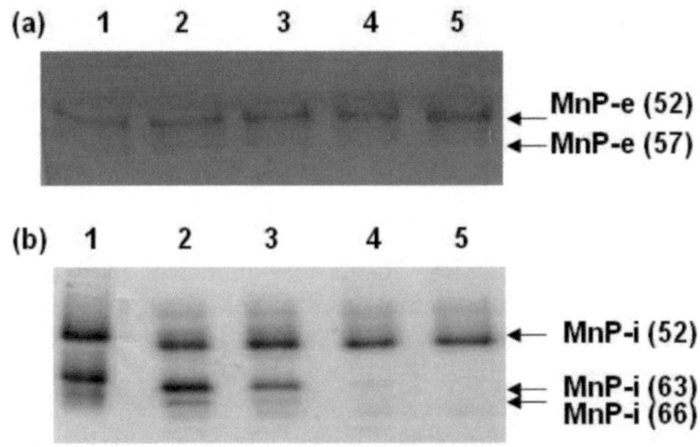

Fig. 8. Effects of glycosidase treatment on enzymatic staining of (**a**) extracellular and (**b**) intracellular MnP isozymes expressed under MnP-induced conditions (at 20 days). Lanes differ according to concentration of glycosidase: lane 1, control (0%); lane 2, 0.25%; lane 3, 0.5%; lane 4, 1.0%; and lane 5, 2.0%, respectively. MnP-i (57) was not detected.

3.6 MnP isozymes in monokaryons

Four monokaryotic strains (#317, #208, #305 and #105), each carrying the mating type factor A_1B_1, A_1B_2, A_2B_1 and A_2B_2, respectively, were derived from basidiospores from dikaryon H600. Although MnP activities of the monokaryons were very weak compared to the MnP activities of H600 (refer Fig. 7), both of the extracellular MnP isozymes (MnP-e (52) and MnP-e (57)) that were detected in the parent dikaryon were also detected in monokaryons, irrespective of their mating-type factors (Fig. 9). This finding suggests that, although these isozymes may be encoded by different loci, the isozymes are not under allelic control (i.e. they are not allozymes).

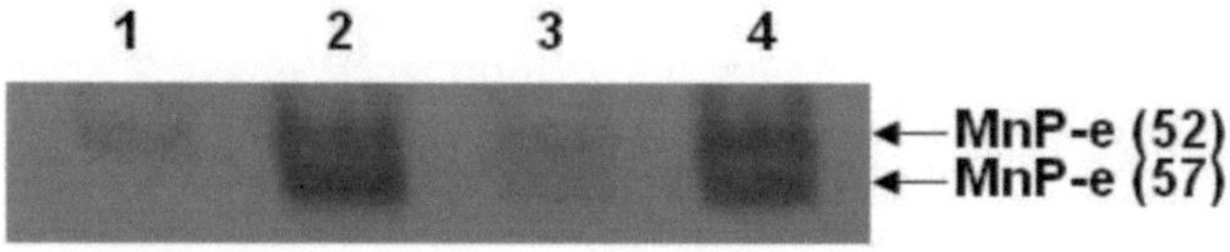

Fig. 9. Extracellular manganese peroxidase isozymes expressed by monokaryotic progenies of H600. Lanes represent strains with mating type factor in parentheses: lane 1, #317 (A_1B_1); lane 2, #208 (A_1B_2); lane 3, #305 (A_2B_1) and lane 4, #105 (A_2B_2).

The results of the Southern blotting experiment of *lemnp2* on *Hin*d III-digested monokaryon genomes are shown in Fig. 10. There was no *Hin*d III restriction site in the amplified region (*lemnp2*S) of the H600 genomic DNA, as expected from the database analysis of different *L. edodes* stock SR-1 (DDBJ Acc. No. AB306944, Sakamoto et al. 2009). Two *lemnp2* hybridization signals appeared at positions between 564-2322 bp in all four of the monokaryotic strains (lanes 1-4 in Fig. 10), and all of the strains exhibited the same two hybridization signals observed in the H600 parent dikaryon (lane 5 in Fig. 10). However, single intense and weak hybridization signals of *lemnp1* were observed using another probe, *lemnp1*S, at different positions between 2322-6557 bp in H600 (lane 6 in Fig. 10), indicating that the two probes did not cross-hybridize with each other. Conversely, it is likely that the weak hybridization signals that appeared between 2322-4631 bp (lane 2 in Fig. 10) were cross-hybridization products between the two probes (see lane 6 in Fig. 10). These observations, combined with the observation of two isozymes being expressed by all of the monokaryons assayed in this study, suggest that there are two copies of *lemnp2* in the haploid genome of *L. edodes*. Nevertheless, to confirm whether the lemnp2 gene is indeed duplicated as proposed here, further analysis will need to be undertaken to assign lemnp2 to a genetic linkage map or on chromosomal DNA which separated by contour-clamped homogeneous electric fields (CHEF) gel electrophoresis. Indeed, such attempts at combining assignments of quantitative trait loci (QTL) related to wood and lignin degradation in fungi would facilitate the identification of new genes involved in another ligninolytic system.

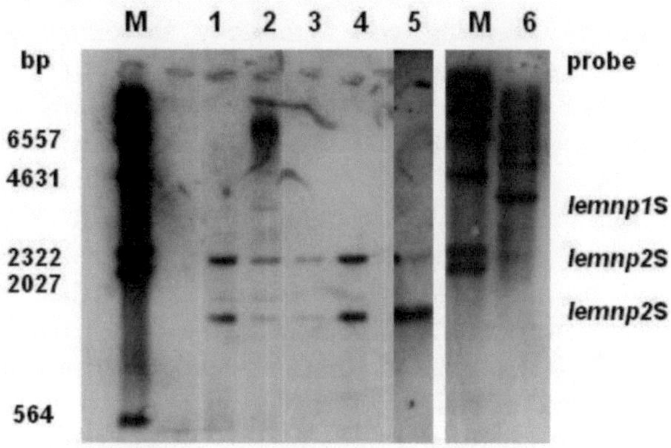

Fig. 10. Southern blot analysis with probe *lemnp2*S on genomic DNAs (digested with *Hin*d III). Lanes represent strains: lane M, size marker (λ/*Hin*d III); lane 1, #317; lane 2, #208; lane 3, #305; lane 4, #105, lane 5, H600; and lane 6, H600 (probed with *lemnp1*S).

3.7 Degradation of β-*O*-4 lignin model compound

We performed preliminarily examinations of the degradation of a β-*O*-4 lignin model compound under MnP- and Lcc-induced conditions (culture experiment) and the degradation of the model compound by incubation with enzyme solutions (incubation experiment) (Table 1).

Culture conditions	Days after inoculation in the culture experiment			Days of incubation in the incubation experiment[1]	
	14	21	42	4	10
MnP-induced	1.4 (nd)[3]	13.4 (32.2)[3]	20.0 (46.8)	16.8 (9.4)	23.8 (6.3)
Lcc-induced	1.1 (nd)	1.8 (47.0)	4.2 (58.5)	3.5 (8.0)	6.9 (3.1)
Enzyme mix[2]				16.0 (12.0 for MnP) (5.0 for Lcc)	22.9 (7.0 for MnP) (3.3 for Lcc)

[1]: Extracellular enzyme solution at an initial activity adjusted to 17 U/ml
[2]: Mixture of the extracellular enzyme solutions (MnP+Lcc), each at an initial activity of 17 U/ml
[3]: Numerals in parentheses represent enzyme activities (U/ml) , nd = not detected

Table 1. Degradation rate (%) of β-*O*-4 lignin model compound under MnP- or Lcc-induced conditions (culture experiment) and after incubation with enzyme solutions prepared from given culture conditions (incubation experiment) (Data from Kochi et al., 2009)

Under MnP-induced conditions, the β-*O*-4 compound was not degraded at all during the initial stages of the culture experiment. Indeed, effective degradation only occurred after day 21 when MnP activities suddenly increased; by day 42, 20.0% of the β-*O*-4 compound had been degraded. Conversely, no degradation of the β-*O*-4 compound was observed under Lcc-induced conditions until day 42 (4.2%). In the incubation experiment with MnP solution, the β-*O*-4 compound was effectively degraded in the initial 4 days of incubation (16.8%), with degradation increasing very gradually thereafter and then decreasing markedly near the end of the experiment; i.e., 23.8% at day 10 and only 7% of the compound was degraded in the latter 6 days. Conversely, degradation of the β-*O*-4 compound incubated with Lcc solution was detectable, but weak, until 10 days after inoculation (6.9%). The change in the degradation rates of the β-*O*-4 compound incubated with a mixture of the MnP and Lcc enzyme solutions (each at an initial activity of 17 U/ml) were similar to the degradation patterns of the MnP solution alone. This similarity indicated that no additive or multiplier effects could be attributed to the interaction of the two enzymes on the degradation of the β-*O*-4 compound. Compared to the initial period of the incubation experiment, the shallow slope of degradation rate in the latter period of the incubation was partly attributable to decreased enzyme activities over the course of the experiment (Table 1). Unfortunately, because we conducted this experiment without a protease-inhibitor, the decrease in enzyme activities was observed in enzyme solutions containing both MnP and Lcc, as well as the mixed MnP+Lcc solutions. In addition, laccase is also capable of degrading non-phenolic lignin model compounds in systems incorporating naturally occurring or synthetic redox mediators (Johannes & Majcherczyk, 2000; Srebotnik & Hamel 2000; Tanaka et al., 2009). Nevertheless, based on the above results, manganese peroxidase (LeMnP2) appears to be more important than laccase (Lcc1) in lignin degradation by *L. edodes*. Prior to the discovery of Lip and MnP, one of the major catabolites formed by the degradation of the β-*O*-4 dimer by *P. chrysosporium*, 2-guaiacoxyethanol (II), was identified (Enoki et al., 1980). The results described above suggest that the assay system developed in this study is well suited for identifying phenol-oxidizing isozymes involved in the degradation of lignin model compounds. Further, these phenol-oxidizing isozymes have been effective for elucidating the mechanisms involved lignin degradation, and this role is likely to extend into the future (Cullen & Kersten, 2004).

Although we attempted to identify other MnP isozymes using another commercial Japanese Shiitake variety, "Bridge 32" (The General Environmental Technos Co. Ltd., Osaka, Japan), which is also used in sawdust cultivation, the variety exhibited the same extracellular MnP isozyme patterns as H600 (data not shown). The estimated heritability (h^2), which is the ratio of the additive genetic components of variance to the phenotypic components of variance, of the variety's wood-degrading ability was relatively low (32.2%) compared to the heritabilities estimated for other traits in crosses of H600 and Bridge 32 (Tanesaka et al., 2007). This low heritability may be attributable to the low allelic variation that exists between the MnP isozymes of the two varieties. The MnP that is produced by *L. edodes* when it is cultured on sawdust media (Buswell et al., 1995; Leatham 1985; Makker et al., 2001; Sakamoto et al., 2009), and which degrades the β-*O*-4 lignin model compound (Kochi et al., 2009), is likely to be critical for mycelial growth and fruit-body development during sawdust cultivation. The system presented here for assaying phenol-oxidizing enzymes under liquid culture conditions could therefore provide a practical screening method for examining isozymes of value in mushroom cultivation, particularly since the assay system targets the wood-degrading ability and the genomic characteristics of the genes involved in lignin-degradation.

4. Conclusions

When cultivated on sawdust-based media, the white-rot basidiomycete *Lentinula edodes* frequently produces the lignin-degrading enzymes MnP and Lcc. In this study, MnP produced by *L. edodes* was induced in a liquid culture supplemented with a sawdust extract of *Castanopsis cuspidata*. Lcc activity was induced by the addition of 2 mM $CuSO_4 \cdot 5H_2O$ into the same media 7 days after initial inoculation. In addition to employing native-PAGE and sequential enzymatic staining to detect the MnP and Lcc secreted by *L. edodes*, we also compared the expression of intra- and extracellular MnP isozymes. To distinguish between the phenol-oxidizing enzymes after native-PAGE, the gel was sequentially stained using an improved enzymatic staining solution (referred to as LccS+EDTA). In addition to containing 0.1 mM acetate buffer (pH 4.0) for Lcc detection, the staining solution contained 1.8 mM *o*-dianisidine as the substrate and 130 mM EDTA to eliminate Mn^{2+} contamination. Subsequently, 0.1 mM H_2O_2 was added to the LccS+EDTA for Per detection (PerS+EDTA), and 0.1 mM $MnSO_4 \cdot 5H_2O$ was added to the PerS, without EDTA, for MnP detection (MnPS). The two extracellular isozyme bands, MnP-e (52) and MnP-e (57), detected in culture medium under MnP-induced conditions, were both identified as manganese peroxidase (LeMnP2). Similarity, the bands Lcc-e (61) and Lcc-e (67), which were detected under Lcc-induced conditions, were both identified as laccase (Lcc1) by Q-TOF mass spectrometry. Four major, intracellular, MnP isozyme bands were detected in mycelial extracts obtained from *L. edodes* cultured under MnP-induced conditions. Of these isozyme bands, two exhibited the same mobilities as extracellular MnP isozymes, while the other two bands, MnP-i (63) and MnP-i (66), were strictly intracellular. The intracellular MnP isozymes were expressed during the initial stage of culture, either several days before, or coincident with, the expression of the extracellular MnP isozymes. Compared to intracellular MnP isozymes, the extracellular MnP isozymes maintained relatively high activities for up to 40 days of culture. While glycosidase treatment of crude enzyme solutions prior to electrophoresis had no effect on the activities of the extracellular MnP isozymes, such treatment completely inactivated the two strictly intracellular MnP isozymes, implying that the intracellular isozymes were active as glycosylated proteins. Both of the extracellular MnP isozymes detected in the dikaryon were also detected in monokaryotic progeny, suggesting that although these isozymes may be encoded by different loci, they are not under allelic control. Southern blot analysis revealed that the probe *lemnp2* region hybridized with the four of the monokaryotic strains used, all of which exhibited the same two hybridization signals that were observed in the parent dikaryon. These observations suggest that there are two copies of *lemnp2* in the *L. edodes* haploid genome. Moreover, degradation assays involving the addition of the β-*O*-4 lignin model compound in cultures under MnP- and Lcc-induced conditions suggest that, rather than laccase (Lcc1), manganese peroxidase (LeMnP2) is a critical enzyme for lignin degradation in *L. edodes*.

In response to the crucial role played by basidiomycetous fungi in the carbon cycle by degrading lignocelluloses, considerable effort has focused on the functional genomics related to the enzymatic systems and mechanisms involved in lignin degradation, particularly in a few model fungus species. Nevertheless, fungal succession on dead logs and leaf litter in nature show that complete degradation of lignocelluloses is a commensal and competitive process affected by numerous fungi. The assay system presented here would be practical and convenient, not only as a method of screening isozymes of value in mushroom breeding and cultivation, but also for evaluating the lignin-degrading abilities of fungi and assessing the antagonistic interactions of different strains under experimental conditions.

5. Acknowledgments

We are grateful to Professor Toshiaki Umezawa from Kyoto University for his kind support and synthesis of the β-*O*-4 lignin model compound. This work was supported by the "Academic Frontier" Project for Private Universities, with a matching fund subsidy from the Ministry of Education, Culture, Sports, Science and Technology (2004–2008), Japan. The Article Processing Charges for this chapter were provided by a fund from The General Environmental Technos Co. Ltd., Osaka, Japan.

6. References

Bonnen, A.; Anton, L. & Orth, A. (1994). Lignin-degrading enzymes of the commercial button mushroom, *Agaricus bisporus. Appl Environ Microbiol*, Vol.60, No.3, (March 1994), pp. 960-965, ISSN 0099-2240

Burke, R. & Cairney, J. (2002). Laccase and other phenol oxidases in ecto- and ericoid mycorrhizal fungi. *Mycorrhiza,* Vol.12, No.3, (June 2002), pp. 105-116. ISSN 0904-6360

Buswell, J.; Cai, Y. & Chang, S-T. (1995). Effect of nutrient and manganese on manganese peroxidase and laccase production by *Lentinula (Lentinus) edodes. FEMS Microbiol Lett,* Vol.128, No.1, (April 1995), pp. 81-88, ISSN 0378-1097

Chang, S-T. & Miles, P. (1989). *Edible mushrooms and their cultivation,* CRC Press, Inc., ISBN 0-8493-6758-X, Boca Raton, Florida, USA

Collins, P. & Dobson, A. (1997). Regulation of laccase gene transcription in *Trametes versicolor. Appl Environ Microbiol,* Vol.63, No.3, (September 1997), pp. 3444-3450, ISSN 0099-2240

Cullen, D. & Kersten, P. (2004). Enzymology and molecular biology of lignin degradation. In: *The Mycota III; Biochemistry and molecular biology. 2nd edition,* Brambl, R., & Marzluf, G. (Eds), pp. 249–273, Springer, ISBN 3-540-42630-2, Berlin, Germany

Enoki, A.; Gwendolyn, P., Goldsby, G. & Gold, M. (1980). Metabolism of the lignin model compounds veratryl glycerol-β-guaiacyl ether and 4-ethoxy-3-methoxyphenylglycerol-β-guaiacyl ether. *Arch Microbiol,* Vol.125, No.3, (April 1980), pp. 227-232, ISSN 0302-8933

Enoki, A.; Tanaka, H. & Fuse, G. (1988). Degradation of lignin-related compounds, pure cellulose, and wood components by white-rot and brown-rot fungi. *Holzforschung,* Vol.42, No.2, (January 1988), pp. 85-93, ISSN 0018-3830

Forrester, I.; Grabski, A., Mishra, C., Kelley, B., Strickland, W., Leatham, G. & Burgess, R. (1990). Characteristics and N-terminal amino acid sequence of a manganese peroxidase purified from *Lentinula edodes* cultures grown on commercial wood substrate. *Appl Microbiol Biotechnol,* Vol.33, No.3, (June 1990), pp. 359-365, ISSN 0175-7598

Galhaup, C.; Goller, S., Peterbauer, C., Strauss, J. & Haltrich, D. (2002). Characterization of the major laccase isozyme from *Trametes pubescens* and regulation of synthesis by metal ions. *Microbiology,* Vol.148, No.7, (July 2002), pp. 2159-2169, ISSN 1350-0872

Galhaup, C. & Haltrich, D. (2001). Enhanced formation of laccase activity by the white-rot fungus *Trametes pubescens* in the presence of copper. *Appl Microbiol Biotechnol,* Vol.56, No.2, (July 2001), pp. 225–232, ISSN 0175-7598

Glenn, J.; Morgan, M., Mayfield, M., Kuwahara, M. & Gold, M. (1983). An extracellular H_2O_2-requiring enzyme preparation involved in lignin biodegradation by the white rot basidiomycete *Phanerochaete chrysosporium. Biochem Biophys Res Commun,* Vol.114, No.3, (August 1983), pp. 1077-1083, ISSN 0006-291X

Glenn, J. & Gold, M. (1985). Purification and characterization of an extracellular manganese (II)-dependent peroxidase from lignin-degrading basidiomycete, *Phanerochaete chrysosporium. Arch Biochem Biophys,* Vol.242, No.2, (November 1985), pp. 329-341, ISSN 0003-9861

Gold, M. & Alic, M. (1993). Molecular biology of the lignin-degrading basidiomycete *Phanerochaete chrysosporium. Microbiological Review,* Vol.57, No.3, (September 1993), pp. 605-622, ISSN 0146-0749

Highley, T.; Murmanis, L. & Palmer, J. (1985). Micromorphology of degradation in western hemlock and sweetgum by the brown-rot fungus *Poria placenta. Holzforschung,* Vol. 39, No.2, (January 1985), pp. 73-78, ISSN 0018-3830

Highley, T. & Murmanis, L. (1987). Micromorphology of degradation in western hemlock and sweetgum by the white-rot fungus *Coriolus versicolor. Holzforschung,* Vol. 41, No.2, (January 1987), pp. 67-71, ISSN 0018-3830

Higuchi, T. (1990). Lignin biochemistry: biosynthesis and biodegradation. *Wood Sci Technol*, Vol.24, No.1, (March 1990), pp. 23-63, ISSN 0043-7719

Iakovlev, A. & Stenlid, J. (2000). Spatiotemporal patterns of laccase activity in interacting mycelia of wood-decaying basidiomycete fungi. *Microbial Ecology*, Vol.39, No. 3, (April 2000), pp. 236-245, ISSN 0095-3628

Johannes, C. & Majcherczyk, A. (2000). Natural mediators in the oxidation of polycyclic aromatic hydrocarbons by laccase mediator system. *Appl Environ Microbiol*, Vol.66, No.2, (February 2000), pp. 524-528, ISSN 0099-2240

JØrgensen, E. & Vejlby, K. (1953). A new polyphenol oxidase test. *Physiologia Plantarum*, Vol.6, No.3, (July 1953), pp. 533-537, ISSN 0031-9317

Kinugawa, K. & Tanesaka, E. (1990). Changes in the rate of CO_2 release from cultures of three basidiomycetes during cultivation. *Trans Mycol Soc, Japan*, Vol.31, No.4, (December 1990), pp. 489–500, ISSN 0029-0289

Kirk, T. & Fenn, P. (1982). Formation and action of the ligninolytic system in basidiomycetes, In: *Decomposer basidiomycetes: their biology and ecology*, Frankland, J., Hedger, J. & Swift, M. (Eds.), pp. 67-90, Cambridge University Press, ISBN 0-521-24634-2, New York, USA

Kirk, T. & Highley, T. (1973). Quantitative changes in structural components of conifer woods during decay by white- and brown-rot fungi, *Phytopathology*, Vol.63, No. 11, (November 1973), pp. 1338-1342, ISSN 0031-949X

Kochi, A.; Saeki, N., Tanesaka, E. & Yoshida, M. (2009). Degradation of β-*O*-4 lignin model compound by Shiitake, *Lentinula edodes*. *J Crop Res* Vol.54, (November 2009) pp. 131−136, ISSN 1882-885X

Kuwahara, M.; Glenn, J., Morgan, M. & Gold, H. (1984). Separation and characterization of two extracellular H_2O_2-dependent oxidases from ligninolytic cultures of *Phanerochaete chrysosporium*. *FEBS Lett*, Vol.169, No.2, (April 1984), pp. 247-250, ISSN 0014-5793

Leatham, G. (1985). Extracellular enzymes produced by the cultivated mushroom *Lentinus edodes* during degradation of a lignocellulosic medium. *Appl Environ Microbiol*, Vol.50, No.4, (October 1985), pp. 859-867, ISSN 0099-2240

Makker, R.; Tsuneda, A., Tokuyasu, K. & Mori, Y. (2001). *Lentinula edodes* produces a multicomponent protein complex containing manganese (II)-dependent peroxidase, laccase and β-glucosidase. *FEMS Microbiol Lett*, Vol.200, No.2, (June 2001), pp. 175–179, ISSN 0378-1097

Mercer, P. (1982). Basidiomycete decay of standing trees, In: *Decomposer basidiomycetes: their biology and ecology*, Frankland, J., Hedger, J. & Swift, M. (Eds.), pp. 143-160, Cambridge University Press, ISBN 0-521-24634-2, New York, USA

Milligan, B. (1998). Total DNA isolation. In: *Molecular Genetic analysis of population, 2nd edition*, Hoelzel, A. (Ed.), pp. 29-64, Oxford University Press, ISBN 0-19-963635-4, New York

Nagai, M.; Sato, T., Watanabe, H., Saito, K., Kawata, M. & Enei, H. (2002). Purification and characterization of an extracellular laccase from the edible mushroom *Lentinula edodes*, and decolorization of chemically different dyes. *Appl Microbiol Biotechnol*, Vol.60, No. 3, (November 2002), pp. 327–335, ISSN 0175-7598

Nagai, M.; Kawata, M., Watanabe, H., Ogawa, M., Saito, K., Takesawa, T., Kanda, K. & Sato, T. (2003). Important role of fungal intracellular laccase for melanin synthesis : purification and characterization of an intracellular laccase from the edible mushroom *Lentinula edodes* fruit bodies. *Microbiology*, Vol.149, No.9, (September 2003), pp. 2455-2462, ISSN 1350-0872

Nagai, M.; Sakamoto, Y., Nakade, K. & Sato, T. (2007). Isolation and characterization of the gene encoding a manganese peroxidase from *Lentinula edodes*. *Mycoscience*, Vol.48, No.2, (April 2007), pp.125-130, ISSN 1340-3540

Ohga, S. & Kitamoto, Y. (1997). Future of mushroom production and biotechnology. *Food Reviews International*, Vol.13, No.3, (October 1997), pp. 461–469, ISSN 8755-9129

Orth, A.; Royse, D. & Tien, M. (1993). Ubiquity of lignin-degrading peroxidases among various wood-degrading fungi. *Appl Environ Microbiol*, Vol.59, No. 12, (December 1993), pp. 4017-4023, ISSN 0099-2240

Palmieri, G.; Giardina, P., Bianko, C., Fontanella, B. & Sannia, G. (2000). Copper induction of laccase isozymes in the ligninolytic fungus *Pleurotus ostreatus*. *Appl Environ Microbiol*, Vol.66, No.3, (March 2000), pp. 920-924, ISSN 0099-2240

Pasteur, N.; Pasteur, G., Bonhomme, F., Catalan, J. & Britton-Davian, J. (1988). *Practical isozyme genetics*, Ellis Horwood Ltd., ISBN 0-7458-0501-9, Chichester, UK

Paszczynski, A.; Crawford, R. & Huynh, V-B. (1988). Manganese peroxidase of *Phanerochaete chrysosporium*: Purification. In: *Methods in enzymology; Vol.161, Biomass, Part B, Lignin, pectin and chitin*, Wood, W. & Kellogg, S. (Eds.), 264-270, Academic Press, ISBN 0-12-182062-9, San Diego, CA, USA

Saeki, N.; Takeda, H., Tanesaka, E. & Yoshida, M. (2011). Induction of manganese peroxidase and laccase by *Lentinula edodes* under liquid culture conditions and their isozyme detection by enzymatic staining on native-PAGE. *Mycoscience*, Vol.52, No.2, (March 2011), pp. 132-136, ISSN 1340-3540

Sakamoto, Y.; Nakade, K., Yano, A., Nakagawa, Y., Hirano, T., Irie, T., Watanabe, H., Nagai, M. & Sato, T. (2008). Heterologous expression of lcc1 from *Lentinula edodes* in tobacco BY-2 cells results in the production an active, secreted form of fungal laccase. *Appl Microbiol Biotechnol*, Vol.79, No.6, (July 2008), pp. 971–980, ISSN 0175-7598

Sakamoto, Y.; Nakade, K., Nagai, M., Uchimiya, H. & Sato, T. (2009). Cloning of *Lentinula edodes* lemnp2, a manganese peroxidase that is secreted abundantly in sawdust medium. *Mycoscience*, Vol.50, No.2, (March 2009), pp. 116-122, ISSN 1340-3540

Sambrook, J.; Fritsch E. & Maniatis, T. (1989). *Molecular cloning: a laboratory manual, Book 2* (2nd edition), Cold Spring Harbor Laboratory Press, ISBN 0-87969-309-6, New York, USA

Saparrat, M.; Guillén, F., Arambarri, A., Martínez, A. & Martínez, M. (2002). Induction, isolation, and characterization of two laccases from the white rot basidiomycete *Coriolopsis rigida*. *Appl Environ Microbiol*, Vol.68, No.4, (April 2002), pp. 1534–1540, ISSN 0099-2240

Scheel, T.; Höfer, M., Ludwig, S. & Hölker, U. (2000). Differential expression of manganese peroxidase and laccase in white-rot fungi in the presence of manganese or aromatic compounds. *Appl Microbiol Biotechnol*, Vol.54, No.5, (November 2000), pp. 686-691, ISSN 0175-7598

Shutova, V.; Revin, V. & Myakushina Y. (2008). The effect of copper ions on the production of laccase by the fungus *Lentinus (Panus) tigrinus*. *Appl Biochem Microbiol*, Vol.44, No.6, (November 2008), pp. 683–687, ISSN 0003-2697

Smith, J.; Fermor, T. & Zadrazil, F. (1988). Pretreatment of lignocellulosics for edible fungi. In: *Treatment of lignocellulosics with white rot fungi*, Zadrazil, F. & Reiniger, P. (Eds.), pp. 3-13, Elsevier Applied Science, ISBN 1-85166-241-3, Essex, England

Soden, D. & Dobson, A. (2001). Differential regulation of laccase gene expression in *Pleurotus sajor-caju*. *Microbiology*, Vol.147, No.7, (July 2001), pp. 1755–1763, ISSN 1350-0872

Srebotnik, E. & Hammel, K. (2000). Degradation of nonphenolic lignin by the laccase/1-hydroxybenzotriazole system. *J Biotechnol*, Vol.81, No.2-3, (August 2000), pp. 179-188, ISSN 0021-9193

Szklarz, G.; Antibus, R., Sinsabaugh, R. & Linkins, A. (1989). Production of phenol oxidase and peroxidases by wood-rotting fungi. *Mycologia*, Vol.81, No.2, (April 1989), pp. 234-240, ISSN 0027-5514

Tanaka, H.; Koike, K., Itakura, S. & Enoki, A. (2009). Degradation of wood and enzyme production by *Ceriporiopsis subvermispora*. *Enzyme Microbial Technol*, Vol.45, No.5, (November 2009), pp. 384-390, ISSN 0141-0229

Tanesaka, E.; Masuda, H. & Kinugawa, K. (1993). Wood degrading ability of basidiomycetes that are wood decomposers, litter decomposers, or mycorrhizal symbionts. *Mycologia*, Vol.81, No.13, (July 1993), pp. 347-354, ISSN 0027-5514

Tanesaka, E.; Nakajima, Y., Uchida, T., Ikeda, Y. & Yoshida, M. (2007). Variation of wood degrading ability among hybrid stocks of Shiitake mushroom, *Lentinula edodes*, and their cultivation traits on sawdust media, *Breeding Research, Proceedings of Japanese Society of Breeding 2007 112th Conference*, Vol.9, No.2, pp. 111, ISSN 1344-7629, Yamagata, Japan, September 22-23, 2007

Thurston, C. (1994). The structure and function of fungal laccases. *Microbiology*, Vol.140, No.1, (January 1994), pp. 19-26, ISSN 1350-0872

Tien, M. & Kirk, T. (1983). Lignin-degrading enzyme from the hymenomycete *Phanerochaete chrysosporium* Burds. *Science*, Vol.221, No. 4611, (August 1983), pp. 661-663, ISSN 0036-8075

Tien, M. & Kirk, T. (1984). Lignin-degrading enzyme from *Phanerochaete chrysosporium*: Purification, characterization and catalytic properties of a unique hydrogen peroxide-requiring oxygenase. *Proc Natl Acad Sci USA*, Vol. 81, No.8, (April 1984), pp.2280-2284, ISSN 0027-8424

Umezawa, T. & Higuchi, T. (1985). A novel C_α-C_β cleavage of lignin model dimer with rearrangement of the β-aryl group by *Phanerochaete chrysosporium*. *FEBS Lett*, Vol.192, No.1 (November 1985), pp. 293-298, ISSN 0014-5793

White, N. & Boddy, L. (1992). Extracellular enzyme localization during interspecific fungal interactions. *FEMS Microbiol Lett*, Vol.98, No.1 (November 1992), pp.75-79, ISSN 0378-1097

Wood, D.; Matcham, S. & Fermor, T. (1988). Production and function of enzymes during lignocellulose degradation, In: *Treatment of lignocellulosics with white rot fungi*, Zadrazil, F. & P. Reiniger, (Eds.), pp. 43-49, Elsevier Applied Science, ISBN 1-85166-241-3, Essex, England

Yoshikawa, T.; Tanesaka E. & Yoshida M (2004) Determination of ligninolytic culture conditions in expression of *Lentinula edodes* manganese peroxidase, *Breeding Research, Proceedings of Japanese Society of Breeding 2004 105th Conference*, Vol.6, No.1, pp. 15, ISSN 1344-7629, Tokyo, Japan, March 30-31, 2004

21

Protection Studies by Antioxidants Using Single Cell Gel Electrophoresis (Comet Assay)

Pınar Erkekoglu
Hacettepe University, Faculty of Pharmacy, Department of Toxicology, Ankara Turkey

1. Introduction

Oxidation-reduction reactions, simply referred as "redox" reactions, describe all the chemical reactions in which atoms have their oxidation state changed. This can either be a simple redox process like the oxidation of carbon (C) to carbon dioxide (CO_2) or the reduction of C by hydrogen (H) to yield methane (CH_4). However, in biology redox reactions are rather complex and 'redox biology' is fundamental to aerobic life **(Peters et al., 2008; Baliga et al., 2007)**. The simplest example to give is the oxidation of glucose to CO_2 and water in photosynthesis **(Halliwell, 2006)**.

Aerobes are constantly subject to free radicals, but modulate their actions by synthesizing antioxidants. Free radicals are atoms, molecules, or ions with one or more unpaired electrons on an open shell configuration **(Gutteridge & Halliwell, 2000)**. The simplest form is the atomic H. There are many types of free radicals in living systems, but both nitrogen (N) and oxygen (O) radicals are the main concern for the researchers of several fields as they are suspected to be the underlying factors of several conditions and diseases **(Halliwell, 2006)**. O_2 toxicity was suggested to be due to the inactivation of a variety of enzymes (particularly of antioxidant enzymes) by targeting the thiol group of cysteine residues. In the last decades, molecular biology techniques established that the toxic effects of O_2 are directly linked to its reactive forms, the reactive oxygen species (ROS), acting on cellular components. Oxidative stress is a serious imbalance between the generation of ROS and antioxidant protection in favor of the former, causing excessive oxidative damage **(Dröge, 2002; Halliwell, 2011)**. Oxidative stress and ROS can account for changes that may be detrimental to the cells **(Dröge, 2002)**. ROS are shown to contribute to cellular damage, apoptosis and cell death **(Dalton et al., 1999; Finkel, 1998)**. The link between O_2 toxicity and many pathologies, e.g. pulmonary diseases, **(Frankl, 1991)**, and its effect on swelling of the blood–gas barrier **(Drath et al., 1981)**, retina defects **(Geller et al., 2006)**, bowel disease **(Grisham, 1994)** neurodegeneration **(Wang et al., 2006)**, cancer **(Cerutti, 1994)**, diabetes **(Seet et al., 2010)** and ageing **(Irminger-Finger, 2007)** is very well-established. Besides, in the last decade a relationship between obesity and ROS was demonstrated **(Seet et al., 2010; Halliwell, 2011)**.

Antioxidant is a molecule that protects a biological target against oxidative damage **(Halliwell, 2011)**. Accumulating data implicate that both low antioxidant status and genetics may contribute to the risk of several types of malignancies **(Peters et al., 2008;**

Baliga et al., 2007). The field of antioxidants and free radicals is often perceived as focusing around the use of antioxidant supplements to prevent human disease. Currently, there is a growing interest in environmental chemicals that can cause oxidative stress. The genotoxic effects of some compounds are of particular interest for researchers as humans are exposed to these chemicals abundantly. Exposure to such chemicals may result in disturbances of several physiological processes and may lead to wide variety of degenerative diseases including cancer **(Soory, 2009)**.

First described by **Östling & Johanson (1984)**, and then modified by **Singh et al. in 1988,** the single cell gel electrophoresis assay (also known as Comet assay) is an uncomplicated and sensitive technique for the detection of DNA damage at the level of the individual eukaryotic cell. It has since gained in popularity as a standard technique for evaluation of DNA damage/repair, biomonitoring and genotoxicity testing **(Singh et al., 1988)**.

2. Why comet assay is a suitable tool for antioxidant research?

Comet assay can easily detect the *in vitro* toxicity of environmental chemicals on different cell types, as well as *in vivo* toxicity in tissue samples obtained from animals. Besides, it is also a valid technique to evaluate whether antioxidants/micronutrients are able to protect the integrity of the genetic material **(Anderson et al., 1997; Heaton et al., 2002; Novotna et al., 2007)**.

The benefits of Comet assay can be summarized as below:

- Sensitivity for detecting low levels of DNA damage: The limit of sensitivity is approximately 50 strand breaks per diploid mammalian cell and will lose sensitivity above about 10,000 breaks per cell **(Olive & Banáth, 2006)**.
- Requirement for small number of cells per sample: <10,000 cells are enough to perform the assay.
- Flexibility: Comet assay is applicable to virtually any type of cell, as long as a single cell suspension is obtained. Besides, different combinations of unwinding and electrophoresis conditions and lesion-specific enzymes can be used to detect different types and levels of DNA damage **(Wong et al., 2005)**.
- Low cost and ease of application **(Anderson et al., 1997).**
- Studies can be conducted using relatively small amounts of a test substance **(Anderson et al., 1997)**
- A relatively short time is needed to complete an experiment.

The advantages and disadvantages of Comet assay are shown in **Figure 1**.

3. Technical information on comet assay

DNA damage can simply be evaluated using Comet assay that allows the measurement of DNA single- and double-strand breaks (frank strand breaks and incomplete excision repair sites) together with alkali labile sites and crosslinking. By choosing different pH conditions for electrophoresis and the preceding incubation, different levels of damage can be assessed. The degree of DNA migration can be correlated to the extent of DNA damage occurring in each single cell. *In vitro* studies can be performed on virtually with any cell type; however, the cell-type-of-choice in biomonitoring is mostly the lymphocyte because blood is easily

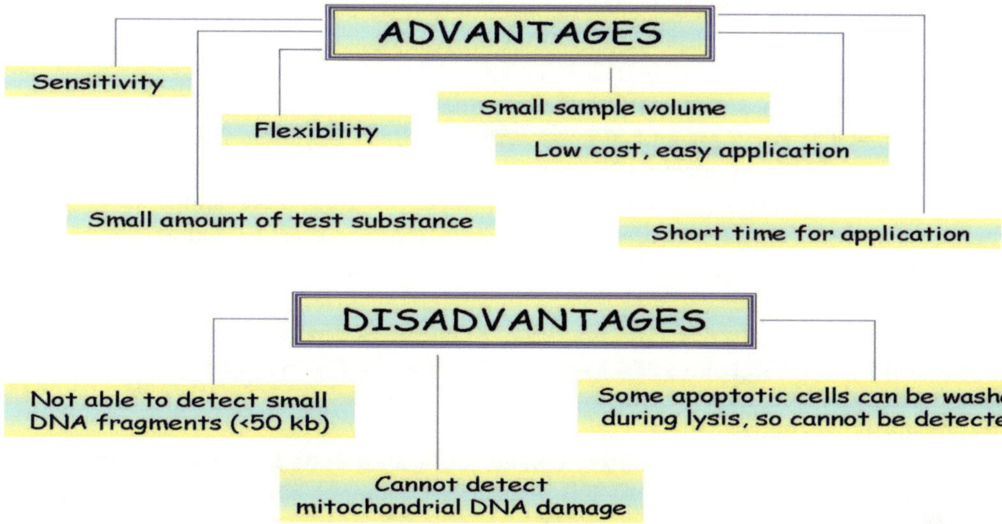

Fig. 1. Advantages and disadvantages of Comet assay

collected and lymphocytes have proved to be good surrogate cells. For example, lymphocytes exhibited genotoxicity caused by anticancer agents targeting several different organs **(Faust et al., 2004)**.

There are differences between laboratories in the isolation of lymphocytes, cells from organs/tissues or other specimens, or in the solutions used for electrophoresis. A simple alkaline Comet assay protocol can be performed in the following steps:

a. The slides that will be used in the study should be covered with agarose (1%) the day before the experiment.
b. In the basic alkaline Comet assay, for primary and other cell cultures, after exposing small number of cells to a physical or chemical agent, the cells are trypsinized, centrifuged, washed, and resuspended in PBS. Because of the flexible application of the technique, the cells used can be isolated lymphocytes, cells isolated from bone marrow, cells isolated from solid organs or tissues or cells from primary or other cell cultures. Lymphocytes can be isolated from whole blood using different isolation solutions and centrifugation. Cells from bone marrow can be obtained by perfusing femur in cold mincing solutions and centrifugation. Solid organs or tissues must be minced into fine pieces, later be suspended in cold mincing solutions and centrifugated. Blood-rich organs like liver and kidney have to minced into larger pieces, the mincing solution can be aspirated and fresh mincing solution should be added. Mincing solution can be Hank's Buffered Salt Solution (HBSS, with 20 mM EDTA and 10% DMSO).
c. Usually 50 µl of the cells obtained from either cell cultures blood or organs/tissues should be mixed with 450 µl solution of low melting point agarose (0.6% in PBS), and 100 µl of the solution is spread on microscope slides covered with agarose.
d. Cells are lysed (in 2.5 M NaCl, 0.5 MNa_2-EDTA, 10 mM Tris, 1% sodium lauryl sulfate, 1% Triton X-100, 10% DMSO, pH 10) at 4°C in dark for 1 h. After lysis, cells were

immersed in freshly prepared alkaline electrophoresis buffer (300 mM NaOH, 1 mM Na_2-EDTA, pH 13) for 30 min to allow DNA unwinding.

e. Electrophoresis is then performed at 25 V/300 mA for 30 min.

f. After electrophoresis, slides are rinsed three times for 5 min with neutralization buffer (0.4 M Tris-HCl, pH 7.4), and stained with ethidium bromide (20 µg/ml) in PBS. Ethidium bromide is an intercalating agent commonly used as a fluorescent nucleic acid stain in molecular biology. There are a number of alternative stains to ethidium bromide, including acridine orange, propidium iodide, YOYO-1 iodide stain, SYBR Gold nucleic acid gel stain, SbYR Green I stain, TOTO-3 stain and silver (for non-fluorescent staining).

g. For quantification, a fluorescence microscope can be used which can be connected to a charge-coupled device (CDC) and a computer-based analysis system.

h. The extent of DNA damage was determined after electrophoretic migration of DNA fragments in the agarose gel.

i. For each condition randomly selected comets (50/100/200) on each slide can be scored, and % head DNA, % tail DNA, tail length, tail moment and comet length can be determined. Usually, % tail DNA and tail moment are preferred for assessing the DNA damage.

Rather than making use of the cell's own repair enzymes to reveal damage, we can achieve greater specificity and higher sensitivity by treating the DNA with purified repair enzymes which will convert particular lesions into breaks. Thus, Comet assay protocol can also be performed using different base or nucleotide excision repair enzymes **(Collins et al., 1997)**. The most commonly used repair enyme is formamidopyrimidine DNA glycosylase (Fpg) which recognizes and removes 8-oxodeoxyguanosine (8-oxoGua) and other oxidized purines. 8-oxoguanine glycosylase (OGG1) also recognizes 8-oxoGua. Endonuclease III (Endo III) deals with oxidized pyrimidines; and T4 endonuclease V is able to incise at sites of pyrimidine dimers. Digestion with these enzymes is carried out after the initial lysis step. The excision repair pathways act more slowly than strand break rejoining **(Collins & Horvathova, 2001)**, and samples should be taken over a period of a few hours.

Different versions of Comet assay are also used for different puposes. Neutral Comet assay is usually used for assessing double strand DNA breaks in sperm cells. On the other hand, a "Comet Chip" protocol, first introduced by Massachusetts Institute of Technology (MIT) Engelward Lab, is nowadays gaining significant importance as a high throughput DNA damage analysis platform. This new method is also used for evaluating DNA strand breaks, sites of DNA modification and interstrand crosslinks. A limitation of the traditional assay is that each sample requires a separate glass slide and image analysis is laborious and data is intensive, thus reducing throughput. This new technique uses microfabrication technologies to enable analysis of cells within a defined array, resulting in a >200 fold reduction in the area required per condition. Each well of a 96-well plate contains patterned microwells for single cell capture and DNA damage quantification. The "CometChip" can be used to analyze dozens of conditions on a single chip. The newly developed automated image analysis software is used for detection of DNA damage, thus greatly reducing analysis time. This new technology will enable the researchers to conduct both large scale epidemiological and clinical studies **(Engelward Lab, 2011)**.

A new technique "Comet fluorescence in situ hybridization (Comet FISH)" combines two well-established methods. The Comet assay comprises the basis of Comet-FISH and allows separation of fragmented from nonfragmented DNA and quantification of DNA damage and repair. FISH enables detection of specifically labeled DNA sequences of interest, including whole chromosomes. The combined technique of Comet-FISH is a modification of the Comet assay that inserts a hybridization step after unwinding and electrophoresis and permits the labeling of specific gene sequences or telomeres. Comet-FISH has been applied for detection of site-specific breaks in DNA regions that are relevant for development of various diseases, and has also been used to study the distribution of DNA damage and repair in the complete genome. Moreover, DNA sequence modifications can be detected in individual cells using Comet-FISH. The results from the Comet assay alone are only reflections of overall DNA damage. However, the addition of the FISH technique allows the assignment of the probed sequences to the damaged or undamaged part of the comet (tail or head, respectively) **(Schlörmann & Glei, 2009)**.

A spesific illustration for alkaline Comet assay methodology is shown in **Figure 2**. Different protocols of Comet assay in research field are given in **Figure 3**.

In this chapter, I will mainly focus on the genotoxicity of different environmental chemicals and both *in vivo* and *in vitro* protection studies by several selenocompounds, vitamins, and isothiocyanates (ITCs) against the toxicity of these compounds.

4. Protection studies using comet assay

4.1 Prevention of genotoxicity by selenocompounds

There is considerable interest in developing strategies that prevent genotoxicity and cancer with minimal risk or toxicity. Trace elements like selenium (Se) are of particular interest as it is the key component of antioxidant enzyme systems.

The requirement for Se and its beneficial role in human health have been known for several decades. Se is an essential trace element commonly found in grains, nuts, and meats and many years of research showed that that low, non-toxic supplementation with either organic and inorganic forms could reduce cancer incidence following exposure to a wide variety of carcinogens **(El-Bayoumy, 2004)**.

Along with its important role for the cellular antioxidant defense, Se is also essential for the production of normal spermatozoa and thus plays a critical role in testis, sperm, and reproduction **(Flohé, 2007)**. In the physiological dosage range, Se appears to function as an antimutagenic agent, preventing the malignant transformation of normal cells and the activation of oncogenes (**Schrauzer, 2000**). Although most of its chemopreventive mechanisms still remain unclear, the protective effects of Se seem to be primarily associated with its presence in the glutathione peroxidases (GPxs), which are known to protect DNA and other cellular components from damage by oxygen radicals **(Negro, 2008)**. Low activity of another important peroxidase, GPx4, can lead to reduction in reproduction **(Flohé, 2007)**.

Selenoenzymes are known to play roles in carcinogen metabolism, in the control of cell division, oxygen metabolism, detoxification processes, apoptosis induction and the functioning of the immune system oncogenes (**Schrauzer, 2000**). Several studies have determined the low activity of Se-containing cytosolic GPx, known as GPx1, as a substantial

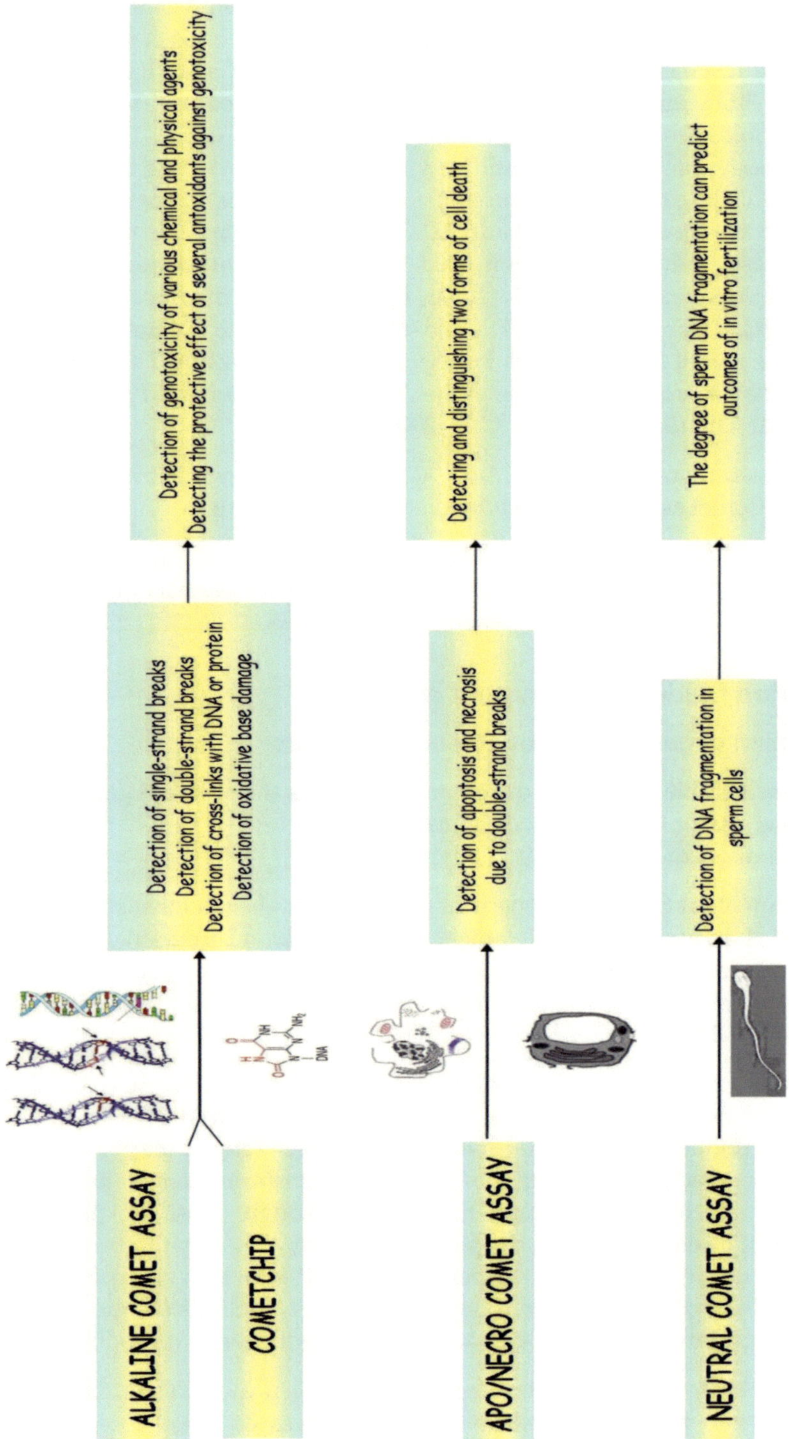

Fig. 2. Different protocols of Comet assay in research field

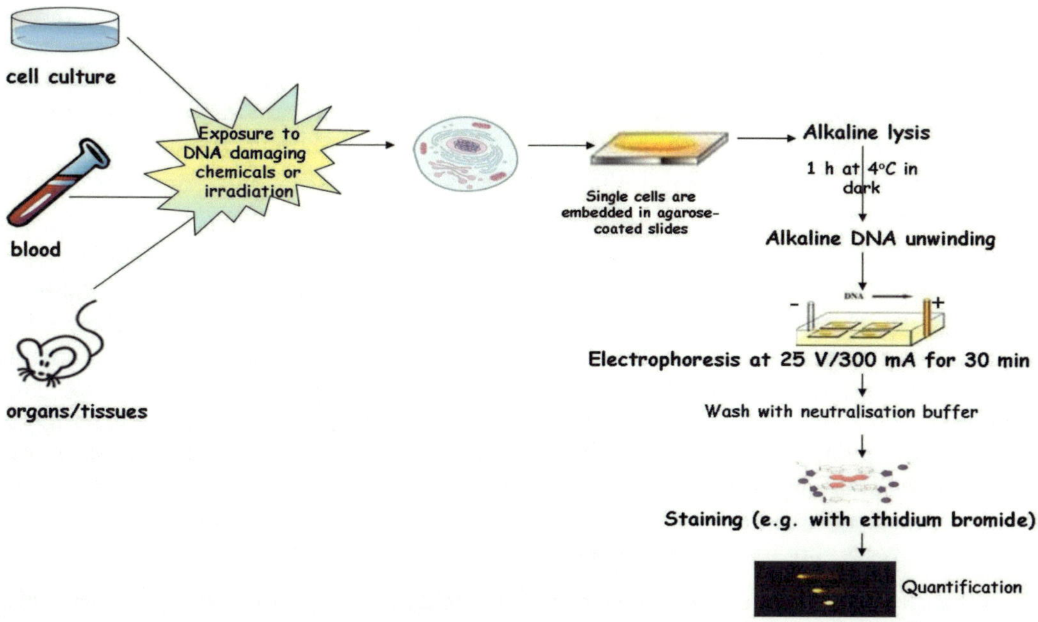

Fig. 3. Alkaline Comet assay methodology

factor in cancer risk **(Esworthy et al., 1985)**. Other modes of action, either direct or indirect, may also be operative, such as the partial retransformation of tumor cells and the inactivation of oncogenes. However, the effects of Se in the physiological dosage range are not attributable to cytotoxicity, allowing Se to be defined as a genuine nutritional cancer-protecting agent **(Yu et al., 1990)**. On the other hand, selenocompounds such as selenodiglutathione, methylselenol, selenomethionine (SM), and Se-methylselenocysteine might affect the metabolism of carcinogens, thus preventing initiation of carcinogenesis **(Gopalakrishna & Gundimeda, 2001)**. These compounds might also restrict cell proliferation by inhibiting protein kinases and by halting phases of the cell cycle that play a central part in cell growth, tumor promotion, and differentiation **(Brinkman et al., 2006)**. A further possible mechanism of action is enhancement of the immune system by stimulating the cytotoxic activities of natural killer cells and lymphokine activated killer cells to act against cancer cells **(Combs, 1998)**. The anticarcinogenic effects of Se are counteracted by Se-antagonistic compounds, and elements **(Schrauzer, 2000)**.

For maximal utilization of its cancer-protective potential, Se supplementation should start early in life and be maintained over the entire lifespan **(Schrauzer & White, 1978; Persson-Moschos et al., 1998; Schrauzer, 2000)**. In addition, exposure to Se antagonists and carcinogenic risk factors should be minimized by appropriate dietary and lifestyle changes **(Schrauzer, 1976; Schrauzer, 1977)**. Because geographical studies done in the 1970s reported a possible inverse association between Se and cancer mortality, epidemiological studies have focused on investigating the anticarcinogenic properties of this nutrient **(Brinkman et al., 2006)**. Two key findings that emerged from these early studies were the inverse association between Se and cancer seemed to be both sex and organ specific **(Li et al., 2004)**.

A larger difference in the reduced death rates was reported for men than for women in regions with high levels of Se, and mortality was significantly lower for some types of cancer **(Shamberger et al., 1976; Clark et al., 1991)**. Higher blood levels of Se have been associated with a lower risk of many types of neoplasia, including prostate, lung, colorectal, and possibly bladder, although the data are inconsistent. A significant 39% decreased risk of bladder cancer associated with high levels of Se by combining results from seven epidemiologic studies, conducted in different populations, which applied individual levels of Se measured in serum or toenails **(Brinkman et al., 2006)**.

Supra-physiological levels of sodium selenite (SS) in the presence of polythiols have oxidative properties that might have an anticancer effect by increasing the vulnerability of cancer cells to destruction. It was stated that Se, independent of type (organic/inorganic), can alter several genes to prevent cancer. High doses of Se might upregulate phase II detoxification enzymes, some Se-binding proteins, and some apoptotic genes, and downregulate phase I activating enzymes and cell proliferation genes **(El Bayoumy & Sinha, 2005)**. Inhibition of carcinogen–DNA adducts formation and induction of apoptosis by high doses of Se suggests that protection occurs at both the initiation and post-initiation phases of carcinogenesis **(El Bayoumy & Sinha, 2005)**. However, at lower physiological doses, Se prevents apoptosis, and induces DNA repair **(Longtin, 2003)**.

The literature agrees on the protective effect of Se evaluated with the Comet assay towards a variety of chemical or physical toxic agents. However, it remains inconclusive which is/are the most suitable Se compound/s to prevent DNA damage and which doses should be used to observe protection. In this chapter, the protective effects of both inorganic and organic selenocompounds, against phthalate and radiation toxicity will be discussed.

4.1.1 Prevention of phthalate genotoxicity by selenocompounds

Phthalate esters are a widespread class of peroxisome proliferators (PPs) and endocrine disruptors. They have attracted substantial attention due to their high production volume and use in a variety of polyvinyl chloride (PVC)-based consumer products **(Akingbemi et al., 2001; Grande et al., 2006)**.

Uses of the various phthalates mainly depend on their molecular weight (MW). Higher MW phthalates, such as di(2-ethylhexyl) phthalate, (DEHP), are used in construction materials and in numerous PVC products including clothing (footwear, raincoats), food packaging, children products (toys, grip bumpers), and medical devices **(Heudorf et al., 2007)**, while relatively lower MW phthalates like di-methyl phthalate (DMP), di-ethyl phthalate (DEP), and di-n-butyl phthalate (DBP) are mainly used as odor/color fixatives or as solvents and in cosmetics, insecticides and pharmaceuticals, but are also used in PVC **(Heudorf et al., 2007)**.

Phthalate migrate out from PVC-containing items into food, air, dust, water, and soils and create human exposure in various ways **(Clark et al., 2003)**. Increasing number of studies on human blood and urine have revealed the ubiquitous phthalate exposure of consumers in industrialized countries **(Wormuth et al., 2006, Frederiksen et al., 2008; Frederiksen et al., 2010; Janjua et al., 2011 , Durmaz et al., 2010)**.

DEHP is the most important phthalate derivative with its high production, use and occurrence in the environment. It is mainly used in PVC plastics in the form of numerous consumer and personal care products and medical devices **(Doull et al., 1999)**. The biological effects of DEHP are hence of major concern but so far elusive. Although, the main mechanism underlying hepatocarcinogenicity of phthalates is not fully elucidated, ROS are thought to be associated with the mechanism of tumorigenesis by PPs, including DEHP. This assumption is based to a fact that various proteins that are induced by DEHP in liver parenchymal cells (peroxisomes, mitochondria and microsomes) are prone to formation of H_2O_2 and other oxidants. Besides, activation of metabolizing enzymes and peroxisome proliferator-activated receptor α (PPARα) might be other substantial factors leading to high intracellular ROS production **(O'Brien et al., 2005; Gazouli et al., 2002)**. However, the mechanisms by which phthalates and particularly DEHP exert toxic effects in reproductive system are not yet fully elucidated. Irreversible and reversible changes in the development of the male reproductive tract like vimentin collapse of Sertoli cells as well as apoptosis of germ cells, effects on sex hormones (mainly on testosterone) as well as follicle stimulating hormone (FSH) and luteinizing hormone (LH), histopathological changes in testis and sperm anomalies were observed with phthalate exposure **(Corton & Lapinskas, 2005; Foster et al., 2001; Erkekoglu et al., 2011a; Erkekoglu et al., 2011b; Kasahara et al, 2002; Noriega et al., 2009)**. Most of the toxic effects were related to its antiandrogenic potential **(Ge et al., 2007)**. A PPARα-mediated pathway based on its peroxisome proliferating (PP) activity **(Gazouli et al., 2002)**, and activation of metabolizing enzymes have also been suggested **(O'Brien et al., 2005)**. While the induction of an oxidative stress may represent a common mechanism in endocrine disruptor–mediated dysfunction, especially on testicular cells **(Latchoumycandane et al., 2002)**, recent studies are also providing supporting evidences for such an effect with DEHP and its major metabolite, mono(2-ethylhexyl)phthalate (MEHP) **(Erkekoglu et al. 2010a; Erkekoglu et al. 2010b; Erkekoglu et al. 2011c; Fan et al. 2010)**. Thus, the primary targets for the DEHP and MEHP are the Sertoli and Leydig cells of testis. In several studies, it was shown that DEHP caused disruption in the function of both cell types. In fact, **Richburg and Boekelheide (1996)** demonstrated histopathological disturbances and alterations of cytoplasmatic distribution of vimentin in Sertoli cells in testis of 28-day-old Fisher rats after a single oral dose of MEHP (2000 mg/kg). Administration of MEHP to Wistar rats at a single oral dose (400 mg/kg bw) was toxic to Sertoli cells and caused detachment of germ cells **(Dalgaard et al., 2000). Tay et al. (2007)** reported vimentin disruption in MEHP-treated C57Bl/6N mice, and gradual disappearance of vimentin in Sertoli cell cultures as time and dose increased. We have also reported that in DEHP-treated rats, significant disruption and collapse of vimentin filaments and disruption of seminiferous epithelium in Sertoli cells was observed **(Erkekoglu et al., 2011b)**. Among several others, an earlier data has demonstrated the increase of ROS generation and depletion in antioxidant defenses by DEHP treatment in rat testis **(Kasahara et al., 2002)**. Our recent studies on MA-10 Leydig **(Erkekoglu et al., 2010b)** and LNCaP human prostate cells **(Erkekoglu et al., 2010a)** have also produced comprehensive data suggesting that at least one of the mechanisms underlying the reproductive toxicity of DEHP is the induction of intracellular ROS. The data of **Fan et al. (2010)** have also suggested oxidative stress as a new mechanism of MEHP action on Leydig cells steroidogenesis *via* CYP1A1-mediated ROS stress. On the other hand, in rats exposed to 1000 mg/kg DEHP for 10 days, we observed that this particular phthalate induced oxidative stress in rat testis, as evidenced by significant decrease in GSH/GSSG redox ratio (>10-fold) and marked increase in TBARS levels **(Erkekoglu et al., 2011d)**.

Several strategies have been attempted to prevent the oxidative stress caused by toxic chemicals and the use of antioxidant vitamins has been the most common approach. **Ishihara et al. (2000)** showed that supplementation of rats with vitamin C and E protected the testes from DEHP-gonadotoxicity. **Fan et al. (2010)** reported that the increase in ROS generation with MEHP exposure in MA-10 cells was inhibited by N-acetylcysteine (NAC). In the above mentioned *in vitro* studies **(Erkekoglu et al., 2010a; Erkekoglu et al., 2010b)**, we demonstrated that Se supplementation in either organic form (SM, 10 μM) or in inorganic form (SS, 30 nM) was highly protective against the cytotoxicity, ROS producing and antioxidant status-modifying effects of DEHP and MEHP in both MA-10 Leydig and LNCaP cells.

Concerning LNCaP cells, we observed that DEHP had a flat dose–cell viability response curve while MEHP showed a very steep dose–response curve and the cytotoxicity of the MEHP was much higher than that of the parent compound. On the other hand, we determined that both organic and inorganic Se supplementation increased resistance to DEHP and MEHP cytotoxicity. From these data, the doses of DEHP and MEHP to be used for the antioxidant status measurements and Comet assay were chosen as close to IC_{50} values and were 3 mM for DEHP and 3 μM for MEHP. We demonstrated that MEHP was the main active form in LnCAP cells with an almost ~1000- fold higher cytotoxicity than the parent compound. Intracellular ROS production showed marked increases with both DEHP and MEHP treatment; however the effect of MEHP was much higher. Both selenocompounds were partially effective in reducing intracellular ROS production. For the antioxidant enzymes, both DEHP and MEHP caused substantial decreases in GPx1 activity (~3-fold, and ~4-fold, respectively) compared to control cells. However, there was no significant difference between the effects of the two phthalate derivatives. Se supplementation with either SS or SM effectively countered the effect of DEHP by completely restoring the activity up to the control level (NT-C) or even higher. In the case of MEHP treatments, both SS and SM supplementations significantly restored the effect of 3 μM MEHP on GPx1 activity, providing ~2-fold increase. For thioredoxin reductase (TrxR) activity, DEHP did not cause a change compared to control; however, MEHP caused a marked increase. Se supplementation in both organic and inorganic forms increased the TrxR activity almost up to the levels of SS and SM supplemented cells alone. However, no changes were observed with both of the phthalates in glutathione S-transferase (GST) activity and total glutathione (GSH) levels. On the other hand, using alkaline Comet assay, we have demonstrated that in LnCAP cells both DEHP and MEHP produced significant DNA damage as evidenced by increased tail % intensity (~2.9-fold and ~3.2-fold, respectively), and tail moment (~2.4-fold and ~2.6-fold, respectively) compared to NT LNCaP cells. The overall difference between the DNA damaging effects of the parent compound and the metabolite was insignificant. Se supplementation itself did not cause any alteration in the steady-state levels of the biomarkers of DNA damage in LNCaP cells, whereas the presence of Se either in SS or SM form reduced the genotoxic effects of DEHP and MEHP as evidenced by significant (~30%) decreases in tail % intensity. These results thus indicated that the Se with the doses and forms used in this study was not genotoxic, but showed antigenotoxic activity against the genotoxicity of DEHP and MEHP. However, the protective effect of Se with the doses used in this study was not complete. Tail intensity remained ~90% and ~80% higher than that of NT-C in SS/DEHP-T and SM/DEHP-T cells, respectively. Similarly, in SS/MEHP-T and SM/ MEHP-T cells, tail intensities were still ~95% and ~120% high compared to NT-C cells. On the other hand, the extent of tail moment increase induced by DEHP was reduced ~30% with SS and ~18% with SM supplementations, and the tail moment induced by MEHP was reduced ~24% with SS supplementation; however, none of these were statistically significant. Only SM supplementation provided a significant (~34%) reduction in the tail moment induced by MEHP. But again, tail moments remained ~64 and ~95% higher than that of NT-C in SS/DEHP-T and SM/DEHP-T cells, respectively; similarly in SS/MEHP-T and SM/MEHP-T cells, tail moments were still ~94 and ~69% high compared to NT-C cells. In all cases, protective effects of SS and SM were not significantly different than each other **(Erkekoglu et al., 2010a)**.

For Leydig MA-10 cells, The IC_{50} values for DEHP and MEHP were again found to be ~3 mM and ~3 µM, respectively. Se supplementation of the cells with either SS (30 nM) or SM (10 µM) was protective against the cytotoxic effects of DEHP, and MEHP. Intracellular ROS production showed substantial increases with both of the phthalates where the effect of MEHP was much more pronounced. SS and SM showed partial protection against the ROS increment for both the phthalates. In cells exposed to DEHP or MEHP, GPx1 and TrxR activities decreased significantly. Se supplementation either with SS or SM in DEHP-exposed cells was able to enhance the both of the selenoenzyme activities. Moreover, GST activity also decreased significantly with both of the phthalates. However, Se supplementation in both of the forms was not effective in restoring GST activity. GSH levels also decreased significantly in DEHP and MEHP treated Leydig cells while Se supplementation in both forms provided significant restoration in both groups. On the other hand, both DEHP and MEHP produced high level of DNA damage as evidenced by significantly increased tail % intensity (~3.4-fold and ~3.8-fold, respectively), and tail moment (~4.2-fold and ~3.8-fold, respectively) compared to non-treated MA-10 cells. The difference between the DNA damaging effects of the parent compound and the metabolite was insignificant. Se supplementation itself did not cause any alteration on the steady state levels of the DNA damage biomarkers of MA-10 cells. But Se was highly effective to decrease the genotoxic effects of phthalate esters. Increased tail % intensities by DEHP and MEHP exposure were lowered ~50–55% with SS supplementation, whereas SM treatment provided ~30–40% protection. SS decreased the tail moments of the DEHP- or MEHP-exposed cells by ~55–65%, whereas the protective effect of SM on tail moments was significantly lower than SS as being ~45% and ~34% for the effects of DEHP and MEHP, respectively. However, both SS and SM reduced the tail moments of the DEHP- and MEHP-exposed cells down to the levels that were not significantly different than that of control cells **(Erkekoglu et al., 2010b)**.

4.1.2 Prevention of radiation genotoxicity by selenocompounds

Ultraviolet (UV) light is electromagnetic radiation with a wavelength shorter than that of visible light, but longer than X-rays, in the range 10-400 nm, and energies from 3 -124 eV. UV light is found in sunlight, can be emitted by electric arcs and specialized lights such as black lights. It can cause chemical reactions, and it causes many substances to glow or fluoresce. Most UV is classified as non-ionizing radiation **(Müller et al., 1998, Griffiths et al., 1998; Grossman et al., 1988)**.

The toxic effects of UV from natural sunlight and therapeutic artificial lamps are a major concern for human health. The major acute effects of UV irradiation on normal human skin comprise sunburn inflammation erythema, tanning, and local or systemic immunosuppression. On the other hand, UV irradiation present in sunlight is an environmental human carcinogen. There is considerable evidence that UV is implicated in skin carcinogenesis and the risk of cutaneous cancers has increased during the last decade due to increase of sun exposure. For a long time, ultraviolet B radiation (UVB: 290-320 nm) have been considered to be the more efficient wavelength in eliciting carcinogenesis in human skin. It is today clear that ultraviolet A (UVA, 320-400 nm), especially UVA_1 (340-400 nm) also participate to photo-carcinogenesis. It penetrates deeply, but it does not cause sunburn. One of molecular mechanisms in the biological effects of UV is the induction of ROS directly or through endogenous photosensitization reactions. UVA radiation mainly acts *via* this production of ROS and the subsequent oxidative stress seems to play a crucial role in the deleterious effects of UVA. UVA does not damage DNA directly like UVB and UVC, but it can generate highly reactive chemical intermediates, such as hydroxyl and oxygen radicals, which in turn can damage DNA and lead to the formation of 8-oxoGua **(Ridley et al., 2009)**. UVB light can cause direct DNA damage. The radiation excites DNA molecules in skin cells, causing aberrant covalent bonds to form between adjacent cytosine bases, producing a dimer. When DNA polymerase comes along to replicate this strand of DNA, it reads the dimer as "AA" and not the original "CC". This causes the DNA replication mechanism to add a "TT" on the growing strand. This mutation can result in cancerous growths, and is known as a "classical C-T mutation". The mutations caused by the direct DNA damage carry a UV signature mutation that is commonly seen in skin cancers. The mutagenicity of UV radiation can be easily observed in bacterial cultures. This cancer connection is one reason for concern about ozone depletion and the ozone hole. UVB causes some damage to collagen, but at a very much slower rate than UVA. Fortunately, the skin possesses a wide range of inter-linked antioxidant defense mechanisms to protect itself from damage by UV-induced ROS. However, the capacity of these systems is not unlimited; they can be overwhelmed by excessive exposure to UV and then ROS can reach damaging levels. An interesting strategy to provide photoprotection would be to support or enhance one or more of these endogenous systems **(Béani, 2001)**.

There is limited number of studies in literature concerning the protective effect of selenocompounds on UV-caused genotoxicity. In a study by **Emonet-Piccardi et al. (1998)**, the researchers determined the protective effects of NAC (5 mM), SS (0.6 μM) or zinc chloride ($ZnCl_2$, 100 μM) against UVA radiation in human skin fibroblasts using Comet assay. The cells were incubated with NAC, SS or $ZnCl_2$ and then UVA was applied as 1 to 6 J/cm2 to the cells. The tail moment increased by 45% (1 J/cm^2) to 89% (6 J/cm^2) in non-supplemented cells ($p<0.01$). DNA damage was significantly prevented by NAC, SS and $ZnCl_2$, with similar efficiency from 1 to 4 J/cm^2. For the highest UVA dose (6 J/cm^2), SS and $ZnCl_2$ were more effective than NAC.

In a study assessing the effects of pretreatment of primary human keratinocytes with Se on UV-induced DNA damage, cells were irradiated with UVB from FS-20 lamps and were subjected to Comet assay. Comet tail length due to UVB-induced T4 endonuclease V-sensitive sites (caused by cyclopyrimidine dimers, CPDs) increased to 100% immediately after irradiation (time 0). After 4 h, 68% of the damage remained and after 24 h, 23% of the damage was still present. Treatment with up to 200 nM SM or 50 nM SS had no effect on CPD formation or rates of repair, or on the number of excision repair sites as measured by cytosine arabino furanoside and hydroxyurea treatment. However, both SS and SM protected against oxidative damage to DNA as measured by formation of formamidopyrimidine (FaPy) glycosylase-sensitive sites, which are indicative of 8-oxoGua photoproduct formation. Preincubation for 18 h with 50 nM SS or with 200 nM SM abolished the UVB-induced increase in comet length. The researchers concluded that both of selenocompounds were protective against UVB-induced oxidative damage in human keratinocytes; however they did not protect from formation of UVB-induced excision repair sites **(Rafferty et al., 2003)**.

Diphenyl diselenide (DPDS) is an electrophilic reagent used in the synthesis of a variety of pharmacologically active organic Se compounds. Studies have shown its antioxidant, hepatoprotective, neuroprotective, anti-inflammatory, and antinociceptive effects. In a study by **Rosa et al. (2007)**, the researchers used a permanent lung fibroblast cell line derived from Chinese hamsters and investigated the antigenotoxic and antimutagenic properties of DPDS. In the clonal survival assay, at concentrations ranging from 1.62 to 12.5 µM, DPDS was not cytotoxic, while at concentrations up to 25 µM, it significantly decreased survival. The treatment with this DPDS at non-cytotoxic dose range increased cell survival after challenge with H_2O_2, methyl-methanesulphonate, and UVC radiation, but did not protect against 8-methoxypsoralen plus UVA-induced cytotoxicity. In addition, the treatment prevented induced DNA damage, as verified in the Comet assay. The mutagenic effect of these genotoxic agents, as measured by the micronucleus test, similarly attenuated or prevented cytotoxicity and DNA damage. Treatment with DPDS also decreased lipid peroxidation levels after exposure to H_2O_2, MMS, and UVC radiation, and increased GPx1 activity in the cells. The results of this study demonstrated that DPDS at low concentrations presents antimutagenic properties, which are most probably due to its antioxidant properties **(Rosa et al., 2007)**.

4.2 Prevention of genotoxicity by vitamins

4.2.1 Ascorbic acid

Diet should include components such as vitamins and flavonoids and the antioxidant capacity of body is directly linked to the diet. Vitamins like ascorbic acid (vitamin C, AA) are important antioxidants. About 90% of AA in the average diet comes from fruits and vegetables **(Vallejo et al., 2002)**.

AA is a water soluble dietary antioxidant that plays an important role in controlling oxidative stress **(Vallejo et al., 2002)**. Most importantly, AA is a mild reducing agent. For this reason, it degrades upon exposure to oxygen, especially in the presence of metal ions and light. It can be oxidized by one electron to a radical state or doubly oxidized to the stable form called "dehydroascorbic acid". Typically it reacts with oxidants such as ROS, such as the •OH formed from H_2O_2. Hydroxyl radical is the most detrimental species, due to its high interaction with nucleic acids, proteins, and lipids. AA can terminate these chain radical reactions by electron transfer. AA is special because it can transfer a single electron, owing to the stability of its own radical ion called "semidehydroascorbate". The oxidized forms of AA are relatively unreactive, and do not cause cellular damage. However, being a good electron donor, high concentrations of AA in the presence of free metal ions can not only promote, but also initiate free radical reactions, thus making it a potentially dangerous pro-oxidative compound in certain metabolic contexts **(Choe and Min, 2006; Blokhina et al., 2003)**.

AA is able to suppress ROS efficiently *in vivo;* thus, reducing DNA damage to tumor suppressor genes which might explain its anticancer properties **(Crott et al., 1999)**. *In vitro,* AA acts in conjunction with vitamin E, present in lipid membranes, to quench free radicals and prevent lipid peroxidation **(Niki et al., 1995)**.

In the Comet assay, evidence of protection was seen against the effects of H_2O_2 when AA was present at low concentrations (up to 1 mM); by contrast, there was exacerbation at higher doses (>5 mM) **(Harréus et al., 2005; Anderson et al., 1994; Anderson and Phillips, 1999)**. After 2–4 h after intake, AA provided significant protection to the DNA of isolated lymphocytes when challenged with H_2O_2 **(Panayiotidis and Collins, 1997)**. Besides, AA was found to be protective against H_2O_2-induced DNA damage (DNA strand breaks and oxidized purines/pyrimidines) in human hepatoma cells (HepG2 cells) **(Arranz et al., 2007a, Arranz et al., 2007b)**. In intervention studies, supplementation of 100 mg/day to 50–59 year-old men led to a decrease in oxidative base damage and enhanced resistance against oxidative damage **(Duthie et al., 1996)**. In a long-term study, the antioxidant effect of AA was studied by measuring oxidative DNA damage and DNA repair in blood cells with the Comet assay. Male smokers were given AA (2 × 250 mg) daily in the form of plain or slow release tablets combined with plain release vitamin E (2 × 91 mg), or placebo for 4 weeks. The results of this study suggested that long-term AA supplementation at a high dose, i.e. 500 mg, together with vitamin E in moderate dose, i.e. 182 mg, decreased the steady-state level of oxidative DNA damage in lymphocytes of smokers **(Møller et al., 2004)**. In a study performed on gastric epithelial cells SGC-7901, both AA and SS were found to be protective against Helicobacter pylori-induced oxidative stress and genotoxicity **(Shi and Zheng, 2006)**.

AA was also tested for its protective effects against the genotoxicity of several toxic chemicals, drugs and metals. Using peripheral blood lymphocytes, AA as well as vitamin E were found to be protective against benzo(a)pyrene [B(a)P]-induced DNA damage **(Gajecka et al., 1999)**. In rats, using Comet assay, the genotoxicity of p-dimethylaminoazobenzene (DAB), a hepatocarcinogen, was found to be decreased by AA administration. Besides, vitamin A, vitamin E and combination of these three vitamins were also found be effective against the toxicity **(Velanganni et al., 2007)**. A significant increase in the levels of protein oxidation, DNA strand breaks, and DNA-protein cross-links was observed in blood, liver, and kidney of rats exposed to arsenic (100 ppm in drinking water) for 30 days. Co-administration of AA and vitamin E in the form of α-tocopherol to arsenic-exposed rats showed a substantial reduction in the levels of arsenic-induced oxidative products of protein and DNA **(Kadirvel et al., 2007)**. For anti-cancer drugs there are inconclusive results. AA was protective against epirubicin- and adriamycin-induced genotoxicity in cancer patients **(Mousseau et al., 2005; Shimpo et al., 1991)**. However, there was no evidence of a protective effect of AA against the damage caused by bleomycin **(Anderson & Phillips, 1999)**. Moreover, results were also inconclusive when oestrogenic compounds were co-incubated with AA (0.5 and 1 mM) in isolated lymphocytes showing no common pattern in the responses **(Anderson et al., 2003)**.

Nitrosamines (NOCs) can be formed endogenously from nitrate and nitrite and secondary amines under certain conditions such as strongly acidic pHs of the human stomach **(Jakszyn and Gonzalez, 2006; Bofetta et al., 2008; Tricker, 1997)**. Humans are exposed to a wide range of NOCs from diet (cured meat products, fried food, smoked preserved foods, foods subjected to drying, pickled and salty preserved foods), tobacco smoking, work place and drinking water **(Bartsch and Spiegelhalder, 1996; Bofetta et al., 2008; Jakszyn & Gonzalez, 2006; Tricker, 1997)**.

In several studies, AA was found to be protective against NOC-induced genotoxicity using Comet assay. In a study by **Robichová et al. (2004)**, the researchers used three cell lines (HepG2, V79 and VH10) to determine the genotoxic effect of N-Nitrosomorpholine (NMOR). NMOR was found to induce DNA damage in a dose-dependent manner but the extent of DNA migration in the electric field was unequal in the different cell lines. Although the results obtained by Comet assay confirmed the genotoxicity of NMOR in all cell lines studied, the number of chromosomal aberrations was significantly increased only in HepG2 and V79 cells, while no changes were observed in VH10 cells. In HepG2 cells pre-treated with vitamin A, vitamin E and AA the researchers found a significant decrease of % tail DNA induced by NMOR. The reduction of the clastogenic effects of NMOR was observed only after pretreatment with Vitamins A and E. AA did not alter the frequency of NMOR-induced chromosomal aberrations under the experimental conditions of this study. In a study by **Arranz et al. (2007)**, HepG2 cells were simultaneously treated with AA and the genotoxic effects of the N-nitrosamines, namely, N-nitrosodimethylamine (NDMA), N-nitrosopyrrolidine (NPYR), N-nitrosodibutylamine (NDBA) or N-nitrosopiperidine (NPIP) were reduced in a dose-dependent manner. At concentrations of 1-5 μM AA, the protective effect was higher towards NPYR-induced oxidative DNA damage (78-79%) than against NDMA (39-55%), NDBA (12-14%) and NPIP (3-55%), in presence of Fpg enzyme. However, a concentration of 10 μM AA led to a maximum reduction in NDBA (94%), NPYR (81%), NPIP (80%) and NDMA (61%)-induced oxidative DNA damage, in presence of Fpg enzyme. The greatest protective effect of AA (10 μM) was higher towards NDBA-induced oxidative DNA damage. The authors concluded that one feasible mechanism by which AA exerted its protective effect could be that it might interact with the enzyme systems catalyzing the metabolic activation of the N-nitrosamines, blocking the production of genotoxic intermediates.

In our previous studies performed using Comet assay, we have shown that AA was highly protective in HepG2 cells against the genotoxicity of both nitrite and three important NOC, namely NDMA, Nitrosodiethylamine (NDEA) and NMOR **(Erkekoglu et al., 2010c)**. Nitrite was added as 20 μM, NDMA as 10 mM, NDEA as 10 mM and NMOR as 3 mM to the medium for 30 min with or without AA (10 μM). When compared to untreated cells, nitrite ($p>0.05$), NDMA ($p<0.05$), NDEA ($p<0.05$), and NMOR ($p<0.05$) raised the tail intensity up to 1.18-, 3.79-, 4.24-, and 4.16-fold, respectively. AA was able to reduce the tail intensity caused by nitrite, NDMA, NDEA, and NMOR to 34%, 59%, 44%, and 44%, respectively, and these reductions were statistically significant when compared to each individual toxic compound applied group (all, $p<0.05$). Besides, nitrite, NDMA, NDEA, and NMOR increased the tail moment up to 1.94, 6.04, 6.05, and 5.70, respectively. AA (10 μM) enabled a reduction of 27%, 30%, 23%, and 22% in the tail moment in nitrite, NDMA, NDEA, and NMOR-treated cells, respectively, and these reductions were statistically significant when compared to each individual toxic compound applied group (all, $p<0.05$) **(Erkekoglu et al., 2010c)**.

In an experiment performed on multiple organs of mice, the genotoxicity of endogenously formed N-nitrosamines from secondary amines and sodium nitrite was evaluated in, using Comet assay. Dimethylamine, proline, and morpholine were simultaneously with sodium nitrite and the stomach, colon, liver, kidney, urinary bladder, lung, brain, and bone marrow were sampled 3 and 24 h after these compounds had been ingested. DNA damage was observed mainly in the liver following simultaneous oral ingestion of these compounds **(Ohsawa et al., 2003)**.

4.2.2 Vitamin E

Vitamin E refers to a group of fat-soluble compounds that include both tocopherols and tocotrienols **(Brigelius- Flohé and Traber, 1999)**. Naturally occurring vitamin E exists in eight chemical forms (alpha-, beta-, gamma-, and delta-tocopherol and alpha-, beta-, gamma-, and delta-tocotrienol) that have varying levels of biological activity. Alpha- (or α-) tocopherol is the only form that is recognized to meet human requirements. γ-tocopherol is the most common in the North American diet **(Traber, 1998)**. γ-tocopherol can be found in corn oil, soybean oil, margarine and dressings **(Bieri and Evarts, 1974; Brigelius-Flohé & Traber, 1999)**. The most biologically active form of vitamin E, α-tocopherol, is the second most common form of vitamin E in the North American diet and perhaps the common form in European and Mediterranean diet. This variant of vitamin E can be found most abundantly in wheat germ oil, sunflower, and safflower oils **(Reboul et al., 2006)**. Serum concentrations of α-tocopherol depend on the liver, which takes up the nutrient after the various forms are absorbed from the small intestine. The liver preferentially resecretes only α-tocopherol *via* the hepatic α-tocopherol transfer protein **(Traber, 2006)**. As a result, blood and cellular concentrations of other forms of vitamin E are lower than those of α -tocopherol and have been the subjects of less research **(Sen et al., 2006; Dietrich et al., 2006)**.

Vitamin E is an important vitamin for preventing lipid peroxidation and it has many reported health effects and is recognized as the most important lipid-soluble, chain-breaking antioxidant in the body **(Fenech & Ferguson, 2001)**. This vitamin might have a protective role against chromosomal damage, DNA oxidation and DNA damage. Vitamin E has also been reported to play a regulatory role in cell signaling and gene expression. Epidemiological studies showed that high blood concentrations of vitamin E were associated with a decreased risk of certain cancers. This effect might emerge in part, by enhancing immune function **(Frank, 2005; Claycombe & Meydani, 2001, Salobir et al., 2010)**. Vitamin E might also block the formation of carcinogenic NOCs formed in the stomach from nitrite and secondary amines **(Weitberg and Corvese, 1997)**.

Vitamin E was shown to prevent the genotoxicity of several environmetal chemicals and several drugs. Nitrosamine toxicity was shown to be protected by vitamin E. Hepatocytes freshly isolated from rats fed with a common diet or a vitamin A- or vitamin E-supplemented diet were assayed for sensitivity to DNA breakage and cytogenetic changes induced by several carcinogens including NMOR. NMOR was the only agent that induced DNA breaks, chromosomal aberrations, and micronuclei. Both vitamin A and vitamin E were able to reduce these effects, and the protection by vitamin A was more pronounced **(Slamenová, 2001)**. On the other hand, vitamin E was also found to be protective against the genotoxic properties of one of the most commonly used herbicides, atrazine, in male rats. Atrazine caused a significant increase in tail length of comets from blood and liver cells compared to controls. Co-administration of vitamin E (100 mg/kg bw) along with atrazine resulted in decrease in tail length of comets as compared to the group treated with atrazine alone. Besides, micronucleus assay revealed a significant increase in the frequency of micro-nucleated cells (MNCs) following atrazine administration. In the animals administrated vitamin E along with atrazine, there was a significant decrease in percentage of micronuclei as compared to atrazine treated rats. The increase in frequency of micronuclei in liver cells and tail length of comets confirm genotoxicity induced by atrazine in blood and liver cells. In addition, the findings clearly demonstrated protective effect of vitamin E in attenuating atrazine-induced DNA damage **(Singh et al., 2008)**. In mouse retina, both vitamin E and AA were shown to markedly reduce the cell apoptosis, lipid peroxidation and DNA damage caused by the organophosphorus insecticide chlorpyrifos **(Yu et al, 2008)**. Vitamin E supplementation was also protective against pyrethroid (both cypermethrin and permethrin), induced lymphocyte DNA damage **(Gabbianelli et al., 2004)**.

Vitamin E was also shown to reduce the genotoxic effects of the anti-HIV drug stavudine **(Kaur & Singh, 2007)** and the antibiotic, ciprofloxacin **(Gürbay et al., 2006)**. In a study performed on primary culture of rat astrocytes, the researchers incubated the cultured cells with various concentrations of ciprofloxacin, and DNA damage was monitored by Comet assay. The results showed a concentration-dependent induction of DNA damage by ciprofloxacin. Pretreatment of cells with Vitamin E for 4 h provided partial protection against this effect **(Gürbay et al., 2006)**.

Vitamin E was also found to be protective against the toxicity of anesthesics. In a study performed with sevoflurane on rabbits, vitamin E and SS were administered 15 days before the anesthesia treatment and blood samples were collected after 5 days of treatment with sevoflurane. Both vitamin E and SS administration prevented the sevoflurane induced genotoxicity in the lymphocytes **(Kaymak et al., 2004)**.

Several supplementation studies have also been performed both vitamin E and AA. Supplementation of the diet for 12 weeks with AA and vitamin E resulted in a significant decrease in the DNA damage in diabetic patients **(Sardaş et al., 2001)**. Vitamin E supplementation was also shown to reduce oxidative DNA damage in both hemodialysis and peritoneal dialysis patients **(Domenici et al., 2005)**. In another study performed on 26 healthy subjects, a daily drink including 1.8 mg vitamin E was administered for 26 days and blood samples were obtained. The DNA damage was measured in the lymphocytes subjected to oxidative stress and genotoxicity was found to be significantly lower (42%, $p<0.0001$) **(Porrini et al., 2005)**.

There are few protection studies with vitamin E against radiation toxicity using Comet assay. An *in vitro* study on dermal microvascular endothelial cells by the same research group, gamma- irradiated cells at 3 and 10 Gy, and 0.5 mM of pentoxifylline (PTX) and trolox (Tx, 6-hydroxy-2,5,7,8-tetramethylchroman-2-carboxylic acid, a water-soluble derivative of vitamin E), were added either before (15 min) or after (30 min or 24 h) irradiation. ROS measured by the dichlorodihydrofluorescein diacetate assay, and DNA damage, assessed by the Comet and micronucleus assays, were measured at different times after exposure (0 - 21 days). The PTX/Tx treatment decreased the early and delayed peak of ROS production by a factor of 2.8 in 10 Gy-irradiated cells immediately after irradiation and the basal level by a factor of 2 in non-irradiated control cells. Moreover, the level of DNA strand breaks, as measured by the comet assay, was shown to be reduced by half immediately after irradiation when the PTX/Tx treatment was added 15 min before irradiation. However, unexpectedly, DNA strand breaks was decreased to a similar extent when the drugs were added 30 min after radiation exposure. This reduction was accompanied by a 2.2- and 3.6-fold higher yield in the micronuclei frequency observed on days 10 and 14 post-irradiation, respectively. These results suggest that oxidative stress and DNA damage induced in dermal microvascular endothelial cells by radiation can be modulated by early PTX/Tx treatment. These drugs acted not only as radical scavengers, but they were also responsible for the increased micronuclei frequency in 10 Gy-irradiated cells. Thus, these drugs may possibly interfere with DNA repair processes **(Laurent et al., 2006)**.

In another study, the effects of vitamin E supplementation were evaluated in cultured primary human normal fibroblasts exposed to UVA. Cells were incubated in medium containing α-tocopherol, α-tocopherol acetate or the synthetic analog Trolox for 24 h prior to UVA exposure. DNA damage in the form of frank breaks and alkali-labile sites, collectively termed single-strand breaks (SSB), was assayed by Comet assay, immediately following irradiation or after different repair periods. The generation of H_2O_2 and superoxide ion was measured by flow cytometry through the oxidation of indicators into fluorescent dyes. Pretreatment of cells with any form of vitamin E resulted in an increased susceptibility to the photo-induction of DNA SSB and in a longer persistence of damage, whereas no significant change was observed in the production of H_2O_2 and superoxide, compared to controls. The researchers indicated that in human normal fibroblasts, exogenously added vitamin E exerted a promoting activity on DNA damage upon UVA irradiation and might lead to increased cytotoxic and mutagenic risks **(Nocentini et al., 2001)**.

In an *in vivo* study by **Konopacka at al. (1998),** the modifying effects of treatment with vitamin E, AA and vitamin A in the form of β-carotene on the clastogenic activity of gamma rays were investigated in mice. Damage *in vivo* was measured by the micronucleus assay in bone marrow polychromatic erythrocytes and exfoliated bladder cells. The vitamins were administered orally, either for five consecutive days before or immediately after irradiation with 2 Gy of gamma rays. The results showed that pretreatment with vitamin E (100-200 mg/kg/day) and β-carotene (3-12 mg/kg/day) were effective in protecting against micronucleus induction by gamma rays. AA depending on its concentration enhanced the radiation effect (400 mg/kg/day), or reduced the number of micro-nucleated polychromatic erythrocytes (50-100 mg/kg/day). Such effect was weekly observed in exfoliated bladder cells. The most effective protection in both tissues was noted when a mixture of these vitamins was used as a pretreatment. Administration of the all antioxidant vitamins to mice immediately after irradiation was also effective in reducing the radiation-induced micronucleus frequency. The data from the *in vitro* experiments based on the Comet assay show that the presence of the vitamins in culture medium influences the kinetic of repair of radiation-induced DNA damage in mouse leukocytes.

4.3 Prevention of genotoxicity by thiocyanates

Human cancer can be prevented by changing the dietary habits **(Kelloff , 2000; Vallejo et al., 2002; Hecht, 1996; Milner , 2004; Davis & Milner, 2006)**. Studies show that antioxidant-rich diets are associated with low risk of cancer and whole diet plays a more important role than the individual components. The protective effects of vegetables and fruits may be attributed to the combined effect of various phytochemicals, vitamins, fibers, and allium compounds rather than the effect of a single component **(Lee et al., 2003)**. There is powerful evidence in literature for a cancer-protective effect of the vegetables of the family *Cruciferae* that includes broccoli, watercress, cabbage, kale, horseradish, radish, turnip, and garden cress **(Verhoeven et al., 1996; Hecht, 1999)**. This effect is attributed to ITCs, which occur naturally as thioglucoside conjugates (glucosinolates). They are hydrolysis products of glucosinolates and are generated through catalytic mediation of myrosinase, which is released upon processing (cutting or chewing) of cruciferous vegetables from a compartment separated from glucosinolates. Evidence exists for conversion of glucosinolates to ITCs in the gut. At least 120 different glucosinolates have been identified. ITCs have a common basic skeleton but differ in their terminal R group, which can be an alkyl, an alkenyl, an alkylthioalkyl, an aryl, a β-hydroxyalkyl, or an indolylmethyl group. The widely studied ITCs include phenethyl isothiocyanate (PEITC), benzyl isothiocyanate (BITC), indole-3-carbinol (I_3C) and allyl isothiocyanate (AITC) **(Fahey et al., 2001; Arranz et al., 2006)**.

The most important biological property discovered about ITCs is their ability to inhibit carcinogenesis, induced by several chemicals including nitrosamines in the lung, stomach, colon, liver, esophagus, bladder and mammary glands in animal models **(Hecht, 1999; Zhang et al., 2003; Zhang and Talalay, 1994; Hecht et al., 1995; Munday et al., 2003)**. Two mechanisms can be suggested for the protective effect of ITCs against nitrosamine-induced DNA damage:

a. Blocking the production of genotoxic intermediates by inhibiting Phase I enzymes: PEITC was shown to reduce p-nitrophenol hdroxylase (CYP2E1), ethoxyresorufin O-deethylase (CYP1A1) and coumarin hdroxylase (CYP2A6) activities **(García et al., 2008)**.
b. Enhancement of detoxification pathways through the induction of Phase II enzymes **(Arranz et al., 2006)**.

Furthermore, ITCs may have ROS scavenging capacity, alter cell proliferation, stimulate DNA-repair, and induce NAD(P)H: quinine oxidoreductase activity as also mentioned for AA before **(Gamet-Payrastre et al., 2000; Chaudière and Ferrari-Iliou et al., 1999; Surh, 2002; Surh et al., 2001; Roomi et al., 1998)**.

ITCs were shown to be effective in the inhibition of lung tumorigenesis in mice and rats induced by the tobacco-specific carcinogen 4-(methylnitrosamino)-1-(3-pyridyl)-1-butanone (NNK). Because NNK is believed to play a significant role as a cause of lung cancer in smokers, PEITC is being developed as a chemopreventive agent, which is presently in Phase I a clinical trial in healthy smokers **(Hecht, 1996; Stoner et al., 1991)**. PEITC is a potent inhibitor of rat esophageal tumorigenesis induced by NBMA **(Stoner et al., 1991)**. A comparative study demonstrates that phenylpropyl isothiocyanate (PPITC) is even more potent, whereas BITC and 4-phenylbutyl isothiocyanate (PBITC) have little effect on tumorigenesis **(Wilkinson et al., 1995)**. However, phenylhdroxyl isothiocyanate (PHITC) enhances tumorigenesis in the same model **(Stoner et al., 1995)**. Mechanistic studies clearly show that PEITC inhibits the metabolic activation of NBMA in the rat esophagus, probably through inhibition of a cytochrome P450 (CYP450) enzyme **(Morse et al., 1997)**. Concomitant with this inhibition, inhibition of O^6-methylguanine formation in rat esophageal DNA was observed. The inhibitory effects on tumorigenicity correlate with their inhibitory effects on O^6-methylguanine formation **(Wilkinson et al., 1995; Stoner & Morse, 1997)**. Inhibition of

N'- nitrosonornicotine (NNN) tumorigenicity in the rat esophagus by PEITC also appears to be due to inhibition of its metabolic activation **(Stoner et al., 1998)**.

The antimutagenic properties of ITCs have been reported towards NDMA and NPYR-induced oxidative stress before. In studies performed by **Knasmüller et al. (1996, 2003)** using PEITC as a chemopreventive agent, the researchers observed a reduction in NDMA- and NPYR-induced DNA damage in *Escherichia coli* K-12 and a considerable reduction in NDMA-induced micronuclei in HepG2 cells. The results of several studies demonstrated that ITCs exhibited strong antimutagenic effects against NDMA and NPYR in a dose dependent manner. In a study by **Smerák at al. (2009)**, the researchers investigated the effect of PEITC on the mutagenic activity of indirect-acting mutagens and carcinogens like aflatoxin B1 (AFB1) and 2-amino-3-methylimidazo[4,5-f]quinoline (IQ) using the *Ames* bacterial mutagenicity test, the Comet assay, an *in vivo* micronucleus test, and direct-acting mutagen and carcinogen N-nitroso-N-methylurea (MNU). In the Ames test, the antimutagenic activity of PEITC was studied in the concentration range 0.3-300 μg/plate. PEITC at concentrations of 0.3, 3 and 30 μg/plate reduced dose-dependently mutagenicity of AFB1 and IQ in both *Salmonella typhimurium* TA98 and TA100 strains. In the case of the direct mutagen MNU, the antimutagenic effect of PEITC was detected only at concentration of 30 μg/plate in the strain TA100. The PEITC concentration 300 μg/plate was toxic in the Ames test. The 24 h pre-treatment of HepG2 cells with PEITC at concentration 0.15 μg/ml resulted in a significant decrease of DNA breaks induced by MNU at concentrations 0.25 and 0.5 mM. Although a trend towards reduced strand break level were determined also at PEITC concentrations 0.035 and 0.07 μg/ml, it did not reach the statistical significance. No effect, however, of PEITC on IQ-induced DNA breaks was observed. Chemopreventive effect of PEITC was revealed also *in vivo*. Pretreatment of mice with PEITC concentrations of 25 and 12.5 mg/kg bw administered to mice in three daily doses resulted in reduction of micronucleus formation in mice exposed to all three mutagens under study, with statistically significant effect at concentration of 25 mg/kg. Results of this study indicated that the strong PEITC antimutagenic properties may have an important role in the prevention of carcinogenesis and other chronic degenerative diseases that share some common pathogenetic mechanisms. In a recent study by **Tang et al. (2011),** PEITC was shown to induce a dose-dependent decrease in cell viability through induction of cell apoptosis and cell cycle arrest in the G_2/M phase of DU 145 human prostate cells. Besides, PEITC induced morphological changes and DNA damage in DU 145 cells. The induction of G_2/M phase arrest was mediated by the increase of p53 and Wee1 and it reduced the level of M-phase inducer phosphatase 3 (CDC25C) protein. The induction of apoptosis was mediated by the activation of caspase-8-, caspase-9- and caspase-3-depedent pathways. Results of this study also demonstrated that PEITC caused mitochondrial dysfunction, increasing the release of cytochrome c and Endo G from mitochondria, and led cell apoptosis through a mitochondria-dependent signaling pathway. The researchers concluded that PEITC might exhibit anticancer activity and become a potent agent for human prostate cancer cells in the future.

There are a few studies on ITCs against nitrosamine-induced genotoxicity in literature. In a study by **Arranz et al. (2006)**, the protective effect of three ITCs was tested. ITC were highly protective against NPYR-induced oxidative DNA damage than against NDMA. The greatest protective effect towards NPYR-induced oxidative DNA damage was shown by I_3C (1 μM,

79%) and by PEITC (1 μM, 67%) and I_3C (1 μM, 61%) towards NDMA (in presence of Fpg enzyme). However, in absence of Fpg enzyme, AITC (1 μM, 72%) exerted the most drastic reduction towards NPYR-induced oxidative DNA damage, and PEITC (1 μM, 55%) towards NDMA. These results indicated that ITCs protect human-derived cells against the DNA damaging effect of NPYR and NDMA, two carcinogenic compounds that occur in the environment. Another study performed by **García et al. (2008)** aimed to investigate the protective effect of ITCs alone or in combination with AA towards NDBA or NPIP-induced oxidative DNA damage in HepG2 cells by Comet assay. PEITC and I_3C alone showed a weak protective effect towards NDBA (0.1 μM, 26-27%, respectively) or NPIP (1 μM, 26-28%, respectively)-induced oxidative DNA damage. AITC alone did not attenuate the genotoxic effect provoked by NDBA or NPIP. In contrast, HepG2 cells simultaneously treated with PEITC, I_3C and AITC in combination with AA showed a stronger inhibition of oxidative DNA-damage induced by NDBA (0.1 μM, 67%, 42%, 32%, respectively) or NPIP (1 μM, 50%, 73%, 63%, respectively) than ITCs alone. One feasible mechanism by which ITCs alone or in combination with AA exert their protective effects towards N-nitrosamine-induced oxidative DNA damage could be by the inhibition of their CYP450 dependent bioactivation. PEITC and I_3C strongly inhibited the p-nitrophenol hydroxylation (CYP2E1) activity (0.1 μM, 66-50%, respectively), while the coumarin hydroxylase (CYP2A6) activity was slightly reduced (0.1 μM, 25-37%, respectively). However, the ethoxyresorufin O-deethylation (CYP1A1) activity was only inhibited by PEITC (1 μM, 55%). The results indicated that PEITC and I_3C alone or PEITC, I_3C and AITC in combination with AA protect human-derived cells against the oxidative DNA damaging effects of NDBA and NPIP.

In our study performed on HepG2 cells, we tested AITC (0.5 μM) against the nitrite and nitrosamine toxicity. Nitrite was added as 20 μM, NDMA as 10 mM, NDEA as 10 mM and NMOR as 3 mM to the medium for 30 min with or without AITC. When compared to untreated cells, nitrite, NDMA, NDEA and NMOR raised the tail intensity up to 17 %, 279 %, 324 % and 288 %, respectively (all, $p<0.05$). AITC was able to reduce the tail intensity caused by nitrite 36 %, by NDMA 36 %, by NDEA 49 % and by NMOR 32 %, respectively. These reductions were statistically significant when compared to each individual toxic compound applied group (all, $p<0.05$). Besides, when compared to untreated cells, nitrite, NDMA, NDEA and NMOR raised the tail intensity up to 94%, 126%, 157% and 207%, respectively (all, $p<0.05$). AITC was able to reduce the tail moment caused by nitrite 16 %, by NDMA 32 %, by NDEA 41 % and by NMOR 19 %, respectively and these reductions were statistically significant when compared to each individual toxic compound applied group **(Erkekoglu & Baydar, 2010d)**.

5. Conclusion

The protective effect of antioxidants is universally accepted. However, as also seen in AA, the mode of action of antioxidants particularly with dual behavior (prooxidant and antioxidant) remain unclear and more research must be conducted on these compounds. For instance, the elucidation of how antioxidant properties operate *in vitro* can provide a better understanding of the *in vivo* situation. On the other hand, Comet assay can be an important tool for the determining of the genotoxic effect of several environmental chemicals, as well as the antioxidant properties of several compounds.

Most of these chemicals exert their toxicity over their ability of producing ROS. ROS can be balanced by the antioxidant action of non-enzymatic antioxidants as well as antioxidant enzymes and it was shown that the genotoxicity of several environmental chemicals can be reversed by proper doses of antioxidants *in vitro*. More *in vitro* studies are needed to prove the beneficial antioxidant effects of trace elements and vitamins. Medicine might benefit from current investigations demonstrating the properties of a vast number of antioxidants as well as studying the effects of different diets. Modest antioxidant supplementation might help prevent chemical-induced carcinogenesis in healthy individuals. On the other hand, antioxidant applications might be beneficial in individuals who may have polymorphisms in genes, including those for antioxidant enzyme. Additionally, populations deficient in several trace elements and vitamins might exhibit modest DNA-repair defects that could be functionally rescued by dietary antioxidants. The future interest of several researchers as well as ours is to understand the pathways underlying the genotoxicity of several agents, particularly phthalates and to determine the antioxidant effect of trace elements and vitamins against the toxic effects of such agents *in vitro* and *in vivo* systems.

6. References

Akingbemi, B.T.; Youker, R.T.; Sottas C.M.; Ge R.; Katz E.; Klinefelter G.R.; Zikrin, B.R. & Hardy, M.P. (2001). Modulation of rat Leydig cell steroidogenic function by di(2-ethylhexyl)phthalate. *Biology of Reproduction*, Vol. 65, No.4, (October 2001), pp. 1252-1259.

Anderson, D. & Phillips, B.J. (1999). Comparative in vitro and in vivo effects of antioxidants. *Food and Chemical Toxicology*, Vol. 37, No. 9-10, (September-October 1999), pp. 1015-1025.

Anderson, D.; Schmid, T.E.; Baumgartner, A.; Cemeli-Carratala, E.; Brinkworth, M.H. & Wood, J.M. (2003). Oestrogenic compounds and oxidative stress (in human sperm and lymphocytes in the Comet assay). *Mutation Research*, Vol. 544, No. 2-3 (November 2003), pp. 173-178.

Anderson, D.; Phillips B.J. & Schmezer P. (1994) The effect of various antioxidants and other modifying agents on oxygen-radical-generated DNA damage in human lymphocytes in the COMET assay. *Mutation Research*, Vol. 307, No. 1, (May 1994), pp. 261-271.

Arranz, N.;, Haza, A.I.; García, A.; Rafter, J. & Morales P. (2007a). Protective effect of vitamin C towards N-nitrosamine-induced DNA damage in the single-cell gel electrophoresis (SCGE)/HepG2 assay. *Toxicology In Vitro*, Vol. 27, No. 1, (October 2007), pp. 1311-1317.

Arranz, N.; Haza, A.I.; García, A.; Delgado, E.; Rafter, J. & Morales, P. (2007b). Effects of organosulfurs, isaothiocyanates and vitamin C towards hydrogen peroxide-induced oxidative DNA damage (strand breaks and oxidized purines/pyrimidines) in human hepatoma cells. *Chemical and Biological Interactions*, Vol. 169, No. 1, (August 2007), pp. 3-71.

Novotna, B.; J. Topinka, I.; Solansky, I.; Chvatalova, Z. & Lnenickova, R.J. (2007). Impact of air pollution and genotype variability on DNA damage in Prague policemen, *Toxicology Letters*, Vol.172, No. 1-2, (July 2007), pp. 37–47.

Baliga, M.S.; Wang, H.; Zhuo, P.; Schwartz, J.L. & Diamond, AM. (2007). Selenium and GPx-1 overexpression protect mammalian cells against UV-induced DNA damage. *Biological Trace Element Research*, Vol. 115, No. 3, (March 2007), pp. 227-242.

Béani, J.C. (2001). Enhancement of endogenous antioxidant defenses: a promising strategy for prevention of skin cancers. *Bulletin de l'Académie Nationale de Médecine*, Vol. 185, No. 8, pp. 507-525.

Bieri, J.G. & Evarts, R.P. (1974). Gamma tocopherol: metabolism, biological activity and significance in human vitamin E nutrition. *American Jotrnal of Clinical Nutrition*, Vol. 27, No. 9, (September 1974), pp. 980-986.

Blokhina, O.; Virolainen, E.; & Fagerstedt, K.V. (2003). Antioxidants, oxidative damage and oxygen deprivation stress: a review. *Annuals of Botanics*, (January 2003), pp. 179-194.

Boffetta, P., Hecht, S., Gray, N., Gupta, P. & Straif, K. (2008). Smokeless tobacco and cancer. *Lancet Oncology*, Vol. 9, No. 7, (July 2008), pp. 667-675.

Brigelius-Flohé, R. & Traber, M.G. (1999). Vitamin E: function and metabolism. *FASEB Journal*, Vol. 13, No. 10, (July 1999), pp. 1145-1155.

Brinkman, M., Buntinx, F., Muls, E. & Zeegers MP. Use of selenium in chemoprevention of bladder cancer. *Lancet Oncology*, Vol. 7, No. 9, (September 2006), pp. 766-774.

Chaudière, J. & Ferrari-Iliou, R. (1999). Intracellular antioxidants: from chemical to biochemical mechanisms, *Food and Chemical Toxicology*, Vol. 37, No. 9-10, (September-October 1999), pp.949-962.

Choe, E. &Min, D.B. (2006). Chemistry and reactions of reactive oxygen species in foods. *Critical Reviews in Food Science and Nutrition*, Vol. 46, No. 1, pp. 1-22.

Clark, L.C.; Cantor, K.P.; & Allaway, W.H. (1991). Selenium in forage crops and cancer mortality in U.S. counties. *Archieves of Environmental Health*. Vo. 46, No. 1, (January-February 1991), pp. 37-42.

Claycombe, K.J. & Meydani SN. (2001).Vitamin E and genome stability. *Mutation Research*, Vol. 475, No. 1-2, (April 2001), pp. 37-44.

Collins, A.R.; Dobson, V.L.; Dusinska, M.; Kennedy, G.; & Stětina, R. (1997). The comet assay: what can it really tell us? *Mutat Research*, Vol. 375, No. 2, (April 1997), pp. 183-193.

Collins, A.R. & Horváthová, E. (2001). Oxidative DNA damage, antioxidants and DNA repair:applications of the comet assay. *Biochemical Society Transactions*, Vol. 29, No. Pt 2, (May 2001), pp. 337-341.

Combs, G.F. Jr. & Gray, W.P. (1998). Chemopreventive agents: selenium. *PharmacologicalTherapy*, Vol. 79, No. 3, (September 1998), pp. 179-192.

Corton, J.C. & Lapinskas, P.J. (2005). Peroxisome proliferator-activated receptors: mediators of phthalate ester-induced effects in the male reproductive tract? *Toxicological Sciences*, Vo. 83, No. 1, (January 2005), pp. 4-17.

Crott, J.W. & Fenech, M. (1999). Effect of vitamin C supplementation on chromosome damage, apoptosis and necrosis ex vivo. *Carcinogenesis*, Vol. 20, No. 6, (June 1999), pp. 1035-1041.

Drath, D.B.; Shorey, J.M. & Huber, G.L. (1981). Functional and metabolic properties of alveolar macrophages in response to the gas phase of tobacco smoke, *Infection and Immunity*, Vol. 34, No. 9-10, (September-October1981), pp. 11–15.

Anderson, D.; Phillips B.J.; Yu T.W.; Edwards A.J.; Ayesh R. & Butterworth K.R. (1997). The effects of vitamin C supplementation on biomarkers of oxygen radical generated damage in human volunteers with low or high cholesterol levels. *Environmental and Molecular Mutagenesis*, Vol. 30, No. 2, pp. 161–174.

Dalgaard, M.; Ostergaard, G.; Lam, H.R.; Hansen, E.V. & Ladefoged, O. (2000). Toxicity study of di(2-ethylhexyl)phthalate (DEHP) in combination with acetone in rats. *Pharmacological Toxicology*, Vol. 86, No. 2, (February 2000), pp. 92-100.

Dalton, T.P.; Shertzer, H.G. & Puga, A. (1999). Regulation of gene expression by reactive oxygen. *Annual Review of Pharmacology and Toxicology*, Vol. 39, pp. 67-101.

Davis, C.D. & Milner, J.A. (2006). Nutritional health: strategies for disease prevention In: *Diet and Cancer Prevention,* N.J.Temple; T. Wilson & D.V. Jacobs, (Eds). 151-171, Humana Press; Totowa, NJ, USA.

Dietrich, M.; Traber, M.G.; Jacques, P.F.; Cross, C.E.; Hu, Y & Block, G. (2006). Does gamma-tocopherol play a role in the primary prevention of heart disease and cancer? A review. *Journal of American College of Nutrition*, Vol. 25, No. 4, (August 2006), pp. 292-299.

Domenici, F.A.; Vannucchi, M.T.; Jordão, A.A. Jr.; Meirelles, M.S. & Vannucchi, H. (2005). DNA oxidative damage in patients with dialysis treatment. *Renal Failure*. Vol. 27, No. 6, pp. 689-694.

Doull, J.; Cattley, R.; Elcombe, C.; Lake, B.G.; Swenberg, J.; Wilkinson, C.; Williams, G. & van Gemert, M. A cancer risk assessment of di(2-ethylhexyl)phthalate: application of the new U.S. EPA Risk Assessment Guidelines. *Regulatory Toxicology and Pharmacology*, Vol. 29, No. 3, (June 1999), pp. 327-357.

Dröge, W. (2002). The plasma redox state and ageing. *Ageing Research Reviews*, Vol.1, No.2, (April 2022), pp. 257-78.

Durmaz, E.; Ozmert, E.N.; Erkekoglu, P.; Giray, B.; Derman, O.; Hincal, F. & Yurdakök, K. (2010). Plasma phthalate levels in pubertal gynecomastia. *Pediatrics*, Vol. 125, No. 1, (January 2010), pp. e122-e129.

Duthie, S.J.; Ma, A.; Ross, M.A. & Collins, A.R. (1996). Antioxidant supplementation decreases oxidative DNA damage in human lymphocytes. *Cancer Research*, Vol. 56, No. 6, (March 1996), pp. 1291-1295.

El-Bayoumy, K. & Sinha, R. Mechanisms of mammary cancer chemoprevention by organoselenium compounds. (2004). *Mutation Research*, Vol. 551, No. 1-2, (July 2004), pp. 181-197.

El-Bayoumy, K. & Sinha, R. (2005). Molecular chemoprevention by selenium: a genomic approach. *Mutation Research*, Vol. 591, No. 1-2, (December 2005), pp. 224-236.

Emonet-Piccardi, N.; Richard, M.J.; Ravanat, J.L.; Signorini, N.; Cadet, J. & Béani JC. (1998). Protective effects of antioxidants against UVA-induced DNA damage in human skin fibroblasts in culture. *Free Radical Research*, Vol. 29, No. 4, (October 1998), pp. 307-313.

Engelward Lab. (2011). High throughput DNA damage analysis. Available at:http://web.mit.edu/engelward-lab/comet.htm.

Erkekoglu, P.; Rachidi, W.; De Rosa, V.; Giray, B.; Favier, A & Hincal, F. (2010a). Protective effect of selenium supplementation on the genotoxicity of di(2-ethylhexyl)phthalate and mono(2-ethylhexyl)phthalate treatment in LNCaP cells. *Free Radical Biology anf Medicine*, Vol. 49, No. 4, (August 2010), pp. 559-566.

Erkekoglu, P.; Rachidi, W.; Yuzugullu, O.G.; Giray, B.; Favier, A.; Ozturk, M. & Hincal F. (2010b). Evaluation of cytotoxicity and oxidative DNA damaging effects of di(2-ethylhexyl)-phthalate (DEHP) and mono(2-ethylhexyl)-phthalate (MEHP) on MA-10 Leydig cells and protection by selenium. *Toxicology and Applied Pharmacology*, Vol. 248, No. 1, (October 2010), pp. 52-62.

Erkekoglu, P. & Baydar, T. (2010c). Evaluation of the protective effect of ascorbic acid on nitrite- and nitrosamine-induced cytotoxicity and genotoxicity in human hepatoma line. *Toxicology Mechanisms and Methods*, Vol. 20, No. 2, (February 2010), pp. 45-52.

Erkekoglu, P. & Baydar, T. (2010d). Effect of allyl isothiocyanate (AITC) in both nitrite- and nitrosamine-induced cell death, production of reactive oxygen species, and DNA damage by the single-cell gel electrophoresis (SCGE): does it have any protective effect on HepG2 cells? *International Journal of Toxicology*, Vol. 29, No. 3, (May-June 2010), pp. 305-312.

Erkekoglu, P.; Zeybek, N.D.; Giray, B.; Asan, E.; Arnaud, J. & Hincal F. (2011a) Reproductive toxicity of di(2-ethylhexyl) phthalate in selenium-supplemented and selenium-deficient rats. *Drug and Chemical Toxicology*, Vol. 34, No. 4, (October 2011), pp. 379-389.

Erkekoglu, P.; Zeybek, N.D.; Giray, B.; Asan, E. & Hincal F. (2011b). The Effects of Di(2-Ethylhexyl)Phthalate Exposure and Selenium Nutrition on Sertoli Cell Vimentin Structure and Germ-Cell Apoptosis in Rat Testis. *Archives of Environmental Contamination and Toxicology*. doi 10.1007/s00244-011-9712-9.

Erkekoglu, P.; Rachidi, W.; Yüzügüllü, O.G.; Giray, B.; Oztürk, M.; Favier, A & Hıncal F. (2011c). Induction of ROS, p53, p21 in DEHP- and MEHP-exposed LNCaP cells-protection by selenium compounds. *Food and Chemical Toxicolgy*, Vol. 49, No. 7, (July 2011), pp. 1565-1571.

Erkekoglu, P.; Giray, B.; Rachidi, W; Hininger-Favier, I.; Roussel, A.M.; Favier, A. & Hincal F. (2011d). Effects of di(2-ethylhexyl)phthalate on testicular oxidant/antioxidant status in selenium-deficient and selenium-supplemented rats. *Environmental Toxicology*, doi: 10.1002/tox.20776.

Esworthy, R.S.; Baker, M.A. & Chu FF. (1995). Expression of selenium-dependent glutathione peroxidase in human breast tumor cell lines. *Cancer Research* Vol. 55, No. 4, (February 1995), pp. 957-962.

Faust, F.; Kassie, F.; Knasmuller S.; Boedecker, R.H.; Mann M. & Mersch-Sundermann, V. (2004). The use of the alkaline comet assay with lymphocytes in human biomonitoring studies, *Mutation Research*, Vol. 566, No. 3, (May 2004),pp. 209–229.

Fahey, J.W.; Zalcmann, A.T. & Talalay P. (2001) The chemical diversity and distribution of glucosinolates and isothiocyanates among plants. *Phytochemistry*. Vol. 56, No 1, pp. 5-51.

Fan, J.; Traore, K.; Li, W.; Amri, H.; Huang, H.; Wu, C.; Chen, H.; Zikrin, B. & Papadopoulos, V. (2010). Molecular mechanisms mediating the effect of mono-(2-ethylhexyl) phthalate on hormone-stimulated steroidogenesis in MA-10 mouse tumor Leydig cells. *Endocrinology*. Vol. 151, No. 7, (July 2010), pp. 3348-3362.

Fenech, M. & Ferguson, L.R. (2001). Vitamins/minerals and genomic stability in humans. *Mutation Researh*, Vol, 475, No. 1-2, (April 2001), pp. 1-6.

Finkel, T. (1998). Oxygen radicals and signaling. *Current Opinions in Cell Biology*, Vol. 10, No. 2, (April 1998), pp. 248-253.

Flohé, L. (2007). Selenium in mammalian spermiogenesis. *Biological Chemistry*, Vol. 288, No. 10, (October 2007), pp. 987-995.

Foster, P.M.; Mylchreest, E.; Gaido, K.W. & Sar M. (2001). Effects of phthalate esters on the developing reproductive tract of male rats. *Human Reproduction Update*, Vol. 7, No. 3, (May-June2001), pp. 231-235.

Frank J. (2005). Beyond vitamin E supplementation: an alternative strategy to improve vitamin E status. *Journal of Plant Physiology*, Vol. 162, No. 7, (July 2005), pp. 834-843.

Frederiksen, H.; Aksglaede, L.; Sorensen, K.; Skakkebaek, N.E.; Juul, A. &Andersson AM. (2011). Urinary excretion of phthalate metabolites in 129 healthy Danish children and adolescents: estimation of daily phthalate intake. *Environmental Research*, Vol. 111, No. 5, (July 2011), pp. 656-663.

Frederiksen, H.; Jørgensen, N. & Andersson, A.M. (2010). Correlations between phthalate metabolites in urine, serum, and seminal plasma from young Danish men determined by isotope dilution liquid chromatography tandem mass spectrometry. *Journal Analytical Toxicology*, Vol. 34, No. 7, (September 2010), pp. 400-410.

Gabbianelli, R.; Nasuti, C.; Falcioni, G. & Cantalamessa F. (2004). Lymphocyte DNA damage in rats exposed to pyrethroids: effect of supplementation with Vitamins E and C. *Toxicology*. Vol. 203, No. 1-3, (October 2004), pp. 17-26.

Gajecka, M.; Kujawski, L.M.; Gawecki, J. & Szyfter K. (1999). The protective effect of vitamins C and E against B(a)P-induced genotoxicity in human lymphocytes. *Journal of Environmental Pathology, Toxicology and Oncology*, Vol. 18, No. 3, pp. 159-167.

Gamet-Payrastre, L.; Li, P.; Lumeau, S.; Cassar, G.; Dupont, M.A.; Chevolleau, S.; Gasc, N.; Tulliez, J. &Tercé, F. (2000). Sulforaphane, a naturally occurring isothiocyanate, induces cell cycle arrest and apoptosis in HT29 human colon cancer cells, *Cancer Research*, Vol 60, No. 5, (March 2000), pp. 1426-1433.

García, A.; Haza, A.I.; Arranz, N.; Rafter, J. & Morales, P. (2008). Protective effects of isothiocyanates alone or in combination with vitamin C towards N-nitrosodibutylamine or N-nitrosopiperidine-induced oxidative DNA damage in the single-cell gel electrophoresis (SCGE)/HepG2 assay. *Journal of Applied Toxicology*, Vol. 28, No. 2, (March 2008), pp. 196-204.

Gazouli, M.; Yao, Z.X.; Boujrad, N.; Corton, J.C.; Culty, M. & Papadopoulos V. (2002). Effect of peroxisome proliferators on Leydig cell peripheral-type benzodiazepine receptor gene expression, hormone-stimulated cholesterol transport, and steroidogenesis: role of the peroxisome proliferator-activator receptor alpha. *Endocrinology*, Vol. 143, No. 7, (July 2002),pp. 2571-2583.

Gopalakrishna, R. & Gundimeda, U. (2001). Protein kinase C as a molecular target for cancer prevention by selenocompounds. *Nutrition and Cancer*. Vol. 40, No. 1, pp. 55-63.

Grande, S.W.; Andrade, A.J.; Talsness, C.E.; Grote, K.; Golombiewski, A.; Sterner-Kock, A. & Chahoud, I. (2007). A dose-response study following in utero and lactational exposure to di-(2-ethylhexyl) phthalate (DEHP): reproductive effects on adult female offspring rats. *Toxicology*, Vol. 229, No. 1-2, (January 2007), pp. 114-122.

Griffiths, H.R.; Mistry, P.; Herbert, K.E. & Lunec, J. (1998). Molecular and cellular effects of ultraviolet light-induced genotoxicity. *Critical Reviews in Clinical Laboratory Sciences*, Vol. 35, No. 3, (June 1998), pp. 189-237.

Grossman, L.; Caron, P.R.; Mazur, S.J. & Oh, E.Y. (1998). Repair of DNA-containing pyrimidine dimers. *FASEB Journal*, Vol. 2, No. 11, (August 1988), pp. 2696-2701.

Gutteridge, J.M. & Halliwell, B. (2000). Free radicals and antioxidants in the year 2000. A historical look to the future. *Annuals of New York Academy of Sciences*, Vol. 899, pp. 136-147.

Gürbay, A.; Gonthier, B.; Signorini-Allibe, N.; Baret, L.; Favier, A. & Hincal, F. (2006). Ciprofloxacin-induced DNA damage in primary culture of rat astrocytes and protection by Vitamin E. *Neurotoxicology*, Vol. 27, No. 1, (January 2006), pp. 6-10.

Halliwell, B. (2006). Reactive species and antioxidants. Redox biology is a fundamental theme of aerobic life. *Plant Physiology*, Vol. 141, No. 2, (June 2006), pp. 312-322.

Halliwell B. Free radicals and antioxidants - quo vadis? (2011). *Trends in Pharmacological Sciences*, Vol. 32, No. 3, (March 2011), pp. 125-130.

Harréus, U.; Baumeister, P.; Zieger, S. & Matthias C. (2005). The influence of high doses of vitamin C and zinc on oxidative DNA damage. *Anticancer Research*, Vol. 25, No. 5, (September-October 2005), pp. 3197-3201.

Hecht, S.S. (1996). Recent studies on mechanisms of bioactivation and detoxification of 4-(methylnitrosamino)-1-(3-pyridyl)-1-butanone (NNK), a tobacco-specific lung carcinogen. *Critical Reviews in Toxicology*, Vol. 26, No. 2, pp. 163-181.

Hecht, S.S. (1999). Chemoprevention of cancer by isothiocyanates, modifiers of carcinogen metabolism. *Journal of Nutrition*, Vol. 129, No. 3, pp. 768-774.

Hecht, S.S.; Chung, F.L.; Richie, J.P. Jr; Akerkar, S.A.; Borukhova, A.; Skowronski, L. & Carmella, S.G. (1995). Effects of watercress consumption on metabolism of a tobacco-specific lung carcinogen in smokers. *Cancer Epidemiology Biomarkers and Prevention*, Vol. 4, No. 8, (December 1995), pp. 877-884.

Heudorf, U.; Mersch-Sundermann, V. & Angerer, J. (2007). Phthalates: toxicology and exposure. *International Journal of Hygiene and Environmental Health*, Vol. 210, No. 5, (October 2007), pp. 623-634.

Irminger-Finger, I. (2007). Science of cancer and aging, *Journal of Clinical Oncology*, Vol. 25, No. 14, (May 2007), pp. 1844–1851.

IARC (International Agency for Research on Cancer). (2000). Some industrial chemicals. Vol. 77. Available at: http://monographs.iarc.fr.

Ishihara, M.; Itoh, M.; Miyamoto, K.; Suna, S.; Takeuchi, Y.; Takenaka, I. & Jitsunari F. (2000). Spermatogenic disturbance induced by di-(2-ethylhexyl) phthalate is significantly prevented by treatment with antioxidant vitamins in the rat. *International Journal of Andrology*, Vol. 23, No. 2, (April 2000), pp. 85-94.

Wang, J.Y.; Wen, L.L.; Huang, Y.N.; Chen, Y.T. & Ku M.C. (2006). Dual effects of antioxidants in neurodegeneration: direct neuroprotection against oxidative stress and indirect protection via suppression of glia-mediated inflammation, *Current Pharmaceutical Design*, Vol. 12, No. 27, pp. 3521–3533.

Jakszyn, P. & Gonzalez, C.A. (2006). Nitrosamine and related food intake and gastric and oesophageal cancer risk: a systematic review of the epidemiological evidence. *World Journal of Gastroenterology*, Vol. 12, No. 27, (July 2006), pp. 4296-4303.

Janjua, N.R.; Frederiksen, H.; Skakkebaek, N.E.; Wulf, H.C. Andersson, A.M. (2008). Urinary excretion of phthalates and paraben after repeated whole-body topical application in humans. *Internatonal Journal of Andrology*, Vol. 31, No. 2, (April 2008), pp. 118-30.

Kadirvel, R; Sundaram, K.; Mani, S.; Samuel, S.; Elango, N. & Panneerselvam C. (2007). Supplementation of ascorbic acid and alpha-tocopherol prevents arsenic-induced protein oxidation and DNA damage induced by arsenic in rats. *Human and Experimental Toxicology*, Vol. 26, No. 12, (December 2007), pp. 939-946.

Kasahara, E.; Sato, E.F.; Miyoshi, M.; Konaka, R.; Hiramoto, K.; Sasaki, J.; Tokuda, M.; Nakano, Y. & Inoue M. (2002). Role of oxidative stress in germ cell apoptosis induced by di(2-ethylhexyl)phthalate. *Biochemical Journal*, Vol. 365, No. Pt 3, pp. 849-856.

Kaur, P. & Singh, R. (2007). In vivo interactive effect of garlic oil and vitamin E against stavudine induced genotoxicity in Mus musculus. *Indian Journal of Experimental Biology*, Vol. 45, No. 9, (September 2007), pp. 807-811.

Kaymak, C.; Kadioglu, E.;Basar, H. & Sardas, S. (2004). Genoprotective role of vitamin E and selenium in rabbits anaesthetized with sevoflurane. *Human and Experimental Toxicology*, Vol. 23, No. 8, (August 2004), pp. 413-419.

Kelloff, G.J. (2000). Perspectives on cancer chemoprevention research and drug development. *Advances in Cancer Research*, Vol. 78, pp. 199-334.

Knasmüller, S.; Huber, W.W.; Kienzl, H.; & Schulte-Hermann, R. (1992). Inhibition of repairable DNA-damage in Escherichia coli K-12 cells recovered from various organs of nitrosamine-treated mice by vitamin A, phenethylisothiocyanate, oleic acid and triolein. *Carcinogenesis*, Vol. 13, No. 9, (September 1992), pp. 1643-1650.

Knasmüller, S.; Mersch-Sundermann, V.; Kevekordes, S.; Darroudi, F.; Huber, W.W., Hoelzl, C.; Bichler, J. & Majer, B.J. (2004). Use of human-derived liver cell lines for the detection of environmental and dietary genotoxicants; current state of knowledge. *Toxicology*, Vol. 198, No. 1-3, pp. 315-328.

Konopacka, M.; Widel M. & Rzeszowska-Wolny J. (1998). Modifying effect of vitamins C, E and beta-carotene against gamma-ray-induced DNA damage in mouse cells. *Mutation Research*, Vol. 417, No. 2-3, (September 1998), pp. 85-94.

Frankl, L. (1991). Developmental aspects of experimental pulmonary oxygen toxicity. *Free Radical Biology and Medicine*, Vol. 11, No. 5, pp. 463–494.

Laurent, C.; Pouget, J.P. & Voisin P. (2005). Modulation of DNA damage by pentoxifylline and alpha-tocopherol in skin fibroblasts exposed to Gamma rays. *Radiation Research*, Vol. 164, No. 1, (July 2005), pp. 63-72.

Laurent, C.; Voisin, P. & Pouget, J.P. (2006). DNA damage in cultured skin microvascular endothelial cells exposed to gamma rays and treated by the combination pentoxifylline and alpha-tocopherol. *International Journal of Radiation Biology*, Vol. 82, No. 5, (May 2006), pp. 309-321.

Lee, K.W.; Lee, H.J.; Surh, Y.J. & Lee, C.Y. (2003). Vitamin C and cancer chemoprevention: reappraisal. *American Journal of Clinical Nutrition*, Vol. 78, No. 6, pp. 1074-1078.

Li, H.; Stampfer, M.J.; Giovannucci, E.L.; Morris, J.S.; Willett, W.C.; Gaziano, J.M. & Ma, J. (2004). A prospective study of plasma selenium levels and prostate cancer risk. *Journal of National Cancer Institute*, Vol. 96, No. 9, (May 2004), pp. 696-703.

Longtin, R. (2003). Selenium for prevention: eating your way to better DNA repair? *Journal of National Cancer Institute* Vol. 95, No. 2, (January 2003), pp. 98-100.

Grisham, M.B. (1994). Oxidants and free radicals in inflammatory bowel disease, *Lancet*, Vol. 344, No. 8926, (September 1994), pp. 859–861.

Milner, J.A. (2004). Molecular targets for bioactive food components. *Journal of Nutrition*, Vol. 134, No. 9, pp. 2492S-2498S.

Møller, P.; Viscovich, M.; Lykkesfeldt, J.; Loft, S.; Jensen, A. & Poulsen H.E. (2004). Vitamin C supplementation decreases oxidative DNA damage in mononuclear blood cells of smokers. *European Journal of Nutrition*, Vol. 43, No. 5, (October 2004), pp. 267-274.

Morse, M.A.; Lu, J.; Gopalakrishnan, R.; Peterson, L.A.; Wani, G.; Stoner, G.D.; Huh, N. & Takahashi, M. (1997). Mechanism of enhancement of esophageal tumorigenesis by 6-phenylhexyl isothiocyanate. *Cancer Letters*, Vol 12, No. 1, (January 1997), pp. 119–125.

Mousseau, M.; Faure, H.; Hininger, I.; Bayet-Robert, M. & Favier, A. (2005). Leukocyte 8-oxo-7,8-dihydro-2'-deoxyguanosine and comet assay in epirubicin-treated patients. *Free Radic Research*, Vol. 39, No. 8, (August 2005), pp. 837-843.

Munday, R.; Munday, J.S. & Munday, CM. (2003). Comparative effects of mono-, di-, tri-, and tetrasulfides derived from plants of the Allium family: redox cycling in vitro and hemolytic activity and Phase 2 enzyme induction in vivo. *Free Radical Biology and Medicine*, Vol. 34, No. 9, pp. 1200-1211.

Müler, L.; Kasper, P.; Kertsen, B. & Zhang, J. (1998). Photochemical genotoxicity and photochemical carcinogenesis--two sides of a coin? *Toxicology Letters*, Vol. 102-103, (December 19989, pp. 383-387.

Negro, R. (20089. Selenium and thyroid autoimmunity. *Biologics*. Vol. 2, No. 2, (June 20089, pp. 265-273.

Niki, E.; Noguchi, N.;, Tsuchihashi, H. & Gotoh, N. (1995). Interaction among vitamin C, vitamin E, and beta-carotene. *American Journal of Clinical Nutrition*, Vol. 62, No. 6 Supplement, (December 1995), pp. 1322S-1326S.

Nocentini, S.; Guggiari, M.; Rouillard, D. & Surgis, S. (2001). Exacerbating effect of vitamin E supplementation on DNA damage induced in cultured human normal fibroblasts by UVA radiation. *Photochemical Photobiology*, vol. 73, No.4, (April 2001), pp. 370-377.

Noriega, N.C.; Howdeshell, K.L.; Furr, J.; Lambright, C.R.; Wilson, V.S. & Gray, L,E, Jr. (2009). Pubertal administration of DEHP delays puberty, suppresses testosterone production, and inhibits reproductive tract development in male Sprague-Dawley and Long-Evans rats. *Toxicological Sciences*, Vol. 111, No. 1, (September 2009), pp. 163-178.

O'Brien, M.L.; Spear, B.T. & Glauert HP. (2005). Role of oxidative stress in peroxisome proliferator-mediated carcinogenesis. *Critical Reviews in Toxicology*, Vol. 35, No. 1, (January 2005), pp. 61-88.

Ohsawa, K.; Nakagawa, S.Y.; Kimura, M.; Shimada, C.; Tsuda, S.; Kabasawa, K.; Kawaguchi, S. & Sasaki, Y.F. (2003). Detection of in vivo genotoxicity of endogenously formed N-nitroso compounds and suppression by ascorbic acid, teas and fruit juices. *Mutation Research*, Vol. 539, No.1-2, (August 2003), pp. 65-76.

Olive, P.L.; Banáth, J.P. (2006). The comet assay: a method to measure DNA damage in individual cells. *Nature Protocols*, Vol. 1, No. 1, pp. 23-29.

Östling, O. & Johanson, K.J. (1984). Microelectrophoretic study of radiation-induced DNA damages in individual mammalian cells, *Biochemical and Biophysical Research Communications*, Vol. 123, No. 1, (August 1984), pp. 291–298.

Cerutti, P.A. (1994). Oxy-radicals and cancer. *Lancet*, Vol. 344, No. 8926 , (September 1994), pp. 862–863.

Heaton, P.R.; Reed, C.F.; Mann, S.J.; Ransley R.; Stevenson J.; Charlton C.J.; Smith, B.H.; Harper, E.J. & Rawlings, J.M. (2002). Role of dietary antioxidants to protect against DNA damage in adult dogs. *Journal of Nutrition*, Vol. 132, No. 6 Supplement 2 , (June 2002), pp. 1720S–1724S.

Panayiotidis, M. & Collins, A.R. (1997). Ex vivo assessment of lymphocyte antioxidant status using the comet assay. *Free Radical Research*, Vol. 27, No. 5, (November 1997), pp. 533-537.

Persson-Moschos, M.; Alfthan, G. & Akesson, B. (1998). Plasma selenoprotein P levels of healthy males in different selenium status after oral supplementation with different forms of selenium. *European Journal of Clinical Nutrition*, Vol. 52, No. 5, (May 1998), pp. 363-367.

Peters, U. & Takata, Y. (2008). Selenium and the prevention of prostate and colorectal cancer. *Molecular Nutrition and Food Research*, Vol. 52, No. 11, (November 2008), pp. 1261-1272.

Porrini, M.; Riso, P.; Brusamolino, A.; Berti, C.; Guarnieri, S. & Visioli, F. (2005). Daily intake of a formulated tomato drink affects carotenoid plasma and lymphocyte concentrations and improves cellular antioxidant protection. *British Journal of Nutrition*, Vol. 93, No. 1, (January 2005), pp. 93-99.

Rafferty, T.S.; Green, M.H.; Lowe, J.E.; Arlett, C.; Hunter, J.A.; Beckett, G.J. & McKenzie, R.C. (2003). Effects of selenium compounds on induction of DNA damage by broadband ultraviolet radiation in human keratinocytes. British *Journal of Dermatology*, Vol. 148, No. 5, (May 2003), pp. 1001-1009.

Reboul, E.; Richelle, M.; Perrot, E.; Desmoulins-Malezet, C.; Pirisi, V. & Borel, P. (2006). Bioaccessibility of carotenoids and vitamin E from their main dietary sources. *Journal of Agricultural Food Chemistry*, Vol. 54, No. 23, (November 2006), pp. 8749-8755.

Richburg, J.H. & Boekelheide, K. (1996). Mono-(2-ethylhexyl) phthalate rapidly alters both Sertoli cell vimentin filaments and germ cell apoptosis in young rat testes. *Toxicology and Applied Pharmacolgy*, Vol. 137, No. 1, (March 1996), pp. 42-50.

Ridley, A.J.; Whiteside, J.R.; McMillan, T.J. & Allinson, S.L. (2009). Cellular and sub-cellular responses to UVA in relation to carcinogenesis. *International Journal of Radiation Biology*, 85, No. 3, (March 2009), pp. 177-195.

Robichová, S.; Slamenová, D.; Chalupa, I. & Sebová, L. (2004). DNA lesions and cytogenetic changes induced by N-nitrosomorpholine in HepG2, V79 and VH10 cells: the protective effects of Vitamins A, C and E. *Mutation Research*, Vol. 560, no. 2, (June 2004), pp. 91-99.

Roomi, M.W.; Netke, S. & Tsao, C. (1998). Modulation of drug metabolizing enzymes in guinea pig liver by high intakes of ascorbic acid. *International Journal for Vitamin and Nutrition Research*, Vol. 68, No. 1, pp. 42-47.

Rosa, R.M.; Moura, D.J.; Romano, E.; Silva, A.C.; Safi, J. & Pêgas Henriques, J.A. (2007). Antioxidant activity of diphenyl diselenide prevents the genotoxicity of several mutagens in Chinese hamster V79 cells. *Mutation Research*, Vol. 631, No. 1, (July 2007), pp. 44-54.

Geller, S.; Krowka, R.; Valter, K. & Stone, J. (2006). Toxicity of hyperoxia to the retina: evidence from the mouse. *Advances in Experimental Medicine and Biology*, Vol. 572, 425–437.

Salobir, J.; Zontar, T.P.; Levart, A. & Rezar, V. (2010). The comparison of black currant juice and vitamin E for the prevention of oxidative stress. *International Journal for Vitamin and Nutrition Research*, Vol. 80, No. 1, (January 2010), pp. 5-11.

Sardaş, S.; Yilmaz, M.; Oztok, U.; Cakir, N. & Karakaya, A.E. (2001). Assessment of DNA strand breakage by comet assay in diabetic patients and the role of antioxidant supplementation. *Mutation Research*, Vol. 490, No. 2, (February 2001), pp. 123-129.

Schlörmann, W. & Glei, M. (2009). Comet fluorescence in situ hybridization (Comet-FISH):detection of DNA damage. *Cold Spring Harbor Protocols*, No.5, (May 2009), pdb.prot5220.

Schrauzer, G.N. (2000). Anticarcinogenic effects of selenium. *Cellular and Molecular Life Sciences*, Vol. 57, No. 13-14, (December 2000), pp. 1864-1873.

Schrauzer, G.N. (1976). Selenium and cancer: a review. *Bioinorgic Chemistry*, Vol. 5, No. 3, pp. 275-281.

Schrauzer, GN. (1977). Trace elements, nutrition and cancer: perspectives of prevention. *Advances in Experimental Medicine and Biology*, Vol. 91, pp. 323-344.

Schrauzer, G.N. & White, D.A. (1978). Selenium in human nutrition: dietary intakes and effects of supplementation. *Bioinorgic Chemistry*, Vol. 8, No.,4, (April 1978), pp. 303-318.

Seet, R.C.; Lee, C.Y.; Lim, E.C.; Quek, A.M.; Huang, S.H.; Khoo, C.M. & Halliwell, B. (2010) Markers of oxidative damage are not elevated in otherwise healthy individuals with the metabolic syndrome. *Diabetes Care*, Vol. 33, No. 5, (May 2010), pp. 1140–1142.

Sen, C.K.; Khanna, S. &Roy, S. (2006). Tocotrienols: Vitamin E beyond tocopherols. *Life Sciences*, Vol. 78, No. 18, (March 2006), pp. 2088-2098.

Shamberger, R.J.; Tytko, S.A. & Willis, CE. (1976). Antioxidants and cancer. Part VI. Selenium and age-adjusted human cancer mortality. *Archieves of Environmental Health*, Vol. 31, No. 5, (September-October 1976), pp. 231-235.

Shi, L.Q. & Zheng, R.L. (2006). DNA damage and oxidative stress induced by Helicobacter pylori in gastric epithelial cells: protection by vitamin C and sodium selenite. *Pharmazie*, Vol. 61, No. 7, (July 2006), pp. 631-637.

Shimpo, K.; Nagatsu, T.; Yamada, K.; Sato, T.; Nimi, H; Shamoto, M.; Takeuchi, T.; Umezawa, H. & Fujita, K. (1991). Ascorbic acid and adriamycin toxicity. *American Journal of Clinical Nutrition*, Vol. 54, No. 6 Supplement, (December 1991), pp. 1298S-1301S.

Singh, M.; Kaur, P.; Sandhir, R. & Kiran, R. (2008). Protective effects of vitamin E against atrazine-induced genotoxicity in rats. *Mutation Research*, Vol. 654, No. 2, (July 2008), pp. 145-149.

Singh, N.P.; McCoy, M.T.; Tice, R.R. & Schneider, E.L. (1988). A simple technique for quantitation of low levels of DNA damage in individual cells. *Experimental Cell Research*, Vol. 175, No. 1, (March 1988), pp. 184–191.

Slamenová, D. (2001). Contemporary trends in in vivo and in vitro testing of chemical carcinogens. *Neoplasma*, Vol. 48, No. 6, pp. 425-434.

Smerák, P.; Polívková, Z.; Stetina, R. & Bártová, J. Bárta I. (2009). Antimutagenic effect of phenethyl isothiocyanate. *Central European Journal of Public Health*, Vol. 17, No.2, (June 2009), pp. 86-92.

Soory, M. (2009). Redox status in periodontal and systemic inflammatory conditions including associated neoplasias: antioxidants as adjunctive therapy? *Infectious Disorders Drug Targets*, Vol. 9, No. 4, (August 2009), pp. 415-427.

Spiegelhalder, B. & Bartsch H. (1996). Tobacco-specific nitrosamines. *European Journal of Cancer Prevention*, Vol. No. 5 Supplement 1, (September 1996), pp. 33-38.

Stoner, G.D.; Morrissey, D.T.; Heur, Y.H.; Daniel, E.M.; Galati, A.J. & Wagner, S.A. (1991). Inhibitory effects of phenethyl isothiocyanate on N-nitrosobenzylmethylamine carcinogenesis in the rat esophagus. *Cancer Research*, Vol. 51, No. 8, (April 1991), pp. 2063-2068.

Stoner, G.D.; Siglin, J.C.; Morse, M.A.; Desai, D.H.; Amin, S.G.; Kresty, L.A.; Toburen, A.L.; Heffner, E.M&. Francis, D.J. (1995). Enhancement of esophageal carcinogenesis in male F344 rats by dietary phenylhexyl isothiocyanate. *Carcinogenesis* Vol. 16, No. 10, (October 1995), pp. 2473–2476.

Stoner, G.D. & Morse, M.A. (1997). Isothiocyanates and plant polyphenols as inhibitors of lung and esophageal cancer. *Cancer Letters*, Vol. 114, No. 1-2, (March 1997), pp. 113-119.

Stoner, G.D.; Adams, C.; Kresty, L.A.; Hecht, S.S.; Murphy, S.E. & Morse, M.A. (1998). Inhibition of *N*-nitrosonornicotine-induced esophageal tumorgenesis by 3-phenylpropyl isothiocyanate. *Carcinogenesis*, Vol. 19, No. 12, (December 1998), pp. 2139–2143.

Surh, Y.J. (2002). Anti-tumor promoting potential of selected spice ingredients with antioxidative and anti-inflammatory activities: a short review. *Food and Chemical Toxicology*, Vol. 40, No. 8, (August 2002), pp. 1091-1097.

Surh, Y.J., Chun, K.S., Cha, H.H., Han, S.S., Keum, Y.S., Park, K.K., Lee, S.S. (2001). Molecular mechanisms underlying chemopreventive activities of anti-inflammatory phytochemicals: down-regulation of COX-2 and iNOS through suppression of NF-kappa B activation. *Mutation Research*, Vol. 480-481, pp. 243-268.

Svensson, K.; Hernández-Ramírez, R.U.; Burguete-García, A.; Cebrián, M.E.; Calafat, A.M.; Needham, L.L.; Claudio, L. & López-Carrillo L. (2011). Phthalate exposure associated with self-reported diabetes among Mexican women. *Environmental Research*, Vol. 111, No. 6 (August 2011), pp. 792-796.

Tay, T.W.; Andriana, B.B.; Ishii, M.; Tsunekawa, N.; Kanai, Y. & Kurohmaru, M. (2007). Disappearance of vimentin in Sertoli cells: a mono(2-ethylhexyl) phthalate effect. *International Journal of Toxicology*, Vol. 26, No. 4, (July-August), pp. 289-295.

Traber, M.G. (1998). The Biological Activity of Vitamin E. http://lpi.oregonstate.edu/sp-su98/vitamine.html. Last updated May, 1998.

Traber, M.G. (2006). Vitamin E. In: *Modern Nutrition in Health and Disease,* M.E. Shils; M. Shike; A.C. Ross; B. Caballero; R. Cousins, (Eds), 396-411, Lippincott Williams & Wilkins, Baltimore, MD, USA.

Traber, M.G. (2007). Vitamin E regulatory mechanisms. *Annual Review of Nutrition*, Vol.27, pp. 347-362.

Tricker, A.R. (1997). N-nitroso compounds and man: sources of exposure, endogenous formation and occurrence in body fluids. *European Journal of Cancer Previews*, Vol. 6, No. 3, (June 1997), pp. 226-268.

Vallejo, F.; Tomas-Barberan, F.A. & Garcia-Viguera, C. (2002). Glucosinolates and vitamin C content in edible parts of broccoli florets after domestic cooking. *European Food Research and Technology*, Vol. 215, No. 4, pp. 310-316.

Vallejo, F.; Tomas-Barberan, F.A. & Garcia-Viguera, C. (2002). Potential bioactive compounds in health promotion from broccoli cultivars grown in Spain. *Journal of the Science of Food and Agriculture,* Vol. 82, No. 11, pp. 1293-1297.

Velanganni, A.A.; Dharaneedharan, S.; Geraldine, P. & Balasundram, C. (2007). Dietary supplementation of vitamin A, C and E prevents p-dimethylaminoazobenzene induced hepatic DNA damage in rats. *Indian Journal of Biochemistry and Biophysics,* Vol. 44, No. 3, (June 2007), pp. 157-163.

Verhoeven, D.T.H.; Goldbohm, R.A.; van Poppel, G.; Verhagen, H. & van den Brandt PA. (1996). Epidemiological studies on Brassica vegetables and cancer risk. *Cancer Epidemiology, Biomarkers and Prev*ention, Vol. 5, No.9, pp. 733-748.

Weitberg, A.B. & Corvese, D. (1997). Effect of vitamin E and beta-carotene on DNA strand breakage induced by tobacco-specific nitrosamines and stimulated human phagocytes. *Journal of Exprimental and Clinical Cancer Research,* Vol. 16, No. 1, (March 1997), pp. 11-14.

Wilkinson, J.T., Morse, M.A., Kresty, L.A. Stoner, G.D. (1995). Effect of alkyl chain length on inhibition of *N*-nitrosomethylbenzylamine-induced esophageal tumorigenesis and DNA methylation by isothiocyanates. *Carcinogenesis*, Vol. 16, No. 5, (May 1995), pp. 1011–1115.

Wong, V.W.C.; Szeto, Y.T.; Collins, A.R. & Benzie, I.F.F. (2005). The comet assay: a biomonitoring tool for nutraceutical research. Current Topics in Nutraceutical Research Vol. 3, No. 1, pp. 1-14.

Wormuth, M.; Scheringer, M.; Vollenweider, M. & Hungerbühler K. (2006). What are the sources of exposure to eight frequently used phthalic acid esters in Europeans? *Risk Analaysis*, Vol. 26, No. 3, (June 2006), pp. 803-824.

Yu, F.; Wang, Z.; Ju, B.; Wang, Y.; Wang, J. & Bai, D. (2008). Apoptotic effect of organophosphorus insecticide chlorpyrifos on mouse retina in vivo via oxidative stress and protection of combination of vitamins C and E. *Experimental Toxicology and Pathology*, Vol. 59, No. 6, (April 2008), pp. 415-423.

Zhang, Y. & Talalay, P. (1994). Anticarcinogenic activities of organic isothiocyanates: chemistry and mechanisms. *Cancer Research,* Vol. 54, No. 7S, pp. 1976s-1981s.

Zhang, Y.; Tang, L. & Gonzalez, V. (2003). Selected isothiocyanates rapidly induce growth inhibition of cancer cells. *Molecular Cancer Therapy,* Vol. 2, No. 10, pp. 1045-1052.

22

Gel Electrophoresis as Quality Control Method of the Radiolabeled Monoclonal Antibodies

Veronika Kocurová
Nuclear Physics Institute, Academy of Sciences of the Czech Republic,
Řež near Prague
Czech Republic

1. Introduction

Neurodegeneration is the leading term for the progressive loss of the neuron structure, including death of neurons. Many neurodegenerative diseases including the specific diseases - such as Parkinson's, Alzheimer's, and Huntington's occur as a result of the neurodegenerative processes. As research progresses, many similarities appear which relate these diseases to one another on a sub-cellular level. Discovering these similarities offers hope for therapeutic advances that could ameliorate many diseases simultaneously. There are many parallels between different neurodegenerative disorders including atypical protein assemblies as well as induced cell death followed by an apoptosis. Apoptosis is a form of the programmed cell death in the multicellular organisms. It is one of the main types of the programmed cell death, and, last but not least, involves a series of the biochemical reactions leading to a characteristic cell morphology changes, and, finally, death. In according to the previously mentioned knowledge, there is a necessity to develop an imaging method which describes these cellular changes. The principal goal of the investigation monoclonal antibodies and their fragments is to examine the possibility of developing of an imaging radiotracer that would be specific for cytoskeleton of destructed dendrites and neuronal bodies. One of the suitable fitting marker, specific for neuronal tissue, performs anti III β-tubulin (bTcIII) antibody - TU-20 with molecular weight 150 kDa and its scFv fragment with molecular weight 27.7 kDa. The scFv fragment of TU-20 was synthesized for its higher mobility through tissue and vascular barriers. Biochemical characteristics (especially immunoaffinity) of the specific binding substance - anti III β-tubulin scFv fragment - is preserved, and, moreover, the biological availability is much better than in case of the whole antibody. See the structure in the Fig. 1.

To examine this hypothesis, it is necessary to radiolabel both substances with ^{125}I and ^{123}I. The next step is chemical analysis and, furthermore, biochemical properties are extensively investigated. The quality control, performed by gel filtration, electrophoresis, ELISA testing determines adequate properties of the radiolabeled substances for further studies.

Affinity coupling and RIA analytic methods occur under development with focusing on specifics of the antibody and its fragment behavior. *In vitro* experiment shows an extent of the preserved binding specifity of the species by incubation of the both radiolabeled substances with mice brain slices followed by an autoradiography.

The *in vivo* biodistribution confirmes behavior of elimination of the radiolabeled TU-20 and scFv from mice. The bi-exponential model for two-phase clearance to determine short phase half-life $t_{1/2\alpha}$ and long phase half-life $t_{1/2\beta}$ values is used. For comparative study, a transgene population G93A1 Gur was chosen to show different behavior of the substances in normal mouse and in modified organism with amyotrophic lateral sclerosis (ALS).

The main objective of this work is to develop a method for direct imaging of the structural degradation of peripheral neurones by various types of neuropathies.

2. Methods and materials

The monoclonal antibody TU-20 and its scFv was purchased from Exbio, CZ. The antibody recognizes the peptide sequence ESESQGPK. ScFv TU-20 is a recombinant protein expressed in *E. coli*. (Dráberová et al., 1998)

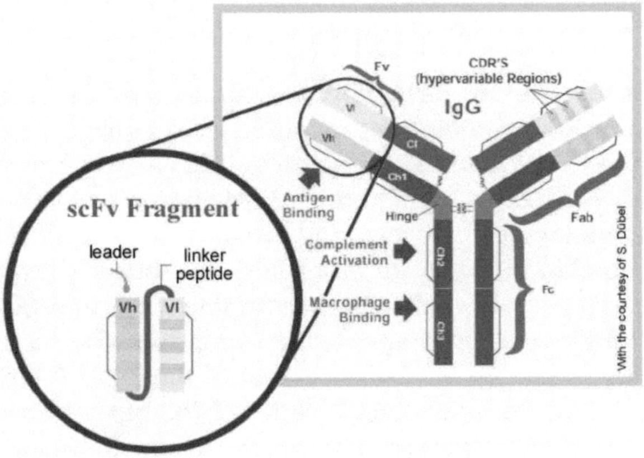

Fig. 1. The structure of the monoclonal antibody on the base of IgG and its scFv fragment.

2.1 Radioiodination of the antibody

^{125}I ($T_{1/2}$ = 59,4 h) radioiodination of TU-20 and scFv TU-20 was performed via chloramine-T with or without stopping reaction with sodium thiosulfate agent. The ratio of an amount of TU-20 to radioactivity was 1 μg to 5.5-7.0 MBq of ^{125}I. The ratio of an amount of the fragment to radioactivity was 1 μg to 1.5-2.0 MBq of ^{125}I. ^{123}I ($T_{1/2}$ = 13,3 h) radioiodination of the fragment scFv TU-20 was performed via chloramine-T with stopping reaction with sodium thiosulfate. The ratio of an amount of the fragment to radioactivity was 1 μg to 3-5 MBq ^{123}I (Švecová et al., 2008). The structure of the radiolabeled antibody is shown in the Fig. 2.

The monoclonal antibody TU-20 was radioiodinated by using either chloramine-T or iodogen as an oxidizing agent. Iodination via chloramine-T was provided in two alternative ways: either with or without stopping a reaction by a reducing agent (Dráberová et al., 1998).

The reaction was performed under following conditions: 10 μl of TU-20 (1 mg/ml) was transferred to 10 μl phosphate buffer (PBS, 0,01 M, pH 7,4) in a reaction vessel and ^{125}I radioactivity (approximately 5,36 MBq) was added. Finally, the solution of chloramine-T in PBS (0,1 mg/ml) was added to the reaction vessel. The amount of chloramine-T ranged from 0,5 to 6 μg per 10 μg of the antibody. After the reaction time (60 seconds), during which the

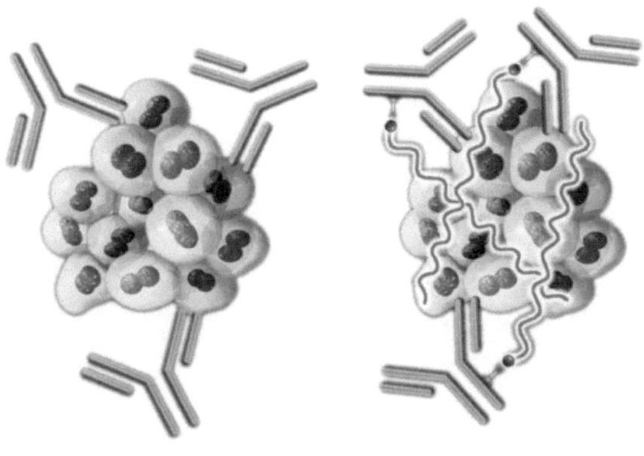

Fig. 2. Radiolabeled monoclonal antibody. Radiotracer is bound to the antibody structure via -OH group of the tyrosine.

reaction mixture was gently agitated, the reaction alternatively might be or not stopped with 100 μl of the solution sodium thiosulfate in water (4 mg/ml) (Chizzonite et al., 1991).

Iodination tubes, for both methods, were prepared in the same way. 100 μl of iodogen dissolved in chloroform (10 - 500 μg/ml) was given in a glass tube and chloroform was evaporated under a slow stream of nitrogen. The prepared iodination tubes were used immediately. The procedure for the direct method consisted in adding 10 μl of TU-20 (1 mg/ml) into the reaction tube with 50 μl of phosphate buffer (PB, 0,05 M, pH 8,5) and an equal amount of $Na^{125}I$ around 5,4 MBq. Reaction time was 15 minutes.

The indirect method was performed in two steps. Firstly, radioactivity in PB was added into the tube coated with iodogen. After 15 minutes an activated iodide was withdrawn, transferred into the vessel containing 10 μl of the antibody and the mixture was agitated for 20 minutes (Švecová et al., 2008).

Radioiodination of the fragment scFv TU-20 was performed via chloramine-T without stopping reaction with thiosulfate as described previously for TU-20. In both cases, at the end of labeling, the reaction mixture was loaded on the top of a BSA-blocked polyacrylamide desalting column with an exclusion limit 6 kDa. Fractions were eluted with 0,1 % BSA in PBS and measured for radioactivity. (Hamilton, 2002), (Katsetos, 2003).

2.2 Immunoreactivity testing by enzyme linked immunosobent assay (ELISA)

The immunoreactivity of the radiolabeled monoclonal antibody TU-20 was determined by an enzyme linked immunosorbent assay (ELISA) using the commercial set for detection of mouse anti - β III tubulin antibodies from VIDIA, CZ. One of the most useful of the immunoassays is the two antibody sandwich ELISA. This assay is used to determine the antigen concentration in unknown samples. This ELISA is fast and accurate, and if a purified antigen standard is available, the assay can determine the absolute amount of antigen in an unknown sample. The principle of ELISA testing is shown in the Fig. 3.

The sandwich ELISA requires two antibodies that bind to epitopes that do not overlap on the antigen. This can be accomplished with either two monoclonal antibodies that recognize discrete sites or one batch of affinity-purified polyclonal antibodies. To utilize this assay, one

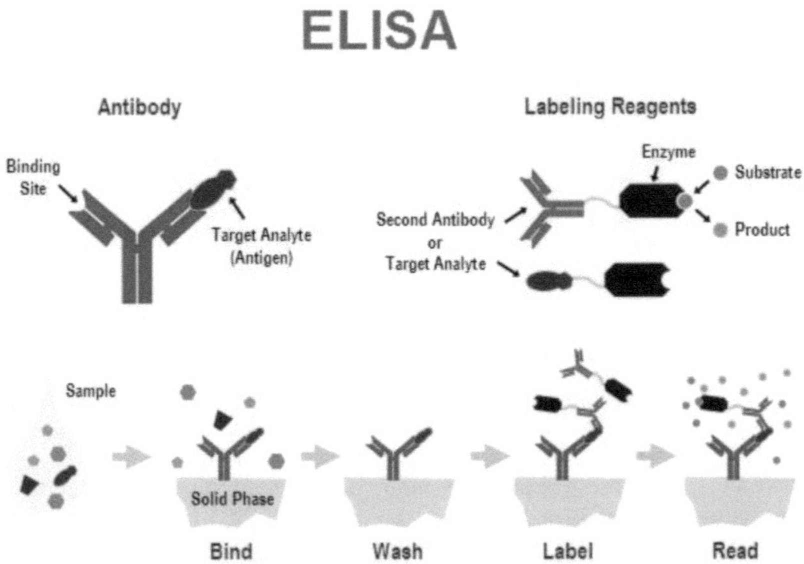

Fig. 3. ELISA principle. A specific antigen (an antibody plays the role of the "antigen" in the case of the antibody ELISA detection) is bound to the specific antibody coated on the solid carrier (microtitration plate). Subsequently, another specific antibody (labeled by an apropriate enzyme which catalyzes the coloured and easily detectable reaction) is added to the previously bound antigen.

antibody (the 'capture' antibody) is purified and bound to a solid phase typically attached to the bottom of a plate well.

Afterwards, an antigen is added, and, allowed to complex with the bound antibody. Unbound products are then removed with a wash, and a labeled second antibody (the 'detection' antibody) is allowed to bind to the antigen, and, therefore, the setting is described as the sandwich. The assay is then quantified by measuring the amount of labeled second antibody bound to the matrix, through the use of a colorimetric substrate.

Major advantages of this technique are that the antigen does not need to be purified prior to use, and that these assays are very specific. However, one disadvantage is that not all antibodies can be used. Monoclonal antibody combinations must be qualified as "matched pairs", meaning that they can recognize separate epitopes on the antigen so they do not hinder each other's binding.

ELISA procedures utilize substrates that produce soluble products. Ideally the enzyme substrates should be stable, safe and inexpensive. Popular enzymes are those that convert a colorless substrate to a colored product, e.g., pnitrophenylphosphate (pNPP), which is converted to the yellow p-nitrophenol by alkaline phosphatase. Substrates used with peroxidase include 2,2'-azo-bis(3-ethylbenzthiazoline-6-sulfonic acid) (ABTS), o-phenylenediamine (OPD) and 3,3'5,5'-tetramethylbenzidine base (TMB), which yield green, orange and blue colors, respectively. In our case, TMB was used for colorimetric visualization. The settlement of the procedure see in the Fig. 4.

Fig. 4. Equipment for optical density measurement in the ELISA settings.

2.3 Immunoaffinity testing by radioimmunoassay (RIA)

In radioimmunoassay, a fixed concentration of radio-labeled antigen in trace amounts is incubated with a constant amount of antiserum such that the total antigen binding sites on the antibody are limited such that the only 30–50 % of the total radio-labeled antigen may be bound in the absence of the antigen. When unlabeled antigen, either as standard or test sample, is added to this system, there is competition between radio-labeled antigen and unlabeled antigen for the limited constant number of binding sites on the antibody.

The amount of radio-labeled antigen bound to antibody decreases as the concentration of unlabeled antigen increases. Following optimal incubation condition e.g. buffer, pH, time and temperature, radio-labeled antigen bound to antibody is separated from unbound radio-labeled antigen.

RIA analytic method was developed in two modifications of surface of the reactive vessel.

2.4 Immunoaffinity separation affinity coupling (AC)

Affinity coupling was develop by use the basic matrix activated Sepharose 4 Fast Flow by Pierce which was modified specific binding octapeptide (Vijayalakshmi, 1992). Activated media enable successful, convenient immobilization of ligands without the need for complex chemical syntheses or special equipment. The Sepharose matrix provides a wide range of high-capacity media with a variety of coupling chemistries for fast, easy, and safe immobilization through a chosen functional group. The principle is to immobilize the antibodies or other large proteins containing -NH2 groups by coupling them to the matrix without the need for an intermediate spacer arm.

The correct choice of an activated medium is dictated by both the group available in the ligand molecule, and by the nature of the binding reaction with the substance to be purified. To ensure minimal interference with the normal binding reaction, immobilization should be attempted through the least critical region of the ligand (Haugland, 1995).

2.5 Stability testing by electrophoresis

Mostly used variation of the electrophoresis for the intention of the quality control of the radiolabeled substances is SDS-PAGE formation of the electrophoresis. It concerns of zone electrophoresis in gel in surface placement. The mixture of the substances is analyzed by division in accordance to the molecular weight.

2.5.1 Polymerization of the polyacrylamide gel

Polyacrylamide gel is prepared to the form by polymerization of the basic monomer acrylamide (CH2=CH-CO-NH2; abbrev. AA) and N,N'-methylen-bis-akrylamid (CH2=CH-CO-NH-CH2-NH-CO=CH-CH2; abbrev. BIS) which is implemented to the polymere randomly and might covalently bind two linear chains of the polyacrylamide. Ammonium persulfate (abbrev. APS) is used as the initiative reactant and N,N'-tetramethylendiamine (abbrev. TEMED) as the catalyzer, see the Fig. 5.

The inhibitor of the reaction is oxygen, and, therefore, the gel must be protected against the oxygen atmosphere. The polymerization has the radical and exothermic process, and, therefore, the cooling is neccessary during the whole polymerization. The ratio of AA:BIS is crucial for the gel mechanic and separation characteristics. The suggested ratio is ranging of about 40:1 (from 20:1 up to 100:1) (Jones, 2004).

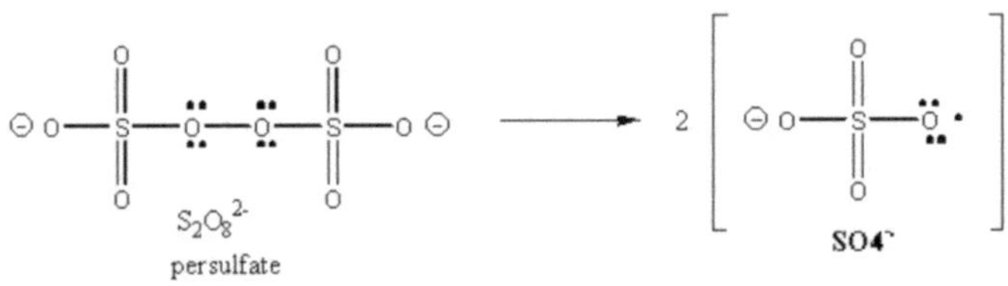

Fig. 5. The Free radical Polymerization of the Acrylamide Initiated on the Addition of the Ammonium Persulfate which Forms the Free Reactive Radicals in the Water.

2.5.2 PAGE electrophoresis

PAGE separation could be conducted in the gel with the same content of the acrylamide in two different following gels, so called Laemmli electrophoresis, when the first gel contents lower percentage of the acrylamide and it is intended to the concentration of the sample at the begining of the separation (so called the concentration gel). The bigger sharpness of the zones in the gel is provided by means of the lower pH (of about two degree) against the surrounding setting. The itself separation takes place in the following part of the gel with the higher density (so called the separation gel). (Laemmli, 1970) The structure of the polymerization process see in the Fig. 6.

Other variation performs the creation of the gradient gel, where the concentration gradient of the polyacrylamide (from the part with lowe density to the part of higher density, in the direction of the separation) is created. The bigger sharpness of the gel zones of the molecules of the similar size is ensured in this arrangement.

The choice of pH of the used buffer by polymerization process, and, also the division of the molecules by the classical PAGE, because, the suitable buffer ensures the sufficient differences in the specific charge of the assorted parts of the protein mixture. The acid proteins require slightly alkaline or neutral pH (the molecules moves to the anode) and alkaline proteins require a slightly acid pH (the molecules migrate to the cathode) (Bernard et al., 1979).

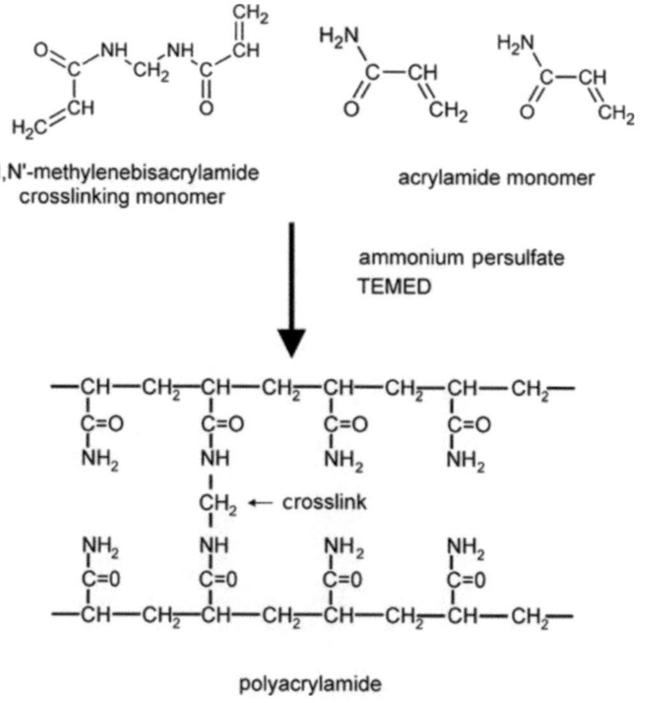

Fig. 6. Polymerization Process of the Structure Networking.

2.5.3 SDS-PAGE electrophoresis

The perfectly suitable modification of the PAGE electrophoresis is an arrangement in the –sodium dodecyl sulfate (abbrev. SDS, or NaDS), which makes an ability of the proteins to bind the SDS in amount of abouti 1,4 mg per 1 mg of the protein by means of the hydrofobic reaction. SDS carries a huge negative charge which enables to equalize the charge of the molecules, and, those, move in one direction in the electrophoretic gel in accordance of the molecular size. The complex SDS-protein unifies either the charge density, or, conformation on the surface of the complex, see the structure in the Fig. 7.

The mobility of the SDS-protein complex in the polyacrylamide gel is proportional to the logarithm of the molecular weight of an appropriate protein, which enables the gel calibration (Rédei, 2008). It is quiet convenient that the examined samples are adjusted before the whole process.

First, an appropriate buffer is added (e.g. TrisHCL) and SDS so that we have the same homogennous reaction setting.

Second, the glycerol is added, because it makes the settings in the gel more dense, so that the samples fill the sample holes properly and do not swirl. Glycerol also decreases the electroendoosmosis and makes the movement and distribution of the proteins even better.

Third, the bromophenol blue is added as the protein movement indicator. Fourth, dithiothreitol (abbrev. DTT) could be added to cleave the proteins to make an analysis more suitable. The samples could be also denatured in the hot water by the temperature of about 65 °C.

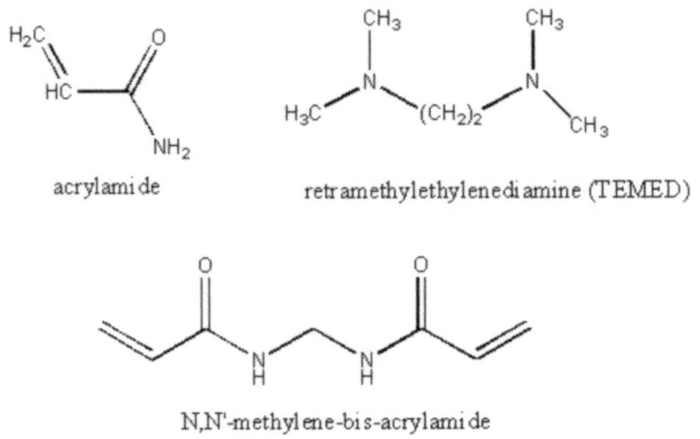

Fig. 7. The Structural Formulae of the Substances in the SDS-PAGE.

2.5.4 Visualization and radiodetection in electrophoresis

The proteins can be visualized directly in gel after electrophoresis proceeding, or, subsequently Western Blot technique could be processed and detection is performed in the membrane where the proteins are transferred from gel. Adsorption of the pigment is used for visualization.

2.5.5 Staining in electrophoresis

A Silver Staining shows another alternative for dying of the proteins in gel. The silver ion is insoluble and colourless, and, distinguishes the places with protein and without proteins in the polyacrylamide gel (formation of the silver complexes with alkaline or sulphuric proteins).

After this procedure, the silver ions are reduced by formaldehyde into the form of the metal silver which is perfectly visible and insoluble. The amount of proteins, which could be visualized by this procedure, ranges from the hundreds of picograms to the units nanograms.

Another staining, which si possible for this purpose of detection, is dying by means of Coomassie Blue which is less sensitive (of about 50 times), but it has another advantage that Coomassie Blue is bound to the protein in the stechiometry ratio, and, therefore, it represents a quantitative densitometry detection (maximum absorbance ranges from 560 nm to 575 nm), see the Fig. 8. An autoradiography may be used as an alternative for detection in gel of the radiolabeled compounds. The differences between electrophoresis by non-reductive (see Fig. 9 and Tab. 1) and reductive conditions (see Fig. 10 and Tab. 2) are shown below.

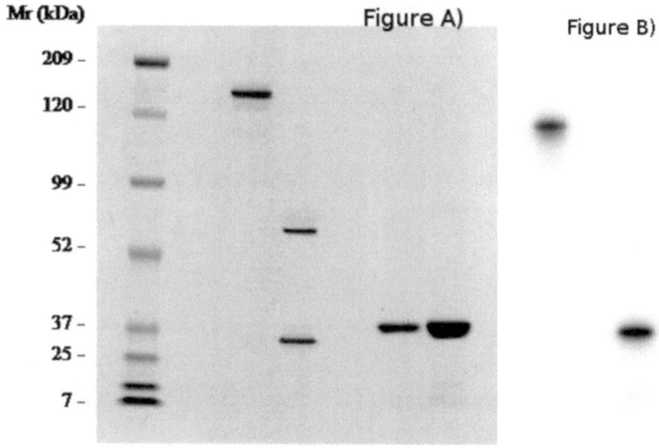

Fig. 8. Electrophoresis of the TU-20 and scFv TU-20 in the Gradient Gel by the Non-Reductive (–) and Reductive (+) Conditions Figure A) Gel Coloured by Coomassie Blue: 1. Molecular Marker; 2. TU-20 (–); 3. TU-20 (+); 4. scFv TU-20 (–); 5. scFv TU-20 (+). Figure B) Autoradiography: 1. [^{125}I]TU-20 (–); 2. [^{125}I]scFv TU-20 (–). .

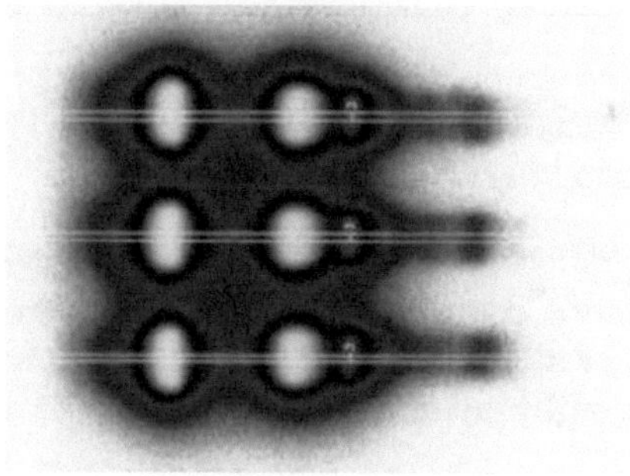

Fig. 9. Autoradiography of electrophoresis SDS-PAGE ^{125}I-TU-20 (all lines) by non-reductive conditions

Peak	Integral Density in PSL	% Ratio of Peak
^{125}I-TU-20	59237,2	46,7
BSA(I)	53228,4	42,0
BSA(II)	14319,4	11,3

Table 1. Autoradiographical interpretation of SDS-PAGE of ^{125}I TU-20 by non-reductive conditions. An autoradiographical visualization of the SDS-PAGE gel (which contains the radiolabeled antibody by non-reductive conditions) after developing on the luminiscent plate by means of the AIDA software.

2.5.6 Immunoblotting

Western Blot transfers the proteins, closed into the gel matrix, into the nitrocellulose membrane for further purposes of investigation after finishing of electrophoresis. Western Blot (Immunoblotting), used for the protein detection, transferes the proteins from the gel into the membrane by means of electrophoresis.

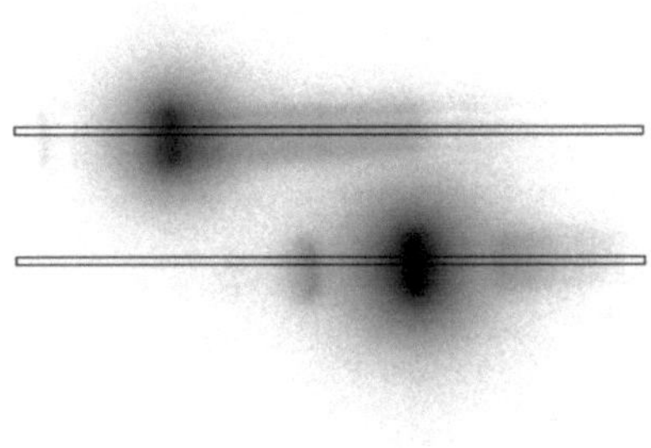

Fig. 10. Autoradiography of electrophoresis SDS-PAGE of ^{125}I-TU-20 and ^{125}I-scFv TU-20 by reductive conditions. An autoradiographical visualization of the SDS-PAGE gel (which contains the radiolabeled antibody and fragment by reductive conditions) after developing on the luminiscent plate by means of the AIDA software.

Peak	Integral Density in PSL	% Ratio of Peak
^{125}I-TU-20	63445,2	98,1
BSA(I)	1228,8	1,9
^{125}I-scFv TU-20	77988,6	97,6
BSA(I)	1909,7	2,4

Table 2. Autoradiographical interpretation of SDS-PAGE of ^{125}I-scFv TU-20 (upper line) and ^{125}I TU-20 (bottom line) by reductive conditions.

The particular proteins are subsequently indentified by the appropriate radiolabbeled antibodies (labeled by enzymatic reaction, or, by the radiolabeling reaction with ^{125}I). The proteins bound into the membrane could be submitted to the non-specific staining, or, as an alternative, to the autoradiography. After drying, the membrane is stored with much better results than dried gel.

When the electrophoresis with all its instruments and alternatives is used as a quality control method of the radiolabeled antibodies, the following parametres were proved and chosen for this setting as the most suitable. Stability of the radiolabeled TU-20 and its scFv TU-20 was investigated on 4 - 12 % Bis-Tris gel electrophoresis.

Protein bands were visualized by staining the gels with Silver Stain Plus. ^{125}I-labeled scFv fragment was processed by autoradiography exposing plate BAS-SR 2025, and finally developed by BAS-1800II. Autoradiographs were evaluated by AIDA 2.0 software, see the Fig. 11.

2.6 Immunohistochemistry testing

Preserved binding properties of the radiolabeled MAb or scFv for neuronal tissue were confirmed by the method of double labeling. It is based on the immunohistochemistry and autoradiography of the brain tissue slices. The 50 μm thick brain slices from the wild type mouse (C57B/6/J) were incubated with the radiolabeled TU-20. The second incubation was performed with anti-mouse IgG polyclonal antibody conjugated with horseradish peroxidase (Sigma-Aldrich, USA). Afterwards, the immunohistochemistry was finalized by staining with 3,3′ – diaminobenzidine (DAB) that revealed the neuronal structure, see the Fig. 12 and 13.

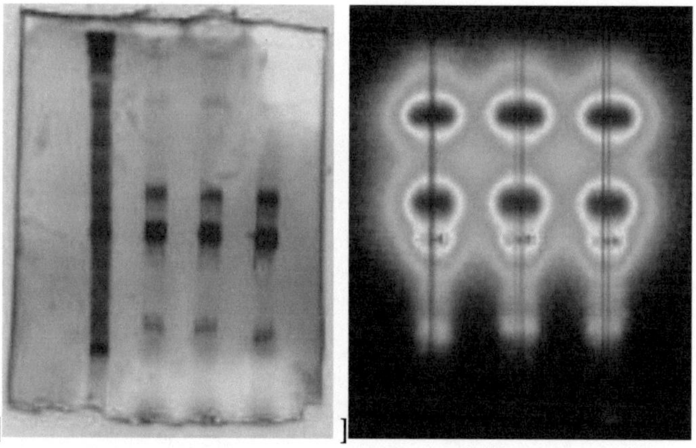

Fig. 11. Gel electrophoresis analysis of [^{125}I]TU-20 – autoradiography (a) and silver staining (b).

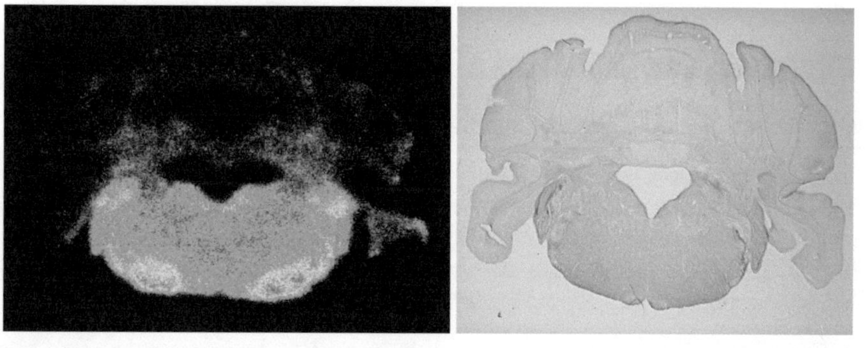

Fig. 12. [^{125}I]TU-20 autoradiographical - Figure A) ,and, immunohistochemical visualization of the bound radiolabeled antibody in the mice brain slice - Figure B) image of the coronal mice brain slice.

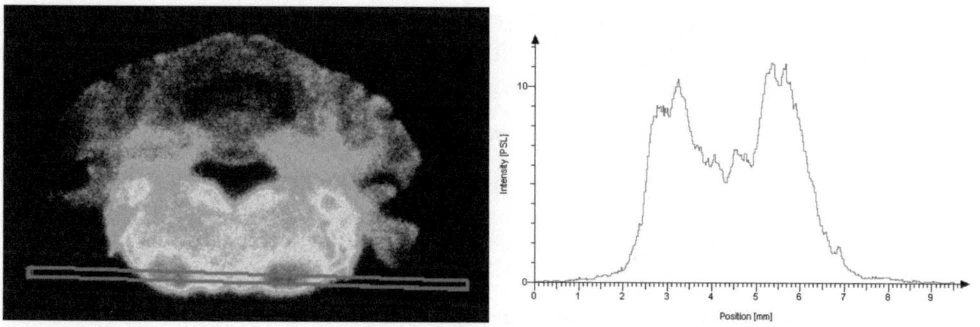

Fig. 13. Autoradiography visualization - Figure A), and, visualization of the 1D-interpretation of the bound radiolabeled antibody in the mice brain slice - Figure B) of the labeled mice brain slices by means of the software AIDA.

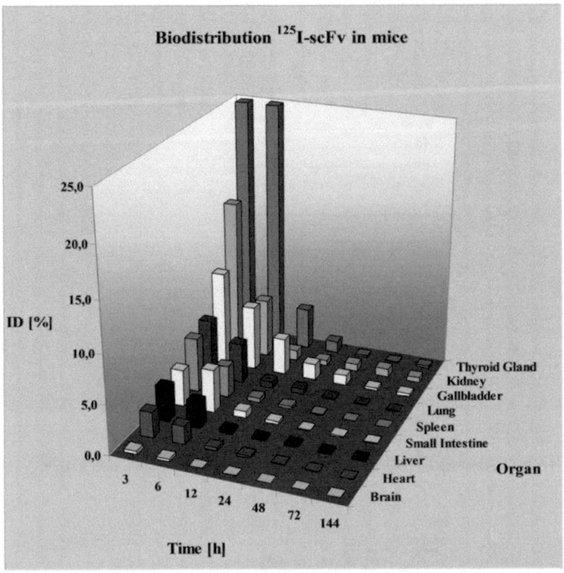

Fig. 14. [^{125}I]scFv biodistribution in normal mice.

2.7 In vivo preparative biodistribution testing in normal mice

The in vivo biodistribution was carried out with the male normal mice - wild type C57B/6/J. Biodistribution studies were performed following an i.v. injection. The main focus is intended for scFv fragment due to its better mobility in organism. ^{125}I-labeled scFv fragment, for comparison with the biodistribution of Na^{125}I, was applied in amount of 50 kBq/50 μl. ^{123}I-labeled scFv fragment was injected in amount of 200 kBq/50 μl.

Mice were sacrificed at designated times points in groups by 3 animals. The kinetic time intervals were: 3, 6, 12, 24, 48, 72, 144 hours for ^{125}I-labeled scFv TU-20 fragment and 0,5, 1, 2, 3, 6, 12 hours for ^{123}I-labeled scFv fragment.

Blood and major organs (included thyroid gland, kidneys, lung, heart, brain, spleen, muscle, fat, skin, gallbladder, testicles, stomach, liver, small intestine, and colon) were removed, weighed, and counted in a gamma scintillation counter to determine the % ID/g (percentage of injected dose per gram) for each radiolabeled substance.

The biodistribution figures are shown below, see the Fig. 14 and 15.

Blood clearance data for ^{125}I-labeled scFv fragment were obtained by analyzing blood samples by using a bi-exponential model for two-phase clearance to determine short phase half-life $t_{1/2\alpha}$ and long phase half-life $t_{1/2\beta}$ values.

2.8 In vivo SPECT imaging biodistribution testing

[^{123}I]scFv TU-20 and [^{123}I]TU-20 behavior in mice (wild type C57B/6/J) was observed by use of the SPECT camera. Kinetic intervals were 0.5, 1, 2, 3 h by [^{123}I]scFv TU-20 - see the Fig. 16 and 1, 2, 3, 6 h by [^{123}I]TU-20 - see the Fig. 17.

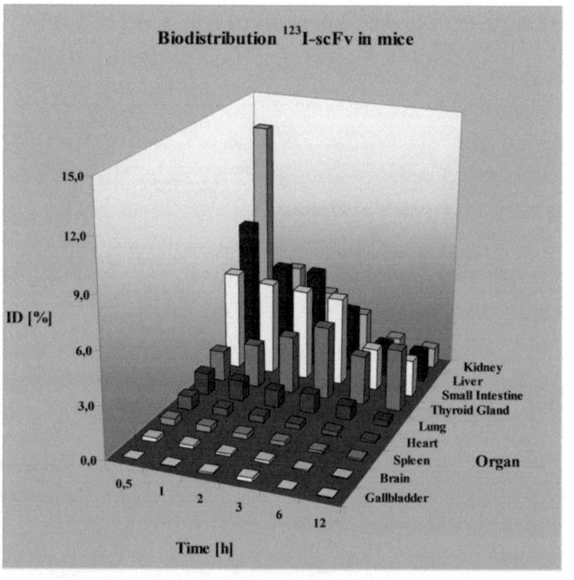

Fig. 15. [^{123}I]scFv biodistribution in normal mice.

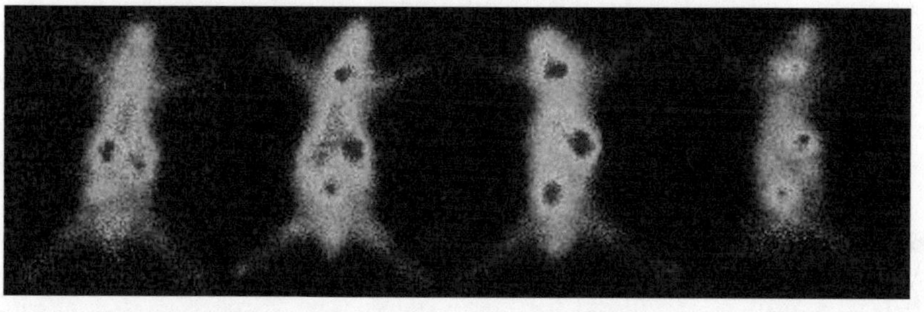

Fig. 16. [^{123}I]scFv TU-20 SPECT camera images –biodistribution study in kinetic intervals 0,5, 1, 2 and 3 h.

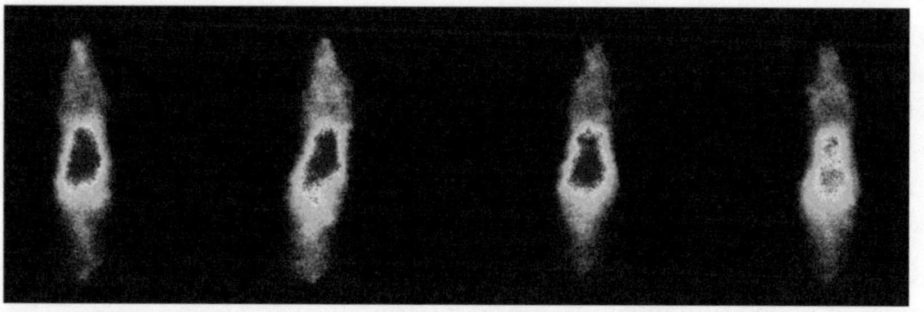

Fig. 17. ^{123}I]-TU-20 SPECT camera images –biodistribution study in kinetic intervals 1, 2, 3, 6 h.

2.9 In vivo biodistribution testing in genetically modified mice

Transgene population G93A1 Gur was used for comparative study to show different behavior of the substances in normal mouse and in modified organism with amyotrophic lateral sclerosis (ALS). Biodistribution kinetic intervals were 3 h (^{125}I-scFv) and 6 h (^{125}I-TU-20). (Heiman-Patterson T.D., 2005) (Naini A., 2007)

3. Conclusion

TU-20 and its scFv were labeled with ^{125}I and ^{123}I by chloramine-T (with average yield 0,72 and 0,50, resp.). Radiochemical purity and stability was revealed by gel filtration (decrease to 80 % and 50 % in two months, resp.) Fragmentation of the labeled antibody and its fragment was estimated by bis-tris gel electrophoresis followed by silver staining and autoradiography (over 95 % of radioactivity bound in the substances).

Affinity coupling and RIA adaptation for the specific conditions showed 10-30 % preserved immunoreactivity of the labeled compounds. Otherwise, these methods carry out quite high discrepancy and it will be neccessary to provide further optimising search.

In vitro studies performed on mice brain slices confirmed several important assumptions. The antibody is preferentially bound in the layer of Purkinje cells in the cerebellum. SPECT camera in vivo experiment deals with these results: activity bound in scFv is primarily distributed to the thyroid gland and digestive tract, then passes quickly through kidneys.

Distribution images of the labeled TU-20 provides ambigous because the substance is accumulated in the chest and ventral part and image resolution do not afford more detailed biodistribution identification. However, it is known from previous biodistribution preparative study that activity is distributed in lung, heart, liver, stomach and colon in first 6 h.

In vivo experiments were focused on investigation of the blood clearance and organ distribution of the radiolabeled TU-20 and scFv fragment in mice. Let´s show especially the results from scFv biodistribution study in preference. It was verified that the major part of activity, according to the amount of the labeled scFv fragment, was eliminated from blood during 2-3 hours. Minor part of activity, according to the amount of the labeled scFv fragment (0,5 - 1,0 %), was kept in the blood for some days. The value $t_{1/2\alpha}$ for ^{125}I-labeled scFv fragment was calculated as 2,3 h and the $t_{1/2\beta}$ was estimated as 62,4 h. The half-life for overall elimination of Na^{125}I from blood was 4,5 h.

In comparison, we found that the ^{125}I-labeled scFv fragment uptake in thyroid gland appeared much lower than for Na^{125}I, as expected. The $t_{1/2\alpha}$ value for ^{123}I-labeled scFv fragment was calculated as 1,4 h, but the long phase elimination half-life $t_{1/2\beta}$ was not estimated due to short half-life of the isotope ^{123}I. The radiolabeled scFv fragment passed in general through the digestive tract (stomach and intestine) and finally was eliminated through kidneys in preference.

TU-20 and ScFv TU-20 showed suitable properties for further investigation in animals which are genetically modified mutants with the ALS (Amyotrophic Lateral Sclerosis). Comparing biodistribution experiments in modified organism confirmed expected behavior. The most significant biodistribution differences occured in the area of the limbs and caudal part of spinal cord and spine.

Finally, as I can summarize, TU-20 and its scFv fragment were successfully labeled with radioiodine ^{123}I and ^{125}I, and, subsequently, the biochemical and analytical characteristics were investigated. Biological properties of the radiolabeled TU-20 and its scFv were evaluated in vivo by biodistribution studies.

The expected behavior of biomolecules during their elimination was observed. Furthermore, the elimination parametres were calculated. ^{125}I-labeling of the TU-20 and its scFv is very suitable for investigation of the radiolabeled antibody fragment behavior and properties due to the long ^{125}I half-life. On the other hand, ^{123}I-labeling of the scFv fragment TU-20 is intended for practical imaging at SPECT camera.

In summary, TU-20 shows better immunospecific behavior in organism together with slower kinetics, on the other hand, scFv TU-20 reveales worse immunospecific characteristics in combination with much faster kinetics.

4. Acknowledgement

This work was supported by the projects No. E!3177 - DIAGIM (1P040E167) and E!2510 - NEUROTUB (0E91) of EUREKA, and by the project No. IBS1048301 of Grant Agency CAS.

5. References

Bernard, A., Goret, A., Buchet, J. P., Roels, H. & Lauwerys, R. (1979). Comparison of sodium dodecyl sulfate-polyacrylamide gel electrophoresis with quantitative methods for the analysis of cadmium-induced proteinuria, *International Archives of Occupational and Environmental Health* 44: 139–148. 10.1007/BF00381129.
URL: *http://dx.doi.org/10.1007/BF00381129*

Chizzonite, R., Truitt, T., Podlaski, F., Wolitzky, A., Quinn, P., Nunes, P., Stern, A. & Gately, M. (1991). Il-12: monoclonal antibodies specific for the 40-kda subunit block receptor binding and biologic activity on activated human lymphoblasts, *The Journal of Immunology* 147(5): 1548–1556.
URL: *http://jimmunol.org/content/147/5/1548.abstract*

Dráberová, E., Lukáš, Z., Ivanyi, D., Viklický, V. & Dráber, P. (1998). Expression of class iii β-tubulin in normal and neoplastic human tissues, *Histochemistry and Cell Biology* 109: 231–239. 10.1007/s004180050222.
URL: *http://dx.doi.org/10.1007/s004180050222*

Hamilton, S., O. J. W. G. K. J. (2002). Reducing renal accumulation of single-chain fv against melanoma-associated proteoglycan by coadministration of l-lysine, *Melanoma Res.* 12: 373–379.

Haugland, R. P. (1995). Coupling of monoclonal antibodies with enzymes, *in* W. C. Davis & J. M. Walker (eds), *Monoclonal Antibody Protocols*, Vol. 45 of *Methods in Molecular Biology*, Humana Press, pp. 235–243. 10.1385/0-89603-308-2:235.
URL: *http://dx.doi.org/10.1385/0-89603-308-2:235*

Heiman-Patterson T.D., Deitch J.S., B. E. (2005). Background and gender effects on survival in the tgn(sod1-g93a)1gur mouse model of als, *J. Neurol. Sci.* 236: 1–7.

Holliger, P., H. P. (2005). Engineered antibody fragments and the rise of single domains, *Nat. Biotechnol* 23: 1126–1136.

Jones, N. A. (2004). Two-dimensional polyarcylamide gel electrophoresis for proteome analyses, *in* P. Cutler (ed.), *Protein Purification Protocols*, Vol. 244 of *Methods in Molecular Biology*, Humana Press, pp. 353–359. 10.1385/1-59259-655-X:353.
URL: *http://dx.doi.org/10.1385/1-59259-655-X:353*

Katsetos, C.D., H. M. M. S. (2003). Class iii β-tubulin in human development and cancer. cell. motil. cytoskeleton, *Cell. Motil.* 55: 77–96.

Laemmli, U. (1970). Cleavage of structural proteins during the assembly of the head of bacteriophage t4, *Nature* 227: 680–685.

Naini A., Mehrazin M., L. J. G. P. M. H. (2007). Identification of a novel d109y mutation in cu-zn superoxide dismutase (sod1) gene associated with amyotrophic lateral sclerosis, *J. Neurol. Sci.* 254: 17–21.

Rédei, G. P. (2008). Electrophoresis, *Encyclopedia of Genetics, Genomics, Proteomics and Informatics*, Springer Netherlands, pp. 591–592. 10.1007/978-1-4020-6754-99_5192. URL: *http://dx.doi.org/10.1007/978-1-4020-6754-9_5192*

Robles, A. M., B. H. O. P. W. M. P. E. (2001). Improved radioiodination of biomolecules using exhaustive chloramine-t oxidation, *Nucl. Med. Biol.* 28: 999–1008.

Švecová, H., Kleinová, V., Seifert, D., Chaloupková, H., Bäurle, J., Kranda, K., Král, V. & Fišer, M. (2008). Radioiodination of mouse anti-iii β-tubulin antibodies and their evaluation with respect to their use as diagnostic agents for peripheral neuropathies, *Applied Radiation and Isotopes* 66(3): 310 – 316. URL: *http://www.sciencedirect.com/science/article/pii/S0969804307002813*

Vijayalakshmi, M. A. (1992). Histidine ligand affinity chromatography, *in* A. Kenney, S. Fowell & J. M. Walker (eds), *Practical Protein Chromatography*, Vol. 11 of *Methods in Molecular Biology*, Humana Press, pp. 33–44. 10.1385/0-89603-213-2:33. URL: *http://dx.doi.org/10.1385/0-89603-213-2:33*

23

Gel Electrophoresis as a Tool to Study Polymorphism and Nutritive Value of the Seed Storage Proteins in the Grain Sorghum

Lev Elkonin, Julia Italianskaya and Irina Fadeeva
Agricultural Research Institute for South-East Region
Russia

1. Introduction

Seed storage proteins of cereals constitute the basis of mankind nutrition. However, climate changes, especially, increased droughts that are distinctly observed in many regions all over the globe, hamper sustainable production of traditional cereals, such as wheat, maize and barley, and dictates necessity to cultivate drought resistant and heat tolerant crops. Among these crops, the grain sorghum, owing to its ability for sustainable grain production in conditions of minimal level of precipitation, takes one of the leading places. However, application of sorghum grain for food and feed purposes is limited by its relatively low nutritive value in comparison with other cereals.

One of the reasons of poor nutritive value of sorghum grain is the resistance of its seed storage proteins (kafirins) to protease digestion. Kafirins are alcohol-soluble prolamin proteins making up to 80% of endosperm sorghum proteins (Hamaker et al., 1995). As well as other prolamins, sorghum kafirins contain high levels of proline and glutamine and are deposited in protein bodies of endosperm cells during kernel development. According to differences in solubility in aqueous *tert*-butanol solutions, molecular weight, structure and immunochemical similarity to zeins (maize prolamins) the kafirins were classified into α-, β- and γ-kafirins (Shull et al., 1991; for review, see: Belton et al., 2006). The α-kafirins are highly hydrophobic prolamin proteins (soluble in 40-90% aqueous *tert*-butanol solutions), they comprise 66-84% of total kafirins, depending on the endosperm type (vitreous or opaque). By SDS-PAGE the α-kafirins usually are resolved into two proteins, 25 kDa and 23 kDa. The γ-kafirin accounts for 9-21% of total kafirins depending on the endosperm type (Waterson et al., 1993). According to immunochemical data, the γ-kafirin is a protein with molecular mass of 28 kDa (Shull et al., 1991) although the sequence of the γ-kafirin gene corresponds to the protein with molecular mass of about 20 kDa (De Barros et al., 1991). The β-kafirin, in different endosperm types, accounts for about 7-13% of the total kafirins, and is resolved by the SDS-PAGE into three bands of 20, 18 and 16 kDa (Shull et al., 1991; 1992) or produced one band of 20kDa (El Nour et al., 1998); such variability, perhaps, is due to genotype differences.

One of the main characteristic features of kafirin proteins is their ability to form olygo- or polymers of high molecular weight. These oligomers comprise α- and γ-kafirins linked together by disulphide (S-S) bonds, which are formed by sulphur-containing amino acids (Nunes et al., 2005). In the native state, both mono- and oligomers are present, while in 'reduced' extracts (i.e. with addition of 5% 2-mercaptoethanol that destroys S-S bonds) only monomers were detected (El Nour et al., 1998).

The causes of the poor kafirin digestibility appear to be multi-factorial (Duodu et al., 2003). Among these factors are chemical structure of kafirin molecules, some of which (γ- and β-kafirins) are abundant with sulfur-containing amino acids that are capable to form S-S bonds, resistant to protease digestion; interactions of kafirins with non-protein components such as polyphenols and polysaccharides; and spatial organization of different kafirins in the protein bodies of endosperm cells.

Among the methods that were developed for investigation of sorghum protein digestibility (Pedersen & Eggum, 1983; Mertz et al., 1984; Aboubacar et al., 2003), pepsin digestion of the flour proteins with subsequent gel electrophoresis is the most informative. This method, originally applied by B. Hamaker and co-workers (Weaver et al., 1998; Aboubacar et al., 2001) has been used in a number of studies (Nunes et al., 2004; Wong et al., 2010). Application of this method allowed to isolate sorghum lines with high protein digestibility (Weaver et al., 1998) and to find out that γ-kafirin plays an important role in resistance of sorghum seed storage proteins to protease digestion, namely, γ-kafirin forms a disulfide-bound enzyme-resistant layer at the periphery of protein bodies that restricts access of proteases to the inferior-located and more easily digested α-kafirins (Oria et al., 2000).

In our investigations (Italianskaya et al., 2009), we studied the protein digestibility in different sorghum lines and hybrids using this method and revealed significant polymorphism for *in vitro* kafirin digestibility as well as the strong genetic bases of this trait. In this paper, we summarize the results of these studies, which allowed isolating sorghum lines and F_1 hybrids with increased nutritive value. In addition, we demonstrate that kafirin polymorphism may be used in genetic experiments, namely, in determination of genetic structure of endosperm in sorghum.

2. Material and methods

In vitro protein digestibility was studied in 10 lines and seven F_1 hybrids of the grain sorghum (*Sorghum bicolor* (L.) Moench) (Table 1).

To study *in vitro* protein digestibility the modified method of whole-grain flour pepsin treatment was used (Oria et al. 1995). For each variety 25 mg of flour was treated with 5 ml of 0.15% pepsin solution (P7000 Sigma-Aldrich) in the 0.1 M potassium-phosphate buffer (pH 2.0) for 120 min at 37 ^{0}C with repeated shaking. Analysis of seed storage protein (kafirin) spectra was performed before and after pepsin treatment by SDS-PAG electrophoresis (SDS-PAGE) in reducing conditions. SDS-PAGE was carried out in the 12.5% (w/v) acrylamide separating gel (0.375 M TRIS·HCl, pH 8.8) and 4% stacking gel (0.125 M TRIS, pH 6.8) according to modified Laemmli method (Laemmli, 1970). SDS-reducing buffer: 62.5 mM TRIS·HCl, pH 6.8, 20% glycerol, 2% SDS, 5% β-mercaptoethanol; running buffer: 25.0 mM TRIS·HCl, 192 mM glycine, 0.1% SDS, pH 8.3; spacer thikness 1.00 mm. Gels were electrophoresed at 20-23 ma for about 5 hr. Gels were stained with Coomassie Brilliant Blue G-250 or R-250 (Diezel et al, 1972).

Line, F_1 hybrid[1]	Grain color	Endosperm type
VIR-120	white	floury
Pishchevoe-614 (P-614)	light-brown	semi-vitreous
Volzhskoe-4	light-brown	floury
Volzhskoe-4 waxy (V-4w)	pink	semi-vitreous
Karlikovoe beloe (KB)	white	semi-vitreous
Milo-10	yellow	floury
KVV-45	white	semi-vitreous
KVV-97	white	vitreous
KVV-3	white	semi-vitreous
KP-70	creamy	semi-vitreous
Topaz	creamy	semi-vitreous
O-1237	white	semi-vitreous
Sudzern svetlyi (Sud)	creamy	semi-vitreous
F_5 [M35-1A] Pishchevoe-614/KVV-45	white	semi-vitreous
A2 Karlikovoe beloe/Pishchevoe-614 (A2 KB/P-614)	light-brown	semi-vitreous
A2 Karlikovoe beloe/KP-70 (A2 KB/P-614)	white-yellowish	semi-vitreous
M35-1A Karlikovoe beloe /KVV-45 (M35-1A KB/KVV-45)	white	semi-vitreous
A2 KVV-97/Pishchevoe-614	light-brown	semi-vitreous
A2 Sudzern svetlyi/Topaz (A2 Sud/Topaz)	creamy	semi-vitreous
A2 O-1237/ Pishchevoe-614 (A2 O-1237/P-614)	light-brown	semi-vitreous

In parenthesis: brief designation used in the paper. F_1 hybrids were obtained using male-sterile counterparts of fertile lines; they are designated as A2 or M35-1A depending on the type of male sterility-inducing cytoplasm.

Table 1. The grain sorghum entries used in this investigation

For quantitative estimation of kafirin digestibility the SDS-PAGE banding patterns were scanned by laser densitometer ULTROSCAN XL (LKB-Pharmacia) with wavelength 633 nm. The protein quantity in each fraction was expressed as the area (mm^2) of the appropriate peak on densitogram, which was calculated by Software LKB 2222 (Version 3.00). In some experiments, the SDS-PAGE banding patterns were analyzed by Scangel program (developed by Dr. A.F. Ravich). The protein quantity in each fraction and in each lane of electrophoregramm was expressed as the amount of dots in the appropriate protein band. Experiments were performed in two replications. The data on digestibility of kafirins (the ratio of protein peak area before and after pepsin digestion) were subjected to variance analysis using the program Agros (Version 2.09; Dr. S. Martynov, Wheat Genetic Resources Department, N.I. Vavilov Institute of Plant Production, St. Petersburg, Russia).

In some lines and hybrids, the dependence of *in vitro* protein digestibility from *in vitro* starch digestibility was studied. In this experiment, the flour, firstly, was subjected to amylolitic enzyme treatment according to the method of B.V. McCleary (McCleary et al., 2002) using Megazyme Resistant Starch Kit (Megazyme Co, Ireland). The pellet remained after removal of solubilised starch was used for pepsin treatment according to the method described above, and, after that, the protein spectrum of the sample was studied by SDS-PAGE.

In order to use kafirins as markers of genetic structure of endosperm the modified technique of SDS-PAGE was applied. In these experiments, AS-1a line of the grain sorghum, which is characterized by a low frequency of parthenogenic embryo formation (Elkonin et al., 2012)

was used. Emasculated panicles of this line were pollinated with the pollen of the line Volzhskoe-4w homozygous for dominant gene *Rs*, conditioning purple color of coleoptiles, seedling leaves and stem. To study the origin of the kernels (apomictic or sexual) with the aid of the kafirin polymorphism, the kernels were split into two parts. The part with an embryo was put in a tray on a moisture filter paper to study the phenotypic traits of a seedling (expression of the *Rs* gene). Another part was used in SDS-PAGE to study its kafirin spectrum. In these experiments, gels were electrophoresed at constant voltage (70 V) for about 15 hr. Gels were stained with $AgNO_3$ solution.

3. *In vitro* kafirin digestibility

SDS-PAGE spectra of the seed storage proteins of a number of lines used in our investigations, before and after pepsin digestion, are shown on Figures 1 and 2.

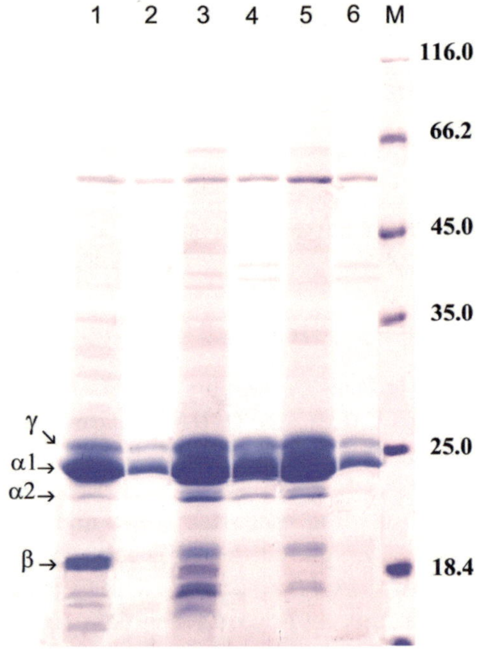

Fig. 1. Electrophoretic patterns of sorghum seed storage proteins before (1, 3, 5) and after (2, 4, 6) pepsin digestion. Lanes 1, 2 – Volzhskoe-4; 3,4 – Pishchevoe-614; 5,6 – F_5 [M35-1A] Pishchevoe-614/KVV-45; M – molecular weight markers (kDa).
α, β, γ – individual kafirin fractions. Gels were stained with Coomassie Brilliant Blue R-250.

In electrophoretic spectra of sorghum lines subjected to pepsin digestion, one could clearly distinguish the γ- (28 kDa), α1 (25 kDa) and α2 (23 kDa) kafirins and one or several bands of β-kafirin fractions (Fig. 1). These electrophoretic patterns correspond to kafirin spectra previously described in the literature (Shull et al., 1991; El Nour et al., 1998; Nunes et al., 2004). In our previous investigations (Table 2) we determined the relative content of different kafirin fractions and observed significant variation among different cultivars. The α1 and γ-kafirins were the most abundant in all lines and hybrids tested: 24-37% and 10-13% of all endosperm proteins, respectively; β-kafirins represent relatively small fractions

(4-10%) that is in concordance with the literature data (Shull et al., 1991; Waterson et al., 1993).

Line, F_1 hybrid	Protein fraction, % [1]			
	γ	α1	α2	β
KVV-45	13.2	37.3	2.0	3.9
Milo-10	12.8	30.7	5.3	7.2
A2 KVV-97	13.1	24.3	3.2	5.7
A2 KVV-97/P-614	13.3	31.4	4.4	5.3
P-614	10.7	26.9	5.5	5.5
A2 KVV-114	10.8	26.0	4.1	6.9
A2 KVV-114/V-4w	9.5	24.9	4.6	8.9
V-4w	10.3	33.1	3.4	10.2

[1] Relative content of each fraction is expressed as percentage of its peak area from the total endosperm proteins peak area sum. Mean data of two replications.

Table 2. Relative content of different kafirin fractions in some sorghum lines and F_1 hybrids (Italianskaya et al., 2009)

After pepsin digestion the amount of protein in kafirin fractions substantially reduced (Figs. 1; 2). Different sorghum lines and cultivars differed significantly by this trait. For example, among the entries presented in Figure 2 the highest digestibility level had VIR-120 - 90.8% (lanes 1 and 2), while the kafirins of line KVV-3 (lanes 9 and 10) were the most resistant to pepsin digestion (54.5% digestibility level) (Table 3).

In our previous study (Italianskaya et al., 2009), we observed significantly higher variation among the lines. For example, in the cultivar Volzhskoe-4 (V-4, registered standard), the amount of undigested γ- and α-kafirins after pepsin digestion was 80% and 73% from their initial contents, respectively. The total amount of undigested kafirins in cv. V-4 was 70% (digestibility level was 30%). At the same time, in the line KVV-45, the total amount of undigested proteins was 37% (digestibility level was 63%). Percentage of undigested α1 and γ-kafirins in the line KVV-45 was only 25% and 30%, respectively. The differences in kafirin spectra between this line and cv. V-4 before and after pepsin treatment are clearly seen in the Figure 3. Further investigation confirmed a high level of protein digestibility in this line (78.4%) (Table 3). Perhaps, the line KVV-45 contains mutation(s) in the genes encoding structure or deposition of kafirin molecules and, therefore, is of a great interest for future experiments.

Remarkably, in subsequent investigation it was found that in the line Topaz the digestibility level was even higher than in the KVV-45 and reached 89% (see chapter 4). This value is sufficiently high; it corresponds to digestibility level of whole grain flour protein of the best condenced-tannin-free sorghum entries (Axtell et al., 1981, and other reports, as cited in Duodu et al., 2003). One should expect that this line would have high nutritive value.

One should note high digestibility of the β-kafirin fractions in majority of lines. This fact contradicts to hypothesis that explains poor kafirin digestibility by formation of S-S bonds because β-kafirins as well as γ-kafirins contain a high amount of cystein, a sulfur-containing amino acid (Belton et al, 2006). In addition, in all lines, the polypeptides with molecular weight approx. 42 and 46 kDa were prominent in electrophoretic spectra after pepsin digestion. These polypeptides, perhaps, represent kafirin dimers, which were formed as a result of association of kafirin monomers. Earlier, the formation of similar polypeptides (45 kDa) was observed after the cooking process (Duodu et al., 2003; Nunes et al., 2004).

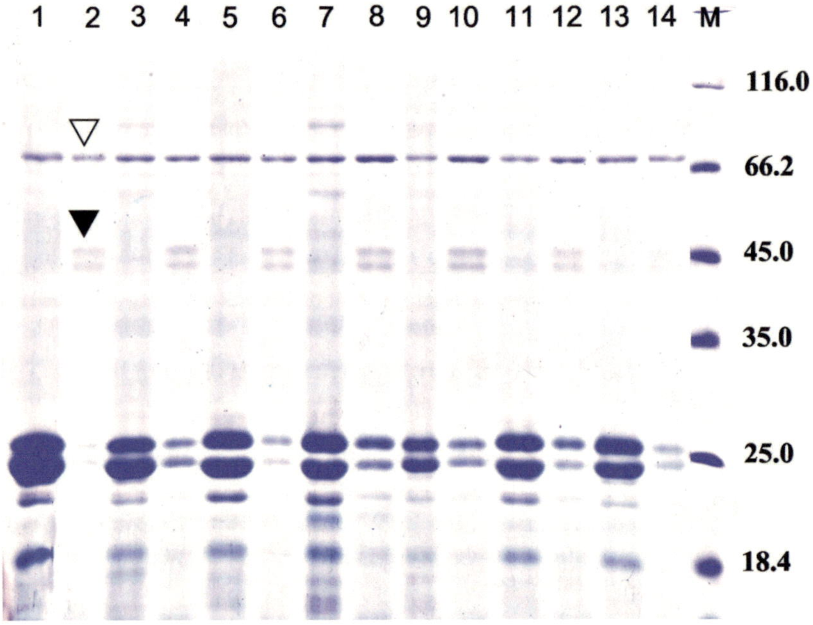

Fig. 2. Electrophoretic patterns of sorghum seed storage proteins before (1, 3, 5, 7, 9, 11, 13) and after (2, 4, 6, 8, 10, 12, 14) pepsin digestion. Lanes 1, 2 - VIR-120; 3, 4 - Volzhskoe-4w; 5, 6 - KVV-45; 7, 8 - KVV-97; 9, 10 - KVV-3; 11, 12 - Karlikovoe beloe; 13, 14 - KP-70; M – molecular weight markers (kDa). di- and trimers of kafirins are indicated by arrows, ◀ and ◁, respectively. Gels were stained with Coomassie Brilliant Blue R-250.

Lane number	Line	Total amount of dots in the lanes		Amount of undigested protein, %	Digestibility, %
		control	after pepsin digestion		
1,2	VIR-120	9769124	897710	9.2	90.8
3,4	Volzhskoe-4w	7285338	2692241	37.0	63.0
5,6	KVV-45	16465667	3554046	21.6	78.4
7,8	KVV-97	26995517	9483915	35.1	64.9
9,10	KVV-3	12242662	5571704	45.5	54.5
11,12	Karlikovoe beloe	13897393	4335642	31.2	68.8
13,14	KP-70	14462063	3651537	25.2	74.8

Table 3. Densitometry of electrophoretic patterns of seed storage proteins shown in Figure 2. The SDS-PAGE banding patterns were scanned and analyzed by Scangel program (developed by Dr. A.F. Ravich)

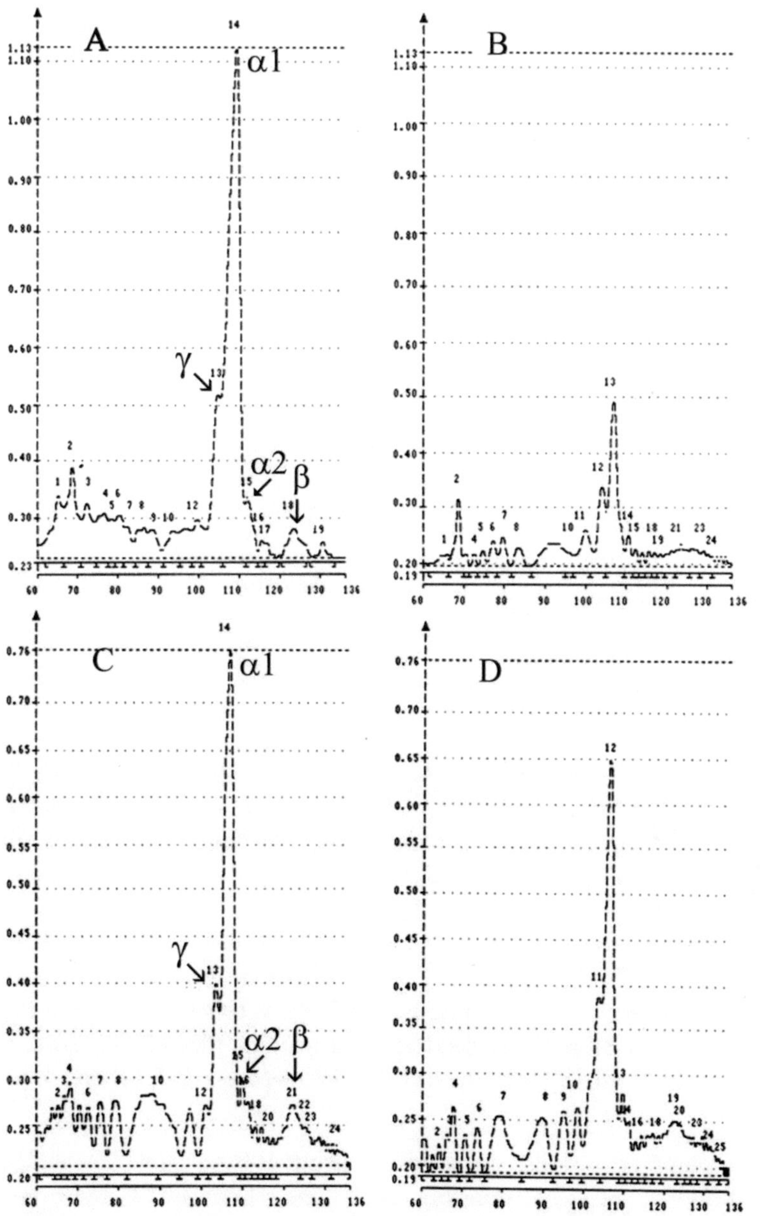

Fig. 3. Densitograms of electrophoretic spectra of endosperm proteins of sorghum line KVV-45 (a, b) and cultivar Volzhskoe-4 (c, d) before (a, c) and after (b, d) pepsin digestion. α1, α2, β and γ-kafirin fractions are indicated.

In order to explore the genetic basis of kafirin digestibility, we studied the expression of this trait in the F_1 hybrids between parental lines differing by resistance to pepsin digestion. Comparison of kafirin digestibility in the F_1 hybrids and their parental lines showed that different hybrid combinations had different mode of inheritance of resistance to pepsin affect (Table 4).

Line, F_1 hybrid[1]	Amount of undigested protein, percent from untreated sample[1]			
	γ	α1	β	Total proteins
KVV-45	24.4	24.6	32.2	24.5 a
M35-1A Karlikovoe beloe /KVV-45	36.2	33.9	34.2	26.8 ab
Karlikovoe beloe	21.3	37.2	26.0	32.1 bcd
A2 Karlikovoe beloe /KP-70	39.5	51.5	42.3	41.6 g
KP-70	22.4	29.5	22.3	26.1 a
A2 Karlikovoe beloe/Pishchevoe-614	41.1	51.3	42.3	40.4 efg
Pishchevoe-614	53.4	64.5	34.7	33.7 cd
A2 KVV-97/Pishchevoe-614	48.9	55.7	44.3	40.5 fg
KVV-97	40.4	30.3	20.4	34.2 d
$F_{0.05}$				14.76*
$LSD_{0.05}$				5.4

[1] Mean from two replications. Data followed by the same letter did not differ significantly ($p<0.05$) according to Duncan Multiple Range Test.
* Significant at $p<0.05$.

Table 4. *In vitro* protein digestibility of endosperm proteins in F_1 sorghum hybrids and their parental lines

The F_1 hybrids A2 KB/P-614, A2 KB/KP-70 and A2 KVV-97/P-614 had significantly lower kafirin digestibility than parental lines, which were characterized by its relatively high level. The reasons of such negative heterosis are unclear. Perhaps, genetic factors conditioning relatively high kafirin digestibility of KP-70, KB and P-614 are recessive and locate in different loci. At the same time, the F_1 hybrid M35-1A KB/KVV-45 did not differ from parental lines and retained high level of kafirin digestibility of the line KVV-45. Perhaps, high digestibility of KVV-45 contrary to other lines may be controlled by any dominant gene(s). This hybrid as well as the line KVV-45, is of great importance for fundamental investigation of factors influencing seed storage protein digestibility in sorghum (kafirin gene structure, structural organization of protein bodies and others) and for practical breeding.

Strong effect of genotype was also found on spectrum of high-molecular weight kafirins that were observed after pepsin digestion (Fig. 4). In some lines and F_1 hybrids two peaks di- and trimers) were found (Fig. 4, A-C), while in others only one peak (trimers) was seen (Fig. 4, D-F). Remarkably, densitograms of the F_1 hybrids in the peak area clearly resembled parental ones. One should note that while the peaks corresponding to trimers were observed in electrophoretic spectra already before pepsin treatment and their amount usually reduced after that, the dimers (45 kDa) were observed only after pepsin action. In some entries kafirin polymers were highly resistant to pepsin digestion, as in the KVV-45, while in others, as in the line P-614 and F_1 hybrid A2 KVV-97/P-614 (Fig. 5, A,B), these peaks were faint or almost absent. These data point on the genetic bases of formation of these molecules, which affect nutritive value of sorghum grain.

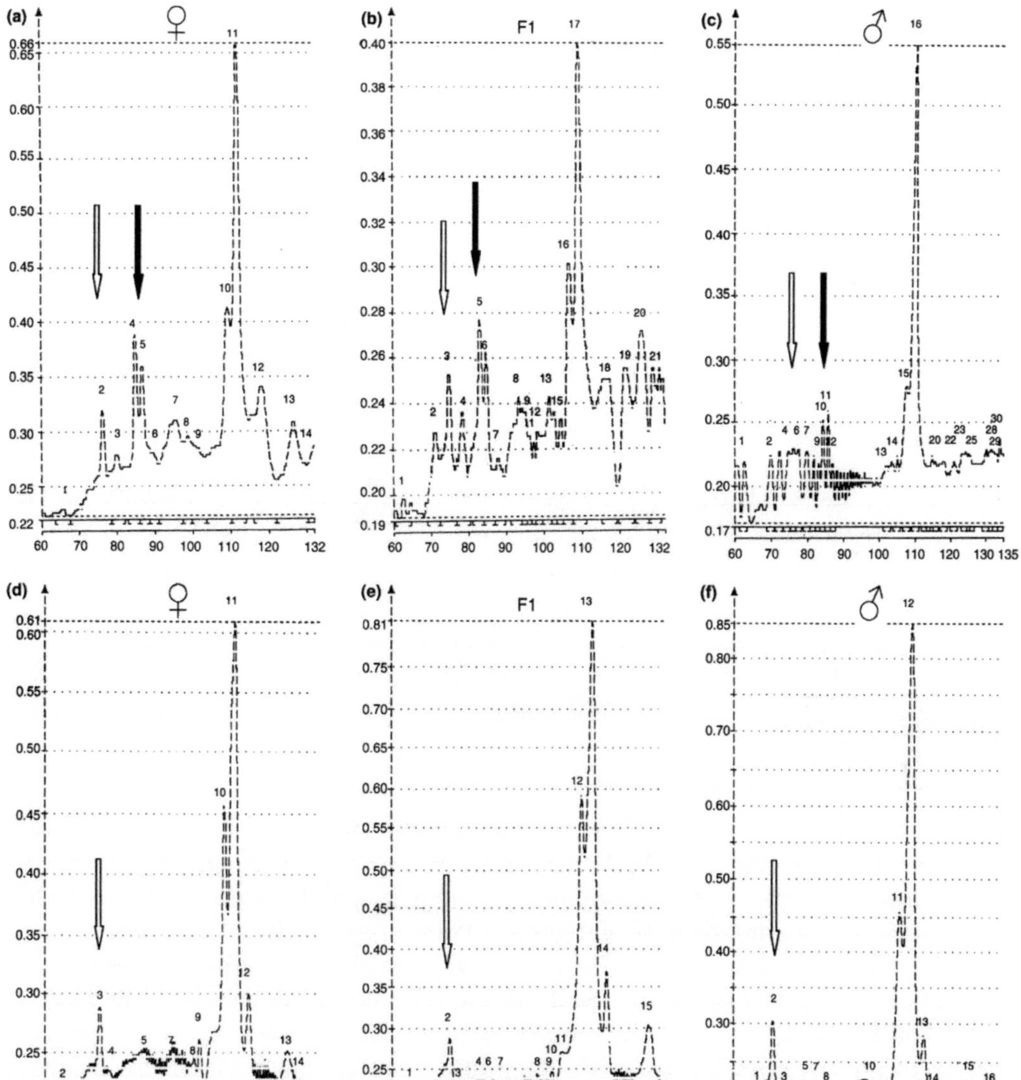

Fig. 4. Densitograms of endosperm proteins electrophoretic spectra of F_1 hybrids and their parental lines after pepsin digestion: a - A2 KVV-114, b - F_1 A2 KVV-114/V-4w, c - V-4w, d - A2 KB, e - F_1 A2 KB/KP-70, f - KP-70. Fractions of di- and trimers of kafirin proteins (45kDa and 66 kDa) are shown by arrows, ➡ and ⇨, respectively.

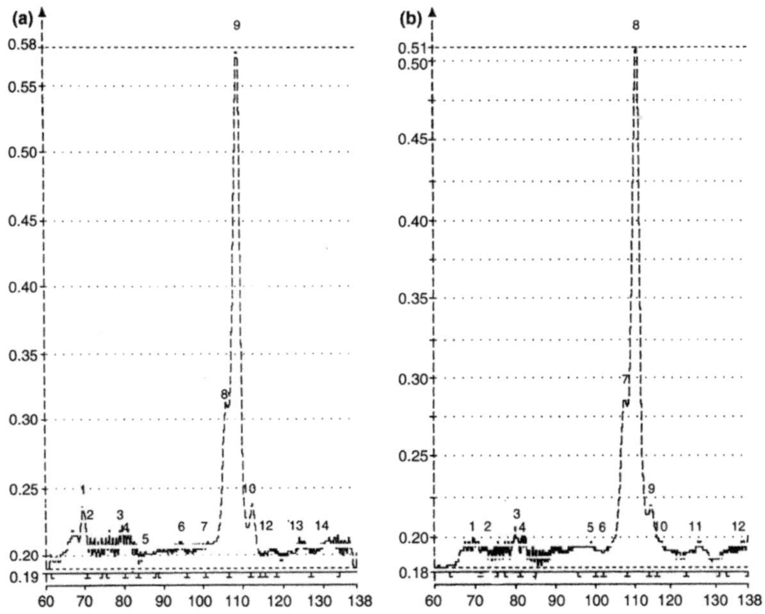

Fig. 5. Densitograms of endosperm proteins electrophoretic spectra of the line P-614 (a) and F_1 hybrid A2 KVV-97/P-614 (b) after pepsin digestion.

4. Interaction of starch and protein digestibility

In order to found out dependence of sorghum protein digestibility on starch digestibility the flour of several lines and F_1 hybrids was subjected to pepsin action after removal of digestible starch by the amylolytic enzymes treatment, and then was studied by SDS-electrophoresis for the presence of undigested proteins. It was found that after action of amylolitic enzymes the amount of protein in the kafirin fractions significantly increases (Fig. 6): in the lanes 3, 7 and 11 (samples after amylolitic enzyme action) almost all the protein is concentrated in the kafirin fractions, in comparison with the lanes 1, 5 and 9 (samples without amylolitic enzyme action). However, contrary to expectation that removal of starch will favor to kafirin digestion, the pepsin treatment of the samples treated before it with amylolitic enzymes (lanes 4, 8 and 12) were digested significantly fewer than samples digested by pepsin only (lanes 2, 6 and 10). Gel densitometry confirmed this visual conclusion (Table 5). Such phenomenon was observed in all F_1 hybrids studied (A2 Sud/Topaz, A2 O-1237/P-614, M35-1A KB/KVV-45) and their parental lines. Perhaps, partially digested starch molecules may interact with kafirin molecules by any physical or, probably, chemical way and prevent their protease digestion. One should not exclude that similar process might take place in *in vivo* conditions and thus decrease sorghum protein digestibility and reduce its nutritive value.

In addition, it was found that after amylolitic enzyme treatment the amount of di- and trimer fractions significantly reduced in comparison with the non-fermented control samples. In the F_1 hybrid A2 Sud/Topaz their amount was significantly fewer even in comparison with pepsin treatment only. Such a reduction of kafirin oligomers may be also responsible in increase of the level of kafirin monomers. These data testify that starch molecules might participate in formation of kafirin oligomer molecules. They are important for understanding the factors influencing kafirin and starch interactions in sorghum endosperm and their digestibility.

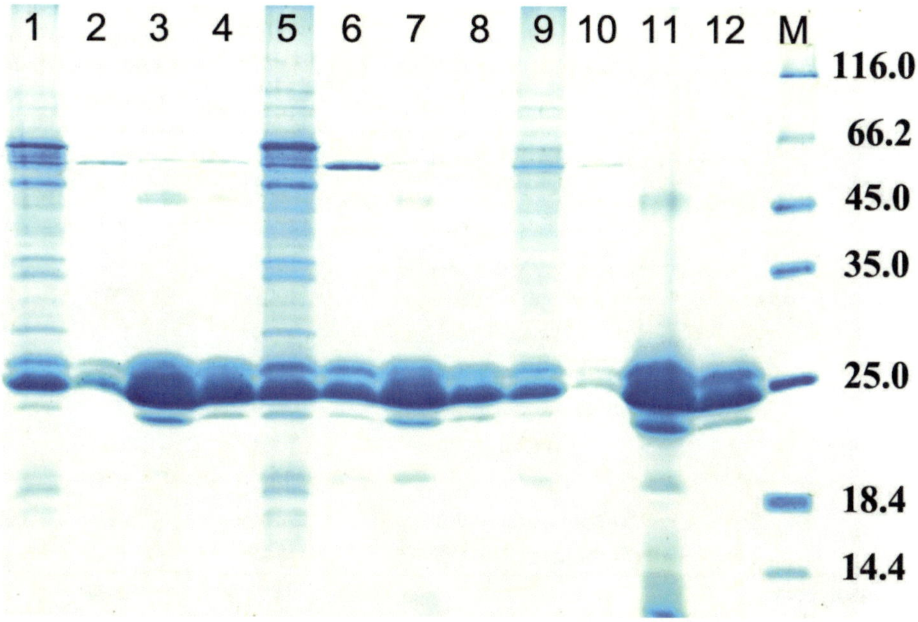

Fig. 6. Electrophoretic patterns of sorghum seed storage proteins from the flour before (1, 3, 5, 7, 9, 11) and after (2, 4, 6, 8, 10, 12) pepsin digestion; lanes 3, 4, 7, 8, 11, 12 - after removal of soluble starch by amylolitic enzymes before pepsin digestion; lanes 1, 2, 5, 6, 9, 10 - without this procedure. Lanes 1-4 - Sudzern svetlyi; 5-8 - F_1 A2 Sudzern svetlyi/Topaz; 9-12 - Topaz; M - molecular weight markers (kDa). Gels were stained with Coomassie Brilliant Blue G-250.

5. Kafirins as the markers of endosperm genetic structure

In addition to variation of a number of β-kafirin fractions in different sorghum entries described above, we have revealed polymorphism of the α-kafirins. The line Volzhskoe-4w (V-4w) that is used as a tester line to distinguish the hybrid seedlings from the maternal ones, possessed specific kafirin spectrum, which was rarely observed in other sorghum lines and cultivars. The α1 fraction was composed from three polypeptides: α1-1, α1-2, and α1-3; α2 fraction was composed from two polypeptides: α2-1 and α2-2 (Fig. 7, lanes 1-3). We hypothesized that this polymorphism could be used in studies of genetic structure of endosperm in apomixis research in sorghum.

To test this possibility we used the AS-1a line, which is characterized by ability for development of aposporous embryo sacs and parthenogenetic embryos (Elkonin et al., 2012). Gel electrophoresis showed that kafirin spectrum of this line differs from V-4w (Fig. 7). Two polypeptides were observed in the α1 fraction (α1 and α1-2), the α1-2 was in trace amount, and α1-3 was absent; the α2 fraction did not subdivide into two polypeptides (Fig. 7, lanes 4-6).

Genotype	Experimental treatment	Amount of protein in different kafirin fractions		Total proteins percent to the control
		Individual fractions (α+β+γ), percent to the control	Olygomers percent to the control	
Sudzern svetlyi	Control	100.0	100.0	100.0
	Pepsin	48.2	7.4	23.0
	Amylolitic enzymes	177.0	8.9	76.5
	Amylolitic enzymes, pepsin	102.9	5.5	44.2
A2 Sudzern / Topaz	Control	100.0	100.0	100.0
	Pepsin	66.5	19.1	35.6
	Amylolitic enzymes	111.2	4.5	52.4
	Amylolitic enzymes, pepsin	60.9	1.2	27.1
Topaz	Control	100.0	100.0	100.0
	Pepsin	18.1	10.2	11.1
	Amylolitic enzymes	225.6	24.9	139.4
	Amylolitic enzymes, pepsin	124.7	6.4	64.7
F_A (genotypes)		0.858	3.834	1.338
F_B (treatment)		6.340*	836.245***	7.945*
F_{AB}		0.995	5.964*	1.245
	In average for treatment			
	Control	100.0 a	100.0 c	100.0 b
	Pepsin	44.3 a	12.2 b	23.2 a
	Amylolitic enzymes	171.2 b	12.8 b	89.4 b
	Amylolitic enzymes + pepsin	96.1 a	4.4 a	45.3 a

Mean data of two replications; data followed by the same letter did not differ significantly ($p<0.05$) according to Duncan Multiple Range Test;
*, and *** significant at $p<0.05$, and $p<0.001$, respectively.

Table 5. Densitometry of seed storage proteins electrophoretic patterns of F_1 A2 Sudzern/Topaz and its parents after treatment with pepsin and/or α-amylase and amyloglucosidase

In order to use this polymorphism for identification of seeds formed via apomixis, the kernels obtained by pollination of emasculated panicles of AS-1a with the pollen of V-4w were split into two parts. The part with an embryo was used to study the phenotypic traits of a seedling. Another part was used in SDS-PAGE to study its kafirin spectrum. In the case of autonomous endosperm development, no V-4w proteins should be found in the kafirin spectra of the kernels yielded maternal seedlings, while in the case of pseudogamous endosperm development, in the electrophretic spectra of these kernels, the SDS-PAGE must reveal V-4w proteins. It was found that kafirin spectra of kernels, which yielded maternal seedlings (Fig. 7, lanes 11,12) did not differ from the spectrum of AS-1a line (Fig. 7, lanes 4-6), while in the spectra of the kernels, which yielded hybrid seedlings the α1-3 protein was clearly distinguished (Fig. 7, lanes 7-10). These data support the results of our cyto-embryological observations of autonomous endosperm development in the AS-1a line (Elkonin et al., 2012) and are in accordance with the literature data on other sorghum lines with apomictic potentials (Rao et al., 1978; Wu et al., 1994; Ping et al., 2004).

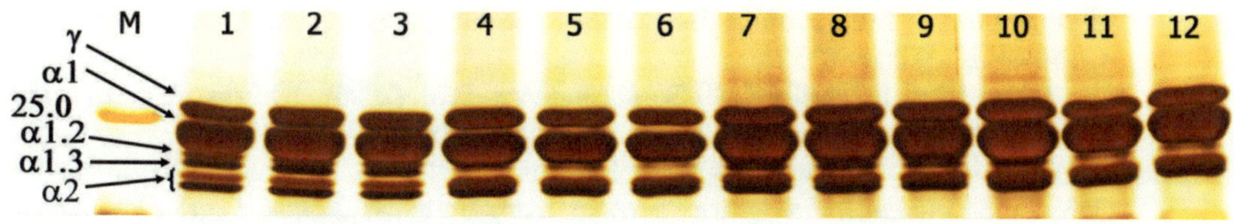

Fig. 7. Kafirin spectra of Volzhskoe-4w (lanes 1-3), AS-1a (4-6) and of the kernels, which were set on emasculated panicles of AS-1a pollinated with the Volzhskoe-4w pollen and yielded the F_1 hybrid seedlings (7-10) and maternal plants (11-12); M - molecular weight marker (kDa). Gels were stained with $AgNO_3$.

6. Conclusion

Summarizing, the results of our investigation demonstrate that gel electrophoresis of the seed storage proteins is a powerful instrument in researches on sorghum genetics and breeding that have both fundamental and applied orientation. It allowed to isolate of sorghum lines with individual kafirin fractions more sensitive to protease action, and, therefore, with increased protein digestibility - one of the main trait characterizing the nutritive value of sorghum grain. These lines may be used in breeding programs for developing new CMS-lines and F_1 hybrids. In addition, these lines (for example, KVV-45) may be used in future investigations on molecular organization of genes encoding structure and/or deposition of kafirins, their cloning and transfer into other sorghum lines by methods of classical genetics or genetic engineering.

Gel electrophoresis of the flour subjected to amylolitic enzyme action has demonstrated that starch digestion decreases content of kafirin polymers and reduces subsequent kafirin digestion by pepsin. This finding may explain the reduced nutrient value of sorghum grain, in comparison with other cereals. These data point on the complex mode of interactions of storage proteins and starch in sorghum endosperm.

Gel electrophoresis of the seed storage proteins allowed to determine genetic structure of endosperm in sorghum kernels with parthenogenic embryos developing in the line AS-1a with apomictic potentials and may be used in development of sorghum lines with high frequency and stable expression of this trait.

7. Acknowledgement

Authors are grateful to Dr. Alexander Ravich for the Software Scangel. This work was funded partly by the Russian Foundation for Basic Researches, grant 10-04-00475.

8. References

Aboubacar, A.; Axtell, J. D.; Huang, C.P. & Hamaker, B.R. (2001). A rapid protein digestibility assay for identifying highly digestible sorghum lines. *Cereal Chemistry,* Vol. 78, pp. 160-165, ISSN 0009-0352

Aboubacar, A.; Axtell, J. D., Nduulu, L. & Hamaker, B. R. (2003). Turbidity assay for efficient identification of high protein digestibility sorghum lines. *Cereal Chemistry,* Vol. 80, pp. 40-44, ISSN 0009-0352.

Belton, P.S.; Delgadillo, I.; Halford, N.G. & Shewry, P.R. (2006). Kafirin structure and functionality. *J Cereal Science,* Vol. 44, pp. 272-286, ISSN 0733-5210

De Barros, E.G.; Takasaki, K.; Kirleis, A.W. & Larkins, B.A. (1991). Nucleotide sequence of a cDNA clone encoding γ-kafirin protein from *Sorghum bicolor. Plant Physiolology,* Vol. 97, pp. 1606–1607, ISSN 0032-0889

Diezel, W.; Kopperschlader, G. & Hofman, E. (1972). An improved procedure for protein staining in polyacrylamide gels with a new type of Coomassie Brilliant Blue. *Annals of Biochemistry,* Vol. 48, pp. 617-620, ISSN 0003-2697

Duodu, K.G.; Taylor, J.R.N.; Belton, P.S. & Hamaker, B.R. (2003). Factors affecting sorghum protein digestibility. *J Cereal Science,* Vol. 38, pp. 117-131, ISSN 0733-5210

El Nour, I.N.A.; Peruffo, A.D.B. & Curioni, A. (1998). Characterisation of sorghum kafirins in relation to their cross-linking behaviour. *J Cereal Science,* Vol. 28, pp. 197–207, ISSN 0733-5210

Elkonin, L.A.; Belyaeva, E.V. & Fadeeva, I.Yu. (2012). Expression of apomictic potentials and selection for apomixis in sorghum line AS-1a. *Russian Journal of Genetics,* Vol. 48, No 1, pp. 32–40, ISSN 1022-7954

Hamaker, B.R.; Mohamed, A.A. & Habben, J.E. (1995). Efficient procedure for extracting maize and sorghum kernel proteins reveals higher prolamin contents than the conventional method. *Cereal Chemistry,* Vol. 72, pp. 583–588, ISSN 0009-0352

Italianskaya, J.V.; Elkonin, L.A. & Kozhemyakin, V.V. (2009). Characterization of kafirin composition and *in vitro* digestibility in CMS-lines, fertility restorers and F_1 hybrids with the new types of CMS-inducing cytoplasms of sorghum. *Plant Breeding,* Vol. 128, pp. 624-630, ISSN 0179-9541

Laemmli, U.K. (1970). Cleavage of structural proteins during the assembly of the head of bacteriophage T4. *Nature,* Vol. 227, pp. 680-685, ISSN 0028-0836

Mertz, E.T.; Hassen, M.M.; Cairnms-Wittern, C.; Kirleis, A.W.; Tu, L. & Axtell, J.D. (1984). Pepsin digestibility of proteins in sorghum and other major cereals. *Proceedings of National Academy of Science of USA*, Vol. 81, pp. 1-2, ISSN 0027-8424

McCleary, B.V.; McNally, M. & Rossiter, P. (2002). Measurement of resistant starch by enzymic digestion in starch samples and selected plant materials: Collaborative Study. *Journal of AOAC International*, Vol. 85, №5, pp. 1103-1111, ISSN 1060–3271

Nunes, A.; Correia, I.; Barros, A. & Delgadillo, I. (2004). Sequential *in vitro* pepsin digestion of uncooked and cooked sorghum and maize samples. *Journal of Agricultural and Food Chemistry*, Vol. 52, pp. 2052-2058, ISSN 0021-8561

Nunes A.; Correia, I.; Barros A. & Delgadillo, I. (2005). Characterization of kafirin and zein oligomers by preparative sodium dodecyl sulfate-polyacrylamide gel electrophoresis. *Journal of Agricultural and Food Chemistry,* Vol. 53, pp. 639-643, ISSN 0021-8561

Oria, M.P.; Hamaker, B.R. & Shull, J.M. (1995). Resistance of Sorghum α-, β- and γ-kafirins to pepsin digestion. *Journal of Agricultural and Food Chemistry*, Vol. 43, pp. 2148–2153, ISSN 0021-8561

Oria, M.P.; Hamaker, B.R.; Axtell, J.D. & Huang, C.P. (2000). A highly digestible sorghum mutant cultivar exhibits a unique folded structure of endosperm protein bodies. *Proceedings of National Academy of Science of USA*, Vol. 97, pp. 5065-5070, ISSN 0027-8424

Pedersen, B. & Eggum, B.O. (1983). Prediction of protein digestibility by an *in vitro* enzymatic pH-stat procedure. *Zeitschrift fur Tierphysiologie, Tierernaehrung und Futtermittelkunde*, Vol. 49, pp.265-277, ISSN 1439-0396

Ping, J.-A.; Zhang, F.-Y.; Cui, G.-M.; Cheng Q.-J.; Du Z.-H.; Zhang Y.-X. (2004). A study of the properties of autonomous seed setting and embryology in Sorghum apomictic line 2083. *Acta Agronomica Sinica.* Vol. 30, № 7, pp. 714-718, ISSN 0496-3490

Rao, N.G.P.; Narayana, L.L. & Reddy, V.R. (1978). Apomixis and its utilization in grain Sorghum. I. Embryology of two apomictic parents. *Caryologia.* Vol. 31, № 4, pp. 427-433, ISSN 0008-7114

Shull, J.M.; Watterson, J.J. & Kirleis, A.W. (1991). Proposed nomenclature for the alcohol-soluble proteins (kafirins) of *Sorghum bicolor* (L. Moench) based on molecular weight, solubility and structure. *Journal of Agricultural and Food Chemistry*, Vol. 39, pp. 83–87 ISSN 0021-8561

Shull, J.M.; Watterson, J.J. & Kirleis, A.W. (1992). Purification and immunocytochemical localization of kafirins in *Sorghum bicolor* (L. Moench) endosperm. *Protoplasma*, Vol. 171, pp. 64–74, ISSN: 0033-183X

Watterson, J.J.; Shull, J.M., & Kirleis, A.W. (1993). Quantitation of α-, β- and γ-kafirins in vitreous and opaque endosperm of *Sorghum bicolor*. *Cereal Chemistry*, Vol. 70, pp. 452–457, ISSN 0009-0352

Weaver, C.A.; Hamaker, B.R. & Axtell, J.D. (1998). Discovery of grain sorghum germplasm with high uncooked and cooked *in vitro* protein digestibility. *Cereal Chemistry*, Vol. 75, pp. 665-670, ISSN 0009-0352

Wong, J.H.; Marx, D.B.; Wilson, J.D.; Buchanan, B.B.; Lemaux, P.G. & Pedersen, J.F. (2010). Principal component analysis and biochemical characterization of protein and starch reveal primary targets for improving sorghum grain. *Plant Science*, Vol. 179, pp. 598-611, ISSN 0168-9452

Wu, S.-B.; Shang, Y.-J. & Han, X.-M. (1994). Embryological study on apomixis in a sorghum line SSA-1. *Acta Botanica Sinica.*, Vol. 36, №11, pp. 833-837, ISSN 0577-7496

24

Extraction and Electrophoresis of DNA from the Remains of Mexican Ancient Populations

Maria de Lourdes Muñoz et al*
Department of Genetics and Molecular Biology, Centro de Investigación y de Estudios Avanzados del Instituto Politécnico Nacional, Mexico D. F.
Mexico

1. Introduction

Ten years ago, the first reports of human genome sequencing were published in Nature and Science (Venter et al., 2001; Sachidanandam et al., 2001; Lander, 2011). This was very exciting and expectations for the application of genome sequencing technology were high. In the past decade, the cost of sequencing has gone down several orders of magnitude, making it a more accessible technology for research studies. The medical value of comprehensive genome sequencing is now becoming apparent: for example, the genetic cause of a rare and debilitating vascular disorder was solved by genome sequencing at NIH (Jasny and Zahn, 2011; Lander, 2011). It is also possible to solve the genetics of individual Mendelian disorders thereby relating phenotype to genotype. In addition, better treatments for diseases such as cancer, metabolic disorders, inflammation, neurodegeneration or diabetes are expected to be found through studies involving genome sequencing (Lander, 2011). Sequencing also has been used to query variation in populations worldwide, and sequences are now available from extinct hominids as well as from thousands of other species (Rasmussen et al., 2010; Krause et al., 2010; Reich et al., 2010; Balter, 2010; Rasmussen et al., 2010). We expect to know very soon what variation exists among individuals at almost all sites in the genome. This is a great opportunity for population genetics to reconstruct the entire genealogical and mutational history of humans (Callaway, 2011), to understand the evolutionary and genetic forces that affected every region of the genome, to determine disease mutations present in human populations, to elucidate the genetic bases of cognitive and physiological adaptations, and/or to determine the demographic events that led to the colonisation of the earth.

The question remains: what is the relationship between morphological features and ancient deoxyribonucleic acid (aDNA)? The evolutionary processes that generated modern species

* Mauro Lopez-Armenta[1,2], Miguel Moreno-Galeana[1], Alvaro Díaz-Badillo[1], Gerardo Pérez-Ramirez[1], Alma Herrera-Salazar[1], Elizabeth Mejia-Pérez-Campos[3], Sergio Gómez-Chávez[4] and Adrián Martínez-Meza[5]

[1]*Department of Genetics and Molecular Biology, Centro de Investigación y de Estudios Avanzados del Instituto Politécnico Nacional; Mexico*

[2]*Posgrado en Ciencias Genómicas, Universidad Autónoma de la Ciudad de México, Mexico*

[3]*Instituto Nacional de Antropología e Historia, Querétaro, Mexico D.F., Mexico*

[4]*Teotihuacan, Mexico*

[5]*Mexico City, Mexico*

and populations are commonly inferred through the analysis of morphological and genetic markers in addition to analyses of contemporary organisms to create tentative reconstructions. To confirm this indirect evidence, it is necessary to check the reconstructions against the fossil records. Nevertheless, the comparison has been made possible now by analysing morphological characters, and the application of recent advances in deoxyribonucleic acid (DNA) sequencing technologies for aDNA are now allowing the genetic record to be generated. This new technology let us focus not only on single genetic loci, such as mitochondrial DNA (mtDNA), but it made possible to obtain whole genome sequences of extinct species and populations (Lander et al., 2011), our closest extinct relatives the Neanderthal (Green et al., 2010), and the extinct hominid group from Siberia, the Denisovans (Reich et al., 2010).

The field of aDNA was initiated more than twenty years ago (Higuchi et al., 1984; Cooper et al., 1992; Greenwood et al., 1999) and research efforts continue to grow and expand into new areas (Stoneking and Krause, 2011). The first aDNA studies demonstrated the inefficiency of bacterial cloning to amplify small sequences recovered from the skins of animals and human mummies (Higuchi et al. 1984; Pääbo, 1985) and showed that DNA was at very low concentrations of short damaged fragments. However, these studies are considered very important because they will elucidate population origins, migrations, relationships, admixture and changes in population size, essentially revealing the demographic history of the human population.

It is now accepted that DNA is preserved in ancient samples under a wide range of depositional environments (Willerslev and Cooper, 2005). Although the DNA of a deceased organism degrades rapidly, part of it may survive for more than 100,000 years under favourable conditions, such as cold, stable temperatures and a dry environment (Pääbo et al., 2004). Fortunately, the development of new technologies has made possible the recovery and manipulation of these molecules as well as the genetic characterisation of these samples. Because this DNA is degraded the analysis is complicated, nevertheless, the new sequencing technology makes it possible to obtain historical information. In addition, the presence of polymerase inhibitors makes DNA amplification exceedingly difficult. Research in this area shares a common problem with forensics and other approaches requiring analyses of museum and non-invasively collected specimens; the amount of endogenous DNA available in the samples is limited. In addition, when working with human samples it is also possible to have contamination from contemporary human DNA. Careful adherence to currently established procedures is necessary to avoid such contamination (Deguilloux et al., 2011).

Because aDNA contains the information of our past its analysis is of high importance. Here, we will review a variety of methods for extraction, purification, amplification and sequencing of aDNA segments informative for genetic population studies. Future prospects for the potential direction of ancient DNA research will be discussed. Furthermore, contributions to migratory theories will also be analysed based on population diversity, taking into account ancient mtDNA studies.

Although there is new technology to determine the sequence of nuclear DNA, we will focus on mtDNA analysis. mtDNA analysis has been very useful to extensively examine human population history throughout the world because of its relatively rapid rate of mutation, lack of recombination and maternal inheritance. Mitochondrial DNA sequence variations at the hypervariable regions HVI and HVII will be described and their importance in population genetic studies will be discussed. Technical differences between DNA extraction procedures for ancient bones and mummy tissue will also be described. Molecular phylogenetic analysis, haplotype and haplogroup determination through software will also be defined and examined.

2. Procedures to study ancient DNA

There have been several aDNA extraction protocols suggested over the years. The first method was purification based on phenol/chloroform extraction, alcohol precipitation (Kalmár et al., 2000; Munoz et al., 2003; Hagelberg and Clegg, 1991; Hänni et al., 1995) and silica binding (Höss and Pääbo, 1993; Yang et al., 1998). In addition, other methods have been suggested, such as using Chelex (Faerman et al., 1995), centricon filters (Anzai et al., 1999), Dextran Blue (Kalmár et al., 2000), decalcifying bone with EDTA (Hagelberg and Clegg, 1991; Hänni et al., 1995; Yang et al., 1998) and hybridisation and magnetic separation (Anderung et al., 2008). The methods most commonly used now combine EDTA decalcification and silica purification (Yang et al., 1998; Krings et al., 1997; Anzai et al., 1999). It is evident that many different techniques have been used, demonstrating that no single procedure has clear advantages. Based on our experience, the selected method is a function of the sample characteristics, including considerations for the origin of the sample, from the skeleton or a mummy.

2.1 Samples

Samples from this study include bones pertaining to pre-Hispanic populations from different periods of time (200 to 1500 years before present). Bone samples of two individuals from Monte Albán, Oaxaca, one from Teotihuacán and a tissue portion from the mummy Pepita were used in the examples presented in this study. To work with the ancient Mexican samples, we made a written agreement with the "Intituto Nacional de Antropologia e Historia" (Mexico). Research on ancient unidentifiable human remains is excluded from the requirement of ethics review by the Research Ethics Boards.

Sampling should be conducted as soon as the bones appear in excavation, and gloves, mask and coat must be used to prevent contamination from excavators. This is not always possible because some samples were collected before these studies were initiated. Samples also have to be deposited directly in hermetic sterile tubes and frozen at -70°C. These practices prevent the introduction of contaminant DNA during the sample collection. In addition, it is also very important to manipulate the sample in a sterile clean room, to use bleach and ultraviolet light to degrade potential contaminants and to keep strict physical separation of modern DNA work from aDNA (Miller et al., 2008, Cooper and Poinar, 2000).

2.2 Ancient DNA extraction

All DNA purification and PCR experiments were carried out under sterile conditions in separate dedicated rooms. Samples were handled wearing protective clothing from collection to DNA isolation, and the laboratory equipment and reagents are maintained DNA-free. The laboratory managing the ancient samples has a high-pressure system to filter the incoming air and a laminar flow hood as well as UV light irradiation and bleach were used to clean of every surface to avoid contamination (Knapp et al., 2011).

DNA purification was preceded by a decontamination step to eliminate surface exogenous DNA when samples were collected and manipulated by unknown people. Each sample was washed with bleach followed by a water rinse and UV light irradiation for 30 min on each face. Some authors suggest removing the surface of the bone, however this procedure may also contaminate the inside of the bone if it has some kind of porosity. When bones were collected as soon as they appeared during the exaction with the necessary equipment to prevent contamination from excavators (i.e., gloves, mask and coat), it was not necessary to treat the sample with the decontamination steps (Deguilloux et al., 2011) and the potential to damage template or impede the efficiency of PCR was therefore avoided. Overall, there is no way to guarantee complete removal of contaminant DNA through decontamination procedures, but these practices are used to eliminate as much contamination as possible.

Bone powder was generated by grinding in a mortar with pestle until a fine powder was obtained, when bone quantities were around 1 g. When the weight of the samples was ≤ 0.5 mg, the bone sample was ground under liquid nitrogen with a sterile screw cylinder modified from those suggested by Thomas M. G. and Moore L. J., (1997). The powder (0.250-0.500 g) was transferred into a sterile 15 ml tube and was suspended in 2 ml of extraction buffer (0.01 M Tris-HCl, 0.1 M EDTA and 0.2% SDS pH 8.0), and the tubes were capped and sealed with Parafilm. After incubation with gentle agitation for 1 h at 37ºC, 1 mg/ml proteinase K was added, and the sample was incubated at 50ºC for 2 h. A blank extraction treated identically to the experimental samples throughout the procedure was included to monitor for contamination during the DNA extraction process. Finally, the samples were centrifuged at 5,000×g for 5 min, and the supernatants were extracted using phenol-chloroform-isoamyl alcohol (24:24:1) organic extraction (Maniatis, et al., 1989; Munoz et al., 2003; Hughes et al., 2006). Subsequently, the aqueous phase was concentrated by precipitation by the addition of 0.1 volumes of 3 M sodium acetate at pH 5.0 and 2.5 volumes of ethanol. After mixing, the sample was incubated at -78ºC overnight and centrifuged at 15,000 rpm for 10 min at 4°C. The supernatant was decanted, and the precipitate was rinsed with 70% ethanol. After drying the pellet at ambient temperature in a sterile area, the pellet was resuspended in 100 µl of high quality sterile water. Alternatively, the aqueous phase can be concentrated using Amicon® Ultra-0.5 30 kDa columns (Millipore, Billerica, USA), in a final volume of 40 µl.

Another method to extract the aDNA is by binding to silica: the powered sample (0.250 g) was suspended in 1 ml of extraction buffer (0.01 M Tris-HCl, 0.5 M EDTA pH 8.0) and after incubation at 37ºC for 16 h, the suspension was incubated at 56ºC for 3 h and centrifuged at 5,000xg for 2 min. The supernatant was transferred into 3 ml of binding buffer (5 M GuSCN, 0.025 M NaCl, 0.010 M Tris-HCl pH 8.0) in a 15 ml sterile conical tube and adjusted to pH 4.0 by adding 30% HCl in 25 µl aliquots. Then, the solution is passed through a QIAquick (Qiagen) silica column. The column was rinsed twice with the washing buffer (50% ethanol, 0.125 M NaCl and 0.010 M Tris and 0.001 M EDTA, pH 8.0) and dried for 15 min. Finally, aDNA was eluted from the column with 100 µl of TE buffer (0.01 M Tris-HCl, 0.001 M EDTA, pH 8).

Extracted DNA was kept in aliquots of 25 µl at -70 ºC.

Ancient DNA can also be extracted by the Chelex-100 method: Extraction of DNA using Chelex1-100 (Bio-Rad Laboratories, CA, USA) was performed with 5% Chelex-100 in sterile H_2O using the protocol described by Walsh et al. (1991). Briefly, 200 µl of DNA extracted by phenol-chloroform-isoamyl alcohol (24:24:1) was boiled at 94 °C for 10 min with 5% Chelex-100 and centrifuged, and an aliquot of the supernatant was taken as the template for the PCR experiment.

2.3 Amplification of DNA from pre-Hispanic samples

Analysis in Native Americans of mtDNA by PCR amplification and high-resolution restriction analysis with 14 endonucleases (Torroni, et al., 1992; Torroni, et al., 1993; Torroni, et al., 1994a,b; Richards et al., 1996) identified four major mtDNA lineages or haplogroups (A-D). These haplogroups of Asian ancestry, each defined by specific polymorphisms, together encompass 96.9 % of the mtDNA observed in modern Native Americans. Each lineage is characterized by specific mtDNA marker: the 9-bp deletion in the COII/tRNAlys region (haplogroup B); a HaeIII restriction site gain at nucleotide position 663 of the reference sequence (haplogroup A) (Anderson et al., 1981); a HincII restriction site loss at nucleotide 13259 (haplogroup C); and an AluI restriction site loss at nucleotide 5176 (haplogroup D) (Wallace et al., 1985; Schurr et al., 1990; Torroni et al., 1992; Wallace and Torroni, 1992). Sequence data indicate a correspondence between each marker and particular hypervariable region I (HVI) mutations (Horai et al., 1993; Bailliet et al. 1994). Consequently, the mtDNA amplification of the specific region has to be performed to characterize the Native Americans (ancient and contemporary) populations. Primers to amplify HVRII were also included, although we did not included any example, because analysing the HVRII region is not as informative as the HVRI.

Enzymatic amplification by PCR was performed as described previously (Munoz et al., 2003; Campos, et al., 2011) using heat-resistant *Thermus aquaticus* (Taq) DNA polymerase (FINNZYMES), or Platinum® Taq High Fidelity (Invitrogen). The PCR parameters were as follows: 2.5 U of hot start DNA polymerase, 1X buffer, 2.5 mM $MgCl_2$, 200 µM of each dNTP, 0.25 mg/ml bovine serum albumin (BSA) and 0.2 µM of each primer in a total volume of 25 µl, and 5 µl of the aDNA template. The primers used to amplify and sequence human mitochondrial DNA were as follow:

HVR I:
L15975-15996 5′-CTCCACCATTAGCACCCAAAGC-3′;
H16401-16420 5′-TGATTTCACGGAGGATGGTG-3′ (Vigilant et al., 1989);
L16140-16159 5′-TACTTGACCACCTGTAGTAC-3′;
H16236-16255 5′-CTTTGGAGTTGCAGTTGATG-3′ (Wilson et al., 1995);
L15989-16008 5′-CCCAAAGCTAAGATTCTAAT-3′;
H16130-16152 5′-AGGTGGTCAAGTATTTATGGTAC-3′ (Eichmann and Parson, 2008);
L16094-16122 5′-TCGTACATTACTGCCAGYC-3′;
H16228-16248 5′-GTTGCAGTTGATGTGTGATAG-3′ (Eichmann and Parson, 2008);
L16190-16209 5′-CCCCATGCTTACAAGCAAGT-3′;
H16380-16398 5′-CAAGGGACCCCTATCTGAG-3′ (Poinar et al., 2001);

HVR II:
L8-29 5′-GGTCTATCACCCTATTAACCAC-3′;
H408-429 5′-CTGTTAAAAGTGCATACCGCC-3′ (Vigilant et al., 1989)

Haplogroup A:
L610-633 5′-TGAAAATGTTTAGACGGCCTCACA-3′;

H712-730 5′-CCAGTGAGTTCACCCTCTA-3′ (Parr et al., 1996).
Haplogroup B:
L8196-8215 5′-ACAGTTTCATGCCCATCGTC-3′;
H8297-8316 5′-CTGTAAAGCTAACTTAGCAT-3′ (Wrischnik et al., 1987);
Haplogroup C:
L13198-13213 5′- GCAGCAGTCTGCGCCC -3′;
H13384-13403 5′- ATATCTTGTTCATTGTTAA -3′ (Lorenz and Smith, 1996)
(1996)
Haplogroup D:
L5101-5120 5′-TAACTACTACCGCATTCCTA-3′;
H5230-5249 5′-TGCCCCCGCTAACCGGCTTT-3′ (Stone and Stoneking, 1993)

All amplifications were carried out in a GeneAmp® PCR System 9700 thermocycler with the following profile: 5 min at 94°C, followed by 40 cycles of 1 min at 94°C, 1 min at 59°C (haplogroups A, D and RHVs) or 55 °C (haplogroups B and C), and 1 min at 72°C, with a final extension of 10 min at 72°C. At least one PCR blank was amplified alongside each batch.

The PCR products were visualised on 2% agarose gels with ethidium bromide, and all positive products were purified using the QIAquick kit (Qiagen) and sequenced using the BigDye® Terminator v3.1 kit (Applied Biosystems) in an ABI PRISM 310 genetic analyser.

2.4 Data analysis

Phylogenetic analysis. The sequences of the pre-Hispanic PCR products from the HVI segment were aligned with representative Amerindian mtDNA control-region sequences (GenBank accession numbers: AY195760, EU719927, EU719811, EU719679, EU720004, EU720308, EU720078, EU719797, AY195749, EU720177, EU720123, EU719764, HQ012155, EU720242, HQ012184, HQ012164, HQ012134, AY195772, EU720073, EU720339, AY195759, EU720071, EU720336, EU720102, EU720202, HQ012188, HQ012198, EU720029, HQ012255, HQ012254, HQ012253, GQ449339, EU034320, AY195748, AF214088, DQ973581, AF478614) and two ancient sequences from a prehistoric Oneota population (Stone and Stoneking, 1998) using the Clustal W program (Thompson et al., 1994). Then, the phylogenetic tree was constructed with the Jukes-Cantor method, and the distances were obtained from a neighbour joining algorithm. Finally, the tree was optimised for maximum likelihood, using Hy-Phy software (Kosakovsky-Pond et al., 2005).

Haplotype network analysis. The median-joining (ε=0) networks (Bandelt et al., 1999) of haplotypes were constructed using the Network package, v4.5.1.0 (Fluxus Engineering). Sequences used were those described for phylogenetic analysis. This method is for constructing networks from recombination-free population data.

3. Examples of ancient DNA extraction procedures

An example of DNA extraction from a pre-Hispanic sample is depicted in Figures 1 and 2. Figure 1 displays the contaminants with different colours during the phenol-chloroform-isoamyl alcohol procedure. Figure 2 shows the DNA extracted by the phenol-chloroform-isoamyl alcohol technique from 0.25 g of two powdered bone samples from the same individual. In this figure, we observe the sample contaminants that are one of the major obstacles to these studies because they inhibit the Taq polymerase. The contaminants, such as Maillard products of reducing sugars (Pääbo, 1989) and humic acids with phenolic

groups, were observed by fluorescent stain in blue while the DNA degraded (which results in a smear pattern) is stained in pink by ethidium bromide (Figure 2, panel A). Figure 2, panel B shows the second sample in lane 3, which displays only contaminants. The DNA was not apparent. These compounds can be partially eliminated using kits such as the Amicon® Ultra-0.5 30 kDa columns. These results show variation in DNA yields between extracts taken from different samples of the same bone, even when using the same extraction method. We attribute such differences to heterogeneity within the bones.

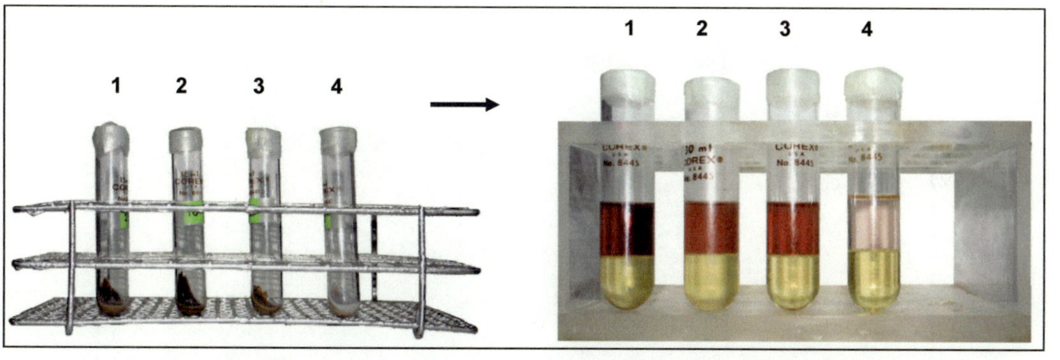

Fig. 1. Extraction of aDNA of bone samples from four different pre-Hispanic samples using the phenol-chloroform-isoamyl alcohol technique.

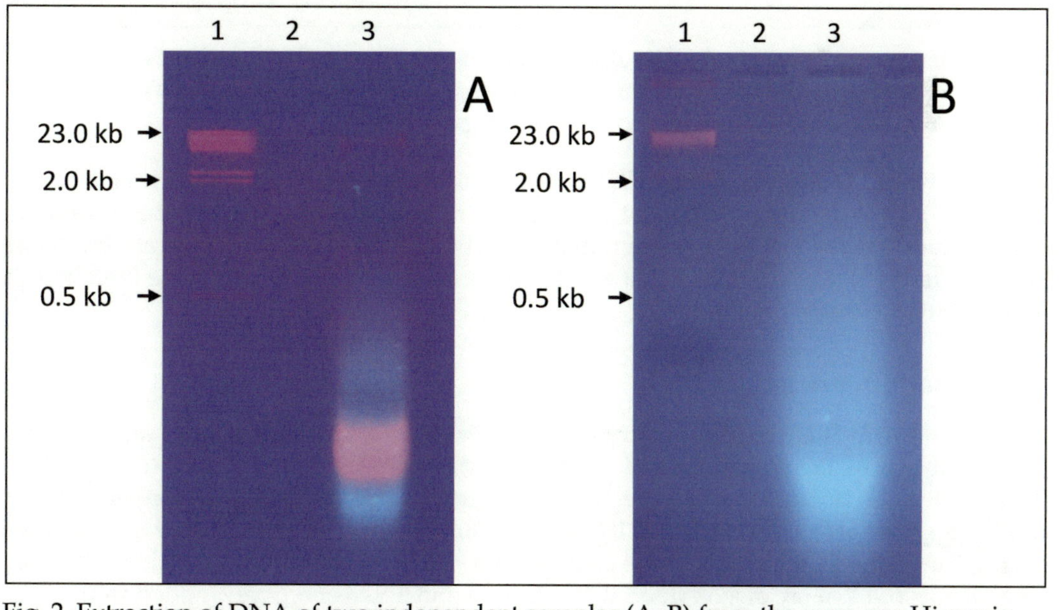

Fig. 2. Extraction of DNA of two independent samples (A, B) from the same pre-Hispanic individual by the technique of phenol-chloroform-isoamyl alcohol and ethanol precipitation. Lanes 1, molecular weight markers of *HindIII*; Lanes 2, no-sample; lane 3, DNA extracted from sample 1 of the Mexican pre-Hispanic population from Monte Alban.

Because DNA concentration is only possible with limited precision and concentrations of standard dilution series change over time in storage, we evaluated the relative performance of the DNA during PCR amplification using serial dilutions of the extracted DNA starting from 5 µl. Using this method, we were able to dilute the inhibitors of the Taq DNA polymerase.

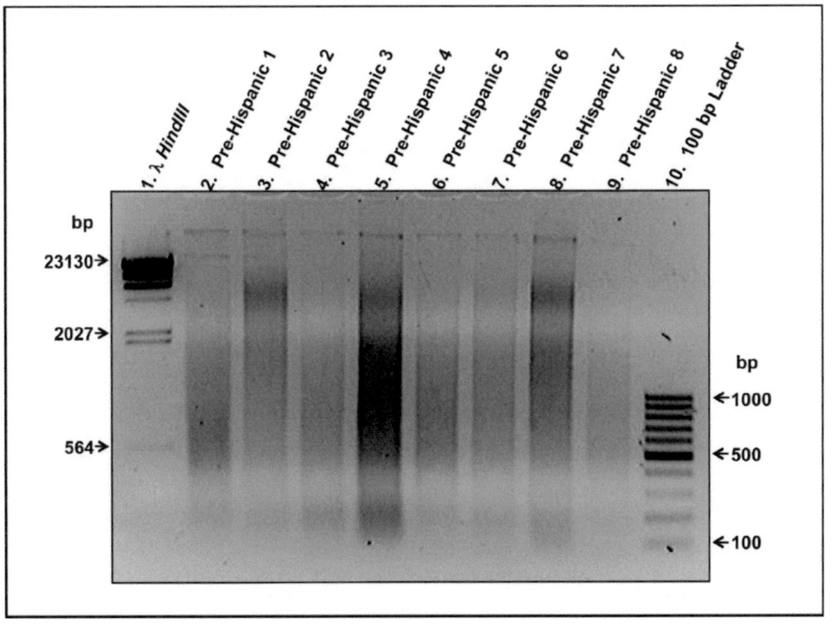

Fig. 3. Extraction of DNA of pre-Hispanic samples by the silica technique. Lanes 1 and 10, molecular weight markers of 23 kbp and 100 bp, respectively. Lanes 2 to 9, DNA extracted from different samples of Mexican pre-Hispanic populations.

Silica gel was also used to purify aDNA, results are shown in Figure 3. Each lane of this figure (2-9) displays aDNA extracted from 0.25 g of different pre-Hispanic bone samples. Lanes 1 and 10 show molecular weight markers. The quantity of Taq polymerase inhibitors is not evident, although we know that all ancient samples contain some of these inhibitors in different concentrations.

4. PCR performance

To study the effects of the Taq polymerase inhibitors, we added decrease quantities of the DNA extracted by the phenol-chloroform-isoamyl alcohol technique (shown in Figure 1) to the amplification reaction of the hypervariable segment I (15975-16420) using contemporary DNA. Figure 4, lane 1 displays the 100 bp molecular weight marker, lane 2 the negative control, lane 3 the PCR product of the contemporary DNA with the aDNA without dilution, lanes 4, 5, 6 and 7 show aDNA diluted 1:1, 1:2, 1:4 and 1:8, respectively, added to the PCR reaction mix and lane 8 contemporary DNA without any addition (positive control). Contaminated DNA allowed positive amplifications when aDNA was diluted at least 1:4 (Figure 4, lane 6). Therefore, an additional method to obtain the PCR product from the aDNA is by sample dilution.

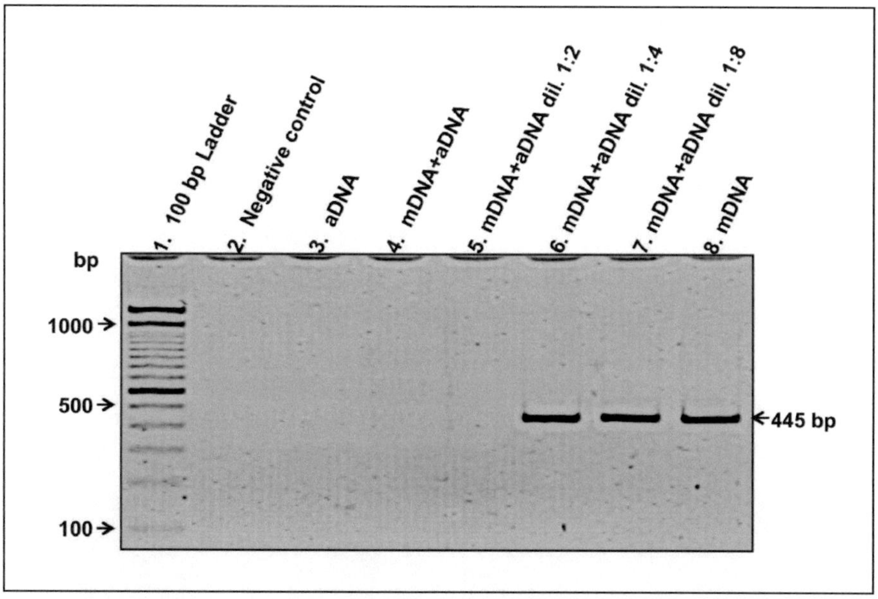

Fig. 4. Inhibition of the mitochondrial DNA hypervariable segment I amplification via inhibition of Taq polymerase by aDNA contaminants.

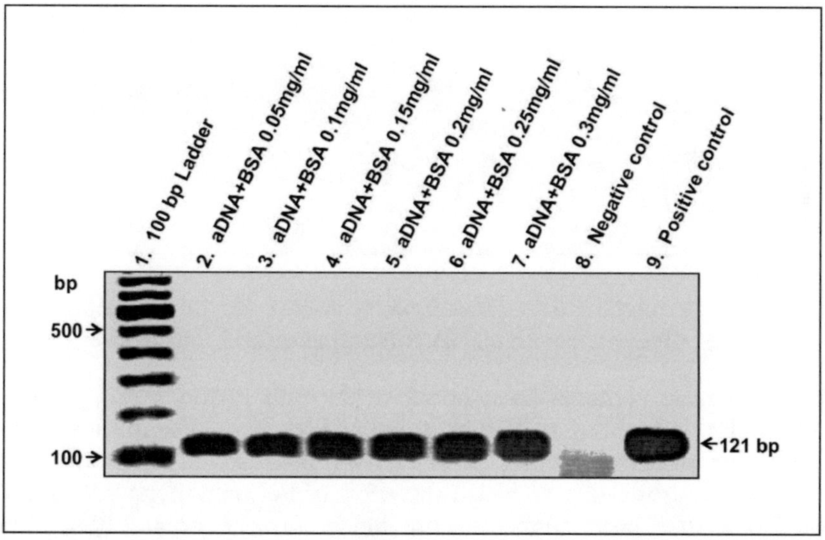

Fig. 5. Positive effect of bovine serum albumin (BSA) on the PCR performance of aDNA extracted by the phenol-chloroform-isoamyl alcohol procedure.

Sometimes aDNA dilution is not enough to obtain the PCR products, so we tested the effect of BSA addition by increasing the concentration of BSA in the reaction from 0.1 to 0.25 mg/ml. Figure 5 shows the positive effect of BSA on the PCR of aDNA extracted by the phenol-chloroform-isoamyl alcohol procedure. Increased amplification was observed in the PCR experiments when BSA was added in increasing concentrations (Figure 5, lanes 2-7, BSA at a concentration of 0.05, 0.1, 0.15, 0.2, 0.25 and 0.3 mg/ml, lane 8, negative control and lane 9, positive control). Based on these results, we added 0.25 mg/ml BSA to all PCR experiments.

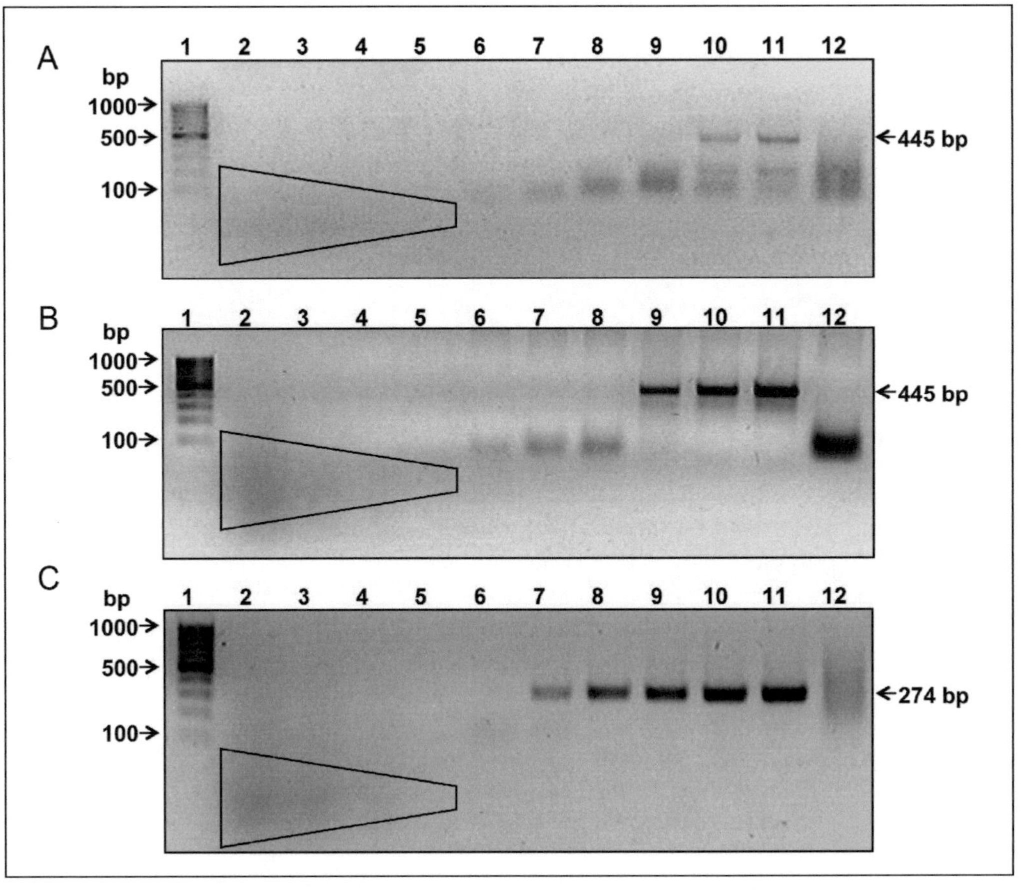

Fig. 6. Inhibition of the mitochondrial DNA hypervariable segment I amplification by inhibition of the Taq polymerase with aDNA contaminants and the effect of PVP.

Soils with high organic contents have humic acids with phenolic groups that denature biological molecules by bonding to N-substituted amides or oxidise to form a quinone that bonds to DNA or proteins (Young et al., 1994). Because aDNA contains these Taq polymerase inhibitors from soil, we tested the effect of Polyvinylpyrrolidone (PVP) during DNA extraction as has been suggested previously (Young et al., 1994; Rohland and Hofreiter, 2007). In addition, to make sure that PVP did not inhibit the PCR experiment, the reagent was added directly to the amplification mix. Figure 6 shows that 2% PVP added during the DNA extraction had a positive effect on DNA amplification. PCR amplification of contemporary DNA containing different dilutions of aDNA (1:1, 1:2, 1:3, 1:4, 1:8, 1:16, 1:32, 1:64, 1:128; Figure 6, lanes 3 to 10, respectively) and 0.25 mg/ml BSA is shown in Figure 6, panel A, and amplifications using the same conditions in the presence of 2% PVP during aDNA extraction is shown in Figure 6, panel B. The use of 2% PVP during aDNA extraction resulted in amplification at an aDNA dilution of 1:64, in contrast to the aDNA sample without PVP in which amplification is only observed at 1:128 dilution or beyond.

The HVI mtDNA segment of 445 bp was amplified in Figure 6, panels A and B with the primers L15975-15996 and H16401-16420. When the primers L15975-15996 and H16228-16248 were used, the PCR product is shorter (273 bp), and the presence of PVP makes evident the PCR fragment at a dilution of 1:16 (Figure 6, panel C). In addition, PVP at a concentration of 0.4% in the PCR experiment did not inhibit amplification, as was previously published (Young et al., 1993). The molecular weight marker is in lane 1; positive control with contemporary DNA alone is in lane 11; and the negative control with no DNA is in lane 12. Nevertheless, the positive effect was not evident in all aDNA bone samples, likely because the amount of Taq polymerase inhibitors is different in each sample.

Contemporary DNA is very easy to amplify. However, when working with aDNA, the PCR reaction efficiency is greatly reduced. For example, in Figure 7, we show a PCR-amplified fragment of mDNA using the human-specific primers L15975-15996 and H16236-16255 where 4 of the 7 bone samples from the pre-Hispanic populations displayed the PCR product (Figure 7, lanes 2 to 8). Positive and negative controls are shown in lanes 9 and 10, respectively, and the molecular weight marker is shown in lane 1.

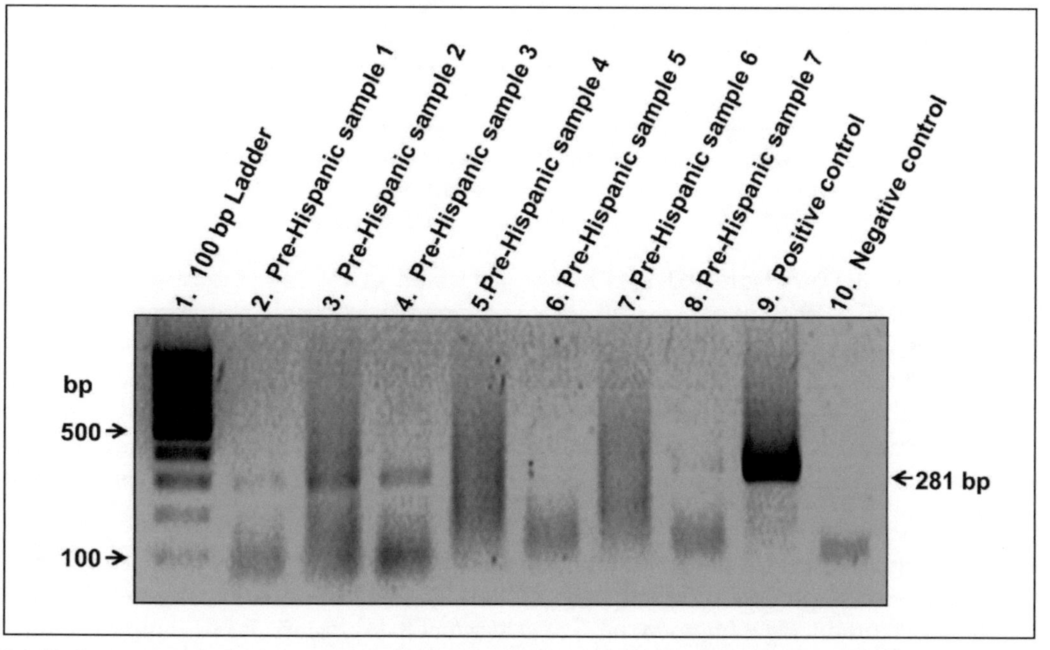

Fig. 7. Agarose gel showing the amounts of PCR-amplified product obtained after DNA extraction with the phenol-chloroform-isoamyl alcohol procedure.

Using the procedure indicated in the methods section, we purified and amplified the DNA from 14 bone samples of pre-Hispanic Native Americans to type them for haplogroup A described for Amerindians. The PCR products obtained were digested by the restriction enzyme *HaeIII*. Haplogroup A was detected in the 10 samples typed (Figure 8, lanes 2-11). Partial restriction digestion was observed in all of the ancient PCR products. This finding suggests the presence of the Amerindian polymorphism; however, we must sequence these amplification fragments or use real-time PCR to confirm the presence of the specific polymorphism because the partial restriction observed.

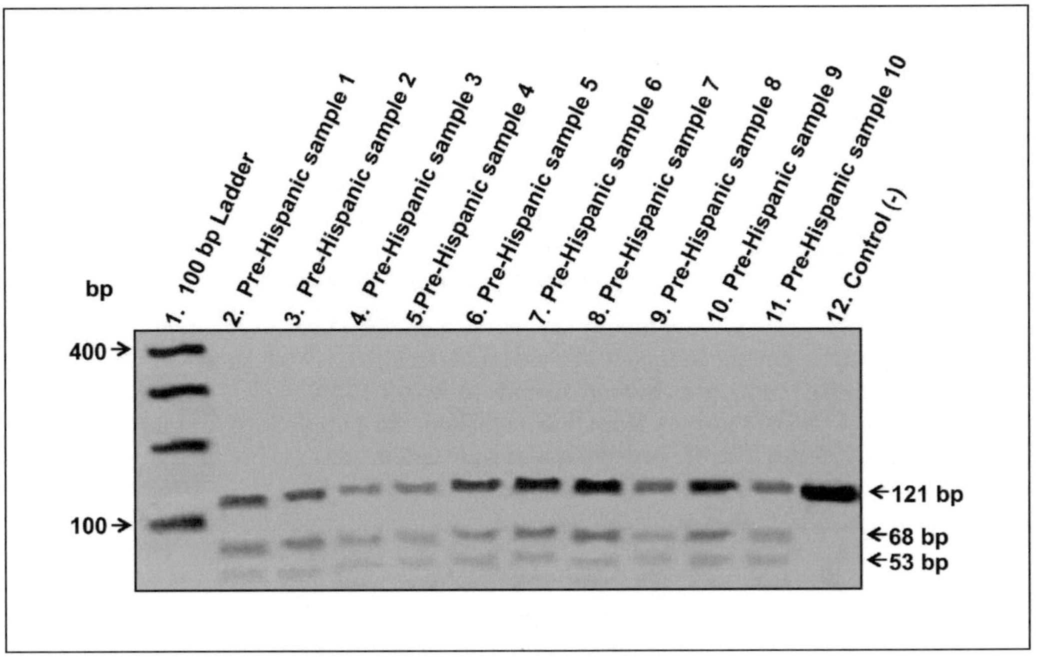

Fig. 8. Agarose gel showing the amounts of PCR-amplified product obtained after DNA extraction with the phenol-chloroform-isoamyl alcohol procedure and digested with the restriction enzyme *HaeIII*. Primers for amplification were specific to type haplogroup A.

Amplification of DNA extracted by Chelex was tested in the samples from pre-ceramic bones (Figure 9). Although experiments that compared the phenol-chloroform and Chelex method concluded that the Chelex method was simple and fast, inhibitory substances had not been eliminated in most of the cases (Kalmár et al., 2002). In our experience, DNA extracted by the phenol-chloroform method followed by Chelex treatment may improve DNA purification. Nevertheless, the silica method was better overall in our experience. The amplification products are observed at DNA dilutions of 1:30 in all samples, as shown in Figure 9.

Figure 10 displays the PCR amplification fragments using the specific primers L15975-15996 and H16236-16255 producing a fragment of 281 bp in panel A and L16140-16159 and H16380-16398 producing a fragment of 259 bp in panel B. We compared the amplification of aDNA extracted by the silica procedure and phenol-chloroform-isoamyl alcohol. Our results showed that aDNA extraction with the silica procedure was better than the phenol-chloroform-isoamyl alcohol method in this specific sample from pre-Hispanic populations because the amplification was observed exclusively in samples in which the DNA was extracted by the silica method. However, this may not be the case for all types of samples, and it is important to consider that when one method does not give good results, other methods may be useful.

Figure 10 shows aDNA from pre-Hispanic samples extracted by the silica gel method (lanes 2 and 3) compared with the phenol-chloroform-isoamyl alcohol procedure followed by concentration of aDNA with filter units (Centricon®) (lanes 4-5).

Ancient DNA was added to the PCR mix without any dilution (lane 2), diluted 1:10 (lane 3); aDNA phenol-chloroform-isoamyl alcohol extracted (lane 4); same procedure but diluted 1:10 (lane 5); washing buffer of the Centrifugal filter units (Centricon®) that were used to purify and concentrate aDNA in the phenol-chloroform-isoamyl alcohol purification procedure (lane 6) and diluted 1:10 (lane 7); negative control without DNA (lane 8); positive control with contemporary DNA (lane 9); and no sample (lane 10).

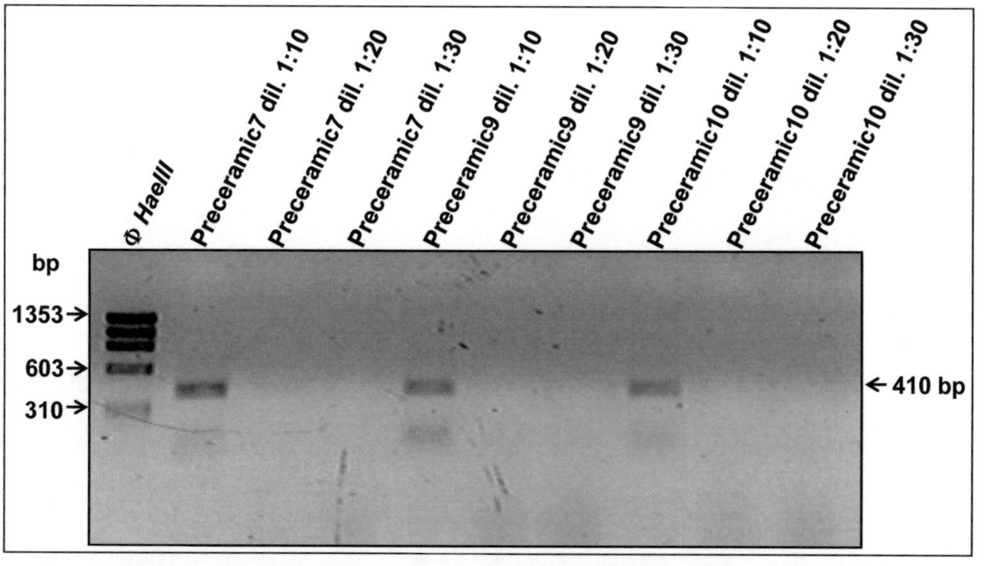

Fig. 9. Amplification of mtDNA HVI segment of 410 bp with specific primers (L15975-15996 and H16380-16398) using aDNA from pre-ceramic samples extracted with phenol-chloroform-isoamyl alcohol procedure followed by treatment with Chelex. Lanes 1 and 10, molecular weight markers φ. Lanes 2-4 pre-ceramic 7 diluted 1:10; 1:20 and 1:30; lanes 1-7, pre-ceramic 9 same dilutions as in lanes 2-4; lanes 8-10, pre-Ceramic 10, same dilutions as lanes 2-4.

When extracting DNA from small, degraded forensic samples or degraded ancient samples, the final concentration of DNA is usually too low for subsequent amplification. Consequently, we concentrated the aDNA samples extracted by the phenol-chloroform-isoamyl alcohol with filter units (lanes 2-3 and 4-5). Figure 10 shows clearly how aDNA was amplified using the DNA that was concentrated with the filter units. Furthermore, the washing buffer did not show any amplification, confirming in part that we do not have DNA contaminating our assays.

Next, we wanted to test all these methods with tissue from different mummies and determine the differences in using internal tissue and skin. From our results, we observe that when the mummy tissue is compact and from an internal organ the quantity of aDNA is very high compared with that obtained from bone samples. In addition the aDNA from the internal tissue was better as far as content is concerned. We had the opportunity to obtain DNA from the internal tissue of the mummy called Pepita that was intact and had no contamination by contemporary DNA. We were able to amplify the HVI segment using the specific primers for a mtDNA fragment of 445 bp (L15975-15996 and

H16401-16420, Figure 11, lane 8) and a fragment of 281 bp (L15975-15996 and H16236-16255, Figure 11, lane 1) or to amplify the specific second segment of HVI (L16140-16159 and H16401-16420, lane 5). The aDNA from this mummy was very well conserved. We have previously published DNA extraction from Mexican mummies with different origin and age (López-Armenta et al., 2008; Bustos-Ríos et al., 2008; Herrera-Salazar et al., 2008).

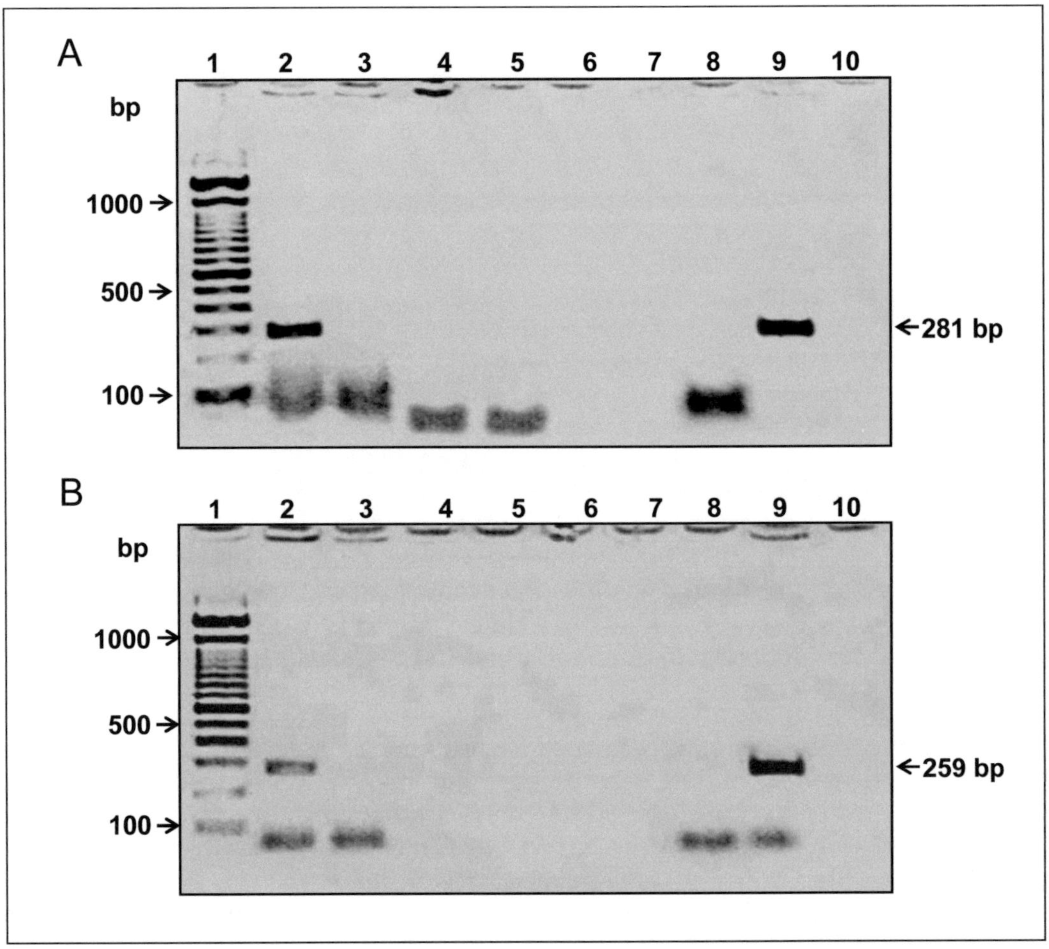

Fig. 10. Amplification of mtDNA HVI segments with the specific primers L15975-15996 and H16236-16255 producing a fragment of 281 bp (A); and L16140-16159 and H16380-16398 producing a fragment of 259 bp (B).

To examine the relationships between mtDNA lineages found in ancient and contemporary Native Americans, phylogenetic trees were constructed with the Jukes-Cantor method, and the distances were obtained from a neighbour joining algorithm and optimised for maximum likelihood, using Hy-Phy software (Kosakovsky-Pond et al., 2005). A total of 290 bp (nucleotides 16104–16394) of the HVI common to all sequences were used for these analyses. Sequences from Monte Alban and Teotihuacán from this study as well as those from the Oneota population were clustered in the haplogroup D linage.

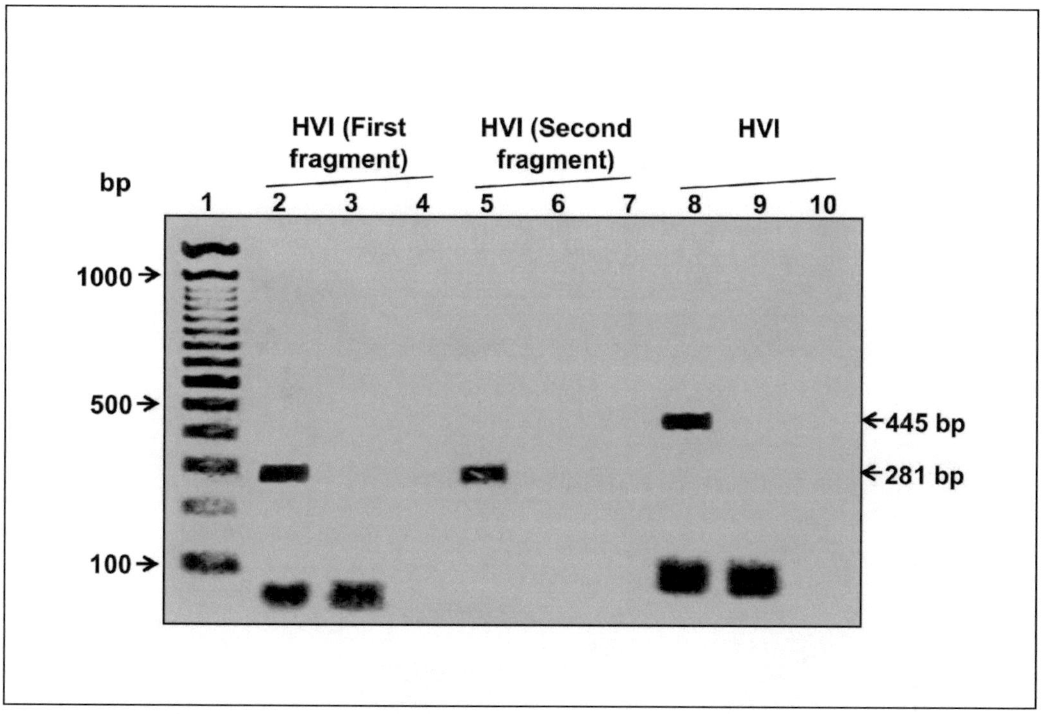

Fig. 11. Amplification of aDNA extracted from a Mexican mummy.

Haplotype network analyses were carried out on 290 bp of the mtDNA HVI from nucleotides 16104 to 16394. These networks were constructed using the Network package, v4.5.1.0 (Fluxus Engineering). These analyses included sequences from our own work and from other authors. The accession numbers of the sequences included in this network analyses were mentioned in the data analysis section. The pre-Hispanic DNA sequences included two ancient sequences from the prehistoric Oneota population (Stone and Stoneking, 1998), two sequences from Monte Alban, Oaxaca, Mexico and one from Teotihuacán, Mexico. Interestingly, the haplotype from the Oneota sequence may be derived from the Teotihuacán haplotype. The sequences from Monte Alban were grouped in the same haplotype as the more frequent haplotype from Native American populations. These results showed the potential to know the relationship among all Mexican pre-Hispanic populations or other populations as well as some haplotypes that were lost through the time.

It is important to mention that we never observed contaminant fragments with the specific HVR-1 mutations carried by the excavators or the geneticists. Therefore we are confident that following the procedures recommended by previous authors and our laboratory generates authentic sequences. Problems arise when the samples come from museums or collections where the researcher does not know how they were managed. In these conditions, additional controls are recommended for all of the procedures.

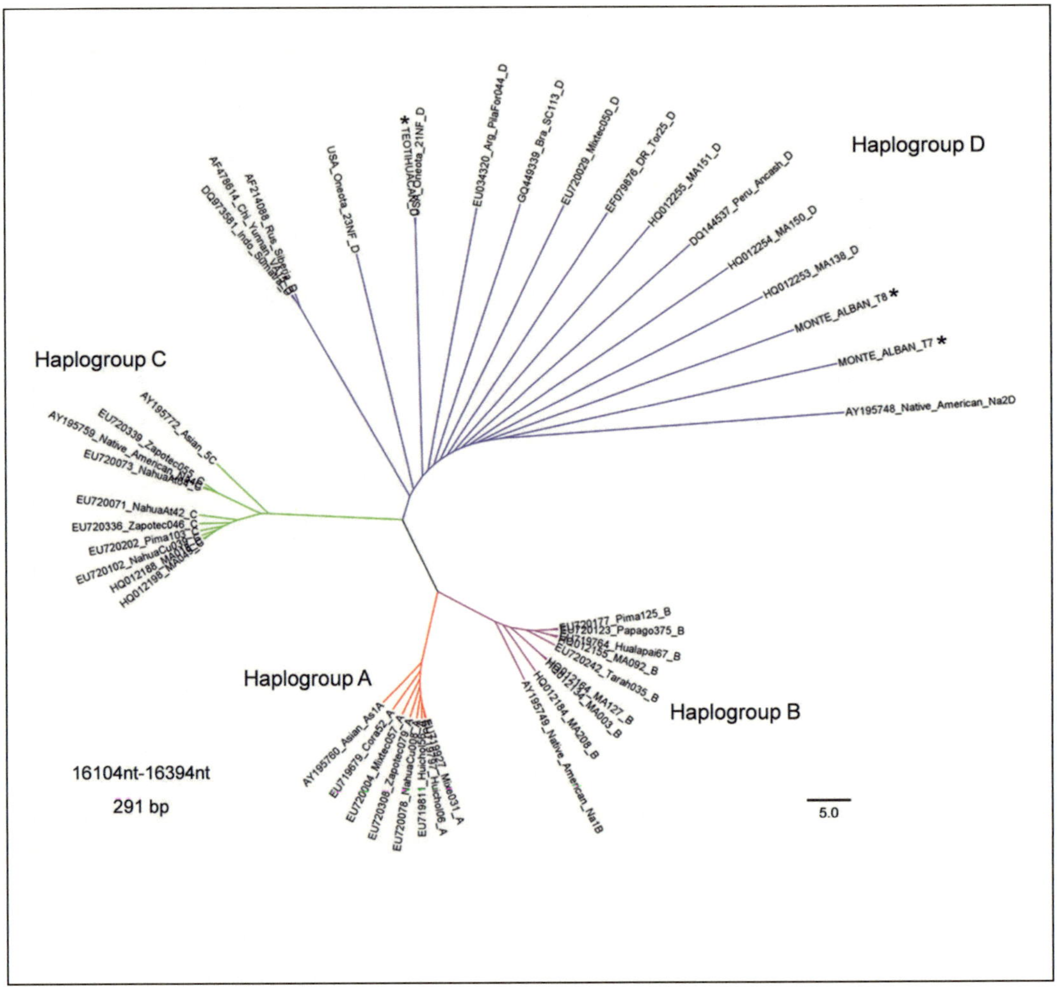

Fig. 12. Phylogenetic analyses of American native populations including five sequences from samples of pre-Hispanic populations. Tree of Native American and ancient pre-Hispanic Amerindian, constructed with the Jukes-Cantor method, and the distances were obtained from a neighbour joining algorithm and optimised for maximum likelihood using Hy-Phy software (Kosakovsky-Pond et al., 2005). The lanes in different colours indicate the haplogroup designation of lineages. Sequences of this study are marked with an asterisk.

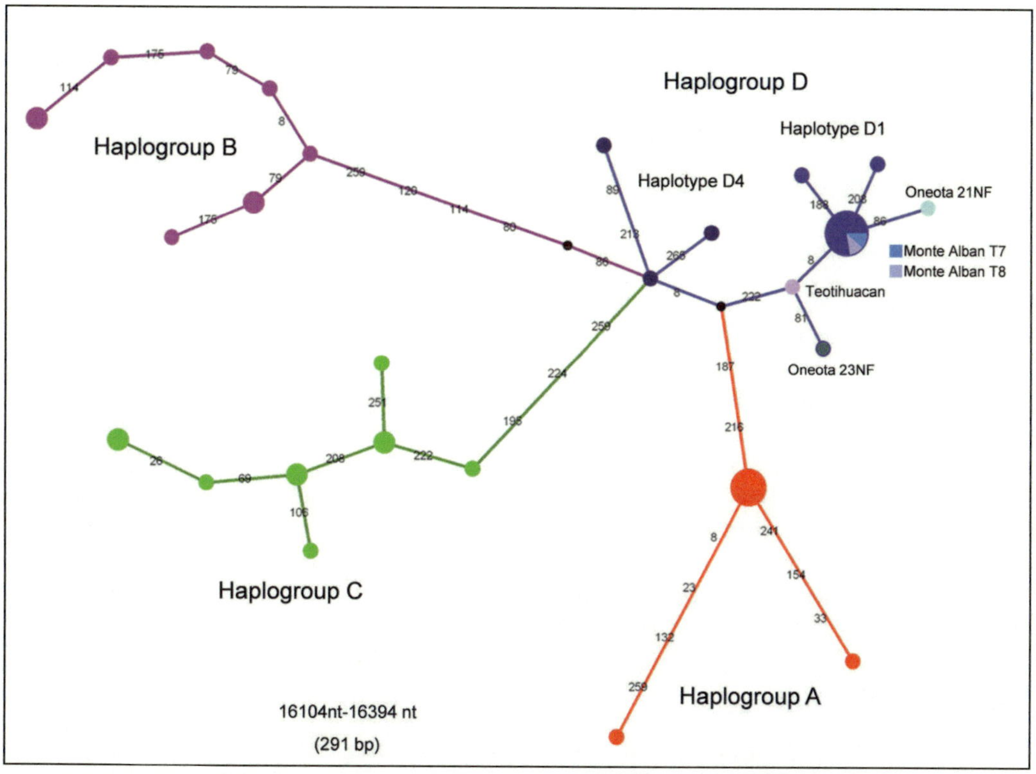

Fig. 13. Haplotype network of American native populations including five sequences from samples of pre-Hispanic populations. Each haplotype is represented by a circle in which the square radius (surface) is proportional to its population frequency. Circle colours show the site location as represented by the indicated colours. Dark circles without number in the network indicate mutational steps between haplotypes (theoretically extinct or unrepresented in the sample). Numbers between haplotypes represent mutational steps.

5. Conclusion

This review offers a direct overview of the different methods of aDNA extraction, including all special conditions needed in the laboratory to avoid contamination by contemporary DNA. It reveals the complexity involved in demonstrating the authenticity of human aDNA because the risk of contamination is very high. However, exogenous DNA contamination can be avoided if the necessary care is taken. In our experience and the experience of other laboratories, obtaining the ancient sample with coat, gloves and mask, and maintaining it in sterile conditions without human contact reduces the chances of sample contamination. It is also very important to test all reagents to verify that they are free of contemporary DNA. In addition, we also recommend performing negative control PCR experiments with at least 45 cycles to convincingly demonstrate the absence of contemporary DNA contamination. In our experience the best method to purified aDNA is phenol-chloroform-isoamyl alcohol with concentration using Amicon® Ultra-0.5 30 kDa columns (Millipore, Billerica, USA) or the Silica gel method using the QIAquick (Qiagen) columns. We also prefer to include the EDTA in the extraction buffer to optimise the aDNA extraction. This is supported by recent publications that have demonstrated that some DNA may be lost during decalcification (Campos et al., 2011). It is also important to keep DNA at -70°C in aliquots to maintain its integrity. Maintaining bone tissue samples at -70°C during aDNA extraction is useful to avoid additional DNA degradation. In our point of view, the best method will be that containing the least sample manipulation because this will avoid DNA contamination. Finally, there will be always risk of contamination by contemporary human DNA; however, next generation sequencing methods do provide a greatly improved means of measuring the degree of contamination in a sample.

Sequencing of the PCR products from aDNA as well as phylogenetic and network analyses of remains from America would allow testing of the hypotheses concerning single versus multiple waves of migration to the New World. This analysis will also reveal new haplotypes that were lost through time because not all migrations were successful in terms of leaving descendants among contemporary populations. Furthermore, the development of next generation sequencing is revolutionising aDNA research. The examples presented in Figure 13 and 14 display the relationship between the Oneota sample and that from Teotihuacán showing different haplotypes. There were also two ancient samples from Monte Albán that were grouped with the more frequent haplotype in the D1 haplogroup. Further analysis of more pre-Hispanic human samples will give us more detailed information about the history of these populations.

6. Acknowledgment

This work was supported by Instituto de Ciencia y Tecnología del Distrito Federal, México grant Clave: PICSA10-189 (01/11/2010-2011).

7. References

Anderung, C., Persson, P., Bouwman, A., Elburg, R., & Götherström, A. (2008). Fishing for ancient DNA. *Forensic Sci. Int. Genet.* Vol. 2, pp. 104-107.

Balter, M. (2010). Human evolution. Ancient DNA from Siberia fingers a possible new human lineage. *Science.* Vol. 327, No. 5973, pp. 1566-1567.

Bailliet G, Rothhammer F, Carnese FR, Bravi CM, & Bianchi NO. (1994). Founder mitochondrial haplotypes in Amerindian populations. *Am. J. Hum. Genet.* Vol. 54, No. 1, pp. 27-33.

Bandelt, H. J., Forster, P., & Röhl, A. (1999). Median-Joining Networks for Inferring Intraspecific Phylogenies. *Mol. Biol. Evol.* Vol. 16, No. 1, pp. 37-48.

Bustos-Ríos, D., López-Armenta, M., Moreno-Galeana, M. A., Herrera-Salazar, A., Pérez-Campos, E. M., Chávez-Balderas, X., & Muñoz, M. L. (2008). Purification of DNA from an ancient child mummy from Sierra Gorda, Queretaro. Mummies and Science. *World Mummies Research/Proceediñngs of the VI World Congress on Mummy Studies (Teguise, Lanzarote=; Pablo Atoche,* Conrado Rodríguez and Ma. Angeles Ramírez (eds.) Santa Cruz de Tenerife: Academia Canaria de la Historia (etc.), p 700. ISBN: 978-84-612-5647-1 Edition: Academia Canaria de la Historia, Ayuntamiento de Teguise, Cabildo Insular de Lanzarote, Caja Canarias, Fundción Canaria Mapfre Guanarteme, Universidad de las Palmas de Gran Canaria. First Edition, pp. 259-266.

Callaway, E. (2011). Ancient DNA reveals secrets of human history. *Nature*. Vol. 476, No. 7359, pp. 136-137. doi:10.1038/476136a

Campos, P. F., Craig, O. E., Turner-Walker, G., Peacock, E., Willerslev, E., & Gilbert, M. T. (2011). DNA in ancient bone - Where is it located and how should we extract it? *Ann. Anat.* [Epub ahead of print]

Cooper, A., Mourer-Chauviré, C., Chambers, G.K., van Haeseler, A. Wilson, A.C., & Pääbo, S. (1992). Independent origin of New Zealand Moas and kiwis. *Proc. Natl. Acad. Sci. USA*. Vol. 89, pp. 8741-8744.

Cooper, A., & Poinar, H. N. (2000). Ancient DNA: do it right or not at all. *Science*. Vol. 289, No. 5482, pp. 1139.

Deguilloux, M. F., Ricaud, S., Leahy, R., & Pemonge, M. H. (2011). Analysis of ancient human DNA and primer contamination: one step backward one step forward. *Forensic Sci. Int.* Vol. 210 No. 1-3, pp. 102-109.

Eichmann, C., & Parson, W. (2008). 'Mitominis': multiplex PCR analysis of reduced size amplicons for compound sequence analysis of the entire mtDNA control region in highly degraded samples. *Int. J. Legal. Med.* Vol. 122, No. 5, pp. 385-388.

Faerman, M., Filon, D., Kahila, G., Greenblatt, C. L., Smith, P., & Oppenheim, A. (1995). Sex identification of archaeological human remains based on amplification of the X and Y amelogenin alleles. *Gene*. Vol. 167, No. 1-2, pp. 327-332.

Green, R. E., Krause, J., Briggs, A.W., Maricic, T., Stenzel, U., Kircher, M., Patterson, N., Li, H., Zhai, W., Fritz, M. H., Hansen, N. F., Durand, E. Y., Malaspinas, A. S., Jensen, J. D., Marques-Bonet, T., Alkan, C., Prüfer, K., Meyer, M., Burbano, H. A., Good, J. M., Schultz, R., Aximu-Petri, A., Butthof, A., Höber, B., Höffner, B., Siegemund, M., Weihmann, A., Nusbaum, C., Lander, E. S., Russ, C., Novod, N., Affourtit, J., Egholm, M., Verna, C., Rudan, P., Brajkovic, D., Kucan, Z., Gusic, I., Doronichev, V. B., Golovanova, L. V., Lalueza-Fox, C., de la Rasilla, M., Fortea, J., Rosas, A., Schmitz, R. W., Johnson, P. L., Eichler, E. E., Falush, D., Birney, E., Mullikin, J. C., Slatkin, M., Nielsen, R., Kelso, J., Lachmann, M., Reich, D., & Pääbo S. (2010). A draft sequence of the Neandertal genome. *Science*. Vol. 328, No. 5979, pp. 710-722.

Greenwood, A. D., Capelli, C., Possnert, G., & Pääbo, S. (1999). Nuclear DNA sequences from late pleistocene megafauna. *Mol Biol Evol.* Vol. 16, No. 11, pp. 1466-1473.

Hagelberg, E., & Clegg, J. B. (1991). Isolation and characterization of DNA from archaeological bone. *Proc Biol Sci.* Vol. 244, No. 1309, pp. 45-50.

Hänni, C. T., Brousseau, V., Laudet, D. & Stehelin, D. (1995). Isopropanol precipitation removes PCR inhibitors from ancient bone extracts. *Nucleic Acids Res.* Vol. 23, No. 5, pp. 881-882.

Herrera-Salazar, A., Bustos, Ríos, D., López-Armenta, M., Moreno-Galeana, M. A., Martínez-Meza, A., & Muñoz, M. L. (2008). Mitochondrial DNA analysis of mummies from the North of México. *World Mummies Research/Proceediñngs of the VI World Congress on Mummy Studies (Teguise, Lanzarote; Pablo Atoche,* Conrado Rodríguez and Ma. Angeles Ramírez (eds.) Santa Cruz de Tenerife: Academia Canaria de la Historia (etc.), p 700. ISBN: 978-84-612-5647-1. Edition: Academia Canaria de la Historia, Ayuntamiento de Teguise, Cabildo Insular de Lanzarote, Caja Canarias, Fundción Canaria Mapfre Guanarteme, Universidad de las Palmas de Gran Canaria. First Edition, pp. 417-422.

Higuchi R, Bowman B, Freiberger M, Ryder OA, & Wilson AC. (1984). DNA sequences from the quagga, an extinct member of the horse family. *Nature*. Vol. 312, No. 5991, pp. 282-284.

Horai, S., Murayama, K., Hayasaka, K., Matsubayashi, S., Hattori, Y., Fucharoen, G., Harihara, S., Park, K. S., Omoto, K., & Pan, I.H. (1996). mtDNA polymorphism in East Asian populations, with special reference to the peopling of Japan. Am. J. Hum. Genet. Vol. 59, No. 3, 579-590.

Höss, M. & Pääbo, S. (1993). DNA extraction from Pleistocene bones by a silica-based purification method. *Nucleic Acids Res.* Vol. 21, No. 16, pp. 3913-3914.

Hughes, S., Hayden, T. J., Douady, C. J., Tougard, C., Germonpré, M., Stuart, A., Lbova, L., Carden, R. F., Hänni, C., & Say, L. (2006). Molecular phylogeny of the extinct giant deer, Megaloceros giganteus. *Mol. Phylogenet. Evol.* Vol. 40, pp. 285-291.

Jasny, B. R., & Zahn, L. M. (2011). Genome-sequencing anniversary. A celebration of the genome, part I. *Science*. Vol. 331, No. 6017, pp. 546.

Kalmár, T., Bachrati, C. Z., Marcsik, A., & Raskó, I. (2000). A simple and efficient method for PCR amplifiable DNA extraction from ancient bones. *Nucleic Acids Res.* Vol. 28, No. 12, E67.

Knapp, M., Clarke, A. C., Horsburgh, K. A., & Matisoo-Smith, E. A. (2011). Setting the stage -building and working in an ancient DNA laboratory. *Ann. Anat.* [Epub ahead of print].

Kosakovsky-Pond, S. L., Frost, S. D. W., & Muse, S. V. (2005). HyPhy: hypothesis testing using phylogenies. *Bioinformatics.* Vol. 21, pp. 676-679.

Krause, J., Fu, Q., Good, J. M., Viola, B., Shunkov, M. V., Derevianko, A. P., & Pääbo, S. (2010). The complete mitochondrial DNA genome of an unknown hominid from southern Siberia. *Nature.* Vol. 464, No. 7290, pp. 894-7.

Krings, M., Stone, A., Schmitz R. W., Krainitzki, H., Soneking, M., & Pääbo, S. (1997). Neandertal DNA sequences and the origin of modern humans. *Cell.* Vol. 90, No. 1, pp. 19-30.

Lander, E. S. (2011). Initial impact of the sequencing of the human genome. *Nature.* Vol. 470, No. 7333, pp. 187-197.

Lorenz, J., & Smith, D. G. (1996). Distribution of four founding mtDNA haplogroups among native North Americans. Am. J. Phys. Anthropol. Vol. 101, No. 1, pp. 307-323.

López-Armenta, M., Bustos, Ríos, D., Moreno-Galeana, M. A., Herrera-Salazar, A., Pérez-Campos, E. M., Chávez-Balderas, X., & Muñoz, M. L. (2008) Genetic origin of a mommy from Queretaro (Pepita). Mummies and Science. *World Mummies Research/Proceediñngs of the VI World Congress on Mummy Studies (Teguise, Lanzarote=; Pablo Atoche,* Conrado Rodríguez and Ma. Angeles Ramírez (eds.) Santa Cruz de Tenerife: Academia Canaria de la Historia (etc.), p 700. ISBN: 978-84-612-5647-1. Edition: Academia Canaria de la Historia, Ayuntamiento de Teguise, Cabildo Insular de Lanzarote, Caja Canarias, Fundción Canaria Mapfre Guanarteme, Universidad de las Palmas de Gran Canaria. First edition, pp. 251-258.

Maniatis, T., Fritsch, E. F., & Sambrook, J. (1989). *Molecular Cloning. Cold Spring Harbor Laboratory Press.*

Miller, W., Drautz, D. I., Ratan, A., Pusey, B., Qi, J., Lesk, A. M., Tomsho, L. P., Packard, M. D., Zhao, F., Sher, A., Tikhonov, A., Raney, B., Patterson, N., Lindblad-Toh, K., Lander, E. S., Knight, J. R., Irzyk, G. P., Fredrikson, K. M., Harkins, T. T., Sheridan, S., Pringle, T., & Schuster, S. C. (2008). Sequencing the nuclear genome of the extinct woolly mammoth. *Nature.* Vol. 456, No. 7220, pp. 387-390.

Muñoz, M. L., Moreno-Galeana, M., Díaz-Badillo, A., Loza-Martínez, I., Macías-Juárez, V. M., Márquez-Morfin L., Jiménez-López, J. C. & Martínez-Meza, A. (2003).Análisis de DNA mitocondrial de una población prehispánica de Monte Albán, Oaxaca, México. In: *Antropología y Biodiversidad, Vol. 2, Ediciones Bellaterra S.L., Nava de Tolsa,* 289 bis. 08026 Barcelona, España. M. Pilar Aluja, Asunción Malgosa y Ramon M.a Nogués (eds.) ISBN: 84-7290-206-4. P.p. 170-182.

Parr, R. L., Carlyle, S. W., & O'Rourke, D. H. 1996. Ancient DNA analysis of Fremont Amerindians of the Great Salt Lake Wetlands. *Am. J. Phys. Anthropol.* Vol. 99, No. 4, pp. 507-518.

Pääbo, S. (1985). Molecular cloning of ancient Egyptian mummy DNA. *Nature.* Vol. 314, No. 6012, pp. 644-645,

Pääbo, S., Poinar, H., Serre, D., Jaenicke-Despres, V., Hebler, J., Rohland, N., Kuch, M., Krause, J., Vigilant, L., & Hofreiter, M. (2004). Genetic analyses from ancient DNA. *Annu. Rev. Genet.* Vol. 38, pp. 645-679.

Poinar, H. N., Kuch, M., Sobolik, K. D., Barnes, I., Stankiewicz, A. B., Kuder, T., Spaulding, W. G., Bryant, V. M., Cooper, A., & Pääbo, S. (2001). A molecular analysis of dietary diversity for three archaic Native Americans. *Proc Natl. Acad. Sci. U S A.* Vol. 98, No. 8, pp. 4317-22.

Rasmussen, M., Li, Y., Lindgreen, S., Pedersen, J. S., Albrechtsen, A., Moltke, I., Metspalu, M., Metspalu, E., Kivisild, T., Gupta, R., Bertalan, M., & Nielsen, K. (2010). Ancient human genome sequence of an extinct Palaeo-Eskimo. *Nature.* Vol. 463, No. 7282, pp. 757-62.

Reich, D., Green, R. E., Kircher, M., Krause, J., Patterson, N., Durand, E. Y., Viola, B., Briggs, A. W., Stenzel, U., Johnson, P. L., Maricic, T., Good, J. M., Marques-Bonet, T., Alkan, C., Fu, Q., Mallick, S., Li, H., Meyer, M., Eichler, E. E., Stoneking, M., Richards, M., Talamo, S., Shunkov, M. V., Derevianko, A. P., Hublin, J. J., Kelso, J., Slatkin, M., & Pääbo, S. (2010). Genetic history of an archaic hominin group from Deniscova Cave in Siberia. *Nature.* Vol. 468, No. 7327, pp. 1053-1060.

Richards, M., Côrte-Real, H., Forster, P., Macaulay, V., Wilkinson-Herbots, H., Demaine, A., Papiha, S., Hedges, R., Bandelt, H. J., & Sykes, B. (1996). Paleolithic and Neolithic lineages in the European mitochondrial gene pool. *Am. J. Hum. Gent.* Vol. 59, No. 3, pp. 185-203.

Rohland, N., & Hofreiter, M. (2007). Comparison and optimization of ancient DNA extraction. *Bio Techniques.* Vol. 42, No. 3, pp. 343-352.

Sachidanandam, R., Weissman, D., Schmidt, S. C., Kakol, J. M., Stein, L. D., Marth, G., Sherry, S., Mullikin, J. C., Mortimore, B. J., & Willey, D. L. (2001). International SNP Map Working Group. A map of human genome sequence variation containing 1.42 million single nucleotide polymorphisms. *Nature.* Vol. 409, No. 6822, pp. 928-933.

Schurr, T. G., Ballinger, S. W., Gan, Y.-Y., Hodge, J. A., Merriwether, D. A., Lawrence, D. N., Dnower, W. C., Weiss, K. M., & Wallace, D. C. (1990). Amerindian mitochondrial DNAs have rare Asian mutations at high frequencies, suggesting they derived from four primary maternal lineages. *Am. J. Hum. Genet.* Vol. 46, No. 3, pp. 768-765.

Stone, A. C., & Stoneking, M. (1993). Ancient DNA from a pre-Columbian Amerindian population. *Am J Phys Anthropol.* Vol. 92, No. 4, pp. 463-471.

Stone, A. C., & Stoneking, M. (1998). mtDNA analysis of a prehistoric oneota population: Implications for the Peopling of the New World. *Am. J. Hum. Genet.* Vol. 62, No. 5, pp. 1153-1170.

Stoneking M. & Krause, J. (2011). Learning about human population history from ancient and modern genomes. *Nat. Rev. Genet.* Vol. 12, No. 9, pp. 603-614. doi: 10.1038/nrg3029.

Thomas, M. G., & Moore, L. J. (1997). Preparation of bone samples for DNA extraction: a nuts and bolts approach. *Biotechniques*. Vol. 22, No. 3, pp. 402

Thompson, J. D., Higgins, D. G. & Gibson, T. J. (1994). CLUSTAL W: improving the sensitivity of progressive multiple sequence alignment through sequence weighting, positions-specific gap penalties and weight matrix choice. *Nucleic. Acids Res*. Vol. 22, pp. 4673-4680.

Torroni, A., Chen, Y.-S., Semino, O., Santachiara-Beneceretti, A. S., Scott, C. R., Lott, M.T., Winter, M., & Wallace, D. C. 1994a. mtDNA and Y-chromosome polymorphisms in four Native American Populations from Southern Mexico. *Am. J. Hum. Gene*. Vol. 54, No. 2, pp. 303-318.

Torroni, A., Neel, J. V., Barrantes, R., Schurr, T. G., & Wallace, D. C. 1994b. Mitochondrial DNA "clock" for the Amerinds and its implications for timing their entry into North America. *Proc. Natl. Acad. Sci. U. S.A*. Vol. 91, No. 3, pp. 1158-1162.

Torroni, A., Schurr, T. G., Cabell, M. F., Brown, M. D., Neel, J. V., Larsen, M., Smith, D. G., Vullo, C. M., & Wallace, D. C. 1993. Asian affinities and continental radiation of the four founding Native American mtDNAs. *Am. J. Hum. Gene*. Vol. 53, No. 3, 563-590.

Torroni, A., Schurr, T. G., Yang, C.-C., Szathmary, E. J., Williams, R. C., Schanfield, M. S., Troup, G. A., Knowler, W. C., Lawrence, D. N., Weiss, K. M., & Wallace, D. C. 1992. Native American mitochondrial DNA analysis indicates that the Amerind and the NaDene populations were founded by two independent migrations. *Genetics*. Vol. 130, pp. 153-162.

Venter, J. C., Adams, M. D., Myers, E. W., Li, P. W., Mural, R. J., Sutton, G. G., & Smith, H. O. (2001).The sequence of the human genome. *Science*. Vol. 291, No. 5507, pp. 1298-1302.

Vigilant, L., Pennington, R., Harpending, H., Kocher, T. D., & Wilson, A. C. (1989). Mitochondrial DNA sequences in single hairs from a southern African population. *Proc Natl Acad Sci U S A*. Vol. 86, No. 23, pp. 9350-9354.

Wallace, D. C., Garrison, K., & Knowler, W. C. (1985). Dramatic founder effects in Amerindian mitochondrial DNAs. *Am. J. Phys. Anthropol*. Vol. 68, No. 2, pp. 149-155.

Wallace, D. C, & Torroni, A. (1992). American Indian prehistory as written in the mitochondrial DNA: A review. *Hum. Biol*. Vol. 64, No. 3, pp. 403-416.

Walsh, P. S., Metzger, D. A., & Higuchi, R. (1991). Chelex 100 as a medium for simple extraction of DNA for PCR-based typing from forensic material. *Biotechniques*. Vol. 10, No. 4, pp. 506–513.

Willerslev, E., & Cooper, A. (2005). Ancient DNA. *Proc. R. Soc. B*. Vol. 272, pp. 3-16.

Wilson, M. R., DiZinno J. A., Polanskey, D., Replogle, J., & Budowle, B. (1995). Validation of mitochondrial DNA sequencing for forensic casework analysis. *Int. J. Legal. Med*. Vol. 108, No. 2, pp. 68-74.

Wrischnik, L. A., Higuchi, R. G., Stoneking, M., Erlich, H. A., Arnheim, N., & Wilson, A. C. (1987). Length mutations in human mitochondrial DNA: direct sequencing of enzymatically amplified DNA. *Nucleic Acids Res*. Vol. 15, No. 2, pp. 529-42.

Young, C. C., Burghoff, R. L., Keim, L. G., Minak-Bernero, V., Lute, J. R., & Hinton, S. M. (1993). Polyvinylpyrrolidone-Agorose Gel Electrophoresis Purification of Polymerase Chain Reaction-Amplifiable DNA from Soils. *Appl Environ Microbiol*. J. Vol. 59, No. 6, pp.1972-1974.